SOUTH-WESTERN

MATH MATTERS BOOK 1

An Integrated Approach

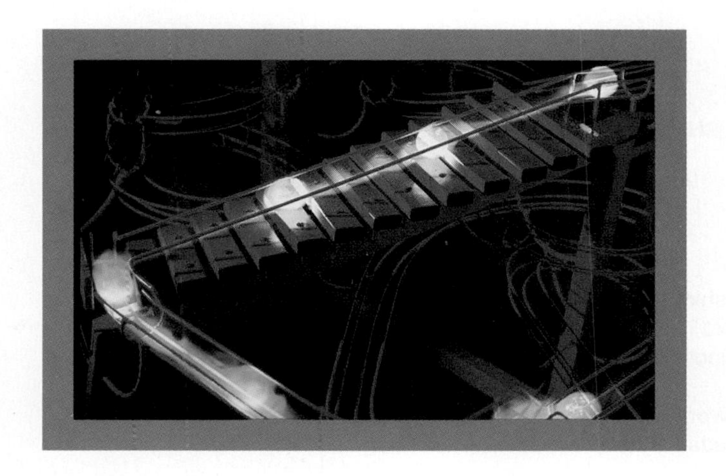

CHICHA LYNCH
Capuchino High School
San Bruno, California

EUGENE OLMSTEAD
Elmira Free Academy
Elmira, New York

SOUTH-WESTERN PUBLISHING CO.

Managing Editor: Eve Lewis
Mathematics Consultant: Carol Ann Dana
Production Editor: Thomas N. Lewis
Editorial Production Manager: Carol Sturzenberger
Cover Photo Photographer: Wayne Sorce © June 1988
Audio Kinetic Sculptures by George Rhoads

About the Math Matters Cover
Artist George Rhoads' "audio kinetic sculptures" appear in shopping centers, malls, terminals, and museums throughout North America. The mechanisms in the sculptures are visible and easy to understand, but the mathematical design provides infinite variety. Crowds gather to watch the acrobatic activity and hear the clanging noises triggered by randomly propelled balls traveling down twisting skeletal ramps.

International Thomson Publishing
South-Western Publishing Co. is an ITP Company.
The ITP trademark is used under license.

Authorized adaptation of a work first published by Nelson Canada,
A Division of Thomson Canada Limited,
1120 Birchmount Road,
Scarborough, Ontario
M1K 5G4

ISBN: 0-538-63951-2

Library of Congress Number: 91-62724

2 3 4 5 6 7 8 9 KI 99 98 97 96 95 94

Printed in the United States of America

About the **Math Matters** Authors

Chicha Lynch

Ms. Lynch is mathematics department head at Capuchino High School, San Bruno, California. As a specialist with the California Mathematics Project, Ms. Lynch helped her school district redesign courses to meet the state framework for Math A. Since 1988 she has served the state of California as a Mentor Teacher for Math A/B. Ms. Lynch is a graduate of the University of Florida and received the LaBoskey Award in 1988 from Stanford University for her contribution to its teacher education program. Most recently Ms. Lynch has been a research associate at the Far West Educational Laboratory, San Francisco, developing assessment for secondary mathematics teachers for certification by the National Board of Professional Teaching Standards. She was a state finalist in 1988 for the Presidential Award for Excellence in Mathematics Teaching.

Eugene Olmstead

Mr. Olmstead is a mathematics teacher at Elmira Free Academy, Elmira, New York. He has had 20 years of public school experience teaching courses from general mathematics through calculus. Mr. Olmstead earned his B.S. in mathematics at State University College at Geneseo, New York, and received his M.S. in mathematics education at Elmira College, Elmira, New York. Mr. Olmstead has worked as a counselor at the Calculator and Computer Precalculus Institutes at The Ohio State University and frequently gives workshops on teaching mathematics using technology. He was a state finalist in 1991 for the Presidential Award for Excellence in Mathematics Teaching.

Program Consultants

Bert K. Waits
Professor of Mathematics
The Ohio State University
Columbus, Ohio

Henry S. Kepner, Jr.
Professor of Mathematics
and Computer Education
University of Wisconsin-Milwaukee
Milwaukee, Wisconsin

Tommy Eads
Mathematics Teacher
North Lamar High School
Paris, Texas

Ann M. Farrell
Assistant Professor of
Mathematics and Statistics
Wright State University
Dayton, Ohio

Reviewers

Lucy Duffley
Mathematics Department Chair
Canyon Springs High School
Moreno Valley, California

David D. Molina
Assistant Professor
Department of Education
Trinity University
San Antonio, Texas

Valmore Guernon
Mathematics Teacher
Lincoln Junior/Senior High School
Lincoln, Rhode Island

David M. Otte
Principal
St. Henry High School
Erlanger, Kentucky

Gerald A. Haber
Assistant Principal
Winthrop Intermediate School
Brooklyn, New York

Janice Udovich
Mathematics Teacher
Washington High School
Milwaukee, Wisconsin

Sonja Hubbard
Mathematics Teacher
Milton High School
Alpharetta, Georgia

Mary Weber
Mathematics Teacher
Washington High School
Milwaukee, Wisconsin

Joan C. Lamborne
Supervisor of Mathematics
and Computers
Egg Harbor Township High School
Pleasantville, New Jersey

Peter Westergard
Mathematics Teacher
Nicolet High School
Glendale, Wisconsin

Contents

THEME
Earth Science

THEME
Recreation

THEME
Road Signs and Other Signs

THEME Sports

THEME Patterns

Credits

TEXT ACKNOWLEDGMENTS

The authors and editors have made every effort to trace the ownership of all copyrighted selections found in this book and to make full acknowledgments of their use. Grateful acknowledgment is made to the following authors, publishers, agents, and individuals for their permission to reprint copyrighted materials.

Marshall Cavendish Limited: "Cases of Worldwide Smallpox" and "Distribution of Blood in the Human Body" in *The Marshall Cavendish Illustrated Encyclopedia of Family Health.* Copyright © 1988.
D. C. Heath and Company: "How Many Calories Are Used In Ten Minutes?" in *Houghton Mifflin Health* by Bud Getchell, Rusty Pippin, Jill Varnes. Copyright © 1987. Reprinted by permission of D. C. Heath and Company.
St. Martin's Press, Inc.: "Top Speeds of Athletes in Various Sports." Copyright © 1980 by The Diagram Group. From the book *Comparisons* and reprinted with permission from St. Martin's Press, Inc.
U.S. News & World Report: "On a Roll," Copyright © 1989. August 21, 1989 issue.
USA TODAY: "Le Tour de France." Copyright © 1991, USA TODAY. Reprinted with permission.

PHOTO ACKNOWLEDGMENTS

CONTENTS

p. v: Robert E. Daemmrich/Tony Stone Worldwide; p. vi: Sheila Turner/Monkmeyer Press Photo Service, Inc. (top); Viesti Associates, Inc. (bottom); p. vii: Joe Viesti/Viesti Associates, Inc.; p. ix: Joe Viesti/Viesti Associates, Inc.; p. x: Diane Johnson/Tony Stone Worldwide; p. xi: Boris Berranger/Viesti Associates, Inc.; p. xii: Luis Villota/The Stock Market

CHAPTER 1

p. 3: Bill Stanton/International Stock Photography, Ltd. (top); Agence Vandysladt, Paris, France/Allsport Photography, Inc. (center); p. 4: Don and Pat Valenti/Tony Stone Worldwide (left); Peter Frank/Tony Stone Worldwide (middle); David Brownell/The Image Bank (right); p. 5: Courtesy of University of California at Berkeley Public Information Office; p. 7: Gerhard Gscheidle/Peter Arnold, Inc.; p. 8: Obremski/The Image Bank; p. 9: Manny Millan/Sports Illustrated (right); Wayne Sproul/International Stock Photography, Ltd. (left); p. 10: Wayne Sproul/International Stock Photography, Ltd.; p. 11: David R. Frazier/Photo Researchers, Inc. (top); Jan Halaska/Photo Researchers, Inc. (bottom); p. 12: Chip Henderson/Woodfin Camp & Associates; p. 13: Gabe Palmer/The Stock Market; p. 16: Jim Pickerell/Tony Stone Worldwide; p. 17: Robert Isear/Photo Researchers, Inc. (top); Salmoiraghi/The Stock Market (bottom); p. 18: Sports Illustrated; p. 19: Bill Stanton/International Stock Photography, Ltd. (top); Bob Firth/International Stock Photography, Ltd. (middle); Ken Levinson/International Stock Photography, Ltd. (bottom); p. 22: Michael Salas/The Image Bank; p. 24: Tom McCarthy/The Stock Market; p. 26: Courtesy of Barbara and Steven Griffel (both); p. 28: Mimi Cotter/International Stock Photography, Ltd.; p. 29: Jose Azel/Woodfin Camp & Associates; p. 30: Elyse Lewin/The Image Bank (left); Dick Luria/FPG International Corp. (right); p. 32: Sports Illustrated; p. 33: Paul Solomon/Woodfin Camp & Associates

CHAPTER 2

p. 40: Barry Seidman/The Stock Market (top); Ed Bock/The Stock Market (center); Pete Salantos/The Stock Market (bottom); p. 41: Gabe Palmer/ Mugshots/The Stock Market; p. 43: Jerry Cooke/Photo Researchers, Inc.; p. 44: Ray Shaw/The Stock Market; p. 45: Joanne Savio/Falletta Associates; p. 47: Joanne Savio/Falletta Associates (top); Craig Hammell/The Stock Market (bottom); p. 48: Roy Morsch/The Stock Market; p. 50: John G. Zimmerman/Sports Illustrated; p. 52: Joanne Savio/Falletta Associates; p. 53: Michael Simpson/FPG International Corp. (top); William D. Adams/FPG International Corp. (bottom); p. 54: Stephen Johnson/Tony Stone Worldwide; p. 55: Guido Alberto Rossi/The Image Bank; p. 59: Chip Henderson/Tony Stone Worldwide; p. 60: Robert H. McNaught/SPL/Photo Researchers, Inc.; p. 63: Obremski/The Image Bank; p. 64: Dr. William C. Keel/SPL/Photo Researchers, Inc.; p. 66: NASA; p. 67: Art Wilkinson/Woodfin Camp & Associates; p. 68: Don Johnson/The Stock Market; p. 69: Richard Steedman/The Stock Market; p. 70: Joanne Savio/Falletta Associates; p. 71: Carroll Seghers/Photo Researchers, Inc.; p. 72: Jose Fuste Raga/The Stock Market; p. 73: Stephanie Stokes/The Stock Market; p. 75: Joanne Savio/Falletta Associates; p. 77: Eric Curry/H. Armstrong Roberts, Inc. (top); Sonya Jacobs/The Stock Market (bottom); p. 79: Barbara Filet/Tony Stone Worldwide; p. 80: Martin Chaffer/Tony Stone Worldwide; p. 81: John P. Kelly/The Image Bank (top); Snowdon/Hoyer/Woodfin Camp & Associates (bottom)

CHAPTER 3

p. 88: Angus McBean Estate (top left); EMI Film Distributors Ltd./Movie Still Archives (top center); Courtesy of Paramount Pictures Release (top right); Snark International/Art Resource (bottom left); Weber-Sipa Press/Art Resource (bottom right); p. 89: Culver Pictures (left); Wide World Photos (right); p. 94: Joanne Savio/Falletta Associates (top); CBS (bottom); p. 98: Travelpix/FPG International Corp.; p. 100: M. Lyons/The Image Bank; p. 102: Bob Rashid/Monkmeyer Press Photo Service, Inc.; p. 103: Michael Chua/The Image Bank; p. 104: Roy Morsch/The Stock Market (left); Focus on Sports (right); p. 105: Larry Busacca/Retna Ltd.; p. 107: Sam Griffith/Tony Stone Worldwide; p. 109: David Walberg/Sports Illustrated

CHAPTER 4

p. 116: Jon Riley/Tony Stone Worldwide (top); Gabe Palmer/Mugshots/The Stock Market (right); Joe Sohm/Chromosohm/Nawrocki Stock Photo, Inc. (bottom); p. 117: Will McIntyre/Photo Researchers, Inc. (top); Joe Sohm/Chromosohm/Nawrocki Stock Photo, Inc. (bottom); p. 120: David Woods/The Stock Market; p. 121: Don Smetzer/Tony Stone Worldwide; p. 124: Photri, Inc.; p. 125: Photri, Inc. (left); Photri, Inc. (center); Chris Jones Photo/The Stock Market (right); p. 126: Joanne Savio/Falletta Associates (both); p. 129: Suzanne Murphy/Tony Stone Worldwide; p. 133: Manfred Kage/Peter Arnold, Inc. (both); p. 135: Joe Sohm/Chromosohm/Nawrocki Stock Photo, Inc.; p. 136: Joseph Nettes/Photo Researchers, Inc.; p. 137: Roy Morsch/The Stock Market; p. 139: Russ Kinne/Photo Researchers, Inc.; p. 141: Lawrence Migdale Photography; p. 142: Charles Gupton/The Stock Market; p. 143: Don Mason/The Stock Market; p. 145: Comnet/Nawrocki Stock Photo, Inc. (top); Michael A. Keller/The Stock Market (bottom)

CHAPTER 5

p. 152: Melanie Carr/Nawrocki Stock Photo, Inc. (top); Candee Prod./Nawrocki Stock Photo, Inc. (middle); Jose Fernandez/Woodfin Camp & Associates (bottom); p. 153: Jose Fernandez/Woodfin Camp & Associates (right top); Nick Nicholson/The Image Bank (left); N. Ferguson/Photri, Inc. (middle); Jeffrey Apoian/Nawrocki Stock Photo, Inc. (bottom); p. 154: Bachman/Photri, Inc.; p. 157: Joanne Savio/Falletta Associates (top); Al Clayton, International Stock Photography, Ltd. (bottom); p. 158: H. Armstrong Roberts, Inc. (top left); Sepp Seitz/Woodfin Camp & Associates (top center); Joanne Savio/Falletta Associates (all others); p. 161: Joanne Savio/Falletta Associates; p. 162: Robert E. Daemmrich/Tony Stone Worldwide; p. 165: Bettye Lane/Photo Researchers, Inc.; p. 166: Joanne Savio/Falletta Associates; p. 167: Peter Russell Clemens/International Stock Photography, Ltd.; p. 170: Hugh Sitton/Tony Stone Worldwide; p. 172: from *The Man Who Knew Infinity* by Robert Kanigel, published by Simon & Schuster; p. 173: Superstock; p. 175: Courtesy of Oldsmobile; p. 176: Chuck Keeler/Tony Stone Worldwide (top left and right);) Don Smetzer/Tony Stone Worldwide (bottom left); Brent Jones/Tony Stone Worldwide (bottom right)

CHAPTER 6

p. 184: Steve Vidler/Nawrocki Stock Photo, Inc. (top); David Weintraub/Photo Researchers, Inc. (bottom); Ed Zirkle/Woodfin Camp & Associates (left); p. 185: Del Mulkey/Photo Researchers, Inc. (left); Rob Goldman/FPG Inernational, Corp. (right); p. 186: Photri, Inc. (left); Andrew Conway/ACE/Nawrocki Stock Photo, Inc. (right); p. 190: Melanie Carr/Nawrocki Stock Photo, Inc.; p. 193: Will/Deni McIntyre/Photo Researchers, Inc.; p. 196: Photo Library International, Ltd./Nawrocki Stock Photo, Inc. (left); Steve Vidler/Nawrocki Stock Photo, Inc. (right); p. 203: Vito Palmisano/Nawrocki Stock Photo, Inc.; p. 204: Courtesy of Ruth Gonzales (top); Robert J. Bennett/Photri, Inc. (bottom); p. 205: C. P. George/H. Armstrong Roberts, Inc.; p. 207: Peter Russell Clemens/ International Stock Photography, Ltd.; p. 212: Richard Shock/Nawrocki Stock Photo, Inc. (left); David Lissy/Nawrocki Stock Photo, Inc. (right); p. 213: Photri, Inc.

CHAPTER 7

p. 220: Ken Lax/Photo Researchers, Inc.; p. 221: Ken Lax/Photo Researchers, Inc.; p. 222: David Vance/The Image Bank; p. 227: Photri, Inc.; p. 229: John D. Luke/Index Stock Photography, Inc.; p. 230: Photri, Inc. (top); Richard Hutchings/Photo Researchers, Inc. (bottom); p. 235: Sepp Seitz/Woodfin Camp & Associates; p. 238: H. Abernathy/H. Armstrong Roberts, Inc.; p. 239: Courtesy of Chevrolet; p. 240: Addison Geary/Nawrocki Stock Photo, Inc.; p. 242: Chris Hackett/The Image Bank; p. 243: FPG International Corp.(left); Timothy Eagan/Woodfin Camp & Associates (right); p. 244: Lawrence Migdale Photography; p. 247: Mark Scott/FPG International Corp.; p. 248: Photri, Inc.; p. 249: Peter B. Kaplan/Photo Researchers, Inc.; p. 251: Photri, Inc.; p. 252: Photri, Inc.; p. 257: Tony Stone Worldwide; p. 259: Guy Gillette/Photo Researchers, Inc.; p. 261: Jose Fernandez/Woodfin Camp & Associates

CHAPTER 8

p. 268: Mike Kullen/Sports Illustrated (left); Jerry Wachter/Sports Illustrated (right); p. 269: Chuck Solomon/Sports Illustrated (top and bottom); Bill Smith/Sports Illustrated (center); p. 271: Photri, Inc. (top); Emily Harste/Bruce Coleman, Inc. (bottom); p. 273: John Iacono/Sports Illustrated (left); WVF/Soenar Chamid/Nawrocki Stock Photo, Inc. (right); p. 274: Jerry W. Myers/FPG International Corp.; p. 275: Maggi & Maggi/The Stock Market; p. 276: Steve Whalen/Nawrocki Stock Photo, Inc.; p. 277: Photri, Inc. (left); Murray Alcosser/The Image Bank (right); p. 279: Les Riess/Photri, Inc.; p. 281: Arnold J. Kaplan, APSA-AFIAP/Photri, Inc.; p. 282: Catherine Ursillo/Photo Researchers, Inc.; p. 283: Chuck Solomon/Sports Illusrated; p. 284: Billy E. Barnes/FPG International Corp.; p. 285: Robert E. Daemmrich/Tony Stone Worldwide; p. 286: Dale E. Boyer/Photo Researchers, Inc. (left); Hugh Rogers/Monkmeyer Press Photo Service, Inc. (right); p. 287: Pedro Coll/The Stock Market (left); Spencer Grant/Monkmeyer Press Photo Service, Inc. (right).; p. 290: Joanne Savio/Falletta Associates; p. 293: Joanne Savio/Falletta Associates

CHAPTER 9

p. 300: David Bentley/Nawrocki Stock Photo, Inc. (top); Richard Hackett/International Stock Photography, Ltd. (bottom); p. 301: Fotex/Drechsler/Nawrocki Stock Photo, Inc. (top); Bachman/Photri, Inc. (bottom); p. 303: Nancy J. Pierce/Photo Researchers, Inc.; p. 305: Mark Bolster/International Stock Photography, Ltd.; p. 306: Photri, Inc.; p. 308: Larry Mulvehill/Photo Researchers, Inc.; p. 309: Melanie Carr/Nawrocki Stock Photo, Inc.; p. 311: Bill Stanton/International Stock Photography, Ltd.; p. 312: Bachmann/Photri, Inc.; p. 313: Tom Carter/Photri, Inc.; p. 315: Photri, Inc.; p. 317: Ken Lax/Photo Researchers, Inc.; p. 318: Freda Leinwand/Monkmeyer Press Photo Service, Inc.; p. 321: Fotex/R. Wolloner/Nawrocki Stock Photo, Inc.; p. 323: David R. Frazier/Photo Researchers, Inc.; p. 325: Blair Seitz/Photo Researchers, Inc.; p. 326: Richard Hutchings/Photo Researchers, Inc.; p. 327: Comnet/Nawrocki Stock Photo, Inc.; p. 329: Mike Kagan/Monkmeyer Press Photo Service, Inc.; p. 330: Roy Morsch/The Stock Market; p. 331: Susan McCartney/Photo Researchers, Inc.; p. 333: Jeff Apoian/Nawrocki Stock Photo, Inc.

CHAPTER 10

p. 340: George Hall/Woodfin Camp & Associates (both); p. 341: George Hall/Woodfin Camp & Associates; p. 342: Lowell Georgia/Photo Researchers, Inc.; p. 345: David Weintraub/Photo Researchers, Inc.; p. 349: Bob Firth/International Stock Photography, Ltd. (left); Richard Steedman/The Image Bank (center); Joanne Savio/Falletta Associates (right); p. 350: Lee Foster/Bruce Coleman, Inc.; p. 354: Joanne Savio/Falletta Associates; p. 363: Chris Tortora/Woodfin Camp & Associates; p. 365: Donald Miller/Monkmeyer Press Photo Service, Inc.; p. 369: Ted Clutter/Photo Researchers, Inc.(left); Carl Purcell/Photo Researchers, Inc. (right)

CHAPTER 11

p. 384: Heinz Kluetmeier/Sports Illustrated (left top); Wachter/Photri, Inc. (left bottom); Mimi Forsythe/Monkmeyer Press Photo Service, Inc. (center); Michael Lichter Photography/International Stock Photography, Ltd. (far right); p. 385: Steve Proehl/The Image Bank (top); Steve Dunwell/The Image Bank (bottom); p. 386: Robert Alan Soltis/Nawrocki Stock Photo, Inc.; p. 395: Lawrence Migdale Photography; p. 398: Gary Russ/The Image Bank; p. 399: Bernard Roussel/The Image Bank; p. 403: Ira Lipsky/International Stock Photography, Ltd.; p. 405: Renate Hiller/Monkmeyer Press Photo Service, Inc.; p. 406: Keith Gunnar/Bruce Coleman, Inc.; p. 407: Marcel Ivy-Schwart/The Image Bank; p. 411: Michal Heron/Woodfin Camp & Associates; p. 414: John McDonough/Sports Illustrated; p. 415: Lionello Fabbri/Photo Researchers, Inc.; p. 417: Joanne Savio/Falletta Associates; p. 418: Steve Allen/Peter Arnold, Inc.; p. 419: Art Wilkinson/Woodfin Camp & Associates (top); James Sugar/Woodfin Camp & Associates (bottom); p. 422: Hugh Sitton/Tony Stone Worldwide; p. 427: Jeffrey W. Myers/Nawrocki Stock Photo, Inc.; p. 429: Robert E. Daemmrich/Tony Stone Worldwide

CHAPTER 12

p. 436: John Sanford/SPL/Photo Researchers, Inc. (left); Robin Scagell/SPL/Photo Researchers, Inc. (right); p. 439: NRAD/SPL/Photo Researchers, Inc.; p. 443: Dr. J. Lorre/SPL/Photo Researchers, Inc.; p. 451: Julian Baum/SPL/Photo Researchers, Inc. (top); NASA/Photo Researchers, Inc. (middle); Dr. Steve Gull & John Fielden/Photo Researchers, Inc. (bottom); p. 453: Julian Baum/SPL/Photo Researchers, Inc. (left); UPI/Bettman Newsphotos (right); p. 454: ARCHIV/Photo Researchers, Inc. (left); Photo Researchers, Inc. (center); U.S. Geological Survey/NASA/Photo Researchers, Inc. (right); p. 455: Denny Tillman/The Image Bank; Eunice Harris/Photo Researchers, Inc.; p. 463: John Sanford/SPL/Photo Researchers, Inc.; p. 465: Astron Society of the Pacific/Photo Researchers, Inc.

CHAPTER 13

p. 472: Ken Biggs/Photo Researchers, Inc. (top); Ray Ellis/Photo Researchers, Inc. (bottom); p. 473: Fulvio Roiter/The Image Bank; (left) Bill C. Kenny/The Image Bank (right); p. 474: Joanne Savio/Falletta Associates; p. 476: Joanne Savio/Falletta Associates; p. 477: Ted Kawalerski/The Image Bank; p. 478: Blair Seitz/Photo Researchers, Inc. (top); Joanne Savio/Falletta Associates (bottom); p. 481: Joanne Savio/Falletta Associates; p. 482: Joanne Savio/Falletta Associates; p. 483: Palmer/Kane, Inc./The Stock Market; p. 484: Courtesy of Oldsmobile; p. 485: Steve Dunwell/The Image Bank (top); Mel Digiacomo/The Image Bank (bottom); p. 486: Photri, Inc.; p. 487: Brock May/Photo Researchers, Inc.; p. 489: Jeffrey W. Myers/FPG International Corp.; p. 490: Joanne Savio/Falletta Associates; p. 491: Joanne Savio/Falletta Associates; p. 492: Joanne Savio/Falletta Associates; p. 494: Grant LeDuc/Monkmeyer Press Photo Service, Inc. (left); Michael Heron/Woodfin Camp & Associates (right); p. 495: Roger A. Clark, Jr./Photo Researchers, Inc.; p. 497: Roger Dollarhide/Monkmeyer Press Photo Service, Inc.

CHAPTER 14

p. 504: Viviane Moos/The Stock Market (left); Masahiro Sano/The Stock Market (right); p. 505: Paul Jablonka/International Stock Photography Ltd. (top left); Masahiro Sano/The Stock Market (left center); Viviane Moos/The Stock Market(right center and bottom right); p. 509: George Hall/Woodfin Camp & Associates; p. 511: ABY/Photri, Inc.; p. 515: Dennis Hallinan/FPG International Corp.; p. 516: David Lissy/Nawrocki Stock Photo, Inc.; p. 518: Steven Burr Williams/The Image Bank (left); Jürgen Vogt/The Image Bank (right); p. 521: Renate Hiller/Monkmeyer Press Photo Service, Inc. (top); Alexas Urba/The Stock Market (bottom); p. 522: Joanne Savio/Falletta Associates; p. 526: Sybil Shackman/Monkmeyer Press Photo Service, Inc.; p. 528: Photri, Inc.; p. 529: Ch. Petit/Photo Researchers, Inc.; p. 530: James F. Palka/Nawrocki Stock Photo, Inc.; p. 531: Courtesy of Rosemary Chang; p. 533: Hungerfond/Photri, Inc.; p. 534: Joanne Savio/Falletta Associates

DATA FILES

p. 548: G. Galicz/Photo Researchers, Inc.; p. 549: Bob Firth/International Stock Photography, Inc. (top); Art Montes de Oca/FPG International Corp. (bottom); p. 550: Guy Gillette/Photo Researchers, Inc. (top); David J. Maenzo/The Image Bank (bottom); p. 551: Camerique/H. Armstrong Roberts, Inc. (top); Cyril Toker/Photo Researchers, Inc. (center); p. 552: Van Bucher/Photo Researchers, Inc. (top); David R. Frazier/Photo Researchers, Inc. (bottom); p. 553: Ron Sherman/Tony Stone Worldwide (top); William Caram/Index Stock Photography, Inc. (bottom left); Menschenfreund/The Stock Market (bottom right); p. 554: Rev. Ronald Royer/SPL/Photo Researchers, Inc. (top); Hale Observatories/Photo Researchers, Inc. (bottom); p. 555: Bill W. Marsh/Photo Researchers, Inc. (top); Hale Observatories/Photo Researchers, Inc. (left); p. 556: Photri, Inc. (top); William Strode/Woodfin Camp & Associates (bottom); p. 557: Jeff Friedman Photography (top); Fotex/Drewa/Nawrocki Stock Photo, Inc. (bottom); p. 558: J. Barry O'Rourke/The Stock Market (top); Richard L. Carlton/Photo Researchers, Inc. (right bottom); p. 559: David J. Cross/ Peter Arnold, Inc. (top left); John V.A. F. Neal/International Stock Photography, Ltd. (top right); Simon Fraser/SPL/Photo Researchers, Inc. (center); p. 560: Verna Brainard/Photri, Inc. (top); R. Kord/H. Armstrong Roberts, Inc. (bottom); p. 561: Joanne Savio/Falletta Associates (top); Index Stock Photography, Inc. (center); p. 562: Dr. C. W. Biedel/Photri, Inc.; p. 563: Suzanne Szasz/Photo Researchers, Inc. (top); Jon Riley/Tony Stone Worldwide (bottom left); Alfred Gescheidt/The Image Bank (bottom right); p. 564: Walter Iooss, Jr./Sports Illustrated (top); Lawrence Migdale Photography (bottom); p. 565: Wayne Sproul/International Stock Photography; p. 566: Tom McHugh/Photo Researchers, Inc.; p. 568: Gregory Sams/SPL/Photo Researchers, Inc. (top left); Character House/Photo Researchers, Inc. (bottom right); p. 571: Gregory Sams/SPL/ Photo Researchers, Inc. (bottom left); p. 570: Gregory Sams/SPL/Photo Researchers, Inc. (top); Paul Jablonica/International Stock Photorgraphy, Ltd. (bottom); p. 571: Gregory Sams/SPL/Photo Researchers, Inc. (top); Dale Boyer/USCF/Photo Researchers, Inc. (center); John Wells/SPL/Photo Researchers, Inc. (bottom)

All statistical information that appears in the Data Files is in the public domain unless otherwise noted. We gratefully acknowledge the following sources:

Playing with Fire — Hazardous Waste Incineration, © 1990 Greenpeace U.S.A., Washington, DC, "Quantities of Hazardous Waste Burned in U.S.," p. 559.

Acknowledgments are continued on page 586.

Preface

Welcome to *Math Matters*. You may find this textbook different from other math books you have used. Unlike more traditional textbooks which focus on one specific subject, *Math Matters* combines mathematic topics in an integrated program. Number sense, algebra, geometry, statistics, and logic are presented as tools for investigating phenomena and exploring new math concepts.

As you work through *Math Matters*, you will find new ways to use estimation and approximation skills to use computers and calculators more effectively. You will discover that algebraic concepts can enhance your critical thinking skills. You will see how geometry relates to reasoning and problem solving. You will find that you can evaluate the meaning of statistics that are presented on TV and in newspapers. In short, you will learn how to use mathematics to your advantage in school, at home, and at work.

To help you achieve success with this program, we suggest you take time now to learn how to use this textbook. Here are a few ideas you may find worthwhile.

GETTING STARTED WITH NOTEBOOKS AND LEARNING JOURNALS

Before you begin using *Math Matters*, you will find that keeping an organized notebook or learning journal is essential to successful learning throughout the school year. It is important that the organization of your notebook or learning journal reflect your particular needs and learning style. Begin by thinking about the kinds of information you will want to record and how you might easily get access to this information. Below are some ideas to help you organize your notebook or journal. Be sure, however, to listen carefully to what your teacher requires of you, since he or she may wish to add or change the kinds of information you record.

Vocabulary List

For each unfamiliar word, definition, or property, write the correct definition as given in the text or Glossary. Also include an example if that will help. Then rewrite the definition and example in your own words. Review your vocabulary list periodically to make sure that you maintain a clear idea of what each term means. A few minutes of vocabulary review each week will go a long way to improving your opportunities for success.

Notes and Questions

The contents of this part of your notebook or learning journal will depend somewhat on your teacher's method of instruction. This section should include any important problems, exercises, and solutions that your teacher puts on the chalkboard for class discussion. It also might include highlights of key points that your teacher makes during a class or lecture. You should write any questions, examples, or exercises that need further explanation. You may want to mark your questions in some way—perhaps, with an asterisk (*) or a question mark in the margin. Then, when you get a satisfactory explanation or you are able to answer your own question, you can cross out or circle the symbol you used to indicate that you no longer need to be concerned about that problem.

Summaries and Conclusions

In this section, you can reflect in your own words on what you have learned, which topics interest you the most, and which topics have been difficult. Summarizing and drawing conclusions is often the last step in mastering what you have learned. These skills are also helpful in clarifying areas for further study.

Formulas, Properties, and Drawings

You may wish to put this section at the back of your notebook or learning journal so that you can quickly find what you need. Here you can record commonly used formulas, properties, and drawings that explain or summarize key information.

Homework and Assignments

Have a special place in your notebook or learning journal where you can write the complete assignment and make a regular habit of recording all the assignments you receive. Doing these things will prevent last-minute "panic" phone calls to classmates, asking, "Do you know what the math homework is?" Be sure that you have some special way of indicating long-term projects so that the due day doesn't sneak up on you unexpectedly. Your teacher will tell you whether you should keep your homework in your notebook or turn it in each day.

LEARN HOW TO USE THIS BOOK

Math Matters has a consistent organization and many recurring features. It is wise to get to know how the book is organized and what the purpose of each feature is. Knowing what to expect will allow you to focus efficiently on what you need to learn. Take time now to become familiar with these features of your text, and you will save valuable time later on.

Skills Preview

Each chapter begins with a Skills Preview that will help you to discover which topics of the chapter you may already know and which will require careful study. Don't be concerned by what you get wrong on the Skills Preview. By identifying what you already know, you will be able to spend more time on those topics that are new to you and require your full attention.

Chapter Opener

The Chapter Opener helps to set the stage for the topics you are about to study. It tells you about the subject matter and theme of the chapter and presents some fascinating data and visual information for you to examine. Each opener requires that you use the data presented to make decisions and that you complete a group project or exploration tied to the mathematical content and theme of the chapter.

Every chapter is divided into a group of numbered sections, each of which is either four pages or two pages long. Every four-page section touches on a group of related math concepts or skills and follows a consistent organization.

Explore and Explore/Working Together

Each four-page section begins with an activity that provides a stimulating introduction to the mathematical concepts or skills to be studied. These activities often involve hands-on investigations that lead to meaningful mathematical insights. Many of the Explore activities require you to work cooperatively with other students. Look over each Explore activity before you begin working on it. In this way you will understand what is really expected of you, and you can gather together what you will need to complete the activity.

Skills Development

This section explains the concepts and skills you are expected to master. Here you will find helpful discussions, along with essential definitions. The Skills Development presents the key points of the lesson through model Examples and worked-out Solutions for you to study and apply later on.

Try These

These are questions and problems that you will likely discuss in your classroom. The items in Try These are similar to the ones presented in the Examples and Solutions of the Skills Development, so they are an excellent way for you to determine if you really have absorbed the key points of each lesson. If you have trouble with any Try These items, it is worth your while to take a second look at the relevant model Examples and Solutions in the Skills Development.

Exercises

The Exercises provide numerous opportunities to practice and apply the mathematical concepts and skills that you have studied and to make connections with other areas of mathematics and with the use of mathematics in real-world situations and in other subject areas. Each section contains three levels of Exercises, which provide increasingly challenging practice and problem solving. These items appear under the headings Practice/Solve Problems, Extend/Solve Problems, and Think Critically/Solve Problems.

From time to time in each chapter a four-page section ends with a special full-page feature called Problem Solving Applications. These pages make connections between the mathematics presented in the section and the practical application of those mathematical skills and concepts within a particular career or professional field.

Problem Solving Strategies

Each two-page Problem Solving Strategies section focuses on one important strategy. Each Problem Solving Strategies section has a brief introduction, model Problems and Solutions, and practice Problems. Problem Solving Strategies, introduced and applied throughout the book, include, for example, Using Logical Reasoning, Finding a Pattern, Working Backward, and Making an Organized List. From time to time throughout the book there will be a special Choose a Strategy lesson that presents a variety of problems involving any of the strategies that have been presented up to that point in the book.

Problem Solving Skills

These are special two-page sections that introduce a skill essential to successful problem solving. Each Problem Solving Skills section has a brief introduction, model Problems and Solutions, and practice Problems to help you apply the problem solving skill being taught. Problem Solving Skills, introduced and applied throughout the book, include Choosing an Operation, Choosing an Appropriate Scale, Using Venn Diagrams, and Making a Conjecture.

Within the sections of each chapter you will encounter a variety of special features, each of which is identified by a special visual symbol, or icon. The following is a description of the various features and their corresponding icons.

Check Understanding

This feature helps you to evaluate your understanding of key concepts and skills by posing questions about model Examples or Exercises.

Talk It Over

This feature provides thought-provoking questions that will spark interesting group discussions between you and your classmates.

Reading Mathematics

This feature provides an opportunity to become familiar with the vocabulary of mathematics and with the special meanings that ordinary words sometimes have within the discipline.

Writing About Math

In this feature you will be asked to demonstrate your knowledge of mathematical concepts and skills by writing about them in a variety of meaningful ways, including notes, advertisements, and brief essays.

Connections

This feature helps you to make connections between the mathematics you are learning and other areas of mathematics, other disciplines, and real-world problem solving situations.

Math in the Workplace

Here you will learn how the math you are studying is applied in the everyday work world.

Math: Who, Where, When

In this feature you will learn about mathematical ideas from different cultures around the world and the contributions that people have made to the development of mathematics.

Computer and Computer Tips

These features provide useful computer programs for solving math problems, along with helpful tips to make them work effectively and to adapt them to other uses.

Calculator and Calculator Tips

These features explain how to solve problems using a calculator, as well as tips to improve your efficiency and accuracy when using a calculator.

Problem Solving, Estimation, and Mental Math Tips

These features will help you to use a variety of strategies to solve problems that are presented in the text.

Mixed Review

This feature presents a quick way to refresh and maintain your mastery of previously learned mathematical skills and concepts.

At the end of each chapter you will find several features to help you maintain your mastery of concepts and skills and to help you prepare for tests.

Chapter Review

Each chapter ends with a two-page Chapter Review that highlights all the major rules and definitions in the chapter and provides additional examples and practice items. It is very helpful to look at the Chapter Review even before beginning a new chapter since the Review will help to focus your attention on the chapter's key concepts and skills.

Chapter Test

The one-page Chapter Test is very much like the Skills Preview at the beginning of the chapter. Try comparing your answers on both these features to learn which concepts and skills you have mastered and which you should review more carefully before taking any out-of-book examinations.

Cumulative Review

The Cumulative Review feature provides on-going opportunities to maintain math concepts and skills learned earlier in the course of study.

Cumulative Test

The Cumulative Test feature provides additional opportunities to practice test-taking skills for concepts and skills taught throughout the year. The test items are in multiple-choice format similar to the type encountered on standardized tests.

Your teacher may ask you to do other things to evaluate how well you are learning and using the math ideas that have been presented. You may be asked to build a portfolio of your work or to work with other students to complete extended projects that pull together the math ideas with which you have worked. You also may be asked to reflect on your experiences in mathematics class and your attitudes toward mathematics throughout the school year. In any case, remember that you are responsible for your own learning and for producing your best work.

Data Files

This section provides fascinating data organized around major themes, such as Health and Fitness, Economics, Ecology, Useful Math References, and so on. The tables, charts, graphs, and other visual and verbal data provided in this section are needed to answer questions that appear from time to time under the heading Using Data.

Glossary

This is an alphabetical listing of all the key terms introduced and defined in the text. Use the Glossary entries as you develop your vocabulary lists. Note that after each term there is a reference indicating the page of the chapter section where the term is first introduced and defined.

Selected Answers

This section provides answers to selected odd-number items of lessons and features. Using the Selected Answers on a regular basis is a good way to assess your own learning progress in an on-going way.

CHAPTER 1 SKILLS PREVIEW

A survey was taken in a town to find out how people felt about the new water-conservation program. The interviewer polled people as they boarded the 7:05 train one morning.

1. What kind of sampling does this represent?
2. What is an advantage of this kind of sampling?

The pictograph shows the number of books read last month by the students in four grades.

3. The students in which grade read the most books?
4. How many books did the juniors read?
5. How many books were read in all?

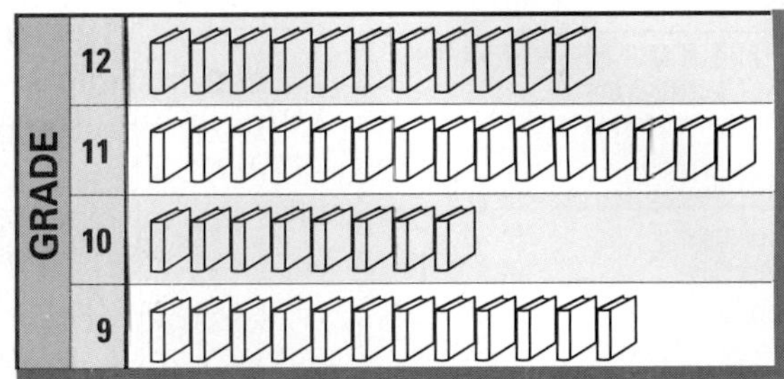

Key: [book] = 10 books

At a large family gathering, the heights of the family members were recorded in inches. The data are shown in this stem-and-leaf plot:

4|7 represents 47 in.

Stem	Leaves
3	9
4	6 7 7 8
5	0 0 2 4 5 5 8
6	1 1 7 8 8 8
7	0 2 2 4 5

The heights of how many people fall within each interval?

6. 30–39 in. 7. 60–69 in. 8. 70 in. or more

How many people are there of each height?

9. 47 in.
10. 54 in.
11. 68 in.

The bar graph gives the number of aluminum cans brought to a recycling center one week.

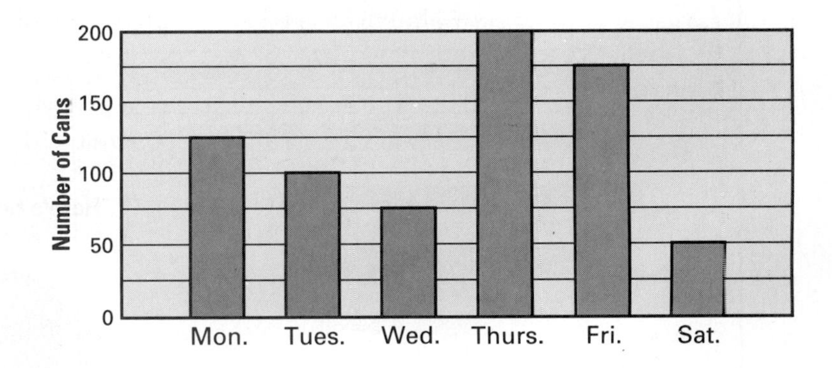

On which days were more cans brought to the recycling center?

12. Tuesday or Wednesday
13. Monday or Thursday
14. Friday or Saturday

The number of books read during the summer by eight students were: 3, 6, 4, 4, 5, 2, 8, 9.

15. Find the mean, median, mode, and range of the data.

EXPLORING DATA

THEME Health and Fitness

Pieces of information are called **data. Statistics** is the branch of mathematics that involves collecting data and organizing them in such a way that they can be used as a basis for making decisions. In this chapter you will have a chance to collect and organize data and display it in stem-and-leaf plots, pictographs, bar graphs, and line graphs. You will also work with statistical measures that help you describe a set of data.

AMERICANS WHO REGULARLY PARTICIPATE IN SPORTS

Sport	Number of Participants
Swimming	105,400,000
Bicycling	69,800,000
Bowling	43,300,000
Jogging/running	35,700,000
Tennis	32,300,000
Softball	28,500,000
Roller skating	25,400,000
Basketball	24,000,000
Ice skating	18,900,000
Golf	15,900,000
Skiing	15,400,000
Football	14,300,000
Racquetball	10,700,000
Soccer	6,500,000
Archery	5,500,000
Ice hockey	1,700,000

DECISION MAKING

Using Data

1. How do you suppose the data were collected for the sports table? Do you think each of the people represented in the table was polled individually?
2. If you wanted to be sure to drink enough juice to take in 200 mg of vitamin C today, what would you drink?
3. If you could have just two days to videotape Le Tour de France, which dates would you choose? In which cities would you be taping?

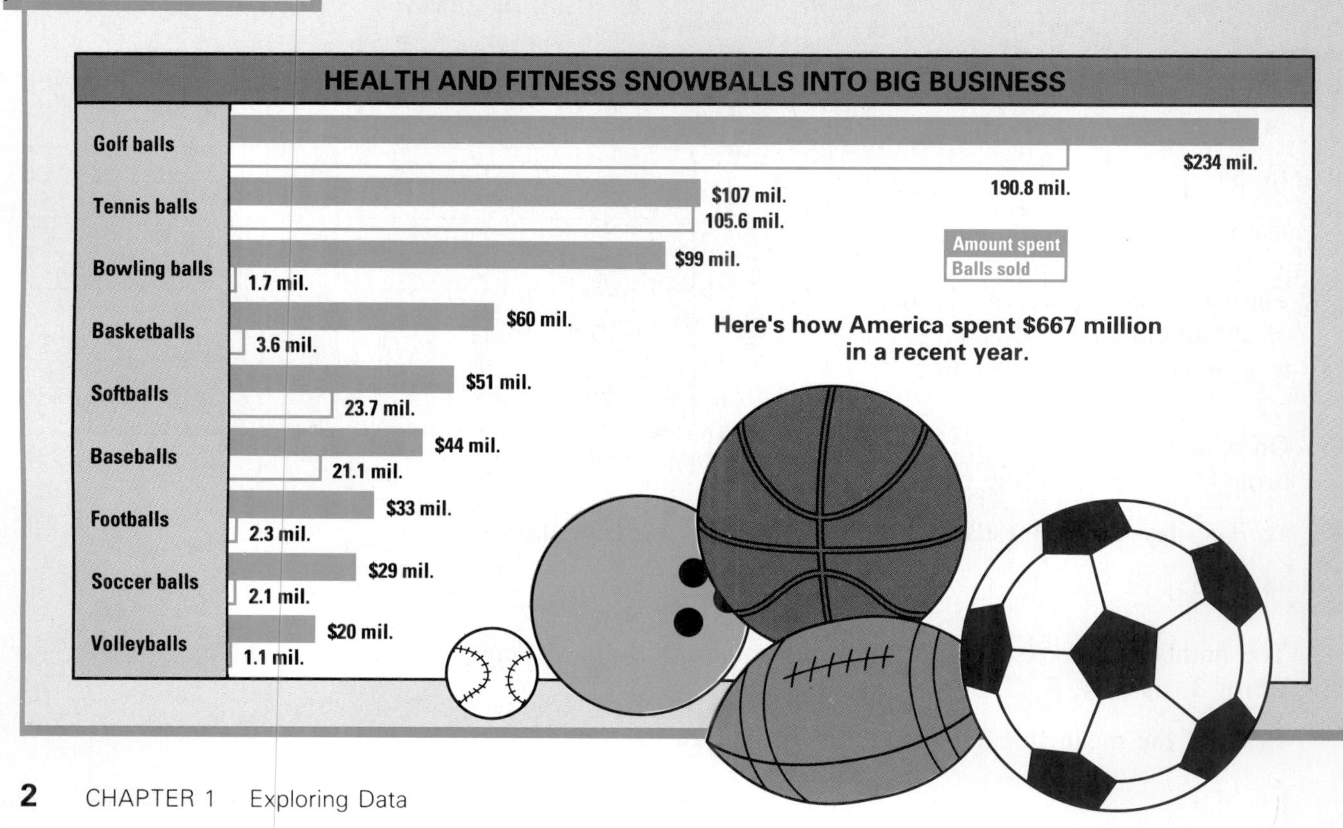

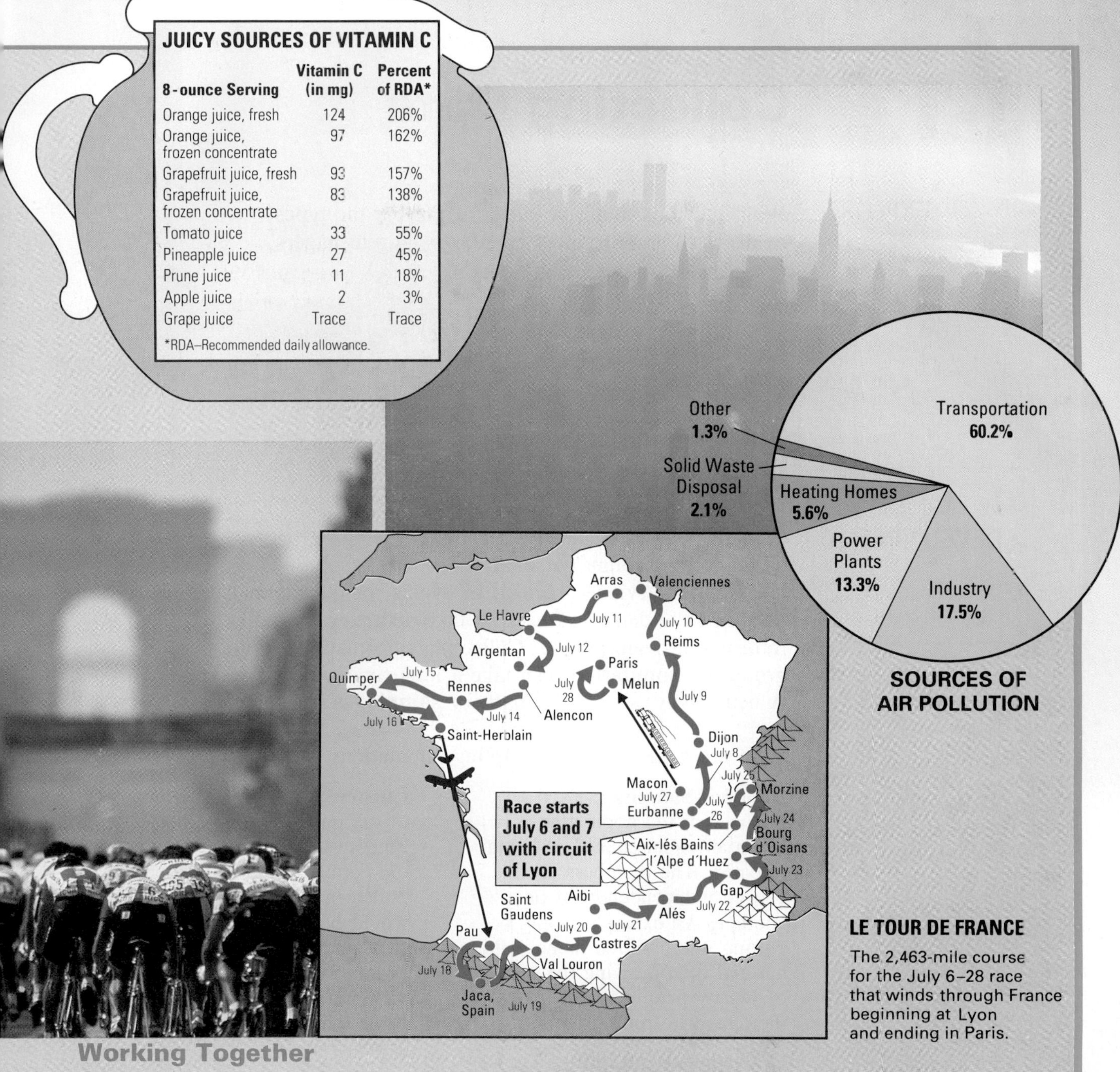

JUICY SOURCES OF VITAMIN C

8-ounce Serving	Vitamin C (in mg)	Percent of RDA*
Orange juice, fresh	124	206%
Orange juice, frozen concentrate	97	162%
Grapefruit juice, fresh	93	157%
Grapefruit juice, frozen concentrate	83	138%
Tomato juice	33	55%
Pineapple juice	27	45%
Prune juice	11	18%
Apple juice	2	3%
Grape juice	Trace	Trace

*RDA–Recommended daily allowance.

LE TOUR DE FRANCE

The 2,463-mile course for the July 6–28 race that winds through France beginning at Lyon and ending in Paris.

Working Together

Find examples of statistical data about aspects of health, fitness, sports, business, science, or other areas of interest to you. You can collect these data from magazines, newspapers, TV, or books or from talking to people. Discuss how each set of data is presented—in a chart, list, graph, or other form. Then, write a question for each set of data that could be answered by analyzing it. Give an example of why someone might need these data.

1-1 Collecting Data

EXPLORE

What methods might you use to find out the types of exercise that are most popular among teenagers?

SKILLS DEVELOPMENT

READING MATH

The word *data* is the plural form of the Latin word *datum.*

Whenever you say or write the word *data,* be sure to follow it with the plural form of the verb *to be.* Say, for example,

The data *are* correct.
These data *were* collected.

Data are information from which decisions can be made. Data can be gathered through personal or telephone interviews, records of events, or questionnaires. In order to obtain the data, you need to take a **survey**, or **poll**, to collect information from people. Collecting information from an entire group, or **population**, can take a long time and can be very costly. A better way is to poll a part, or **sample**, of the population.

Here are some ways of sampling populations.

Random Sampling: Each member of the population is given an equal chance of being selected. The members are chosen independently of one another. An example would be putting one hundred names into a hat and then drawing the names of twenty people to ask about their favorite rock groups.

Convenience Sampling: The population is chosen only because it is readily available. An example would be polling the ten students who happen to be sitting near you in the cafeteria to find out teenagers' favorite styles of jeans.

Cluster Sampling: The members of the population are chosen at random from a particular part of the population and are then polled in clusters. An example would be choosing parts of town at random, visiting the pet shops in these areas, and then asking *all* the pet owners you meet to name their pets' favorite brands of pet food.

Systematic Sampling: After a population has been ordered in some way, its members are chosen according to a pattern. An example would be pulling every tenth item off a production line for quality-control testing.

In order to be able to make decisions about an entire population based on a sample alone, you must first be sure to choose an appropriate sampling method.

Example 1

Which of these reflects the *random sampling* method that a health club owner might use to identify the most popular exercise machine in the club?

a. Ask the first twenty members who enter the club one morning.
b. Ask the members whose phone numbers end with the digit 7.
c. Ask the members who live on the six busiest streets in town.

Solution

Choice **b.** If phone numbers are assigned at random, those members whose phone numbers end in 7 represent a random sample. ◄

Example 2

Every twentieth rowing machine in a run of 500 was quality tested as it came off the assembly line. Two were found to be defective.

a. What kind of sampling does this situation represent?
b. What might be an advantage of this kind of sampling?
c. What might be a disadvantage of this kind of sampling?

Solution

a. This example represents *systematic* sampling.
b. An advantage is that the sample comes from the whole population.
c. If there were additional runs, then a sample taken from the last run might differ in quality from a sample taken from the first run. ◄

MATH: WHO, WHERE, WHEN

David Blackwell (1919-), a professor emeritus of statistics at the University of California, Berkeley, originally planned to become an elementary school teacher. Often portrayed today as a theoretician who presents his ideas in beautiful and elegant ways, Professor Blackwell has made contributions to Bayesian probability, game theory, set theory, dynamic programming, and information theory. At the outset of his teaching career, he taught at Howard University in Washington, D.C.

TRY THESE

1. Which of these reflects the *cluster sampling* method that a committee of high school students might use to find out how many adults would attend a school fair?

   a. Ask all the adults who leave the local supermarket on Tuesday between 5:00 and 6:00 p.m.
   b. Ask all the members of the men's and women's softball leagues.
   c. Ask all the adults who live in houses chosen at random on some of the streets surrounding the school.

In a telephone survey of the first fifty people who said that they earned their living by working in their home offices, it was found that nine out of ten people eat breakfast at least some of the time.

2. What kind of sampling does this situation represent?
3. What might be an advantage of this kind of sampling?
4. What might be a disadvantage of this kind of sampling?

EXERCISES

PRACTICE/ SOLVE PROBLEMS

Suppose you need to determine the most popular video arcade game. Name the advantages and disadvantages of choosing a sample in each of the following ways.

1. Ask every twelfth person who enters the local video arcade.
2. Ask a randomly chosen sample of twelve students in your school.
3. Ask the first twelve people you meet in a video-rental store.

EXTEND/ SOLVE PROBLEMS

Use the following information for Exercises 4–7. The owners of a shopping mall have to fill two vacant stalls in their food court. They want to find out what kinds of fast food shoppers like best. Suppose they station interviewers outside the food court between 12:00 noon and 1:00 p.m. and have them ask the first 100 people who pass about their food preferences.

4. What kind of sampling is represented by this situation?
5. How could this method be changed to produce a systematic sample?
6. How could the survey be conducted to produce a random sample?
7. Which kind of sample do you think would be of most use to the owners of the mall? Explain.

THINK CRITICALLY/ SOLVE PROBLEMS

Conduct your own poll! Work alone or in a group. Ask forty students about a subject of your choice.

8. Decide on a subject whose popularity or frequency you would like to know more about. Then write a question about the subject that can be answered "yes" or "no."
9. Decide on the population from which you will take your sample.
10. Give a pretest! Ask your question of a few students in your class to see if others understand the question as you mean it to be understood. (If necessary, change the wording of your question.)
11. Choose forty students for your sample. Decide on a way to record your data. Then take the poll.
12. Write an article for the school or community newspaper in which you describe your poll. Tell why you decided on the question, tell why you chose the particular population to poll, and describe the data that resulted.

Problem Solving Applications:

FREQUENCY TABLES AND LINE PLOTS

Brad interviewed people about their favorite type of TV program. He used this code to record his data.

S	Sports	Q	Quiz shows
M	Mysteries	V	Music videos
So	Soap operas	A	Adventure
N	News	C	Comedies

These are the data Brad collected.

V	C	So	N	Q	A	M	S	M	So	A	So
Q	C	C	N	A	C	N	M	Q	S	So	
A	S	Q	A	C	S	N	Q	A	Q	S	
A	A	M	S	So	V	N	C	A	C	S	

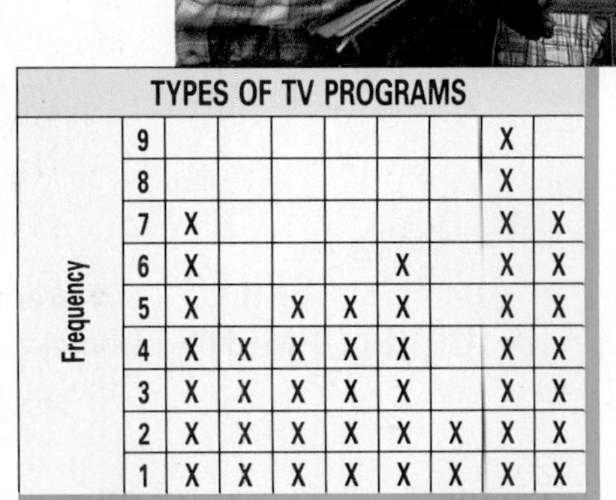

Brad displayed the data in a frequency table and in a line plot.

In a **frequency table**, a tally mark is used to record each response. The total number of tally marks for a given response is the frequency of that response. In a **line plot**, an X is made to record each response. The X's may be stacked one on top of another until all the data are recorded. Read the scale along the side, then follow it across to the X at the top of each stack to find the frequency of each response.

TYPES OF TV PROGRAMS

Program	Tally	Frequency
Sports	卌 II	7
Mysteries	IIII	4
Soap operas	卌	5
News	卌	5
Quiz shows	卌 I	6
Music videos	II	2
Adventures	卌 IIII	9
Comedies	卌 II	7

TYPES OF TV PROGRAMS

Frequency	Sports	Mysteries	Soap operas	News	Quiz shows	Music videos	Adventures	Comedies
9							X	
8							X	
7	X						X	X
6	X				X		X	X
5	X		X	X	X		X	X
4	X	X	X	X	X		X	X
3	X	X	X	X	X		X	X
2	X	X	X	X	X	X	X	X
1	X	X	X	X	X	X	X	X

Use Brad's frequency table or line plot for Exercises 1–3.

1. How many more people chose sports programs than chose news?
2. How many fewer people chose music videos than chose comedies?
3. As many people chose adventures as chose which two possible pairs of programs together?

Gather your own data about popular types of exercise among teenagers.

4. Decide on how to word a survey question about exercising. Then decide on a method of sampling and the number of people that will make up your sample.
5. Ask your question, record the data, then display them in a frequency table or in a line plot.
6. Use your data display in a chart or poster that could serve as an ad for fitness and good health.

1-2 Stem-and-Leaf Plots

EXPLORE/ WORKING TOGETHER

MENTAL MATH TIP

Here is a shortcut for checking your pulse. Instead of counting to find the number of beats in a minute, count the number of beats in 10 seconds, then multiply this number by 6. Explain why this works.

You can find your heart rate by checking your pulse. Find the beat at the side of your neck or at your wrist. Have a partner watch the second hand of a clock for exactly 1 minute and tell you when to start counting the number of beats and when to stop. Record the pulse rates for the class.

a. Which pulse rate was the fastest?
b. Which pulse rate occurred most frequently?
c. How did you record the data?
d. Suggest other ways to record the data so that they are easy to analyze.

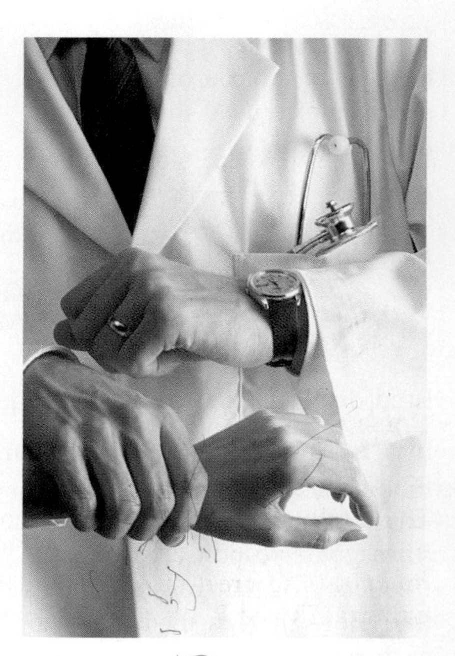

SKILLS DEVELOPMENT

A **stem-and-leaf plot** can be used to organize data. In the example below, the digits in the tens places make up the **stems.** The digits in the ones places make up the **leaves.**

CHECK UNDERSTANDING

Once the leaf data have been entered on a stem-and-leaf plot in the order in which they appear, the leaves may be rewritten in order, from least to greatest, on a second plot.

Use the pulse-rate plot at the right to make another plot in which the leaves are arranged in order, from least to greatest.

Example 1

Organize these pulse rates into a stem-and-leaf plot.

51	73	79	90	86	66	71	86	64
67	81	67	53	59	91	73	87	86
61	66	68	58	93	76	73	82	76

Solution

To start making a stem-and-leaf plot, first identify the least and greatest values in the set of data. The digit in the tens place in each of these values forms the least and greatest extremes of the stems. Write the stems in a column, from least to greatest. Draw a vertical line to the right of the stems. Write the ones digits of the values, in the order in which the values appear, to the right of their stems. Write an explanation of the plot to the left of the stems.

	Stems	Leaves	
	5	1 3 9 8	Least pulse rate is 51.
	6	6 4 7 7 1 6 8	
7|3 represents a pulse rate of 73	7	3 9 1 3 6 3 6	
	8	6 6 1 7 6 2	
	9	0 1 3	Greatest pulse rate is 93.

Analyze the data in a stem-and-leaf plot by looking for the greatest and least values and outliers, clusters, and gaps. **Outliers** are data values that are much greater than or much less than most of the other values. **Clusters** are isolated groups of values. **Gaps** are large spaces between values.

Example 2

This stem-and-leaf plot shows the heights, in inches, of twenty-two students. Write a description of the data. Be sure to note any outliers, clusters, and gaps.

5|1 represents 51 in. tall

Stem	Leaves
4	7 8 9 9
5	1 5 7 7 8 9 9
6	2 2 2 2 3 3 5
7	1 1 2 7

Solution

This set of data indicates that the shortest of the students is 47 in. tall and the tallest is 77 in. tall. The value 77 is an outlier, since it is considerably separated from the other values. There are several clusters of values—those in the high 40s, those in the high 50s, and those in the low 60s. The greatest gaps in values fall between 65 and 71 and between 72 and 77. ◄

TRY THESE

1. Organize these basketball scores into a stem-and-leaf plot.

61	70	36	52	49	77	62
37	55	32	44	82	96	73
55	61	43	38	37	56	59
62	32	46	58	63	72	84

2. This stem-and-leaf plot shows the number of points scored by a high school basketball team during a recent season. Write a description of the data, noting any outliers, clusters, and gaps.

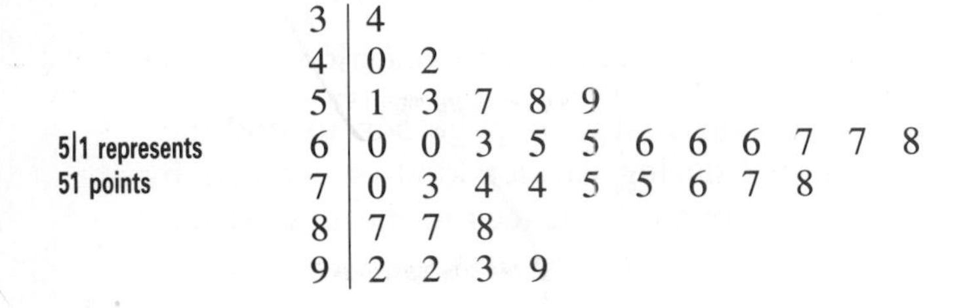

5|1 represents 51 points

Stem	Leaves
3	4
4	0 2
5	1 3 7 8 9
6	0 0 3 5 5 6 6 6 7 7 8
7	0 3 4 4 5 5 6 7 8
8	7 7 8
9	2 2 3 9

CHECK UNDERSTANDING

What are the *extremes* of the stems for these data?

EXERCISES

PRACTICE/ SOLVE PROBLEMS

These data represent the numbers of baskets sunk by basketball players, each taking 50 shots.

27	34	41	35	48	36
12	32	42	50	39	38
42	17	8	42	28	37
3	15	23	6	25	43
9	4	14	41	21	29
11	17	9	26	31	34

1. How many players are represented?
2. Organize the data into a stem-and-leaf plot.
3. How many players sank exactly 17 baskets? 21 baskets? 46 baskets?
4. Write a description of the data. Note the greatest and least numbers of baskets sunk and any outliers, clusters, and gaps.

EXTEND/ SOLVE PROBLEMS

USING DATA Use the Data Index on page 546 to find the statistics for the numbers of miles walked by workers in various jobs during the course of their workdays. Use these data for Exercises 5 and 6.

5. Make a stem-and-leaf plot for the data. (Hint: Think of the vertical line as replacing the decimal points.)
6. Write a paragraph telling which of the jobs might be of interest to you and whether or not the amount of walking involved is important to your decision.

THINK CRITICALLY/ SOLVE PROBLEMS

These data represent the number of voters at forty polling stations.

300	580	390	520	460	290	270	440	200	290
230	320	240	620	350	400	420	600	340	620
670	380	290	290	690	410	300	410	280	530
530	520	350	490	410	500	230	150	290	150

7. Make a stem-and-leaf plot for these data. Use the explanation that 2|3 represents 230 voters.
8. Write two questions based on the data. Then answer them.
9. Ask your questions of others. Answer their questions.
10. ***USING DATA*** Use the Data Index on page 546 to find the number of calories burned during various kind of exercise by a person weighing 110 lb. Round the values to the nearest ten. Then make a stem-and-leaf plot for the data.

Problem Solving Applications:

BACK-TO-BACK STEM-AND-LEAF PLOTS

A **back-to-back stem-and-leaf plot** can be used to organize two sets of data so that they can be compared. The stems form the "backbone" of this kind of plot. This back-to-back stem-and-leaf plot shows age data for the first three U.S. presidents.

Age at Inauguration		Age at Death
	9	0
	8	3
	7	
1	6	7
7 7	5	
	4	

AGES OF UNITED STATES PRESIDENTS AT INAUGURATION AND AT DEATH

Name	Term	Age at Inaug.	Age at Death
1. Washington	1789–1797	57	67
2. J. Adams	1797–1801	61	90
3. Jefferson	1801–1809	57	83
4. Madison	1809–1817	57	85
5. Monroe	1817–1825	58	73
6. J. Q. Adams	1825–1829	57	80
7. Jackson	1829–1837	61	78
8. Van Buren	1837–1841	54	79
9. W. H. Harrison	1841	68	68
10. Tyler	1841–1845	51	71
11. Polk	1845–1849	49	53
12. Taylor	1849–1850	64	65
13. Fillmore	1850–1853	50	74
14. Pierce	1853–1857	48	64
15. Buchanan	1857–1861	65	77
16. Lincoln	1861–1865	52	56
17. A. Johnson	1865–1869	56	66
18. Grant	1869–1877	46	63
19. Hayes	1877–1881	54	70
20. Garfield	1881	49	49
21. Arthur	1881–1885	50	56
22. Cleveland	1885–1889	47	71
23. B. Harrison	1889–1893	55	67
24. Cleveland	1893–1897	55	—
25. McKinley	1897–1901	54	58
26. T. Roosevelt	1901–1909	42	60
27. Taft	1909–1913	51	72
28. Wilson	1913–1921	56	67
29. Harding	1921–1923	55	57
30. Coolidge	1923–1929	51	60
31. Hoover	1929–1933	54	90
32. F. D. Roosevelt	1933–1945	51	**63**
33. Truman	1945–1953	60	**88**
34. Eisenhower	1953–1961	62	**78**
35. Kennedy	1961–1963	43	**46**
36. L. B. Johnson	1963–1969	55	**64**
37. Nixon	1969–1974	56	**81**
38. Ford	1974–1977	61	—
39. Carter	1977–1981	52	—
40. Reagan	1981–1989	69	—
41. Bush	1989–	64	—

1. Copy and complete the back-to-back stem-and-leaf plot to include the data for all forty-one presidents.
2. How many presidents were inaugurated when they were in their fifties?
3. Who was the youngest president to be inaugurated? How old was he? Who was the oldest president to be inaugurated? How old was he?
4. Which president died the youngest? How old was he?
5. Describe any outliers, clusters, and gaps in the data.
6. Research two sets of data that can be compared. You might use, for example, the numbers of games won by the teams in two leagues or the amounts of fat and cholesterol found in various foods. Organize the data into a back-to-back stem-and-leaf plot.
7. Exchange the plot you made for Exercise 6 for a classmate's plot. Write a description of the data in your classmate's plot.

CHECK UNDERSTANDING

Why are there dashes in the Age-at-Death column for the last four presidents?

Why is there a dash in the Age-at-Death column for the twenty-fourth president, Cleveland?

1-3 Pictographs

EXPLORE

This graph shows the results of a survey taken of joggers one day in a big city park.

Use this clue to complete the key:
One hundred twenty-five joggers ran a distance of from 2 to 4 km on the day of the survey.

a. Did more joggers run 0–2 km or over 10 km?
b. The fewest joggers ran how many kilometers?
c. How many joggers ran 4–6 km?
d. How many joggers were surveyed?
e. Write a title for this graph.

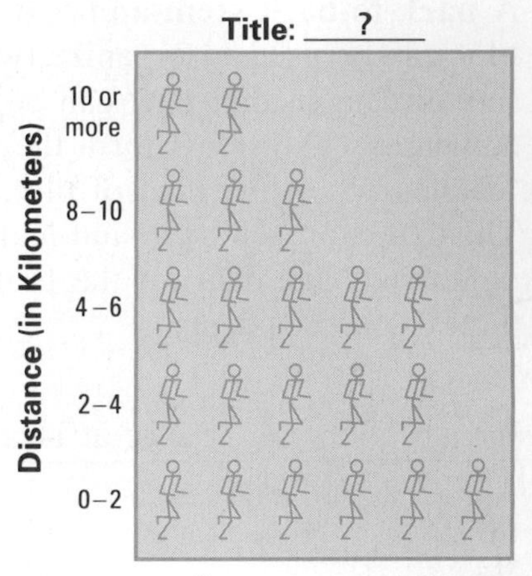

Key: represents ? joggers

SKILLS DEVELOPMENT

A picture graph, or **pictograph**, is a means of displaying data by using graphic symbols. The **key** identifies the number of data items represented by each symbol. The symbols often represent rounded, rather than exact, amounts. To read a pictograph, start by interpreting the key. Then multiply the value of one symbol by the number of symbols in a row to find the value for the row.

Example 1

Use this pictograph.

FISH CONSUMPTION AROUND THE WORLD (in Pounds Eaten, Per Person, Per Year)
U.S.A.
France
Japan
India
Brazil

Key: represents 10 lb of fish

a. In which country is the most fish eaten? About how many pounds of fish does the average person in that country eat in a year?
b. How much fish does the average person in the U.S. eat per year?
c. The average Norwegian eats about 110 lb of fish per year. How many fish symbols would be needed to represent this information on the pictograph?

Solution

a. The average person in Japan eats about 190 lb of fish per year.
b. The average person in the U.S. eats about 35 lb of fish per year.
c. Eleven fish symbols would be needed to represent the amount of fish eaten by the average Norwegian. ◄

To construct a pictograph, choose a symbol for the key, then determine a value for the symbol. Draw the symbols to represent the data. Write a title for the graph.

Example 2

Construct a pictograph for these data.

Number of Students Holding Part-Time Jobs:

Convenience-store worker	25	Newspaper deliverer	10
Supermarket cashier	15	Waiter	30
Short-order cook	25	Recycling-center worker	15
Babysitter	40		

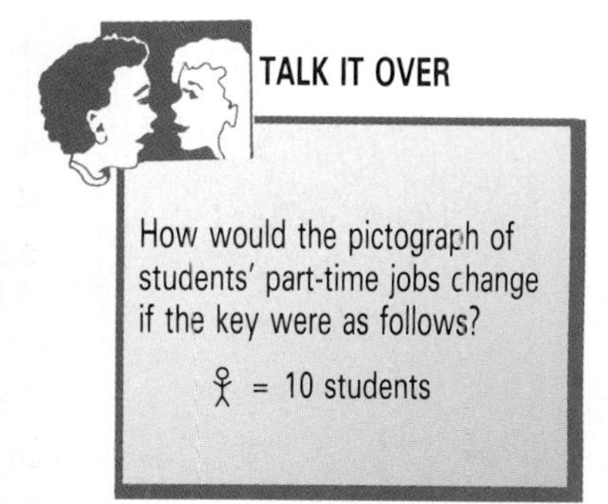

Solution

NUMBER OF STUDENTS HOLDING PART-TIME JOBS

Job	Symbols
Convenience-store worker	5
Supermarket cashier	3
Short-order cook	5
Babysitter	8
Newspaper deliverer	2
Waiter	6
Recycling-center worker	3

Key: (symbol) represents 5 students ◄

TRY THESE

Use this pictograph.

AMOUNT OF SODIUM IN SOME FOODS

Food	Symbols
Pizza (1 slice)	4
Tomato juice (1 cup)	9
Pita bread (1)	2
Potato salad (1 cup)	13
Carrot cake (1 slice)	3½

Key: (symbol) represents 100 mg of sodium

1. Which food contains the most sodium? About how much sodium does it contain?
2. About how much sodium is there in two slices of pizza?
3. There are 1,075 mg of sodium in 1 cup of cream of mushroom soup. How many saltshaker symbols would be needed to represent this information on the pictograph?
4. Construct a pictograph for these data.

   Number of Students in After-School Activities:

Writer's Workshop	13	Softball Team	10	Math Team	18
Soccer Team	11	Crafts Club	8	Runner's Club	30

EXERCISES

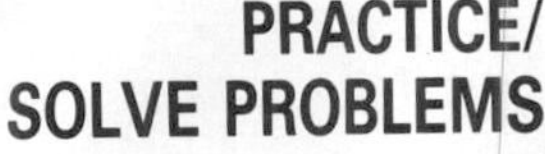

PRACTICE/ SOLVE PROBLEMS

Use this pictograph for Exercises 1–5.

1. Which type of music is most popular?
2. Which type of music is least popular?
3. How many students prefer country and western?
4. How many more students prefer blues than prefer classical music?
5. What kinds of music could be represented by "other"? How many students might prefer each of the "others"?

MIXED REVIEW

Add or subtract.

1. 4,685 + 1,709
2. 17,505 − 8,936
3. 30,710 − 19,666
4. 23,006 + 10,084

Find the fractional part.

5. $\frac{1}{3}$ of 360
6. $\frac{1}{4}$ of 360
7. $\frac{1}{9}$ of 360
8. $\frac{3}{4}$ of 360

Solve.

9. A dump truck contains 1,440 lb of sand that must be moved by wheelbarrow to a construction site. If the wheelbarrow can move 90 lb in one trip from the dump truck to the site, how many trips must be made to move all the sand?

Roberto has scored the most points on the basketball team. Here are his scores for a recent season's game.

Game	1	2	3	4	5	6	7	8
Number of points	12	20	16	12	20	4	16	24

6. Construct a pictograph for the above data. For your key use a basketball symbol to represent 4 points.
7. In which game did Roberto score the greatest number of points?
8. In which game did he score the least number of points?
9. What was Roberto's total number of points in the first half of the season?
10. What was Roberto's total in the second half of the season?

The number of points scored by the hockey players on one team are shown in the chart.

Player	Kelly	Green	Tookey	Currie	Smith	Charron
Points	60	74	45	35	40	48

11. Construct a pictograph to show the numbers of points scored. For your key use a hockey puck, , to represent 10 points and to represent 5 points.
12. How many more points did Green score than Charron?
13. Kelly scored half again as many points as which player?

EXTEND/ SOLVE PROBLEMS

This pictograph shows the number of situps done by the students in one school. Use this pictograph for Exercises 14–19.

NUMBER OF SITUPS
0 to 35 (poor)
36 to 45 (satisfactory)
46 to 60 (good)
61 or more (excellent)

Key: represents 25 students

14. How many students are in the excellent category?

15. There are 75 more students in the satisfactory category than in which other category?

16. How many students do these data represent?

In the second half of the school year, 20 more students enrolled in school. They were all tested. Each did between 36 and 45 situps.

17. How many students are now in the satisfactory category?

18. How many students could now be represented by the pictograph?

19. How would you change the graph to include the new students?

THINK CRITICALLY/ SOLVE PROBLEMS

Here are the greatest numbers of home runs hit by the top American and National League players for three recent years. Use these data to help you make decisions for Exercises 20–22.

American League players: 42, 36, 51
National League players: 39, 47, 40

This graph shows the number of home runs hit by the top players on two high school baseball teams for five recent years.

HOME RUNS HIT BY TOP HIGH SCHOOL PLAYERS
1992
1991
1990
1989
1988

Key: represents _?_ Crain H.S. Home Runs
represents _?_ St. Clair H.S. Home Runs

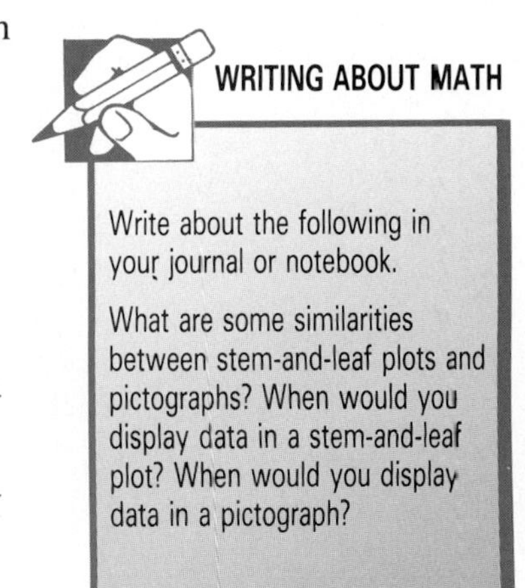

20. Which of these is a reasonable number with which to complete the key for the high school baseball data—2, 3, 5, or 10?

21. The players on which team have the better five-year record? According to the number you determined for the key, by how many home runs is this team ahead of the other team?

22. Write a paragraph explaining your answer to Exercise 20.

1-4 Problem Solving Strategies:

INTRODUCING THE 5-STEP PLAN

You know that people use computers as powerful problem solving tools. Computers can store data and then organize and compute them at breakneck speeds. However, computers still cannot solve problems the way people can—by thinking, or reasoning!

PROBLEM

A computer magazine published a survey for home-computer users. Of those who responded to the survey, 5,410 users said they used only daisy-wheel printers; 11,068 said they used only dot-matrix printers; and 3,709 said they used only laser printers. How many users responded to the survey?

SOLUTION

5-STEP PROBLEM SOLVING PLAN

1. READ . . . Read the entire problem slowly and carefully. Then go back and read the question again.

How many users responded to the survey?

2. PLAN . . . Decide what you can do to solve the problem.

Add to find the number of users that responded.

3. SOLVE . . . Put your plan into action.

$5{,}410 + 11{,}068 + 3{,}709 = 20{,}187$

4. ANSWER . . . State your answer in terms of the original question.

A total of 20,187 users responded to the survey.

5. CHECK . . . Determine whether or not your answer is *reasonable*.

Round each addend to the nearest thousand. Then add.

$5{,}000 + 11{,}000 + 4{,}000 = 20{,}000$

So, the answer is reasonable.

PROBLEMS

Use the 5-step plan to solve these problems.

This stem-and-leaf plot shows the number of students in the computer classes in a school system. Use it for Exercises 1 and 2.

	Stem	Leaves
	0	7 9 9
	1	0 3 5 5 7 8 9
0/7 represents	2	1 4 5 8 9
7 students	3	3 6 7 7 8
	4	0 1 1 2

1. If the classes of 40 or more students were disbanded, with the students redistributed among the other classes, how many computer classes would remain?
2. What is the total student enrollment in the computer classes?

For the cost of an airline ticket, a passenger may check two suitcases. Checking a third suitcase costs an additional $35. Checking the fourth to the sixth suitcases costs $45 each. Checking a seventh suitcase or more costs $55 each.

3. How much money would the airline collect for additional suitcases if nine passengers each checked a third suitcase?
4. How much would a passenger pay for checking a total of five suitcases?
5. If a family of three moving to another state paid $289 each for their airline tickets and checked eight suitcases, what would be their total cost?

6. If a doctor-recommended weight-loss diet calls for consuming a total of 1,800 calories a day, what is a reasonable day's allotment of calories for each of three meals and two snacks?

PROBLEM SOLVING TIP

To determine the numbers of calories for the meals and the snacks in Exercise 6, try *working backward.*

1-5

- READ
- PLAN
- SOLVE
- ANSWER
- CHECK

Problem Solving Skills:

CHOOSE AN APPROPRIATE SCALE

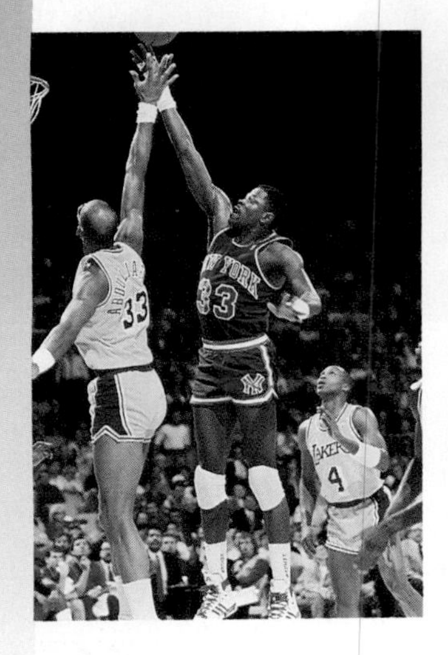

In the lessons that follow you will be constructing various types of graphs. For each type of graph, you will need to determine an appropriate scale. When determining the scale, you will find that it helps to first write the data in a simpler form.

BASKETBALL PLAYERS' TOP SALARIES FOR A RECENT SEASON		
"Hot Rod" Williams	Cleveland	$5,000,000
Patrick Ewing	New York	$4,250,000
Hakeem Olajuwon	Houston	$4,020,000
Chris Mullin	Golden St.	$3,300,000
David Robinson	San Antonio	$3,200,000
Magic Johnson	L.A. Lakers	$3,142,000
Danny Ferry	Cleveland	$3,100,000
Charles Barkley	Philadelphia	$3,000,000
Michael Jordan	Chicago	$2,950,000
Sam Perkins	L.A. Lakers	$2,800,000

PROBLEM

Assume you have to graph the data in this table. What size interval would you use for the scale of the graph?

SOLUTION

The data are given in millions of dollars. Simplify the data. Express the greatest and least salaries in simpler forms.

You can write $5,000,000 as 5.0 million dollars. You can write $2,800,000 as 2.8 million dollars. So, all the data lie between 5.0 and 2.8 million dollars.

Now, determine the size of the intervals for your scale. If you use intervals of 0.2, you would need 25 intervals. If you use intervals of 0.5, you would need 10 intervals. If you use intervals of 1.0, you would need 5 intervals.

You might decide that 9 intervals of 0.6 would best reflect the data.

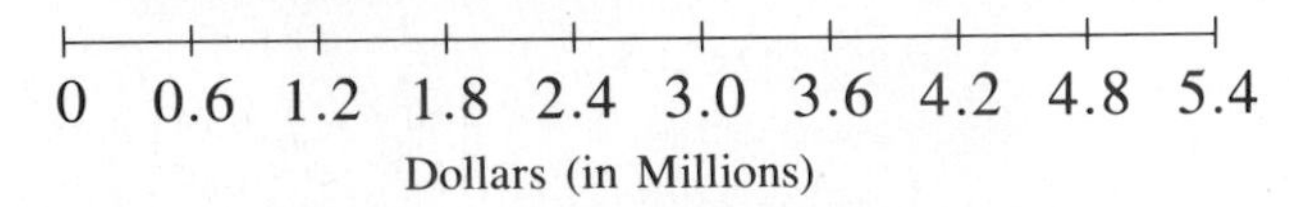

MIXED REVIEW

Round each number to the nearest thousand, nearest hundred, and nearest ten thousand.

1. 85,461
2. 712
3. 1,034,510
4. 151,151

Round each number to the nearest million, nearest ten million, and nearest tenth of a million.

5. 57,306,200
6. 636,489,000
7. 659,201
8. A large box holds twice as many nails as a medium box, which holds twice as many as a small box. If 3,150 nails are used to fill one box of each size, how many nails are in each box?

PROBLEMS

Determine the intervals of the scale you would use to graph each set of data.

1. ***USING DATA*** Use the Data Index on page 546 to find statistics on the shows that have run for the longest periods of time on Broadway.

2. While bananas are often considered to be the best source of potassium, many other foods actually have more.

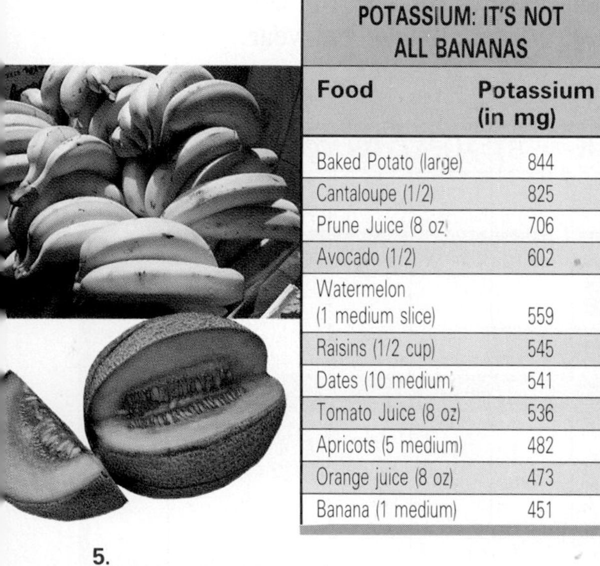

POTASSIUM: IT'S NOT ALL BANANAS	
Food	**Potassium (in mg)**
Baked Potato (large)	844
Cantaloupe (1/2)	825
Prune Juice (8 oz)	706
Avocado (1/2)	602
Watermelon (1 medium slice)	559
Raisins (1/2 cup)	545
Dates (10 medium)	541
Tomato Juice (8 oz)	536
Apricots (5 medium)	482
Orange juice (8 oz)	473
Banana (1 medium)	451

3.

Most Popular Programs at Summer Camp

Program	Camps Offering Program
Swimming	3,426
Art	1,986
Cycling	1,679
Horseback riding	1,573
Canoeing	1,540
Tumbling, gymnastics	1,266
Archery	1,256
Drama	807
Tennis	766
Sailing	733
Hockey	693
Physical fitness	668
Softball	655
Basketball	619
Rowing	554
Backpacking	483
Music	442
Nutrition	434
Hiking	379

4. A recent survey showed the time, to the nearest 15 minutes, spent watching TV each week by an average household in selected locations.

Market	Time (h:min)
Dallas-Fort Worth	56:45
Detroit	56:45
Houston	55:30
Atlanta	54:15
Philadelphia	53:00
Boston	50:30
Chicago	50:30
Sacramento-Stockton	50:30
Hartford-New Haven	49:15
Los Angeles	49:15
Memphis	49:15
New Orleans	49:15
New York	49:15
Seattle-Tacoma	49:15
Washington, D.C.	49:15
Cleveland-Akron	48:00
Denver	48:00
Miami-Ft. Lauderdale	48:00
Minneapolis-St. Paul	46:30
Nashville	46:30
Pittsburgh	46:30
Raleigh-Durham	46:30
Cincinnati	45:15
Oklahoma City	45:15
San Francisco-Oakland-SanJose	45:15
St. Louis	45:15
Portland, Oregon	44:30
Buffalo, N.Y.	44:00
Charlotte	44:00
Indianapolis	44:00
Orlando-Daytona Beach-Melbourne	44:00
Tampa-St. Petersburg-Sarasota	44:00
Baltimore	42:45
Columbus, Ohio	42:45
Kansas City, Mo.	42:45
Greenville-Spartanburg, S.C.-Asheville	42:45
Milwaukee	41:30
Grand Rapids-Kalamazoo-Battle Creek	41:30
Phoenix	40:15
San Diego	37:45

5.

Money Spent on Concessions

Event	Money Spent on Concessions
Amusement park	$5.25
Tractor pull	$5.00
Boxing	$4.75
NFL football game	$4.00
Major-league baseball game	$3.90
Wrestling match	$3.75
Ice-hockey game	$3.00
Jai alai match	$3.00
NBA basketball game	$3.00

6.

Major League Lifetime Records	
Player	**Average**
Ty Cobb	.366
Rogers Hornsby	.358
Joe Jackson	.356
Pete Browning	.347
Ed Delahanty	.346
Willie Keeler	.345
Billy Hamilton	.344
Ted Williams	.344
Tris Speaker	.344
Dan Brouthers	.342
Jesse Burkett	.342
Babe Ruth	.342
Harry Heilmann	.342
Bill Terry	.341
George Sisler	.340
Lou Gehrig	.340

7.

Some of the World's Largest Dams		
Largest[1]	Location	yd^3
New Cornelia	Arizona	274,026,000
Tarbela	Pakistan	186,000,000
Fort Peck	Montana	125,600,000
Oahe	S. Dakota	92,000,000
Mangla	Pakistan	85,870,000
Gardiner	Canada	85,740,000
Oroville	California	78,000,000

[1]All earthfill except Tarbela (earth and rockfill).

1-6 Bar Graphs

EXPLORE/ WORKING TOGETHER

The data in the table show the number of injuries for which emergency-room treatment was needed in one recent year.

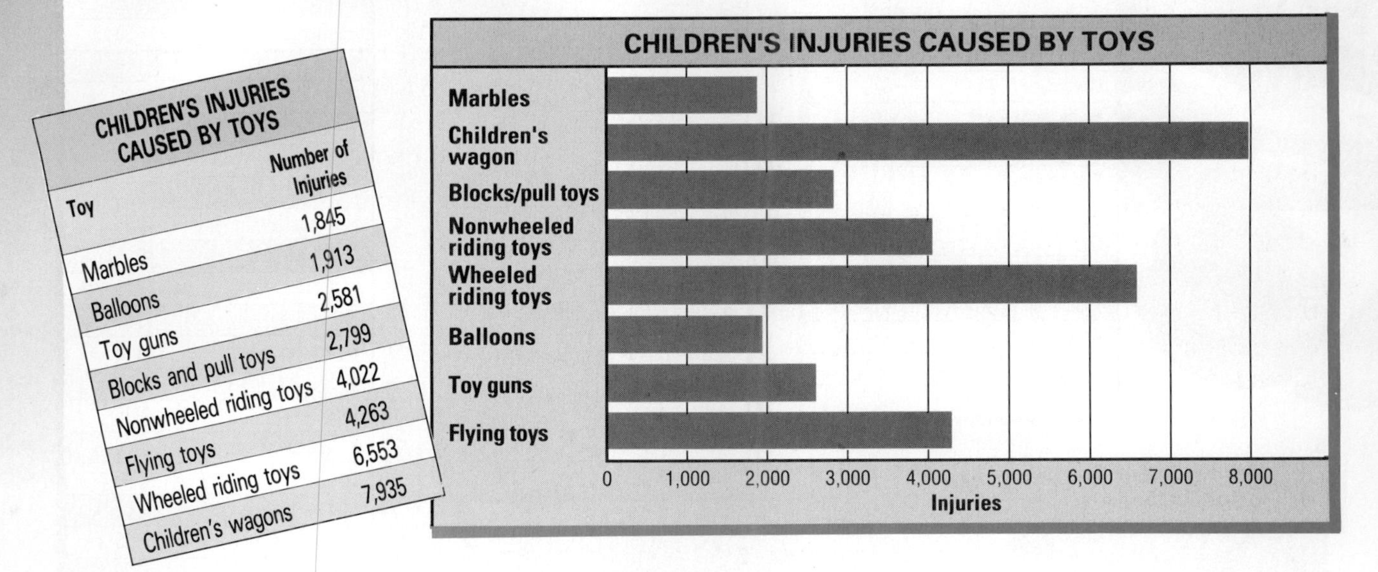

CHILDREN'S INJURIES CAUSED BY TOYS

Toy	Number of Injuries
Marbles	1,845
Balloons	1,913
Toy guns	2,581
Blocks and pull toys	2,799
Nonwheeled riding toys	4,022
Flying toys	4,263
Wheeled riding toys	6,553
Children's wagons	7,935

Use the bar graph of the data in the chart to answer these questions.

a. The scale at the bottom of the graph is marked in intervals of what number?

b. The left edges of all the bars represent what number on the scale?

c. Find the bar for injuries caused by flying toys. Imagine a vertical line drawn from the right edge of this bar downward to meet the scale. Between which two numbers on the scale would the vertical line fall?

d. What does the shortest bar tell you?

e. What does the longest bar tell you?

SKILLS DEVELOPMENT

In a **bar graph**, either horizontal or vertical bars are used to display data. A scale is used to show number values.

To read a horizontal bar graph, look at the right edge of each bar. Match that edge with the number on the scale at the bottom of the graph to find the value for that bar. To read a vertical bar graph, match the top edge of each bar to the scale at the side of the graph to find the value of the bar.

Example 1

Use the bar graph at the right.

a. Which bar represents the longest river? About how long is that river?

b. About how much longer is the Nile than the Yangtze?

c. Which rivers are over 5,400 km long?

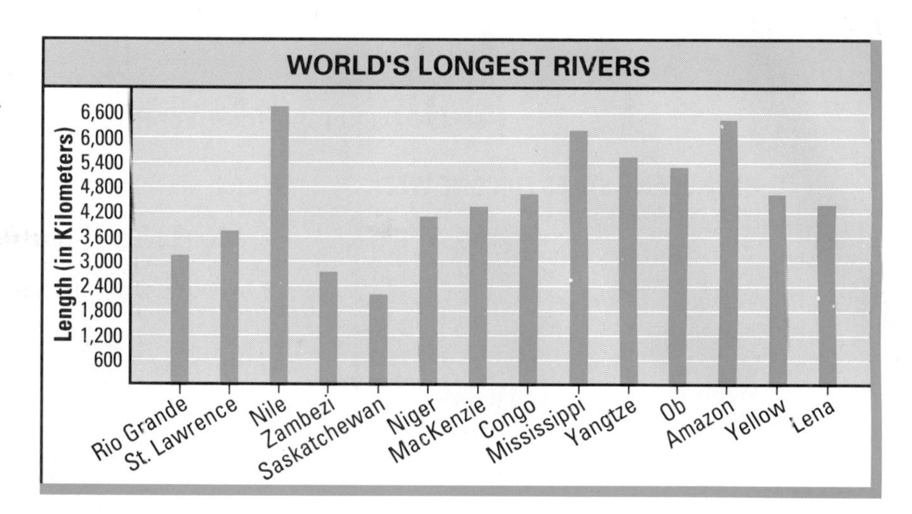

Solution

a. The longest bar represents the Nile River. The top of the bar is very close to the 6,600-km mark on the scale. The Nile is about 6,600 km long.

b. The Nile is about 1,200 km longer than the Yangtze.

c. The Nile, the Mississippi, the Yangtze, and the Amazon are each over 5,400 km long. ◄

When you construct a bar graph, you may have to estimate the lengths of some of the bars.

Example 2

Construct a bar graph for these data.

NATIONAL HOCKEY LEAGUE STANDINGS			
Team	**Points**	**Team**	**Points**
N.Y. Islanders	65	Philadelphia	59
Detroit	44	N.Y. Rangers	53
Montreal	56	Vancouver	26
Toronto	31	Washington	14
Boston	47	Edmonton	14

Solution

To construct a bar graph, choose an appropriate scale, then label it. Draw and label the bars. Write a title for the graph.

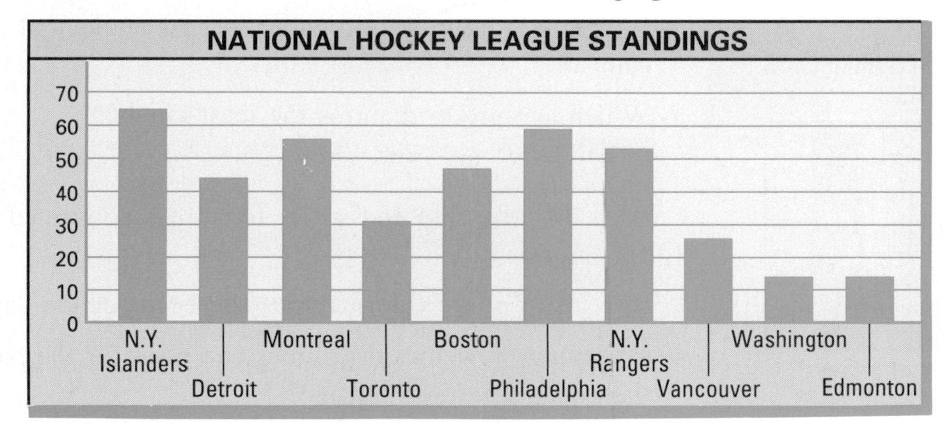

◄

TRY THESE

Use the bar graph for Exercises 1–11.

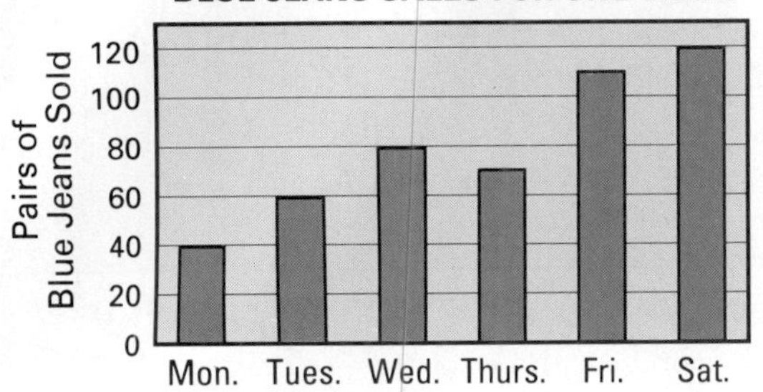

Which day had the lower number of sales?

1. Monday or Thursday
2. Thursday or Friday
3. Wednesday or Saturday
4. Saturday or Tuesday

About how many pairs of blue jeans were sold on each day?

5. Thursday
6. Friday
7. Saturday
8. On which days were more than 60 pairs of blue jeans sold?
9. On which days were more than 80 pairs of blue jeans sold?
10. About how many more pairs of blue jeans were sold on Thursday than on Monday?
11. About how many pairs of blue jeans were sold on Wednesday and Friday together?
12. Construct a bar graph for the video-rental data.

Student Preferences for Video Rentals

Types of Videos	Number of Videos Rented
Westerns	32
Musicals	17
Comedies	41
Action/Adventure	53
Science Fiction	70
Horror	50
Other	20

EXERCISES

PRACTICE/ SOLVE PROBLEMS

Use the bar graph below for Exercises 1–6.

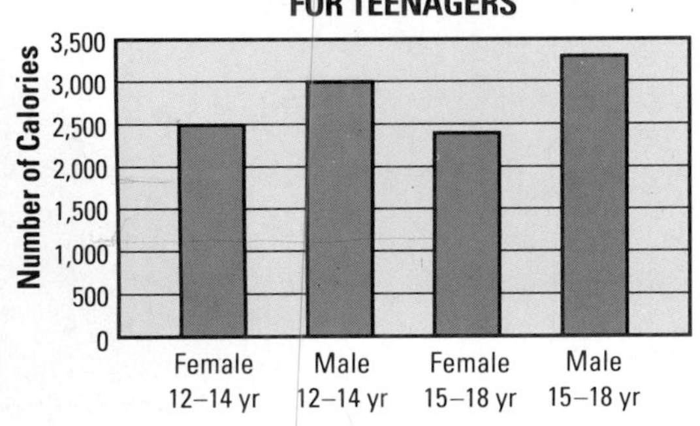

1. Which age group requires the greatest number of calories?
2. Which age group requires the least number of calories?

Carla is 14 years old and so far today has consumed 1,700 calories.

3. How many more calories does she need for the day?
4. How many fewer calories does she need for the day than does a boy of her age?

Michael is 16 years old. He consumed 1,100 calories at breakfast.

5. How many more calories does he need for the day?

6. How many more calories does he need for the day than does a girl of his age?

EXTEND/ SOLVE PROBLEMS

7. ***USING DATA*** Use the Data Index on page 546 to find the heights of the tallest buildings in the world. Construct a bar graph for these data. Show either the numbers of stories or the heights of the buildings, in meters or in feet.

Use the table below.

DEPTHS OF LARGE BODIES OF WATER			
Body of Water	**Average Depth (m)**	**Body of Water**	**Average Depth (m)**
Hudson Bay	90	Atlantic Ocean	3,740
Mediterranean Sea	1,500	Pacific Ocean	4,190
Arctic Ocean	1,330	Black Sea	1,190

8. Construct a horizontal bar graph for the data.

9. The average depth of the Caribbean Sea is 2,575 m. Which of the bodies of water listed here are shallower?

10. The average depth of the Red Sea is 540 m. Which of the bodies of water listed here are deeper than the Red Sea?

11. Create a question about the bar graph. Then answer it.

THINK CRITICALLY/ SOLVE PROBLEMS

A double-bar graph allows you to compare two sets of data.

USING DATA Use the Data Index on page 546 to find the data showing the links between vegetable consumption and lung cancer. For Exercises 12–14, compare the data within each double-bar graph. Then compare the two graphs.

12. About how much greater is the risk of lung cancer in women who consume few tomatoes than in women who consume many tomatoes?

13. About how much lower is the risk of lung cancer in men who consume many cruciferous vegetables than in men who consume few of these vegetables?

14. For whom, men or women, do all vegetables have the greater effect on the risk of lung cancer?

15. ***USING DATA*** Use the Data Index on page 546 to find the sleep-patterns data. About how much sleep does a 1-year-old child get daily?

CONNECTIONS

Beta carotene, found in orange, yellow, red, and dark-green vegetables, has long been known to lower the risk of lung cancer. Scientists have recently identified *lutein* in dark-green vegetables, *lycopene* in tomatoes, and *indoles* and *phenols* in vegetables in the cabbage family, that also protect against lung cancer.

1-7

► READ
► PLAN
► SOLVE
► ANSWER
► CHECK

Problem Solving Skills:

ESTIMATION

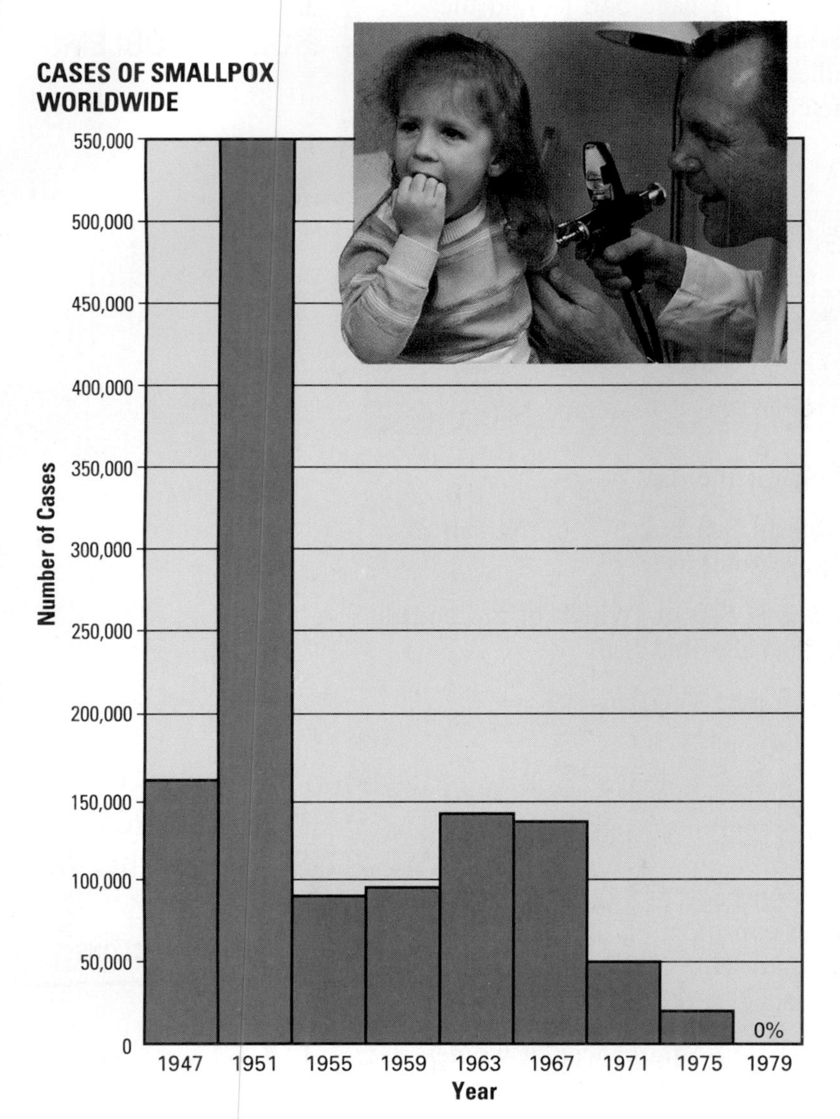

Some graphs display greater numbers in rounded terms, instead of in exact terms. You have to make estimates in order to interpret such graphs.

PROBLEM

Use the graph to estimate about how many times as many smallpox cases there were in 1951 as compared to the number of cases in 1971.

SOLUTION

In 1951, there were about 550,000 cases of smallpox. So, the top of the bar for 1951 is at the 550,000 mark. There were about 50,000 cases in 1971. So, the top of the bar for 1971 is at the 50,000 mark. So there were about 11 times as many smallpox cases in 1951 as there were in 1971.

PROBLEMS

Use the bar graph above for Exercises 1–2.

1. Estimate in which years the number of smallpox cases was about twice the number that occurred in 1971.
2. Estimate in which year the number of smallpox cases was about one-third the number of cases that occurred in 1947.

Use these graphs for Exercises 3–4. Estimate each answer.

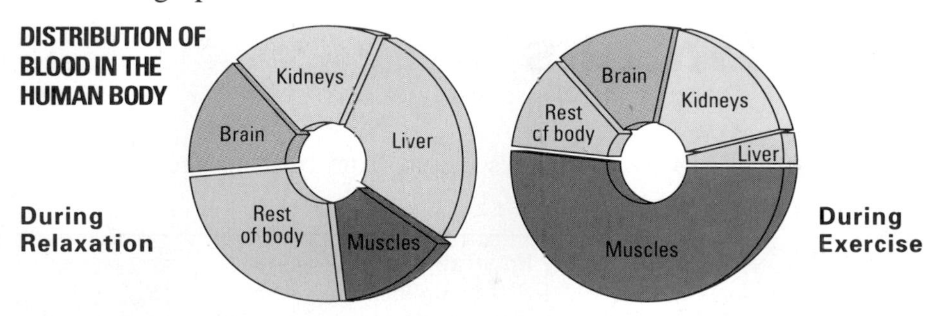

3. How many times as much blood is in the muscles during exercise as during relaxation?
4. How many times as much blood is in the liver during relaxation as during exercise?

Use this bar graph for Exercises 5–6.

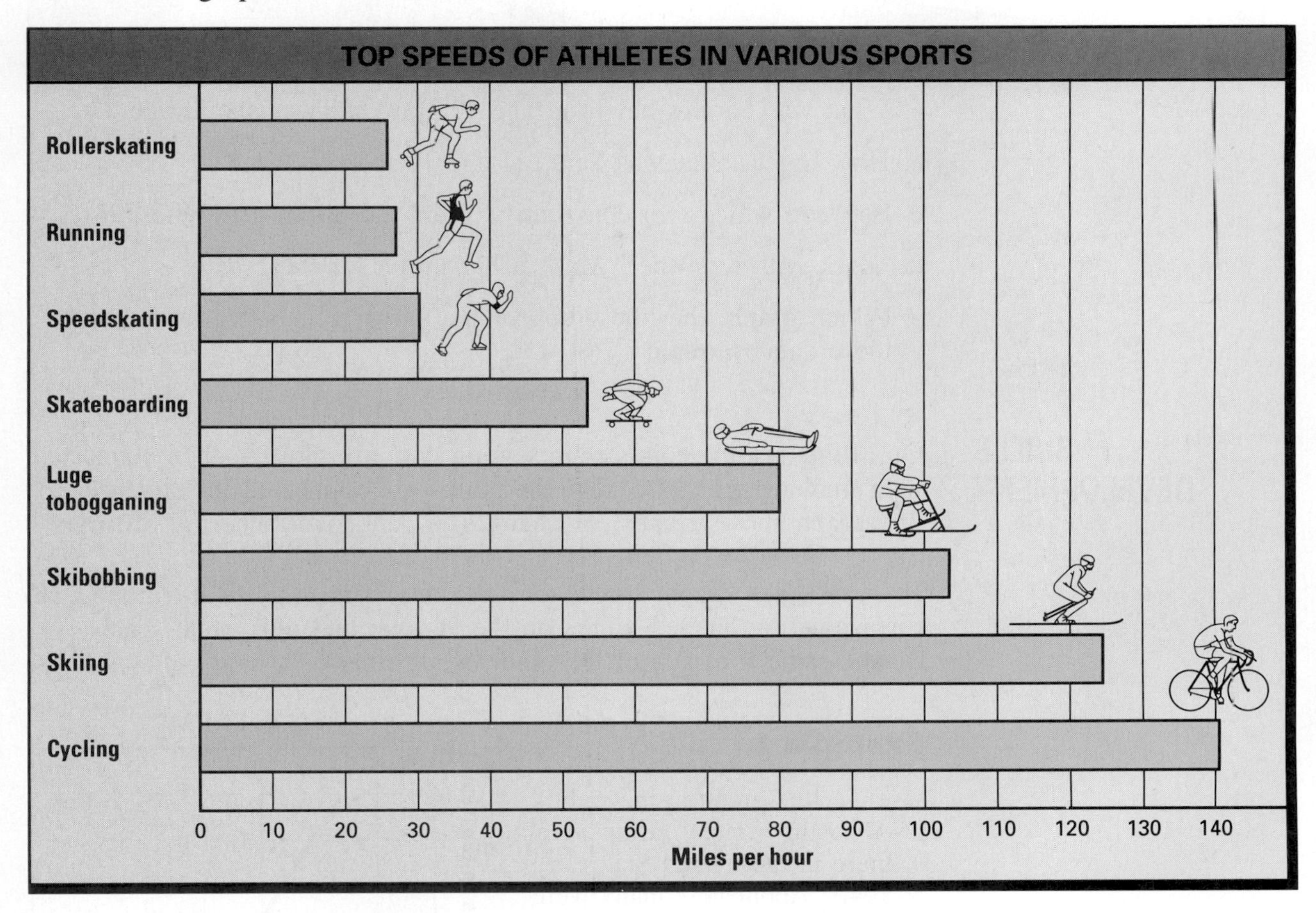

5. The record speed for cycling is about twice as fast as the record speed for which sport?
6. The record speed for roller skating is about one-third the record speed for which sport?

1-8 Line Graphs

EXPLORE/ WORKING TOGETHER

This line graph shows one child's height from birth to age 10.

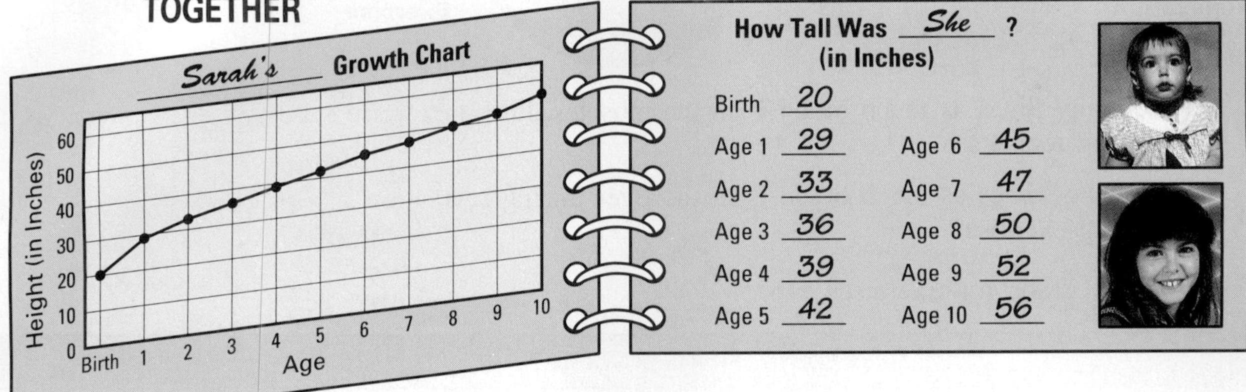

a. What was Sarah's height at age 4? How tall was she at age 8?

b. How much taller was Sarah at age 10 than at age 5?

c. Between which two consecutive years did Sarah grow the most?

d. Work with a partner. Make a bar graph for these data.

e. Which graph, the line graph or the bar graph, better represents these data? Explain.

SKILLS DEVELOPMENT

On a **line graph**, points representing data are plotted, then connected with line segments. Because the points are connected in sequence, a line graph shows trends, or changes in data, over a period of time.

To read a line graph, locate the first data point. Relate it to the corresponding labeled points on the vertical and horizontal scales. Do the same for each of the other data points.

Example 1

Use the line graph at the right.

a. About how many farms were there in the United States in 1940? About how many were there in 1990?

b. About how many fewer farms were there in 1960 than in 1950?

c. Describe the trend of the data as *increasing* or *decreasing*.

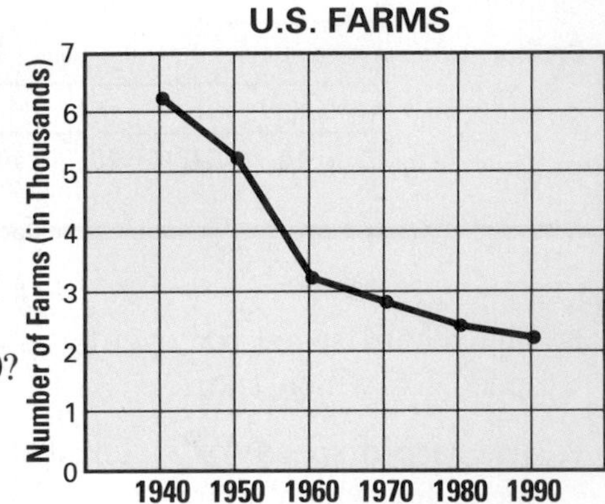

Solution

a. In 1940, there were about 6,000 farms. In 1990, there were about 2,000 farms.
b. There were about 2,000 fewer farms in 1960 than in 1950.
c. The trend is decreasing. ◄

Example 2

These data show the prices paid to farmers for a bushel of corn. Construct a line graph for the data.

PRICE OF CORN (per bushel)							
Year	1930	1940	1950	1960	1970	1980	1990
Price	$0.60	$0.62	$1.52	$13.00	$1.33	$3.11	$2.25

Solution

To construct a line graph, draw and label the horizontal and vertical axes. Locate and mark the data points. Draw straight lines to connect the points. Write a title for the graph.

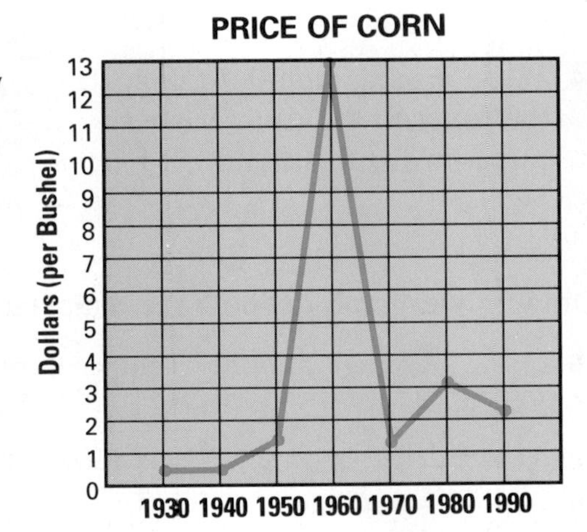

◄

CHECK UNDERSTANDING

Why do you think the price of corn in 1960 was so much higher than in each of the other years?

TRY THESE

Use the line graph below for Exercises 1–3.

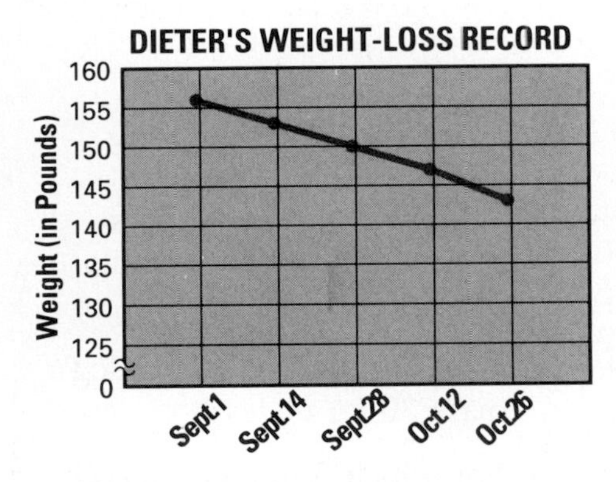

1. About how much did the dieter weigh on Sept. 1? On Oct. 26?
2. About how many pounds were lost over the eight-week period?
3. Describe the trend of the data as *increasing* or *decreasing*.

4. These data show the prices paid by consumers for a pound of fish. Construct a line graph for the data.

PRICE OF FISH (per pound)							
Year	1960	1965	1970	1975	1980	1985	1990
Price	$0.85	$0.95	$1.29	$1.89	$3.59	$4.99	$6.99

JUST FOR FUN

What is the greatest amount of money you can have in coins and still not have change for a dollar?

EXERCISES

PRACTICE/ SOLVE PROBLEMS

The line graph shows the changes in the price of the best-selling hand-held calculator from 1965 to 1990.

1. What was the price of the calculator in 1965?
2. Between which two years did the price of a calculator drop the most?
3. By which year was the price of the calculator the lowest?
4. Write a paragraph describing the trend in the data. Suggest a reason for this trend.

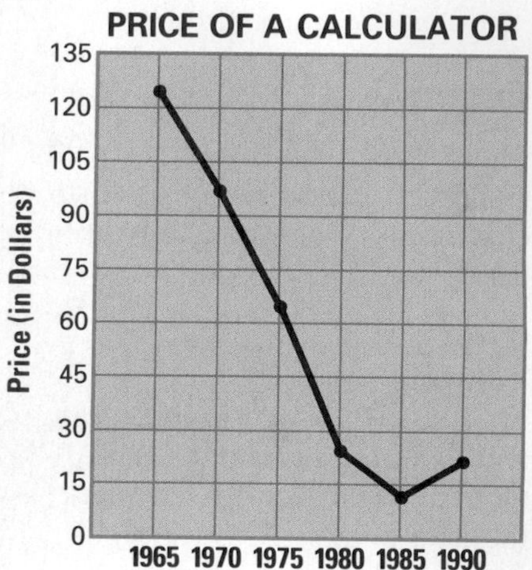

Draw a line graph to show the temperature data for each city.

5. **Average Monthly Temperature in Seattle, Washington**

Month	J	F	M	A	M	J	J	A	S	O	N	D
Temp. (°C)	4	4	6	9	13	15	17	17	14	10	7	4

6. **Average Monthly Temperature in Miami, Florida**

Month	J	F	M	A	M	J	J	A	S	O	N	D
Temp. (°C)	22	23	24.5	26.5	27.5	28	29.5	30.5	29	27.5	23.5	22

EXTEND/ SOLVE PROBLEMS

Use this line graph for Exercises 7–10.

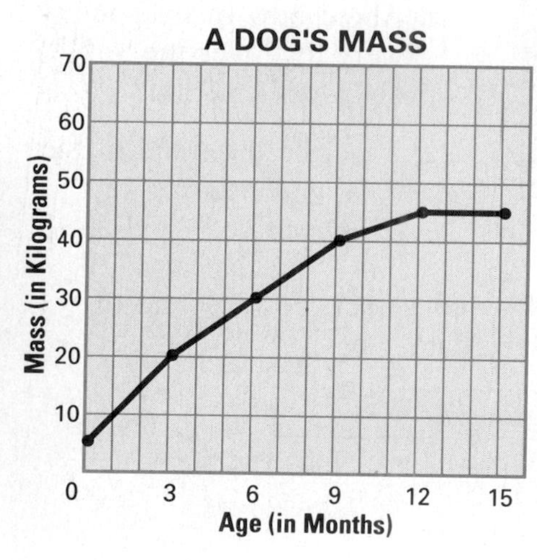

7. What was the dog's mass at 6 months? What was its mass at 12 months?
8. Until what age was the dog's mass 30 kg or less?
9. For which 3-month interval did the dog's mass increase the most? By how much did it increase?
10. Predict the dog's mass at 18 months. Explain your prediction.

11. Beginning at age 8, Dana kept track of how much she weighed on each birthday. Construct a line graph for her weight data.

Dana's Weight

Age	8	9	10	11	12	13	14
Weight (lb)	56	62	69	77	86	98	107

12. By how much did Dana's weight increase between her eighth and fourteenth birthdays?

13. Between which two birthdays did Dana's weight increase the most?

14. Can you predict what Dana's weight will be on her fifteenth birthday? Can you predict her weight on her eigtheenth birthday? Explain your answers.

THINK CRITICALLY/ SOLVE PROBLEMS

Use this multiple-line graph.

15. If you weighed 120 lb, about how many more calories would you use swimming for 10 min than walking for 10 min?

16. About how many calories would a person weighing 140 lb use by walking for 10 min and then playing tennis for 10 min?

17. Of what use could these data be to a diet-conscious person?

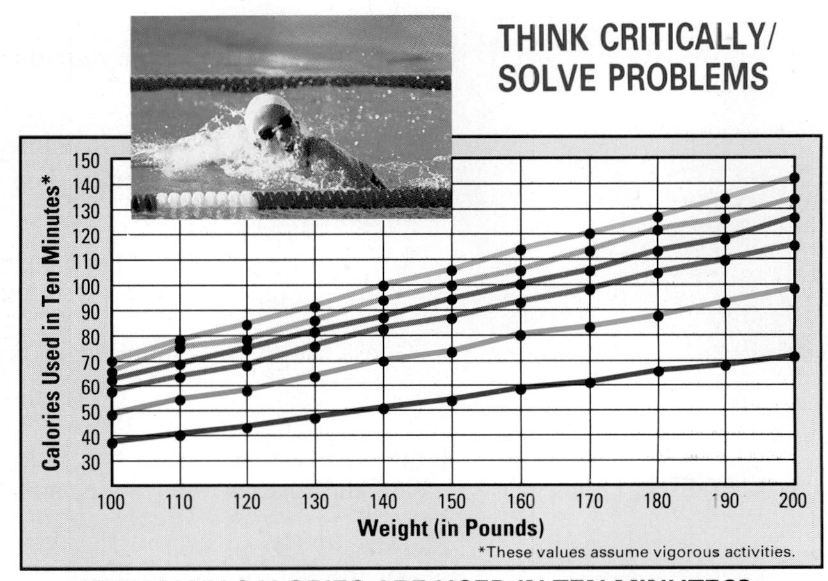

USING DATA Use the Data File on page 546 to find life-expectancy information.

18. What was the projected life expectancy of a baby boy born in 1960?

19. What is the approximate projected life expectancy of a baby girl born today?

20. Predict what the life expectancies will be for boys and for girls born in the year 2010. Give reasons for your predictions.

1-9 Mean, Median, Mode, Range

EXPLORE/ WORKING TOGETHER

What do you think is the average number of books that a high school student carries to school each day?

Count the number of books you brought with you today. (Remember to include this math book in your count.) Record the number of books for each student in the class.

a. Is there one number in your data that occurs more frequently than the others?

b. If you arrange the data in numerical order, what is the middle number? (If the number of data is even, there are two middle numbers.)

c. What is the difference between the greatest and the least numbers of books per student?

d. Guess the average number of books for the students in your class.

SKILLS DEVELOPMENT

Four statistical measures that help you describe a set of data are the mean, the median, the mode, and the range.

- The **mean**, or arithmetic average, is found by dividing the sum of the data by the number of data.
- The **median** is the middle number when the data are arranged in numerical order. (If the number of data is even, the median is the average of the *two* middle numbers.)
- The **mode** is the number that occurs most frequently in the set of data. Some sets of data have no mode; some have more than one mode.
- The **range** is the difference between the greatest and the least values in a set of data.

The mean, median, mode, and range enable you to analyze data and to determine how the data are grouped. Because the mean, median, and mode help you to locate the center of a set of data, these terms are sometimes called **measures of central tendency**.

When you need to analyze a set of data, it usually helps to begin by arranging the data in order, from least to greatest.

COMPUTER

Calculating the mean, especially for a large set of data, can be tedious. In this program when you enter the number of data items and the data, the computer will calculate the sum and the mean for you.

```
10 INPUT "HOW MANY
   DATA ITEMS? ";N
20 LET S = 0
30 FOR K = 1 TO N
40 PRINT : INPUT "DATA "; D
50 LET S = S + D
60 NEXT K
70 LET M = S / N
80 PRINT : PRINT "SUM = ";
    S; "MEAN = ";M
90 PRINT : INPUT "RUN
    AGAIN? Y OR N";X$:
    PRINT : PRINT
100 IF X$ = "Y" GOTO 10
```

Example

The seven students at one table in the cafeteria had these numbers of books with them: 4 3 6 5 5 7 5

a. Find the mean.
b. Find the median.
c. Find the mode.
d. Find the range.

Solution

a. The mean is 5 because the sum of the data is 35 and 35 divided by 7 is 5.
b. The median is 5 because when the data are arranged in order—3, 4, 5, 5, 5, 6, 7—the middle number is 5.
c. The mode is 5 because it is the number that occurs most frequently.
d. The range is 4, because the difference between 7 and 3 is 4. ◄

TRY THESE

Sports Shoes Unlimited had 11 pairs of men's running shoes on sale in these sizes: 10 7 7 9 10 8 11 12 10 8 10

1. Find the mean to the nearest tenth.
2. Find the median.
3. Find the mode.
4. Find the range.

EXERCISES

PRACTICE/
SOLVE PROBLEMS

Here are one cyclist's competition times in seconds for eight tries in a season: 9.1 8.7 9.2 9.0 8.7 8.9 15.0 9.2

1. Find the mean, median, and mode.

2. Which measure of central tendency is the best indication of the cyclist's ability?

Tell which measure of central tendency would best represent each of the following.

3. The average mass of an adult polar bear.

4. The most requested song at a radio station.

5. The usual number of goals scored by a hockey team in one game.

6. The average height of a female student in your grade.

The table shows the salaries for the executives of a large company.

Position	Salary
President	$125,000
Vice President	$ 86,000
Plant Manager	$ 71,500
Accounting Manager	$ 53,000
Personnel Manager	$ 41,250
Research Manager	$ 40,975

7. Find the mean, median, and mode of the salaries.
8. Which measure of central tendency best describes the average executive salary?

During a gymnastics competition, the judges awarded these scores:

7.0 7.0 9.6 9.4 10.0 8.8 9.2 9.8 9.0 8.5

9. Find the mean, median, and mode of the scores.
10. Which of the measures of central tendency do you think best describes the data?

EXTEND/ SOLVE PROBLEMS

The stores at a mini-mall recorded the number of customers they had each day for a week.

Find each of the following for Jeans City:

	Mon.	Tues.	Wed.	Thurs.	Fri.	Sat.
Keon's Electronics	206	184	212	253	267	184
Computer Store	212	241	208	279	296	197
Jeans City	198	147	164	196	281	162
Record Shop	146	162	180	234	294	184
A and M Dept. Store	553	607	692	756	914	833

11. the mean number of customers per day
12. the median number of customers per day
13. How many customers should Jeans City expect in one 30-day month?
14. On which three days of the week might Jeans City have to hire extra part-time help?
15. Find the mean number of customers for the five stores on Thursday.
16. Which store would you go to if you were looking for a part-time job on Thursdays?
17. Find the total number of customers in all five stores for the entire week.
18. Find the mean number of customers for the week for all the stores.

19. Create a problem involving measures of central tendency based on the information in the table. Then, write a solution for your problem.

The number of minutes that a sample of customers spent in a record store is shown in the table at the right. Use the table for Exercises 20–21.

Person	Time (in min)
Ravi	80
Agnes	45
Rosalie	75
Liz	30
Cathy	50
Ian	80
George	25
Phil	40

20. Find the mean, median, and mode for these data.

21. How would your answers to Exercise 20 be affected if each person had spent 5 minutes more in the store?

MATH: WHO, WHERE, WHEN

As a child, the German mathematician Karl Friedrich Gauss (1777–1855) astounded his teacher with his almost instantaneous response to a challenging problem.

The teacher had asked the class to find the sum of the numbers from 1 to 100. Karl found the sum mentally. How did he do this? What was Karl's trick?

Eight crates were loaded onto a delivery truck. The mass of each crate was recorded in kilograms as follows:

40.2 50.0 39.8 46.1 34.8 39.8 40.2 44.7

22. Find the mean, medain, and mode for the data. Which measure best represents the data?

23. If four larger crates with masses 80.4 kg, 91.6 kg, 86.2 kg, and 90.5 kg were added to the load in the truck, which measures of central tendency would change for these loads?

24. Which measure best represents the data after the four crates were loaded?

25. Lorimer High School's wrestling team has six wrestlers with a mean weight of 153 lb. If the weights of five of the wrestlers are 158 lb, 143 lb, 140 lb, 162 lb, and 156 lb, what is the weight of the sixth wrestler?

THINK CRITICALLY/ SOLVE PROBLEMS

26. Find the mean of the first one hundred counting numbers.

27. Calculate the mean and median of any number of consecutive whole numbers. What conclusion can you draw?

28. In 1990, the minimum wage was raised from \$3.80 to \$4.25 per hour. How do you think this affected the median income?

Create sample data to illustrate your answers to Exercises 29–31.

29. If the median of a set of data is 15, must one of the data be 15?

30. If the mean of a set of data is 15, must one of the data be 15?

31. If the mode of a set of data is 15, must one of the data be 15?

CHAPTER 1 • REVIEW

	a. cluster
	b. mode
	c. convenience
	d. systematic
	e. data
	f. random
	g. median
	h. mean

1. If each member of a population being polled has an equal chance of being selected, the sample is ___?___ .
2. Information from which decisions can be made is called ___?___ .
3. A random sample chosen from a particular part of the population is a ___?___ sample.
4. The number that occurs most often in a set of data is the ___?___ .
5. If a population has been ordered and the members are sampled according to a pattern, the sample is ___?___ .
6. A ___?___ sample is data obtained from a readily available population.
7. Adding the data, then dividing the sum by the number of data, gives the ___?___ of the data.
8. When data are arranged in numerical order, the middle number is the ___?___ .

SECTION 1–1 COLLECTING DATA (pages 4–7)

- **Data,** or information from which decisions can be made, can be collected by sampling a population. Some ways of sampling are *random, convenience, cluster,* and *systematic.*
- Data may be organized in a *frequency table* or a *line plot.*

A population of homeowners is to be surveyed to determine the benefits of various kinds of home-heating systems. Name the kind of sampling that is represented.

9. All the homeowners on various streets in four neighborhoods are surveyed.
10. A telephone survey is conducted of names chosen from the phone book.
11. Which of the samples outlined in Exercises 9 and 10 would be the more valuable? Explain.

SECTION 1–2 STEM-AND-LEAF PLOTS (pages 8–11)

- Numerical data may be organized in a **stem-and-leaf plot.**
- In a stem-and-leaf plot for two-digit numbers, the digits in the tens places make up the *stems.* The digits in the ones places make up the *leaves.*

12. These data represent one class's grades on an English exam. Make a stem-and-leaf plot for the data.

53 68 82 61 91 69 72 67 72 67
66 55 79 87 67 90 84 62 98 67
55 81 75 79 55 79 82 67

SECTIONS 1–3, 1–6, 1–8 GRAPHS (pages 12–15, 20–23, 26–29)

- **Pictographs** and **bar graphs** are used to compare data.
- **Line graphs** are used to show trends, or changes over time.

13. Decide which kind of graph—pictograph, bar graph, or line graph—would best represent the exam-grade data in Exercise 12. Explain.

SECTIONS 1–4, 1–5, 1–7 PROBLEM SOLVING (pages 16–19, 24–25)

- To graph a set of data involving greater numbers, it helps to first simplify the numbers in order to determine the interval for the scale.
- Some graphs present data in rounded, instead of in exact, terms. To read these graphs, you have to make estimates.

14. What intervals would you use on a graph of these annual salaries for five supermarket employees?

$23,900 $24,000 $19,570 $22,550 $21,050

About how many books are there in each library system?

15. Chicago

16. Dallas

17. Detroit

18. San Francisco

NUMBER OF BOOKS IN LIBRARY SYSTEM	
Chicago	
Dallas	
Detroit	
San Francisco	

Key: = 1 million books

SECTION 1–9 MEAN, MEDIAN, MODE, AND RANGE (pages 30–33)

- The **mean** of a set of data is found by dividing the sum of the data by the number of data.
- The **median** is the middle number in a set of data. If there are two middle numbers, the mean of the middle numbers is the median.
- The **mode** is the most frequently occurring number in a set of data. Some sets of data have no mode. Some have more than one mode.
- The **range** is the difference between the greatest and least values in a set of data.

USING DATA For Exercises 19–22, use the pulse-rate data on page 8.

19. What is the mean?

20. What is the median?

21. What is the mode?

22. What is the range?

23. ***USING DATA*** Use the data on page 2 to find out how much more money was spent in a year on bowling balls than on baseballs.

CHAPTER 1 • TEST

To test the quality of a shipment of light bulbs, a sample of 500 was chosen randomly from different lots, then tested. In the sample, 14 bulbs were defective.

1. What kind of sampling does this represent?
2. What is an advantage of this kind of sampling?

The pictograph below shows how students travel to Laurier High School.

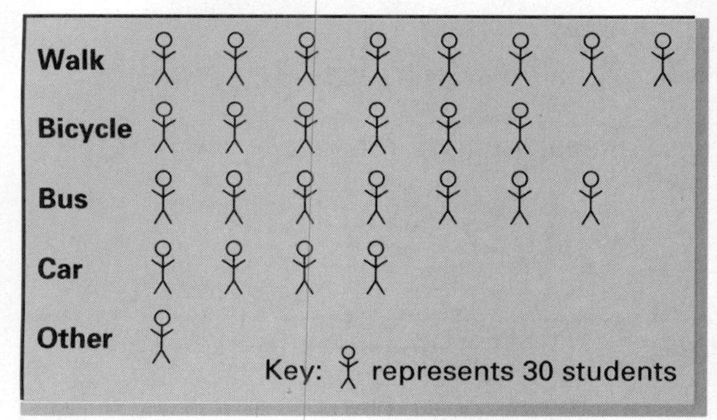

3. How many students were surveyed?
4. Which method of travel is used by the greatest number of students?
5. How many students walk to school?

Linda took a survey of people's ages and recorded the data in this stem-and-leaf plot.

4|1 represents an age of 41

Stem	Leaves
0	7 9 9
1	1 4 5 5 6 6 6 6 8 9
2	2 2 3 4 5 5 6 7 7 8
3	1 1 1 2 3 3 5 8
4	1 2 2 5 5 6 7 9 9
5	3 5 6 6 7
6	0 2 4 6 6 6
7	4 5 8
8	6

How many people were in each age group?

6. 0–9 7. 10–19 8. 30–39

How many people were there of each age?

9. 66 10. 15 11. 57

The bar graph gives the number of tapes sold in one store in a week. On which day were more tapes sold?

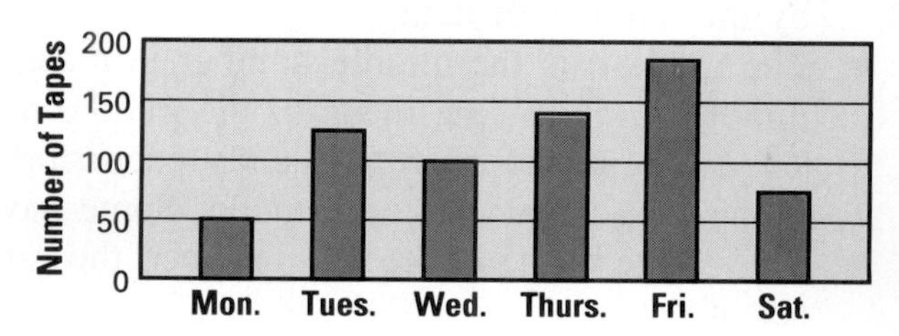

12. Tuesday or Wednesday 13. Monday or Thursday 14. Friday or Saturday

The number of repairs needed for 20 TVs during the first five years were as follows:

3	2	0	0	4	5	7	3	2	1
1	3	6	4	8	3	10	4	0	2

15. Find the mean, median, mode, and range of the data.

The following data show the numbers of pairs of blue jeans sold during one week by various clothing stores.

6|0 represents 60 pairs of blue jeans

Stem	Leaves
0	8 9
1	5 6 8 8
2	2 2 4 5 5 5
3	4 4 6 6 9 9 9
4	1 1 2 3 3 5 5 6 7 7
5	2 3 3 4
6	0 0 1
7	1

How many stores sold the following numbers of pairs of jeans?

1. 43 **2.** 25 **3.** 67

4. How many stores sold 46 or fewer pairs of jeans during the week?

5. How many stores sold more than 50 pairs of jeans during the week?

Students were asked the following question: "How far is it from your home to your school?"

DISTANCE FROM HOME TO SCHOOL (mi)						
1.7	4.3	5.6	0.5	2.5	3.1	3.7
1.9	5.0	6.8	3.4	3.7	2.1	4.3
6.2	0.8	2.4	3.7	2.5	0.9	1.9
5.6	7.4	4.3	3.6	13.6	8.1	5.0

6. Make a stem-and-leaf graph for these data.

7. How many students were interviewed?

8. How many students travel 10 mi or more?

9. How many students travel less than 2.6 mi?

10. What is the average number of miles traveled from home to school?

11. Make a bar graph to show the data.

Lake	Depth (in feet)
Erie	210
Huron	751
Michigan	922
Ontario	777
St. Clair	33
Superior	1,332

Use the line graph for Exercises 12–15.

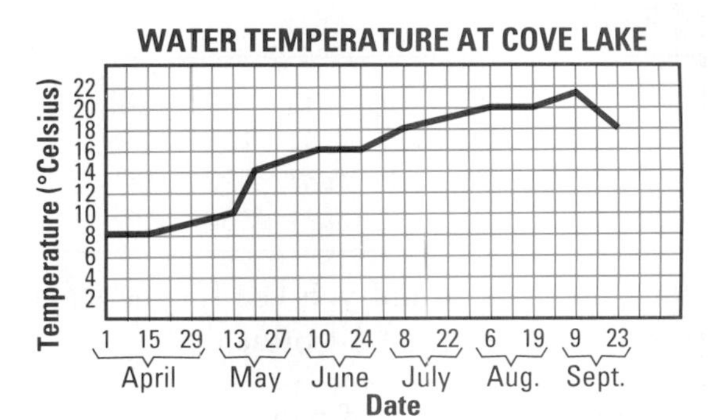

12. Approximately when was the water the warmest? What was the temperature?

13. At what time intervals was the lake temperature measured?

14. During which two intervals did the temperature increase the most?

15. Estimate the temperature of the lake water in October.

The following numbers of boats going through a strait were recorded on ten days.

36 41 58 45 36 39 52 40 43 47

16. Find the mean, median, and mode for the data.

17. Which measure of central tendency in Exercise 16 best represents the data?

18. If on the last day the number was 94 instead of 47, which measure of central tendency would change? Which measure would then best represent the data?

1. A radio newscaster asked every fifth caller to answer the question "Do you think the President is doing a good job?" Which term describes the sample population?

A. random B. clustered
C. systematic D. convenience

2. These data represent the number of hamburgers sold each day for one month in a school cafeteria.

98	86	112	94	133	89	138	111
102	124	130	126	94	105	93	107
114	126	83	104	137	122	100	
135	132	101	116	121	136	108	

What numbers would you use for the stems of a stem-and-leaf plot of this data?

A. 6–4 B. 1–9
C. 8–13 D. 4–14

3. Use this pictograph. What are the *most common* and the *least common* shoe sizes in the group sampled?

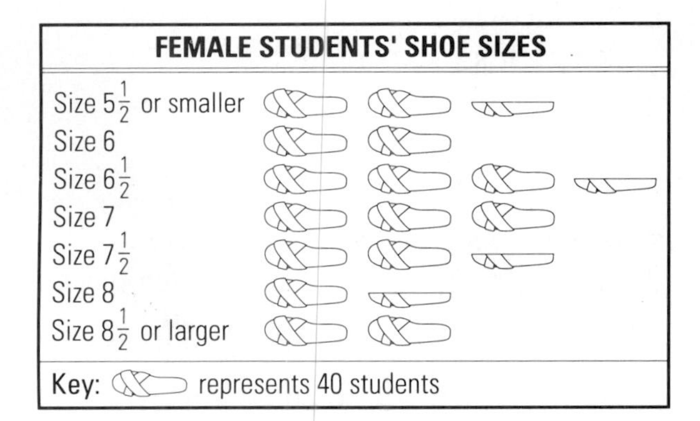

A. $6\frac{1}{2}$ and 8 B. 7 and $8\frac{1}{2}$
C. $5\frac{1}{2}$ and $8\frac{1}{2}$ D. 7 and 8

The bar graph shows the number of animals in some zoos, according to species. Use the graph for Exercises 4–7.

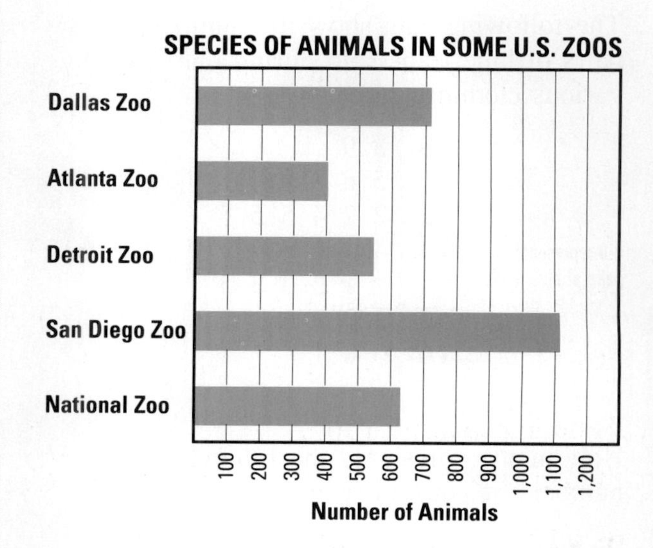

4. The zoo with the greatest number of species is

A. San Diego B. Detroit
C. Dallas D. Atlanta

5. Which zoo has about two times as many species as Detroit?

A. Atlanta B. Detroit
C. San Diego D. Dallas

6. The zoo with about 700 species of animals is

A. Atlanta B. Detroit
C. San Diego D. Dallas

7. The zoo with the least number of species is

A. Atlanta B. National
C. San Diego D. Detroit

8. The weekly wages of 10 students working part time are as follows:

\$35	\$40	\$34	\$75	\$36
\$39	\$41	\$38	\$42	\$37

Which of the following best represents the typical wage?

A. mean B. median
C. mode D. range

CHAPTER 2 SKILLS PREVIEW

Find each answer. Estimate and compare to see if your answer is reasonable.

1. $3{,}876 - 1{,}624$
2. 68×49
3. $98 \times 4{,}152$
4. $7{,}332 \div 78$
5. $6.38 + 11.3 + 0.367$
6. $19.3 - 0.782$
7. 0.75×0.27
8. 8×8.6
9. $6{,}930 \div 35$

Evaluate each expression. Let $a = 2$ and $b = 8$.

10. $a + 15$
11. $b \div 4$
12. $b \div 4a$
13. $12a \div b$

Write in exponential form.

14. $4 \times 4 \times 4$
15. $9 \times 9 \times 9 \times 9 \times 9$
16. $7 \times 7 \times 7 \times 7$

Write in standard form.

17. 8^3
18. 2^7
19. 3^3
20. 4^2

Write in scientific notation.

21. 5,483
22. 228
23. 549,761
24. 1,475,265

Write each answer in exponential form.

25. $7^4 \times 7^3$
26. $9^{10} \div 9^5$
27. $x^9 \div x^2$
28. $y^3 \times y^7$

Simplify.

29. $5 \times 6 - 27 \div 3$
30. $2 \times (8 - 4) + 3$
31. $(5 + 3)^2 \times 2$

Complete.

32. $7(■ - 3) = (7 \times 2) - (7 \times 3)$
33. $■(4 + 8) = (3 \times 4) + (3 \times 8)$
34. $9.6 + (■ + 2.4) = (9.6 + 5) + 2.4$
35. $■ \times 2{,}158 = 2{,}158$
36. $4.89 \times ■ = 0$
37. $2.1 \times 3.92 = 3.92 \times ■$

Simplify.

38. $3(4 - a)$
39. $2(b + 3)$
40. $6(8 - x)$
41. $5(n + 9)$

Solve.

42. Keisha put $15 in her savings account in January, $22 in February, $29 in March, and $36 in April. Following this pattern, find how much money she put into the account in October.

CHAPTER 2 EXPLORING WHOLE NUMBERS AND DECIMALS

THEME Money Around the World

Numbers are an important part of everyday life. We use numbers daily as we check the time, run a specific distance, or pay for lunch. In many countries, as in the United States, the monetary system is a decimal system, in which the basic monetary unit is equivalent to 100 smaller units. For example, in the United States the dollar is the basic monetary unit, and one dollar is equivalent to 100 cents. In this chapter, you will use whole numbers and decimals to solve problems involving money. You will also work with numbers expressed in exponential form.

Most countries of the world issue their own currency in coins and paper notes. The table shows the values of the currencies of five countries as compared with the currency of the United States. The column headed "Late Wed" shows the exchange rate for one Wednesday in a recent year. Note that the table gives the equivalent U.S. dollar values for one British pound and for one Canadian dollar, but gives the value of one U.S. dollar in Swiss francs, in Japanese yen, and in German marks.

CURRENCY	Late Wed	Day's High	Day's Low	12-Month High	12-Month Low
British pound (in U.S. dollars)	1.6215	1.6255	1.6015	2.0040	1.6010
Canadian dollar (in U.S. dollars)	0.8707	0.8720	0.8697	0.8859	0.8504
Swiss franc (per U.S. dollar)	1.5715	1.5690	1.5955	1.2325	1.5903
Japanese yen (per U.S. dollar)	138.62	138.10	138.75	124.33	150.85
German mark (per U.S. dollar)	1.8135	1.8080	1.8370	1.4475	1.8356

DECISION MAKING

Using Data

Use the information in the table above to answer each question. Use the exchange rates in the "Late Wed" column for Exercises 1 and 2.

1. What U.S. dollar amount was equivalent to one Canadian dollar?
2. Would you have received more German marks or more Swiss francs for one U.S. dollar? Explain.
3. What was the difference in the exchange rates for the 12-month high and the 12-month low for the British pound?
4. Why was the 12-month high a lesser amount than the 12-month low for the Swiss franc, the Japanese yen, and the German mark?

Working Together

Your group's task is to research the currency exchange rates for five countries during the past twelve months. Choose a date, such as the third of the month, and find the high and low exchange rates on that date for each of the previous twelve months. Find these data for five countries. You may refer to newspapers, business magazines, or other sources of business information for your data. Assemble the results of your research in a table. Compare your group's table with those made by other groups. Did you choose the same or different countries? Which country had the greatest fluctuation in the exchange rate? Which country had the least fluctuation in the rate? What are some reasons for changes in the exchange rate?

2-1 Whole Numbers

EXPLORE

Napier's rods can be used to multiply and divide. Use them to multiply 3 × 425.

a. Position the rods so that the digits in the top row form the factor 425.

b. From the top of the rods, count down 3 boxes.

c. Add the numbers inside each colored diagonal section as shown in the diagram. The product is 1,275.

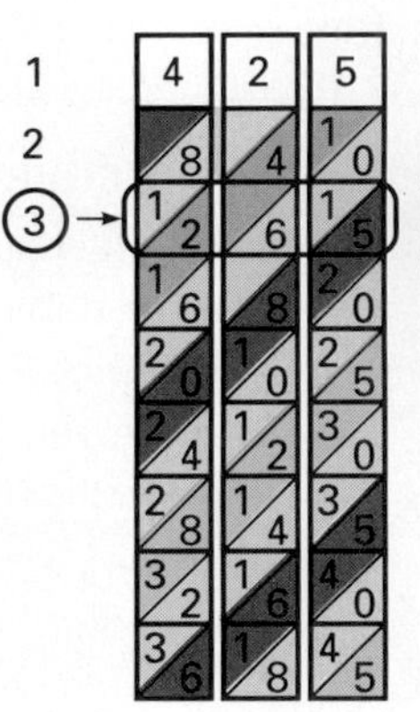

Add the numbers inside each colored diagonal section.

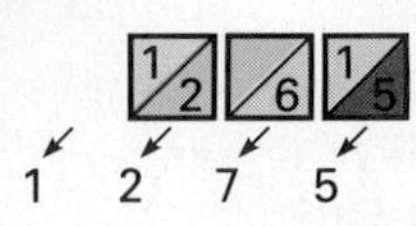

3 × 425 = 1,275

Explain how you could use the rods to find the product 7 × 254.

SKILLS DEVELOPMENT

MATH: WHO, WHERE, WHEN

John Napier, born in the sixteenth century in Scotland, is credited with many discoveries and inventions. While some of his discoveries had military uses, many others had mathematical applications.

Napier is best known for devising the logarithmic tables. But Napier's rods were his most unusual invention. (They are sometimes called Napier's bones, since animal bones were first used for this purpose.)

Answers to computations can be checked for reasonableness by **estimating**. One technique to help you estimate is **rounding.** To round a number follow these steps.

- Locate the digit in the place to which you are rounding.
- Increase this digit by 1 if the next digit to the right is 5 or greater.
- Leave the digit unchanged if the next digit to the right is less than 5.

Example 1

Round 1,538 to the indicated place-value position.

a. the nearest ten b. the nearest hundred c. the nearest thousand

Solution

a. 1,538 → 1,540 The digit to the right of 3 is greater than 5. So, increase 3 to 4.

b. 1,538 → 1,500 The digit to the right of 5 is less than 5. So, leave 5 unchanged.

c. 1,538 → 2,000 The digit to the right of 1 is 5. So, increase 1 to 2. ◄

Example 2

Find the actual answer. Use rounding to estimate the answer and compare to see if your actual answer is reasonable.

a. 3,234 + 2,578 b. 796 − 308 c. 93 × 198 d. 836 ÷ 22

Solution

a.

Actual		Estimate
3,234	⟶	3,000
+2,578	⟶	+3,000
5,812		6,000

b.

Actual		Estimate
796	⟶	800
−308	⟶	−300
488		500

c.

Actual		Estimate
198	⟶	200
× 93	⟶	× 90
18,414		18,000

d.

Actual	Estimate
$22\overline{)836}$ = 38	$20\overline{)800}$ = 40

Since each estimate is close to the actual answer, the answers are all reasonable. ◄

Compatible numbers, or numbers with which you can compute mentally, are often used in estimating quotients.

Example 3

Find the quotient 2,535 ÷ 13. Then use compatible numbers to estimate the answer. Compare the estimate with the actual answer.

Solution

Actual: $13\overline{)2,535}$ = 195

13 is close to 12 and 2,535 is close to 2,400.
12 and 24 are compatible numbers.
So, mentally divide 24 hundred by 12.

Estimate: $12\overline{)2,400}$ = 200

Since 195 is close to 200, the answer is reasonable. ◄

Example 4

Friday's attendance at a football game was 21,904. Saturday's attendance was 38,053.

a. Estimate the total attendance to the nearest ten thousand.

b. Estimate how many more people attended on Saturday than on Friday.

Solution

a.

21,904	⟶	20,000
+38,053	⟶	+40,000
		60,000

The total attendance was about 60,000 people.

b.

38,053	⟶	40,000
−21,904	⟶	−20,000
		20,000

About 20,000 more people attended on Saturday. ◄

In BASIC, the function INT finds the greatest INTeger in a number. All digits to the right of the decimal are dropped. Only the whole number is given. For example:

INT(8.32) = 8
INT(8.95) = 8

You can round in BASIC by adding 0.5 to the number and then finding the INT of the sum.

```
10 INPUT "ENTER THE
   NUMBER: ";N
20 R = INT (N + 0.5)
30 PRINT : PRINT
   "ROUNDED TO THE
   NEAREST INTEGER: ";R:
   PRINT
40 INPUT "ENTER T TO
   TRY AGAIN OR Q TO
   QUIT ";X$
50 PRINT : PRINT : IF X$ =
   "T" GOTO 10
```

TRY THESE

Round each of the following.

1. 68 to the nearest ten
2. 21 to the nearest ten
3. 358 to the nearest hundred
4. 844 to the nearest hundred
5. 7,333 to the nearest thousand
6. 2,724 to the nearest thousand

Find the actual answer. Use rounding to estimate the answer. Then compare to see if your actual answer is reasonable.

7. 345 + 953 **8.** 28 × 325 **9.** 3,127 − 1,984

Find the actual answer. Use compatible numbers to estimate the answer. Compare your estimate with your actual answer.

10. 9,834 ÷ 33 **11.** 2,432 ÷ 32 **12.** 1,820 ÷ 52

Solve.

13. On Friday, there were 5,146 students at the Ice Palace. On Saturday, 3,821 students were there.
   a. Estimate the total attendance to the nearest thousand.
   b. Estimate how many more students were at the Ice Palace on Friday than on Saturday.

EXERCISES

PRACTICE/ SOLVE PROBLEMS

Round each of the following.

1. 34 to the nearest ten **2.** 12 to the nearest ten

3. 684 to the nearest hundred **4.** 8,237 to the nearest hundred

5. 1,255 to the nearest thousand **6.** 4,328 to the nearest thousand

Find the actual answer. Use rounding to estimate the answer. Then compare to see if your actual answer is reasonable.

7. 2,168 + 3,804 **8.** 8,796 − 4,521 **9.** 24 × 352

10. 1,932 ÷ 21 **11.** 8,142 − 3,490 **12.** 5,942 + 1,298

Find the actual answer. Use compatible numbers to estimate the answer. Compare your estimate with your actual answer.

13. 2,175 ÷ 75 **14.** 5,712 ÷ 68 **15.** 7,452 ÷ 92

16. 1,888 ÷ 59 **17.** 4,392 ÷ 72 **18.** 2,895 ÷ 15

Solve.

19. There were 6,758 people at a local theater production on Friday and 9,056 people on Saturday.
   a. Estimate the total attendance to the nearest thousand.
   b. Estimate how many more people were at the theater on Saturday than on Friday.

20. XYZ Corporation sold 64,918 cassette tapes in November and 75,985 cassette tapes in December.
   a. Estimate, to the nearest thousand, the total number of tapes sold in November and December.
   b. Estimate how many more tapes were sold in December than in November.

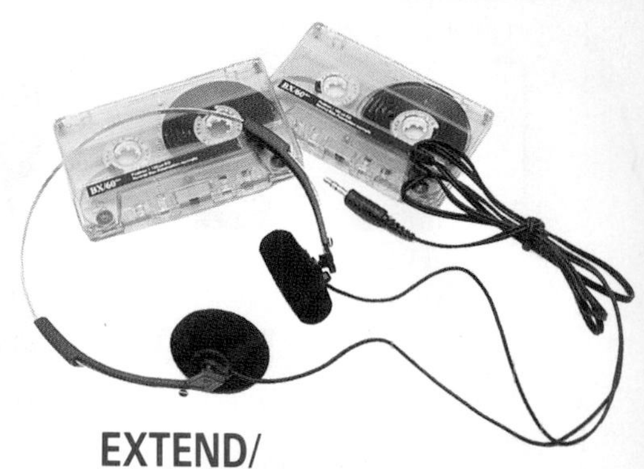

EXTEND/ SOLVE PROBLEMS

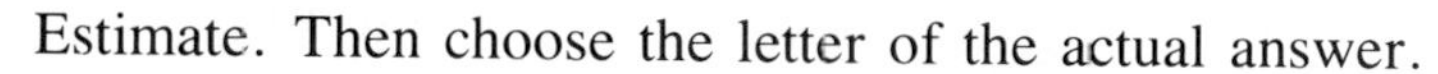

Estimate. Then choose the letter of the actual answer.

	A	B	C
21. $27 \times 32 \times 49$	423,360	4,233	42,336
22. $83{,}978 \div 398$	21	211	441
23. $1{,}594 + 375 + 8{,}946$	10,915	9,695	18,904
24. $96{,}350 + 235 + 2{,}145$	100,845	98,730	141,300
25. $21{,}914 - 13{,}450$	45,816	18,917	8,464

USING DATA Use the Data Index on page 546 to find the seigniorage chart. Round the seigniorage of coin and silver to the nearest million dollars for the given year.

26. 1974   27. 1984   28. 1988

29. A student gave the sum of the seigniorage for 1970, 1980, and 1986 as 1,629,478,350.10. Is this a reasonable answer? Explain.

COMPUTER TIP

Data base software is a powerful application tool of the computer. It allows you to organize, store, and retrieve information in a variety of ways. A computer that is equipped with a CD ROM can search very large data bases, for example, all the volumes of an encyclopedia. It enables you to retrieve information, complete with pictures, very quickly.

Solve.

30. There are 12 rows of seats in an auditorium with 48 seats in each row. It was reported that 850 students were seated in the auditorium on Monday. Is this a reasonable answer? Explain.

31. A student had these five test scores: 92, 75, 84, 89, 75.
   a. Estimate to see if a mean score of 83 is reasonable.
   b. Find the median and the mode.

THINK CRITICALLY/ SOLVE PROBLEMS

Round to estimate.

32. Three numbers are given as the missing factor for the following: $56 \times \blacksquare = 4{,}536$. Which one is reasonable?
   a. 903   b. 81   c. 93

33. Three numbers are given as a divisor for $1{,}862 \div \blacksquare = 38$. Which one seems reasonable?
   a. 4   b. 49   c. 490

34. It was reported that about 80,000 people attended a football game. If this number was rounded to the nearest ten thousand:
   a. What is the least number of people that could have attended?
   b. What is the greatest number of people that could have attended?

2-2 Variables and Expressions

EXPLORE

Try the following number trick four times, each time starting with a different number. Suppose you choose 3 and then add 5. Adding 5 to 3 leads to the statement $3 + 5 = 8$. Write a statement for each step from b–g.

Predict the Number!

a. Choose any number.
b. Add 5.
c. Multiply by 2.
d. Subtract 4.
e. Divide by 2.
f. Subtract the original number.
g. Record your answer.

What conclusions can you draw?

SKILLS DEVELOPMENT

Often it is necessary to use a **variable,** a letter such as a, n, x, or y, to represent an unknown number. Expressions with at least one variable, such as $n + 5$ and $x \div 9$, are **variable expressions.** You can write variable expressions to describe word phrases.

Example 1

Write a variable expression. Let n represent "a number."

a. 8 more than a number
b. 9 less than a number
c. a number divided by 3
d. 3 times a number

Solution

a. $n + 8$
b. $n - 9$
c. $n \div 3$ or $\frac{n}{3}$ A bar indicates division.
d. $3 \times n$ or $3n$ $3n$ means multiply n by 3. ◄

When you **evaluate** a variable expression, you replace the variable with a number. (The number is called a **value** of the variable.) Then you perform any indicated operations, keeping in mind these important rules.

- ► First do all multiplications and divisions in order from left to right.
- ► Then do any additions and subtractions in order from left to right.

Example 2

Evaluate each expression.

a. $9 + m$, if $m = 12$
b. $a - 7$, if $a = 15$
c. $3y - 4$, if $y = 9$
d. $m \div 7 + 1$, if $m = 21$

MIXED REVIEW

Find the mean, median, and mode for each set of data.

1. 178, 235, 679, 384, 384
2. 13, 98, 45, 67, 67, 98, 17, 28, 98
3. 31, 56, 94, 79, 85, 62, 31, 92, 39, 31
4. Give the difference between the value of the mean and the value of the median of these numbers: 7, 9, 2, 3, 8, 9, 5, 1, 9, 4, 3, 6, 8, 0, 1

Solution

a. $9 + m$
$9 + 12 = 21$

b. $a - 7$
$15 - 7 = 8$

c. $3y - 4$
$3 \times 9 - 4$ Multiply first.
$27 - 4 = 23$ Then subtract.

d. $m \div 7 + 1$
$21 \div 7 + 1$ Divide first.
$3 + 1 = 4$ Then add. ◄

Some expressions have more than one variable.

Example 3

Evaluate the expression $2x + 5y$, if $x = 2$ and $y = 3$.

Solution

$2x + 5y$
$2 \times 2 + 5 \times 3$ Multiply first.
$4 + 15 = 19$ Then add. ◄

You can write and evaluate expressions to describe real-world situations.

Example 4

Write an expression for each real-world situation.

a. c pens shared equally by 4 boys
b. s dollars of savings decreased by \$27
c. d dollars more than \$7
d. 5 times p points
e. t movie tickets at \$7 each

Solution

a. $c \div 4$ or $\frac{c}{4}$
b. $s - 27$
c. $7 + d$
d. $5 \times p$ or $5p$
e. $7 \times t$ or $7t$ ◄

TRY THESE

Write a variable expression. Let n represent "a number."

1. 7 greater than a number
2. 5 less than a number
3. the product of 3 and a number
4. a number divided by 3

Evaluate each expression. Let $a = 8$.

5. $13 - a$
6. $a + 12$
7. $a \div 2$
8. $5a$

Evaluate each expression. Let $a = 3$ and $b = 6$.

9. $a + b$
10. $b \div a$
11. $b - 2a$
12. $2b \div a$

Write an expression to describe each situation.

13. b baseballs less 7
14. s dollars of savings equally deposited in 5 accounts
15. s dollars increased by \$12
16. three times d dollars

Solve.

17. A carton holds c cans of food. The cans were shared equally by 8 people.
    a. Write an expression to describe the situation.
    b. Evaluate the expression. Let $c = 48$.

EXERCISES

PRACTICE/ SOLVE PROBLEMS

Write a variable expression. Let n represent "a number."

1. half of a number
2. 4 less than a number
3. six times a number
4. a number increased by 9
5. 7 greater than a number
6. a number divided by 7
7. the product of 10 and a number
8. 16 decreased by a number

Evaluate each expression. Let $a = 9$.

9. $a - 2$
10. $2a$
11. $27 \div a$
12. $21 + a$
13. $a \div 3$
14. $7a$
15. $12 - a$
16. $a + 13$
17. $3a$
18. $3a - 2$
19. $3a + 4$
20. $\frac{4a}{9}$

Evaluate each expression. Let $a = 2$ and $b = 8$.

21. $a + b$
22. $b - a$
23. $b \div a$
24. $9a - b$
25. $b - 4a$
26. $3b \div a$
27. $7a \times 4b$
28. $9a - 2b$
29. $12a \div b$
30. $b \div 2a$
31. $b - 3a$
32. $5a + 6b$

Write an expression to describe each situation.

33. 9 more than w wallets
34. d dollars shared equally by 3 people
35. p party favors at 25 cents each
36. e erasers decreased by 2

Solve.

37. A box holds c pencils. The pencils were shared equally by 8 students.
    a. Write an expression to describe the situation.
    b. Evaluate the expression if c is 24.
38. Jenny has c cassette tapes. She gives 5 tapes away.
    a. Write an expression to describe the situation.
    b. Evaluate the expression if c is 38.

39. The temperature in the morning was 65°F. It rose d degrees before noon.
   a. Write an expression to describe the situation.
   b. Evaluate the expression if d is 14.

EXTEND/ SOLVE PROBLEMS

Evaluate each expression. Let $a = 2$, $b = 3$, and $c = 6$.

40. $a + c - b$	41. $c - 2a + 1$	42. $a + \frac{12}{b} - 1$
43. $a + b + c$	44. $c + a - b$	45. $\frac{c}{a} + b$
46. $b - a + c$	47. $c - b + a$	48. $b + c - 2a$
49. $3c + 2a + b$	50. $2c - 4a + b$	51. $12b \div 6 - 1$

PROBLEM SOLVING TIP

For Exercise 52, remember to count Ian as one of the people who share.

Solve.

52. Ian shares d dollars equally with 5 friends.
   a. Write an expression to describe the situation.
   b. Evaluate the expression if d is 30.

53. In May, Chang had d dollars in his bank account. Three months later, he had 4 more than 3 times that amount.
   a. Write an expression to describe the situation.
   b. Evaluate the expression if d is 25.

54. Ramona's test scores—96, 88, x, 88, and 85—are given in order from greatest to least.
   a. The median is represented by x. Find its value.
   b. Find the mean score.

THINK CRITICALLY/ SOLVE PROBLEMS

55. Let x represent an odd number. Then write an expression that describes each of the following.
   a. next odd number
   b. preceding odd number
   c. next whole number
   d. preceding whole number
   e. sum of the odd number and the next whole number
   f. sum of the odd number and the next odd number

56. Try this number trick. Pick a number. Add 20, multiply by 6, divide by 3, subtract 40, then divide by 2.
   a. Write a conclusion about your result.
   b. Change this number trick so that your result is twice the number you started with.

57. Write a number trick in which the result is your age. Try it out with a classmate.

2-3 Problem Solving Skills:

CHOOSE THE OPERATION

► READ
► PLAN
► SOLVE
► ANSWER
► CHECK

Remember to use the 5-step problem solving plan to help you organize your work when you solve any problem.

PROBLEM

A professional athlete earned the same amount of money for each game played. The athlete earned a total of \$67,536 for 36 games. How much did the athlete earn for each game?

SOLUTION

These steps can help you solve the problem.

READ *What information do you know?*

Amount earned: \$67,536
Number of games played: 36
Equal amounts of money earned for each game.

What information are you asked to find?

How much was earned for each game?

PLAN *Decide what operation(s) you need to use.*

Because the same amount of money was earned for each game, divide the total amount by the number of games to find how much was earned for each game.

$67{,}536 \div 36 = n$

SOLVE *Carry out the plan.*

$67{,}536 \div 36 = 1{,}876$

ANSWER *Give the answer in a sentence.*

The athlete earned \$1,876 per game.

CHECK *Check to see if the answer is reasonable.*

Use compatible numbers to estimate.

$70{,}000 \div 35 = 2{,}000$

Since 1,876 is close to 2,000, the answer is reasonable.

PROBLEMS

Choose the operation. Do not solve.

1. Willie Mays of the Giants hit 660 career home runs. Joe DiMaggio hit 361 home runs. How many more home runs did Mays hit than DiMaggio?

2. A Ford V8 needs spark plugs at a cost of $2 each. What is the total cost of the spark plugs?

Choose the operation(s). Then solve.

3. There are 174 people who work at a tourist attraction. During the summer, 37 more people are hired. How many people in all are working at the attraction during the summer?
4. Twelve couples bought a load of 288 lb of potatoes. How many pounds will each of the couples receive if they share equally?
5. Ron bought 14 stickers at 3 cents each. What was the total cost?
6. Three brothers shared a birthday gift of $135 equally. How much money did each brother receive?
7. Mandy raises worms for bait. One year she raised 23,167 worms. The next year she raised 15,154 more worms than the year before. How many worms did she raise the second year?
8. Kevin bought 8 boxes of erasers. Each box contained 44 erasers. How many erasers did he buy in all?
9. There were 175 benches in the auditorium, each holding the same number of people. There were 1,400 people in all. How many people were seated on each bench?
10. The combined height of two acrobats, one standing on the other's head, was 128 inches. The height of one acrobat was 68 inches. What was the height of the other acrobat?
11. How many cartons of 12 eggs could be filled with 144 eggs?
12. On election night, it was reported that out of 1,545 polling stations that were closed, 698 had completed counting their votes and 150 stations were still counting. How many stations had not yet begun counting?
13. A sleeping person breathes an average of 15 times per minute. How many breaths would someone take during 8 h of sleep?
14. Kesper Auto Limited carries spark plugs in boxes of 10. They have 24 boxes of plugs in stock. How many cars requiring 6 spark plugs each could they service with this amount?

READING MATH

When you read a problem, how do you decide which operation is involved in the solution? There are no hard-and-fast rules, but certain words in the problem may suggest which operation(s) you should consider. Consider these words and word phrases, for example. Then see if you can think of others to add to the list.

Addition
plus
sum
total
altogether
combined
in all

Subtraction
minus
difference
how much more
how much less

Multiplication
times
product
doubled, tripled, and so on
total
altogether
in all

Division
quotient
shared equally
equal parts

USING DATA Use the Data Index on page 546 to find the listing of the early monetary system used in the Pacific Islands.

15. How many coconuts would have to have been gathered to equal the monetary value of 1 dog's tooth?
16. What is the value in dog's teeth of 500 coconuts?
17. What is the value in porpoise teeth of 100 strings of white teeth?

2-4 Adding and Subtracting Decimals

EXPLORE/ WORKING TOGETHER

Choose a sum of money between 25 cents and one dollar. Take turns in your group showing various combinations of coins that are equivalent in value to this sum. For example, to show 47 cents, someone might choose the following coin combination:

	Number of Coins	Coins	Number of Cents	Decimal Form
	1	Quarter	25	$0.25
	2	Dimes	20	$0.20
	2	Pennies	2	$0.02
Totals	5		47	$0.47

How many combinations of coins did your group find for the sum you chose? Repeat for four more sums, determining for each sum all the possible combinations of coins that are equivalent in value. Make a table like the one above to record all the combinations.

For each sum, compare the totals in the last two columns.

SKILLS DEVELOPMENT

What you learn in one part of mathematics often can be used to guide you as you learn other skills. For example, you can use what you know about adding and subtracting whole numbers to help you add and subtract decimals.

- To add or subtract whole numbers, line up the digits in the ones, tens, and hundreds columns and within any other columns that represent values greater than 1. Then add or subtract in each column.
- To add or subtract decimals, line up the digits in the tenths, hundredths, and thousandths columns and within any other columns that represent values less than 1. Aligning the decimal points helps you align the digits correctly. Then add or subtract in each column. Place the decimal point in the answer directly below the other decimal points.

Example 1

Find each answer. Use rounding to estimate the answer and compare to see if the answer is reasonable.

a. $8.6 + 3.89 + 6.2$ **b.** $86.2 - 19.76$

Solution

Align the decimal points. If necessary, annex zeros so all numbers have the same number of decimal places. Then add or subtract.

a.

Actual		Estimate
8.60	→	9
3.89	→	4
+ 6.20	→	+ 6
18.69		19

Since 18.69 is close to 19, the answer is reasonable.

b.

Actual		Estimate
86.20	→	86
− 19.76	→	− 20
66.44		66

Since 66.44 is close to 66, the answer is reasonable. ◄

You also can use **front-end estimation.**

Example 2

Find the total cost of these purchases, including tax: tape, $6.99; CD, $15.99; earphones, $19.99; sales tax, $2.58. Use front-end estimation and compare your estimate with the actual answer.

Solution

Actual	Add front-end digits.		Adjust remaining digits.
$ 6.99	6		0.99 → 1
15.99	10		5.99 → 6
19.99	10		9.99 → 10
+ 2.58	+ 2		0.58 → 1
$45.55	28	+	18 = 46

The total, $45.55, is close to 46, so the answer is reasonable. ◄

TRY THESE

Compute. Round to estimate. Compare to check your answer.

1. $6.29 + $16.29 + $4.28
2. 294.8 − 124.7

Compute. Use front-end estimation to see if the answer is reasonable.

3. 6.94 + 41.77 + 5.2 + 8.4
4. $2,456.69 − $1,297.70

Solve. Estimate and compare to see if the answer is reasonable.

5. Maria works part time. One week, her gross pay was $34.55. She paid $8.75 in taxes. What was her take-home pay that week?

EXERCISES

PRACTICE/ SOLVE PROBLEMS

Compute. Round to estimate. Compare to check your answer.

1. $16.52 + $5.03 + $166.29
2. $3.23 + $17.58 + $100.01
3. 1,568.33 − 422.88
4. 24,367.1 − 14,368.9

Compute. Use front-end estimation to see if the answer is reasonable.

5. 356.5 + 3,496.80 + 524.6
6. \$525.49 − \$137.84
7. 6.66 + 66.66 + 666.66 + 66.666
8. \$1,489.65 − \$1,211.15

Solve. Estimate and compare to see if your answer is reasonable.

9. Bill earned \$15.60, \$21.75, and \$16.80 for baby-sitting one week. He also earned \$12.50 doing yard work. How much money did he earn in all that week?
10. Anwar had a balance of \$157.92 in his checking account. He wrote a check for \$62.39. What was his new balance?

EXTEND/ SOLVE PROBLEMS

USING DATA Use the Data Index on page 546 to find the table that shows how the dollar has "shrunk."

11. Find the total cost of the four foods for each of the six years.
12. Find the change in the total cost of the four foods between the following pairs of years.
    a. 1890 and 1910 b. 1910 and 1930 c. 1930 and 1950
    d. 1950 and 1970 e. 1970 and 1975 f. 1890 and 1975

Solve. Estimate and compare to see if your answer is reasonable.

13. Allison purchased a sweater for \$72.21 and a jacket for \$102.99. She gave the clerk \$200. How much change did she receive?
14. Lynne and Rico each loaned Damon \$5.00. Damon used the money to buy a bouquet of flowers for his mother. How much money did Damon have left after spending \$8.37 for the flowers?

THINK CRITICALLY/ SOLVE PROBLEMS

Complete. Estimate and compare to see if your answers are reasonable.

15. $$\begin{array}{r} 43.25\blacksquare \\ +\ 0.352 \\ \hline 4\blacksquare.\blacksquare 11 \end{array}$$

16. $$\begin{array}{r} 5.004 \\ -1.9\blacksquare\blacksquare \\ \hline \blacksquare.\blacksquare 05 \end{array}$$

17. $$\begin{array}{r} 9.\blacksquare 16 \\ -3.7\blacksquare 9 \\ \hline \blacksquare.04\blacksquare \end{array}$$

18. $$\begin{array}{r} 2.16 \\ 3.\blacksquare\blacksquare \\ +\blacksquare.05 \\ \hline 9.34 \end{array}$$

19. $$\begin{array}{r} 4.\blacksquare 7 \\ -\blacksquare.3\blacksquare \\ \hline 1.92 \end{array}$$

20. $$\begin{array}{r} 2.1\blacksquare 5 \\ 6.\blacksquare 94 \\ +\blacksquare.058 \\ \hline 8.57\blacksquare \end{array}$$

21. Enrico had some money. He earned \$15 for baby-sitting. Then he bought food for \$18.74 and gasoline for \$21.93. He has \$35.84 left. How much money did he start out with?

Problem Solving Applications:
TRAVEL EXPENSES

After graduation, three students took a 7-day trip across part of the United States. They recorded their expenses.

Day	Food	Motel	Gas
Sunday	$65.00	$35.00	$45.29
Monday	72.16	48.00	56.16
Tuesday	34.91	27.00	18.92
Wednesday	52.38	54.00	14.38
Thursday	43.27	38.00	10.21
Friday	63.64	29.00	54.98
Saturday	38.72		21.37

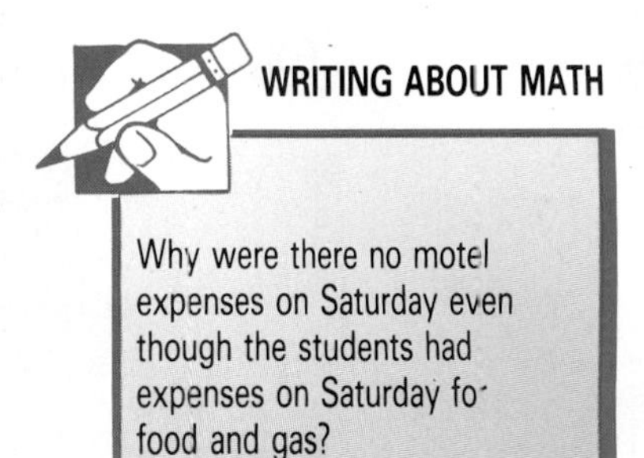

Why were there no motel expenses on Saturday even though the students had expenses on Saturday fo· food and gas?

Use the expense record to solve these problems.

1. Find the total spent during the week for each.
   a. food **b.** motels **c.** gas
2. The students shared the costs equally. Find out each student's cost for each item.
   a. food **b.** motels **c.** gas
3. What was each student's total expense?
4. Each student started the trip with $400. How much did each have left to spend on other items after paying for food, motels, and gas?

Lawanda is a high school senior. In a national contest, she won a "dream trip" for outstanding scholarship.

5. In Mexico, Lawanda bought a pair of maracas for 2,500 pesos. She paid with a 5,000-peso note. How many pesos did she receive in change?
6. In India, she bought a bag for 1,200 rupees, a sari for 2,300 rupees, and a brass dish for 750 rupees. How many rupees did she spend?
7. In Canada, she purchased a coat for $123.45 and a scarf for $11.98. She handed the clerk $150.00. How much change did she receive?
8. While in Canada, Lawanda attended two games in a hockey exhibition. Each ticket cost $25.75. She bought food totaling $12.35 at the snack bar. How much did she spend?

2-5 Multiplying and Dividing Decimals

EXPLORE

Mark off a 10 by 10 grid on graph paper.

a. Shade all the squares that make up four rows. Write the decimal that describes the shaded part.
b. Mark an X in all the squares that make up three columns. Next to your first decimal, write the decimal for the squares with X's.
c. Count the squares that are both shaded and marked with an X. Write a decimal for these squares next to your second decimal.
d. Put a multiplication sign between the first two decimals and an equals sign between the last two.

You have just multiplied two decimals! Experiment with other numbers.

SKILLS DEVELOPMENT

MENTAL MATH TIP

Multiply a decimal by 10, and the decimal point in the product moves 1 place to the right.

10 x 0.24 = 2.4

Multiply a decimal by 100, and the decimal point in the product moves 2 places to the right.

100 x 0.24 = 24

What happens to the decimal point in the product when a decimal is multiplied by 1,000?

1,000 x 0.24 = ?

Multiply and divide decimals just as you multiply and divide whole numbers. Then place a decimal point in the answer.

Study and analyze these examples. See if you can find the pattern that was used to place the decimal points in the products.

64	64	64	6.4	6.4	6.4	0.64	0.64
× 2	×0.2	×0.02	× 2	×0.2	×0.02	× 2	×0.02
128	12.8	1.28	12.8	1.28	0.128	1.28	0.0128

► To find the number of decimal places to put in a product, find the sum of the decimal places in the factors. Count off that number of places in the product, counting from right to left.

Example 1

Find the product. Estimate and compare to see if it is reasonable.

a. 4 × 8.7 b. 6.3 × 9.2 c. 0.98 × 16.4

Solution

a. Actual → Estimate

Actual		Estimate
8.7	→	9
× 4	→	×4
34.8		36

b.

Actual		Estimate
9.2	→	9
×6.3	→	×6
276		54
5520		
57.96		

c.

Actual		Estimate
16.4	→	16
×0.98	→	× 1
1312		16
14760		
16.072		

Each estimate is close to the actual answer, so each product is reasonable. ◄

Study and analyze these examples. See if you can find the pattern that was used to place the decimal points in the quotients.

$$3\overline{)36} = 12 \quad 3\overline{)3.6} = 1.2 \quad 3\overline{)0.36} = 0.12 \quad 3\overline{)0.0036} = 0.0012 \quad 7\overline{)0.49} = 0.07$$

$$0.2\overline{)25.0} = 125 \quad 0.2\overline{)2.50} = 12.5 \quad 0.2\overline{)0.250} = 1.25 \quad 0.08\overline{)0.64} = 8 \quad 0.08\overline{)0.0064} = 0.08$$

- ► To divide a decimal by a whole number, place the decimal point in the quotient above the decimal point in the dividend.
- ► To divide a decimal by a decimal, first move the decimal point in the divisor to the right to get a whole number. Then move the decimal point in the dividend the same number of places to the right. Annex zeros if necessary.

MENTAL MATH TIP

Divide a decimal by 10, and the decimal point in the quotient moves 1 place to the left.

$$10\overline{)22.4} = 2.24$$

Divide a decimal by 100, and the decimal point in the quotient moves 2 places to the left.

$$100\overline{)22.4} = 0.224$$

What happens to the decimal point in the quotient when a decimal is divided by 1,000?

$$1{,}000\overline{)22.4} = ?$$

Example 2

Find the quotient. Estimate and compare to see if it is reasonable.

a. 23.76 ÷ 24 **b.** 11.178 ÷ 2.3

Solution

a. Actual

$$\begin{array}{r} 0.99 \\ 24\overline{)23.76} \\ \underline{21\,6} \\ 2\,16 \\ \underline{2\,16} \\ 0 \end{array}$$

Estimate

$$24\overline{)24} = 1$$

b. Actual

$$\begin{array}{r} 4.86 \\ 2.3\overline{)11.1\,78} \\ \underline{9\,2} \\ 1\,9\,7 \\ \underline{1\,8\,4} \\ 1\,38 \\ \underline{1\,38} \\ 0 \end{array}$$

Estimate

$$2\overline{)10} = 5$$

Each estimate is close to the actual answer, so each quotient is reasonable. ◄

Try These

Write the letter for the product.

	A	B	C
1. 4.6 × 3	138	13.8	1.38
2. 42 × 5.8	24.36	243.6	2.436
3. 1.9 × 2.8	5.32	0.532	53.2

Find each product. Estimate and compare to see if it is reasonable.

4. 10 × 1.3
5. 100 × 0.31
6. 1,000 × 5.63
7. 4.8 × 4
8. 92.3 × 0.35
9. 0.75 × 0.92

Write the letter for the quotient.

	A	B	C
10. 17.28 ÷ 4.8	36	3.6	0.36
11. 39.2 ÷ 14	28	2.8	0.28
12. 56.7 ÷ 6.3	0.09	9	0.9

Find each quotient. Estimate and compare to see if it is reasonable.

13. $10\overline{)12.3}$ 14. $100\overline{)18.96}$ 15. $1{,}000\overline{)489}$

16. $6\overline{)94.56}$ 17. $1.8\overline{)36.36}$ 18. $0.05\overline{)16.75}$

EXERCISES

PRACTICE/ SOLVE PROBLEMS

Write the letter of the product.

	A	B	C
1. 4 × 3.68	1.472	14.72	147.2
2. 5.1 × 38	19.38	193.8	1,938
3. 4.6 × 3.8	1.748	17.48	174.8
4. 3.36 × 4.2	14.112	141.12	1,411.2

Find each product. Estimate and compare to see if it is reasonable.

5. 10 × 2.6 6. 10 × 0.45 7. 100 × 0.86

8. 1,000 × 4.62 9. 4 × 7.2 10. 3.2 × 5

11. 3 × 6.9 12. 13.5 × 1.8 13. 90.5 × 0.26

Write the letter of the quotient.

	A	B	C
14. 29.16 ÷ 12	243.0	24.3	2.43
15. 54.30 ÷ 15	36.2	3.62	0.362
16. 63.18 ÷ 1.3	48.6	4.86	0.486
17. 10.07 ÷ 2.65	0.38	3.8	38

Find each quotient. Estimate and compare to see if it is reasonable.

18. $10\overline{)1.36}$ 19. $100\overline{)12.4}$ 20. $1{,}000\overline{)148.36}$

21. $14\overline{)1.694}$ 22. $22\overline{)16.94}$ 23. $12\overline{)1.44}$

24. $116\overline{)40.6}$ 25. $2.5\overline{)3.75}$ 26. $0.02\overline{)0.354}$

Solve. Estimate and compare to see if the answer is reasonable.

27. A class of 50 students went to a baseball game together. The total cost of admission was $346.00. If the students shared the costs equally, how much did each student pay?

28. Tickets for a school play cost $4.25 each. The attendance at the play was 1,006 people. How much ticket money was collected?

EXTEND/ SOLVE PROBLEMS

Find each answer rounded to the indicated place-value position.

29. 52.7×6.2 to the nearest tenth

30. $484.27 \div 8.14$ to the nearest hundredth

31. $4.799 \div 2.658$ to the nearest thousandth

32. 174.8×16.3 to the nearest hundredth

33. $86.42 \div 26.17$ to the nearest hundredth

Solve.

34. A class of 40 students went to a museum. The cost of renting a bus was $156.40. Admission was $3.45 per student. If they shared the costs equally, how much did each student pay?

35. Tickets to a charity dance cost $45.50 each. The attendance at the dance was 1,262. The cost of refreshments averaged $6.21 per person. What was the profit for the charity?

36. A pet-store owner bought 6 mollies at $0.79 each, 7 guppies at $0.59 each, 12 swordtails at $1.89 per pair, and 14 platies at $1.69 per pair. What was the total cost of the fish?

THINK CRITICALLY/ SOLVE PROBLEMS

USING DATA Use the Data Index on page 546 to find the currency exchange rates. All exchanges were made on Thursday.

37. Find the equivalent number of Finnish markkas you would receive for 30 U.S. dollars.

38. Find the equivalent number of Portuguese escudos you would receive for 75 U.S. dollars.

39. Find the U.S. dollar equivalent for 4,687 Irish punts.

40. Find the U.S. dollar equivalent for 467 Pakistani rupees.

Tell whether the statement is *true* or *false*. Explain.

41. When a decimal is divided by a whole number the quotient always has the same number of decimal places as the dividend.

42. The product of two decimals cannot be a whole number.

2-6 Exponents and Scientific Notation

EXPLORE

Compare these numbers. The top number, 10, can be written as the product 1×10. Write each of the other numbers as a product, using only 10 as a factor. Then write the number of factors beside each number. What do you notice? How many factors of 10 would there be for the number 100,000,000,000? Explain how to use mental math to determine your answer.

10
100
1,000
10,000
100,000
1,000,000
10,000,000
100,000,000
1,000,000,000

SKILLS DEVELOPMENT

You can make large multiples of ten easier to use and to read by writing them in **exponential form**. A number written in exponential form has a base and an exponent.

$$10 \times 10 \times 10 \times 10 = 10^4$$

The **base** tells what factor is being multiplied. The **exponent** tells how many equal factors there are. The expression above, 10^4, is read as "10 to the fourth power."

The exponents zero and one are special.

Any number raised to the first power is that number itself.
$2^1 = 2$, $3^1 = 3$

Any number, except 0, raised to the zero power is equal to one.
$2^0 = 1$, $3^0 = 1$

A number multiplied by itself is said to be **squared**. Read 10^2 as "10 squared." You will learn to express the area of a plane figure in square units. A number multiplied by itself two times is said to be that number **cubed**. Read 10^3 as "10 cubed." You will learn to express the volume of a solid figure in cubic units.

Example 1

Write in exponential form.

a. $4 \times 4 \times 4$ **b.** $7 \times 7 \times 7 \times 7 \times 7$ **c.** 0.4×0.4

Solution

a. 4^3 **b.** 7^5 **c.** $(0.4)^2$ ◄

Example 2

Write in standard form.

a. 3^4 b. 5^2 c. 7^3

Solution

a. $3^4 = 3 \times 3 \times 3 \times 3$
$= 81$

b. $5^2 = 5 \times 5$
$= 25$

c. $7^3 = 7 \times 7 \times 7$
$= 343$ ◄

Frequently, very large or very small numbers are written in **scientific notation**. A number written in scientific notation has two factors. The first factor is a number greater than or equal to 1 and less than 10. The second factor is a power of 10.

These numbers are written in scientific notation:

4×10^3 9.9×10^8 6.001×10^{20}

CHECK UNDERSTANDING

Explain why these numbers are *not* written in scientific notation.

0.621×10^2
1.23×5^4
$4.1 \times 1.6 \times 10^3$
21×10^5
$1.25 + 10^3$

Example 3

Write in scientific notation.

a. 50,000 b. 4,183,000

Solution

a. $50{,}000 = 5 \times 10{,}000$ Ask yourself, *What number times 5 equals 50,000?*
$= 5 \times 10^4$

b. $4{,}183{,}000 = 4.183 \times 1{,}000{,}000$
$= 4.183 \times 10^6$ ◄

Example 4

Write in standard form.

a. 2×10^3 b. 6.47×10^8

Solution

a. $2 \times 10^3 = 2 \times 1{,}000$
$= 2{,}000$

b. $6.47 \times 10^8 = 6.47 \times 100{,}000{,}000$
$= 647{,}000{,}000$ ◄

COMPUTER TIP

The computer memory can store numbers with up to nine digits and a decimal point. For numbers with more digits, the computer uses scientific notation. Here are some examples. Can you figure out what the E stands for?

```
]PRINT 98000000*520000
5.096E+13

]PRINT 82000*560000
4.592E+10
```

TRY THESE

Write in exponential form.

1. $5 \times 5 \times 5$
2. $3 \times 3 \times 3 \times 3$

Write in standard form.

3. 2^2
4. 3^2
5. 3^3
6. 4^2

Write in scientific notation.

7. 7,302
8. 452,968
9. 209
10. 6,000,432

Write in standard form.

11. 5.629×10^3
12. 9.04×10^5

EXERCISES

PRACTICE/ SOLVE PROBLEMS

MENTAL MATH TIP

When you write a number in scientific notation, the exponent of 10 is the same as the number of places that you move the decimal point.

700,000 = 700,000
5 places
= 7×10^5

9.04×10^3 = 9.040
3 places
= 9,040

Write in exponential form.

1. $3 \times 3 \times 3$
2. $2 \times 2 \times 2 \times 2 \times 2 \times 2 \times 2$
3. $6 \times 6 \times 6 \times 6$
4. $10 \times 10 \times 10 \times 10 \times 10$

Write in standard form.

5. 2^4
6. 0.5^2
7. 6^2
8. 1^{10}

Write in scientific notation.

9. 625,984
10. 1,961,048
11. 892
12. 54,203

Write in standard form.

13. 6.39×10^4
14. 8.7231×10^9
15. 3×10^2
16. 4.124×10^3

Solve. Explain your answer.

17. A student said that 3^2 equals 2^3. Is this correct?

EXTEND/ SOLVE PROBLEMS

Evaluate. Let $x = 2$, $y = 3$, and $z = 4$.

18. $x^5 - 15$
19. $z^3 + 21$
20. $x^2 + y^2$
21. $x^3 + y^4$
22. $y^2 - x^2$
23. $z^4 - x^3$
24. $x^2 + y^3 + z^2$
25. $z^3 - y^2 + x^5$

Write in standard form. Use a calculator for the computation.

26. 29^5
27. 13^7
28. 2^{12}
29. $1{,}452^1$

Solve. Explain your answer.

30. One light year equals 9,460,000,000,000 km. The mean distance between Earth and the sun is 1.496×10^8 km. Is the distance between Earth and the sun greater than, less than, or equal to one light year?

THINK CRITICALLY/ SOLVE PROBLEMS

Write each number with the given base or exponent.

31. 32 with base of 2
32. 1,024 with base of 4
33. 343 with exponent of 3
34. 1.0 with exponent of 6
35. A fund-raising drive began with $3.00. Every hour the amount in the fund doubled. The goal was met in 8 hours. How long was it before the goal was half met?

► READ
► PLAN
► SOLVE
► ANSWER
► CHECK

Problem Solving Applications:

USING A SCIENTIFIC CALCULATOR

All calculators are not created equal. A scientific calculator has more keys than most standard calculators and uses a different program to perform calculations.

A scientific calculator may have a key labeled [x^y] or [y^x] to compute numbers in exponential form. How you use a calculator to write a number in standard form depends on the type of calculator.

	Scientific Calculator Key Sequence:	Standard Calculator Key Sequence:
$2^4 \rightarrow$	2 [x^y] 4 [=]	2 [×] [=] [=] [=]
		Press [=] one time less than the number indicated by the exponent.

Sometimes a number is too large to fit in the display of a calculator. Depending on the calculator, the display either will indicate an error or it will display the number in scientific notation.

You may see the complete answer or one of the following displays if you use a calculator to find $600 \times 962{,}431 = 577{,}458{,}600$.

Error Message:	ERROR	or	E5.7745860
Scientific Notation:	5.774586[8]	or	5.774586[08]

CALCULATOR TIP

If your calculator has an [EE] or an [EXP] key, you can use it as a quick way to enter a number that is in scientific notation. For example, to enter 8×10^5, use this key sequence:

8 [EXP] 5

Choose the letter of the key sequence to be used on a scientific calculator to find the number given in exponential form.

1. 6^{15}
   a. 15 [x^y] 6
   b. 6 [x^y] 15
   c. [x^y] 6 15

2. 15^6
   a. 15 [x^y] 6
   b. 6 [x^y] 15
   c. [x^y] 15 6

3. 9^8
   a. [x^y] 8 9
   b. 8 [x^y] 9
   c. 9 [x^y] 8

What is the number for which the answer is given in the calculator display? Write it in scientific notation and then in standard form.

4. [9.6578 10]
5. [1.39 11]
6. [2.5555 15]

Write each number in scientific notation. Then tell which of the numbers has the least value and which has the greatest value.

7. 27^{14}
8. 30^{10}
9. 25^{19}
10. [2.6359 16]
11. [2.39245 20]
12. [1.962 25]

2-7 Laws of Exponents

EXPLORE

Find the product by writing each factor in expanded form. Then rewrite it as a power of ten.

CONNECTIONS

A **googol** is an amount equal to the number 1 followed by 100 zeros, or 10^{100}. A **googolplex** is equal to 10^{googol}. Mathematicians and scientists estimate there to be about 10^{20} grains of sand on Coney Island beach in New York City and about 10^{110} protons and electrons in the universe. If you were to write one googolplex as a 1 with zeros following, you would not have enough room to write if you went from Earth to the farthest star, putting down zeros every inch of the way!

$10^2 \times 10^3 = 10 \times 10 \times 10 \times 10 \times 10 = 10^?$
$10^4 \times 10^5 = \underline{\ ?\ } = 10^?$
$10^{10} \times 10^2 = \underline{\ ?\ } = 10^?$
$10^1 \times 10^2 = \underline{\ ?\ } = 10^?$
$10^0 \times 10^3 = \underline{\ ?\ } = 10^?$

How are the exponents in the factors related to the exponent in the product?

Find the quotient by first writing each number in standard form, then dividing. Then rewrite the quotient as a power of ten.

$10^6 \div 10^2 = 1{,}000{,}000 \div 100 = 10{,}000 = 10^?$
$10^4 \div 10^3 = \underline{\ ?\ } = 10^?$
$10^3 \div 10^0 = \underline{\ ?\ } = 10^?$
$10^5 \div 10^1 = \underline{\ ?\ } = 10^?$
$10^2 \div 10^1 = \underline{\ ?\ } = 10^?$

How are the exponents in the dividend and divisor related to the exponent in the quotient?

SKILLS DEVELOPMENT

A number can be easier to work with if it is written in exponential form than if it is written in standard form. For example, 10^{10} is easier to work with than 10,000,000,000. Several rules govern operations with numbers written in exponential form. These rules are called the **laws of exponents.**

To find the product $2^2 \times 2^3$, you could write out the factors for each term and then write the product using exponents.

$$2^2 \times 2^3 = 2 \times 2 \times 2 \times 2 \times 2 = 2^5$$

A faster method is to use the **product rule** which states:

► To multiply numbers with the same base, add the exponents.
$a^m \times a^n = a^{m+n}$, so $2^2 \times 2^3 = 2^{2+3} = 2^5$

Example 1

Use the product rule to multiply.

a. $10^5 \times 10^2$ b. $a^2 \times a^2$

Solution

a. $10^5 \times 10^2$
$= 10^{5+2}$
$= 10^7$

b. $a^2 \times a^2$
$= a^{2+2}$
$= a^4$ ◄

To find the quotient $3^4 \div 3^3$, you could write out the factors for each term and then write the quotient using exponents.

$$3^4 \div 3^3 \text{ can be written as } \frac{3^4}{3^3} = \frac{3 \times 3 \times 3 \times 3}{3 \times 3 \times 3} = 3^1$$

Another method is to use the **quotient rule** which states:

► To divide numbers with the same base, subtract the exponents.

$$a^m \div a^n = a^{m-n}, \text{ so } 3^4 \div 3^3 = 3^{4-3} = 3^1$$

Example 2

Use the quotient rule to divide.

a. $10^3 \div 10^1$ b. $b^7 \div b^6$

Solution

a. $10^3 \div 10^1$
$= 10^{3-1}$
$= 10^2$

b. $b^7 \div b^6$
$= b^{7-6}$
$= b^1$
$= b$ ◄

A number written in exponential form can be raised to a power. You could first write out the factors and then write the product in exponential form. For example:

$$\begin{aligned}(3^3)^2 &= 3^3 \times 3^3 \\ &= 3 \times 3 \times 3 \times 3 \times 3 \times 3 \\ &= 3^6\end{aligned}$$

Compare the exponents in the original number to the exponents in the final number to determine their relationship. This relationship is described by the **power rule** which states:

► To raise an exponential number to a power, multiply the exponents.

$$(a^m)^n = a^{mn}, \text{ so } (3^3)^2 = 3^{3 \times 2} = 3^6$$

Example 3

Use the power rule.

a. $(4^3)^6$ b. $(y^9)^8$

Solution

a. $(4^3)^6 = 4^{3 \times 6} = 4^{18}$

b. $(y^9)^8 = y^{9 \times 8} = y^{72}$ ◄

The laws of exponents can be used when solving some word problems.

Example 4

It is believed that 10^4 craters form on the moon every 10^9 years. On average, about how many years are there between the formation of one crater and the next?

Solution

Divide the entire span of 10^9 years by 10^4, the number of craters.

$10^9 \div 10^4 = 10^{9-4} = 10^5 = 100{,}000$

A crater forms about once every 100,000 years. ◄

TRY THESE

Use the product rule to multiply.

1. $10^4 \times 10^2$ **2.** $5^8 \times 5^5$ **3.** $m^{21} \times m^7$ **4.** $d^6 \times d^3$

Use the quotient rule to divide.

5. $9^8 \div 9^2$ **6.** $4^{11} \div 4^0$ **7.** $p^{20} \div p^{10}$ **8.** $a^9 \div a^3$

Use the power rule.

9. $(3^6)^8$ **10.** $(7^2)^{10}$ **11.** $(x^4)^4$ **12.** $(a^8)^1$

Solve.

13. The total number of bacteria on a surface was estimated to be $(10^{10})^{50}$. If the same number of bacteria was on each of 1,000 of these surfaces, how many bacteria would there be?

EXERCISES

PRACTICE/ SOLVE PROBLEMS

Use the product rule to multiply.

1. $10^{21} \times 10^8$ **2.** $10^1 \times 10^9$ **3.** $5^6 \times 5^2$ **4.** $6^2 \times 6^3$

5. $n^3 \times n^3$ **6.** $d^{11} \times d^0$ **7.** $m^9 \times m^3$ **8.** $a^6 \times a^9$

Use the quotient rule to divide.

9. $10^4 \div 10^2$ **10.** $5^8 \div 5^5$ **11.** $2^8 \div 2^5$ **12.** $6^{16} \div 6^{15}$

13. $b^9 \div b^1$ **14.** $x^{21} \div x^7$ **15.** $y^7 \div y^5$ **16.** $n^6 \div n^3$

Use the power rule.

17. $(45^6)^{10}$ **18.** $(9^8)^8$ **19.** $(5^7)^9$ **20.** $(11^4)^5$

21. $(a^2)^{15}$ **22.** $(x^{20})^5$ **23.** $(n^7)^7$ **24.** $(p^3)^3$

Solve.

25. There are 10^2 centimeters in a meter and 10^3 meters in a kilometer. How many centimeters are there in a kilometer?

26. A googol is 10^{100}. It has been estimated that there are 10^{79} electrons in the universe. About how many times greater is the googol than the number of electrons?

EXTEND/ SOLVE PROBLEMS

Use the laws of exponents.

27. $(4^{10})^{15}$
28. $1^{16} \div 1^8$
29. $8^{24} \times 8^3$
30. $11^8 \div 11^4$
31. $7^{15} \div 7^3$
32. $6^{20} \times 6^5$
33. $(12^5)^5$
34. $2^{90} \times 2^9$
35. $8^{25} \times 8^5$
36. $(103^4)^6$
37. $10^7 \div 10^1$
38. $(5^7)^{10}$

Replace ● with <, >, or =.

39. $5^8 \div 5^6$ ● 5^4
40. $3^6 \times 3^5$ ● 3^{30}
41. $(7^6)^2$ ● $(7^3)^4$
42. $2^3 \times 2^6$ ● 2^8
43. $(9^{24})^8$ ● $(9^8)^{24}$
44. $8^{12} \div 8^6$ ● 8^3

USING DATA Use the Data Index on page 546 to find the Richter Scale. Find the difference in the ground movement between an earthquake measuring 2.5 and one with each magnitude below.

45. 3.5
46. 4.5
47. 5.5

Solve.

48. The number of atoms of oxygen in an average thimble is 1,000,000,000,000,000,000,000,000,000. Since a googol is equal to 10^{100}, about how many times greater is a googol than the number of atoms of oxygen in an average thimble?

THINK CRITICALLY/ SOLVE PROBLEMS

Find the value of each variable.

49. $6^3 \times 6^a = 6^{18}$
50. $7^4 \div 7^b = 7^2$
51. $(6^c)^3 = 6^{12}$
52. $5^d \div 5^8 = 5^5$
53. $4^e \times 4^2 = 4^{24}$
54. $(9^7)^f = 9^{49}$
55. $(2^g)^2 = 2^2$
56. $3^5 \times 3^h = 3^{45}$
57. $8^i \div 8^{15} = 8^{15}$

Solve.

58. Use the laws of exponents to write two factors that would have each of these products.

    a. a googol (10^{100})
    b. a googolplex (10^{googol})

2-8 Order of Operations

EXPLORE/ WORKING TOGETHER

List the steps needed to perform a simple activity, such as getting lunch. Cut your list apart, with each step of the activity on a separate strip of paper. Mix up the strips and place them face down on your desk. Have a partner choose one strip at a time, turning it face up and placing each succeeding strip below the previous one. Then read the steps in this new order. Explain why the order of the steps is important.

SKILLS DEVELOPMENT

To simplify an expression is to perform as many of the indicated operations as possible. Sometimes, to achieve a goal, you must take steps in a specified order. For instance, to start a car, you first have to put the key in the ignition. To simplify mathematical expressions, certain steps must be followed in a certain order. These rules are called the **order of operations.**

- First, perform all calculations within parentheses and brackets.
- Then, do all calculations involving exponents.
- Next, multiply or divide in order, from left to right.
- Finally, add or subtract in order, from left to right.

Sometimes there are no grouping symbols or exponents.

Example 1

Simplify.

a. $7 - 3 \times 2 + 9$ **b.** $8 \div 4 + 2 \times 3$

Solution

a.
$$\begin{aligned} & 7 - 3 \times 2 + 9 \\ &= 7 - 6 + 9 \\ &= 1 + 9 \\ &= 10 \end{aligned}$$

b.
$$\begin{aligned} & 8 \div 4 + 2 \times 3 \\ &= 2 + 6 \\ &= 8 \end{aligned}$$ ◄

Example 2

Simplify.

a. $6 - (4 + 8) \div 2$ **b.** $(3 \times 4)^2 + 6$

Solution

a.
$$\begin{aligned} & 6 - (4 + 8) \div 2 \\ &= 6 - 12 \div 2 \\ &= 6 - 6 \\ &= 0 \end{aligned}$$

b.
$$\begin{aligned} & (3 \times 4)^2 + 6 \\ &= (12)^2 + 6 \\ &= 144 + 6 \\ &= 150 \end{aligned}$$
◄

The way that you enter an expression on your calculator will depend on the type of calculator you have. Some calculators compute according to the order of operations. On calculators you compute through the use of memory keys.

Example 3

Use a calculator to simplify $25 - 2 \times 7 + 1$.

Solution 1

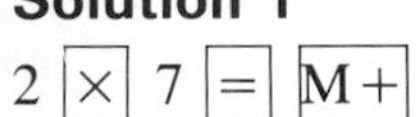

2 [×] 7 [=] [M+]

25 [−] [MR] [+] 1 [=]

Calculate 2 × 7 and enter it into memory.

Solution 2

Use this key sequence on a scientific calculator.

25 [−] 2 [×] 7 [+] 1 [=]

With either calculator, you will find that $25 - 2 \times 7 + 1 = 12$. ◄

You can write and evaluate expressions to solve some word problems.

Example 4

Ed took 100,000 pesos on his trip to Mexico. He bought 3 rings for 20,000 pesos each and 2 sombreros for 10,000 pesos each. Then he exchanged U.S. money for 150,000 pesos. How many pesos did he have then?

Solution

$$\begin{aligned} & 100{,}000 - 3 \times 20{,}000 - 2 \times 10{,}000 + 150{,}000 \\ &= 100{,}000 - 60{,}000 - 20{,}000 + 150{,}000 \\ &= 20{,}000 + 150{,}000 \\ &= 170{,}000 \end{aligned}$$

Ed had 170,000 pesos. ◄

COMPUTER

The computer's computations reflect the order of operations. Here is a list of how a computer "understands" mathematical operations.

Expression	Computer
3 + 2	3 + 2
3 − 2	3 − 2
3 × 2	3 * 2
3 ÷ 2	3/2
3^2	3^2 or 3↑2

Use the PRINT command with the expression you want to evaluate. For example:

```
]PRINT 3*4^2+6
54

]PRINT 24/4-2*3+5
5
```

TRY THESE

Simplify.

1. $8 - 4 \div 2 \times 3$
2. $8 \div 4 - 2 + 3$
3. $16 + 4 \times 3 \times 5$
4. $16 \times 4 + 3 \div 3$

Simplify. Remember to work within parentheses first.

5. $3 \times (8 + 2) \div 6$
6. $(12 - 2) \times 4 - 8$
7. $45 \div (3 + 6) + 5$
8. $6^2 \div 4 \div 2$
9. $6^2 \div (4 \div 2)$
10. $(3 + 3)^2 \times 5$

Simplify using a calculator. Show two different key sequences.

11. $25 \times (18 \div 6) - 5$

12. $(2 \times 16) - 3^3$

Write an expression and solve.

13. Colin ran 6 miles last week. He ran twice that distance this week. How far did he run over the two-week period?

EXERCISES

PRACTICE/ SOLVE PROBLEMS

Simplify. Use the order of operations.

1. $6 \times 4 - 4 + 1$

2. $14 - 4 \times 3 - 2$

3. $72 \div (3 + 5) \times 9$

4. $3 + 2^4 \times 4$

5. $49 - 8 \times (42 \div 7)$

6. $(18 + 7) \times 4 \div 2$

7. $4^3 - (6 \div 3)$

8. $(3 + 2)^4 \times 4$

Simplify using a calculator. Show two different key sequences.

9. $4^2 + 2 \times 6$

10. $9 \times 8 - 6 + 3$

Write an expression and solve.

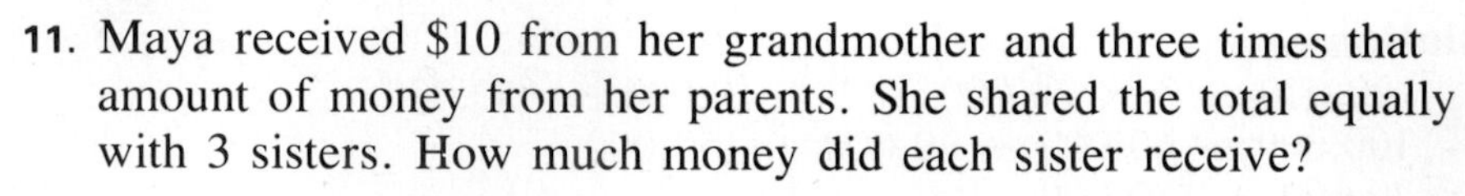

11. Maya received \$10 from her grandmother and three times that amount of money from her parents. She shared the total equally with 3 sisters. How much money did each sister receive?

12. Kirk had \$35.00. He bought 5 tubes of paint for \$2.50 each and 3 canvases for \$5.50 each. How much money did he have left?

EXTEND/ SOLVE PROBLEMS

Simplify.

13. $8 + 6 - 2 \times 2 - 3^2$

14. $4 \times 42 \div (56 \div 8 \times 3)$

15. $27 \div (7 + 2) \times (2 + 3)$

16. $63 \div (7 + 2) - 12 \div 3$

17. $19 + 2 \times (6 - 4)^3$

18. $(65 - 7^2 + 4) \times 9$

Write a numerical expression for each of the following.

19. twenty-five less the product of six and four

20. the sum of eight and nine, multiplied by the difference of ten and two

21. five squared multiplied by the difference of thirty-two and twenty-eight, divided by twenty

Use only one of the symbols $+$, $-$, $\times$, or $\div$, to make each number sentence true.

22. $8 \blacksquare 4 \blacksquare 2 = 1$
23. $12 \blacksquare 9 \blacksquare 3 = 0$
24. $6 \blacksquare 2 \blacksquare 4 = 12$
25. $7 \blacksquare 3 \blacksquare 1 = 11$
26. $8 \blacksquare 3 \blacksquare 6 = 144$
27. $12 \blacksquare 3 \blacksquare 4 = 1$

Solve. Write an expression if one is not given.

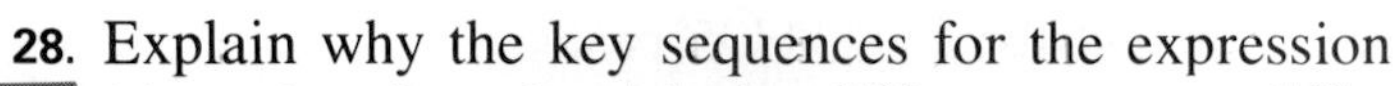

28. Explain why the key sequences for the expression $14 - 6 \times 2 \div 3$ might be different on two different calculators.
29. Jerome ran 4 miles farther than the mean distance of the other runners. The others ran 18 miles, 22 miles, 10 miles, and 26 miles. How far did Jerome run?
30. Lyn has \$60.00. She wants to purchase a hat for \$32.50, two scarves for \$15.75 each, and one hair ribbon priced at 4 for \$1.00. How much more money does she need?

THINK CRITICALLY/ SOLVE PROBLEMS

Write *true* or *false* for each. For any that are false, insert parentheses to make them true.

31. $4 - 3 \times 7 = 7$
32. $7 \times 0 + 8 = 56$
33. $5 \times 24 - 16 = 40$
34. $15 \div 3 - 2 = 3$
35. $16 + 2^2 \div 5 \times 3 = 12$
36. $4^2 \div 0 + 8 = 2$
37. $3^2 - 6 \div 3 = 1$
38. $8^2 + 5 \times 4 \div 2^2 = 69$

Use parentheses and the symbols $+$, $-$, $\times$, and $\div$ to make each number sentence true.

39. $7 \blacksquare 3 \blacksquare 1 \blacksquare 1 = 9$
40. $8 \blacksquare 3 \blacksquare 6 \blacksquare 5 = 4$
41. $10 \blacksquare 5 \blacksquare 5 \blacksquare 2 = 10$
42. $12 \blacksquare 0 \blacksquare 4 \blacksquare 3 = 1$
43. $30 \blacksquare 5 \blacksquare 5 \blacksquare 23 = 30$
44. $12 \blacksquare 3 \blacksquare 4 \blacksquare 5 = 16$

Solve.

45. You can change the value of $3 + 5 \times 4 - 1$ by using parentheses. Show four possible ways to do this. Then find the value of each.
46. Show two different key sequences you could use on two different kinds of calculators to simplify the following expression: $(5 \times 6^3 \div 10) \div 81$

2-9 Properties of Operations

EXPLORE

Describe how the sentences in each pair differ. Then, tell if the order affects the meaning.

A. One peso was exchanged for 0.0003305 dollars.
One dollar was exchanged for 0.0003305 pesos.
B. Jared had one 1,000-peso note and two 500-peso coins.
Jared had two 500-peso coins and one 1,000-peso note.
C. Bo and Jin bought souvenirs from Paco at the Pyramid of the Sun.
Bo bought souvenirs from Jin and Paco at the Pyramid of the Sun.
D. Peso notes and peso coins along with centavos are Mexican monies.
Peso notes along with peso coins and centavos are Mexican monies.

SKILLS DEVELOPMENT

You know that the order of operations always affects the answer to a mathematics exercise. However, the order in which addends or factors are arranged does not affect the answer. Certain patterns, stated as **properties,** can help you understand this.

The **commutative property** states that the order in which numbers are added or multiplied does not affect the sum or product:

$$a + b = b + a, \text{ so } 15 + 12 = 12 + 15$$

$$x \times y = y \times x, \text{ so } 3 \times 2 = 2 \times 3$$

Example 1

Complete.
a. $12 + 7 = ■ + 12$ **b.** $6 \times 12 = 12 \times ■$

Solution
Use the commutative property.

a. $12 + 7 = 7 + 12$
$19 = 19$
The unknown number is 7.

b. $6 \times 12 = 12 \times 6$
$72 = 72$ ◄
The unknown number is 6.

The **associative property** states that the grouping of factors does not affect the sum or product:

$$(a + b) + c = a + (b + c), \text{ so } (3 + 5) + 6 = 3 + (5 + 6)$$

$$(x \times y) \times z = x \times (y \times z), \text{ so } (4 \times 8) \times 2 = 4 \times (8 \times 2)$$

Example 2

Complete.

a. $(10 + 4) + 6 = ■ + (4 + 6)$ b. $(2 \times ■) \times 3 = 2 \times (5 \times 3)$

Solution

Use the associative property.

a. $(10 + 4) + 6 = 10 + (4 + 6)$
$14 + 6 = 10 + 10$
$20 = 20$
The unknown number is 10.

b. $(2 \times 5) \times 3 = 2 \times (5 \times 3)$
$10 \times 3 = 2 \times 15$
$30 = 30$ ◄
The unknown number is 5.

The **identity property of addition** states that the sum of zero and a number is always that number:

$$a + 0 = a, \text{ so } 236 + 0 = 236$$

The **identity property of multiplication** states that the product of 1 and any other factor is always that factor:

$$x \times 1 = x, \text{ so } 863 \times 1 = 863$$

The **property of zero for multiplication** states that any number multiplied by zero has a product of zero:

$$x \times 0 = 0, \text{ so } 213 \times 0 = 0$$

Example 3

Complete. Tell which property you used.

a. $17 + ■ = 17$ b. $1 \times ■ = 36$ c. $45 \times ■ = 0$

Solution

a. $17 + 0 = 17$
identity property of addition
The unknown number is 0.

b. $1 \times 36 = 36$
identity property of multiplication
The unknown number is 36.

c. $45 \times 0 = 0$
property of zero for multiplication ◄
The unknown number is 0.

Example 4

Lina bought a ceramic bowl for 3,325 pesos, a tea set for 2,425 pesos, and mugs for 4,675 pesos. How much did she spend in all?

Solution

$3{,}325 + 2{,}425 + 4{,}675$
$= 2{,}425 + (3{,}325 + 4{,}675)$
$= 2{,}425 + 8{,}000 = 10{,}425$

Use the Commutative and Associative Properties.

Lina spent 10,425 pesos. ◄

TRY THESE

Complete.

1. $6 \times ■ = 7 \times 6$
2. $1{,}588 + ■ = 72 + 1{,}588$
3. $6 \times ■ = 0$
4. $■ + 963 = 963$
5. $2.3 + 6.8 = 6.8 + ■$
6. $23 \times 48 = 48 \times ■$
7. $■ \times 1 = 2{,}586.25$
8. $248 \times ■ = 248$

Complete.

9. $(6 \times 8) \times 5 = 6 \times (8 \times ■)$
10. $(9 + 6) + 8 = 9 + (6 + ■)$
11. $4 + (4 + 6) = (■ + 4) + 6$
12. $(3 \times ■) \times 9 = 3 \times (8 \times 9)$

Simplify. Name the properties you used.

13. $8 \times 9 \times 0$
14. 85×1
15. $0.35 + 0$
16. $8 + 6 + 4$
17. $1.2 + 0.32 + 0.8$
18. $5 \times 16 \times 2$
19. $0.05 \times 1.32 \times 200$
20. $8 + 19 + 2$
21. $9 \times 3 \times 3$

Solve.

22. Wanda spent 36,400 pesos on Wednesday, 106,925 pesos on Thursday, and 65,000 pesos on Friday. How many pesos did she spend on the three days?

EXERCISES

PRACTICE/ SOLVE PROBLEMS

Complete.

1. $8.5 \times 5.8 = 5.8 \times ■$
2. $■ \times 72 = 72$
3. $■ + 0 = 963$
4. $■ + 768 = 768 + 432$
5. $9.36 + 2.58 = ■ + 9.36$
6. $2.5 = ■ \times 1$
7. $713 \times ■ = 0$
8. $19 \times ■ = 25 \times 19$

Complete.

9. $■ \times (3 \times 9) = (2 \times 3) \times 9$
10. $(7 + 8) + 2 = ■ + (8 + 2)$
11. $6 \times (8 \times 5) = (■ \times 8) \times 5$
12. $1 + (■ + 2) = (1 + 7) + 2$

Simplify using mental math. Name the properties you used.

13. $30 + 0 + 30$
14. $0.25 + 72 + 0.75$
15. $13 + 84 + 16$
16. $5 \times 6 \times 2$
17. $0.09 \times 9 \times 100$
18. $4 \times 1 \times 6$
19. $3 \times 0 \times 8$
20. $1 \times 11 \times 5$
21. $5 \times 82 \times 2$

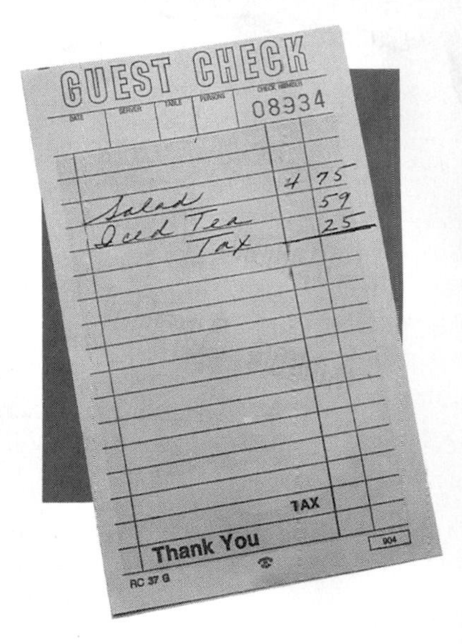

Solve.

22. A server totaled a customer's lunch check. The customer had a salad for \$4.75 and iced tea for \$0.59. Tax on the meal was \$0.25. What was the total?

23. Carla bought 5 packages of socks. Each package had 4 rows of socks with 2 socks in each row. How many socks did she buy?

EXTEND/ SOLVE PROBLEMS

Simplify using mental math.

24. $6 + 9 + 4 + 1$
25. $2 \times 8 \times 5 \times 9$
26. $19 + (2 \times 2) + 1$
27. $14 + 8 + 206 + 92$
28. $0.68 + 0.7 + 0.32 + 1.3$
29. $0.029 \times 100 \times 0 \times 8$

Solve. Use mental math.

30. A student had saved \$4. She earned four times that amount on Friday and six times the original savings on Saturday. How much did she have then?

31. A customer was checking her restaurant bill. She had charges of \$6.72, \$9.01, \$2.28, \$0.75, \$1.25, and \$1.98. The written total, \$22.00, was incorrect. What was the correct total?

32. Janelle rolled coins to take to the bank. She had 6 rolls of 20 dimes each. How much paper money did she receive from the bank in exchange?

THINK CRITICALLY/ SOLVE PROBLEMS

Solve.

33. To simplify $4 \times 16 + 6 + 5 + 4$, can you use the commutative property with $16 + 6$? Explain.

34. To simplify $24 \div 4 \times (3 + 2) \div 6$, can you use the commutative property with 4×3? Explain.

Decide whether each statement is *true* or *false*. Explain.

35. The commutative property cannot be used when dividing.

36. The associative property cannot be used when subtracting.

37. A property can be used only once to evaluate each expression.

2-10 Distributive Property

EXPLORE

Study each diagram and the expression written below it. Then simplify each expression.

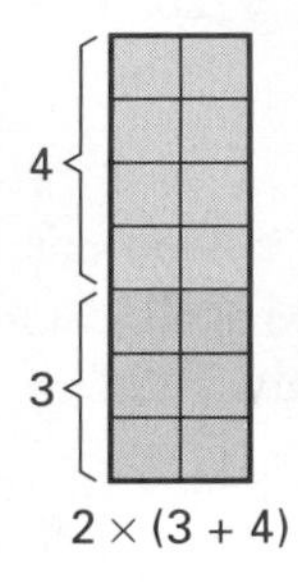

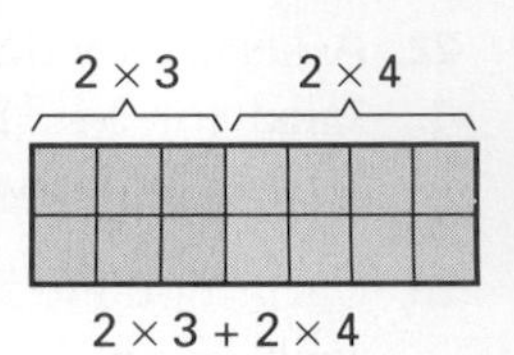

Compare the rows, columns, and the total number of squares in the two diagrams. Then compare the two expressions. How are they alike? How are they different?

Make similar diagrams for numbers of your choice. Then, write two expressions for your diagrams.

SKILLS DEVELOPMENT

There may be more than one expression that will solve a problem. Examine these expressions. Both could be used to answer the same problem and both have the same answer.

$7(12 - 3)$	$7 \times 12 - 7 \times 3$
$= 7 \times 9$	$= 84 - 21$
$= 63$	$= 63$

The **distributive property** states that a factor outside parentheses can be used to multiply *each* term within the parentheses:

$a(b + c) = (a \times b) + (a \times c)$, so $5(3 + 2) = (5 \times 3) + (5 \times 2)$

$x(y - z) = (x \times y) - (x \times z)$, so $6(4 - 1) = (6 \times 4) - (6 \times 1)$

TALK IT OVER

Explain why the term *distributive property* is an appropriate name for this property.

Example 1

Complete.

a. $2(3 + 4) = (2 \times 3) + (\blacksquare \times 4)$

b. $\blacksquare(6 - 1) = (3 \times 6) - (3 \times 1)$

Solution

Use the distributive property.

a. $2(3 + 4) = (2 \times 3) + (2 \times 4)$
$2(7) = 6 + 8$
$14 = 14$ The unknown number is 2.

b. $3(6 - 1) = (3 \times 6) - (3 \times 1)$
$3(5) = 18 - 3$
$15 = 15$ The unknown number is 3. ◄

The distributive property can also be applied in reverse.

$a \times b + a \times c = a(b + c)$, so $8 \times 6 + 8 \times 5 = 8(6 + 5)$

$x \times y - x \times z = x(y - z)$, so $7 \times 5 - 7 \times 3 = 7 \times (5 - 3)$

Example 2

Use the distributive property to rewrite each expression.

a. $6 \times 45 + 6 \times 3$ b. $9 \times 5 - 9 \times 4$

Solution

a. $6 \times 45 + 6 \times 3$
$= 6(45 + 3)$

b. $9 \times 5 - 9 \times 4$
$= 9(5 - 4)$ ◄

You can use the distributive property to break up a number by thinking of it as the sum or difference of two numbers that are easy to work with mentally.

Example 3

Simplify using mental math.

a. 20×130 b. 48×40

Solution

a. 20×130
$= 20 \times (100 + 30)$
$= (20 \times 100) + (20 \times 30)$
$= 2{,}000 + 600$
$= 2{,}600$

b. 48×40
$= (50 - 2) \times 40$
$= (50 \times 40) - (50 \times 2)$
$= 2{,}000 - 100$
$= 1{,}900$ ◄

Use the distributive property to simplify variable expressions.

Example 4

Simplify.

a. $6(4 + x)$ b. $2(y - 7)$

Solution

Use the distributive property to simplify.

a. $6(4 + x)$
$= 6 \times 4 + 6 \times x$
$= 24 + 6x$

b. $2(y - 7)$
$= 2 \times y - 2 \times 7$
$= 2y - 14$ ◄

Example 5

Alicia and her friends crocheted 21 afghans for the homeless shelter. Each afghan was made up of 64 squares. How many squares did Alicia and her friends crochet?

Solution

To find the total number of squares, multiply 21 and 64.

21×64
$= 20 \times 64 + 1 \times 64$ Break up one factor to use the distributive property.
$= 1{,}280 + 64$ Compute, using the order of operations.
$= 1{,}344$

Alicia and her friends crocheted 1,344 squares. 

TRY THESE

MIXED REVIEW

Use the Data Index on page 546 for Exercises 1 and 2.

1. Decide whether a line graph is an appropriate way to express the data in the exchange-rate table. Explain.
2. Explain how you could use a pictograph to show the relative values of U.S. coins to one another.

Find the mean, the median, and the mode for each set of data.

3. 5.4, 2.9, 3.785, 6.04, 2.9
4. 7.3, 8.72, 7.3, 9.085, 4.7

Complete.

1. $5(4 + 2) = (5 \times 4) + (■ \times 2)$
2. $■(3 - 1) = (2 \times 3) - (2 \times 1)$

Rewrite each expression using the distributive property.

3. $9 \times 3 + 9 \times 2$
4. $7 \times 3 + 7 \times 6$
5. $8 \times 8 - 8 \times 4$
6. $4 \times 9 - 4 \times 3$

Simplify using mental math.

7. 9×62
8. 360×2
9. 7×39
10. 82×4
11. 105×3
12. 28×4

Simplify.

13. $2(x - 3)$
14. $18(2 + y)$
15. $21(z + 1)$
16. $9(8 - a)$
17. $11(4 + b)$
18. $8(c - 6)$

Write an expression using the distributive property and solve.

19. The student body filled 11 buses to travel to a football game. Each bus held 36 students. How many students went to the game?

EXERCISES

PRACTICE/ SOLVE PROBLEMS

Complete.

1. $6(3 + 1) = (■ \times 3) + (6 \times 1)$
2. $■(9 + 5) = (4 \times 9) + (4 \times 5)$
3. $9(8 - 5) = (9 \times ■) - (9 \times 5)$

Rewrite each expression using the distributive property.

4. $7 \times 4 + 7 \times 8$
5. $9 \times 7 - 9 \times 3$
6. $2 \times 18 - 2 \times 11$
7. $14 \times 12 + 14 \times 16$

Simplify using mental math.

8. 6×28
9. 5×26
10. 16×9
11. 108×6
12. 7×18
13. 3×198

Simplify.

14. $4(7 - a)$ 15. $6(4 + b)$ 16. $10(c + 10)$

17. $7(d + 1)$ 18. $5(e - 8)$ 19. $3(f - 2)$

Write an expression using the distributive property and solve.

20. There were 17 workers on an assmbly line. In one hour, each worker was expected to put together 9 kits. How many kits in all should be assembled each hour?

EXTEND/ SOLVE PROBLEMS

Find each answer using mental math.

21. 90×62 22. 360×20 23. 70×39

24. 82×40 25. 105×30 26. 25×12

27. 108×60 28. 70×18 29. 30×198

Simplify using the distributive property.

30. 108×210 31. 310×498 32. 961×50

33. $2,005 \times 11$ 34. 603×306 35. 492×201

36. $7,924 \times 80$ 37. $2,900 \times 109$ 38. 815×501

Write an expression using the distributive property and solve.

39. A warehouse contained 198 cases of product. Each case held 25 packages. How many packages were in the warehouse?

40. A gas pipeline pumps 402,000 gallons of crude oil per day. At this rate, how many gallons are pumped in one week?

41. A school homecoming button costs $1.90. How much would it cost if every student in your math class bought a button?

THINK CRITICALLY/ SOLVE PROBLEMS

Write *yes* if the expression is an example of the distributive property. Write *no* if it is not and explain why.

42. $(6 \times 5) \times (6 \times 4) = 30 \times 24$

43. $8 \times 1 + 8 \times 3 = 8(1 + 3)$

44. $17 \times 1 \div 17 \times 1 = 17 \div 17$

Decide whether each is *true* or *false*. Explain.

45. The distributive property cannot be used with division.

46. The distributive property cannot be used when any other properties are used to solve an expression.

47. $a(b - c) = (b - c)a$

2-11 Problem Solving Strategies:
FIND A PATTERN

► READ
► PLAN
► SOLVE
► ANSWER
► CHECK

Sometimes a problem is easier to understand and solve if you can find a pattern in the problem.

CALCULATOR TIP

Whenever you start to work with a calculator, always push [c] first in order to clear the display.

Add 368 + 4,961.

Enter	Calculator Display
[C]	0
368	368
[+]	368
4961	4961
[=]	5329

PROBLEM

Danny's parents put $1 into his savings account on his first birthday, $2 on his second birthday, $4 on his third birthday, $8 on his fourth birthday, and so on. How much money will Danny's parents put into his account on his tenth birthday?

SOLUTION

READ Known: Danny received $1 for his first birthday, $2 for his second, $4 for his third, and $8 for his fourth.

PLAN Find: How much he will receive on his tenth birthday

SOLVE Make a table to organize the information. Then look for a pattern.

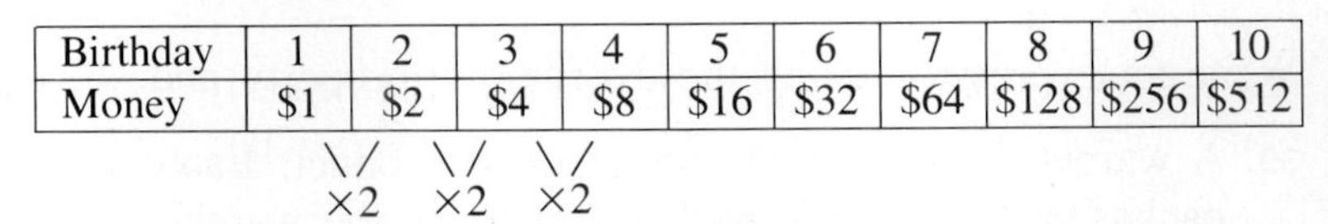

Birthday	1	2	3	4	5	6	7	8	9	10
Money	$1	$2	$4	$8	$16	$32	$64	$128	$256	$512

×2 ×2 ×2

Look at the first four entries in the table. Each amount of money is multiplied by 2 to get the next amount. The pattern continues through Danny's tenth birthday.

ANSWER Danny's parents will put $512 into his account on his tenth birthday.

CHECK Use this key sequence on your calculator to check:

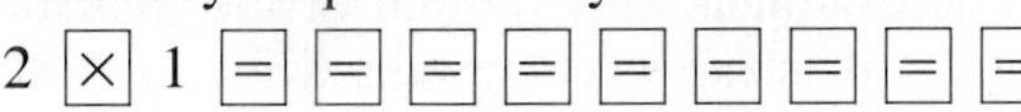

2 [×] 1 [=] [=] [=] [=] [=] [=] [=] [=] [=]

PROBLEMS

Which rule describes the pattern? Copy and complete.

1. 0.08, 0.16, 0.24, ■, ■
   a. Add 0.8.
   b. Subtract 0.8.
   c. Add 0.08.
   d. Subtract 0.08.

2. 2.45, 2.05, ■, 1.25, ■
   a. Subtract 0.5.
   b. Subtract 0.4.
   c. Subtract 0.05.
   d. Subtract 0.04.

3. 7.432, 8.543, 9.654, ■, ■
   a. Add 1.
   b. Add 1.111.
   c. Add 1.11.
   d. Add 1.001.

4. 1.5, 1.0, 2.0, 1.5, ■, ■
   a. Add 0.5, subtract 1.0.
   b. Add 2.0, subtract 0.5.
   c. Subtract 0.5, add 0.5.
   d. Subtract 0.5, add 1.0.

Write the rule. Then copy and fill in the missing numbers.

5. 1.5, 1.3, 1.1, ■, ■, 0.5
6. 0.3, 0.6, 0.9, ■, ■, 1.8
7. 0.9, 4.9, 4.8, 8.8, ■, ■
8. 5.2, 3.2, 5.2, 3.2, ■
9. 9.34, 8.34, 8.54, 7.54, ■, ■
10. 1.005, 1.001, 1.003, ■, 1.001, ■

Solve.

11. Sondra plans to put $5 in her savings account in January, $8 in February, $11 in March, $14 in April, and so on.
    a. How much money will she put into the account in December?
    b. How much money will she put into the account for the entire year.

12. A movie theater plans to give away 2 theater tickets on Monday, 8 tickets on Tuesday, 32 tickets on Wednesday, and so on for Thursday and Friday.
    a. How many tickets will they give away on Friday?
    b. How many tickets will they give away for the five days?

13. Leroy went on a cross-country bike trip. He traveled 9 miles the first day, 15 miles the second day, 21 miles the third day, and so on. At this rate, how many miles will he have traveled by the end of the eighth day?

14. Melinda jogs 1 lap around the track the first week, 3 laps the second week, 6 laps the third week, and 10 laps the fourth week. If she continues this pattern, how many laps will she jog the eighth week?

15. The occurrence of the Fibonacci sequence of numbers appears frequently in nature. The arrangement of the parts of many plants reflects this sequence.

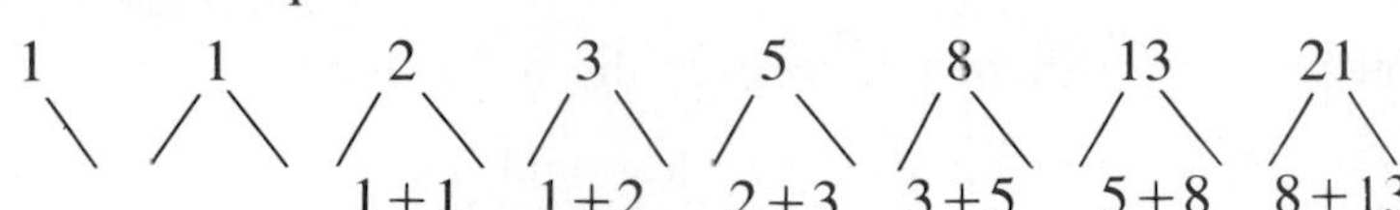

    a. Continue the pattern to find the next three numbers.
    b. Find at least three other instances in which the Fibonacci sequence occurs in nature.

CHAPTER 2 • REVIEW

1. A letter representing an unknown number is a(n) __?__ .
2. 3^4 is written in __?__ .
3. Frequently very large or very small numbers are written in __?__ .
4. Rules that must be followed when evaluating expressions are called the __?__ .
5. According to the __?__ property, the order in which two numbers are added or multiplied does not affect the answer.

a. commutative
b. order of operations
c. exponential form
d. scientific notation
e. variable

SECTION 2–1 WHOLE NUMBERS (pages 42–45)

► To round numbers, locate the digit in the place to which you are rounding.
a. Increase this digit by 1 if the next digit to the right is 5 or greater.
b. Leave the digit the same if the next digit to the right is less than 5.

Find each answer. Round and compare to see if your answer is reasonable.

6. $456 + 938$
7. $1{,}092 - 657$
8. 89×31
9. $1{,}189 \div 29$

SECTION 2–2 VARIABLES AND EXPRESSIONS (pages 46–49)

► When writing an expression to describe a real-world situation, you can use a variable to represent an unknown number.

Evaluate each expression. Let $a = 2$, $b = 4$, and $c = 6$.

10. $a + b$
11. $c - a$
12. $a + c$
13. $b - 2a$
14. $5c \div 3a$
15. $2c \div b$
16. $\frac{3b}{2a}$
17. $6a \times 2c$

SECTIONS 2–3 and 2–11 PROBLEM SOLVING (pages 50–51, 80–81)

► Before you can solve a problem, you need to decide on a strategy and choose the operation needed to solve the problem.

USING DATA Use the table on page 41 to help you answer the following.

18. What is the difference in the exchange rate for the day's low and the 12-month low for the German mark?
19. Use the day's high to find how many U.S. dollars are equal to 15 British pounds.

SECTIONS 2–4 and 2–5 OPERATIONS WITH DECIMALS (pages 52–59)

- When adding or subtracting decimals, align the decimal points.
- When multiplying or dividing decimals, compute just the way you do with whole numbers. In multiplication, the answer has as many decimal places as the sum of the decimal places in the factors.

Find each answer. Estimate and compare to see if your answer is reasonable.

20. $8.679 + 8.543$ **21.** $9.768 - 4.389$ **22.** $4.6 - 2.037$

23. 6×3.954 **24.** 0.32×1.4 **25.** $2.65 \div 0.5$

SECTIONS 2–6 and 2–7 EXPONENTS AND SCIENTIFIC NOTATION (pages 60–67)

- It is easier to work with large numbers when they are written in scientific notation.

Write in scientific notation.

26. 8,409 **27.** 732,037 **28.** 388 **29.** 7,000,849

Write each answer in exponential form.

30. $5^8 \times 5^3$ **31.** $9^8 \div 9^4$ **32.** $x^8 \times x^3$ **33.** $y^5 \div y^2$

SECTION 2–8 ORDER OF OPERATIONS (pages 68–71)

- Follow the **order of operations** to simplify mathematical expressions.

Simplify.

34. $8 \div 4 + 9 \times 3$ **35.** $6 + (2 \times 3) - 7$ **36.** $9 \times 2^2 + 1$

37. $7 \times (3 - 1) + 2$ **38.** $4 \times 5 + 12 \div 4$ **39.** $3^2 \div 9 + 8$

SECTIONS 2–9 and 2–10 NUMBER PROPERTIES (pages 72–79)

- Some of the ways in which we perform operations on numbers are governed by the **commutative, associative, distributive,** and **identity properties** and by the **property of zero for multiplication.**

Complete. Identify the property.

40. $(8 + 3) + 5 = 8 + (■ + 5)$ **41.** $3,492 \times ■ = 0$

42. $16 + 9 + ■ = 9 + 43 + 16$ **43.** $(a + b) + ■ = (a + b)$

CHAPTER 2 • TEST

Find each answer. Estimate and compare to see if your answer is reasonable.

1. $2,154 + 3,895$
2. 38×49
3. $97 \times 3,125$
4. $5,551 \div 61$
5. $3.45 + 12.3 + 0.967$
6. $48.2 - 0.345$
7. 0.34×0.34
8. 5×2.4
9. $36.54 \div 18$

Evaluate each expression. Let $a = 2$ and $b = 6$.

10. $a + 19$
11. $b \div 2$
12. $2b \div 2a$
13. $6a \div b$

Write in exponential form.

14. $8 \times 8 \times 8 \times 8$
15. $5 \times 5 \times 5$
16. $6 \times 6 \times 6 \times 6 \times 6 \times 6$

Write in standard form.

17. 7^3
18. 2^6
19. 3^4
20. 4^3

Write in scientific notation.

21. 6,893
22. 349
23. 435,982
24. 4,358,231

Write each answer in exponential form.

25. $8^2 \times 8^9$
26. $7^{12} \div 7^3$
27. $x^8 \div x^3$
28. $y^5 \times y^4$

Simplify.

29. $7 \times 8 - 36 \div 4$
30. $4 \times 2^3 + 10$
31. $(2 + 2)^2 \times 3$

Complete.

32. $(7 \times 9) \times 4 = ■ \times (9 \times 4)$
33. $■ \times 3,458 = 3,458$
34. $3.75 \times ■ = 0$
35. $7.6 \times 8.4 = 8.4 \times ■$
36. $873 + ■ = 954 + 873$
37. $87.3 + ■ = 87.3$

Simplify.

38. $5(8 - x)$
39. $4(a - 3)$
40. $7(5 - y)$
41. $6(n + 5)$

Solve.

42. Joan put $25 in her savings account in January, $28 in February, $31 in March, $34 in April, and so on. Following this pattern, find how much money she put into the account in December.

CHAPTER 2 ● CUMULATIVE REVIEW

1. A car dealer hired students to determine the ages of the cars owned by the people in one neighborhood. The students recorded the following ages (in years).

   3 5 1 2 6 8 2 4 5 5 6 2 1 2 3 4 4 5
   4 2 3 7 7 1 3 3 1 6 7 7 6 8 2 2 3 1

   Make a pictograph for these data.

Find the mean, median, mode, and range for each of the following sets of data. Round your answers to the nearest tenth.

2. 43 49 56 61 45 52 74
3. 46 48 50 50 52 53 49
4. 6.6 5.9 6.8 9.6 5.2 6.9 9.9 8.8 9.6
5. 30 30 45 28 31 31 45 30
6. What number must be included in the following data for the median to be 14?

   18 11 22 11 10
7. What number must be included in the following data for the mean to be 6.44?

   4.25 3.14 8.72 9.19

Use the following data for Exercises 8–12. The number of basketball free throws sunk by 40 students each taking 50 free throws was recorded as follows:

17	8	29	32	15	31	22	4	35	11
23	39	38	6	42	17	50	34	48	25
3	37	27	43	28	36	41	9	41	29
26	12	42	34	14	9	27	37	21	35

8. Construct a stem-and-leaf plot for these data.

How many students had the following numbers of successful throws?

9. 16–24
10. 40–48
11. less than 16
12. 32 or more

Use the bar graph for Exercises 13–15.

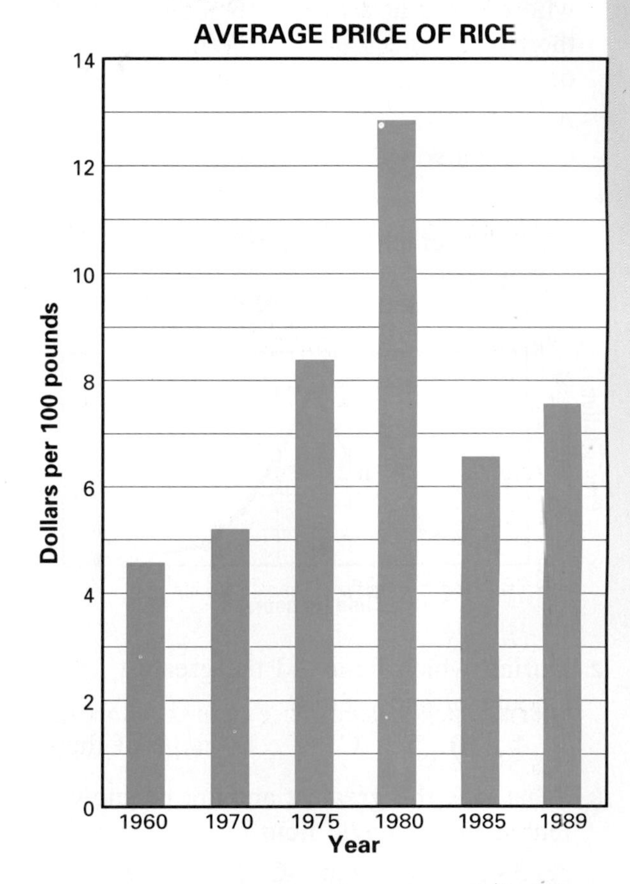

13. In which year was the price of rice the greatest?
14. In which years was the price under \$6?
15. About how much of a price difference was there between 1970 and 1975?
16. In which year was the price about one-half of what it was in 1980?

Find each answer. Estimate to see if the answer is reasonable.

17. $8.349 + 13.83$
18. $18.2 - 0.375$
19. 4.8×0.31
20. $4.941 \div 6.1$
21. $24.96 \div 1.3$
22. 0.02×0.08
23. $83.5 - 9.678$
24. $1.836 \div 1.2$

1. A soap manufacturer sampled people who crossed at a busy intersection about their preferences of soap. What type of sampling did the manufacturer use?
   A. systematic B. convenience
   C. random D. cluster

Use the line graph for Exercises 2–3.

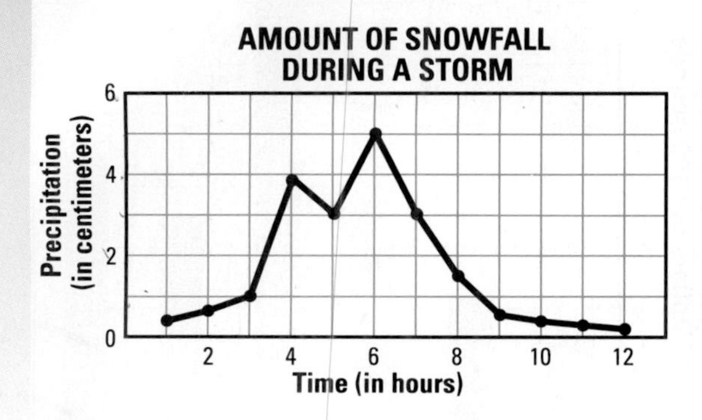

2. During which hour did the greatest amount of snow fall?
   A. 4 B. 5 C. 8 D. none of these

3. What was the greatest amount of snow that fell in a single hour?
   A. 3 cm B. 4 cm C. 5 cm D. 6 cm

4. If a car cost $15,000 and the expected life of the car is 10 years, what is the average cost per year?
   A. $15,000 B. $1,500
   C. $1,250 D. $150

5. Here is the amount of rainfall for a year recorded in centimeters.

J	F	M	A	M	J	J	A	S	O	N	D
63	67	56	43	19	0	0	21	45	57	59	60

   If a stem-and-leaf plot were made for these data, what would be the extremes of the stems?
   A. 6 and 6 B. 1 and 6
   C. 0 and 6 D. 0 and 9

6. A statistic in the newspaper stated that in 1987 the average family size was 3.19 people. What measure of central tendency does 3.19 represent?
   A. mean B. median
   C. mode D. range

7. The median age in the United States in 1987 was 32.1. What can you conclude from this?
   A. Children made up about half of the population.
   B. Half of the population was older and half was younger than 32.1 years.
   C. The mean age of the population was 32.1 years.
   D. More people were 32.1 years old than any other age.

Use this data for Exercises 8–10.

8	19	34	5	39	41	58	63	12

8. What is the mean of the data?
   A. 41 B. 39 C. 31 D. none of these

9. What is the range of the data?
   A. 34 B. 31 C. 58 D. none of these

10. What is the median of the data?
    A. 34 B. 63 C. 31 D. none of these

11. Find 16.3 − 0.361.
    A. 16.061 B. 15.939
    C. 16.661 D. 15.949

12. Find 2.7 × 0.49.
    A. 0.1323 B. 13.23
    C. 1.323 D. 0.01323

13. Find 2.91 ÷ 1.5
    A. 1.94 B. 194
    C. 19.4 D. 1,940

CHAPTER 3 SKILLS PREVIEW

1. Refer to the figure below. Make a statement about line segments HJ and JK. Then check to see whether your statement is true or false.

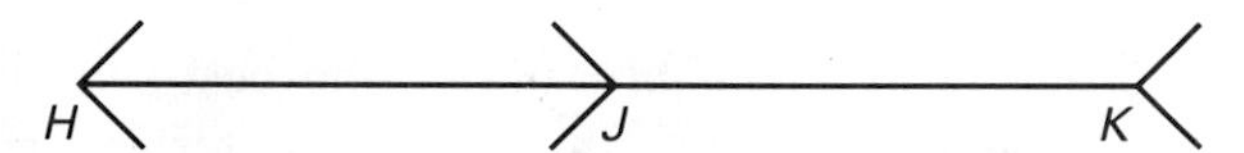

2. Choose three single-digit numbers. Multiply the first by 2. Add 1. Multiply by 5. Add the second number. Multiply by 10. Add the third number and subtract 50. What number do you get? Repeat this process several times with different sets of digits. Make a conjecture about the result of the process.

3. Give an example to show why the following conjecture is false:

   The product of two one-digit whole numbers is a two-digit whole number.

4. Write two *if–then* statements, using the following two sentences:

   The temperature of the water is 212°F.
   The water will boil.

For each of the following statements, write *true* or *false*. If false, give a counterexample.

5. If a number is divisible by 6, then it is divisible by 12.

6. If a woman lives in Dallas, then she lives in Texas.

Determine whether the following arguments are *valid* or *invalid*.

7. If two numbers are odd,
   then their sum is even.
   The sum of a and b is even.

   Therefore, a and b are odd.

8. If a plant is a cactus, then
   it has thorns.
   This plant is a cactus.

   Therefore, this plant has thorns.

9. In a survey of 150 students, it was found that 83 participated in sports, 62 participated in chorus, and 70 participated in band. There were 56 who participated in chorus and band, 12 who participated in sports and chorus, and 15 who participated in sports and band. There were 7 who participated in all three. How many students did *not* participate in sports or chorus or band?

10. Max, Bert, and Laura each play one sport. One plays football, one plays basketball, and one plays baseball. Bert is a poor student. Laura is an only child and a cheerleader for the football team. Each basketball player has a brother or sister, and all of them are good students. What sport does each person play?

CHAPTER 3

REASONING LOGICALLY

THEME Investigating, Solving Mysteries, Detective Work

Mystery and detective novels are among the best-selling books of all time. The most popular American mystery writer is Erle Stanley Gardner (1889–1970). So far more than 320,000,000 copies of his books, printed in 37 languages, have been sold.

The top-selling female mystery writer is British author Agatha Christie (1890–1976). Her 78 mystery novels have been printed in over 183 languages. To date, more than 2 billion copies of Christie's books have been sold.

Two mystery writers named Frederic Dannay and Manfred B. Lee wrote as a team. The two men were cousins, and they published their work under the name Ellery Queen. The following list of fictional detectives and their creators was prepared by Ellery Queen for *The Book of Lists*.

ELLERY QUEEN'S 17 GREATEST FICTIONAL DETECTIVES OF ALL TIME (in alphabetical order)

Fictional Detectives	Authors Who Created Them
1. Uncle Abner	Melville Post
2. Lew Archer	Ross Macdonald
3. Father Brown	G. K. Chesterton
4. Albert Campion	Margery Allingham
5. Charlie Chan	Earl Derr Biggers
6. C. Auguste Dupin	Edgar Allan Poe
7. Dr. Gideon Fell	John Dickson Carr
8. Sherlock Holmes	A. Conan Doyle
9. Inspector Maigret	Georges Simenon
10. Philip Marlowe	Raymond Chandler
11. Miss Marple	Agatha Christie
12. Perry Mason	Erle Stanley Gardner
13. Hercule Poirot	Agatha Christie
14. Ellery Queen	Ellery Queen (Frederic Dannay and Manfred B. Lee)
15. Sam Spade	Dashiell Hammett
16. Lord Peter Wimsey	Dorothy L. Sayers
17. Nero Wolfe	Rex Stout

DECISION MAKING

Using Data

Read each statement. Then determine from the information above whether or not the statement is true. Write *true, false,* or *cannot tell*.

1. Agatha Christie created more than one fictional detective.
2. Sherlock Holmes is the best of the detectives listed.
3. Lord Peter Wimsey was created by G. K. Chesterton.
4. Ellery Queen was a real person.
5. There are only three female authors on the list.

Working Together

Your group's task is to read a story by one of the writers on the list above. You might choose A. Conan Doyle's "The Red-headed League," Dorothy L. Sayers's "The Inspiration of Mr. Budd," or Agatha Christie's *Nemesis*. Write a paragraph describing the reasoning the detective in the story uses to solve the mystery. Discuss whether or not you think the detective's conclusions were drawn logically. Then suggest other ways the mystery could have been solved.

3-1 Optical Illusions

EXPLORE A 5-unit by 13-unit rectangle is shown below.

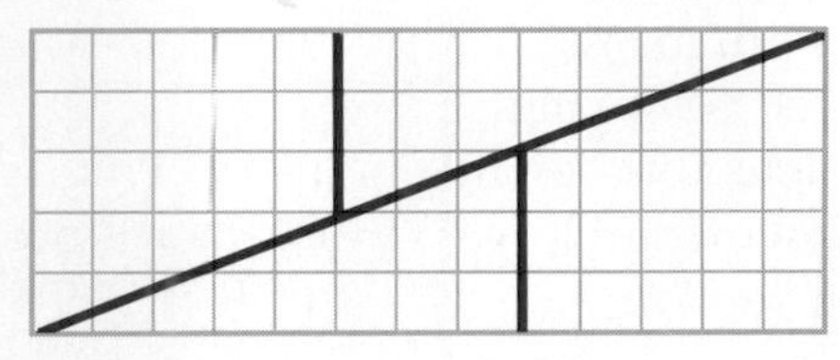

a. Copy the rectangle as shown.

b. Cut it into pieces along the heavy lines shown and rearrange the pieces to form a square. Tape the pieces together.

c. What do you think is the relationship between the area of the rectangle and the area of the figure you made?

d. Count square units to compute the area of the original rectangle. Now count square units to compute the area of the figure that you formed. What do you discover?

e. Discuss the answers to parts c and d. What explains your conclusion?

SKILLS DEVELOPMENT

Look carefully at the set of railroad tracks shown at the right. The two rails appear to get closer and closer together, but they do not actually do so. The distances between the railroad ties seem to get shorter and shorter, but they do not really do so. The trees along the sides of the railroad tracks appear to become smaller and smaller, yet they are all actually about the same height. The picture you see is an **optical illusion.** In an optical illusion, the human eye perceives, or pictures, something to be true that is actually not true.

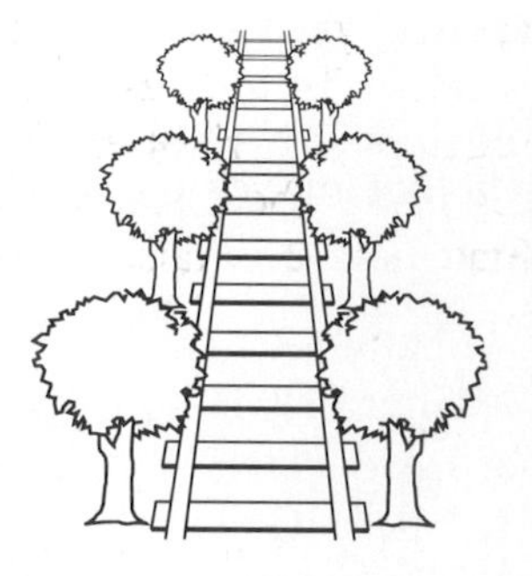

You can make a statement about what you believe to be true of something. A **statement** is a kind of sentence that is either true or false. You can find out if a statement is true or false by testing it.

Example 1

Refer to the figure at the right. Make a statement about the lengths of line segments AB and AC. Then check to see if your statement is true or false.

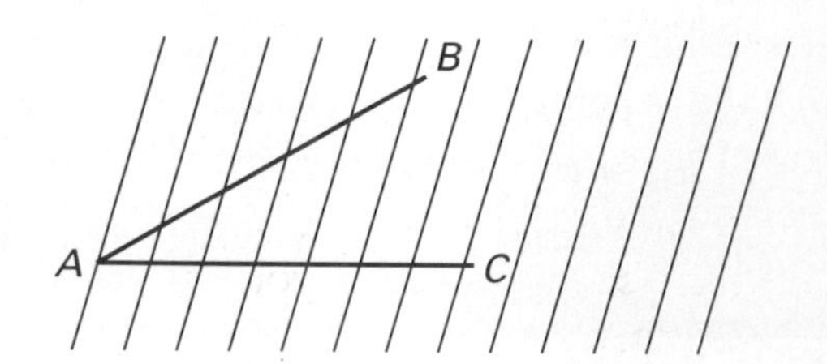

Solution

Line segment *AC* crosses more slant lines than does line segment *AB*.

Statement: *Line segment* AC *is longer than line segment* AB.

Measure both segments with a ruler. You will find that they have the same length. The statement is false. ◄

Example 2

Refer to the figure below. Make statements about the length and direction of *DE* and *FG*. Then check to see if each statement is true or false.

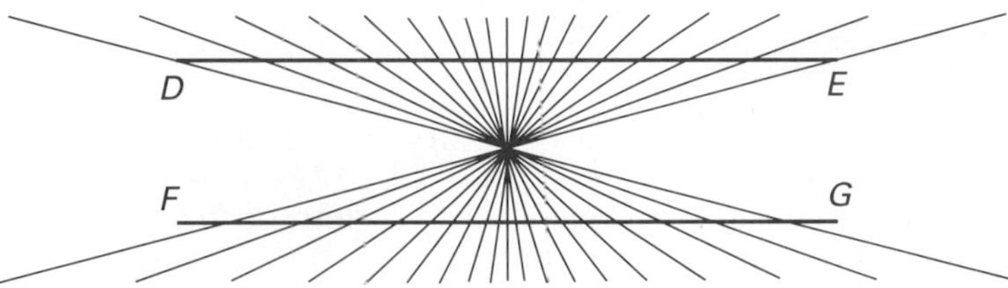

Solution

FG crosses the same number of lines as does *DE*. *FG* extends farther out past the lines, however. *FG* and *DE* seem to bow away from each other in the middle.

Statement 1: FG *is longer than* DE.
Statement 2: FG *and* DE *are not parallel to each other.*

Place a ruler next to *DE*, then next to *FG*. You will find that both are straight line segments, and they have the same length. Statement 1 is false. Now measure the distance between the left end of each segment, the right end of each segment, and the middle of each segment. The distance is the same. The line segments are parallel. Statement 2 is false. ◄

MATH IN THE WORKPLACE

Architects and interior designers use their knowledge of visual perception when they design buildings or interior spaces. For example, to make a building look taller, an architect might use tall, narrow windows. To make a room look wider, an interior designer might use horizontal stripes.

A fashion designer would seldom put horizontally striped clothing on a short person. Why?

TRY THESE

1. Refer to the figure at the right. Write a statement about the lengths of line segments *EF* and *GH*. Then check to see if your statement is true or false.

2. Refer to the figure at the right. Write a statement about *AB*, *AC*, and *BC*. Then check to see if your statement is true or false.

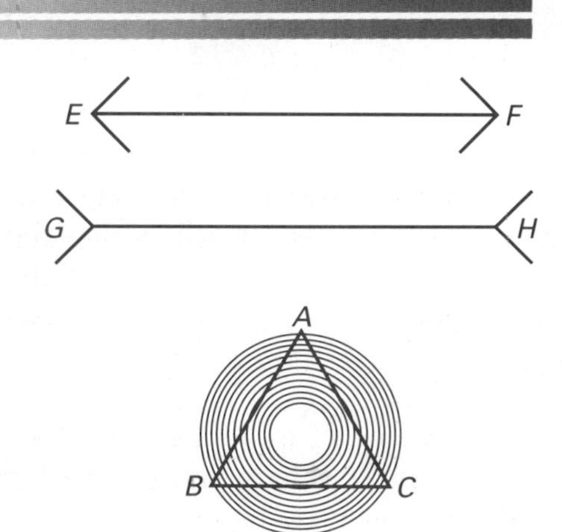

EXERCISES

PRACTICE/ SOLVE PROBLEMS

1. Write a statement about l and m that is suggested by the figure. Check to see if it is true.

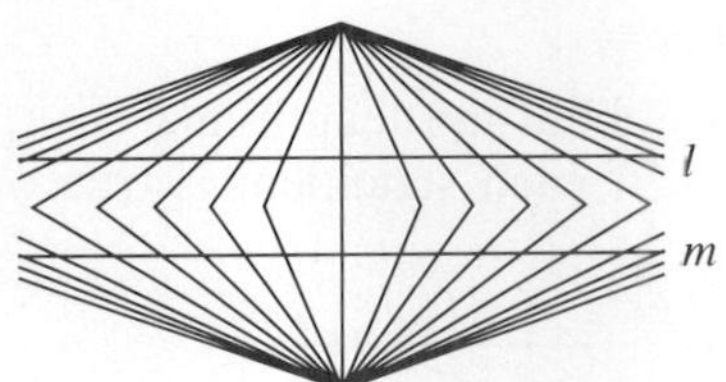

2. Write a statement about line segments AB and BC that is suggested by the picture.

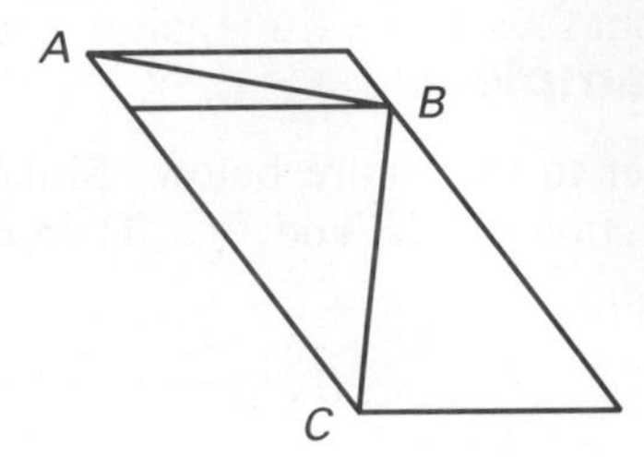

EXTEND/ SOLVE PROBLEMS

3. Write a statement about line segments DE and FG that is suggested by the figure.

4. Write a statement about the dots in the two figures.

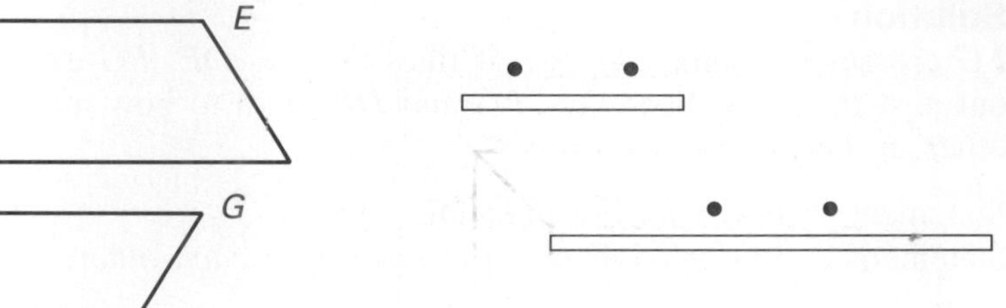

Each drawing below can be viewed in two ways. Write a paragraph that describes what you see in each figure. Justify your statement.

5.

6.

7. ***USING DATA*** Use the Data Index on page 546 to find a reference to two drawings that show optical illusions. Describe what you first see in each drawing. Then look at the figures until you see something different. Describe what you see this time.

THINK CRITICALLY/ SOLVE PROBLEMS

8. Choose one of the optical illusions shown in the lesson. Write a paragraph in which you describe the illusion and explain what it is in the picture that tricks the eye.

► READ
► PLAN
► SOLVE
► ANSWER
► CHECK

Problem Solving Applications:

IMPOSSIBLE FIGURES IN ART

Examine the two-dimensional figures shown. If you were to attempt to make a three-dimensional model using straws, sticks, or other objects, could you do it? Justify your answer.

1.

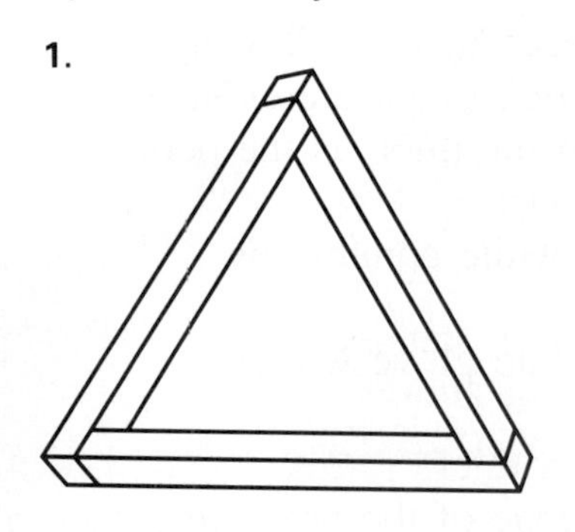

2.

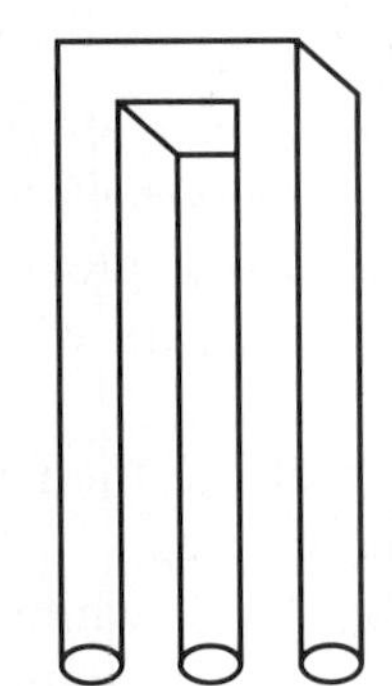

3.

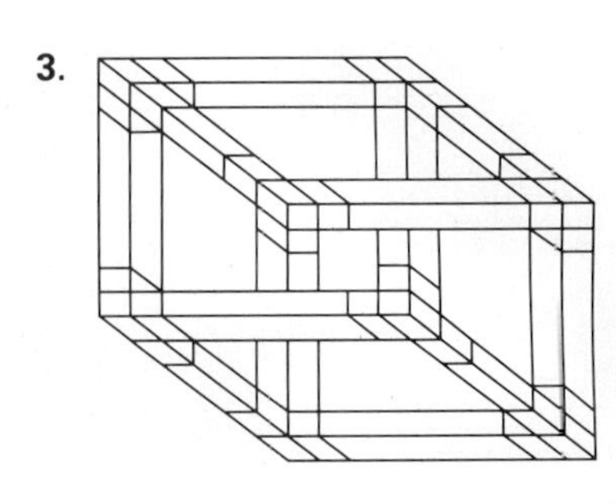

4.

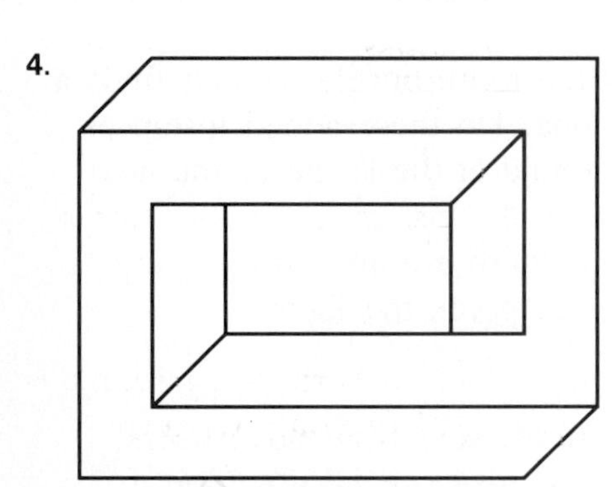

5.

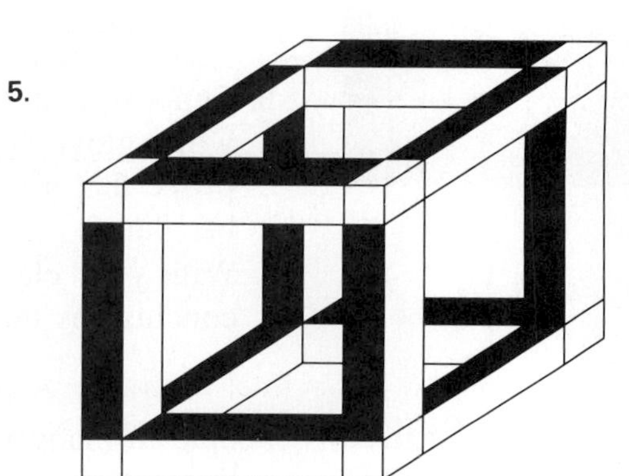

Study the two drawings below carefully. You will notice some strange things. Describe at least one unusual thing in each picture.

6.

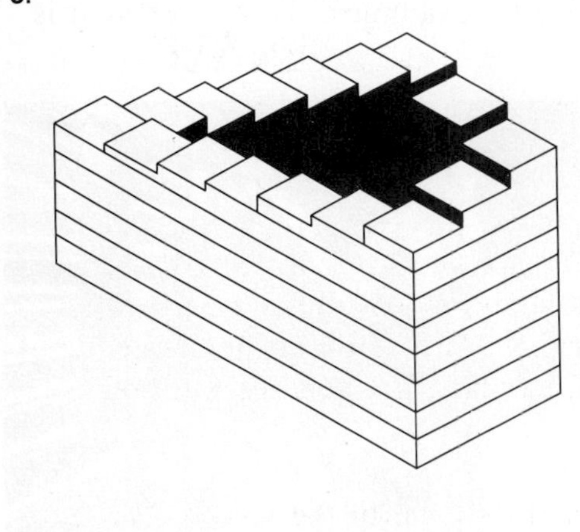

7.

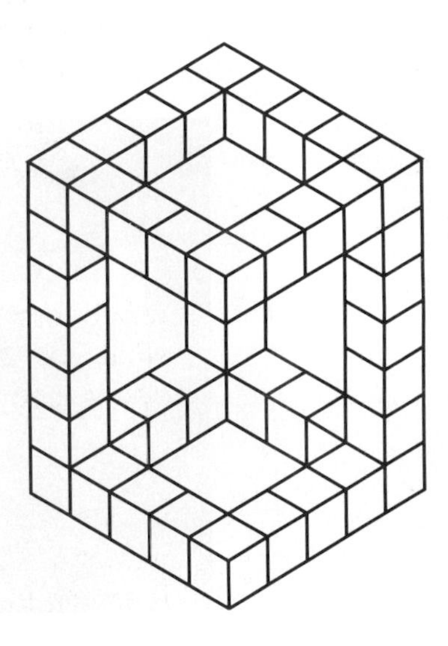

3-2 Inductive Reasoning

EXPLORE/ WORKING TOGETHER

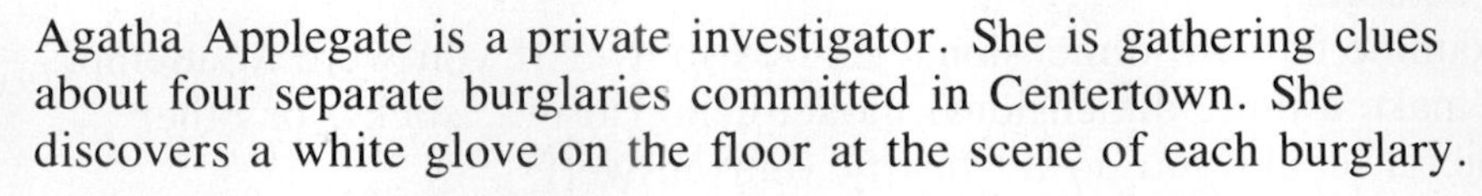

Agatha Applegate is a private investigator. She is gathering clues about four separate burglaries committed in Centertown. She discovers a white glove on the floor at the scene of each burglary.

a. Discuss with your classmates whether each of the following conclusions is *warranted* or *not warranted*, given the evidence.
   - The thief, or someone who is helping the thief, is the person who leaves the white glove on the floor.
   - The thief always wears white gloves while committing burglaries.
   - It is possible that the thief wears a white glove while committing burglaries.
   - Everyone who wears white gloves is a thief.
   - A white glove will be found at the scene of the next burglary committed in Centertown.

b. Suppose that Agatha investigates ten more burglaries and finds a white glove at the scene of each one. Do these added instances prove that a white glove will be found at the scene of the next burglary?

c. With your classmates, determine if there are any other conclusions that might be warranted, given the facts.

SKILLS DEVELOPMENT

People in many walks of life—detectives, scientists, advertisers, politicians, mathematicians, and others—pay special attention to things that happen repeatedly. They hope that by examining some events, they might discover a pattern or rule that applies to all such events. That is, they hope to arrive at a conclusion by using *inductive reasoning*. **Inductive reasoning** is the process of reaching a conclusion based on a set of specific examples. The conclusion is called a **conjecture.**

Example 1

Use inductive reasoning to make a conjecture about the ones digit of 3^{16}.

COMPUTER

The computer program below lets you print the successive powers of 3 whose exponents are multiples of 4, to check your conjecture in Example 1. How would you change line 10 to find the answers that end in 3, 9, or 7?

```
10 FOR X = 4 TO 16 STEP 4
20 V = INT ( 3 ↑ X + 0.5)
30 PRINT : PRINT "3↑";X;
   = ";V
40 NEXT X
```

Solution

First, use a calculator to find several successive powers of 3 and look for a pattern in the ones digits.

$3^1 = 3$	$3^2 = 9$	$3^3 = 27$	$3^4 = 81$
$3^5 = 243$	$3^6 = 729$	$3^7 = 2{,}187$	$3^8 = 6{,}561$
$3^9 = 19{,}683$	$3^{10} = 59{,}049$	$3^{11} = 177{,}147$	$3^{12} = 531{,}441$

In these examples, the ones digits repeat in a pattern:

3 9 7 1   3 9 7 1   3 9 7 1

Next, make a conjecture based on the pattern. For example,

"When the exponent of 3 is a multiple of 4, the ones digit is 1."

Finally, make a specific prediction based on the conjecture and test it. For example,

"Since 16 is a multiple of 4, the ones digit for 3^{16} is 1."

In fact, $3^{16} = 43{,}046{,}721$, a number whose ones digit is 1. ◄

A conjecture may be true, but it is not *necessarily* true. To prove that a conjecture is false, you need to find just one case where the conjecture does not hold true.

Example 2

Pete observed that he could express fractions with two-digit numerators and denominators in lowest terms by this method.

$$\frac{1\not{6}}{\not{6}4} = \frac{1}{4} \qquad \frac{1\not{9}}{\not{9}5} = \frac{1}{5} \qquad \frac{2\not{6}}{\not{6}5} = \frac{2}{5}$$

On the basis of the three examples, he made this conjecture: "Whenever the ones digit in the numerator is the same as the tens digit in the denominator, you can cancel the ones digit in the numerator and the tens digit in the denominator to get an equivalent fraction."

CHECK UNDERSTANDING

From what you have observed about inductive reasoning, why can't you state that if a rule or pattern holds for several examples, then it will hold for all examples?

Is Pete's conjecture always true?

Solution

If possible, find a case in which Pete's conjecture is false.

Use Pete's method for $\frac{12}{24}$.

$$\frac{1\not{2}}{\not{2}4} \rightarrow \frac{1}{4}$$

Use division to write $\frac{12}{24}$ in lowest terms.

$$\frac{12}{24} = \frac{12 \div 12}{24 \div 12} = \frac{1}{2}$$

Since $\frac{1}{4} \neq \frac{1}{2}$, Pete's conjecture is false. ◄

You can adapt the program on page 95 to test your conjecture for Try These Exercises 1 and 2. Change line 10 so $X = 1$ to 5. What change do you need to make to line 20? to line 30? How could you change line 10 to check your answer?

TRY THESE

1. Use inductive reasoning to find the ones digit for 4^6.
2. Use inductive reasoning to find the ones digit for 4^7.
3. Merola found that the sum for each of three pairs of numbers was a two-digit number whose digits are the same:

   $16 + 61 = 77 \qquad 17 + 71 = 88 \qquad 18 + 81 = 99$

   She made this conjecture: Take any two-digit number and reverse the digits. The sum of that number and the original number is a two-digit number whose digits are the same. Is Merola's conjecture always true?

EXERCISES

PRACTICE/ SOLVE PROBLEMS

1. Use inductive reasoning to find the ones digit of any odd-numbered power of 4.
2. Use inductive reasoning to find the ones digit of any even-numbered power of 4.

Use the given examples to complete each conjecture.

3. $3 \times 5 = 15 \quad 7 \times 9 = 63 \quad 13 \times 5 = 65 \quad 23 \times 27 = 621$

   The product of two odd numbers is an __?__ number.
4. $8 \times 12 = 96 \quad 12 \times 16 = 192 \quad 36 \times 74 = 2{,}664$

   The product of two even numbers is an __?__ number.
5. $2 + 8 = 10 \quad 14 + 16 = 30 \quad 6 + 62 = 68 \quad 46 + 76 = 122$

   The sum of two even numbers is an __?__ number.
6. Choose an even number. Add 20. Multiply the sum by 2. Divide the product by 4. Subtract 10 from the quotient. Multiply the difference by 2. What number results? Repeat this experiment with a different starting number. What number results? Make a conjecture.
7. Choose any number. Add 12 to it. Multiply the result by 2. Subtract 4. Divide by 4. Finally, subtract half the number you started with. What number results? Repeat this experiment with a different starting number. What number results? Make a conjecture.
8. Examine the following sequence of numbers. Give a pattern for the sequence and give the next two numbers.

   1, 4, 2, 5, 3, 6, 4, . . .

EXTEND/ SOLVE PROBLEMS

9. Multiply any number by 9. Take the sum of the digits of this product. Keep taking the sum of the digits until the sum has only a single digit. Make a conjecture about the digit that results from this process.

10. Add 5 to your age. Double that sum and then multiply by 50. Add the date of the month on which you were born. Double that sum. Subtract 1,000. Divide by 2. Make a conjecture about the number that results. Test your conjecture with several other people.

11. Pick a three-digit number. Multiply it by 7, then multiply that product by 11, and then multiply that product by 13. Test this process several times. Make a conjecture.

$? \times 7 = ?$

$? \times 11 = ?$

$? \times 13 = ?$

THINK CRITICALLY/ SOLVE PROBLEMS

For each of the following exercises, test the conjecture by finding at least five examples that support it *or* at least one example that disproves it. All the conjectures will be true for at least some cases. Try to explain what is wrong with each false conjecture and modify it if possible.

12. The difference between any two whole numbers is always positive.

13. The sum of consecutive odd whole numbers starting with 1 is always a perfect square. For example, $1 + 3 + 5 = 9$, and 9 is a perfect square.

14. The sum of n consecutive positive whole numbers starting with 1 is always equal to $\frac{n(n + 1)}{2}$. For example, the sum of $1 + 2 + 3 + 4 + \ldots + 99 + 100 = \frac{100\,(100 + 1)}{2} = 5{,}050$.

15. ***USING DATA*** Use the Data Index on page 546 to find the list of prime numbers. A *prime number* is a number greater than 1 having only itself and 1 as factors. Use the formula $n^2 + n + 11$, and replace n with the whole numbers from 1 to 9. Is each resulting number prime? Replace n with 10. Is the result prime?

MATH: WHO, WHERE, WHEN

A mathematician named Christian Goldbach made the following conjecture in 1742

Every even whole number greater than 2 can be written as the sum of two prime numbers.

Examples: $8 = 3 + 5$
$20 = 7 + 13$
$100 = 11 + 89$

This conjecture has never been proved, but no one has ever found an example that disproves it.

3-3 Deductive Reasoning

EXPLORE/ WORKING TOGETHER

Study the map below and read these four statements.

A. If you live in Chicago, then you live in Illinois.
B. If you live in Illinois, then you live in Chicago.
C. If you do not live in Chicago, then you do not live in Illinois.
D. If you do not live in Illinois, then you do not live in Chicago.

Which of the statements are true? Which of the statements are false?

SKILLS DEVELOPMENT

In daily life, we often make statements that involve *if* and *then.* Such statements are called ***if-then* statements,** or **conditional statements.** A conditional statement has two parts, a *hypothesis* and a *conclusion.*

hypothesis ↓   *conclusion* ↓

If **it rains,** then **it is cloudy.**

Example 1

Write two conditional statements that can be made from *The temperature is 0°C* and *Water freezes.*

Solution

Make one statement the hypothesis and the other the conclusion.

If the temperature is 0°C, then water freezes.
If water freezes, then the temperature is 0°C. ◄

Some conditional statements are true and others are false. A **counterexample** shows that a conditional sentence is false by satisfying the hypothesis but not the conclusion.

Example 2

Is the conditional statement *true* or *false*? If false, give a counterexample.

a. If a number is divisible by 10, then it is divisible by 5.

b. If a number is divisible by 5, then it is divisible by 10.

Solution

a. The statement is true, because any number ending in 0 is divisible by 5.

b. The number 25 is a counterexample, because 25 is divisible by 5 but not by 10. So, the statement is false. ◄

When you use a conditional statement along with given information to draw a new conclusion, you use **deductive reasoning,** or make a **logical argument.** The following is an example of deductive reasoning.

If I do my homework by 6:00 p.m., then I may go to the movies.
I *did* my homework by 6:00 p.m.

Therefore, I may go to the movies.

In a **valid** argument, a new statement is obtained from given statements by deductive reasoning. If there is an error in the reasoning, the argument is called **invalid.**

WRITE ABOUT MATH

Suppose that a friend of yours has difficulty understanding the difference between inductive reasoning and deductive reasoning. Write your friend a brief note explaining the difference between the two kinds of thinking. Give an example of each type of thought process.

Example 3

Is this argument *valid* or *invalid*? Use a picture to help you decide.

If a runner wears Speedy Shoes, then the runner will win the race.
Constance, a contestant in the race, wears Speedy Shoes.

Therefore, Constance will win the race.

Solution

Constance is in the set of people who run and wear Speedy Shoes. That group is in the set of people who win. So, Constance is in the set of winners of the race. The argument is valid. ◄

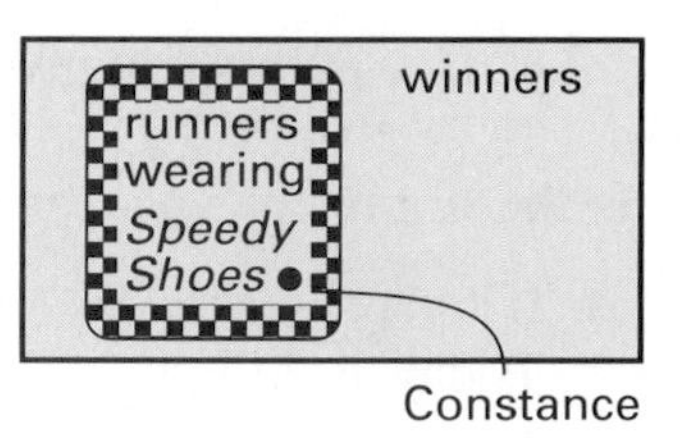

Example 4

Is this argument *valid* or *invalid*? Use a picture to help you decide.

If an animal is a cat, then it loves tuna.
Fido loves tuna.

Therefore, Fido is a cat.

Solution

Draw a shaded region for cats inside the region for tuna lovers. The argument is not valid, because Fido might be a dog that loves tuna. ◄

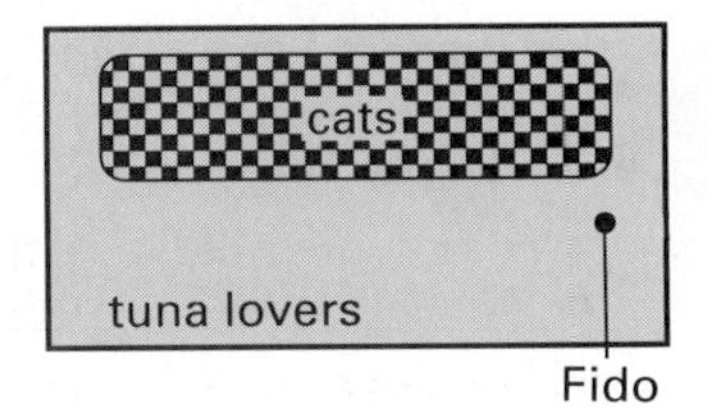

TRY THESE

1. Write two conditional statements that can be made from *There is no gasoline in the tank* and *The car will not run.*

2. Is the conditional statement *true* or *false*? If false, give a counterexample.
   a. If a woman lives in Atlanta, then she lives in Georgia.
   b. If two numbers are even, then their sum is an odd number.

3. Is this argument *valid* or *invalid*? Use a picture to help you decide.
   If a flower is red, then it is pretty.
   This rose is red.
   Therefore, this rose is pretty.

4. Is this argument valid or invalid? Use a picture to help you decide.
   If the battery is dead, then the car will not start.
   The car did not start.
   Therefore, the battery is dead.

EXERCISES

PRACTICE/SOLVE PROBLEMS

1. Write two conditional statements that can be made from *The number is divisible by 9* and *The number is divisible by 3.*

2. Is the conditional statement *true* or *false*? If false, give a counterexample.
   a. If two numbers are odd, then their product is odd.
   b. If a number is divisible by 6, then it is divisible by 12.

Is the argument *valid* or *invalid*? Use a picture to help you decide.

3. If today is Monday, then tomorrow is Tuesday.
   Today is Monday.
   Therefore, tomorrow is Tuesday.

4. If two numbers are even, then their sum is even.
   The sum of 7 and 9 is even.
   Therefore, 7 and 9 are even.

5. If an animal is a bird, then it can fly.
   An ostrich is a bird.
   Therefore, an ostrich can fly.

6. If a fruit is an orange, then it is a citrus fruit.
   This fruit is not an orange.
   Therefore, this fruit is not a citrus fruit.

Write each statement as an *if-then* statement.

7. All even numbers are divisible by 2.
8. All squares are quadrilaterals.
9. All whales are mammals.

EXTEND/SOLVE PROBLEMS

Is the argument *valid* or *invalid*? Use a picture to help you decide.

10. All giraffes have long necks.
    Titus has a long neck.
    Therefore, Titus is a giraffe.
11. All rectangles have four right angles.
    Figure *ABCD* does not have four right angles.
    Therefore, figure *ABCD* is not a rectangle.
12. All multiples of 100 are multiples of 10.
    500 is a multiple of 100.
    Therefore, 500 is a multiple of 10.

Sometimes conditional statements can be strung together. Write an *if-then* statement of your own from the given conditional statements.

13. If a number is a whole number, then it is an integer.
    If a number is an integer, then it is a rational number.
14. If people live in Mexico City, then they live in Mexico.
    If people live in Mexico, then they live in North America.

CONNECTIONS

Advertisements are sometimes misleading. An advertisement might show a young person being invited to a party. A voice simply says "Drink Sizzling Soda." The advertiser wants you to think "If I drink Sizzling Soda, then I will be popular". Find examples of misleading conditional statements like this and discuss them with your classmates.

Write a valid logical argument for each picture.

15.

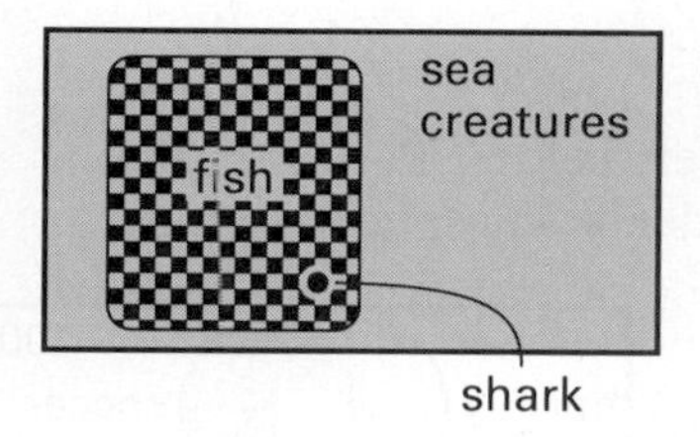

16.

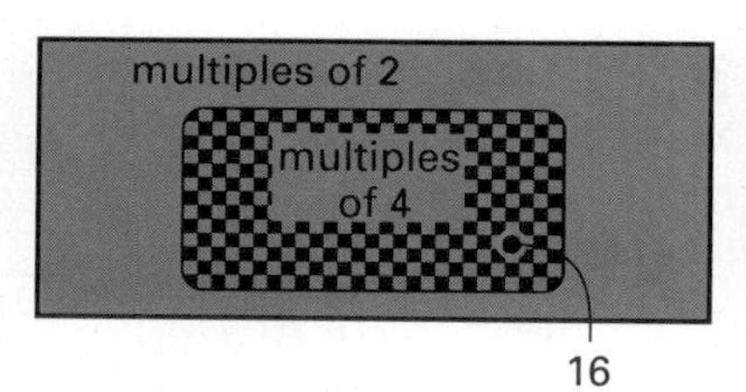

THINK CRITICALLY/ SOLVE PROBLEMS

Lewis Carroll was the nineteenth-century English writer who wrote *Alice in Wonderland.* However, he was also a mathematician, and he wrote a book called *Symbolic Logic.* For the given statements from his book, write a valid argument. (You will need to draw your own conclusion.)

17. All well-fed canaries sing loud.
    No canary is melancholy if it sings loud.
18. All puddings are nice.
    No nice things are wholesome.
    This dish is a pudding.

3-4

► READ
► PLAN
► SOLVE
► ANSWER
► CHECK

Problem Solving Skills:

USING VENN DIAGRAMS

Some problems can be solved with a diagram. In a **Venn diagram,** a collection is represented by a circular region inside a rectangle. The rectangle is called the **universal set.** For instance, the diagram at the right shows the relationships between basketball players and baseball players in a certain school.

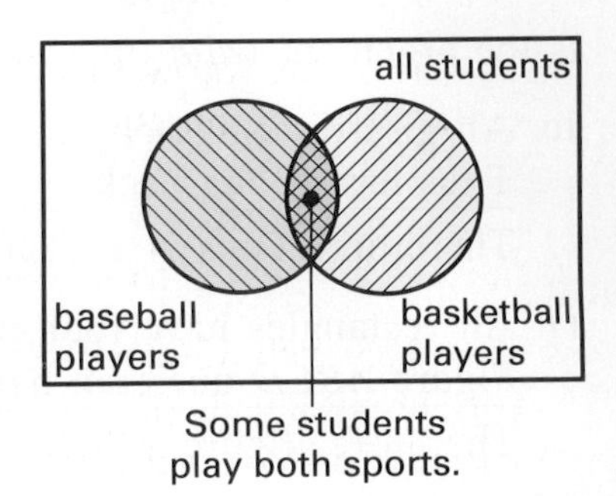

PROBLEM

In a poll of 100 students, the following information was learned.

Number of Students	Sport
66	basketball
50	soccer
48	hockey
22	soccer and basketball
30	hockey and basketball
28	soccer and hockey
12	all three sports

How many play soccer or basketball? How many play no sport?

SOLUTION

Make a Venn diagram like the one at the left below. Put 12 in the region marked *A,* because there are 12 students who play all three sports.

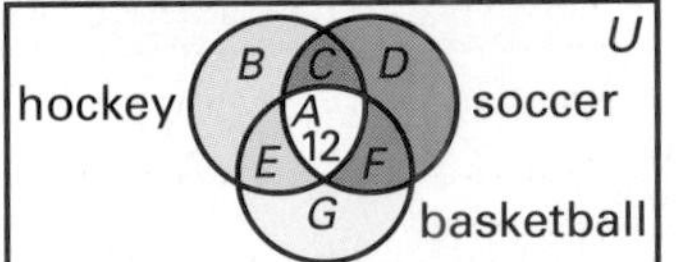

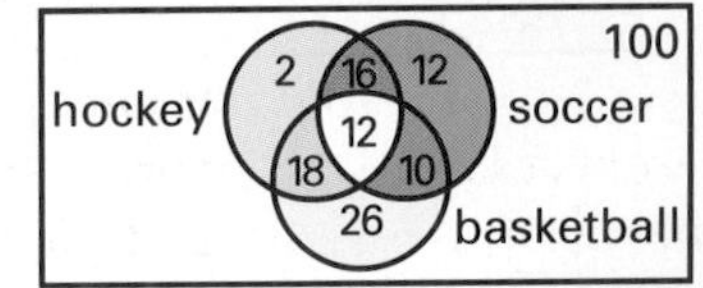

Since 28 students play hockey and soccer (regions A and C together) and region A has 12 people, region C has $28 - 12 = 16$ people. Use addition and subtraction to find the value for each other letter. The completed diagram is at the right above.

The number of students who play soccer or basketball is

$$16 + 12 + 12 + 10 + 18 + 26 = 94.$$

The number of students who play no sport is

$$100 - (2 + 16 + 12 + 12 + 18 + 10 + 26) = 4.$$

PROBLEMS

Refer to the problem on the preceding page.

1. Tell how you know that 18 students play hockey and basketball but not soccer (region *E*).
2. Tell how you know that 2 students play hockey but not soccer or basketball (region *B*). (You will need the answer to Exercise 1.)

Use a Venn diagram to solve each problem.

3. Dan surveyed 120 people at a shopping mall one day. Of the people surveyed, 62 said they would like to see a bookstore included in the mall; 48 would like to see a music store; and 23 would like to see both included. How many did not want either type of store to be included in the mall?
4. In a class of 30 students, 20 were going on family vacations during the summer, 15 were planning on working, and 3 were doing neither. How many were planning on both working and going on a family vacation?

At a school sports banquet, a survey was conducted concerning what sports each person played. The following facts were recorded: 18 students played a fall sport; 16 played a winter sport; 21 played a spring sport. It was also noted that 5 played a fall and a winter sport, 8 played a fall and a spring sport, and 7 played a winter and a spring sport. Only 3 students played a fall, a winter, and a spring sport.

5. How many students played only in fall? in winter? in spring?
6. How many students were surveyed at the banquet?

A survey of 200 commuters produced the following data: 118 used the expressway; 98 wore glasses; 84 listened to the radio; 40 listened to the radio and wore glasses; 58 listened to the radio and used the expressway; 62 used the expressway and wore glasses; 24 did all three.

7. How many wore glasses but did not use the expressway or listen to the radio?
8. How many only used the expressway?
9. How many listened to the radio or used the expressway?
10. How many did none of these things?

3-5 Problem Solving Strategies:

USING LOGICAL REASONING

In order to solve many logic problems, it helps to make a table. Using the *process of elimination,* you can complete the table to answer questions.

PROBLEM

Dexter, Maria, and Phil each play baseball, badminton, and tennis, but not necessarily in that order. Each person plays only one sport. Dexter does not play a sport that requires a ball. Phil does not play baseball.

Which sport does each play?

SOLUTION

Step 1: Make a table listing each person and each sport.

	baseball	badminton	tennis
Dexter			
Maria			
Phil			

Step 2: Since Dexter does not play a sport that requires a ball, place an **X** in the boxes for baseball and tennis. Place a ✓ in the box for badminton. Since Phil does not play baseball, place an **X** in the box for baseball.

	baseball	badminton	tennis
Dexter	X	✓	X
Maria			
Phil	X		

Step 3: Since each person plays only one sport, place an **X** in the badminton boxes for Maria and Phil.

	baseball	badminton	tennis
Dexter	X	✓	X
Maria		X	
Phil	X	X	

Step 4: There is only one choice of sport for Phil. Place a ✓ in the box for tennis for Phil.

	baseball	badminton	tennis
Dexter	X	✓	X
Maria		X	
Phil	X	X	✓

Step 5: By the process of elimination, Maria must be the baseball player. So, Dexter plays badminton, Maria plays baseball, and Phil plays tennis.

PROBLEMS

Make a table to help solve each problem.

1. Three dogs are named Tillie, Spot, and Tippie. Two are long-haired dogs, a collie and a Pekingese. One is a short-haired dog, a basset hound. Tippie does not have short hair. The collie lives next door to Tillie. Tippie lives next door to the Pekingese. Tillie is not a basset hound. Who is the Pekingese?

2. Three singers appear at a concert. One is named Jones, one is named Salazar, and one is named Friedman. One sings only jazz, one sings only opera, and one sings only folk music. Jones dislikes the opera singer, but the folk singer is a friend of Jones. Salazar performs before the opera singer in the concert. Who is the folk singer?

3. My three cousins, Sue, DeeDee, and Betty, just got married, but I have forgotten each husband's name. I remember that my cousins married Tom, Bill, and Harry. I heard Betty say that she hoped Tom didn't argue with her husband. I also remember that DeeDee married Tom's brother and that Harry married DeeDee's sister. Who married whom?

4. Hector Hollingsworth is a private investigator called to the scene of a crime. There are four suspects—the butler, the cook, the chauffeur, and the maid. All four suspects are wearing identical uniforms and refuse to divulge their occupations. Through his investigation, however, Hector is sure that the butler committed the crime. The four suspects are named Alexandra, Beatrice, Cecilia, and Delphine. Hector discovers that Delphine cannot drive. Unknown to Cecilia, her uniform is spattered with kitchen grease. Neither Alexandra nor Delphine can correctly tell Hector where the mops are kept. Who was the butler?

TALK IT OVER

Discuss a mystery or detective story you have seen on TV or in the movies. Were you able to solve the mystery before the story ended or were you surprised at the outcome?

5. Armand, Belinda, Colette, and Dimitri are four artists. One is a potter, one is a painter, one is a violinist, and one is a writer. Armand and Colette saw the violinist perform. Belinda and Colette have modeled for the painter. The writer wrote a story about Dimitri and plans to write a story about Armand. Armand does not know the painter. Who is the writer?

3-6 Brainteasers

EXPLORE/ WORKING TOGETHER

Work in groups of four students each.

1. Three students sit in a triangle facing each other.
2. The fourth student prepares three pieces of tape, marking one or none of them with an **X**. This student then places one piece of tape on the forehead of each student in the triangle.
3. By looking only at the other students, each student decides whether the tape on his or her own forehead is marked with an **X**.

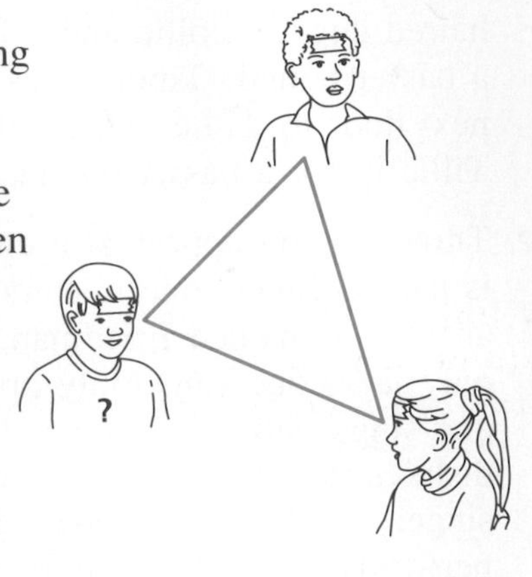

How can each player determine whether his or her tape is blank or marked with an **X**?

SKILLS DEVELOPMENT

Seemingly difficult problems can often be solved if you try a different point of view. In addition to using a different point of view, you will find experimentation, or *trial and error,* helpful.

Example 1

Suppose that you need exactly four quarts of water for a recipe. You have the water available, but you only have 3-quart and 5-quart containers. How can you measure exactly four quarts of liquid by using only the two containers you have on hand?

Solution

It might help to use pictures to help you think of the actions you could take. There are many possible solutions. Here is one of them. In this solution, container A holds 3 quarts and container B holds 5 quarts.

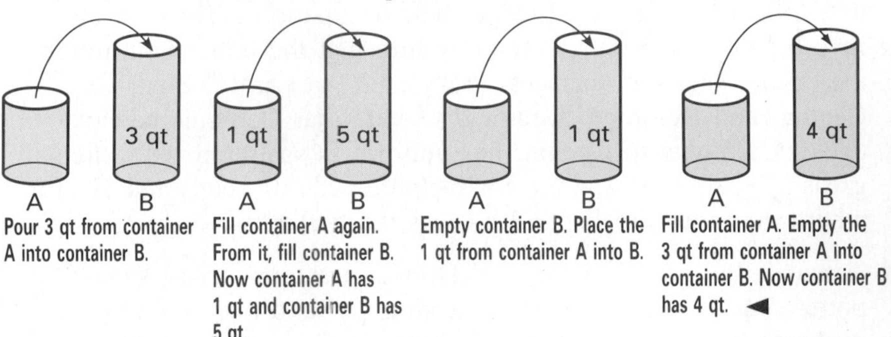

Let's see what happens to container B if we use numbers to symbolize the actions pictured above.

$$3 + 2 - 5 + 1 + 3 = 4$$

Example 2

Divide the figure shown into two pieces having the same size and shape.

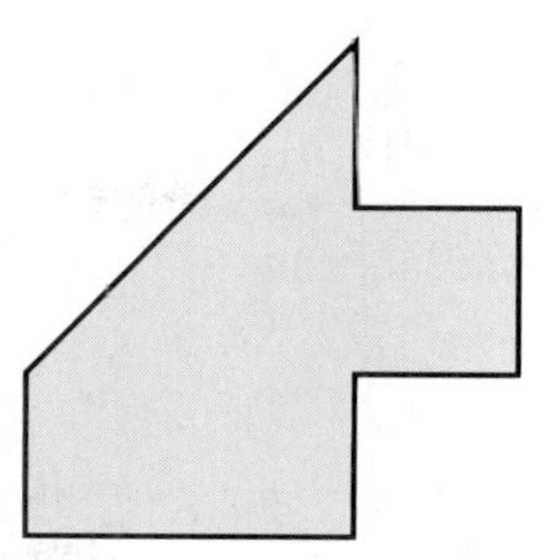

Solution

Suppose that you think about making one cut. Try it. The two figures that result probably won't have the same shape. See the figure at the left below.

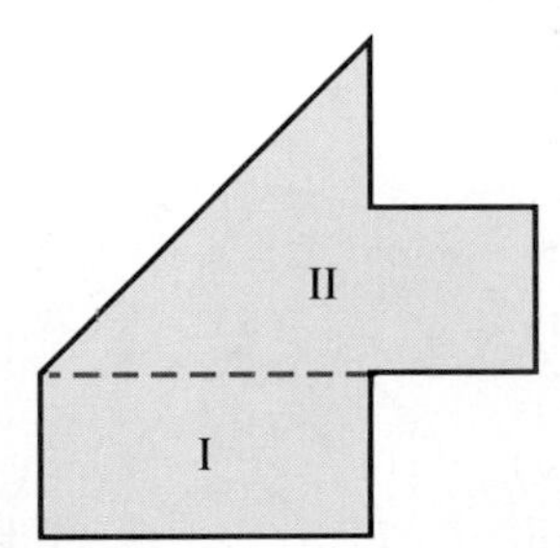

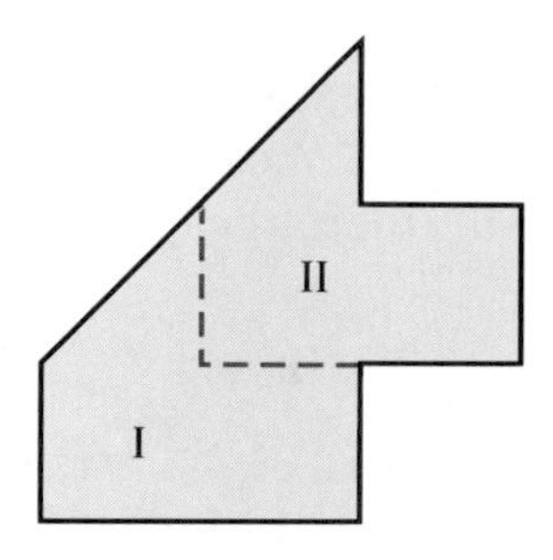

Think about making two cuts. The figure at the right above shows that two cuts will give two figures that have the same size and shape. ◄

TRY THESE

1. A girl put a caterpillar in a glass jar at 8:00 p.m. when she went to bed. Every hour the caterpillar climbed up the wall of the jar 2 inches and slid back 1 inch. If the jar was 6 inches high, at what time did the caterpillar reach the top of the jar?
2. If it takes one minute to make a cut, how long will it take to cut a 10-foot log into ten equal parts?
3. What is the greatest number of pieces into which you could cut a pizza with four straight cuts of a knife?

EXERCISES

PRACTICE/ SOLVE PROBLEMS

1. A pair of sneakers and the laces cost $15.00 and the sneakers cost $10.00 more than the laces. How much do the laces cost?

2. How many triangles are there in the figure?

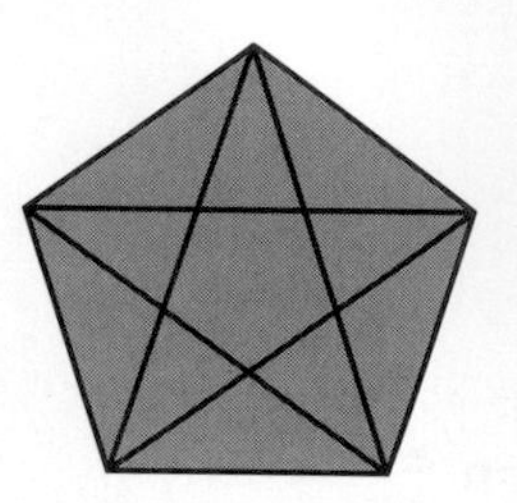

3. A property owner wants to build a fence around a square lot that is 50 feet long on each side. If the fencing comes in 10-foot sections, then how many posts will the owner need?

4. How can you score 100 points with 6 shots if all of them hit the target?

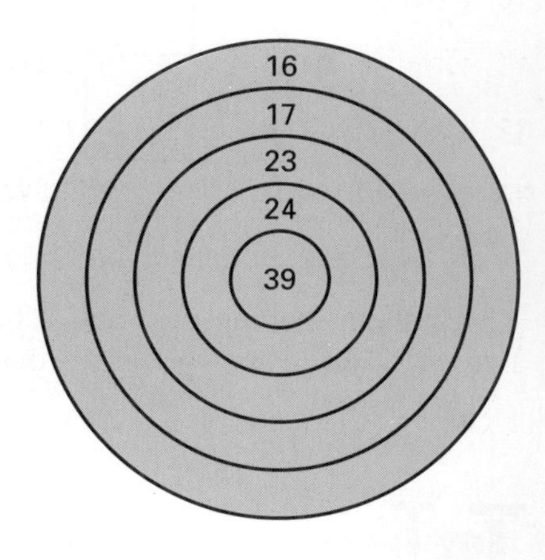

EXTEND/ SOLVE PROBLEMS

5. Suppose you have a 5-gallon, an 11-gallon, and a 13-gallon container. How could you take a 24-gallon container of water and divide it equally into three parts?

6. There are eight marbles exactly alike, except that one is heavier than the others. Using a balance scale, how can you find the heavier marble in two weighings?

7. Trace the four shapes below. Cut them out and arrange the pieces to form an equilateral triangle. Then rearrange them to form a square.

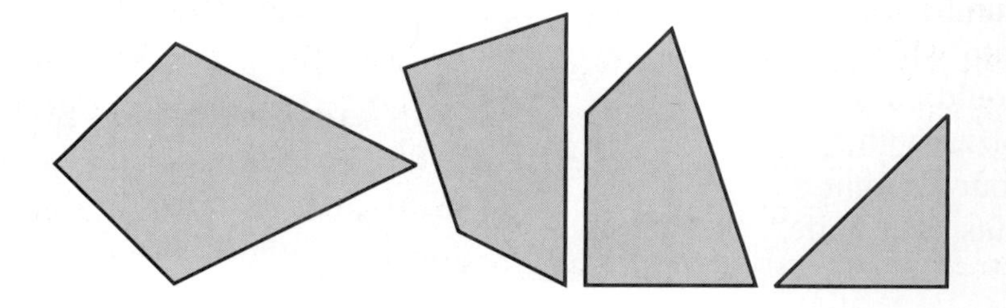

Suppose you have a 3-inch cube painted black.

8. How many cuts would it take to cut the cube into 1-inch cubes?

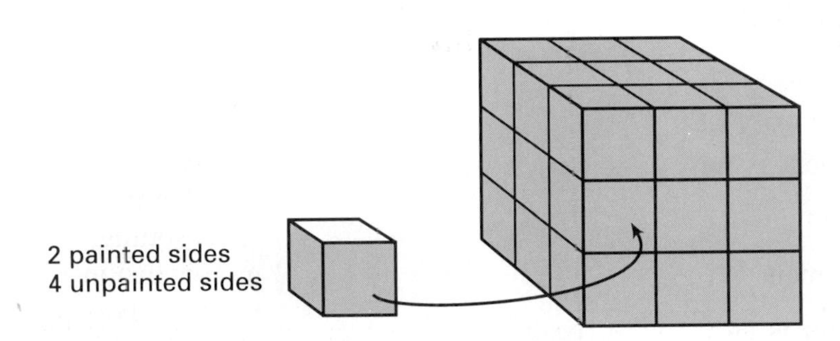

9. How many cubes would you have?
10. How many cubes would have no black sides?
11. How many cubes would have exactly one black side?
12. How many cubes would have exactly two black sides?
13. How many cubes would have exactly three black sides?
14. How many cubes would have exactly four black sides?
15. Add three straight lines to get nine triangles, not counting those that overlap.

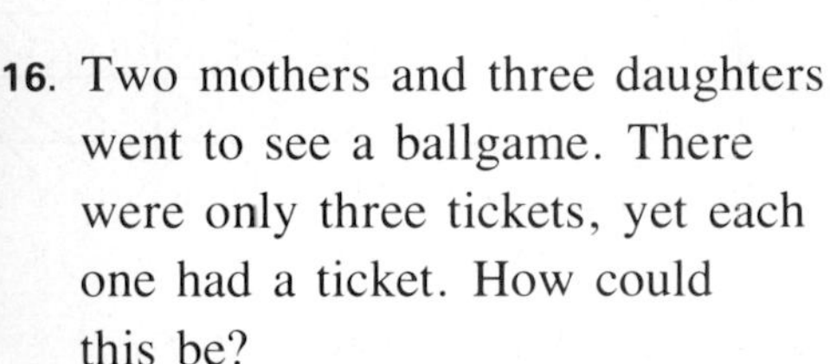

16. Two mothers and three daughters went to see a ballgame. There were only three tickets, yet each one had a ticket. How could this be?

17. A man has to take a wolf, a goat, and some cabbage across a river. His rowboat has enough room for the man plus either the wolf or the goat or the cabbage. If he takes the cabbage with him, the wolf will eat the goat. If he takes the wolf, the goat will eat the cabbage. Only when the man is present is the goat safe from the wolf and the cabbage safe from the goat. How does he get them across? *Hint:* He makes the round trip three times before making his final crossing.

MIXED REVIEW

Find the median, mean, mode, and range for these scores.

1. 82, 87, 90, 77, 65, 81, 82, 88

Add or subtract.

2. 2,159 + 2,607
3. 5,942 − 4,998
4. 6.94 + 128.526
5. 307 − 291.75

Multiply or divide.

6. 1.8×1.8
7. 0.008×0.3
8. 1.96 ÷ 0.7
9. 36 ÷ 1.8

Solve.

10. How much change would you receive if you purchased a sweater that cost $39.99 with a hundred dollar bill?

THINK CRITICALLY/ SOLVE PROBLEMS

CHAPTER 3 • REVIEW

Write the letter for the word at the right that matches each description.

1. Process of reaching a conclusion based on a set of specific examples
2. Trial generalization
3. Shows that a conditional statement is false
4. *If* part of a conditional statement
5. *Then* part of a conditional statement
6. Kind of argument in which there is an error in reasoning
7. Reasoning from a conditional statement and other information, to draw a conclusion
8. A picture that the eye sees as true that is actually not true
9. Shows the relationships between two or three different classes of things

a. hypothesis
b. counterexample
c. Venn diagram
d. invalid
e. conjecture
f. inductive reasoning
g. conclusion
h. deductive reasoning
i. optical illusion

SECTION 3–1 OPTICAL ILLUSIONS (pages 90–93)

► An optical illusion is a picture in which the human eye perceives something to be true that is actually not true.
► A statement is a kind of sentence that is either true or false. You can often make a test to find out whether a statement is true or false.

Refer to the figures below. Make a statement about each figure. Then check to see whether your statement is true or false.

10.

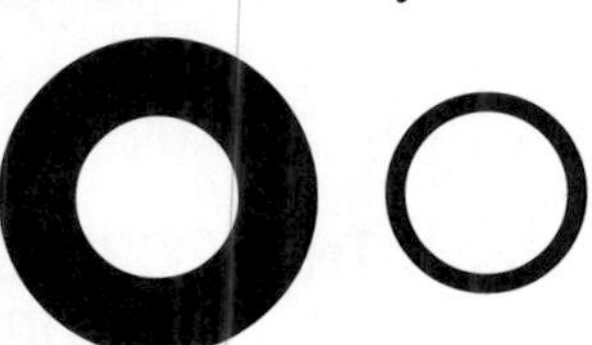

11.

SECTION 3–2 INDUCTIVE REASONING (pages 94–97)

► You can use a large number of examples to support a conjecture, or trial generalization, but the examples do not necessarily prove that the conjecture is always true. Yet you can prove a conjecture false with just one example that contradicts the generalization.

12. Pick a number. Double it. Add 6. Add the original number. Divide by 3. Add 4. Subtract the original number. What do you get? Repeat the process several times using a different original number. Then make a conjecture about the result of the process.

SECTION 3–3 DEDUCTIVE REASONING (pages 98–101)

► An *if-then* statement is called a conditional statement. The statement in the *if* part is called the hypothesis. The statement in the *then* part is called the conclusion. A conditional statement is false if a counterexample satisfies the hypothesis but not the conclusion.

► In a valid argument, a new statement can be derived from given statements by deductive reasoning. If there is an error in the reasoning, the argument is invalid.

13. Write two conditional statements that can be made from these two sentences:

It has been snowing. The streets are slippery.

Determine whether the following arguments are *valid* or *invalid*.

14. If today is Monday, then I will play tennis.
Today is Monday.
Therefore, I will play tennis.

15. If it rains on me, then I shall get wet.
It did not rain on me.
Therefore, I did not get wet.

SECTIONS 3–4 AND 3–5 PROBLEM SOLVING (pages 102–105)

► Some problems involving relationships between classes of things can be solved using a Venn diagram.

► Many problems involving logical relationships are easier to solve if you use a table to keep track of the clues.

16. In a survey of 250 people, 156 said they preferred front-wheel-drive cars, 183 said they preferred bucket seats, and 147 said they preferred white-wall tires. There were 126 who preferred front-wheel drive and bucket seats, 102 who preferred bucket seats and white-wall tires, and 114 who preferred front-wheel drive and white-wall tires. In all, 87 people preferred all three. How many preferred white-wall tires, but not front-wheel drive?

17. Three students named Alfred, Betty, and Charlie receive three different grades on a test. One received an A, one received a B, and one received a C. No student received a grade matching the first letter of the student's name. Alfred did not receive a B. What grade did each receive?

SECTION 3–6 BRAINTEASERS (pages 106–109)

► Solving problems such as brainteasers often requires you to change your point of view toward the problem.

18. Suppose that, 20 years ago, Rhoda was four times as old as Li. Rhoda is only twice as old as Li now. How old is each person now?

CHAPTER 3 ● TEST

1. Refer to the figure below. Make a statement about line segments *MN* and *ST*. Then check to see whether your statement is true or false.

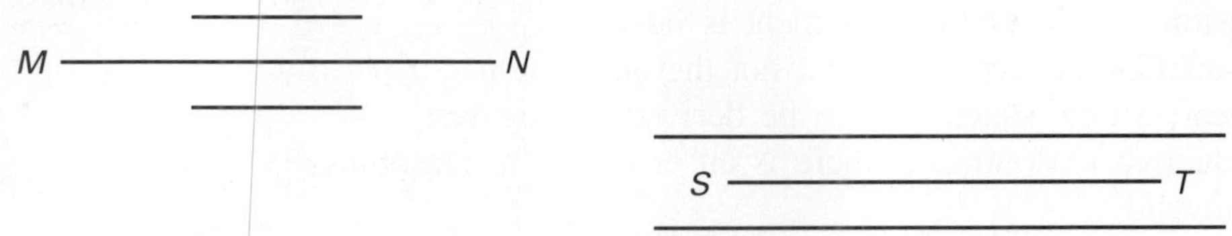

2. Choose any single-digit number. Multiply it by 9. Then multiply by 12,345,679. What number do you get? Repeat this process with several numbers. Then make a conjecture about the result of the process.

3. Experiment with several different sets of numbers and make a conjecture about the product of an even number, an odd number, and an even number.

4. Write two conditional statements, using the following two sentences:
   Something is made of wood. It will burn.

Is each of the following statements *true* or *false*? If false, give a counterexample.

5. If a man is a bachelor, then he is unmarried.

6. If a number is divisible by 7, then it is divisible by 14.

Determine whether the following arguments are *valid* or *invalid*.

7. If today is Friday, then I will go to school.
   I went to school today.
   Therefore, today is Friday.

8. If a horse wins the race, then it runs fast.
   Lightning won the race.
   Therefore, Lightning ran fast.

9. In a survey of employees in an office, it was found that 42 wanted peanuts to be placed in the vending machine, 56 wanted raisins, and 30 wanted granola bars. The survey showed that 18 wanted both peanuts and granola bars, 31 wanted peanuts and raisins, and 12 wanted raisins and granola bars. There were 9 who wanted all three. How many employees wanted either raisins or granola bars in the vending machine?

10. Terri, Jonathan, and Kathy each have different jobs. One is a writer, one is a bus driver, and one is a chef. Terri cannot drive. Kathy cannot cook. Jonathan does not know the chef. Kathy is the writer's sister. What job does each person have?

The following data show the number of pies eaten by each person in a pie-eating contest.

8 4 12 10 3 5 7 9 11
14 6 11 2 1 6 14 8 4
10 13 12 5 7 15 6

1. Construct a stem-and-leaf plot for the data.
2. How many pies did the winner eat?
3. How many contestants ate 10 or more pies?
4. How many contestants ate between 8 and 12 pies?

The table shows the bowling scores for four games played by Jennifer and Ricardo.

Name	Game 1	Game 2	Game 3	Game 4
Jennifer	144	151	136	157
Ricardo	153	146	132	161

5. Find the mean score for Jennifer and for Ricardo.
6. Who had the higher average?

The annual salaries of six people follow:

\$18,600 \$19,400 \$24,000
\$20,000 \$21,000 \$45,000

7. Find the mean salary and the median salary.
8. Is the mean or the median the more representative of the data? Give reasons for your answers.

Use the pictograph for Exercises 9–11.

Key: represents 8 players

9. Which league has the most players? How many players are there?
10. How many more novice players are there than advanced-intermediate?
11. Which group has four times more players than the advanced group?

Evaluate.

12. $6 + 9 \div 3 - 2$
13. $8^2 \div 4 - 10$
14. $3 \times 9 + 12 \div 3$

15. Determine whether the following argument is *valid* or *invalid*.

    If the sun is shining, then
    I will go swimming.
    The sun is shining.
    Therefore, I will go swimming.

16. Three girls named Jo, Myra, and Tawanda have different hobbies. One collects stamps, one collects shells, and one collects bottles. Jo's collection will not fit in a book. Myra does not collect shells or stamps. What does each collect?

CHAPTER 3 ● CUMULATIVE TEST

1. A forester selected one tree out of every 25 along the roadside to test for the effects of pollution. What type of sampling is this?

A. random B. systematic
C. convenience D. cluster

2. A stereo manufacturer tested every twentieth stereo coming off the assembly line. What type of sampling is this?

A. systematic B. convenience
C. cluster D. none of these

Use these data for Exercises 3–6.

7	14	14	8	9	11	8	10	7	6	8

3. What is the mean of the data?

A. 8 B. 9.3 C. 11.3 D. 7

4. What is the median of the data?

A. 8 B. 9 C. 10 D. 9.3

5. What is the mode of the data?

A. 14 B. 8 C. 7 D. 9.3

6. What is the range of the data?

A. 9.3 B. 8 C. 14 D. 7

7. Evaluate. $8 - (4 \div 2) + 2$

A. 8 B. 4 C. 2 D. 0

8. Evaluate. $12 + 3 \div 3 - 4$

A. 1 B. 9 C. 15 D. 14

9. Evaluate. $3 + 6 \div (3 - 1)$

A. 2 B. 4 C. 5 D. 6

10. Evaluate. $6^2 \div 2 - 4$

A. 0 B. 5 C. 14 D. 2

Use the bar graph for Exercises 11–13.

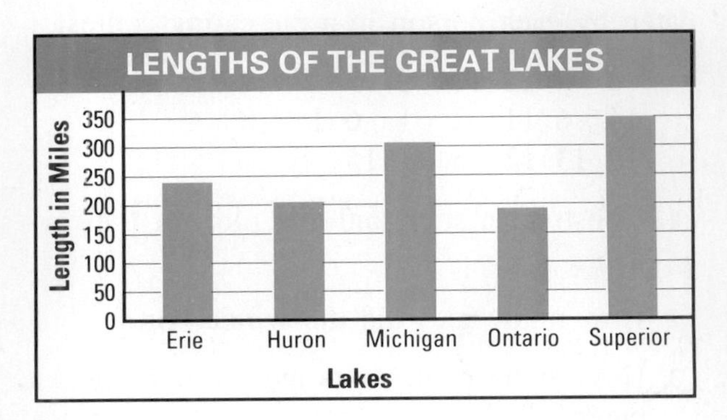

11. Which is the longest lake?

A. Michigan B. Superior
C. Huron D. Ontario

12. Which lakes are over 250 miles long?

A. Superior, Huron
B. Superior, Erie
C. Michigan, Erie
D. Michigan, Superior

13. About how long is Lake Michigan?

A. 350 miles B. 300 miles
C. 250 miles D. 275 miles

14. Which is a counterexample for the following statement?

If a number is prime, then it is not divisible by 7.

A. 11 B. 8 C. 13 D. 7

15. Which conclusions can be drawn from these statements?

If the vase is made of glass, then it is fragile.
The vase is made of glass.

A. The vase is fragile.
B. The vase is not fragile.
C. The vase is made of glass.
D. none of these

CHAPTER 4 SKILLS PREVIEW

1. Find the prime factorization of 280.

Find the GCF of each pair of numbers.

2. 8, 12
3. 4, 10
4. 6, 15
5. 12, 18

Find the LCM of each pair of numbers.

6. 3, 4
7. 9, 15
8. 8, 20
9. 12, 16

10. Multiply to find two fractions that are equivalent to $\frac{11}{15}$.
11. Divide to find two fractions that are equivalent to $\frac{24}{28}$.

Write each fraction in lowest terms.

12. $\frac{15}{25}$
13. $\frac{20}{24}$
14. $\frac{24}{40}$
15. $\frac{14}{21}$

Write each fraction as a decimal.

16. $\frac{31}{100}$
17. $\frac{7}{8}$
18. $\frac{4}{15}$
19. $\frac{7}{9}$

Write each decimal as a fraction in lowest terms.

20. 0.78
21. 0.3
22. 0.001
23. 4.92

Replace ● with <, >, or =.

24. $\frac{5}{8}$ ● $\frac{3}{5}$
25. $\frac{2}{3}$ ● $0.\overline{6}$
26. 0.31 ● $\frac{5}{16}$
27. $\frac{5}{6}$ ● $0.8\overline{3}$

28. Write 6^{-8} as a fraction.
29. Write $\frac{1}{8 \times 8 \times 8 \times 8 \times 8 \times 8 \times 8 \times 8}$ in exponential form.
30. Write 2.813×10^{-6} in standard form.
31. Write 0.00684 in scientific notation.

Find each answer. Write answers in lowest terms.

32. $\frac{1}{2} + \frac{3}{8}$
33. $6\frac{3}{8} - 2\frac{1}{6}$
34. $1\frac{1}{2} \times 2\frac{2}{3}$
35. $\frac{5}{12} \div \frac{2}{3}$

Solve.

36. Yoki has read $\frac{5}{6}$ of a book that is 365 pages long. Estimate to find about how many pages she has read.
37. Rosa plans to hike $1\frac{3}{5}$ mi each day to raise money for a charity. Her sponsor has pledged \$1.25 to the charity for each mile she hikes. If she hikes each day for $6\frac{2}{7}$ weeks, how much money will the charity receive?

CHAPTER 4

EXPLORING FRACTIONS

THEME Environment

The word **fraction** comes from the Latin word *fractio,* which means "act of breaking." A fraction represent a unit broken up into equal parts. Decimals are fractions written as part of the decimal number system. In this chapter, you will compute with fractions, compare decimals and fractions, use negative exponents, and solve problems that involve fractions.

Solid waste is composed of household and industrial wastes. There are three major ways to dispose of solid waste: in landfills, in incinerators, and by recycling. The table shows the amount of solid waste in the United States and how this waste was disposed of over two decades.

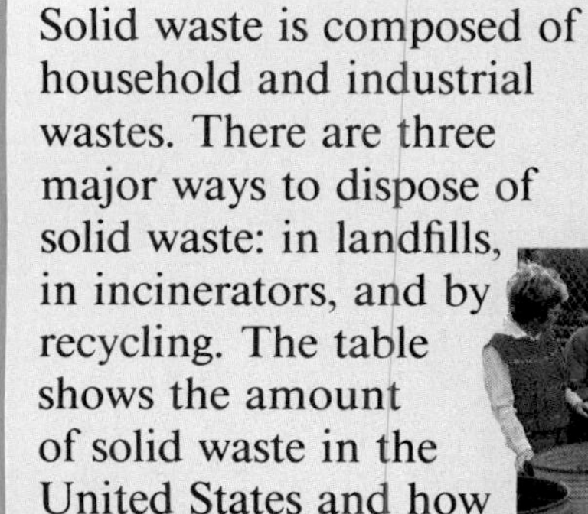

PROFILE OF U.S. WASTE GENERATION, DISPOSAL, AND REUSE

Year	Pounds of waste generated per person per day	Pounds of waste recycled per person per day	Pounds of waste recovered for energy production	Net pounds of waste disposed of per day
1955	2.77	0.17	0.01	2.59
1970	3.16	0.21	0.01	2.94
1975	3.11	0.23	0.02	2.86
1980	3.35	0.32	0.06	2.96
1981	3.36	0.31	0.05	2.99
1982	3.25	0.30	0.08	2.86
1983	3.37	0.32	0.12	2.92
1984	3.43	0.35	0.15	2.93

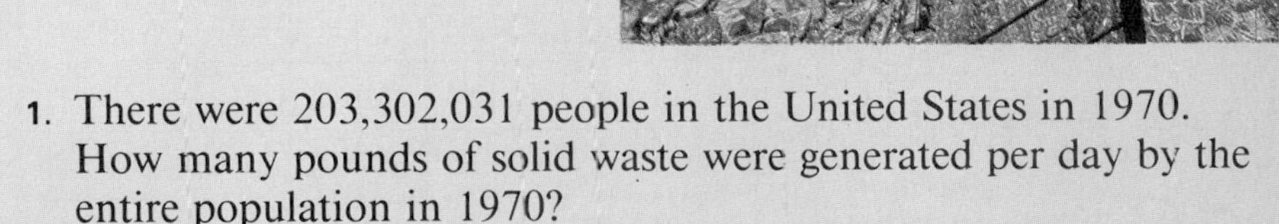

DECISION MAKING

Using Data

Use the information in the table to answer each question.

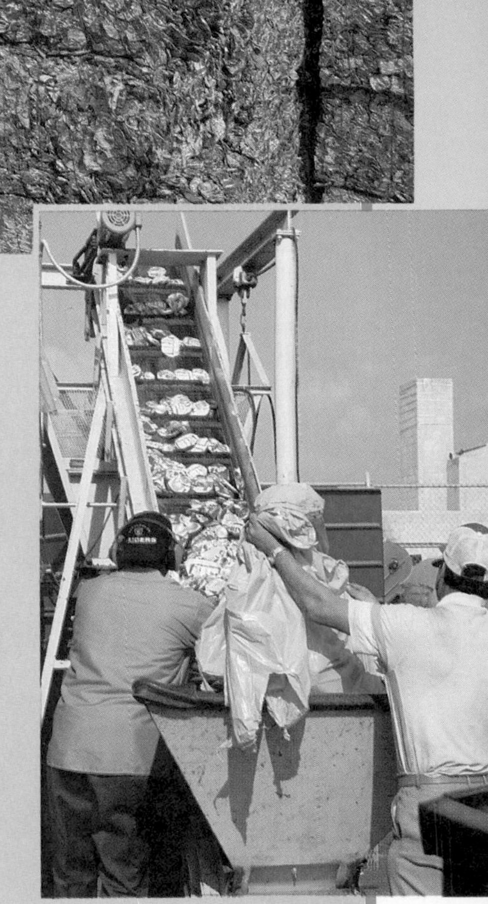

1. There were 203,302,031 people in the United States in 1970. How many pounds of solid waste were generated per day by the entire population in 1970?
2. There were 266,545,805 people in the United States in 1980. How many pounds of solid waste were generated per day by the entire population in 1980?
3. What trend do you see emerging from the data on the total amount of solid waste generated in the United States?
4. What trend do you see emerging from the data on the number of pounds of waste generated per person per day?
5. What trend do you see emerging from the data on the net number of pounds of waste disposed of per person per day?
6. What kind of graph would you choose to display these trends? Explain.
7. What conclusions can you draw about the attitudes of the population that affected these trends?
8. What can you do to encourage recycling at home and in your school?
9. What other conservation methods could you use to save resources?

Working Together

Your group's task is to research the solid-waste management and the recycling programs in your area. Contact the department of waste management in your area, or use newspapers, almanacs, or other reference books. Find the change in the volume of solid waste recycled in your community over the past ten years. If possible, compare the changes in the recycling of different materials, such as newspapers, cans, and glass. Record your results in a chart or graph. Then use the chart or graph to make a poster that will encourage others to recycle.

4-1 Factors and Multiples

EXPLORE

MATH: WHO, WHERE, WHEN

The chart you will complete is called the Sieve of Eratosthenes, named for the Greek scientist who developed it. Eratosthenes of Cyrene lived from approximately 276 B.C. until 194 B.C. He was the first person to calculate the circumference of the earth. He also devised a calendar that included leap years and was known for his writings on the theater and on ethics.

Make a chart like this one.

a. Cross out 1.

b. Cross out all multiples of 2 except 2.

c. Cross out all multiples of 3 except 3.

d. Cross out all multiples of 5 except 5.

e. Cross out all multiples of 7 except 7.

1	2	3	4	5	6	7	8	9	10
11	12	13	14	15	16	17	18	19	20
21	22	23	24	25	26	27	28	29	30
31	32	33	34	35	36	37	38	39	40
41	42	43	44	45	46	47	48	49	50
51	52	53	54	55	56	57	58	59	60
61	62	63	64	65	66	67	68	69	70
71	72	73	74	75	76	77	78	79	80
81	82	83	84	85	86	87	88	89	90
91	92	93	94	95	96	97	98	99	100

The numbers that you did not cross out are called *prime numbers.* How many prime numbers are there between 1 and 100? How many of them, other than 2, are even numbers? Explain how you know.

SKILLS DEVELOPMENT

If one whole number is divisible by a second, the second number is a **factor** of the first. For example, since 6 is divisible by 1, 2, 3, and 6, all these numbers are factors of 6.

$$6 \div 1 = 6 \qquad 6 \div 2 = 3 \qquad 6 \div 3 = 2 \qquad 6 \div 6 = 1$$

A **prime number** is a whole number that has exactly two factors, 1 and itself.

$$2 = 1 \times 2 \qquad 3 = 1 \times 3 \qquad 13 = 1 \times 13$$

A whole number that has more than two factors is called a **composite number.** You can write any composite number as a product of prime numbers. This is called the **prime factorization** of the number.

TALK IT OVER

Is the number 1 prime, composite, or neither? Give an argument to support your answer.

Example 1

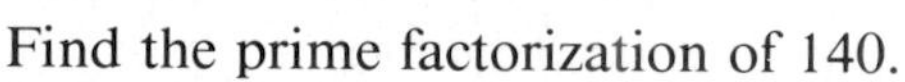

Find the prime factorization of 140.

Solution

Use a *factor tree* to find the prime factors.

- ► Write the composite number as the product of two factors.
- ► Repeat with any remaining composite numbers.
- ► When all numbers are prime, write the prime factorization.

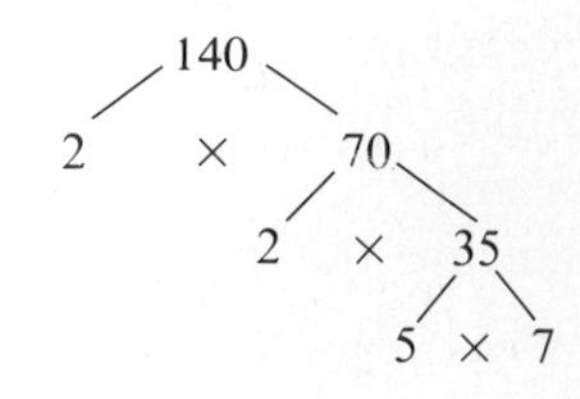

$140 = 2 \times 2 \times 5 \times 7$, or $2^2 \times 5 \times 7$ ◄

Two whole numbers may have some factors that are the same for both numbers. These are called **common factors.** The greatest of these is called the **greatest common factor (GCF)** of the numbers.

CHECK UNDERSTANDING

If the GCF of two numbers is 1, what do you know about the factors of each number?

Example 2

Find the GCF of 18 and 60.

Solution 1
To find the GCF using common factors:

- ► List all the factors of each number. — **factors of 18: 1, 2, 3, 6, 9, 18**
- ► Underline the common factors. — **factors of 60: 1, 2, 3, 4, 5, 6, 10, 12, 15, 20, 30, 60**
- ► Identify the greatest common factor. — The GCF of 18 and 60 is 6. ◄

Solution 2
To find the GCF using prime factors:

- ► Write the prime factorization of each number. — $18 = 2 \times 3^2$; $60 = 2^2 \times 3 \times 5$
- ► Underline the common prime factors.
- ► Choose the least power of each underlined factor. Then multiply. — $2 \times 3 = 6$; The GCF of 18 and 60 is 6. ◄

When you multiply a number by 1, 2, 3, 4, and so on, you obtain **multiples** of that number. Here are some multiples of 3.

$$1 \times 3 = 3 \qquad 2 \times 3 = 6 \qquad 3 \times 3 = 9$$

Two whole numbers may have some multiples that are the same for both numbers. These are called **common multiples.** The least of these is called the **least common multiple (LCM)** of the numbers.

COMPUTER

This program has the computer find the LCM of two numbers by comparing the common multiples of the numbers.

```
10 LET A = 0:B = 0
20 INPUT "FIRST NUMBER?
   ";A
30 INPUT "SECOND NUMBER?
   ";B
40 FOR X = 1 TO B
50 LET M = A * X
60 FOR Y = 1 TO A
70 LET N = B * Y
80 IF M = N THEN GOTO 110
90 NEXT Y
100 NEXT X
110 PRINT : PRINT "THE LCM
    OF ";A;" AND ";B;" IS
    ";M
```

Example 3

Find the LCM of 18 and 60.

Solution 1
To find the LCM using multiples:

- ► List multiples of each number. — **multiples of 18: 18, 36, 54, 72, 90, 108, 120, 144, 163, 180, . . .**
- ► Underline common multiples. — **multiples of 60: 60, 120, 180, . . .**
- ► Write the LCM. — The LCM of 18 and 60 is 180. ◄

Solution 2
To find the LCM using prime factors:

- ► Write the prime factorization of each number. — $18 = 2 \times 3^2$; $60 = 2^2 \times 3 \times 5$
- ► Choose the greatest power of each prime factor. Then multiply. — $2^2 \times 3^2 \times 5 = 180$; The LCM of 18 and 60 is 180. ◄

Example 4

The manager of a popcorn stand gives a free box of popcorn to every 10th customer and a free lemonade to every 15th customer. Which customer will be the first to receive both popcorn and a drink?

Solution

To solve the problem, you need to find the LCM of 10 and 15.

► Write the prime factorization of each number. $10 = 2 \times 5$, $15 = 3 \times 5$

► Choose the greatest power of each prime factor. Then multiply. $2 \times 3 \times 5 = 30$

The 30th customer will receive both popcorn and a drink. ◄

COMPUTER

This computer program will list all the factors of any numbers that you input. You can use the program to help you find the GCF for any two numbers.

```
10 INPUT "FIRST NUMBER?
   ";N
20 FOR X = 1 TO N
30 IF N / X = INT (N / X) THEN
   PRINT " ";N / X;
40 NEXT X
50 PRINT:INPUT "SECOND
   NUMBER? ":A
60 FOR Y = 1 TO A
70 IF A / Y = INT (A / Y)
   THEN PRINT " ";A / Y;
80 NEXT Y
```

TRY THESE

Find the prime factorization of each number.

1. 56
2. 88
3. 117
4. 336

Find the GCF of each pair of numbers.

5. 8, 12
6. 6, 15
7. 6, 8
8. 30, 40

Find the LCM of each pair of numbers.

9. 8, 9
10. 6, 15
11. 18, 24
12. 14, 21

Solve.

13. Every 6th person at a party receives a balloon and every 8th person receives a noisemaker. Which person will be the first to receive both a balloon and a noisemaker?

EXERCISES

PRACTICE/ SOLVE PROBLEMS

Find the prime factorization of each number.

1. 104
2. 108
3. 150
4. 300

Find the GCF of each pair of numbers.

5. 4, 6
6. 8, 12
7. 12, 15
8. 9, 21
9. 10, 22
10. 12, 20
11. 16, 24
12. 24, 32

Find the LCM of each pair of numbers.

13. 4, 7
14. 8, 9
15. 12, 16
16. 9, 15
17. 4, 14
18. 6, 21
19. 18, 20
20. 15, 27

21. Every 4th person in line at the theater will receive a free matinee pass. Every 6th person will receive a free evening pass. Which person in line will be the first to receive both passes?

22. Jane comes to the tennis court every two days. Ellen comes every three days. How many days will pass before the two can play tennis together?

EXTEND/SOLVE PROBLEMS

Write the prime numbers that fall between each pair of numbers.

23. 2^2 and 3^2

24. 2^4 and 4^3

25. 4^3 and 2^5

Find the GCF.

26. 6, 12, 15

27. 4, 6, 8

28. 8, 12, 16

29. $3a^2b$, $6ab^2$

30. $8a^3b^4$, $4a^2b^3$

31. $12a^7b^3$, $15a^6b^5$

Find the LCM.

32. 2, 3, 5

33. 4, 6, 10

34. 3, 7, 12

35. $3ab^2$, $4a^2b$

36. $2a^4b$, $5ab^4$

37. $6ab$, $3a^2b^2$

WRITING ABOUT MATH

In your journal, write a brief paragraph that explains the differences between the GCF and the LCM of two numbers. Then outline the method you prefer to use to find the GCF and the method you prefer to find the LCM. Make a note of any differences in methods.

Solve.

38. On December 1, Jenny, Dan, and Marita went jogging together. Jenny goes jogging every third day, Dan every fourth day, and Marita every fifth day.
    a. On what day will they next go jogging together?
    b. How many more times will Jenny have jogged than Marita by the time they jog together again?

39. At a rummage sale, Christine bought 36 books, 27 cassette tapes, and 15 board games and shared them equally with her friends. With how many friends did she share?

40. Clarence shared $85.50 with each of five friends. What is each person's share?

THINK CRITICALLY/ SOLVE PROBLEMS

Find each answer.

41. If the GCF of 2 and another number is 1, describe what you know about the other number.

42. What is the GCF of two prime numbers?

43. Describe the LCM of two prime numbers.

44. Three road blinkers are turned on simultaneously. One blinks every 6 seconds; another every 10 seconds; and the third every 12 seconds. How many times per hour do they blink at once?

4-2 Equivalent Fractions

EXPLORE

Use fraction bars. Find the bar that is divided into halves. Use one half to cover part of the fourths bar. How many fourths does one half cover?

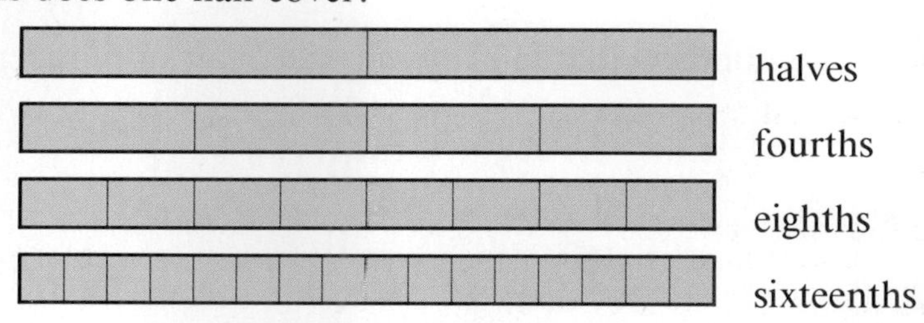

Now use one half to cover the eighths bar, then the sixteenths bar. How many parts of each bar does one half cover?

SKILLS DEVELOPMENT

Fractions that represent the same amount are **equivalent fractions.**

The terms of a fraction, such as $\frac{2}{3}$, tell you what the fraction means.

$$\frac{\text{number of parts being described}}{\text{number of parts that make up the whole}} = \frac{2}{3} \begin{array}{l}\leftarrow \textbf{numerator} \\ \leftarrow \textbf{denominator}\end{array}$$

For example, a fraction, such as $\frac{2}{3}$, can be used to show equal parts of a whole object or a part of a set of objects.

Equal parts of a whole object

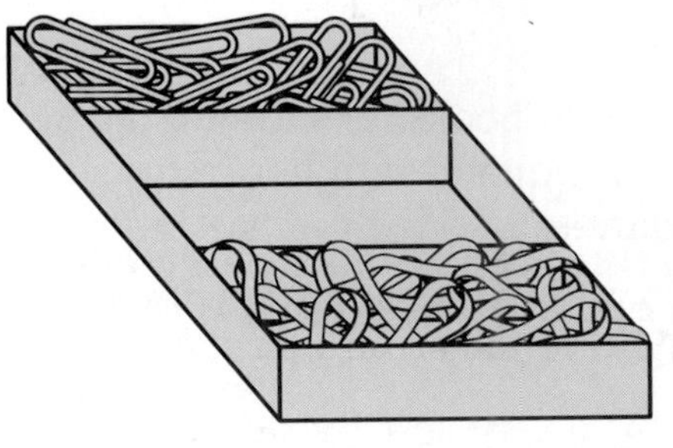

2 out of 3 sections are full.
So $\frac{2}{3}$ of the box is full.

Equal parts of a set of objects.

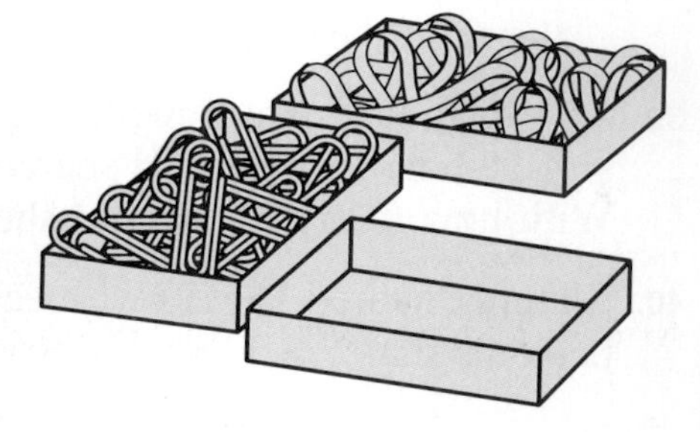

2 out of 3 boxes are full.
So $\frac{2}{3}$ of the boxes are full.

An equivalent fraction can be found for any fraction by multiplying or dividing the numerator and the denominator by the same nonzero number.

Example 1

a. Multiply to find four fractions equivalent to $\frac{1}{2}$.

b. Divide to find four fractions equivalent to $\frac{24}{48}$.

Solution

a. $\frac{1}{2} = \frac{1 \times 2}{2 \times 2} = \frac{2}{4}$ $\quad \frac{1 \times 3}{2 \times 3} = \frac{3}{6}$ $\quad \frac{1 \times 4}{2 \times 4} = \frac{4}{8}$ $\quad \frac{1 \times 5}{2 \times 5} = \frac{5}{10}$

b. $\frac{24}{48} = \frac{24 \div 2}{48 \div 2} = \frac{12}{24}$ $\quad \frac{24 \div 3}{48 \div 3} = \frac{8}{16}$ $\quad \frac{24 \div 4}{48 \div 4} = \frac{6}{12}$ $\quad \frac{24 \div 6}{48 \div 6} = \frac{4}{8}$ ◄

A fraction is said to be in **lowest terms** when the GCF of its numerator and denominator is 1.

Example 2

Write $\frac{24}{48}$ in lowest terms.

Solution

► Find the GCF of the numerator and the denominator. $\quad 24 = 2^3 \times 3$, $48 = 2^4 \times 3$ $\quad$ The GCF is $2^3 \times 3 = 24$.

► Divide both the numerator and the denominator by the GCF. $\quad \frac{24 \div 24}{48 \div 24} = \frac{1}{2}$ ◄

It is easier to compare fractions when they have the same denominator. The **least common denominator (LCD)** of two or more fractions is the least common multiple of their denominators.

Example 3

Compare. Replace ● with <, >, or =: $\frac{5}{6}$ ● $\frac{7}{8}$.

Solution

► Find the LCD. $\quad 6 = 2 \times 3$ and $8 = 2^3$. The LCD is $2^3 \times 3 = 24$.

► Write the fractions using the LCD as the denominator of each. $\quad \frac{5}{6} = \frac{5 \times 4}{6 \times 4} = \frac{20}{24}$ and $\frac{7}{8} = \frac{7 \times 3}{8 \times 3} = \frac{21}{24}$

► Compare. $\quad \frac{20}{24} < \frac{21}{24}$, so $\frac{5}{6} < \frac{7}{8}$. ◄

TRY THESE

Multiply to find two fractions equivalent to the given fraction.

1. $\frac{2}{3}$ 2. $\frac{3}{4}$ 3. $\frac{7}{8}$ 4. $\frac{3}{10}$

Divide to find two fractions equivalent to the given fraction.

5. $\frac{12}{16}$ 6. $\frac{8}{12}$ 7. $\frac{8}{24}$ 8. $\frac{20}{30}$

Write each fraction in lowest terms.

9. $\frac{2}{4}$ 10. $\frac{4}{12}$ 11. $\frac{6}{10}$ 12. $\frac{15}{20}$

Compare. Replace ● with <, >, or = .

13. $\frac{7}{8}$ ● $\frac{6}{8}$ 14. $\frac{3}{5}$ ● $\frac{2}{3}$ 15. $\frac{5}{6}$ ● $\frac{10}{12}$

EXERCISES

PRACTICE/ SOLVE PROBLEMS

Multiply to find two fractions equivalent to the given fraction.

1. $\frac{1}{3}$ **2.** $\frac{3}{8}$ **3.** $\frac{5}{12}$ **4.** $\frac{5}{6}$

Divide to find two fractions equivalent to the given fraction.

5. $\frac{12}{18}$ **6.** $\frac{36}{54}$ **7.** $\frac{30}{36}$ **8.** $\frac{24}{42}$

Write each fraction in lowest terms.

9. $\frac{2}{4}$ **10.** $\frac{6}{8}$ **11.** $\frac{15}{45}$ **12.** $\frac{8}{64}$

13. $\frac{18}{20}$ **14.** $\frac{9}{24}$ **15.** $\frac{8}{12}$ **16.** $\frac{12}{15}$

Compare. Replace ● with <, >, or = .

17. $\frac{3}{8}$ ● $\frac{2}{3}$ **18** $\frac{4}{5}$ ● $\frac{2}{3}$ **19.** $\frac{3}{5}$ ● $\frac{6}{10}$

20. $\frac{2}{3}$ ● $\frac{3}{4}$ **21.** $\frac{5}{6}$ ● $\frac{7}{8}$ **22.** $\frac{7}{8}$ ● $\frac{2}{3}$

Solve. Write the answer in lowest terms.

23. There are 3 people in a minivan that can seat 9 people. What fraction of the seats are occupied?

24. A football team is made up of 30 players. Six players were absent from practice. What fraction of the team was absent?

25. A large glass can hold 16 fl oz of liquid. If it is filled with 12 fl oz of water, what fraction of the glass is filled with water?

EXTEND/ SOLVE PROBLEMS

Order the following sets of fractions from least to greatest.

26. $\frac{1}{8}, \frac{7}{8}, \frac{3}{8}$ **27.** $\frac{1}{3}, \frac{5}{6}, \frac{2}{3}$ **28.** $\frac{1}{2}, \frac{3}{10}, \frac{2}{5}$

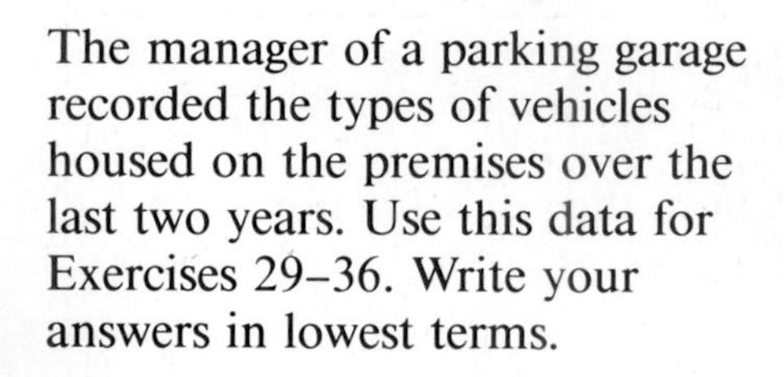

The manager of a parking garage recorded the types of vehicles housed on the premises over the last two years. Use this data for Exercises 29–36. Write your answers in lowest terms.

Vehicles	Last Year	This Year
full-size cars	52	40
compact cars	42	51
vans	6	9

Use the data in the column marked *last year.*

29. What fraction of the total were compact cars?

30. What fraction of the total were full-size cars?

31. What fraction of the total number of vehicles were vans?

32. Compact cars and full-size cars together made up what fraction of the total number of vehicles?

Use the data in the column marked *this year*.

33. What fraction of the total were compact cars?
34. What fraction of the total were full-size cars?
35. What fraction of the total number of vehicles were vans?
36. Compact cars and full-size cars together make up what fraction of the total number of vehicles?
37. A student council consists of eight representatives from each of four classes. Eight representatives were absent from a meeting.
    a. What fraction of the council was absent?
    b. What fraction of the council was present?

ESTIMATION TIP

Fractions that represent everyday situations are sometimes "messy," like $\frac{115}{356}$.

In these cases it helps to know how to write a simple fraction as an estimate.

$\frac{115}{356}$ is close to $\frac{120}{360}$.

So, $\frac{115}{356}$ is about $\frac{120}{360}$, or $\frac{1}{3}$.

THINK CRITICALLY/ SOLVE PROBLEMS

Solve.

Each student is reading a book for a book report. For Exercises 38–41, write a fraction to estimate what part of each student's book has been read so far.

	Student	Pages in Book	Pages Read
38.	Suzanne	464	117
39.	Jack	680	512
40.	Jennifer	720	430
41.	Frank	520	345

42. The denominator of the lowest-terms fraction equivalent to $\frac{48}{84}$ is 5 more than a number. What is the number?
43. The sum when 3 is added to 2 times a number is the numerator of the lowest-terms fraction that is equivalent to $\frac{30}{54}$. What is the number?

4-3 Fractions and Decimals

EXPLORE

Think about the names given to U.S. coins. Why is a 25-cent piece called a *quarter?* (What is it a quarter of?)

None of our other coins in general use today have names that are the same as names of fractions. But suppose they did. Which coins might be called a *tenth,* a *twentieth,* and a *hundredth*?

SKILLS DEVELOPMENT

Since fractions and decimals both show parts of a whole, any fraction can be expressed in decimal form. For example:

1 quarter is $\frac{1}{4}$ of \$1.00, or $\frac{25}{100}$ of \$1.00, which is \$0.25.

1 dime is $\frac{1}{10}$ of \$1.00, or $\frac{10}{100}$ of \$1.00, which is \$0.10.

1 nickel is $\frac{1}{20}$ of \$1.00, or $\frac{5}{100}$ of \$1.00, which is \$0.05.

1 penny is $\frac{1}{100}$ of \$1.00, or $\frac{1}{100}$ of \$1.00, which is \$0.01.

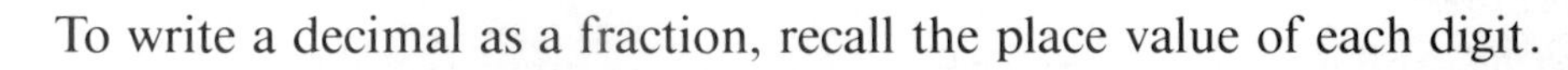

To write a decimal as a fraction, recall the place value of each digit.

Example 1

Write each decimal as a fraction or mixed number in lowest terms.

a. 0.3 **b.** 0.84 **c.** 0.375 **d.** 2.06

Solution

	Decimal	Read	Write fraction with denominator of 10, 100, or 1,000.	Write in lowest terms.
a.	0.3	3 tenths	$\frac{3}{10}$	$\frac{3}{10}$
b.	0.84	84 hundredths	$\frac{84}{100}$	$\frac{84}{100} = \frac{84 \div 4}{100 \div 4} = \frac{21}{25}$
c.	0.375	375 thousandths	$\frac{375}{1,000}$	$\frac{375}{1,000} = \frac{375 \div 125}{1,000 \div 125} = \frac{3}{8}$
d.	2.06	2 and 6 hundredths	$2\frac{6}{100}$	$\frac{6}{100} = \frac{6 \div 2}{100 \div 2} = \frac{3}{50}$; $2\frac{6}{100} = 2\frac{3}{50}$ ◄

Remember that a bar can indicate division. Therefore, you can write a fraction as a decimal by dividing the numerator by the denominator. When the remainder of this division is 0, the decimal is a **terminating decimal.**

Example 2

Write each fraction or mixed number as a decimal.

a. $\frac{3}{8}$

b. $6\frac{1}{2}$

Solution

Divide. Annex zeros if necessary.

a. $\frac{3}{8} \rightarrow 8\overline{)3.000}$ = 0.375

$$\begin{array}{r} 0.375 \\ 8\overline{)3.000} \\ \underline{2\,4} \\ 60 \\ \underline{56} \\ 40 \\ \underline{40} \\ 0 \end{array}$$

Since the remainder is 0, the decimal terminates.

$\frac{3}{8} = 0.375$

b. Write $\frac{1}{2}$ as a decimal.

$$\begin{array}{r} 0.5 \\ \frac{1}{2} \rightarrow 2\overline{)1.0} \\ \underline{1\,0} \\ 0 \end{array}$$

Since the remainder is 0, the decimal terminates.
Now, write $6\frac{1}{2}$ in decimal form.

$6\frac{1}{2} = 6 + \frac{1}{2}$
$= 6 + 0.5$
$= 6.5$ ◄

When the remainder of the division is not zero and a block of digits in the quotient repeats, the decimal is a **repeating decimal.**

Example 3

Write each fraction as a decimal.

a. $\frac{1}{3}$

b. $\frac{1}{11}$

Solution

Divide. Annex zeros if necessary.

a.
$$\begin{array}{r} 0.33\ldots \\ \frac{1}{3} = 3\overline{)1.00} \\ \underline{9} \\ 10 \\ \underline{9} \\ 1 \end{array}$$

The remainder is not zero but the digit 3 repeats in the quotient. Place a bar over the 3.

$\frac{1}{3} = 0.\overline{3}$

b.
$$\begin{array}{r} 0.0909\ldots \\ \frac{1}{11} = 11\overline{)1.0000} \\ \underline{99} \\ 100 \\ \underline{99} \\ 1 \end{array}$$

The remainder is not zero but the digits 0 and 9 repeat as a block in the quotient. Place a bar over 09.

$\frac{1}{11} = 0.\overline{09}$ ◄

Example 4

Replace ● with <, >, or =.

a. $0.3 ● \frac{2}{5}$

b. $\frac{3}{8} ● 0.375$

c. $0.812 ● \frac{7}{8}$

Solution

Before comparing, write each fraction as a decimal.

a. $0.3 ● \frac{2}{5}$
$0.3 ● 0.4$
$0.3 < 0.4$

b. $\frac{3}{8} ● 0.375$
$0.375 ● 0.375$
$0.375 = 0.375$

c. $\frac{7}{8} ● 0.812$
$0.875 ● 0.812$
$0.875 > 0.812$ ◄

COMPUTER TIP

To make a computer act like a calculator, just type ? FRINT followed by the computation you want the computer to do.

MIXED REVIEW

Find each answer. Use rounding to estimate and compare to see if the answer is reasonable.

1. 4,562 − 796
2. $412.36 − $381.53
3. 1,698 + 245
4. $122.19 + 296.34

Solve.

5. Randi bought a bicycle for $263.75, a helmet for $39.95, and a lock for $27.50. There was no tax. How much did she spend in all?

TRY THESE

Write each decimal as a fraction or mixed number in lowest terms.

1. 0.9   2. 0.75   3. 0.512   4. 2.95
5. 0.016   6. 3.8   7. 1.02   8. 7.806

Write each fraction or mixed number as a terminating decimal.

9. $\frac{3}{10}$ 10. $\frac{61}{100}$ 11. $\frac{912}{1,000}$ 12. $\frac{4}{5}$
13. $\frac{3}{4}$ 14. $\frac{3}{15}$ 15. $4\frac{5}{8}$ 16. $2\frac{7}{8}$
17. $1\frac{1}{2}$ 18. $2\frac{1}{4}$ 19. $6\frac{3}{8}$ 20. $4\frac{2}{5}$

Write each fraction as a repeating decimal.

21. $\frac{2}{3}$ 22. $\frac{1}{6}$ 23. $\frac{2}{11}$ 24. $\frac{5}{8}$
25. $1\frac{2}{9}$ 26. $2\frac{11}{12}$ 27. $6\frac{7}{15}$ 28. $4\frac{7}{30}$

Replace ● with <, >, or =.

29. 0.2 ● $\frac{2}{5}$ 30. $\frac{1}{8}$ ● 0.125 31. $\frac{5}{6}$ ● 0.93

EXERCISES

PRACTICE/ SOLVE PROBLEMS

Write each decimal as a fraction or mixed number in lowest terms.

1. 0.7   2. 0.892   3. 0.932   4. 4.62
5. 0.058   6. 4.3   7. 8.05   8. 3.901

Write each fraction or mixed number as a terminating decimal.

9. $\frac{9}{10}$ 10. $\frac{89}{100}$ 11. $\frac{417}{1000}$ 12. $\frac{3}{5}$
13. $\frac{1}{4}$ 14. $\frac{18}{45}$ 15. $\frac{3}{8}$ 16. $\frac{9}{25}$
17. $1\frac{2}{5}$ 18. $8\frac{1}{25}$ 19. $4\frac{7}{8}$ 20. $3\frac{7}{28}$

Write each fraction as a repeating decimal.

21. $\frac{5}{6}$ 22. $\frac{4}{11}$ 23. $\frac{5}{9}$ 24. $\frac{8}{15}$
25. $\frac{8}{9}$ 26. $\frac{1}{15}$ 27. $\frac{1}{3}$ 28. $\frac{21}{22}$

Replace ● with >, <, or =.

29. 0.85 ● $\frac{8}{25}$ 30. 0.625 ● $\frac{5}{8}$ 31. $\frac{7}{12}$ ● 0.578

Solve.

32. Mary Lee jogged $1\frac{4}{5}$ mi. Twana jogged 1.6 mi. Who jogged a greater distance?

33. During basketball practice, Christine shot free throws. She made 60 out of 75 shots. She said she made $\frac{4}{5}$ of her shots. Her coach said she made 0.8 of her shots. Who is correct? Explain.

EXTEND/ SOLVE PROBLEMS

Order each set of numbers from least to greatest.

34. $\frac{1}{8}$, 0.28, $\frac{1}{5}$, 0.75, $\frac{4}{15}$, 10.031

35. 0.561, 0.25, 0.781, $\frac{4}{5}$, $\frac{3}{7}$, $\frac{7}{8}$

Let $a = 1$, $b = 2$, and $c = 3$. Replace ● with <, >, or =.

36. $\frac{b}{c}$ ● 0.45

37. $\frac{a}{c}$ ● 0.125

38. 0.3 ● $\frac{a}{b}$

Solve.
José biked 12.5 mi. Betty biked $12\frac{1}{4}$ mi. Mona biked $12\frac{3}{5}$ mi.

39. Who traveled the greatest distance?

40. Who traveled the least distance?

Shiroh has collected $\frac{8}{9}$ of his goal for a fund-raising drive. Greg has collected 0.83 of his goal. Otis has collected $\frac{16}{17}$ of his goal.

41. Who is closest to meeting his goal?

42. Who is farthest from meeting his goal?

THINK CRITICALLY/ SOLVE PROBLEMS

Find each answer.

43. Name at least four fractions that have the same value as 0.75.

44. What are the greatest and the least decimals that can be written using each of the digits 3, 4, 5, and 7 just once?

Solve.

45. Maria has finished reading $\frac{4}{5}$ of a new novel, while Allison has finished 0.832 of it. Melanie has read an amount that is somewhere in between. Express that amount in decimal form.

4-4 Zero and Negative Exponents

EXPLORE

Describe the patterns formed by the numbers at the right. Then continue the pattern to show 0.00001, 0.000001, and 0.0000001 in exponential form.

Now try to show a similar pattern using 3 as the base.

$10{,}000 = 10^4$
$1{,}000 = 10^3$
$100 = 10^2$
$10 = 10^1$
$1 = 10^0$

$0.1 = \frac{1}{10} = \frac{1}{10^1} = 10^{-1}$
$0.01 = \frac{1}{100} = \frac{1}{10^2} = 10^{-2}$
$0.001 = \frac{1}{1{,}000} = \frac{1}{10^3} = 10^{-3}$
$0.0001 = \frac{1}{10{,}000} = \frac{1}{10^4} = 10^{-4}$

SKILLS DEVELOPMENT

Any nonzero number to the zero power has a value of 1.

$$a^0 = 1$$

Although this fact surprises many people, it is easy to show why it must be true. For instance, this is how you can show that 10^0 must be equal to 1.

You know that $100 \div 100 = 1$, so $10^2 \div 10^2 = 1$.
By the laws of exponents, $10^2 \div 10^2 = 10^{2-2} = 10^0$.
Therefore, $10^0 = 1$.

If a nonzero number has a negative exponent, you can write it as a fraction with a numerator of 1 and a denominator that has a positive exponent.

$$a^{-n} = \frac{1}{a^n}$$

Example 1

Write as a fraction.

a. 4^{-2} b. 6^{-3} c. 2^{-4}

Solution

a. $4^{-2} = \frac{1}{4^2} = \frac{1}{4 \times 4} = \frac{1}{16}$

b. $6^{-3} = \frac{1}{6^3} = \frac{1}{6 \times 6 \times 6} = \frac{1}{216}$

c. $2^{-4} = \frac{1}{2^4} = \frac{1}{2 \times 2 \times 2 \times 2} = \frac{1}{16}$ ◄

Example 2

Write in exponential form.

a. $\frac{1}{11 \times 11 \times 11}$

b. $\frac{1}{3 \times 3}$

Solution

a. $\frac{1}{11 \times 11 \times 11}$

$= \frac{1}{11^3} = 11^{-3}$

b. $\frac{1}{3 \times 3}$

$= \frac{1}{3^2} = 3^{-2}$ ◄

Negative exponents are useful when you need to write very small numbers in scientific notation.

Example 3

Write 3.416×10^{-4} in standard form.

Solution

$3.416 \times 10^{-4} = 3.416 \times \frac{1}{10^4} = 3.416 \times \frac{1}{10{,}000}$

$= 3.416 \times 0.0001$

$= 0.0003416$ ◄

CONNECTIONS

The definition of scientific notation applies to both very large and very small numbers. A number written in scientific notation has one factor that is greater than or equal to 1 and less than 10 and another factor that is a power of ten.

Example 4

Write 0.0016 in scientific notation.

Solution

$0.0016 = 1.6 \times 0.001 = 1.6 \times \frac{1}{1{,}000}$

$= 1.6 \times \frac{1}{10^3}$

$= 1.6 \times 10^{-3}$ ◄

TRY THESE

Write as a fraction.

1. 6^{-2}
2. 3^{-5}
3. 6^{-4}
4. 9^{-3}

Write in exponential form.

5. $\frac{1}{3 \times 3 \times 3 \times 3 \times 3}$
6. $\frac{1}{4 \times 4 \times 4 \times 4}$

Write in standard form.

7. 6.876×10^{-7}
8. 4.2×10^{-6}

Write in scientific notation.

9. 0.00000002195
10. 0.00012
11. 0.000009
12. 0.00000007

MENTAL MATH TIP

Here's a shortcut for writing a number in scientific notation. Count the number of places that the decimal point moves to the left or right. Use that number in the exponent. If the decimal point moves left, the exponent should be positive. If it moves right, the exponent should be negative.

$73{,}000 = 7.3 \times 10^4$

$0.00073 = 7.3 \times 10^{-4}$

EXERCISES

PRACTICE/ SOLVE PROBLEMS

Write as a fraction.

1. 2^{-6}
2. 7^{-3}
3. 3^{-2}
4. 3^{-4}

Write in exponential form.

5. $\frac{1}{8 \times 8 \times 8 \times 8 \times 8 \times 8}$
6. $\frac{1}{5 \times 5 \times 5 \times 5 \times 5 \times 5 \times 5}$
7. $\frac{1}{2 \times 2}$
8. $\frac{1}{3 \times 3 \times 3}$

Write in standard form.

9. 7.1×10^{-3}
10. 8.62154×10^{-8}
11. 9.1302×10^{-6}
12. 4.3276×10^{-7}

COMPUTER TIP

A computer will automatically express any number that has more than nine digits and a decimal point in scientific notation.

Write in scientific notation.

13. 0.0025
14. 0.000000009
15. 0.00000000000196
16. 0.00087
17. Write the number 0.000000000000001 in exponential form.
18. The smallest organisms are the viroids. Each has a diameter of 0.00000000007 in. Write this number using scientific notation.

EXTEND/ SOLVE PROBLEMS

Write as a fraction. Use a calculator for the computation.

19. 2^{-20}
20. 7^{-10}
21. 15^{-8}
22. 3^{-12}

Order the numbers in each row from least to greatest.

23.	6.2×10^{-5}	6.27×10^{-5}	7.3×10^{-4}	$\frac{1}{1{,}000}$
24.	$\frac{1}{10^8}$	1.25×10^{-8}	1.25×10^{-9}	10^{-9}
25.	10^{-7}	8.886×10^{-6}	8.668×10^{-8}	$\frac{1}{10^6}$
26.	$\frac{1}{4^4}$	$\frac{1}{5^4}$	$\frac{1}{6^4}$	$\frac{1}{7^4}$

Write *equal* if all the numbers in each row are equal. If not, indicate which number does not belong and explain why.

27.	$\frac{1}{10 \times 10 \times 10 \times 10}$	0.001	$\frac{1}{10{,}000}$	1^{-4}
28.	$\frac{1}{3 \times 3 \times 3}$	0.0027	$\frac{1}{27}$	3^{-3}
29.	$\frac{1}{2 \times 2 \times 2 \times 2}$	0.0625	$\frac{1}{16}$	2^{-4}
30.	$\frac{1}{20 \times 20 \times 20 \times 20}$	0.000125	$\frac{1}{8{,}000}$	20^{-3}

Solve.

31. The diameter of the nucleus of an atom is measured in femtometers. One femtometer equals 10^{-15} meter. A picometer equals 0.000000000001 meter. Which unit is larger? Explain.

32. One micrometer equals 0.000001 meter. A microsporidian spore can measure from 2 to 20 micrometers. Write this span in meters using scientific notation.

Write in exponential form.

33. $\frac{1}{125}$ 34. $\frac{1}{1,000}$ 35. $\frac{1}{64}$ 36. $\frac{1}{343}$

Use mental math to tell which number is greater.

37. 0.0025 or $\frac{1}{2^2}$ 38. 9 or $\frac{1}{3^3}$

39. 6.2193×10^{-3} or $6.2193 \times 10,000$

40. 7.987×10^3 or 7,987

THINK CRITICALLY/ SOLVE PROBLEMS

Solve.

41. The mass of a proton is 1.673×10^{-27} kg. The mass of a neutron is 1.675×10^{-27} kg.
    a. Which one is heavier?
    b. How much heavier is it?

Replace ● with <, >, or =.

42. 7.192×10^{-2} ● 10^{-2} 43. 3.96×10^4 ● 3.961×10^{-4}

44. x^{-3} ● 10^0, where x is greater than zero but less than one

45. x^{-3} ● 10^0, where x is greater than one

4-5 Multiplying and Dividing Fractions

EXPLORE

Try this activity.

a. Fold a small rectangular piece of paper into thirds using horizontal folds. Shade $\frac{2}{3}$ of the paper yellow.

b. Then fold the paper into fourths using vertical folds and shade $\frac{3}{4}$ of the yellow part blue.

c. There are 6 green sections, so $\frac{6}{12}$, or $\frac{1}{2}$, of the paper is green. You have found that $\frac{3}{4}$ of $\frac{2}{3}$ is $\frac{1}{2}$. You can rewrite this using math symbols as $\frac{3}{4} \times \frac{2}{3} = \frac{1}{2}$.

Now, try folding paper and shading sections to find $\frac{1}{2}$ of $\frac{1}{2}$, $\frac{1}{2}$ of $\frac{3}{4}$, and $\frac{2}{3}$ of $\frac{9}{10}$.

SKILLS DEVELOPMENT

To multiply by a fraction or a mixed number, follow these steps.

Step 1: Write ali whole numbers and mixed numbers as fractions.
Step 2: Multiply the numerators.
Step 3: Multiply the denominators.
Step 4: Write the product in lowest terms.

WRITING ABOUT MATH

Describe in your journal how you would write a whole number as a fraction and how you would write a mixed number as a fraction.

Example 1

a. Multiply: $\frac{1}{2} \times \frac{3}{5}$

b. Multiply: $1\frac{1}{3} \times 3\frac{3}{10}$

Solution

a. $\frac{1}{2} \times \frac{3}{5} = \frac{1 \times 3}{2 \times 5} = \frac{3}{10}$

b. $1\frac{1}{3} \times 3\frac{3}{10} = \frac{\cancel{4}^{2}}{\cancel{3}_{1}} \times \frac{\cancel{33}^{11}}{\cancel{10}_{5}} = \frac{2 \times 11}{1 \times 5} = \frac{22}{5} = 4\frac{2}{5}$ ◄

Divide 4 and 10 by their GCF, 2.
Divide 3 and 33 by their GCF, 3.

Two numbers whose product is 1 are called **reciprocals.** Each pair of numbers is an example of reciprocals.

$\frac{2}{3} \times \frac{3}{2} = 1 \qquad 6 \times \frac{1}{6} = 1 \qquad 1\frac{2}{3} \times \frac{3}{5} = \frac{5}{3} \times \frac{3}{5} = 1$

To divide fractions and mixed numbers, follow these steps.

Step 1: Write all whole numbers and mixed numbers as fractions.
Step 2: Multiply by the reciprocal of the divisor.
Step 3: Write the product in lowest terms.

Give the reciprocal of each number. $\frac{2}{3}, \frac{1}{10}, 3\frac{1}{4}, 7$

Why doesn't zero have a reciprocal?

Example 2

a. Divide: $\frac{3}{8} \div \frac{1}{4}$

b. Divide: $4\frac{1}{2} \div 1\frac{1}{5}$

Solution

The reciprocal of $\frac{1}{4}$ is $\frac{4}{1}$.

a. $\frac{3}{8} \div \frac{1}{4} = \frac{3}{\cancel{8}_2} \times \frac{\cancel{4}^1}{1} = \frac{3}{2} = 1\frac{1}{2}$

Divide 8 and 4 by their GCF, 4.

The reciprocal of $\frac{6}{5}$ is $\frac{5}{6}$.

b. $4\frac{1}{2} \div 1\frac{1}{5} = \frac{9}{2} \div \frac{6}{5} = \frac{\cancel{9}^3}{2} \times \frac{5}{\cancel{6}_2} = \frac{15}{4} = 3\frac{3}{4}$ ◄

Divide 9 and 6 by their GCF, 3.

Example 3

There were 864 glass bottles in the recycling bin. If $\frac{3}{4}$ of them were made from clear glass, how many clear glass bottles were in the bin?

Solution

$\frac{3}{4}$ of $864 = \frac{3}{4} \times 864 = \frac{3}{\cancel{4}_1} \times \frac{\cancel{864}^{216}}{1}$

$= 3 \times 216 = 648$

There were 648 clear glass bottles. ◄

Try These

Multiply. Be sure the product is in lowest terms.

1. $\frac{1}{3} \times \frac{1}{2}$
2. $\frac{3}{8} \times \frac{4}{5}$
3. $\frac{3}{4} \times 40$
4. $\frac{1}{2} \times 1\frac{1}{3}$
5. $1\frac{2}{3} \times 2\frac{1}{2}$
6. $5 \times 2\frac{3}{10}$
7. $3\frac{1}{3} \times 1\frac{1}{5}$
8. $3\frac{1}{2} \times 2\frac{2}{5}$

Divide. Be sure the quotient is in lowest terms.

9. $\frac{3}{8} \div \frac{3}{4}$
10. $\frac{4}{5} \div \frac{3}{10}$
11. $3 \div \frac{3}{10}$
12. $3\frac{1}{2} \div 1\frac{1}{6}$
13. $1\frac{1}{2} \div \frac{2}{3}$
14. $1\frac{1}{2} \div 1\frac{1}{4}$
15. $\frac{8}{15} \div 4$
16. $2\frac{3}{4} \div 1\frac{1}{2}$

Solve.

17. A ride on the roller coaster takes $2\frac{1}{2}$ minutes. Tony rode 7 times. For how many minutes did he ride?

EXERCISES

PRACTICE/ SOLVE PROBLEMS

Multiply. Be sure the product is in lowest terms.

1. $\frac{3}{4} \times \frac{8}{9}$
2. $\frac{2}{5} \times \frac{5}{8}$
3. $\frac{2}{3} \times 27$
4. $\frac{3}{4} \times 1\frac{1}{3}$
5. $1\frac{1}{8} \times 4\frac{5}{7}$
6. $1\frac{3}{5} \times 1\frac{3}{8}$
7. $1\frac{4}{5} \times 3\frac{1}{2}$
8. $5\frac{1}{5} \times 1\frac{1}{2}$

Divide. Be sure the quotient is in lowest terms.

9. $\frac{8}{9} \div \frac{8}{12}$
10. $\frac{3}{10} \div \frac{5}{8}$
11. $14 \div \frac{7}{10}$
12. $1\frac{1}{3} \div 1\frac{1}{6}$
13. $1\frac{1}{5} \div \frac{5}{8}$
14. $3\frac{4}{5} \div 1\frac{9}{10}$
15. $4\frac{1}{2} \div 1\frac{4}{9}$
16. $2\frac{1}{9} \div 12$

Solve.

17. Heather works 4 days each week at the hamburger stand. She works $2\frac{3}{4}$ hours each day. How many hours per week is that?
18. Chris bought 15 yd of fabric from which to cut sashes for the dance team. At $\frac{3}{4}$ yd per sash, how many sashes can he cut?

EXTEND/ SOLVE PROBLEMS

USING DATA Use the Data Index on page 546 to find the carbon dioxide emissions. Then solve each problem.

19. How many tons of carbon dioxide would be emitted by 2 cars that get 60 mi/gal?
20. How many tons of emissions would these be over a $\frac{1}{2}$-year period for a car that gets 26 mi/gal?

Marilyn ordered 24 pizzas for the homeroom party.

21. If each person eats $\frac{2}{3}$ of a pizza, how many students can be fed?
22. Each pizza cost $12.75. If the students share the cost equally, how much will each student pay?

THINK CRITICALLY/ SOLVE PROBLEMS

Use mental math to find each missing number.

23. $1\frac{3}{5} \div \frac{■}{5} = 1$
24. $\frac{4}{■} \times 1\frac{1}{4} = 1$
25. $■ \div \frac{1}{4} = 1$

Solve.

A swim club has 24 members. Which of these statements cannot be true? Explain.

26. One-half are female.
27. Two-thirds are divers.
28. Three-eighths swim daily.
29. Three-fifths are lifeguards.

► READ
► PLAN
► SOLVE
► ANSWER
► CHECK

Problem Solving Applications:

CHANGING RECIPE QUANTITIES

TOMATO AND CHEESE PIZZA Yield: 2 pizzas

DOUGH Combine and set aside.	Combine in food processor.	With the processor running, add steadily:	Process 1 min. Knead dough for 30 s, divide in half, and refrigerate.
$\frac{1}{4}$ c warm water	$2\frac{3}{4}$ c all-purpose flour	yeast mixture	
$\frac{1}{2}$ t sugar	$1\frac{1}{2}$ t salt	$\frac{3}{4}$ c cold water	
1 t dry yeast		3 T olive oil	

SAUCE

1 c tomato sauce
1 6-oz can tomato paste
1 T olive oil
2 T water
1 t oregano
$\frac{1}{2}$ t salt
$\frac{1}{2}$ t sugar
$\frac{1}{8}$ t crushed dried hot red chili peppers

Warm all ingredients in saucepan.

CHEESE TOPPING

$\frac{3}{4}$ lb sliced or grated mozzarella cheese
$\frac{1}{3}$ to $\frac{1}{2}$ c grated Parmesan cheese

DIRECTIONS

Place $\frac{1}{2}$ dough on each baking sheet, top each with $\frac{1}{2}$ sauce and $\frac{1}{2}$ cheese toppings. Bake in 450°F oven for 16–21 min or until cheese is melted.

Ken used this recipe for two pizzas.

Solve.

Suppose Ken cuts the recipe in half to make only one pizza.

1. How much of each ingredient would he use to make the dough?
2. How much of each ingredient would he use to make the sauce?
3. How much of each ingredient would he use to make the cheese topping?

Suppose Ken needs to double the recipe to make four pizzas.

4. How much of each ingredient would he use to make the dough?
5. How much of each ingredient would he use to make the sauce?
6. How much of each ingredient would he use to make the cheese topping?
7. If Ken has $6\frac{3}{4}$ lb mozzarella cheese, how many pizzas can he make with a cheese topping? (Remember that the recipe makes 2 pizzas.)

Explain how to adapt recipes with each number of servings.

8. A recipe serves 12 people. You want to make 8 servings.
9. A recipe serves 4 people. You want to make 6 servings.

Janet wanted to make $\frac{1}{12}$ of a recipe that serves 12 people. The recipe calls for $\frac{1}{2}$ t baking soda.

10. How much baking soda would she have needed for 1 serving?
11. Is it reasonable to think that Janet could make $\frac{1}{12}$ of a serving?

4-6 Adding and Subtracting Fractions

EXPLORE

Model $\frac{4}{5} - \frac{2}{5}$. Cut two identical strips of paper. Divide each strip into 5 equal parts. On one strip, fold back one fifth to show $\frac{4}{5}$. Then, to show how to subtract $\frac{2}{5}$ from $\frac{4}{5}$, fold back two more fifths. Compare the strips. The two parts remaining on the folded strip represent the difference, $\frac{2}{5}$. How would you change the paper strips to model subtraction of other fractions?

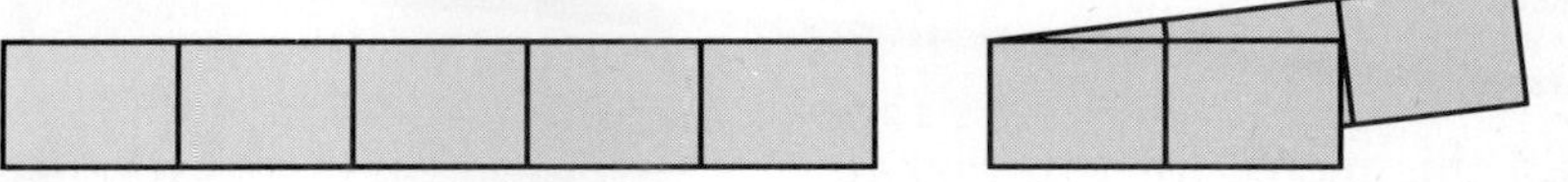

SKILLS DEVELOPMENT

To add or subtract **like fractions,** those fractions with denominators that are alike, follow these steps.

Step 1: Add or subtract the numerators.
Step 2: Keep the same denominator.
Step 3: Write the answer in lowest terms.

Example 1

a. Add: $\frac{3}{8} + \frac{3}{8}$

b. Subtract: $\frac{7}{9} - \frac{1}{9}$

Solution

a. $\frac{3}{8} + \frac{3}{8} = \frac{3+3}{8} = \frac{6}{8} = \frac{3}{4}$

b. $\frac{7}{9} - \frac{1}{9} = \frac{7-1}{9} = \frac{6}{9} = \frac{2}{3}$ ◄

To add or subtract fractions with unlike denominators, first rename them as like fractions so that the denominators are the same. Then add or subtract and write the answer in lowest terms.

Example 2

a. Add: $\frac{1}{2} + \frac{3}{10}$

b. Subtract: $\frac{5}{6} - \frac{2}{3}$

Solution

a.
$$\begin{array}{r} \frac{1}{2} = \frac{5}{10} \\ +\frac{3}{10} = \frac{3}{10} \\ \hline \frac{8}{10} = \frac{4}{5} \end{array}$$

b.
$$\begin{array}{r} \frac{5}{6} = \frac{5}{6} \\ -\frac{2}{3} = \frac{4}{6} \\ \hline \frac{1}{6} \end{array}$$ ◄

To add or subtract mixed numbers with unlike denominators, first rename the mixed numbers using like denominators. Then add or subtract and write the answer in lowest terms.

MIXED REVIEW

Find each answer. Use rounding to estimate and compare to see if the answer is reasonable.

1. 24×86
2. 7.9×3.11
3. $832 \div 13$
4. $15.925 \div 6.5$

Solve.

5. Roberto bought 6 CDs for \$10.99 each and 5 tapes for \$6.78 each. There was no tax. What was the total cost for these purchases?

Example 3

a. Add: $1\frac{11}{12} + 2\frac{5}{6}$

b. Subtract: $4\frac{3}{5} - 3\frac{1}{3}$

Solution

a.
$$\begin{array}{r} 1\frac{11}{12} = 1\frac{11}{12} \\ +\ 2\frac{5}{6} = 2\frac{10}{12} \\ \hline 3\frac{21}{12} = 4\frac{9}{12} = 4\frac{3}{4} \end{array}$$

b.
$$\begin{array}{r} 4\frac{3}{5} = 4\frac{9}{15} \\ -3\frac{1}{3} = 3\frac{5}{15} \\ \hline 1\frac{4}{15} \end{array}$$ ◄

Example 4

On Saturday, Jamie spent 6 hours at a family picnic and $2\frac{3}{4}$ hours at a ball game. How many more hours did Jamie spend at the picnic than at the ball game?

Solution

Subtract.
$$\begin{array}{r} 6 \ = 5\frac{4}{4} \\ -2\frac{3}{4} = 2\frac{3}{4} \\ \hline 3\frac{1}{4} \end{array}$$

Jamie spent $3\frac{1}{4}$ more hours at the picnic than at the ball game. ◄

TRY THESE

Add or subtract. Write each answer in lowest terms.

1. $\frac{1}{3} + \frac{1}{3}$
2. $\frac{7}{8} - \frac{3}{8}$
3. $\frac{8}{12} + \frac{1}{12}$
4. $\frac{7}{9} - \frac{4}{9}$
5. $\frac{1}{3} + \frac{2}{5}$
6. $\frac{5}{6} - \frac{2}{3}$
7. $\frac{1}{5} + \frac{1}{2}$
8. $\frac{1}{2} - \frac{3}{8}$
9. $5\frac{1}{2} - 3\frac{2}{5}$
10. $2\frac{3}{5} + 1\frac{1}{10}$
11. $6\frac{1}{2} + 5\frac{4}{3}$
12. $4\frac{1}{6} - 2\frac{1}{3}$

Solve.

13. Elena found a cone from a Scotch pine that is $1\frac{1}{4}$ in. long and a cone from a white pine that is $4\frac{3}{8}$ in. long. How much longer was the cone from the white pine than the cone from the Scotch pine?

EXERCISES

PRACTICE/ SOLVE PROBLEMS

Add or subtract. Write your answers in lowest terms.

1. $\frac{1}{8} + \frac{5}{8}$
2. $\frac{5}{6} - \frac{1}{6}$
3. $\frac{3}{10} - \frac{1}{10}$
4. $\frac{7}{12} + \frac{1}{12}$
5. $\frac{1}{6} - \frac{1}{8}$
6. $\frac{3}{10} + \frac{1}{2}$
7. $\frac{1}{3} + \frac{2}{5}$
8. $\frac{3}{4} - \frac{1}{3}$
9. $3\frac{1}{5} - 1\frac{1}{10}$
10. $1\frac{1}{2} + 3\frac{1}{5}$
11. $5\frac{3}{5} + 3\frac{3}{5}$
12. $9\frac{1}{2} - 6\frac{2}{3}$
13. $7\frac{3}{8} + 1\frac{1}{6}$
14. $5\frac{1}{6} - 3\frac{1}{6}$
15. $9 - 4\frac{7}{12}$
16. $1\frac{5}{6} + 1\frac{1}{6}$

Solve.

17. Ming caught a trout that weighed $14\frac{3}{4}$ lb and a walleye that weighed $7\frac{7}{8}$ lb. How much more did the trout weigh?

18. Jamie spent $\frac{3}{8}$ of her free time on Tuesday morning and $\frac{1}{6}$ of her free time on Tuesday afternoon at the pool. What fraction of her free time did she spend at the pool on Tuesday?

EXTEND/ SOLVE PROBLEMS

Add. Write each answer in lowest terms.

19. $\frac{1}{2} + \frac{1}{3} + \frac{1}{4}$

20. $\frac{1}{6} + \frac{1}{3} + \frac{1}{2}$

21. $\frac{3}{8} + \frac{1}{6} + \frac{3}{4}$

22. $1\frac{3}{5} + 3\frac{1}{2} + 2\frac{7}{10}$

USING DATA Use the Data Index on page 546 to find the average mix of refuse. Then solve each problem.

23. What other wastes have the same combined weight as the weight of paper and paperboard and wood/fiber combined?

24. What fraction describes the combined weight of household hazardous waste, rubber and leather, and plastics?

The school band practiced the following amounts of time during a two-week period. Express the answers to Exercises 25–27 in hours and minutes.

Week 1: $2\frac{1}{2}$ h, $1\frac{2}{3}$ h   Week 2: $3\frac{1}{2}$ h, $2\frac{3}{4}$ h

25. How long did the band practice during Week 1?

26. How long did the band practice during Week 2?

27. How long did the band practice for the two-week period?

THINK CRITICALLY/ SOLVE PROBLEMS

Complete.

28. $\frac{7}{8} + \frac{\blacksquare}{6} = 1\frac{1}{24}$

29. $1\frac{3}{4} - \frac{\blacksquare}{10} = 1\frac{1}{20}$

30. $3\frac{2}{3} - \blacksquare\frac{5}{\blacksquare} = \frac{5}{6}$

Use the clues to find each fraction in Exercises 31–34.

31. The sum of the digits of the numerator and denominator is 9. Their difference is 7.

32. The sum of the digits of the numerator and the denominator is 11. Their difference is 1.

33. The sum of the digits of the numerator and the denominator is 5. Their difference is 3.

34. Find the sum of the fractions you found for Exercises 31–33.

► READ
► PLAN
► SOLVE
► ANSWER
► CHECK

Problem Solving Applications:

MUSIC FOR FRACTION LOVERS

The notes in a piece of music are given the following names.

𝅝 whole note 𝅗𝅥 $\frac{1}{2}$ note ♪ $\frac{1}{8}$ note

♩ $\frac{1}{4}$ note 𝅘𝅥𝅯 $\frac{1}{16}$ note

Two eighth notes ♪♪ are written as ♫. Two sixteenth notes 𝅘𝅥𝅯𝅘𝅥𝅯 are written as ♬.

The time signature in a piece of music is written in fraction form. The numerator tells the number of beats in one bar. The denominator tells the kind of note that gets one beat.

4 beats to a bar
a quarter note gets one beat

The total value of the notes in each bar must equal 1.

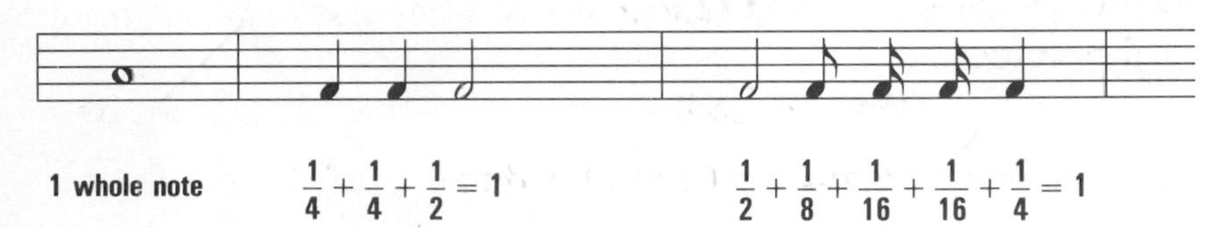

1 whole note $\frac{1}{4} + \frac{1}{4} + \frac{1}{2} = 1$ $\frac{1}{2} + \frac{1}{8} + \frac{1}{16} + \frac{1}{16} + \frac{1}{4} = 1$

Find the missing note that would make each pair equal in value.

1. 𝅗𝅥 = ♩ ♪ ? **2.** ♩ = ♪ 𝅘𝅥𝅯 ? **3.** 𝅗𝅥 = ♬ ♫ ?

4. 𝅝 = 𝅗𝅥 ? **5.** ♪ = 𝅘𝅥𝅯 ? **6.** ♩ = ♬ ?

Each piece of music is written in $\frac{4}{4}$ time. What type of note is missing from each bar?

7. 

8. 

9. 

10. 

11.

12.

13.

14.

4-7 Problem Solving Skills:

ESTIMATING WITH FRACTIONS AND DECIMALS

- ► READ
- ► PLAN
- ► SOLVE
- ► ANSWER
- ► CHECK

You often estimate to see if your answer is reasonable, but sometimes you don't have to find the exact answer—an estimate is all you need.

You have already used some estimation techniques, such as rounding, finding compatible numbers, and front-end estimation, when working with whole numbers and decimals. Now you will learn how to use these techniques when estimating with fractions.

PROBLEM

Garth worked $7\frac{3}{4}$h on Monday, $9\frac{1}{2}$h on Tuesday, $9\frac{1}{3}$h on Wednesday, $7\frac{1}{3}$h on Thursday, and $8\frac{1}{2}$h on Friday. He receives overtime pay for every hour he works over 40 hours. Should he receive overtime pay for this week?

How can you be sure from Garth's answer that he worked *more than* 40 hours?

SOLUTION

Garth and his friends, Jeff and Daryl, estimated to see if he is eligible for overtime pay.

Garth used front-end estimation.	Jeff used numbers close to 0, $\frac{1}{2}$, and 1.	Daryl used rounding.
$7\frac{3}{4} \rightarrow 7$	$7\frac{3}{4} \rightarrow 8$	$7\frac{3}{4} \rightarrow 8$
$9\frac{1}{2} \rightarrow 9$	$9\frac{1}{2} \rightarrow 9\frac{1}{2}$	$9\frac{1}{2} \rightarrow 10$
$9\frac{1}{3} \rightarrow 9$	$9\frac{1}{3} \rightarrow 9\frac{1}{2}$	$9\frac{1}{3} \rightarrow 9$
$7\frac{1}{3} \rightarrow 7$	$7\frac{1}{3} \rightarrow 7\frac{1}{2}$	$7\frac{1}{3} \rightarrow 7$
$8\frac{1}{2} \rightarrow 8$	$8\frac{1}{2} \rightarrow 8\frac{1}{2}$	$8\frac{1}{2} \rightarrow 9$
40	43	43

All the estimation methods indicate that Garth worked more than 40 hours and so is eligible for overtime pay.

PROBLEMS

ESTIMATION TIP

Use compatible numbers to help you estimate the answer to Exercise 10.

Use estimation to check Pilar's math answers. Write *R* for a reasonable answer and *NR* for an unreasonable answer.

1. $4\frac{1}{8} + 5\frac{1}{2} = 9\frac{5}{8}$
2. $8\frac{7}{8} - 2\frac{3}{4} = 6\frac{5}{8}$
3. $1\frac{4}{5} \times 1\frac{2}{9} = 2\frac{1}{5}$
4. $9\frac{1}{8} \div 3\frac{1}{4} = 2\frac{21}{26}$
5. $4 \times 3\frac{9}{10} = 12\frac{3}{5}$
6. $7\frac{1}{2} - 3\frac{11}{12} = 3\frac{7}{12}$
7. $7\frac{1}{2} + 3\frac{5}{8} = 11\frac{1}{8}$
8. $3\frac{2}{3} + 4\frac{1}{12} = 7\frac{1}{12}$
9. $9\frac{1}{4} \div 3\frac{1}{8} = 2\frac{24}{25}$

Estimate to solve each problem.

10. Donna has read about $\frac{2}{9}$ of a book that is 446 pages long. About how many pages has she read?
11. Raúl wants to cut four boards, each $3\frac{5}{12}$-ft long, from a board whose length is 13 ft. Is this possible?
12. Marti wants to buy a loaf of bread for $0.99, 3 cans of beans for $1.25, ground beef for $6.95, apples for $2.38, and frozen items for $5.84. The sales tax was $1.05. She has $20. Does she have enough money to pay for these items?
13. Angela had $50. She purchased 2 cassette tapes at $7.98 each and a sweater for $24.89. The sales tax was $2.45. Does she have enough money left to purchase a compact disc for $19.98?
14. Keith needs fourteen pieces of twine that each measure $1\frac{5}{6}$-ft long. He has a $28\frac{1}{2}$-ft roll of twine. Does he need to buy more?
15. A party balloon costs $0.89. A party favor costs $1.98. Tom has $35. Can he buy 8 balloons and 16 party favors, not including tax? If not, give at least two combinations of items he could buy with his money.
16. A portion of a park trail is 5,885 meters long. Darlene's footstep is 0.62 meter long. She estimates she would take about 1,000 steps if she walked the trail. Do you agree with her estimate? Why or why not?
17. Jeanne is buying a coat that is marked $\frac{1}{4}$-off the original price of $149. The sales tax on the coat comes to $7.45. She has $135. Does she have enough money to pay for the coat?
18. Leon swam $3\frac{1}{10}$ lengths of the pool. Jim swam $1\frac{9}{10}$ times as far. Estimate how many lengths Jim swam.
19. Jackie walked 3.8 km on Monday. She walked half that distance on Tuesday and 5.8 km on Wednesday. Estimate the mean distance she walked per day.

4-8 Problem Solving Strategies:

SOLVE A SIMPLER PROBLEM

Sometimes you can find the answer to a problem by solving another problem that is like it but has simpler numbers. Using simpler numbers can help you focus on choosing the correct operation(s).

PROBLEM

In January, the Willistown Recycling Center paid the Save Our World Club \$0.16 per pound for $95\frac{1}{2}$ lb of newspaper and \$0.32 per pound for $84\frac{3}{4}$ lb of aluminum cans. How much did the recycling center pay the club?

SOLUTION

Solve a simpler problem. Substitute whole numbers for the decimals and fractions.

Simpler Problem

Suppose the recycling center paid \$0.10 per pound for 10 lb of newspaper.

Multiply: $10 \times 10 = 100$ They paid 100 cents, or \$1.00.

Suppose the recycling center paid \$0.10 per pound for 20 lb of aluminum cans.

Multiply: $10 \times 20 = 200$ They paid 200 cents, or \$2.00.

Add the amounts to find the total paid: \$1.00 + \$2.00 = \$3.00

Original Problem

Follow the same operations used to solve the simpler problem.

Multiply: $0.16 \times 95\frac{1}{2} = 15.28 \rightarrow \$15.28$

Multiply: $0.32 \times 84\frac{3}{4} = 27.12 \rightarrow \$27.12$

Add: $15.28 + 27.12 = 42.40 \rightarrow \$42.40$

The Willistown Recycling Center paid the Save Our World Club \$42.40 during the month of January.

Check the answer by estimating.

At \$0.16 per pound, 100 lb of newspaper is \$16; 75 lb is \$12.
At \$0.32 per pound, 100 lb of aluminum cans is \$32; 75 lb is \$24.

Since the actual number of pounds is between 75 and 100, the answer must be between \$36 and \$48. So \$42.40 is reasonable.

READING MATH

Read each problem carefully and describe what each variable represents. Then, choose the expression that tells how to solve the problem.

1. Andy rode his bike *a* miles in April, *b* miles in May, and *c* miles in June. What was the average number of miles he rode each month?
   a. $a + b - c$
   b. $a \times b \times c$
   c. $\frac{a + b + c}{3}$
2. Jenny bought a sweater for *a* dollars, a pair of shoes for *b* dollars, a pocketbook for *c* dollars, a blouse for *d* dollars, and slacks for *e* dollars. She paid the clerk with *f* dollars. How much change did she receive?
   a. $f + a + b + c + d + e$
   b. $f - a + b + c + d + e$
   c. $f - (a + b + c + d + e)$

PROBLEMS

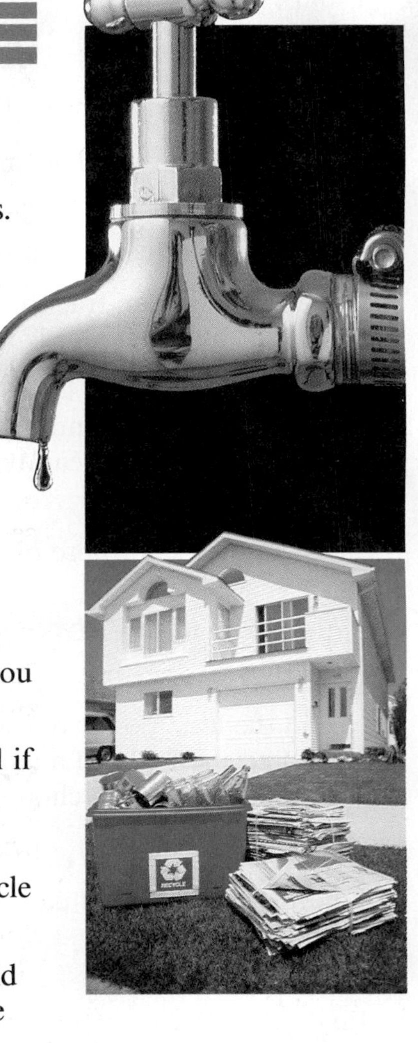

Solve.

1. The recycling center paid \$0.32 for each pound of aluminum cans. The student council received \$18.56 for its returned cans. How many pounds of cans was that?
2. A dripping faucet wasted $\frac{1}{4}$c of water per day. How much water was wasted over $14\frac{2}{3}$ days?

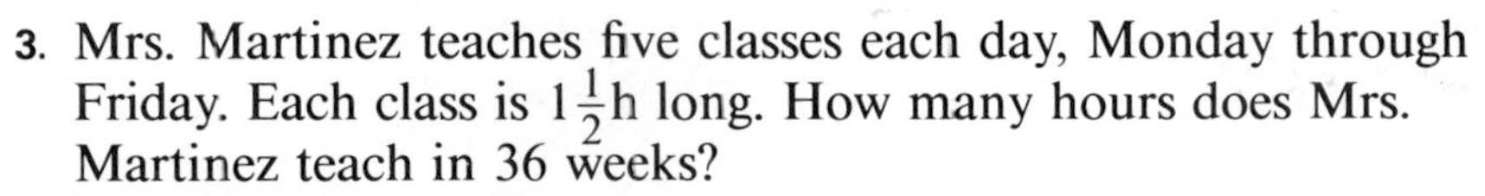

3. Mrs. Martinez teaches five classes each day, Monday through Friday. Each class is $1\frac{1}{2}$h long. How many hours does Mrs. Martinez teach in 36 weeks?
4. The price of cashews is \$7.50 per pound. If you buy 0.46 lb at the grocery and 0.38 lb at the specialty store, how much will you pay altogether?
5. A car averages 31.5 mi/gal. How many miles did the car travel if it used 8.8 gal?
6. A bicycle costs $\frac{4}{5}$ as much as a moped. The moped costs $\frac{1}{6}$ as much as a car. If the car costs \$7,200, how much does the bicycle cost?
7. Recyclers paid \$0.15 per pound for newspapers. How much did a homeowners' group receive for $75\frac{3}{5}$ lb of newspapers in June and $56\frac{1}{2}$ lb of newspapers in July?

A sponsor for the Hike-A-Thon has promised to pay Anita's favorite charity \$0.12 for each mile she hikes. She plans to hike $17\frac{1}{4}$ mi each day for a total of 138 miles.

8. How many days will it take Anita to complete 138 miles?
9. How much money will Anita's sponsor pay to her charity?

A lawn mower costs \$475.98. Next week it will be sold at $\frac{1}{3}$-off. The Garden Club decides to wait until next week to buy the mower at the sale price.

10. How much will the Garden Club save by buying the mower at the sale price?
11. What is the sale price?
12. What is the total amount of money the Garden Club will pay if the sales tax is 0.06 of the sales price?

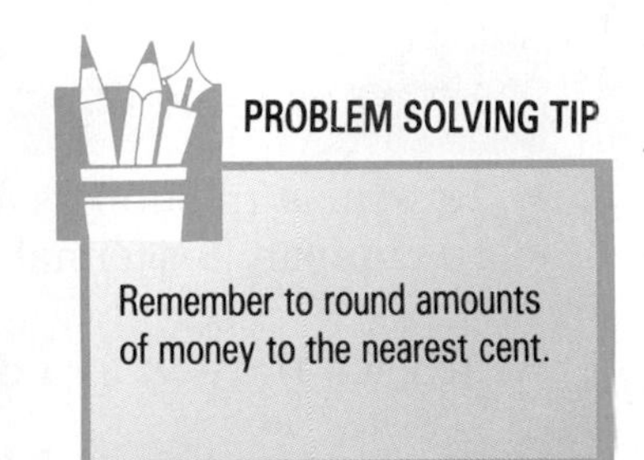

PROBLEM SOLVING TIP

Remember to round amounts of money to the nearest cent.

CHAPTER 4 ● REVIEW

Choose the word from the list that completes each statement.

a. terminating
b. negative
c. prime
d. reciprocals
e. repeating

1. A __?__ number has only two factors, 1 and itself.
2. When writing a fraction as a decimal, the answer is a __?__ decimal when the remainder is zero.
3. When writing a fraction as a decimal, the answer is a __?__ decimal when the remainder is not zero and a block of digits repeats in the quotient.
4. A fraction with a numerator of 1 and a denominator written in exponential form can be written as a whole number with a __?__ exponent.
5. Two numbers whose product is 1 are __?__.

SECTION 4–1 FACTORS AND MULTIPLES (pages 118–121)

► To find the GCF of a pair of numbers, use the prime factorization to find the least power of each common prime factor, then multiply.
► To find the LCM of a pair of numbers, write the prime factorization of each number, then choose the greatest power of each prime factor, and multiply.

Find the GCF of each pair of numbers.

6. 16, 24
7. 12, 18
8. 15, 25
9. 36, 40

Find the LCM of each pair of numbers.

10. 7, 13
11. 10, 25
12. 12, 15
13. 27, 36

SECTION 4–2 EQUIVALENT FRACTIONS (pages 122–125)

► You can multiply or divide the numerator and the denominator of a fraction by the same nonzero number to find an equivalent fraction.

Write each fraction in lowest terms.

14. $\frac{16}{20}$
15. $\frac{7}{14}$
16. $\frac{30}{50}$
17. $\frac{24}{36}$

SECTION 4–3 FRACTIONS AND DECIMALS (pages 126–129)

► To write a fraction as a decimal, divide the numerator by the denominator.
► To compare a decimal and a fraction, write the fraction as a decimal.

Write each fraction as a decimal.

18. $\frac{4}{5}$
19. $\frac{5}{16}$
20. $\frac{7}{9}$
21. $\frac{3}{11}$

SECTION 4–4 ZERO AND NEGATIVE EXPONENTS (pages 130–133)

► To write a number in scientific notation, express it as a product in which one factor is greater than or equal to 1 and less than 10 and another factor is a power of ten.
► A number with a negative exponent can be expressed as a fraction.

Write in exponential form.

22. $\frac{1}{3 \times 3 \times 3 \times 3 \times 3}$ **23.** $\frac{1}{4 \times 4}$ **24.** $\frac{1}{9 \times 9 \times 9 \times 9}$

Write in standard form.

25. 8.63×10^{-5} **26.** 4.11108×10^{-12}

Write in scientific notation.

27. 0.00258 **28.** 0.0000034 **29.** 0.657 **30.** 0.0000009

SECTIONS 4–5 and 4–6 OPERATIONS WITH FRACTIONS (pages 134–141)

► To multiply fractions, multiply the numerators, then the denominators.
► To divide fractions, multiply by the reciprocal of the divisor.
► To add or subtract fractions, rename them, if necessary, so that the denominators are the same; then add or subtract the numerators.

Compute. Write the answer in lowest terms.

31. $\frac{3}{8} \times 4$ **32.** $2\frac{1}{2} \times 2\frac{3}{4}$ **33.** $\frac{3}{8} + \frac{1}{6}$ **34.** $\frac{2}{3} - \frac{1}{2}$

35. $\frac{2}{3} \div 8$ **36.** $1\frac{3}{5} + 2\frac{9}{10}$ **37.** $2\frac{1}{5} \div 1\frac{5}{6}$ **38.** $9\frac{1}{4} - 2\frac{11}{12}$

SECTIONS 4–7 and 4–8 PROBLEM SOLVING (pages 142–145)

► To estimate when working with fractions, it is often helpful to use numbers close to 0, $\frac{1}{2}$, and 1.
► Sometimes you can try simpler numbers to help you focus on choosing the operation(s) needed to solve a problem.

Estimate to solve.

39. Jodie needs 6 pieces of string that each measure $13\frac{1}{2}$ in. She has a roll of string that is 78 in. long. Does she need to buy more?

USING DATA Use the table on page 116 to answer the following question.

40. How many pounds of waste would 150,000,000 people have generated for the years 1980 through 1984? Leap years were 1980 and 1984.

CHAPTER 4 ● TEST

1. Find the prime factorization of 200.

Find the GCF of each pair of numbers.

2. 6, 8
3. 4, 6,
4. 9, 15
5. 24, 36

Find the LCM of each pair of numbers.

6. 7, 8
7. 4, 6,
8. 6, 8
9. 8, 12

10. Multiply to find two fractions that are equivalent to $\frac{11}{12}$.
11. Divide to find two fractions that are equivalent to $\frac{24}{36}$.

Write each fraction in lowest terms.

12. $\frac{25}{75}$
13. $\frac{36}{42}$
14. $\frac{12}{15}$
15. $\frac{18}{30}$

Write each fraction as a decimal.

16. $\frac{71}{100}$
17. $\frac{5}{8}$
18. $\frac{11}{15}$
19. $\frac{5}{9}$

Write each decimal as a fraction in lowest terms.

20. 0.32
21. 0.8
22. 0.004
23. 3.88

Replace ● with <, >, or =.

24. $\frac{7}{8}$ ● $\frac{11}{12}$
25. $\frac{1}{6}$ ● $0.\overline{3}$
26. 0.58 ● $\frac{7}{12}$
27. $\frac{3}{16}$ ● 0.1875

28. Write 7^{-4} as a fraction
29. Write $\frac{1}{4 \times 4 \times 4 \times 4 \times 4}$ in exponential form.
30. Write 7.392×10^{-8} in standard form.
31. Write 0.000138 in scientific notation.

Find each answer. Write answers in lowest terms.

32. $\frac{3}{4} + \frac{1}{8}$
33. $12\frac{3}{8} - 9\frac{1}{3}$
34. $3\frac{1}{2} \times 1\frac{1}{3}$
35. $\frac{2}{3} \div \frac{5}{6}$

Solve.

36. Don has read $\frac{3}{5}$ of a book that is 348 pages long. Estimate to find about how many pages he has read.
37. Mai Ling plans to hike $2\frac{3}{4}$ mi each day to raise money for a charity. Her sponsor has pledged \$1.50 to the charity for each mile she hikes. If she hikes each day for $5\frac{3}{7}$ weeks, how much money will the charity receive?

The following data show math test scores for one class.

80 70 84 95 68 90 76 85 100 65
89 97 71 69 73 88 91 76 85

1. To find out whether you are in the top half of the class, would you use the mean or the median?
2. Would you be in the top half of the class if you had one of the 85's?

Use the bar graph for Exercises 3–6.

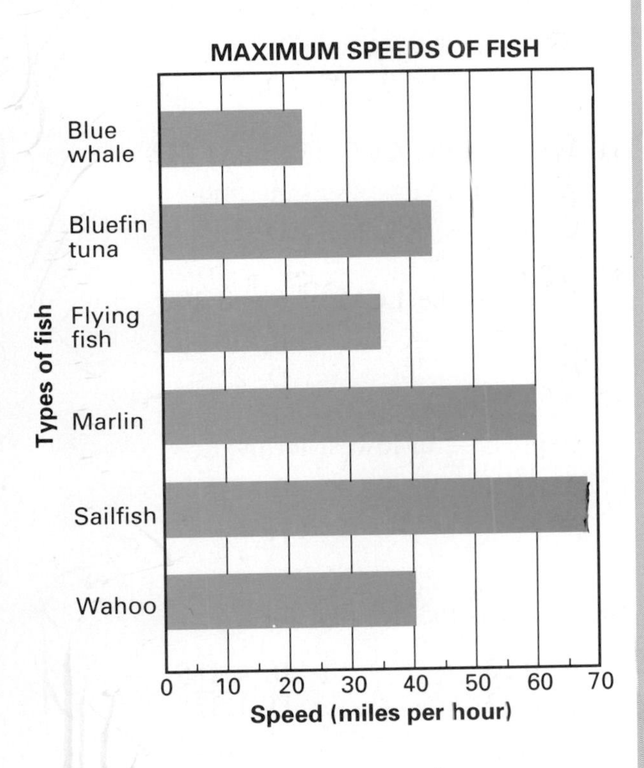

3. Which is the fastest fish?
4. How many miles per hour faster is the marlin than the flying fish?
5. Which fish can swim over 40 miles per hour?
6. The speed of the sailfish is about how many times faster than the speed of the flying fish?

Evaluate each expression.
Let $a = 3$ and $b = 6$.

7. $a + 2b$
8. $4b \div a$
9. $\frac{5b}{2a}$
10. $4a \times 2b$
11. $b - 1a$
12. $\frac{12a}{b}$

13. Determine whether the following argument is *valid* or *invalid*.

    If a number is divisible by 8, then the number is divisible by 4.
    96 is divisible by 8.
    Therefore, 96 is divisible by 4.

14. Pick a number. Double it. Add 3. Then add the original number. Divide by 3. Subtract 1. What number results? Repeat the process several times. Make a conjecture about the result.

15. Write two conditional statements using these two sentences:
    The fruit is a plum.
    The fruit has a pit.

Find the prime factorization for each number.

16. 120
17. 768

Find the GCF of each pair of numbers.

18. 12, 15
19. 14, 24

Find the LCM of each pair of numbers.

20. 6, 8
21. 18, 24

Write in lowest terms.

22. $\frac{42}{48}$
23. $\frac{21}{28}$

CHAPTER 4 • CUMULATIVE TEST

1. To determine the baseball player with the best batting average, which is the best way to collect data?
   A. personal interviews
   B. records of performance
   C. questionnaire
   D. telephone interviews

2. Which is the best method of choosing a sample to gather data about the quality of Trendy Tires?
   A. personal interviews
   B. records of performance
   C. questionnaire
   D. telephone interviews

3. What is the range of this set of data?
   62 16 49 104 30 21
   A. 104　　B. 78
   C. 62　　D. none of these

4. Evaluate $a + 3b$. Let $a = 2$ and $b = 3$.
   A. 9　　B. 8　　C. 11　　D. 12

5. Evaluate $6b \div a$. Let $a = 2$ and $b = 3$.
   A. 4　　B. 9　　C. 6　　D. 8

6. Evaluate $\frac{6b}{3a}$. Let $a = 2$ and $b = 4$.
   A. 4　　B. 1　　C. 2　　D. 3

7. Evaluate $6a \times 3b$. Let $a = 2$ and $b = 4$.
   A. 24　　B. 0　　C. 1　　D. 144

8. Which is a counterexample for the following statement?

   If a whole number x is divisible by 3, then x is divisible by 9.

   A. 27　　B. 81　　C. 9　　D. 39

9. Which conlusion can be drawn from these statements?

   If yesterday was Wednesday, then I went to school.
   Yesterday was Wednesday.

   A. I did not go to school.
   B. I went to school.
   C. Yesterday was Wednesday.
   D. none of these

10. Which is the prime factorization of 288?
    A. $2^3 \times 3^2 \times 4$　　B. $2 \times 3^3 \times 4$
    C. $2^3 \times 4 \times 9$　　D. $2^5 \times 3^2$

11. Which is the GCF of 15 and 24?
    A. 7360　　B. 209
    C. 3　　D. 2

12. Which is the LCM of 6 and 10?
    A. 60　　B. 30
    C. 3　　D. 2

13. Express $\frac{24}{32}$ in lowest terms.
    A. $\frac{12}{16}$　　B. $1\frac{1}{4}$
    C. $\frac{4}{5}$　　D. $\frac{3}{4}$

14. Which is the decimal for $\frac{7}{30}$?
    A. $0.2\overline{3}$　　B. 4.285
    C. 0.32　　D. $1.\overline{37}$

15. Add. $\frac{3}{8} + \frac{1}{6}$
    A. $\frac{2}{3}$　　B. $\frac{4}{14}$
    C. $\frac{2}{7}$　　D. $\frac{13}{24}$

16. Divide. $\frac{1}{2} \div \frac{7}{8}$
    A. $\frac{2}{4}$　　B. $\frac{2}{3}$
    C. $\frac{4}{7}$　　D. $\frac{1}{3}$

CHAPTER 5 SKILLS PREVIEW

Complete.

1. 48 in. = ■ ft
2. 5 gal = ■ qt
3. 8 lb = ■ oz
4. 3 mi = ■ yd
5. 490 cm = ■ m
6. 2 L = ■ mL
7. 958 mg = ■ g
8. 1 km = ■ m

Write each answer in simplest form.

9. 4 lb 8 oz
   +7 lb 9 oz
10. 5 gal
   −3 gal 3 qt
11. 1 c 8 fl oz
   × 5
12. 2 ft 8 in. ÷ 4

Complete.

13. 150 mm + 450 mm = __?__ cm
14. 2 km − 250 m = __?__ km
15. 3.28 cm × 2 = __?__ mm
16. 8.4 m ÷ 6 = __?__ cm

Solve.

17. Will one 40-inch roll of tape be enough tape to make 3 strips of tape, each 1 ft 2 in. long?

Find the perimeter of each figure.

18.

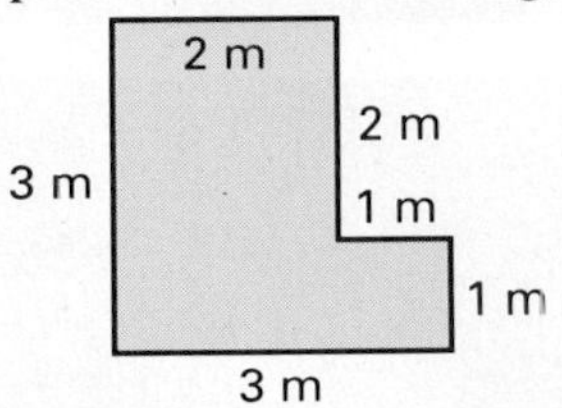

19.

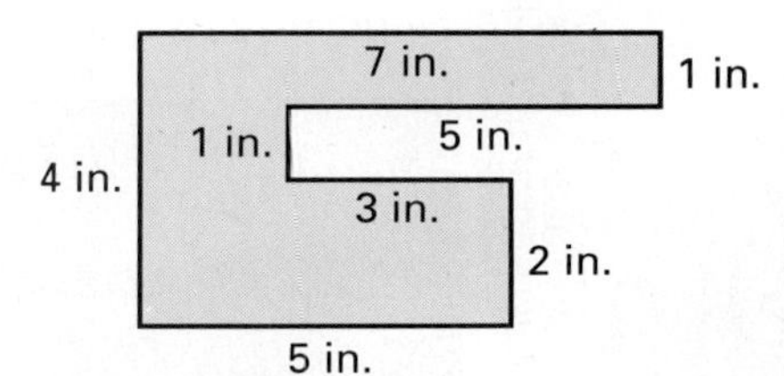

Use the formula to find the perimeter for a rectangular figure with the given dimensions.

20. length: 78.34 m
   width: 29.87 m
21. length: 375.4 mi
   width: 291.3 mi

Find the circumference. Use 3.14 or $\frac{22}{7}$ for π, as appropriate.

22.

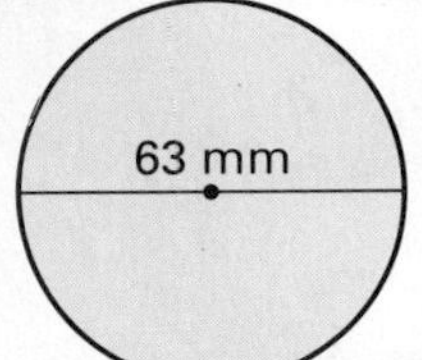

23.

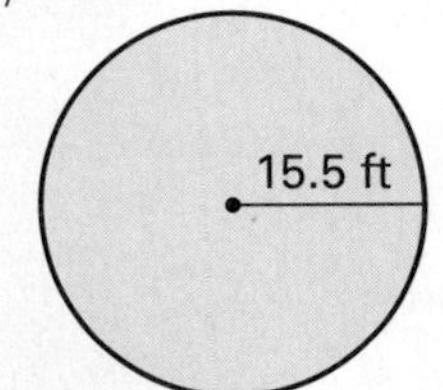

24.

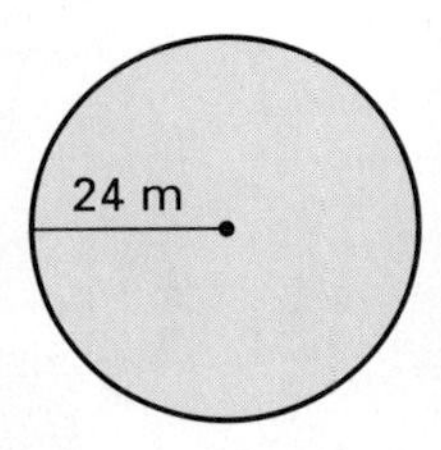

Draw a picture to help you solve.

25. A triangular garden is enclosed by a fence that has 10 posts on each side. How many posts are there in all?

CHAPTER 5

MEASUREMENT AND GEOMETRY

THEME Architecture Around the World

Architecture involves the planning and designing of structures such as buildings, bridges, and towers. Architecture can also involve the designing and positioning of the furniture within a building. Landscape architecture involves arranging the plantings on a piece of property.

The architect seeks to design a structure that is both functional and pleasing to the eye. The style of a structure reflects the cultural preferences of the society that exists at the time during which the structure was produced.

In this chapter, you will solve problems that involve units of measurement, estimating measurements, and finding perimeter and circumference.

Specific styles of architecture became popular during certain periods in history. Some of these styles of Western architecture are listed in the table. The "c." preceding a date is an abbreviation for *circa,* a Latin word meaning "around." This indicates that the date given is an approximation, not an exact date.

STYLES OF WESTERN ARCHITECTURE	
Style	**Date**
Archaic	c. 750–500 B.C.
Classical	c. 500–323 B.C.
Early Byzantine	330–726
Iconoclastic Age	726–843
Byzantine	843–1453
Italian Gothic	c. 1200–1400
Italian Renaissance	1402–1520
Baroque	c. 1600–1750
Classicist	1750–1830
Gothic Revival	c. 1790–1930

DECISION MAKING

Using Data

Use the information in the chart above to answer the following questions.

1. How many years did the Italian Gothic period span?
2. How many years were there between the end of the Classical period and the start of the Baroque period?
3. What name do you think might apply in the future to the style of architecture that will be popular from 2000 to 2100?

Working Together

Research the styles of architecture that have been popular from 1930 to the present. Identify what is meant by the *International Style* and by *post-modernism.* Find out about the work of famous architects, such as Mies van der Rohe, Le Corbusier, and Philip Johnson. Collect pictures to illustrate the work of the style periods and the architects. Decide which style and architect you prefer. Discuss your choices with your group.

5-1 Customary Units of Measurement

EXPLORE

What is the longest bridge you have ever been across? Guess its length, in feet. Write down your guess.

USING DATA Use the Data Index on page 546 to find the list of longest bridge spans in the world. Is the bridge you named listed in the table?

a. If so, find the actual length of the bridge, in feet. How close was your guess to the actual length?

b. If not, compare your guess to the lengths of the longest bridge spans. Was your guess close to any of these lengths? Was it much greater than or much less than these? How do the data suggest that you adjust your guess?

SKILLS DEVELOPMENT

The table below lists some of the **customary units** that are used to measure length, weight, and capacity. Equivalent measurements are also given.

Measure	Units	Equivalents
length	inch (in.)	
	foot (ft)	1 ft = 12 in.
	yard (yd)	1 yd = 3 ft, or 36 in.
	mile (mi)	1 mi = 5,280 ft, or 1,760 yd
weight	ounce (oz)	
	pound (lb)	1 lb = 16 oz
	ton (T)	1 T = 2,000 lb
capacity	fluid ounce (fl oz)	
	cup (c)	1 c = 8 fl oz
	pint (pt)	1 pt = 2 c, or 16 fl oz
	quart (qt)	1 qt = 2 pt, or 4 c
	gallon (gal)	1 gal = 4 qt, 8 pt, or 16 c

To change from a larger unit of measure to a smaller unit, multiply the number of larger units by the number of smaller units equivalent to one larger unit.

Example 1

Change 517 yd to feet.

Solution
1 yd = 3 ft

Multiply 517 by 3 to find how many feet are in 517 yd.

$$517 \times 3 = 1{,}551$$

So 517 yd = 1,551 ft. ◄

To change from a smaller unit of measure to a larger one, divide the number of smaller units by the number of smaller units equivalent to one larger unit.

Example 2

Change 45 c to quarts.

Solution
4 c = 1 qt

Divide 45 by 4 to find how many quarts make 45 cups.

$$45 \div 4 = 11\frac{1}{4}$$

So 45 c = $11\frac{1}{4}$ qt. ◄

Example 3

Ramona has a dress pattern that calls for 7 yd of lace. She already has 4 ft of lace. How many feet of lace will she need to buy?

Solution
1 yd = 3 ft, so **Change 7 yd to the equivalent number of feet.**
7 yd = 21 ft.

21 ft − 4 ft = 17 ft **Subtract 4 ft from 21 ft.**

Ramona needs to buy 17 ft of lace. ◄

JUST FOR FUN

Did you know that a 10-gallon hat can hold only $\frac{3}{4}$ gal of liquid? How much liquid would ten 10-gallon hats hold?

TRY THESE

Change the larger unit of measure to the smaller unit.

1. 5 ft to inches
2. 1 gal to cups
3. $3\frac{1}{4}$ lb to ounces
4. $2\frac{1}{2}$ gal to quarts

Change the smaller unit of measure to the larger unit.

5. 1,760 yd to miles
6. 30 in. to feet
7. 48 fl oz to pints
8. 3,000 lb to tons

Solve.

9. Brooke has jogged 520 yards. How many more yards must she jog before she has jogged 1 mile?

EXERCISES

PRACTICE/ SOLVE PROBLEMS

Complete.

1. 32 oz = ■ lb
2. 4 yd = ■ ft
3. $3\frac{1}{2}$ T = ■ lb
4. 112 in. = ■ ft ■ in.
5. 5 qt = ■ gal ■ qt
6. 40 fl oz = ■ pt ■ c
7. 46 ft = ■ yd ■ ft
8. 46 oz = ■ lb ■ oz

Name the best customary unit for expressing the measure of each.

9. capacity of a pitcher
10. weight of a moving van
11. amount of water needed to fill a bathtub
12. distance from one city to another
13. weight of two plums
14. amount of fabric needed to make a dress

Solve.

15. A living room is 24 ft long. How many yards long is it?
16. A watermelon weighs 5 lb. How many ounces does it weigh?

EXTEND/ SOLVE PROBLEMS

Choose the estimate that best expresses the measure of each.

17. length of a classroom
    a. 30 in.　b. 30 ft　c. 30 yd
18. capacity of a water glass
    a. 12 pt　b. 12 qt　c. 12 fl oz
19. length of a newborn baby
    a. 19 in.　b. 19 ft　c. 19 yd
20. weight of a bag of potatoes
    a. 5 oz　b. 5,000 oz　c. 5 lb

Solve.

21. A cement company has 1,700 lb of sand on its lot. A supplier delivers 4 T of sand. How many pounds of sand are now on the lot?

22. Jeffrey ran 545 yd in the morning and 750 yd in the afternoon. How many more yards must he run later in the day if he wants to have run a total of 3 mi that day?

23. Paul knows that his size $10\frac{1}{2}$ shoe is about 12 in. long. By walking slowly along the length of a room with one foot touching the other, heel to toe, Paul can estimate the length of the room. Paul's son, Davey, wears a shoe that is half as long as Paul's. If Davey uses his father's method of measuring the length of a 24-ft-long room, how many steps will Davey take?

PROBLEM SOLVING TIP

In Exercise 23, how many steps would Paul need to take to measure the room?

THINK CRITICALLY/ SOLVE PROBLEMS

Janice wants to can 18 qt of beans. She already has 9 quart jars and 6 pint jars. She knows she will need to buy more jars.

24. If she buys only quart jars, how many will she need?

25. If she buys only pint jars, how many will she need?

26. List five combinations of quart jars and pint jars that Janice might buy.

PROBLEM SOLVING TIP

For Exercise 26, you can make a table to help you find the combinations of quart jars and pint jars.

5-2 Metric Units of Measurement

EXPLORE

Which metric unit of measurement would you use to express the answer to each of these questions?

a. About how much does a person weigh?

b. About how far might you travel in a day?

c. How much juice would there be in a glass?

SKILLS DEVELOPMENT

The basic unit of *length* for metric measure is the **meter (m).** Other commonly used units include the **kilometer (km),** the **centimeter (cm),** and the **millimeter (mm).**

about 1 km

about 1 m wide

about 1 mm at tip

The most commonly used metric units of *mass* are the **kilogram (kg),** the **gram (g),** and the **milligram (mg).**

about 1 kg

about 1 g

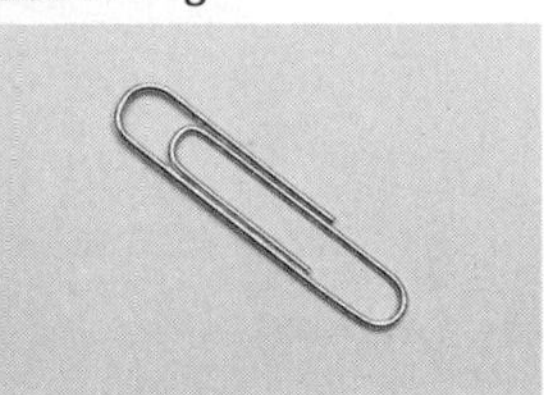

about 100 mg

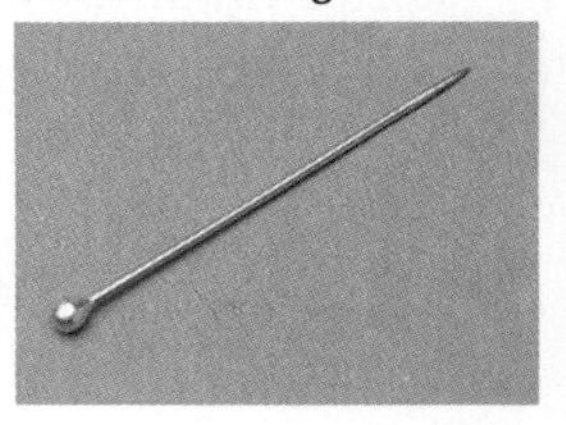

The most commonly used metric units of *capacity* are the **liter (L)** and the **milliliter (mL).**

about 1 L

about 1 mL

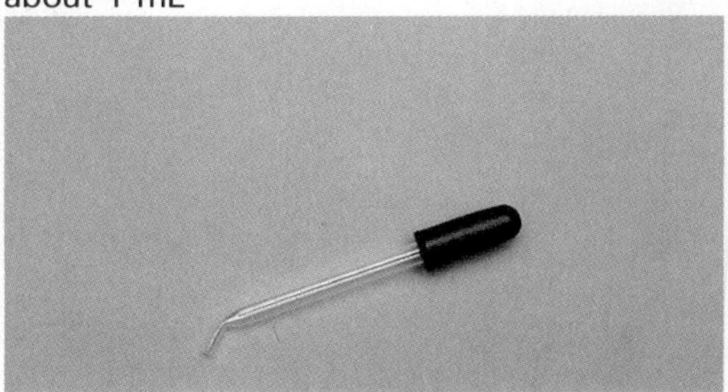

MATH: WHO, WHERE, WHEN

The metric system, developed by French scientists, was adopted in 1799 as the standard system of measurement in France. Most other countries of the world now use the metric system as their standard.

The size of the meter was originally determined as being $\frac{1}{10,000,000}$ (one ten-millionth) of the distance from the equator to either the North Pole or the South Pole. Today, the size is determined differently.

Find out the standard used today for the size of the meter.

This chart can help you see the relationships between the prefixes used to name metric units. It is important to understand these relationships in order to change values from one metric unit to another.

thousands	hundreds	tens	ones	tenths	hundredths	thousandths
kilo-	hecto-	deka-	no prefix	deci-	centi-	milli-
kilometer (km) 1,000 m	hectometer (hm) 100 m	dekameter (dam) 10 m	meter (m) 1 m	decimeter (dm) 0.1 m	centimeter (cm) 0.01 m	millimeter (mm) 0.001 m
kilogram (kg) 1,000 g	hectogram (hg) 100 g	dekagram (dag) 10 g	gram (g) 1 g	decigram (dg) 0.1 g	centigram (cg) 0.01 g	milligram (mg) 0.001 g
kiloliter (kL) 1,000 L	hectoliter (hL) 100 L	dekaliter (daL) 10 L	liter (L) 1 L	deciliter (dL) 0.1 L	centiliter (cL) 0.01 L	milliliter (mL) 0.001 L

Use the chart to help you change units. Just as with customary units, to change from a larger metric unit to a smaller one, multiply the number of larger units by the number of smaller units equivalent to one larger unit.

MENTAL MATH TIP

When you multiply by 10, 100, or 1,000, move the decimal point to the right one place for each zero in the multiplier.

For example:

$10 \times 1.57 = 15.7$

$100 \times 6.8 = 680.$

$1,000 \times 9 = 9000.$

Example 1

Change 5 m to centimeters.

Solution

1 m = 100 cm — **Multiply 5 by 100 to find how many centimeters are in 5 m.**

$5 \times 100 = 500$

So 5 m = 500 cm. ◄

Just as with customary units, to change from a smaller metric unit to a larger one, divide the number of smaller units by the number of smaller units equivalent to one larger unit.

Example 2

Change 5,642 g to kilograms.

Solution

1,000 g = 1 kg — **Divide 5,642 by 1,000 to find how many kilograms make 5,642 g.**

$5,642 \div 1,000 = 5.642$

So 5,642 g = 5.642 kg. ◄

MENTAL MATH TIP

When you divide by 10, 100, or 1,000, move the decimal point to the left one place for each zero in the divisor.

For example:

$15.7 \div 10 = 1.57$

$680 \div 100 = 6.80$

$9,000 \div 1,000 = 9.000$

Example 3

A pharmacist has 4.5 L of cough syrup. How many 150-mL bottles can the pharmacist fill with the cough syrup?

Solution

1 L = 1,000 mL, so
4.5 L = 4,500 mL. — **Change 4.5 L to the equivalent number of milliliters.**

4,500 mL ÷ 150 mL = 30 — **Divide 4,500 mL by 150 mL.**

The pharmacist can fill thirty 150-mL bottles. ◄

READING MATH

In this textbook, you are learning about the most commonly used units in the metric system. You may come across other metric units as you read newspapers, magazines, or other textbooks.

Use a dictionary to find the meaning of each of these metric units. Suggest a situation for which each of these units is likely to be used.

1. nanosecond
2. megahertz
3. micrometer
4. gigameter
5. kilowatt

TRY THESE

Change the larger unit of measure to the smaller one.

1. 7 m to centimeters
2. 2 L to milliliters
3. 12.75 g to milligrams
4. 5.5 km to meters

Change the smaller unit of measure to the larger one.

5. 5,000 m to kilometers
6. 8,000 mL to liters
7. 3,500 g to kilograms
8. 1,254 mg to grams

Solve.

9. The Dixie Luncheonette had 12 L of milk. If 9.25 L were used, how many liters of milk were left? How much is that in milliliters?

EXERCISES

PRACTICE/ SOLVE PROBLEMS

Complete.

1. 8 L = ■ mL
2. 125 cm = ■ m
3. 2 km = ■ m
4. 388 mm = ■ m
5. 4,783 mL = ■ L
6. 307 g = ■ kg
7. 4,000 g = ■ kg
8. 255 mL = ■ L

Name the best metric unit for expressing the measure of each.

9. length of a car
10. mass of a feather
11. capacity of a gasoline tank
12. height of a child
13. mass of a bag of groceries
14. capacity of a teacup

Solve.

15. A container holds 7 L of liquid. How many milliliters does it hold?
16. It is 0.85 km from Jason's house to the library. How far is that distance in meters?

EXTEND/ SOLVE PROBLEMS

Choose the most reasonable unit of measure.
Write *km, m, cm,* or *mm.*

17. The length of an airport landing strip is about 1 ■.

18. The length of a pen is about 14 ■.

19. Janine's height is about 164 ■.

20. The length of a minivan is about 5 ■.

Write *kg, g,* or *mg.*

21. The mass of a nickel is about 5 ■.

22. A box of raisins has a mass of 500 ■.

23. A large pot can hold about 0.75 ■.

24. A grain of rice has a mass of about 1 ■.

Write *L* or *mL.*

25. A fish tank has a capacity of about 60 ■.

26. A baby's bottle can hold about 250 ■.

Solve.

27. Steve began an 8-km hike on Monday by walking 454 m. He walked 2 km on Tuesday and 533 m on Wednesday. How many meters farther must he walk to complete the hike?

28. Maria bought 1,224 cm of ribbon at $0.75 per meter. How much did Maria pay for the ribbon, to the nearest cent?

29. Mike had 2 L of milk. After he used 250 mL to bake muffins and 500 mL to bake bread, how much milk did he have left?

30. The running time on the videotape of a movie that Dina rented was 108 min. Dina began watching the movie at 7:45 p.m. At what time did the movie end?

COMPUTER

This computer program will compute the total value of the gold in Exercise 31 quite quickly. Which line would you change to reflect the change in the price of gold specified in Exercise 32?

```
10 INPUT "WHAT BIRTHDAY?
   ";B: PRINT
20 FOR X = 0 TO B – 16
30 G = 10 * 2 ↑ X
40 V = 20 * G
50 S = S + V
60 NEXT X
70 PRINT : PRINT "GINA'S
   GOLD IS WORTH $";S;"."
```

THINK CRITICALLY/ SOLVE PROBLEMS

Gina's father is a jeweler who cuts gold chain of different sizes for necklaces. On Gina's 16th birthday, she received a 10-g gold necklace. On her 17th birthday, she received a 20-g gold necklace; on her 18th birthday, a 40-g gold necklace; on her 19th birthday, an 80-g gold necklace; and so on.

31. After receiving her necklace for her 21st birthday, Gina decided to determine the value of the gold in all her necklaces. If the price of gold was $20 per gram, find the value of all her necklaces. Make a table to help you.

32. By Gina's 25th birthday, the price of gold had risen to $30 per gram. Find the value of the gold in all of Gina's necklaces, including the one she received for her 25th birthday. (Hint: Make a table or write a calculator sequence to help you find the answer.)

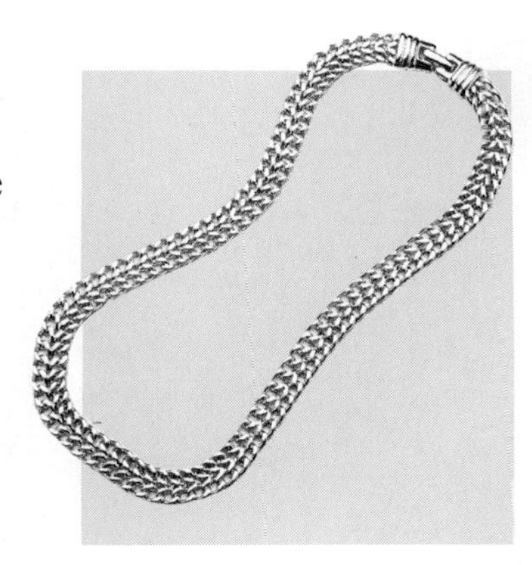

5-3 Working with Measurements

EXPLORE/ WORKING TOGETHER

Work in groups of three. Decide if the measurements in each set are equivalent. If not, determine which is larger.

1. 5,000 cm; 5 m
2. 48 oz; 4 lb
3. 6 ft; 2 yd
4. 3 L; 300 mL
5. 5.3 kg; 5,300 g
6. $7\frac{1}{2}$ qt; 15 pt
7. 3 ft 6 in.; 36 in.
8. 100 oz; 6 lb 4 oz

SKILLS DEVELOPMENT

There are many times that you need to perform operations that involve measurements. In carpentry, for example, cutting wood to an appropriate length often involves adding or subtracting lengths. Similarly, in cooking, adjusting a recipe may involve adding or multiplying weights or liquid capacities. When you add or multiply with customary measurements, you may need to change your answer to its simplest form.

Example 1

Write each answer in simplest form.

a. 3 lb 9 oz
   + 1 lb 9 oz

b. 4 ft 9 in.
   × 3

Solution

a. 3 lb 9 oz
   + 1 lb 9 oz
   4 lb 18 oz   **18 oz is more than 1 lb, so simplify.**

4 lb + 18 oz = 4 lb + (1 lb + 2 oz) = 5 lb 2 oz

b. 4 ft 9 in.
   × 3
   12 ft 27 in.   **Be sure to multiply both the number of inches and the number of feet by 3.**

12 ft 27 in. = 12 ft + (2 ft + 3 in.) = 14 ft 3 in. ◄

When subtracting or dividing with customary measurements, you may need to change a measurement before you can perform the operation.

Example 2

Write each answer in simplest form.

a. $\begin{array}{r} 8 \text{ gal } 1 \text{ qt} \\ -6 \text{ gal } 2 \text{ qt} \\ \hline \end{array}$

b. 7 yd 1 ft ÷ 2

Solution

a. $\begin{array}{r} 8 \text{ gal } 1 \text{ qt} \\ -6 \text{ gal } 2 \text{ qt} \\ \hline \end{array}$ **1 qt is less than 2 qt, so you cannot subtract in this form.**

$8 \text{ gal } 1 \text{ qt} = (7 \text{ gal} + 4 \text{ qt}) + 1 \text{ qt} = 7 \text{ gal } 5 \text{ qt}$

$\begin{array}{r} 8 \text{ gal } 1 \text{ qt} \\ -6 \text{ gal } 2 \text{ qt} \\ \hline \end{array} \rightarrow \begin{array}{r} 7 \text{ gal } 5 \text{ qt} \\ -6 \text{ gal } 2 \text{ qt} \\ \hline 1 \text{ gal } 3 \text{ qt} \end{array}$ ◄

b. 7 yd 1 ft ÷ 2 **1 ft is not divisible by 2, so you need to "borrow" from the number of yards.**

$7 \text{ yd } 1 \text{ ft} = (6 \text{ yd} + 3 \text{ ft}) + 1 \text{ ft} = 6 \text{ yd } 4 \text{ ft}$

$2\overline{)7 \text{ yd } 1 \text{ ft}} \rightarrow \begin{array}{r} 3 \text{ yd } 2 \text{ ft} \\ 2\overline{)6 \text{ yd } 4 \text{ ft}} \end{array}$ ◄

You also may need to change measurements when performing operations with metric measurements.

Example 3

a. Complete: 65 cm + 72 cm = __?__ m

b. 5 L ÷ 8 = __?__ mL

Solution

a. $\begin{array}{r} 65 \text{ cm} \\ +72 \text{ cm} \\ \hline 137 \text{ cm} \end{array}$ $\begin{array}{r} 100 \text{ cm} = 1 \text{ m} \\ 137 \div 100 = 1.37 \\ 137 \text{ cm} = 1.37 \text{ m} \end{array}$ So, 65 cm + 72 cm = 1.37 m. ◄

b. $1 \text{ L} = 1{,}000 \text{ mL}$

$5 \times 1{,}000 = 5{,}000$

$5 \text{ L} = 5{,}000 \text{ mL}$ $\begin{array}{r} 625 \text{ mL} \\ 8\overline{)5{,}000 \text{ mL}} \end{array}$ So, 5 L ÷ 8 = 625 mL. ◄

TALK IT OVER

In Example 3, is it possible to perform the steps in each solution in a different order? Explain.

Example 4

For a craft project, Linnette needs twenty pieces of ribbon that are each 1 ft 4 in. long. She can buy a 25-ft roll of the ribbon for $4.98. Will this be enough ribbon?

Solution

Multiply 1 ft 4 in. by 20 to find the total amount of ribbon that she needs.

$\begin{array}{r} 1 \text{ ft } 4 \text{ in.} \\ \times 20 \\ \hline 20 \text{ ft } 80 \text{ in.} \end{array}$ $\begin{array}{l} 20 \text{ ft } 80 \text{ in.} = 20 \text{ ft} + (6 \text{ ft } 8 \text{ in.}) \\ \qquad = 26 \text{ ft } 8 \text{ in.} \end{array}$

Compare the amount of ribbon on the roll to the amount Linnette needs.

25 ft __?__ 26 ft 8 in.

$25 \text{ ft} < 26 \text{ ft } 8 \text{ in.}$ No, the 25-ft roll is not enough. ◄

MIXED REVIEW

Find the mean, the median, and the mode for each set of data.

1. 178, 235, 679, 384, 384
2. 13, 98, 45, 67, 67, 98, 17, 28, 98
3. 31, 56, 94, 79, 85, 62, 31, 92, 39, 31

Evaluate each expression. Let $a = 4$ and $b = 5$.

4. ab
5. $a^2 + b$
6. $3a - b$

Solve.

7. Bus tickets cost $1.15 each. How much do 8 bus tickets cost?

TRY THESE

Write each answer in simplest form.

1. 4 c 6 fl oz + 1 c 4 fl oz
2. 1 yd 2 ft × 5
3. 7 lb 2 oz − 10 oz
4. 5)6 ft 3 in.

Complete.

5. 200 mL + 750 mL = ___?___ L
6. 3 kg − 145 g = ___?___ g
7. 4.15 cm × 10 = ___?___ mm
8. 7.2 m ÷ 6 = ___?___ cm
9. In her laboratory, Angela has three beakers with a capacity of 750 mL each and a fourth beaker with a capacity of 1.25 L. Do the four beakers together have the capacity to hold 3.5 L of a saline solution?

EXERCISES

PRACTICE/ SOLVE PROBLEMS

Write each answer in simplest form.

1. 6 lb 12 oz + 4 lb 9 oz
2. 9 ft 2 in. − 8 ft 8 in.
3. 8 yd − 2 yd 1 ft
4. 5 gal 2 qt + 1 gal 3 qt
5. 2 c 3 fl oz × 4
6. 1 lb 8 oz × 6
7. 5 yd 1 ft ÷ 2
8. 1 ft 8 in. ÷ 4

Complete.

9. 2.4 kg + 600 g = ___?___ kg
10. 4 × 65 mm = ___?___ cm
11. 1.6 L ÷ 5 = ___?___ mL
12. 4 m − 250 cm = ___?___ cm
13. 1.25 cm × 14 = ___?___ mm
14. 120 mL + 860 mL = ___?___ L
15. 1.9 kg − 1.25 kg = ___?___ g
16. 2,500 mm ÷ 4 = ___?___ cm

17. Don needs two pounds of raisins to make some loaves of bread for a family get-together. He has three 12-oz boxes of raisins. Will this be enough? Explain.

18. Erica is planning to install three new cabinets in her kitchen. She would like to get two cabinets that are each 2 ft 9 in. wide and a third cabinet that is 3 ft 6 in. wide. Will these three cabinets fit side-by-side along a wall space that is 8 ft 6 in. wide?

EXTEND/ SOLVE PROBLEMS

Write each answer in simplest form.

19. 2 gal 2 qt 1 pt
+1 gal 3 qt 1 pt

20. 4 yd 1 ft 3 in.
−1 yd 2 ft 9 in.

21. 2 yd 1 ft 4 in.
−1 yd 1 ft 10 in.

22. 4 gal
−1 gal 3 qt 1 pt

Customary units are often written using mixed numbers. For instance, since a measurement of 6 in. is half of one foot, a measurement of 2 ft 6 in. might be written $2\frac{1}{2}$ ft. In Exercises 23–28, write the answer in mixed number form.

23. 3 ft 4 in. + 1 ft 4 in.

24. 6 lb 2 oz − (1 lb 10 oz)

25. 5 × (2 qt 1 pt)

26. (9 yd 1 ft) ÷ 2

27. 3 lb 4 oz + $1\frac{1}{2}$ lb

28. $4\frac{2}{3}$ ft − 10 in.

THINK CRITICALLY/ SOLVE PROBLEMS

Replace each ■ with the measurement that makes a true statement.

29. 7 lb 6 oz + ■ = 18 lb 2 oz

30. ■ − 3 ft 5 in. = 1 ft 7 in.

31. 5 × ■ = 2 c 4 fl oz

32. ■ ÷ 6 = 1 ft 2 in.

5-4 Perimeter

EXPLORE

Take any book from your desk. Find the distance around the front cover using an eraser or a paper clip as a unit of measure.

a. How many units long is the book's cover? How many units wide?

b. What was the fastest way to find the distance around the cover?

c. If another book's cover measures the same distance around, but has a different length and width, what might be its dimensions?

SKILLS DEVELOPMENT

The distance around a plane figure is called its **perimeter.** You can find the perimeter of a plane figure by adding the lengths of its sides.

Example 1

Find the perimeter of each figure.

a.

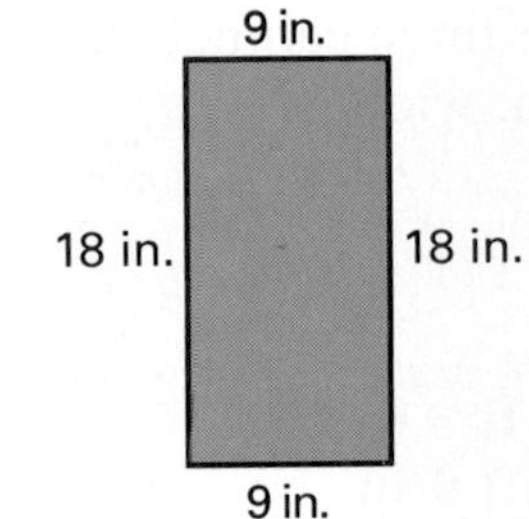

b.

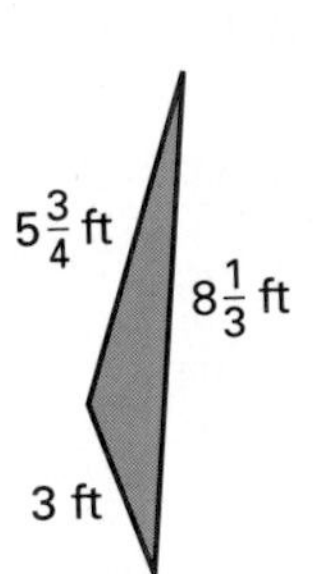

c.

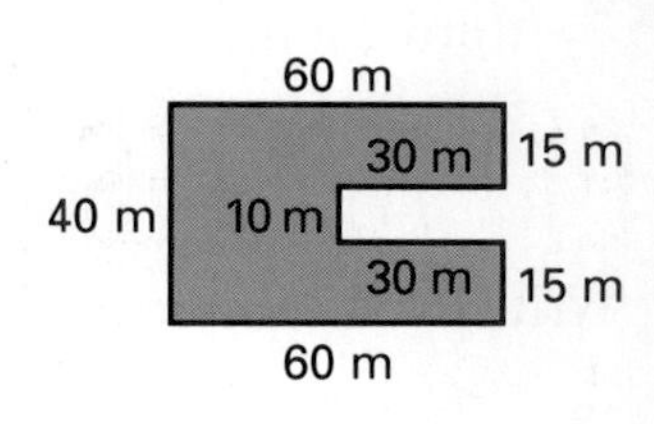

Solution

a. Add the lengths of the four sides.

$$9 + 18 + 9 + 18 = 54$$

The perimeter is 54 in.

b. Add the lengths of the three sides.

$$\begin{array}{rcr} 3 & \rightarrow & 3 \\ 5\frac{3}{4} & \rightarrow & 5\frac{9}{12} \\ +\,8\frac{1}{3} & \rightarrow & +\;8\frac{4}{12} \\ \hline & & 16\frac{13}{12} \rightarrow 17\frac{1}{12} \end{array}$$

The perimeter is $17\frac{1}{12}$ ft.

c. Add the lengths of the eight sides.

$$40 + 60 + 15 + 30 + 10 + 30 + 15 + 60 = 260$$

The perimeter is 260 m. ◄

Example 2

Find the perimeter of a rectangular swimming pool that is 12.4 m long and 5.6 m wide.

Solution

The swimming pool is in the shape of a rectangle, so it has two sides that measure 12.4 m and two sides that measure 5.6 m. Add the lengths of the sides.

$$12.4 + 12.4 + 5.6 + 5.6 = 36$$

The perimeter of the swimming pool is 36 m. ◄

CHECK UNDERSTANDING

In Example 2, the pool is described as being *rectangular.* What do you know about the sides of an object that is described as rectangular? What if it is described as *square?*

TRY THESE

Find the perimeter of each figure.

1. 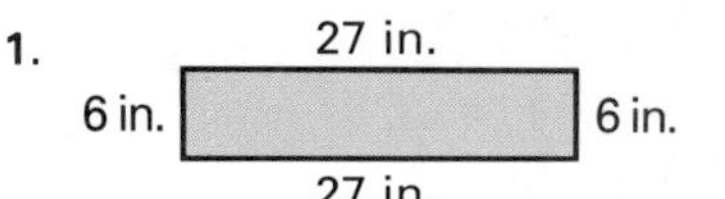

2. 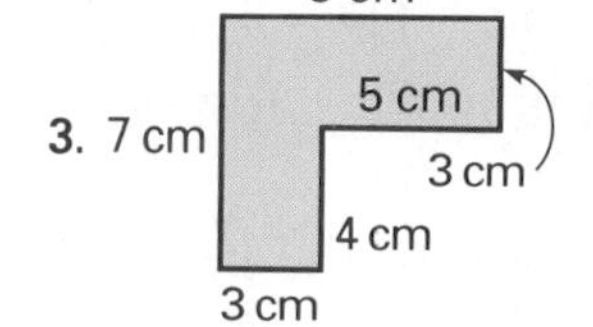

3. 7 cm

4. Find the perimeter of a square courtyard that measures $16\frac{1}{3}$ yd on each side.

EXERCISES

PRACTICE/ SOLVE PROBLEMS

1. Find the perimeter of each figure drawn on the grid if each unit of the grid represents 1 ft.

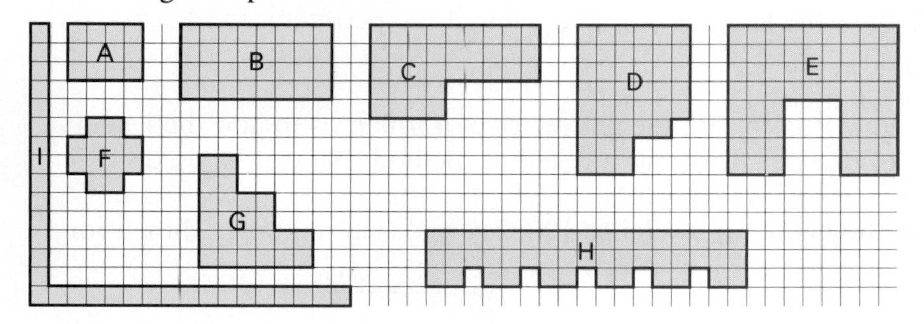

Find the perimeter of each figure.

2. 

3. 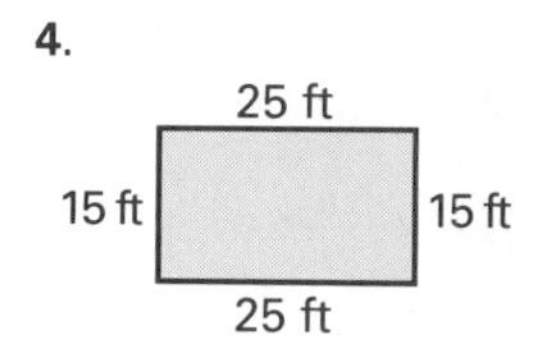

4.

Solve.

5. Find the perimeter of a rectangular building foundation that has two sides that measure 45 ft and two sides that measure $37\frac{1}{2}$ ft.

EXTEND/ SOLVE PROBLEMS

COMPUTER

To find the perimeter of a rectangle, you may want to use this program.

```
10 INPUT "ENTER THE
   LENGTH; ";L : PRINT
20 INPUT "ENTER THE
   WIDTH: ";W: PRINT
30 LET P = 2 * L + 2 * W
40 PRINT "THE PERIMETER IS
   ";P; "."
```

To find the perimeter of a triangle, you could make line 10:

```
10 INPUT "ENTER THE
   LENGTHS: ";A,B,C:?
```

and delete line 20. How would you then have to change line 30 to get the perimeter?

USING DATA Use the Data Index on page 546 to find the table showing dimensions of noted rectangular structures. Find the perimeter, in meters, of each of these structures.

6. Izumo Shrine
7. Step Pyramid of Zosar
8. Kibitsu Shrine (main)
9. Palace of the Governors
10. Bakong Temple, Roluos
11. Kongorinjo hondo

Find the perimeter of each figure.

12.

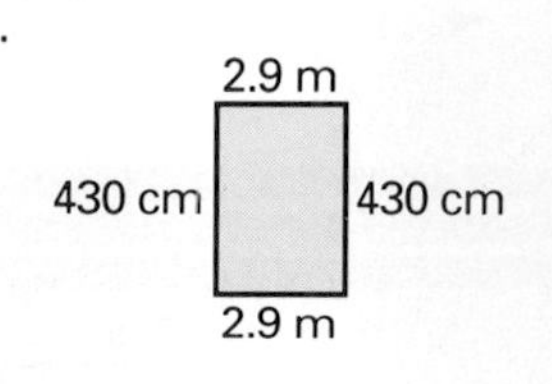

13.

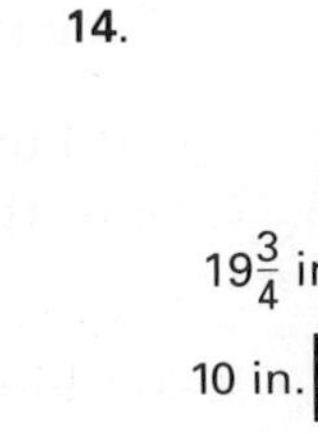

14.

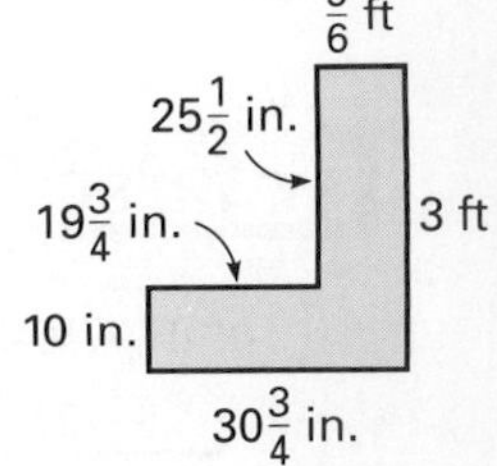

Given the perimeters, find the missing dimensions.

15. The perimeter of this triangle is 35 yd.

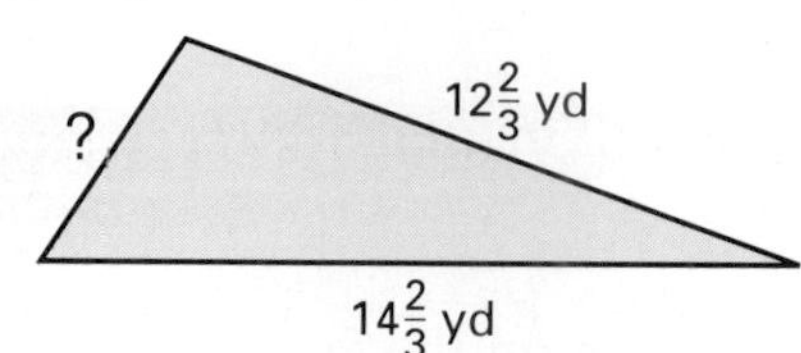

16. The perimeter of this rectangle is 70 m.

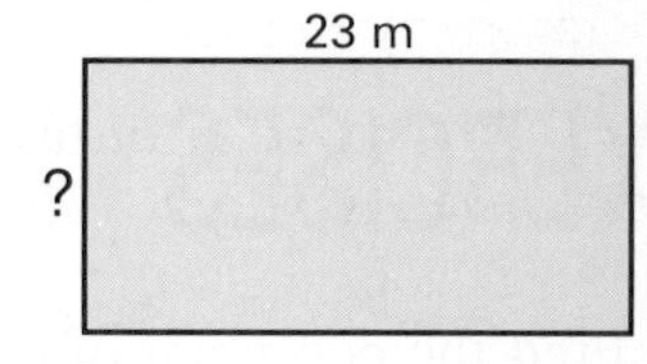

17. A triangular rose garden has two sides that measure 4.8 m each Find the length of the third side if the perimeter is 14.5 m.

THINK CRITICALLY/ SOLVE PROBLEMS

18. One way to find the perimeter of a square is to add the lengths of its sides. Describe a different method that can be used.

WRITING ABOUT MATH

Study the dimensions of the figure used in Exercises 19–22. Write a description of what such a figure could represent. What would be the perimeter of this figure? Why might someone need to know it?

Use the figure for Exercises 19–22. Find each unknown dimension.

19. x
20. y
21. z
22. Find the perimeter.

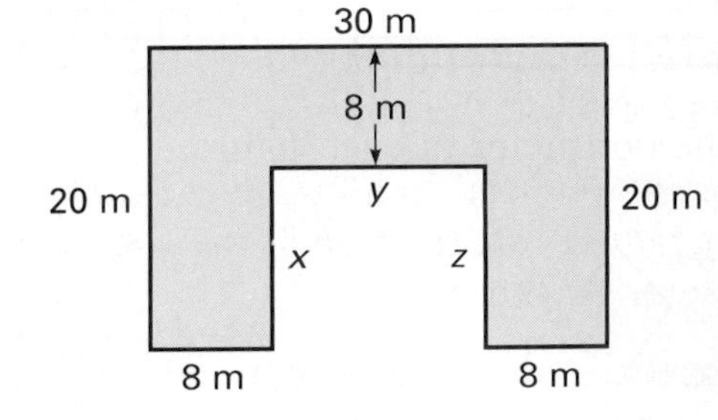

Solve.

23. One side of a rectangle with a perimeter of 45.8 m measures 17 m. Find the lengths of the other three sides.

- ▶ READ
- ▶ PLAN
- ▶ SOLVE
- ▶ ANSWER
- ▶ CHECK

Problem Solving Applications:

SOLVING PROBLEMS THAT INVOLVE PERIMETER

Problems involving perimeter arise in many different practical situations. For example, before you install a fence, you must know how much fencing you need and how much it will cost.

PROBLEM

a. Determine about how much fencing would be needed to fence in the property shown.

b. If the cost of fencing is \$8.92/m, find about how much it would cost to fence in the property shown.

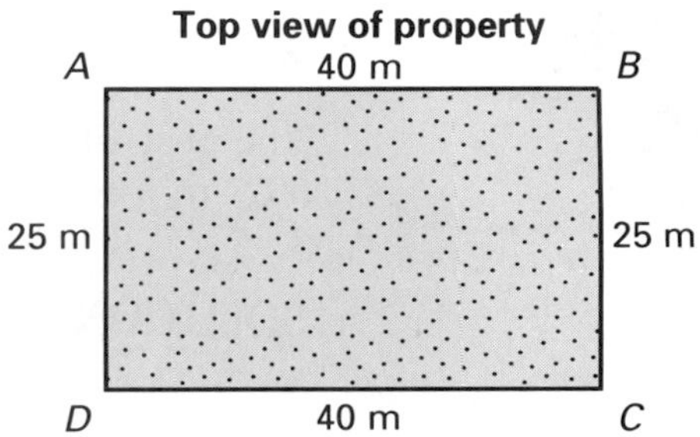

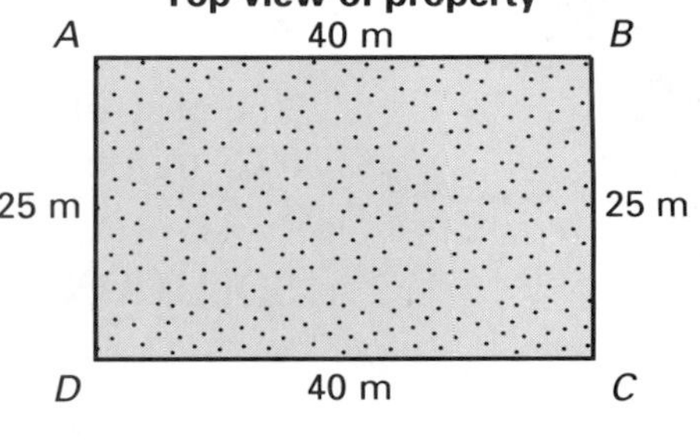

SOLUTION

a. 40 m + 25 m + 40 m + 25 m = 130 m

About 130 m of fencing would be needed to fence in the property shown.

b. Since the perimeter of the property is 130 m, and 1 m of fence costs \$8.92, multiply 130 by \$8.92.

The cost of fencing would be 130 × \$8.92, or about \$1,159.60.

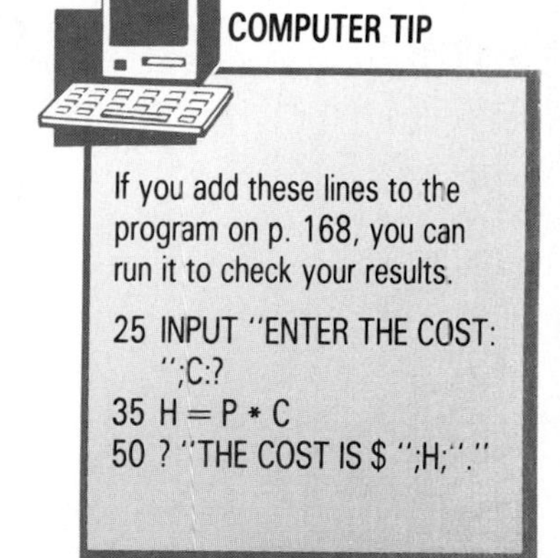

Solve.

1. If the cost of fencing material is \$2.55 per meter, how much fencing would be needed to enclose a rectangular dog run that measures 7.5 m by 3.7 m? What would be the cost of the fencing?
2. A rectangular swimming pool measures 15 m by 20 m. Find the cost of enclosing the pool if fencing costs \$5.25/m.
3. A court is shown in the diagram. What is the cost of placing a railing around the court if railing costs \$6.29/m?
4. The path walked in a golf game is shown in the diagram. How long would it take to walk the course at 5 km/h? Each unit represents 100 m.
5. Find the cost of enclosing a rectangular courtyard that is 12.4 m long and 5.6 m wide with fencing that costs \$7.29/m.
6. A rectangular Japanese garden measures 27.4 m by 43.5 m. Calculate the cost of placing a low decorative fence around it if fencing costs \$2.69/m.

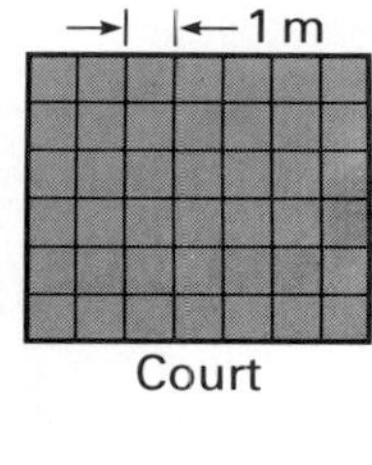

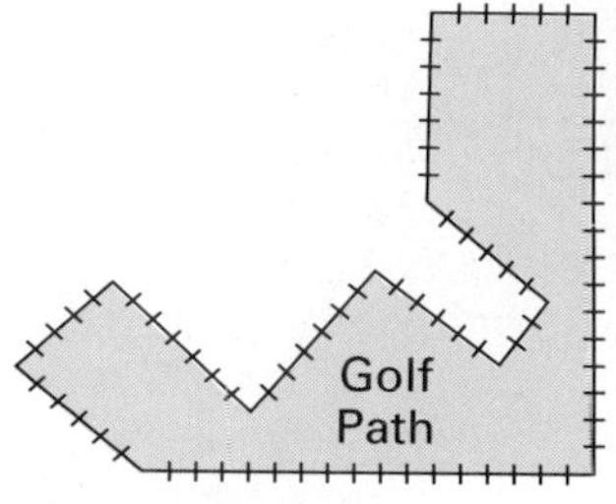

5-5 Problem Solving Skills:

USE A FORMULA

You have solved problems involving perimeter by adding the lengths of the sides to find the distance around a figure. In this lesson, you will learn how to use a formula to find the perimeter of a rectangle.

PROBLEM

The base of the Parthenon in Greece is a rectangular shape about 69.5 m long and 30.9 m wide. Find the perimeter of its base.

SOLUTION

Since a rectangle has two pairs of equal sides, you can find the perimeter by doubling the given length and doubling the given width, then adding the results.

Choose variables: P = perimeter
l = length
w = width

Write a formula: $P = 2l + 2w$. Then replace l and w with the given length and width.

$$P = 2l + 2w$$
$$P = 2(69.5) + 2(30.9)$$
$$P = 139 + 61.8$$
$$P = 200.8$$

The perimeter of the base of the Parthenon is about 200.8 m.

PROBLEMS

1. The base of a building is in the shape of a triangle, with sides that measure a feet, b feet, and c feet.
   a. Write a formula to find the perimeter of the base.
   b. Use your formula to find the perimeter if $a = 65$, $b = 97$, and $c = 46$.
2. The base of a building is shaped like a square with sides of length s meters.
   a. Write a formula to find the perimeter of the base.
   b. Use your formula to find the perimeter if each side of the square is 32 m long.

Architects plan and design many types of buildings. They draw plans for houses, such as the plan shown below. Refer to this plan for Problems 3–13.

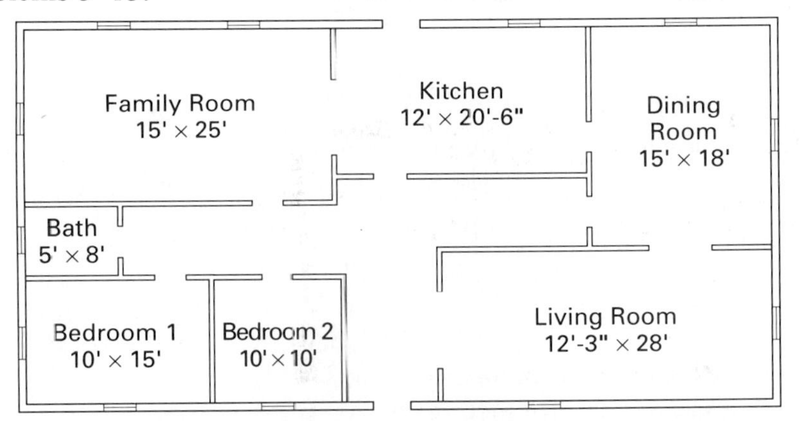

READING MATH

In building plans like the one shown on this page, the symbol ' stands for *feet* and the symbol " stands for *inches*.

Read *12' × 20'6"* as "twelve feet by twenty feet, six inches." This means that the width of the room is 12 ft and the length is $20\frac{1}{2}$ ft.

Suppose the architect wants to install decorative molding around the ceilings in some of the rooms. Use a formula to find how much molding would be needed for each room.

3. family room
4. kitchen
5. dining room
6. bedroom 1
7. bedroom 2
8. living room

Suppose the decorative molding costs $1.79 per foot.

9. How much would the molding cost for each of the six rooms listed above?

10. What would be the total cost of the molding for the six rooms?

The bathroom will be tiled as shown at the right. In Problems 11 and 12, allow a 3-ft opening for the bathroom door.

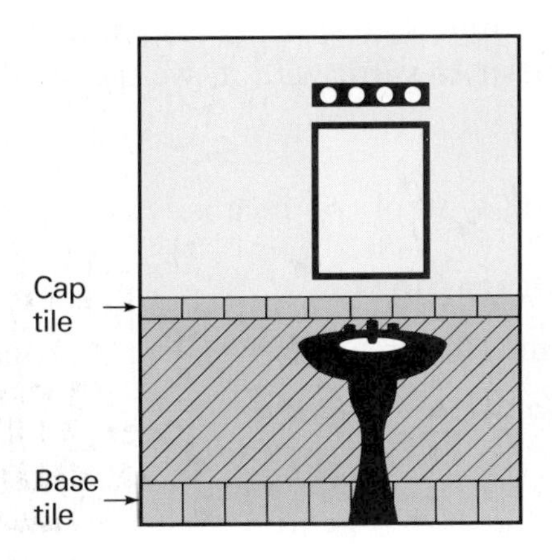

11. How many feet of cap tile are needed for the four walls of the bathroom? How many feet of base tile are needed?

12. At $1.59 per foot for cap tile and $2.89 per foot for base tile, how much will these bathroom tiles cost?

13. ***CRITICAL THINKING*** The new owners of this house want to build a deck that they can walk onto from the kitchen. The deck will be 12 ft wide and will run lengthwise across half the back of the house. How many feet of railing are needed for the deck? Allow a $2\frac{1}{2}$-ft opening for the steps down to the yard.

5-6 Circumference

EXPLORE/ WORKING TOGETHER

How can you find the distance around a circle?

Work with a partner. Carefully fit a piece of string around the circle below. Now, stretch out the string and use a centimeter ruler to measure the amount of string that you fit around the circle.

What is the distance around the circle?

SKILLS DEVELOPMENT

A **circle** is the set of all points in a plane that are a fixed distance from a given point in the plane. The point is called the **center** of the circle. The fixed distance is called the **radius** of the circle. The distance across a circle is twice its radius. This distance is called the **diameter** of the circle. The distance around a circle is called its **circumference.**

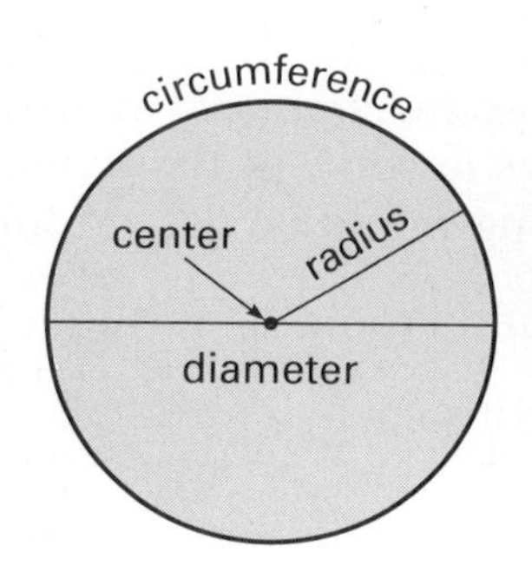

MATH: WHO, WHERE, WHEN

Srinivasa Ramanujan (1887–1920) was a mathematician who was largely self-taught. It was only when he was 27 that he went to England from his native India to study mathematics formally. He is credited with having developed over 6,000 mathematical theorems, one of which was a formula for calculating a decimal approximation for π. Ramanujan's formula was, remarkably, much like that used by modern mathematicians in 1987 when they used high-speed computers to determine the value of π to more than 100 million decimal places!

For a circle of any size, the quotient when the circumference (C) is divided by the diameter (d) is always equal to the same number. This number is represented by the Greek letter π (pi).

$$\frac{C}{d} = \pi, \qquad \text{or} \qquad C = \pi d$$

The number π is an *irrational number.* This means that it is a decimal that never ends, and that its digits never repeat. To make it easier to work with π, we use the following approximations.

$$\pi \approx 3.14 \qquad \text{or} \qquad \pi \approx \frac{22}{7}$$

The symbol $\approx$ means *is approximately equal to.*

Example 1

Find the circumference. Use 3.14 for π.

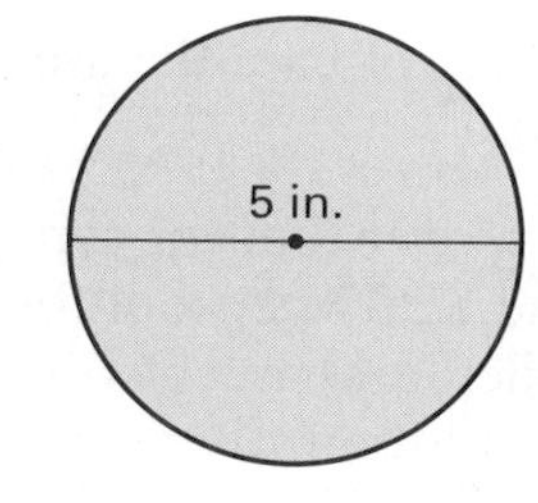

Solution

$C = \pi d$

$C \approx 3.14 \times 5$

$C \approx 15.70$

The circumference is approximately equal to 15.70 in. ◄

To find the circumference of a circle when you know the radius (r), use the formula $C = 2\pi r$.

Example 2

Find the circumference. Use $\frac{22}{7}$ for π.

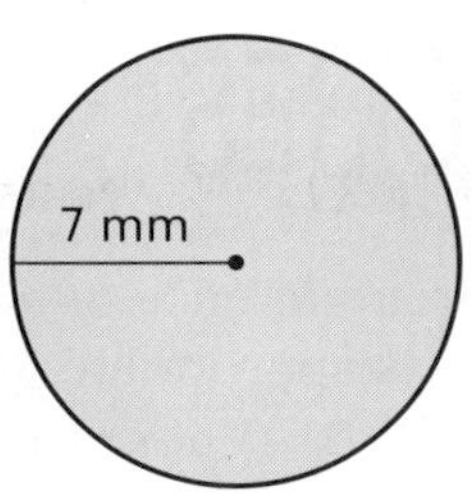

Solution

$C = 2\pi r$

$C \approx 2 \times \frac{22}{\cancel{7}_1} \times \cancel{7}^1$

$C \approx 44$

The circumference is approximately equal to 44 mm. ◄

CHECK UNDERSTANDING

In the solution of Example 2, why is the equal sign in the circumference formula replaced by the symbol $\approx$?

Example 3

A circular trampoline has a radius of 35 in. Find how much rubber stripping is needed to go around the rim of the trampoline.

Solution

The rim of the trampoline is shaped like a circle. The amount of rubber stripping needed is equal to the circumference of this circle. Since $r = 35$, use the formula $C = 2\pi r$.

$C = 2\pi r$

$C \approx 2 \times \frac{22}{\cancel{7}_1} \times \cancel{35}^5$

$C \approx 220$ in., or $18\frac{1}{3}$ ft

About $18\frac{1}{3}$ ft of rubber stripping is needed. ◄

CHECK UNDERSTANDING

In the solution of Example 3, why has $\frac{22}{7}$, and not 3.14, been used for π?

TRY THESE

Find the circumference. Use 3.14 for π. If necessary, round your answer to the nearest tenth.

1.

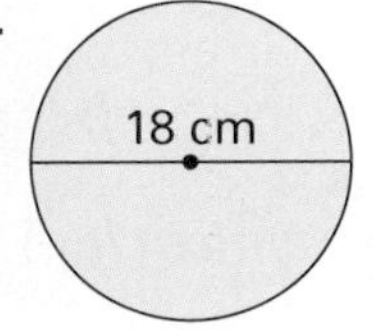

2.

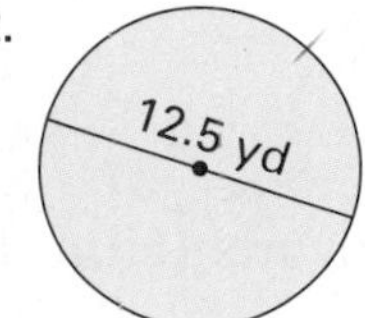

3.

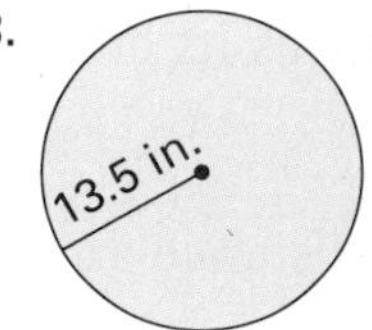

Find the circumference. Use $\frac{22}{7}$ for π.

4.

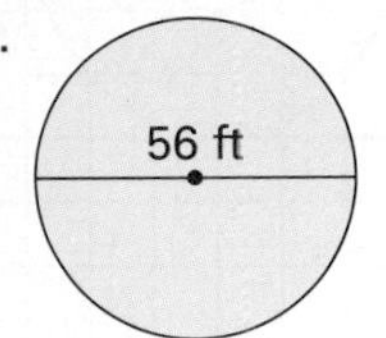

5.

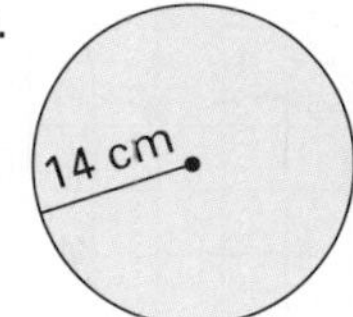

6.

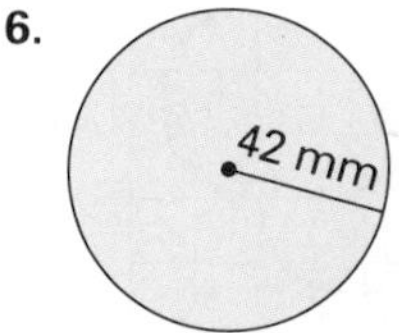

Solve. Use 3.14 or $\frac{22}{7}$ for π, as appropriate. If necessary, round your answer to the nearest tenth.

7. Find the circumference of a patio umbrella with a radius of 1.4 ft.

8. The diameter of the face-off circle used in hockey is 6 m. Find the circumference.

EXERCISES

PRACTICE/ SOLVE PROBLEMS

Find the circumference of a circle with the given diameter. Use 3.14 or $\frac{22}{7}$ for π, as appropriate.

1. $d = 19$ cm 2. $d = 5.4$ mm 3. $d = 13.5$ ft 4. $d = 56$ yd

Find the circumference of a circle with the given radius. Use 3.14 or $\frac{22}{7}$ for π, as appropriate.

5. $r = 21$ in. 6. $r = 5.8$ yd 7. $r = 12.2$ mm 8. $r = 17$ cm

ESTIMATION TIP

Whenever you need to use π in a calculation, an easy way to estimate the answer is to use 3 as an approximate value for π.

For example, to estimate the circumference of a circle with a diameter of 10 cm, think:

$$C = \pi d$$
$$C \approx 3 \times 10$$
$$C \approx 30$$

The circumference is about 30 cm.

Copy and complete the chart.

	radius	diameter	Circumference
9.	15 cm	■	■
10.	■	100 ft	■
11.	500 in.	■	■
12.	■	0.1 mm	■

Solve.

13. The face of the clock known as Big Ben is about 7.1 m in diameter. What is the circumference of the clock face?

14. The first Ferris wheel, named after its designer, George Ferris, was built in 1893. The diameter of that first wheel was 76 m. What was its circumference?

EXTEND/ SOLVE PROBLEMS

READING MATH

What is a *semicircle?* Find out the meaning of the prefix *semi–*. Then, find the meanings of these words.

1. semifinal
2. semiprecious
3. semiannual
4. semiprivate

Find the distance around each figure. (Hint: Look for *semicircles.*)

15.

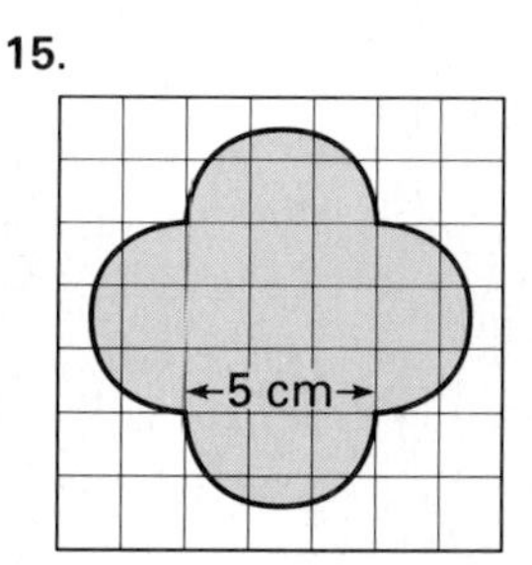

16.

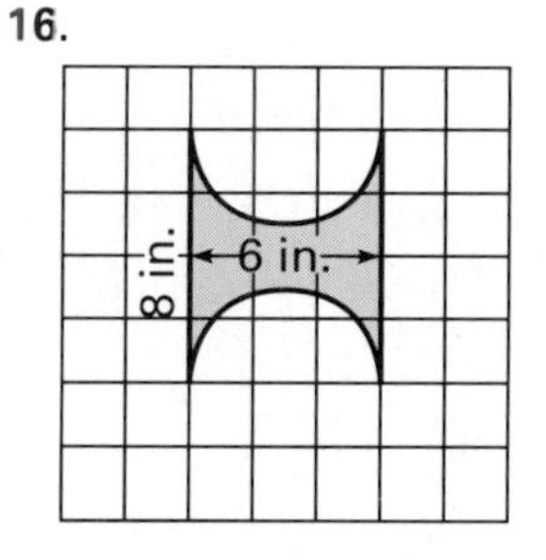

17.

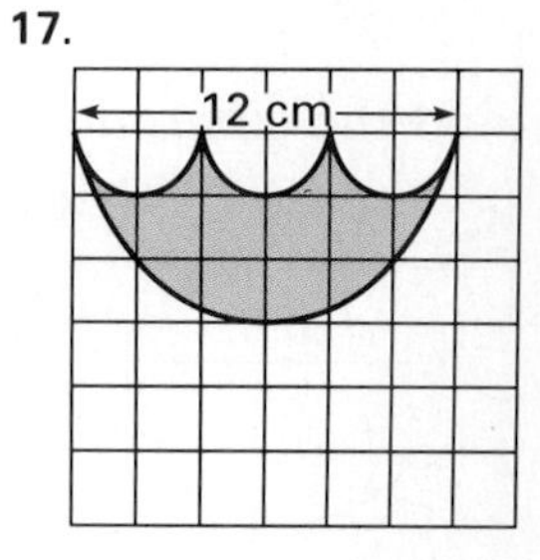

18. Find the distance a runner would cover in running one lap around this track.

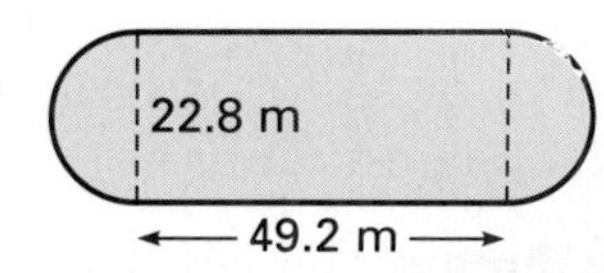

19. Which is greater, the perimeter of a square that measures 55.5 mm on each side or the circumference of a circle with a radius of 55.5 mm?

Use the diagram for Exercises 20 and 21.

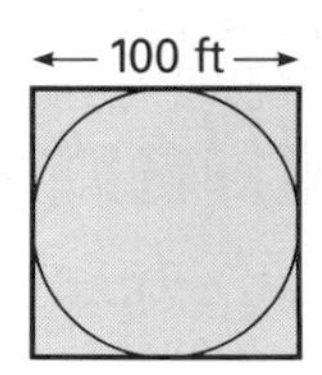

20. Find the perimeter of the square.

21. Find the circumference of the circle.

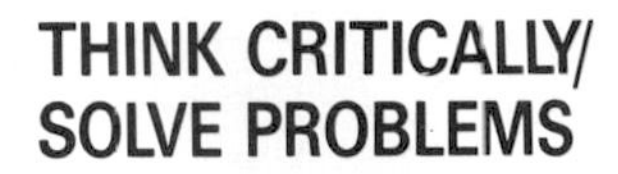

A circular jogging course is shown. Jogger A uses the outer edge of the track. Jogger B uses the inner edge of the track.

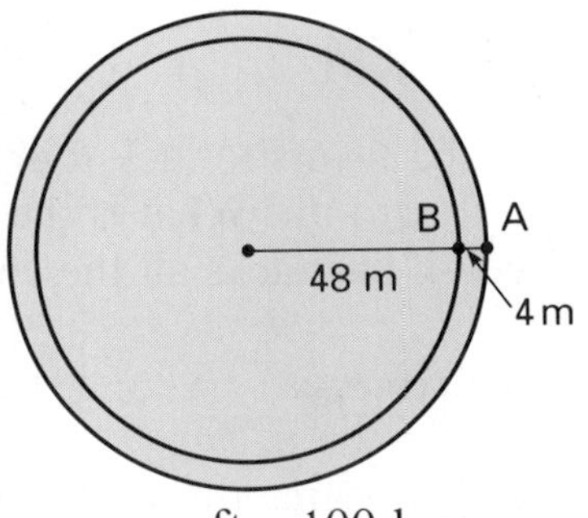

22. How many meters does each jogger cover if each jogs one complete lap?

23. If the two joggers begin at the same time and at the same points on their tracks, how much farther will Jogger A have jogged than Jogger B after each number of laps?

    a. after 5 laps b. after 10 laps c. after 100 laps

24. The diameter of a car tire is 70.4 cm. How many times must the wheel turn to travel 1 km?

The radius of the front wheels of a car is 42.3 cm. The radius of the rear wheels is 47.9 cm. Use these data for Exercises 27 and 28.

25. Find how far the car travels for one turn of the rear wheels.

26. For 100 turns of the front wheels, how many times do the rear wheels turn?

5-7 Problem Solving Strategies:

DRAW A PICTURE

Sometimes a problem is easier to understand and solve if you draw a picture to help you organize and visualize the data.

PROBLEM

Keisha wants to tack four photographs onto the bulletin board in her bedroom. What is the least number of thumbtacks that she will need if she tacks all the corners of every photograph?

SOLUTION

Keisha begins thinking about the problem. She knows that, if she tacks each photograph individually, she will need 4 thumbtacks for each photograph, or 16 in all. She draws this picture to show the arrangement.

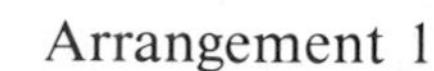

Arrangement 1

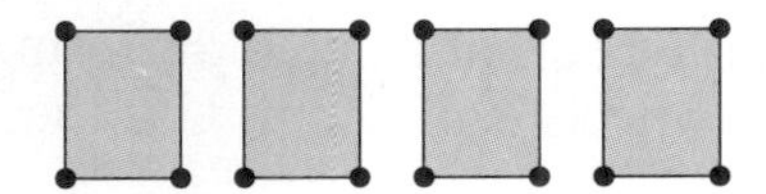

Then Keisha considers overlapping the corners of the photographs in different ways. She draws these three arrangements of the photographs to help her figure out how many thumbtacks she will need.

Arrangement 2

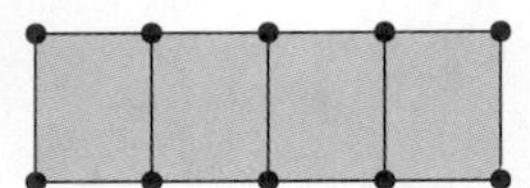

This takes 4 thumbtacks for the first photograph and 2 for each of the other three photographs—10 thumbtacks in all.

Arrangement 3

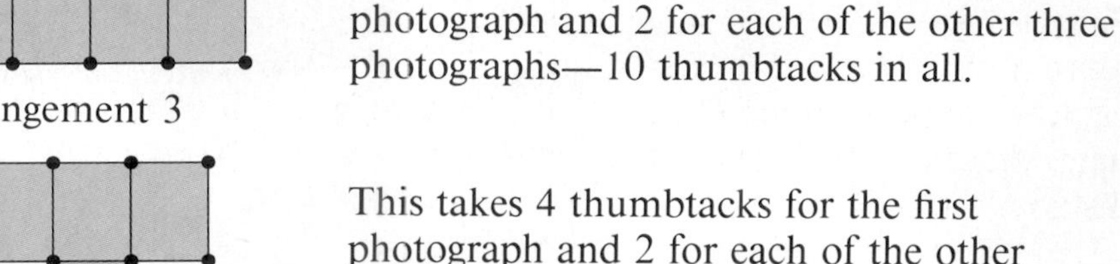

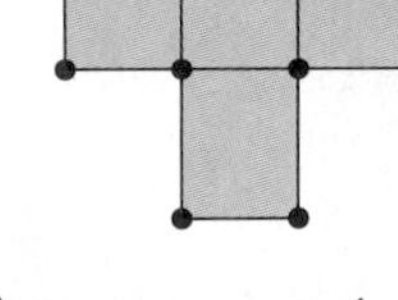

This takes 4 thumbtacks for the first photograph and 2 for each of the other three—10 in all.

Arrangement 4

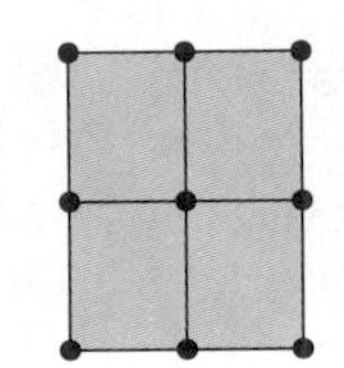

This takes only 9 tacks in all!

TALK IT OVER

Why do you think fewer thumbtacks are needed for Arrangement 4 than for the other arrangements?

PROBLEMS

PROBLEM SOLVING TIP

Try to make arrangements that are shaped like Arrangement 4. (Remember, every corner of every photograph must be tacked.)

What is the least number of thumbtacks that would be needed to tack each number of photographs to a bulletin board?

1. 9 photographs
2. 10 photographs
3. 11 photographs
4. 12 photographs
5. 13 photographs
6. 14 photographs
7. 15 photographs
8. 16 photographs

What is the greatest number of photographs that could be tacked to a bulletin board with each number of thumbtacks?

9. 4 thumbtacks
10. 9 thumbtacks
11. 16 thumbtacks
12. 25 thumbtacks
13. 36 thumbtacks
14. 49 thumbtacks
15. Record your answers for Exercises 9–14 in a table like the one below. Look for a pattern in the numbers of photographs and the numbers of thumbtacks. Then complete the table by predicting the number of photographs that could be displayed using 64 thumbtacks.

Number of Photographs	1	4	■	■	■	■	■
Number of Thumbtacks	4	9	■	■	■	■	■

Suppose you notice that your neighbor, Kathy, has begun to build a fence around her square garden plot. You see that she has installed eight fence posts along each side of the garden. Each of the posts has a diameter of six inches. The posts have been placed so that the distance from one post to the next is one yard.

16. How many posts are there?
17. What is the perimeter of the garden?
18. Suppose Kathy has told you that she would like to plant her square garden so that there are four rows of tomato plants with the same number of plants in each row. Each row will be 100 ft long. If she places the plants four feet apart and starts planting at one corner of the garden, how many plants will she need to buy?

CHAPTER 5 ● REVIEW

a. meters
b. milliliters
c. circumference
d. perimeter
e. radius
f. diameter
g. kilograms

1. The distance across a circle is the circle's ___?___.
2. The distance around a plane figure is its ___?___.
3. The height of a doorway could be expressed in ___?___.
4. The fixed distance from any point on a circle to the circle's center is the ___?___ of the circle.
5. The amount of water in a cup could be expressed in ___?___.
6. The mass of a basket of apples could be expressed in ___?___.
7. The distance around a circle is its ___?___.

SECTIONS 5–1 AND 5–2 CUSTOMARY AND METRIC UNITS (pages 154–161)

► To change from a larger unit of measure to a smaller unit, multiply.
► To change from a smaller unit of measure to a larger unit, divide.

Complete.

8. 4 yd = ■ ft
9. 38 oz = ■ lb ■ oz
10. 5 pt = ■ c
11. 3 mi = ■ yd
12. 28 in. = ■ ft ■ in.
13. 10 ft = ■ yd ■ ft
14. 16 kg = ■ g
15. 900 cm = ■ m
16. 290 mm = ■ m
17. 4 km = ■ m
18. 8 m = ■ cm
19. 4.6 L = ■ mL

SECTION 5–3 WORKING WITH MEASUREMENTS (pages 162–165)

► When you add or multiply with customary measurements, you may need to simplify your answer.
► When you subtract or divide with customary measurements, you may need to change a measurement.
► You may also need to change measurements when performing operations with metric measurements.

Write each answer in simplest form.

20. 6 gal 2 qt
+2 gal 6 qt

21. 4 ft 3 in.
−2 ft 10 in.

Complete.

22. 84 cm + 92 cm = ___?___ m
23. 8 km ÷ 4 = ___?___ m

SECTION 5–4 PERIMETER (pages 166–169)

- The **perimeter** of a plane figure is the distance around it.
- You can find the perimeter of a figure by adding the lengths of its sides.

Find the perimeter of each figure.

24.
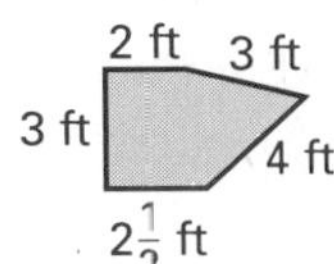

25.
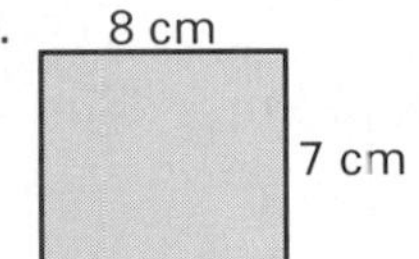

26.
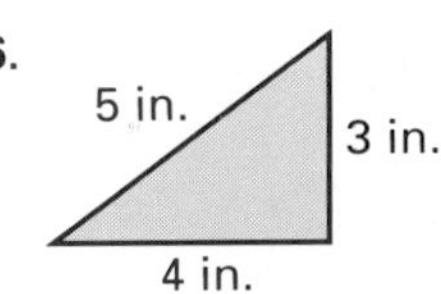

27.
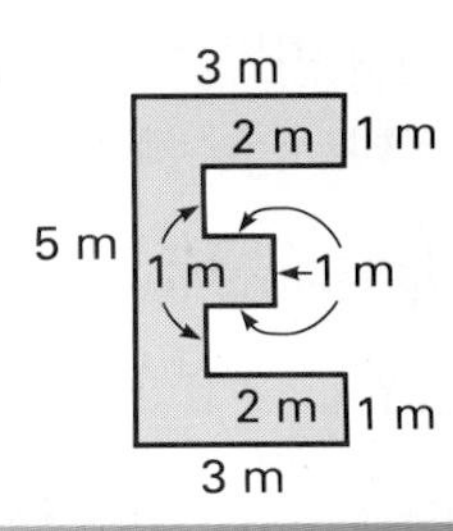

SECTION 5–5 PROBLEM SOLVING SKILLS: USE A FORMULA (pages 170–171)

- You can use the formula $P = 2l + 2w$ to find the perimeter of a rectangle.

Use the formula to find the perimeter of a rectangular figure with the given dimensions.

28. length: 5.6 km
    width: 2.9 km

29. length: 16.5 mi
    width: 12.4 mi

SECTION 5–6 CIRCUMFERENCE (pages 172–175)

- The formulas $C = \pi d$ and $C = 2\pi r$ can be used to find the circumference of a circle.

Find the circumference. Use 3.14 or $\frac{22}{7}$ for π, as appropriate.

30.
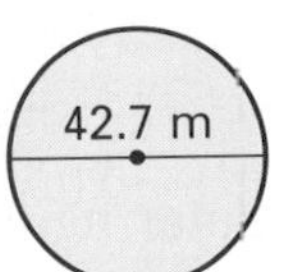

31.
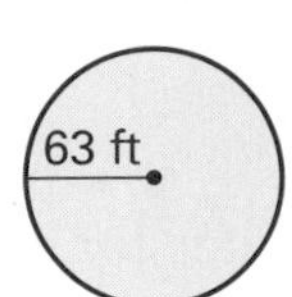

32.
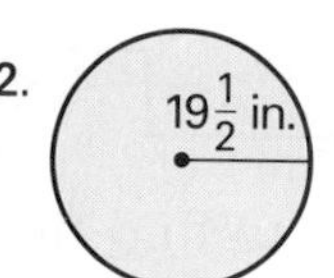

SECTION 5–7 PROBLEM SOLVING STRATEGIES (pages 176–177)

- Sometimes a problem is easier to understand and solve if you draw a picture.

33. A square garden is enclosed by a fence that has 12 posts on each side. How many posts are there in all?

USING DATA Use the Data Index on page 546 to find the table showing dimensions of noted rectangular structures. Find the perimeter, in meters, of the base of each structure.

34. Cleopatra's Needle

35. Ziggurat of Ur

36. Great Pyramid of Cheops

37. Ta Keo Templean.

USING DATA Use the chart on page 153 to answer the question.

38. About how much longer than the Italian Gothic period was the Byzantine period?

CHAPTER 5 ● TEST

Complete.

1. 72 in. = ■ ft
2. 9 gal = ■ qt
3. 10 lb = ■ oz
4. 2 mi = ■ yd
5. 280 mm = ■ m
6. 5 L = ■ mL
7. 589 mg = ■ g
8. 2 km = ■ m

Write each answer in simplest form.

9. 6 c 4 fl oz + 2 c 10 fl oz
10. 6 lb − 4 lb 5 oz
11. 2 ft 6 in. × 3
12. 13 gal 2 qt ÷ 3

Complete.

13. 230 cm + 370 cm = __?__ m
14. 12 m − 750 cm = __?__ m
15. 3.21 cm × 3 = __?__ mm
16. 400 m ÷ 8 = __?__ cm

Solve.

17. Will one 112-inch ball of string be enough to make twelve pieces of string, each $\frac{3}{4}$ ft long?

Find the perimeter of each figure.

18.

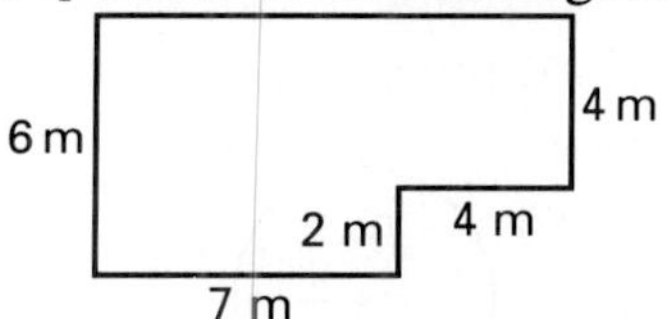

19.

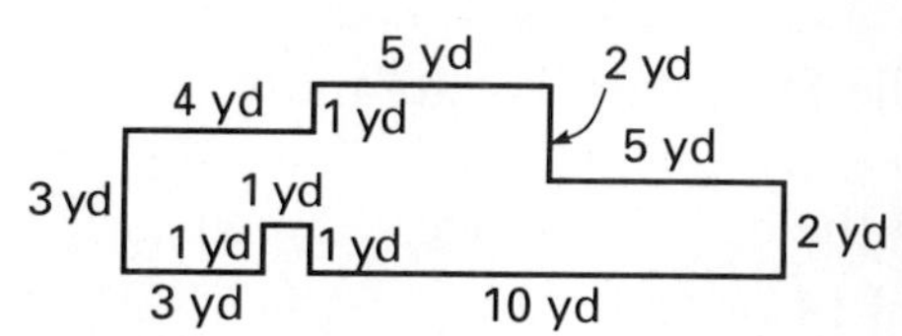

Use the formula to find the perimeter for a rectangular figure with the given dimensions.

20. length: 7.35 m
width: 4.81 m
21. length: 25.4 mi
width: 16.9 mi

Find the circumference. Use 3.14 or $\frac{22}{7}$ for π, as appropriate.

22.

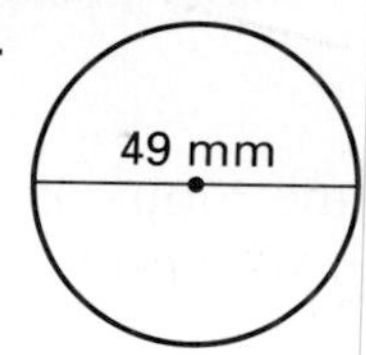

23.

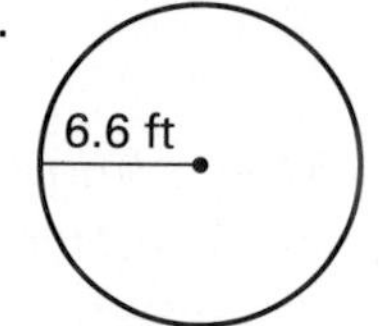

24.

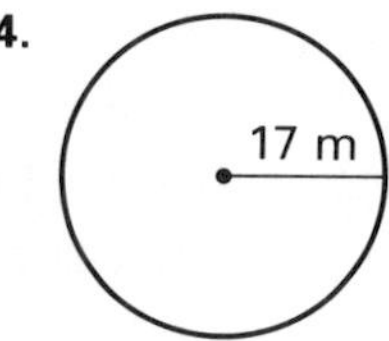

Draw a picture to help you solve.

25. A rectangular garden is enclosed by a fence that has 10 posts on each of its longer sides and 8 posts on each of its shorter sides. How many posts are there in all?

1. Organize these data into a stem-and-leaf plot.

   90 68 78 62 92 73
   87 98 96 70 85 95
   84 67 79 86 64 76

Which measure of central tendency would you use to represent the middle of each set of data?

2. average salary of part-time help
3. salary received by most part-time help
4. speed of vehicles on a highway

Complete.

5. 7.8 = ■ × 1
6. 179 + 612 = ■ + 179
7. 68 × ■ = 24 × 68
8. (8 + 2) + 3 = ■ + (2 + 3)
9. (5 × ■) × 4 = 5 × (9 × 4)
10. 36 × ■ = 0

11. Determine whether the following argument is *valid* or *invalid*.

    If an animal is a snake, then it is a reptile.
    Squeezy is a reptile.
    Therefore, Squeezy is a snake.

12. Is the following conditional statement *true* or *false*? If false, give a counterexample.

    If a number is divisible by 8, then it is divisible by 16.

Find each answer.

13. $\frac{1}{6} \times 18$
14. $\frac{9}{16} \div 3$
15. $19 - 4\frac{5}{6}$
16. $\frac{5}{8} + \frac{2}{3}$
17. $\frac{4}{9} \times \frac{3}{8}$
18. $\frac{3}{4} \div \frac{5}{8}$

Complete.

19. 5 mi = ■ ft
20. 4.5 m = ■ mm
21. 96 oz = ■ lb
22. 1,572 m = ■ km
23. 60 oz = ■ lb ■ oz
24. 6,250 cm = ■ m

Find the perimeter of each figure.

25.

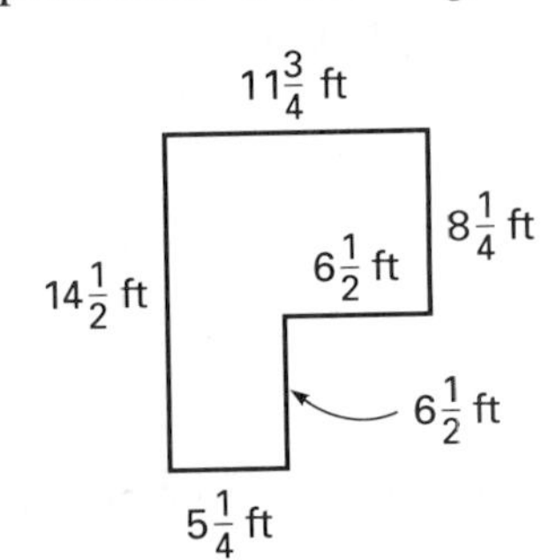

26.

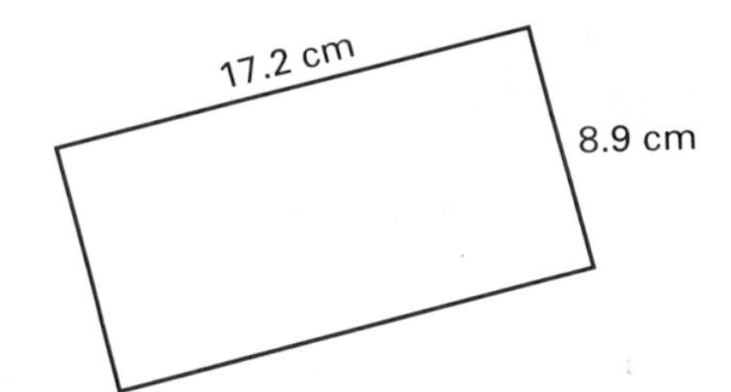

27.

27.2 mm
48.7 mm
37.9 mm
52 mm
31.4 mm

CHAPTER 5 ● CUMULATIVE TEST

In a set of data, the least number is decreased by 4 and the greatest number is increased by 4.

1. How is the range affected?

A. It is increased by 8.
B. It is decreased by 8.
C. It remains the same.
D. none of these

2. How is the mean affected?

A. It is increased.
B. It is decreased.
C. It remains the same.
D. none of these

---

3. Multiply. $5^9 \times 5^4$.

A. 25^{13} B. 5^{13}
C. 5^5 D. 25^5

4. Use the quotient rule to divide. $10^{10} \div 10^7$.

A. 10^{17} B. 17^{10}
C. 100^{17} D. 10^3

5. Complete.
$(9 + 3) + 5 = ■ + (3 + 5)$

A. 17 B. 3
C. 5 D. 9

6. $(7 \times ■) \times 3 = 7 \times (5 \times 3)$

A. 17 B. 5
C. 15 D. 7

---

7. Which conclusion can be drawn from these statements?

If a person lives in Dallas, then that person lives in Texas.
Malerie lives in Dallas.

A. Malerie does not live in Dallas.
B. Malerie lives in Houston.
C. Malerie lives in Texas.
D. none of these

8. Which is a counterexample for the following statement?

If two numbers are odd, then their sum is odd.

A. 3 and 5 B. 3 and 6
C. 4 and 5 D. none of these

---

9. Add. $\frac{5}{6} + \frac{1}{8}$

A. $\frac{6}{14}$ B. $\frac{3}{7}$
C. $\frac{23}{24}$ D. $1\frac{1}{24}$

10. Subtract. $\frac{9}{10} - \frac{2}{5}$

A. $1\frac{2}{5}$ B. $\frac{7}{10}$
C. $\frac{2}{3}$ D. $\frac{1}{2}$

11. Multiply. $\frac{3}{4} \times \frac{8}{15}$

A. $\frac{2}{5}$ B. $\frac{24}{45}$
C. $\frac{15}{32}$ D. $1\frac{13}{32}$

12. Divide. $\frac{3}{8} \div \frac{1}{4}$

A. $\frac{3}{32}$ B. $\frac{1}{2}$
C. $\frac{1}{3}$ D. $1\frac{1}{2}$

---

13. Complete. 5 gal = ■ pt

A. 10 pt B. 20 pt
C. 40 pt D. 50 pt

14. Complete. 4,568 cm = ■ m

A. 0.4568 m B. 456.8 m
C. 45.68 m D. 4.568 m

---

15. A garden has rows of equal numbers of pepper plants. Each row is 40 ft long. The plants are placed 5 ft apart starting at the corner of the garden. How many pepper plants are in the garden?

A. 54 plants B. 60 plants
C. 48 plants D. 240 plants

CHAPTER 6 SKILLS PREVIEW

Find each absolute value.

1. $|3|$ 2. $|-8|$ 3. $|-21|$ 4. $|6|$

Replace each ● with <, >, or =.

5. -16 ● 16 6. -14 ● -12 7. 18 ● -21

8. 0 ● -1 9. -10 ● -1 10. -16 ● 5

Add.

11. $-8 + 4$ 12. $8 + (-12)$ 13. $-6 + (-8)$

14. $5 + (-43)$ 15. $-7 + (-7)$ 16. $12 + (-12)$

Subtract.

17. $7 - (-8)$ 18. $-6 - 4$ 19. $-5 - (-9)$

20. $-4 - 3$ 21. $9 - 12$ 22. $4 - (-8)$

Multiply.

23. 9×6 24. -6×8 25. $-4 \times (-5)$

26. $7 \times (-8)$ 27. $-3 \times (-4)$ 28. -5×9

Divide.

29. $16 \div 4$ 30. $-24 \div 3$ 31. $-56 \div (-8)$

32. $63 \div (-7)$ 33. $-28 \div (-4)$ 34. $-12 \div 4$

Write each rational number as a ratio of two integers.

35. 3 36. 0.9 37. $-3\frac{5}{6}$ 38. 2.6

39. -5.2 40. $4\frac{7}{8}$ 41. 0.17 42. 1

Solve.

43. A number is three less than negative eight. What is the number?

44. A number is nine more than negative twenty-two. What is the number?

45. Six teams will play in a tournament. Each team will play every other team once. How many games will be played?

CHAPTER 6 EXPLORING INTEGERS

THEME Earth Science

Until now, the only numbers you have worked with in this textbook have been positive numbers, numbers greater than zero. The earth science topics in this chapter introduce you to negative numbers, numbers with values less than zero. You can use negative numbers to express measurements such as depths below sea level and temperatures less than zero degrees.

The earth is shaped more or less like a ball. The inside of the ball is made up of layers that differ in composition and in physical properties, such as density and temperature. The very center of the earth is a solid core of iron and nickel.

LAYERS OF THE EARTH	
Layer	**Approximate Thickness (km)**
Core: Inner	1,300
Outer	2,250
Mantle: Lower	2,650
Upper	180
Crust	10–40

DECISION MAKING

Using Data

Use the information in the table to find the answers to the following questions. Give your answers as negative numbers.

1. We live on the surface of the crust. If the surface is at a depth of 0 km, at what depth does the lower mantle begin? (Give your answer as a range of numbers.)
2. What is the greatest depth at which the bottom of the lower mantle lies?
3. At about what depth is the center of the earth? (Round your answer to the nearest hundred kilometers.)

Working Together

Make a model of the earth's interior based on the data in the table. Your model should be a cross section of the earth. Be sure to label each layer.

Use earth science books or encyclopedias to find out the temperatures at the core and at each of the other layers of the earth. Compare the temperatures. Compare them to temperatures you have experienced.

The earth is enclosed by the atmosphere, which begins at the surface of the earth and rises far out into space. Like the interior of the earth, the atmosphere has a layered structure. If you were a space scientist responsible both for launching space vehicles and for the safety of humans aboard, what would you want to know about the structure and composition of the atmosphere? How could a model like the one of the earth's interior help you?

6-1 Exploring Integers

EXPLORE

For most English words that indicate direction, there are antonyms, or words that mean the opposite. *Above* and *below* and *forward* and *backward* are examples of antonyms.

Make a list of antonyms for these words: *in, over, east, north,* and *up.*

Add other related direction words to your list.

SKILLS DEVELOPMENT

An **integer** is any number in the following set.

$$\{\ldots, -3, -2, -1, 0, 1, 2, 3, \ldots\}$$

Integers that are greater than zero are **positive integers.** Integers that are less than zero are **negative integers.** Zero is neither positive nor negative. A positive number can be written with a positive sign (+), but it does not need to have a sign. A negative number must be written with a negative sign (−).

Integers can be shown as points on a number line. On a horizontal number line, positive numbers are to the right of 0; negative numbers are to the left of 0.

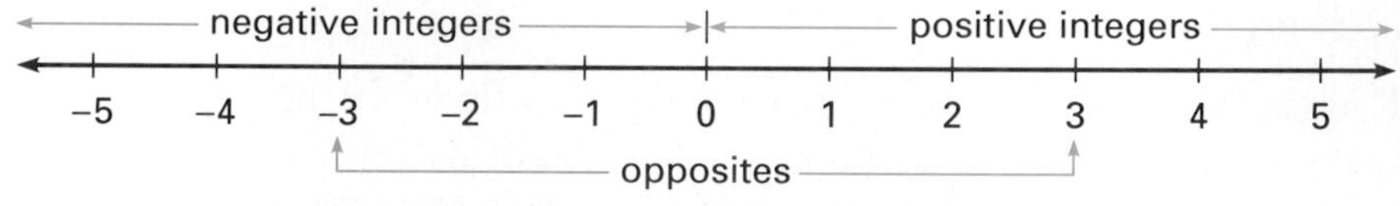

Two numbers that are the same distance from 0, but in opposite directions, are **opposites.** The opposite of a number n is written $-n$.

The distance a number is from zero on a number line is the **absolute value** of the number. The integer 3 is 3 units from 0. The integer −3 is also 3 units from 0. So, opposite integers have the same absolute value. The absolute value of 3 is written $|3|$.

TALK IT OVER

Integers can be used to represent real-life situations. For example, in a football game, 40-yd gain can be represented by +40 and a 40-yd loss by −40.

Discuss other situations that could be represented by positive and negative integers.

Example 1

Write the integer for each lettered point. Then give its opposite.

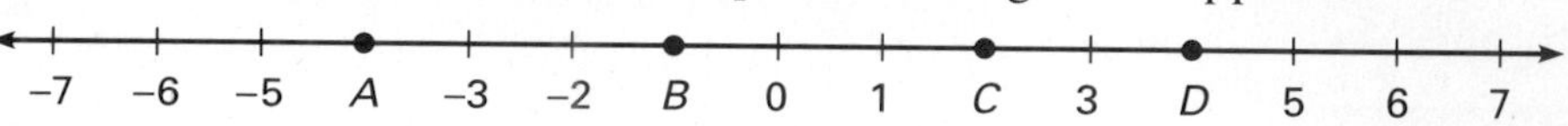

Solution

a. Point A represents -4. Its opposite is 4.
b. Point B represents -1. Its opposite is 1.
c. Point C represents 2. Its opposite is -2.
d. Point D represents 4. Its opposite is -4. ◄

Example 2

Find each absolute value. a. $|-7|$ b. $|7|$ c. $|0|$

Solution

a. Since -7 is 7 units from zero, the absolute value of -7 is 7. $|-7| = 7$
b. Since 7 is 7 units from zero, the absolute of 7 is 7. $|7| = 7$
c. The absolute value of 0 is 0. $|0| = 0$ ◄

COMPUTER TIP

The ABS function will give you the absolute value of any number, N. Just type: ? ABS(N) and press ENTER.

If integers are graphed on a horizontal number line, the number to the right is the greater number.

Example 3

Replace ● with $<$, $>$, or $=$.

a. 3 ● 2 b. -3 ● -4 c. -4 ● 3 d. 2 ● -3

Solution

Use a number line to help you compare integers.

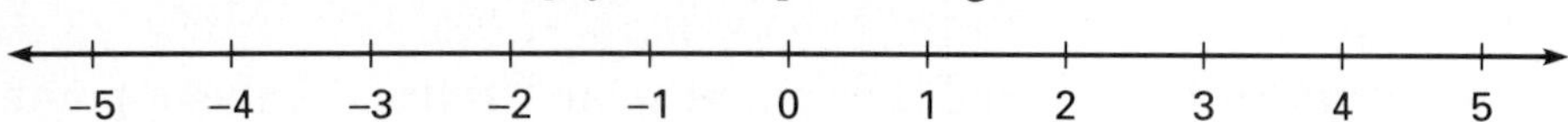

a. Since 3 is to the right of 2, $3 > 2$.
b. Since -3 is to the right of -4, $-3 > -4$.
c. Since -4 is to the left of 3, $-4 < 3$.
d. Since 2 is to the right of -3, $2 > -3$. ◄

TRY THESE

Write the integer for each lettered point. Then give its opposite.

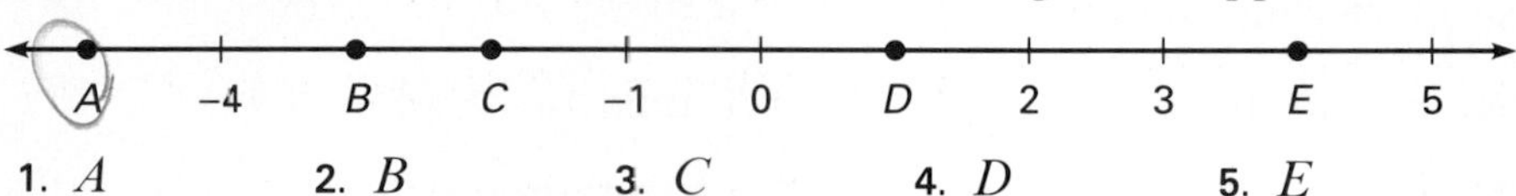

1. *A* 2. *B* 3. *C* 4. *D* 5. *E*

Find each absolute value.

6. $|-8|$ 7. $|3|$ 8. $|-45|$ 9. $|-13|$ 10. $|72|$

Replace each ● with $<$, $>$, or $=$.

11. 0 ● 5 12. -7 ● -9 13. -4 ● 2

MATH: WHO, WHERE, WHEN

The symbols we use for positive and negative first appeared in print about 500 years ago. The positive sign, +, may have been an abbreviated form of the Latin word *et*, meaning "and." The negative sign, −, probably came from the mark called a macron that was over the Latin word *minus*, meaning "less."

EXERCISES

PRACTICE/ SOLVE PROBLEMS

Write the integer for each lettered point. Then give its opposite.

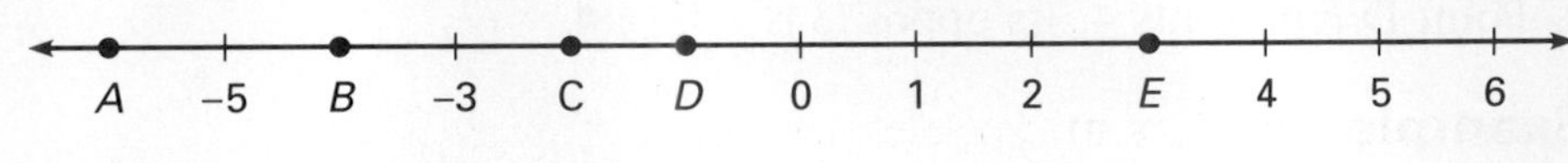

1. A 2. B 3. C 4. D 5. E

COMPUTER TIP

This program will let you check Exercise 26. You can use your knowledge of conditionals in logic to change the program to check Exercises 27–30.

```
10 INPUT "ENTER 2
   INTEGERS: ";A ,B: PRINT
20 IF A > B AND ABS (A) >
   ABS (B) THEN PRINT
   "TRUE": END
30 PRINT "FALSE"
```

Find each absolute value.

6. $|-2|$ 7. $|-34|$ 8. $|0|$ 9. $|17|$ 10. $|-20|$

Replace each ● with $<$, $>$, or $=$.

11. 0 ● -2 12. -4 ● -10 13. -2 ● 2

14. -8 ● -3 15. -9 ● -8 16. 6 ● -7

Write an integer to represent the quantity in each situation.

17. The top of Little Bear, a peak in Colorado, is 4,278 m above sea level.

18. The lowest point on the earth's surface is 10,859 below sea level.

EXTEND/ SOLVE PROBLEMS

Draw a number line. At the zero point, write "Today." Label a point to represent each of the following days on the number line. Write an integer to represent each day.

19. X, yesterday 20. Y, a week from today

21. H, the day after tomorrow 22. Q, the day before yesterday

Write the integers in order from least to greatest.

23. $-9, -13, 14, 0, 7$ 24. $6, -5, 5, 7, -9$ 25. $-8, 14, 8, 9, -15$

THINK CRITICALLY/ SOLVE PROBLEMS

Decide whether each statement is *true* or *false*. Give an example to support your choice.

26. For all integers a and b, if $a > b$, then $|a| > |b|$.

27. For all integers a and b, if $|a| = |b|$, then $a = b$.

28. For all negative integers a and b, if $a > b$, then $|a| > |b|$.

29. For all positive integers a and b, if $|a| > |b|$, then $a > b$.

30. If $a > b$ and $b > c$, decide whether each statement is *true* or *false*.
    a. $a > b > c$ b. $b > c > a$ c. $c > a > b$ d. $a > c$

Problem Solving Applications:

COMPARING TEMPERATURES

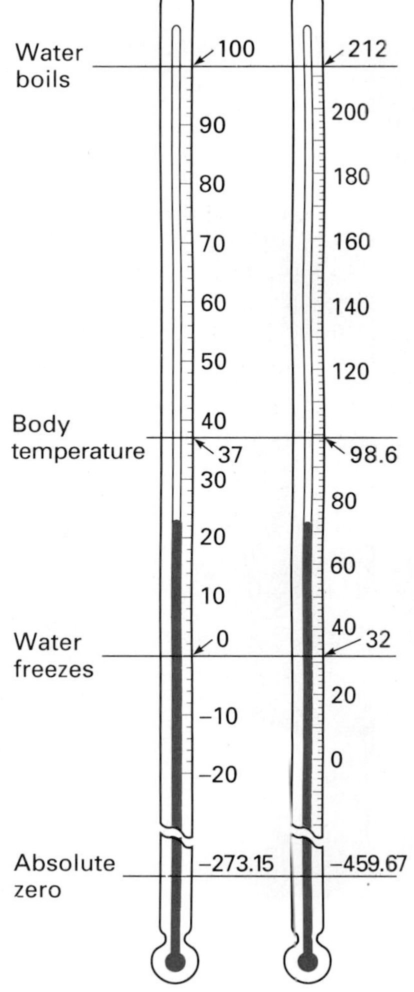

Some thermometers measure temperature in degrees Celsius (°C); others measure temperature in degrees Fahrenheit (°F).

Write a suitable temperature in degrees Celsius for each activity. **Answers will vary.**

1. shoveling snow
2. picnicking in park
3. complaining about heat wave
4. wearing jacket outdoors

Choose the more reasonable temperature for each situation.

5. boiling water: 100°C or 100°F
6. room temperature: 76°C or 76°F
7. swimming in July: 85°C or 85°F
8. removing frost from windshield: 15°C or 15°F

Which temperature is warmer?

9. 0°C or −32°F
10. 92°C or 180°F
11. 32°C or 198°F
12. 100°C or 198°F
13. 20°C or 80°F
14. −10°C or 10°F
15. The highest temperature ever recorded in the United States was 134°F in Death Valley, California. Do you think the integer used to record the equivalent temperature in degrees Celsius is greater or less than 134? Explain.

Use the table for Exercises 16 and 17.

16. Write the cities in order of their temperatures, from lowest to highest.
17. Use an almanac. Find the *highest* recorded temperatures for these cities and order them from lowest to highest.

RECORD LOW TEMPERATURES

	High
Hartford, CT	−26°F
Key West, FL	41°F
Atlanta, GA	− 8°F
Honolulu, HI	53°F
Boston, MA	− 12°F
Helena, MT	− 42°F
Madison, WI	− 37°F

READING MATH

When you read a number like −6, say "negative six" or "the opposite of six." Reading this number as "minus six" could cause confusion with the operation of subtraction. For a similar reason, the number + 6 is best read as "positive six," instead of "plus six."

6-2 Adding Integers

EXPLORE/ WORKING TOGETHER

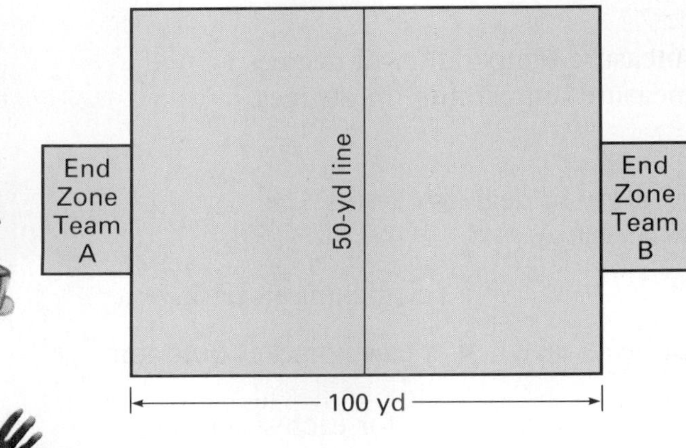

YARDAGE—TEAM A		
Play	Gain (Yd)	Loss (Yd)
1	5	
2	5	
3	7	
4		− 6
5	10	
6	12	
7		− 3
8	8	

In a football game, Team A gained control of the ball on the 50-yd line. Team A's gains and losses of yardage in the next eight plays are shown in the table. At the end of the eight plays, how many yards did Team A still have to cover to reach Team B's end zone and score a touchdown?

Show on a diagram of a football field how Team A's position changed with each of the eight plays.

Discuss with your group how you might use a number line instead of a diagram of a football field to show how Team A's position changed with the eight plays. Hint: Let the 0 on the number line represent the field's 50-yd line.

SKILLS DEVELOPMENT

You can use a number line to show the addition of two integers. Start at 0. Move to the right to add a positive integer. Move to the left to add a negative integer.

Example 1

Use a number line to help you add.

a. $4 + 3$ **b.** $-2 + (-4)$ **c.** $5 + (-4)$ **d.** $-4 + 2$

Solution

a. +4 +3

−1 0 1 2 3 4 5 6 7 8 9

Begin at 0. Move right 4 places, then 3 places more. The stopping place is $+7$.

$$4 + 3 = 7$$

b. −4 −2

−7 −6 −5 −4 −3 −2 −1 0 1 2

Begin at 0. Move left 2 places, then 4 places more. The stopping place is -6.

$$-2 + (-4) = -6$$

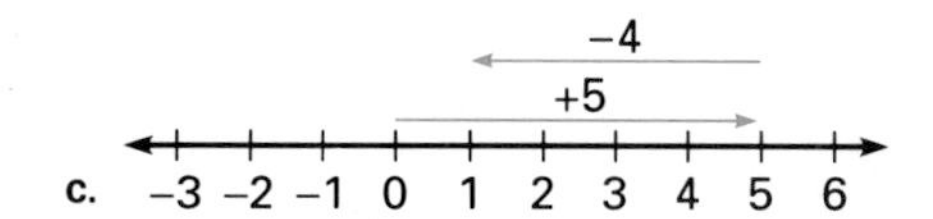

Begin at 0. Move right 5 places. Then move left 4 places. The stopping place is $+1$.

$$5 + (-4) = 1$$

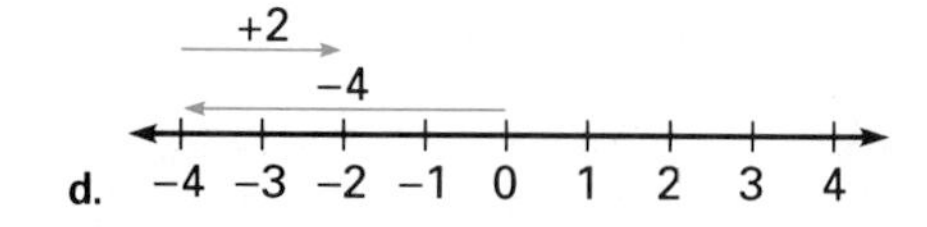

Begin at 0. Move left 4 places. Then move right 2 places. The stopping place is -2.

$$-4 + 2 = -2$$ ◄

You can also add two integers by using some simple rules. The rule that you use depends on whether the integers have the *same* sign or *different* signs. Study these rules to learn how to add integers correctly.

► To add integers with the *same* signs, add the absolute values of the integers. Give the sum of the two integers the sign of the addends.

► To add integers with *different* signs, subtract the absolute values. Give the sum the sign of the addend with the greater absolute value.

Example 2

Add.

a. $-7 + (-3)$ b. $-8 + 5$ c. $9 + (-5)$

Solution

a. Find the absolute values: $|-7| = 7$; $|-3| = 3$
Since the signs are the same, add the absolute values: $7 + 3 = 10$
The addends are both negative. So, the sum is negative:
$-7 + (-3) = -10$

b. Find the absolute values: $|-8| = 8$; $|5| = 5$
Since the signs are different, subtract the absolute values: $8 - 5 = 3$
The negative addend has the greater absolute value. So, the sum is negative: $-8 + 5 = -3$

c. Find the absolute values: $|9| = 9$; $|-5| = 5$
Since the signs are different, subtract the absolute values: $9 - 5 = 4$
The positive addend has the greater absolute value. So, the sum is positive: $9 + (-5) = 4$ ◄

When opposites are added, the sum is always 0. This is called the **addition property of opposites.**

$$a + (-a) = 0 \qquad -a + a = 0$$

CALCULATOR TIP

You can enter a number as a negative integer by pressing the change-sign key ([+/−]). For example, use this key sequence to add $-2 + (-5) + (-7)$.

2 [+/−] [+] 5 [+/−] [+] 7 [+/−] [=] [−14]

The negative symbol appears in the display when you press the change-sign key. (Positive symbols never appear in the display.)

Use the change-sign key in the same way when you subtract, multiply, or divide negative integers.

MENTAL MATH TIP

When you add a string of integers, looking for sums of zero can make your work easier. For example:

$$-10 + \underbrace{2 + 4 + (-2)}_{0}$$
$$= -10 + 4$$
$$= -6$$

Example 3

Add.

a. $7 + (-7)$ b. $-3 + 3$

Solution

Find the absolute values. Then subtract the absolute values.

a. $|7| = 7$
$|-7| = 7$
$7 - 7 = 0$

b. $|-3| = 3$
$|3| = 3$
$3 - 3 = 0$

The integer 0 is neither positive nor negative.

$7 + (-7) = 0$ $-3 + 3 = 0$ ◄

Sometimes you have to add more than two integers.

CHECK UNDERSTANDING

In Example 4c, suppose you added the integers in the order in which they appear. How would this affect the sum of the integers?

What property for addition does this illustrate?

Example 4

Add.

a. $4 + 9 + 16$ b. $-3 + (-6) + (-14)$ c. $12 + (-4) + 8 + (-2)$

Solution

a. $4 + 9 + 16 = 20 + 9 = 29$

b. $-3 + (-6) + (-14) = -3 + (-20) = -23$

The signs are the same. Add the absolute values. Use the sign of the addends.

c. $12 + (-4) + 8 + (-2)$
$= 20 + (-6) = 14$ ◄

The signs are different. Find the sums of integers with the same signs. Then add the sums according to the rule for adding integers with different signs.

TRY THESE

Use a number line to help you add.

1. $6 + 3$
2. $4 + (-3)$
3. $-7 + (-2)$
4. $4 + 2$

Add. Use the rules for adding integers.

5. $-6 + 3$
6. $9 + (-7)$
7. $-8 + (-5)$
8. $-4 + (-4)$
9. $4 + (-4)$
10. $-6 + 6$
11. $-5 + (-5)$
12. $-9 + 9$ **0**
13. $4 + (-12) + (-8)$
14. $-9 + 6 + 12$
15. $14 + (-9) + 3 + 4$
16. $-20 + (-4) + (-7) + 2$

Solve.

17. A football team lost 5 yd, gained 10 yd, gained 22 yd, lost 16 yd, and then lost 15 yd. What was the net loss or net gain in yards?
18. A football team gained 4 yd, lost 10 yd, lost 2 yd, gained 14 yd, and then gained 2 yd. What is the net loss or net gain?

EXERCISES

PRACTICE/ SOLVE PROBLEMS

Add. Use a number line or the rules for adding integers.

1. $5 + 5$	2. $-5 + (-4)$	3. $-6 + (-2)$	4. $6 + (-2)$
5. $-7 + (-3)$	6. $-7 + 3$	7. $3 + (-3)$	8. $-9 + (-9)$

9. $14 + 8 + 6 + (-20)$
10. $-7 + 12 + (-7) + 7$
11. $3 + (-9) + (-6) + 12$
12. $-4 + (-8) + 9 + (-3)$

Solve.

13. The price of a share of stock at Friday's close was \$45. The price rose \$5 on Monday, fell \$2 on Tuesday, rose \$4 on Wednesday, rose \$3 on Thursday, and fell \$6 on Friday. What was the price of a share of stock then?

EXTEND/ SOLVE PROBLEMS

Replace ● with <, >, or =.

14. $-8 + 3$ ● $-3 + (-2)$
15. $4 + (-8)$ ● $-6 + 1$
16. $7 + 2$ ● $-1 + 10$
17. $-6 + 3$ ● $-5 + (-1)$
18. $-7 + 8$ ● $10 + (-8)$
19. $8 + (-6)$ ● $-4 + 10$

On Monday, Jessica had \$264 in her checking account. On Tuesday, she wrote checks for \$69 and \$92. On Thursday, she deposited \$21.

20. Use paper and pencil to show how to find the balance in the account.
21. Use a calculator to find the balance in the account.

THINK CRITICALLY/ SOLVE PROBLEMS

Find the pair of integers.

22. Two negative integers are 6 units apart. Their sum is -8.
23. The sum of a positive integer and a negative integer is 2. They are 14 units apart.
24. A positive integer and a negative integer are 16 units apart. The absolute value of the positive integer is 3 times the absolute value of the negative integer.
25. A positive integer and a negative integer are 48 units apart. The absolute value of the negative integer is half the absolute value of the positive integer.

PROBLEM SOLVING TIP

You may want to use a number line to help you solve Exercises 22–25.

6-3 Subtracting Integers

EXPLORE/ WORKING TOGETHER

You can use integer chips to represent integers. A [+] chip represents +1. A [−] chip represents −1.

[+] [+] [+] = 3 and [−] [−] [−] = −3

a. Use integer chips to model these integers.

6   4   −5   −7

You can use either [+] or [−] integer chips to model addition of integers with the same signs. For example, use [−] chips to model −3 + (−5).

−3   +   −5

[−] [−] [−]   [−] [−] [−] [−] [−]

[−] [−] [−] [−] [−] [−] [−] [−]

−3 + (−5) = −8

b. Use integer chips to model these sums.

−4 + (−3)   6 + 9   −7 + (−5)

You can use both [+] and [−] integer chips to model addition of integers with different signs.

c. The combination of [+] and [−] always equals zero. Explain why.

[+]--[−] = 0

Make combinations of zero to model addition of integers with different signs. For example, make pairs of [+] and [−] chips to add 3 + (−2).

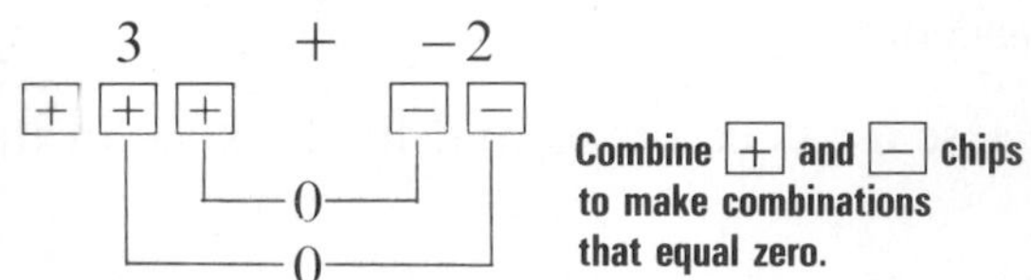

Combine [+] and [−] chips to make combinations that equal zero.

After making all possible combinations, one [+] chip remains.
3 + (−2) = 1

d. Use integer chips to model these sums. Remember that a pair of [+] and [−] chips equals zero.

−7 + 5   6 + (−2)   5 + (−5)

You can use integer chips to model subtraction of integers. For example, use [−] chips to subtract −5 − (−3). Show 5 [−] chips.

Subtract 3 of the chips. [−] [−] [⁄] [⁄] [⁄]

Two [−] chips remain.   −5 − (−3) = −2

Sometimes you must add combinations of zero to get enough [+] or [−] chips to subtract. For example, subtract $3 - 5$.

Show 3.

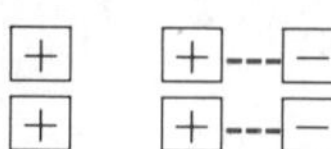

Add two combinations of zero.

Subtract 5 [+] chips.

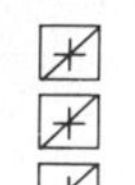

Two [−] chips remain. $3 - 5 = -2$

e. Use integer chips to model these differences.

$-6 - (-4)$ $\quad$ $9 - 5$ $\quad$ $4 - 5$

Using integer chips is one way to subtract integers.

SKILLS DEVELOPMENT

Example 1

Subtract.

a. $-4 - (-3)$ $\quad$ b. $6 - (-2)$

Solution

a. $-4 - (-3)$

Show -4. Then subtract 3 [−] chips. One [−] chip remains.

[−] [−̸] [−̸] [−̸]

$-4 - (-3) = -1$

b. $6 - (-2)$

Show 6. Then add two pairs of [+] and [−] chips.
Subtract 2 [−] chips.
Eight [+] chips remain.

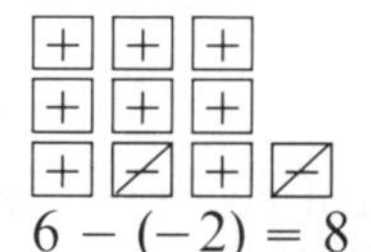

$6 - (-2) = 8$ ◄

Notice how subtraction and addition of integers are related.

$-4 - (-3) = -1$ $\qquad$ $6 - (-2) = 8$

$-4 + 3 = -1$ $\qquad$ $6 + 2 = 8$

These examples illustrate the following rule for subtracting integers.

► To subtract an integer, add its opposite.

Example 2

Subtract.

a. $-7 - (-6)$ $\quad$ b. $-3 - 8$

Solution

a. $-7 - (-6)$

$-7 + 6$ ← **Add the opposite of -6, which is 6.**

-1

b. $-3 - 8$

$-3 + (-8)$ ← **Add the opposite of 8, which is -8.**

-11 ◄

Sometimes you need to subtract integers in order to solve problems.

Example 3

The highest recorded temperature in Colorado is 118°F. The lowest recorded temperature is −61°F. By how many degrees do these temperatures differ?

Solution

Subtract.

$118 - (-61)$

$118 + 61$ ← **Add the opposite of −61, which is 61.**

179

The highest and lowest recorded temperatures differ by 179°F. ◄

TRY THESE

Use integer chips to model these differences.

1. $4 - 1$
2. $-2 - 2$
3. $-1 - 2$
4. $3 - (-6)$
5. $-2 - 1$
6. $-5 - 4$

Subtract. Use the rule for subtracting integers.

7. $6 - 3$
8. $2 - 6$
9. $-8 - (-4)$
10. $-9 - 3$
11. $5 - (-4)$
12. $7 - 8$

Solve.

13. The highest recorded temperature in Texas is 120°F. The lowest is −23°F. By how many degrees do these temperatures differ?

EXERCISES

PRACTICE/ SOLVE PROBLEMS

Use integer chips to model these differences.

1. $12 - (-6)$
2. $-7 - (-10)$
3. $4 - (-2)$
4. $15 - (-6)$

Subtract. Use the rule for subtracting integers.

5. $8 - (-3)$
6. $-17 - (-11)$
7. $-22 - 3$
8. $-8 - (-10)$
9. $40 - (-24)$
10. $-16 - (-6)$
11. $-10 - 13$
12. $17 - 21$
13. $-12 - (-3)$

14. ***USING DATA*** Use the Data Index on page 546 to locate information about the highest elevation and the lowest point on the earth. Find the distance between these two points.

EXTEND/ SOLVE PROBLEMS

Without subtracting, tell whether the answer will be positive or negative.

15. $-16 - (-4)$
16. $5 - 18$
17. $12 - (-7)$
18. $-38 - (-39)$
19. $-72 - 71$
20. $38 - 14$

Replace ● with <, >, or =.

21. $-3 - (-2)$ ● $-8 - (-5)$
22. $-14 - (-5)$ ● $14 - (-5)$
23. $4 - 18$ ● $12 - 5$
24. $5 - 9$ ● $9 - 13$
25. $|-6| - |-9|$ ● $|5| - |-10|$
26. $|12| - |-9|$ ● $|-10| - |-5|$
27. $|-4| - |5|$ ● $|9| - |-9|$
28. $|-8| - |-9|$ ● $|8| - |-9|$

Solve.

29. The highest recorded temperature in Louisiana is 114°F. The high varies from the lowest recorded temperature by 130°F. What is the lowest recorded temperature in Louisiana?

MIXED REVIEW

Find the range, mean, median and mode for each set of data.

1. 63 67 43 40 51 78 89 89
2. 225 275 190 211 263 259 205 204

Express each as a decimal.

3. $\frac{5}{16}$ 4. $\frac{5}{6}$ 5. $\frac{3}{5}$

6. $\frac{1}{8}$ 7. $\frac{7}{8}$ 8. $\frac{9}{25}$

9. Find the circumference of a circle with a diameter of 10 centimeters.

THINK CRITICALLY/ SOLVE PROBLEMS

Replace each ● with + or −.

30. 8 ● $-9 = -1$
31. -8 ● $-9 = 1$
32. -5 ● $-10 = 5$
33. 6 ● $-8 = -2$
34. -9 ● $4 = -5$
35. -16 ● $-4 = -20$

Find the integer represented by ■.

36. $7 -$ ■ $= 2$
37. $-4 -$ ■ $= 5$
38. ■ $- (-3) = -8$
39. ■ $- 7 = -10$
40. $-9 -$ ■ $= 0$
41. ■ $- 2 = 14$
42. $30 -$ ■ $= -40$
43. $0 -$ ■ $= 3$
44. $-6 -$ ■ $= -13$

The highest temperature recorded in Little Rock, Arkansas, is 112°F. The lowest is −5°F. The highest recorded temperature in Concord, New Hampshire, is 102°F. The lowest is −37°F.

45. Find the differences between the high and low temperatures for Little Rock and for Concord.
46. Which of the two cities had the greater temperature difference?

WRITING ABOUT MATH

Would your answer to Exercise 46 be the same if the temperatures were recorded in degrees Celsius instead of in degrees Fahrenheit? Write a paragraph explaining your answer.

6-4 Multiplying Integers

EXPLORE

You have used integer chips to model addition and subtraction of integers. Now you can use integer chips to model multiplication of integers.

Make three groups of four $\boxed{+}$ chips.

Put the three groups of four $\boxed{+}$ chips together to make one large group of $\boxed{+}$ chips.

Write the number sentence that the modeling represents.

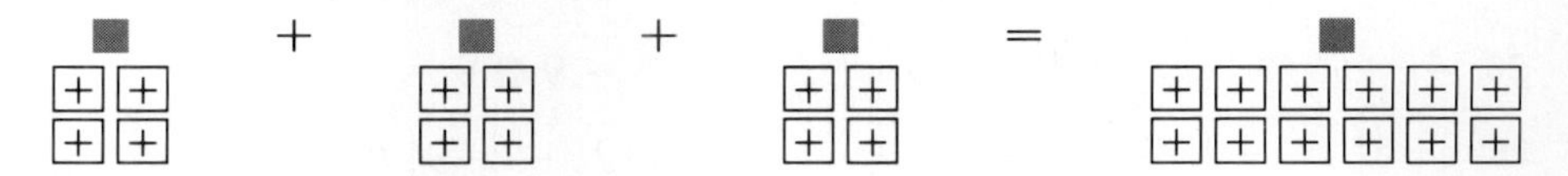

You know that multiplication is repeated addition. So, you can write a multiplication sentence to replace the addition sentence. First, complete this sentence.

■ groups of ■ $\boxed{+}$ chips = 1 group of ■ $\boxed{+}$ chips

Now use integers to complete this multiplication sentence.

■ × ■ = ■

Repeat these steps using $\boxed{-}$ integer chips. Make three groups, each containing four $\boxed{-}$ chips. Then, put the three groups of $\boxed{-}$ chips together to make one large group of $\boxed{-}$ chips. Write the number sentence that the modeling represents.

■ + ■ + ■ = ■

Replace the addition sentence with a multiplication sentence. Use integers to complete the sentence.

■ × ■ = ■

SKILLS DEVELOPMENT

Several techniques can be used to multiply integers. In some cases, you can use integer chips. Sometimes you use the commutative property of multiplication. Sometimes you look for patterns.

Example 1

Multiply.

a. 3×4 **b.** $4 \times (-8)$ **c.** -8×4 **d.** $-4 \times (-3)$

Solution

a. Use integer chips.

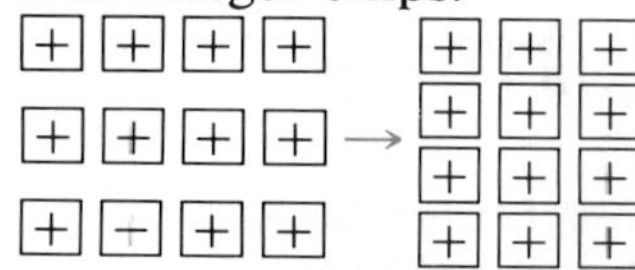

3 groups of 4 $\boxed{+}$ chips = 1 group of 12 $\boxed{+}$ chips

$3 \times 4 = 12$

b. Use integer chips.

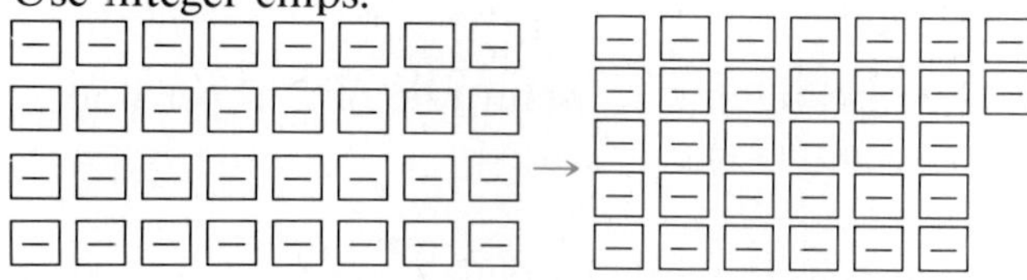

4 groups of 8 $\boxed{-}$ chips = 1 group of 32 $\boxed{-}$ chips

$4 \times (-8) = -32$

c. You know that $4 \times (-8) = -32$. ← **Use the commutative property.**
So, $-8 \times 4 = -32$.

d. Use a pattern. Begin with two factors whose product you know. Then extend the pattern.

$(-4)(3) = -12$ **Look for the pattern. Each product is 4 greater than the preceding product. Extend the pattern.**
$(-4)(2) = -8$
$(-4)(1) = -4$
$(-4)(0) = 0$
$(-4)(-1) = ? \rightarrow (-4)(-1) = 4$
$(-4)(-2) = ? \rightarrow (-4)(-2) = 8$
$(-4)(-3) = ? \rightarrow (-4)(-3) = 12$ ◄

Notice the sign of the product in each part of Example 1. Compare with the signs of the factors. These results illustrate the following rules for multiplying integers.

► The product of two integers with the *same signs* is positive.
► The product of two integers with *different signs* is negative.

Example 2

Multiply.

a. 8×4 b. $-9 \times (-6)$ c. -6×5 d. $4 \times (-2)$

Solution

a. $8 \times 4 = 32$ **The product is positive because the signs of the factors are the same.**

b. $-9 \times (-6) = 54$ **The product is positive because the signs of the factors are the same.**

c. $-6 \times 5 = -30$ **The product is negative because the signs of the factors are different.**

d. $4 \times (-2) = -8$ **The product is negative because the signs of the factors are different.** ◄

What is the sign of the product of any three positive integers? of any three negative integers? Explain.

Example 3

Multiply: $-2 \times 3 \times (-5)$

Solution

$$\begin{aligned} -2 \times 3 \times (-5) &= -2 \times (-5) \times 3 \\ &= 10 \times 3 \\ &= 30 \end{aligned}$$ ◄

You may multiply factors in any order to make it easier to multiply mentally.

Sometimes you need to multiply integers in order to solve problems.

Example 4

The temperature has been falling at a constant rate of 2°F per hour. Describe the temperature 4 hours ago.

Solution

The rate at which the temperature has been falling can be represented by a negative number, -2. The four hours that have passed can be represented by a negative number, -4.

$$-2 \times (-4) = 8$$ **Multiply.**

Four hours ago, the temperature was 8°F higher. ◄

TRY THESE

Multiply. Decide whether to use integer chips, the commutative property, or a pattern.

1. $-4 \times (-5)$
2. $6 \times (-2)$
3. -2×6
4. $-7 \times (-3)$

Multiply. Use the rules for multiplying integers.

5. -7×8
6. $-6 \times (-6)$
7. -4×9
8. $2 \times (-9)$
9. $-4 \times (-3) \times (-2)$
10. $-4 \times 3 \times (-2)$
11. $4 \times 3 \times (-2)$
12. A football team lost 7 yd on each of the last 3 plays. What was the total yardage on the plays?

EXERCISES

PRACTICE/ SOLVE PROBLEMS

Multiply.

1. $4 \times (-8)$
2. $3 \times (-7)$
3. -8×8
4. $-9 \times (-8)$
5. -5×8
6. $-4 \times 5 \times (-2)$
7. $4 \times 5 \times (-2)$
8. $-4 \times (-5) \times (-2)$
9. The temperature has been falling at a constant rate of 5°F per hour. Describe the temperature 3 hours ago.

EXTEND/ SOLVE PROBLEMS

Write the multiplication sentence that describes each word phrase. Then find the product.

10. five times negative eight
11. negative two multiplied by negative six
12. negative seven times four
13. the product when three is multiplied by the sum of negative one and two

Replace each ● with <, >, or = .

14. $-4 \times (-8)$ ● 4×8
15. $2 \times (-5)$ ● -3×3
16. $-4 \times (-6)$ ● -4×6
17. $7 \times (-8)$ ● -9×6
18. -3×1 ● $3 \times (-1)$
19. $3 \times (-9)$ ● -4×7

Multiply.

20. $5 \times (-2) \times (-4) \times (-1)$
21. $5 \times (-7) \times 2 \times (-3)$
22. $10 \times 5 \times (-3) \times 0$
23. $6 \times (-3) \times (-1) \times (-1)$
24. By Friday, the price of a share of stock had fallen $1 for each day of the past work week. How did the price compare to the previous Friday's price?
25. The temperature on Mars drops 15°C each hour at night. Use integers to show the temperature drop for 5 h.

PROBLEM SOLVING TIP

For Exercise 24, a work week consists of five days, Monday through Friday.

THINK CRITICALLY/ SOLVE PROBLEMS

Find the integer represented by ■.

26. $-6 \times$ ■ $= -30$
27. $-15 \times$ ■ $= 75$
28. ■ $\times (-13) = -52$
29. $9 \times$ ■ $= -81$
30. $-24 \times$ ■ $= 144$
31. $5 \times$ ■ $= -60$

Give the next four integers to continue the pattern. Then state the rule.

32. 1, −2, 4, −8, 16, −32, . . .
33. −3, −6, −12, −24, −48, . . .
34. 0, 1, −1, 2, −2, 3, −3, . . .
35. Find two integers with a sum of −7 and a product of 12.
36. Create a rule for predicting the sign of an even power and an odd power of a negative integer.

COMPUTER TIP

A RUN of this program will give you data from which to create a rule for Exercise 36.

```
10 FOR X = 1 TO 8
20 PRINT "(−2) ^";X, (−2) ^
   X
30 NEXT X
```

6-5 Dividing Integers

EXPLORE

You can use what you know about multiplying integers to explore dividing integers.

Recall that multiplication undoes division and division undoes multiplication.

$9 \times 5 = 45$, and $45 \div 5 = 9$
$5 \times 9 = 45$, and $45 \div 9 = 5$

Complete.

a. $-9 \times 5 = -45$, and $-45 \div 5 = ■$

b. $9 \times (-5) = -45$, and $-45 \div (-5) = ■$

c. $-9 \times (-5) = 45$, and $45 \div (-5) = ■$

SKILLS DEVELOPMENT

Division and multiplication are related operations. The rules for finding the sign of the quotient of two integers are the same as the rules for finding the sign of the product of two integers.

► The quotient of two integers with the *same* signs is positive.

► The quotient of two integers with *different* signs is negative.

Because division and multiplication are inverse operations, you can multiply to check your answer to division.

Example 1

Find each quotient. Then check by multiplying.

a. $56 \div 7$ b. $-42 \div (-7)$ c. $48 \div (-6)$ d. $-72 \div 8$

Solution

a. $56 \div 7 = 8$
Check: $8 \times 7 = 56$
The quotient is positive because the signs of the factors are the same.

b. $-42 \div (-7) = 6$
Check: $6 \times (-7) = -42$
The quotient is positive because the signs of the factors are the same.

c. $48 \div (-6) = -8$
Check: $-8 \times (-6) = 48$
The quotient is negative because the signs of the factors are different.

d. $-72 \div 8 = -9$
Check: $-9 \times 8 = -72$
The quotient is negative because the signs of the factors are different. ◄

Example 2

The price of a stock decreased \$20 over four days. If the rate of decrease was spread equally over the four days, how did the price of the stock change in one day?

Solution
The total amount of decrease in the price can be represented by -20.
Divide to find the amount of decrease in one day.
$-20 \div 4 = -5$

The price decreased \$5 in one day. ◄

TRY THESE

Divide. Then check by multiplying.

1. $8 \div (-4)$
2. $-15 \div 3$
3. $24 \div 8$
4. $-21 \div (-7)$
5. $-56 \div 7$
6. $64 \div (-8)$
7. $-48 \div (-6)$
8. $49 \div (-7)$
9. $42 \div 6$
10. $-35 \div (-5)$
11. $28 \div (-4)$
12. $-27 \div 9$

Solve.

13. The price of a share of stock fell \$10 over five days. If the rate of decrease was spread equally over the five days, how much did the price of the stock decrease in one day?
14. A borer was being used to drill a well. It drilled to a depth of 200 ft in four minutes. If the rate of drilling was constant, how far had it drilled in one minute?
15. The temperature fell 12°F over a four-hour period. If the decrease was spread equally over the four hours, how much did the temperature change in each hour?

EXERCISES

PRACTICE/ SOLVE PROBLEMS

Divide. Then check by multiplying.

1. $-36 \div 4$
2. $-35 \div (-5)$
3. $12 \div 3$
4. $40 \div (-8)$
5. $35 \div (-7)$
6. $-54 \div 9$
7. $-56 \div (-8)$
8. $-36 \div 6$
9. $14 \div 2$
10. $18 \div (-9)$
11. $-4 \div (-2)$
12. $-16 \div 4$
13. $-25 \div 5$
14. $72 \div (-6)$
15. $-60 \div (-5)$
16. $-36 \div 12$

17. The temperature decreased 16°F over 8 h. If the decrease was spread equally over the 8 h, how did the temperature change in each hour?

18. A stock portfolio decreased $99 in value. If the decrease was spread equally over 3 days, how did the value of the portfolio change each day?

EXTEND/ SOLVE PROBLEMS

MATH IN THE WORKPLACE

Ruth Gonzalez is a geophysical mathematician. Her work involves developing the mathematics for computer programs used in the exploration and production of oil and gas reservoirs. These programs perform calculations and produce diagrams that help provide information about the earth's subsurface.

Ms. Gonzalez has a bachelor's and master's degree from the University of Texas and a doctorate in mathematical sciences from Rice University.

Copy and complete the table for variables a and b.

	a	b	$a \div b$
19.	■	−4	2
20.	−15	■	−5
21.	■	7	−3
22.	−35	■	−5

	a	b	$a \div b$
23.	■	−8	−7
24.	−72	■	4
25.	■	−16	−4
26.	−100	■	25

USING DATA Use the Data Index on page 546 to find information about record holders from the earth sciences.

27. About how many times deeper is the Marianas Trench than the Grand Canyon?

28. About how many times older are the oldest slime molds than the oldest birds?

Replace each ● with $<$, $>$, or $=$.

29. $-18 \div (-3)$ ● $18 \div 3$

30. $10 \div (-5)$ ● $-15 \div (-3)$

31. $-24 \div (-6)$ ● $-25 \div 5$

32. $54 \div (-6)$ ● $-54 \div 6$

33. $-3 \div 1$ ● $-3 \div (-1)$

34. $49 \div (-7)$ ● $-48 \div 6$

Find the mean of each set of integers.

35. −12, −8, 5

36. −18, 16, −14, −13, 9

37. −9, −7, 15, 16, 12, −15, 14, −2

38. 11, 7, −8, 5, −7, −16, −3, −2, 4, 9

Solve.

39. The temperature decreased 20°F over 4 hours. If the decrease was spread equally over the 4 hours, how did the temperature change in 30 minutes?

40. A stock portfolio had a decrease in price of \$120. If the decrease was spread equally over two work weeks, how much did the price of the portfolio change in one day?

41. The price of a share of stock increased \$8.50. If the increase was spread equally over five days, how much did the price increase in one day?

THINK CRITICALLY/ SOLVE PROBLEMS

Find the integer represented by ■.

42. $18 \div ■ = -3$

43. $-20 \div ■ = -10$

44. $■ \div (-15) = 3$

45. $■ \div (-10) = -5$

46. $-39 \div ■ = -13$

47. $■ \div 11 = -2$

Replace each ● with $+$, $-$, $\times$, or $\div$.

48. $-7 ● -5 ● 10 = -20$

49. $4 ● -3 ● -2 = -14$

50. $-1 ● -1 ● -1 = -3$

51. $5 ● -5 ● 5 = -125$

52. $6 ● -6 ● -6 = -7$

53. $-4 ● 6 ● 3 = -8$

Solve.

54. The average daily temperature decreased 31°F in October. If the decrease had been spread equally over the month, how would the temperature have changed in one week?

55. The sum of two integers is always an integer. Is the quotient of two integers always an integer? Explain.

56. The product of two integers is -36. The quotient is -4. What are the integers?

57. The product of two integers is -16. The quotient is -1. What are the integers?

58. The sum of two integers is 20. The quotient is -5. What are the integers?

59. You can multiply to check your answer to a division problem. Using this knowledge, explain why zero can be divided by a nonzero number, but no number can be divided by zero.

6-6 Problem Solving Skills:

CHOOSE THE OPERATION

In order to solve many problems that involve integers, it is necessary to decide how the numbers in a problem are related. This relationship between the numbers is the key to deciding what operation to use in solving the problem.

PROBLEMS

a. A number is eight more than negative three. What is the number?

b. A number is eleven less than seven. What is the number?

SOLUTIONS

a. The number is greater than negative three. Therefore, add eight to negative three to find the number.

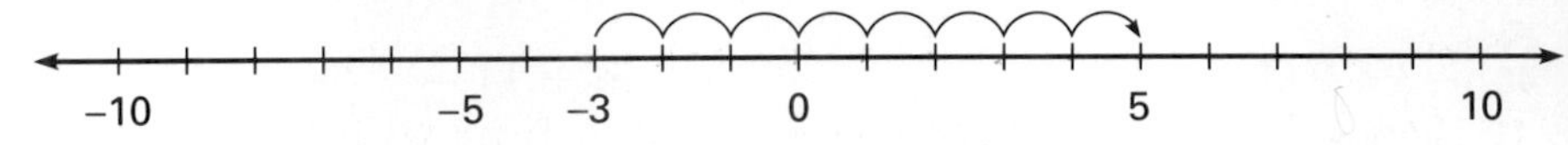

$-3 + 8 = 5$ Check: Five is eight more than negative three.

The number is five.

b. The number is less than seven. Therefore, subtract eleven from seven to find the number.

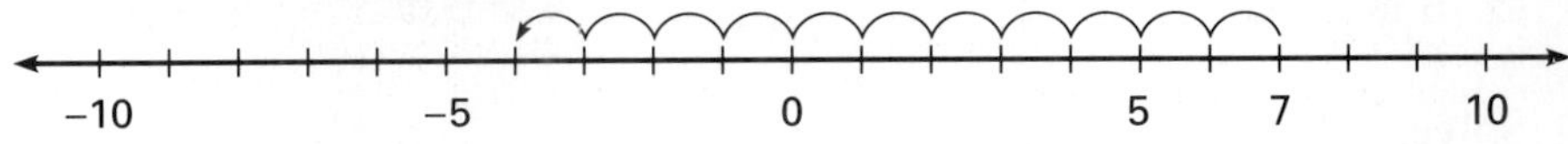

$7 - 11 = -4$ Check: Negative four is eleven less than seven.

The number is negative four.

PROBLEMS

Choose the operation required. Then solve the problem.

1. A number is six less than negative four. What is the number?
2. A number is nine more than negative four. What is the number?
3. Eight is nine more than some number. What is the number?
4. A number is the product of six and negative fourteen. What is the number?

5. A number is eighteen more than the sum of six and negative six. What is the number?

6. Negative seventy-seven is seven times some number. What is the number?

7. A number is the product of negative nine and negative twenty-one. What is the number?

8. A number is four more than negative nine. What is the number?

9. Negative one hundred sixty-eight is fourteen times some number. What is the number?

10. A number is the product of five and negative twelve. What is the number?

11. A number is sixteen more than negative eight. What is the number?

12. A number divided by negative thirteen equals negative twelve. What is the number?

Solve.

13. David has $24 in his checking account. If he writes checks for $6, $13, and $36, what will be the balance in his account?

14. The Falcons received a penalty of -15 yd after they made a 33 yd gain. What was the net gain or loss for the team?

15. Maria ate a breakfast that supplied her with 300 calories of energy. Then she ran for an hour, which used up 14 calories of energy per minute. Give Maria's net calorie intake after running.

16. Swimming uses up 10 calories of energy per minute. If Ivan wants to expend the same number of calories in swimming that he takes in by eating a 350-calorie snack, how long would he have to swim?

17. Juanita had a balance of $16 in her bank account. In the next few days she made a deposit of $8, wrote a check for $9, and made two more deposits of $10 and $15. How much money is in her account now?

18. Carlos played a board game in which a player can move forward or backward from the starting line. In the first round Carlos moved forward 8 squares, then back 6 squares, then back 4 squares, then back 9 squares, and then forward 6 squares. At the end of the round, what was Carlos' position relative to the starting line?

6-7 Rational Numbers

EXPLORE

You already know how to find the quotient of two integers.

$$\frac{10}{2} = 5 \qquad \frac{15}{-3} = -5 \qquad \frac{-18}{3} = -6 \qquad \frac{-24}{-6} = 4$$

Is the quotient of any two integers always an integer?

SKILLS DEVELOPMENT

When you add, subtract, or multiply integers, the result is always another integer. Study these examples.

$4 + (-3) = 1$ $\qquad -15 + 14 = -1$

$-10 + (-4) = -14$ $\qquad 6 - (-5) = 11$

$8 - 9 = -1$ $\qquad -12 - (-14) = 2$

$(6)(5) = 30$ $\qquad (-4)(2) = -8$

$(3)(-2) = -6$ $\qquad (-8)(-2) = 16$

However, the quotient of two integers is not always another integer.

$0 \div 5 = \frac{0}{5} = 0 \qquad 12 \div 3 = \frac{12}{3} = 4 \qquad 3 \div 4 = \frac{3}{4} \qquad 11 \div 8 = \frac{11}{8}$

Quotient is an integer. (first two) **Quotient is a fraction.** (last two)

When an operation on a set of numbers always results in a number that is in the same set, the set of numbers is said to be **closed** under that operation. You have seen that the set of integers is closed under addition, subtraction, and multiplication, but not under division. To obtain a set of numbers that is closed under all four operations, you need to consider the set of *rational numbers.*

A **rational number** is any number that can be expressed in the form $\frac{a}{b}$, where a is any integer and b is any integer except 0. That is, a rational number can be expressed as a ratio of two integers. So all simple fractions are rational numbers, as are all terminating and repeating decimals. Any integer can be expressed as a fraction, such as $4 = \frac{4}{1} = \frac{8}{2}$, so all the integers are rational numbers. The quotient of any two rational numbers is always another rational number.

Example 1

Write each rational number as a ratio of two integers.

a. 4 b. $0.\overline{4}$ c. -0.87 d. $2\frac{2}{5}$ e. -2.3

Solution

a. $4 = \frac{8}{2}$ b. $0.\overline{4} = \frac{4}{9}$ c. $-0.87 = -\frac{87}{100}$

d. $2\frac{2}{5} = \frac{12}{5}$ e. $-2.3 = -\frac{23}{10}$ ◀

Example 2

Name the point that corresponds to each rational number.

a. 0.8 b. $-\frac{3}{5}$ c. -1.2 d. 2.5

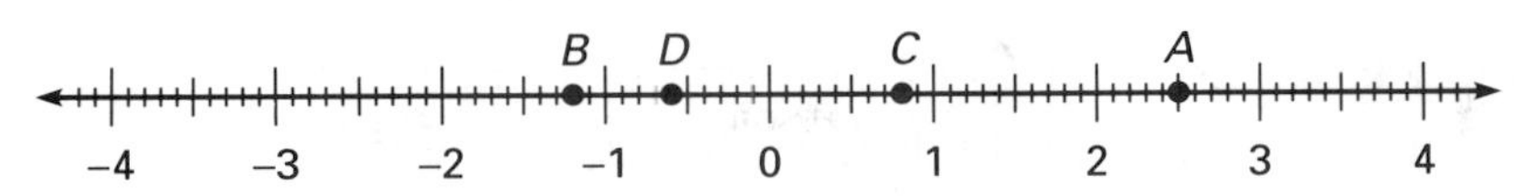

Solution

a. Point C corresponds to 0.8.

b. Point D corresponds to $-\frac{3}{5}$. $\left(-\frac{3}{5} = -\frac{6}{10}\right)$

c. Point B corresponds to -1.2. $\left(-1.2 = -1\frac{2}{10}\right)$

d. Point A corresponds to 2.5. $\left(2.5 = 2\frac{5}{10}\right)$ ◄

Any rational number can be graphed as a point on a number line.

Example 3

Graph each rational number on a number line.

a. $2\frac{1}{6}$ b. $1\frac{2}{3}$ c. $-\frac{5}{6}$ d. $-2\frac{1}{3}$

Solution

The denominator of each rational number is either 3 or 6. Mark the number line in sixths and represent each number by a point on the number line. ◄

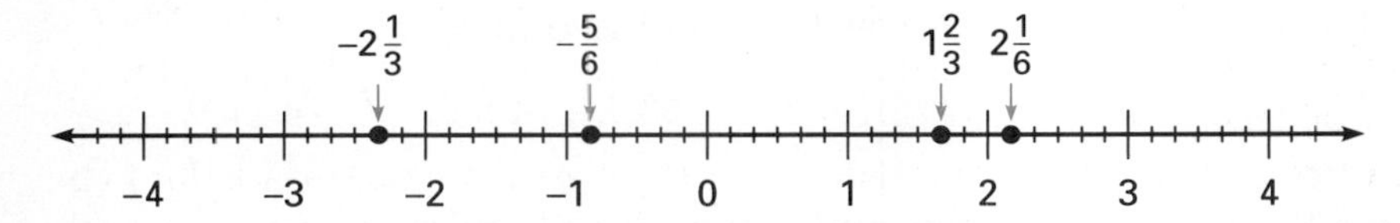

You can use a number line to order rational numbers. If one rational number lies to the right of another on the number line, then it is greater than the other rational number. If one rational number lies to the left of another on the number line, then it is less than the other rational number.

Example 4

Order these rational numbers from least to greatest: $3, -4.1, \frac{1}{2}, 4.5, -2\frac{1}{2}$

Solution

Graph the points on a number line.

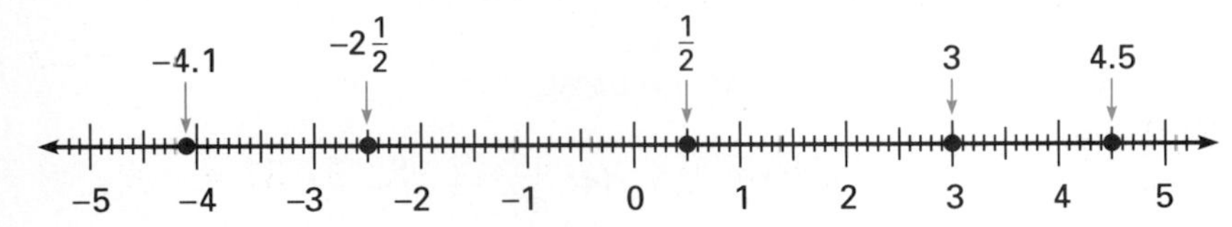

From least to greatest the order is $-4.1, -2\frac{1}{2}, \frac{1}{2}, 3, 4.5$. ◄

TRY THESE

Write each rational number as a ratio of two integers.

1. 6
2. 0.6
3. -0.75 –
4. -1.4

Name the point that corresponds to each rational number.

5. $-\frac{7}{10}$
6. 2.4
7. -2.3
8. 0.7

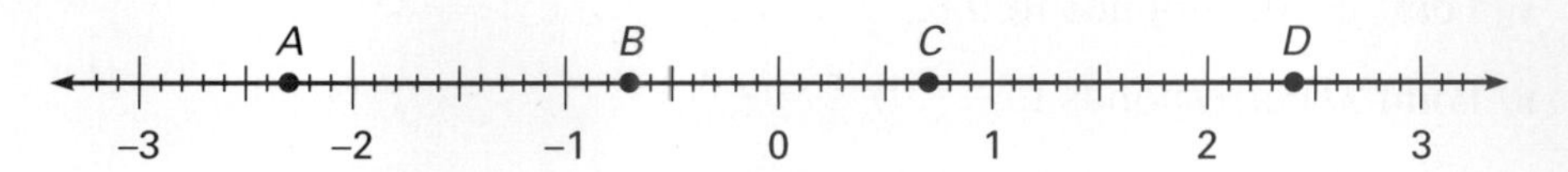

Graph each rational number on a number line.

9. $-1\frac{1}{2}$
10. -0.9
11. $2\frac{1}{5}$
12. Order these rational numbers from least to greatest: 2.7, -2.1, -0.3, 0.3

EXERCISES

PRACTICE/ SOLVE PROBLEMS

Write each rational number as a ratio of two integers.

1. 7
2. 0.7
3. $3\frac{1}{3}$
4. 1.9

Name the point that corresponds to each rational number.

5. $-\frac{3}{10}$
6. 2.5
7. 0.5
8. -2.3

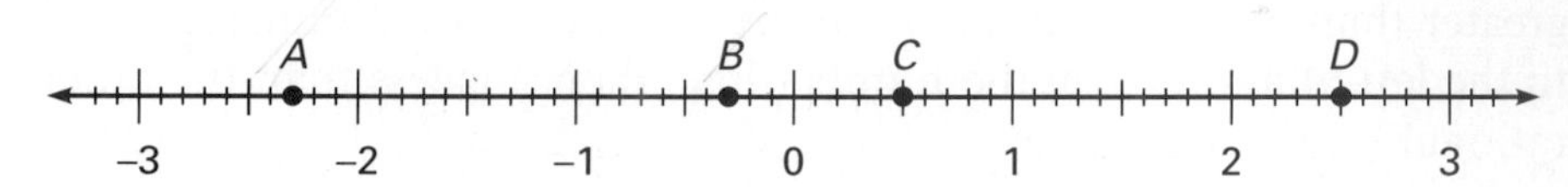

Graph the rational numbers on a number line.

9. $1\frac{1}{3}, -2\frac{2}{3}, -\frac{2}{3}, 3\frac{2}{3}$
10. $-\frac{1}{4}, 2\frac{1}{2}, -1\frac{3}{4}, -2\frac{1}{2}$
11. $\frac{2}{5}, -\frac{1}{5}, 1\frac{3}{5}, -1\frac{4}{5}$
12. $\frac{7}{2}, -\frac{5}{2}, 1\frac{1}{2}, -1\frac{1}{2}$

Order the numbers from least to greatest.

13. $\frac{7}{10}, \frac{7}{20}, \frac{3}{5}$
14. $\frac{15}{40}, \frac{7}{32}, \frac{9}{20}, \frac{5}{16}$
15. $\frac{3}{4}, \frac{2}{3}, \frac{7}{12}$
16. $-0.8, -\frac{13}{16}, -\frac{7}{8}, -\frac{13}{20}$

Solve.

17. In the all-county school chorus, 0.28 of the 200 singers are altos. Express the number of altos as a ratio of two integers.

18. Which of the rational numbers is equivalent to $\frac{7}{4}$?
$\frac{21}{12}$, $\frac{14}{11}$, 1.75, -1.70

19. Which of the rational numbers is equivalent to $-\frac{3}{5}$?
0.8, -0.6, $\frac{3}{10}$, $-\frac{18}{30}$

Replace ● with <, >, or =.

20. $-\frac{4}{5}$ ● $\frac{5}{6}$
21. $-\frac{5}{8}$ ● $-\frac{3}{4}$
22. $\frac{8}{15}$ ● $\frac{2}{3}$
23. $\frac{27}{40}$ ● $-\frac{5}{8}$
24. $\frac{1}{3}$ ● 0.3
25. $\frac{2}{3}$ ● $0.\overline{6}$
26. $-\frac{7}{18}$ ● $-\frac{1}{3}$
27. 0.55 ● $-\frac{11}{20}$

Write a rational number for each description.

28. decrease in price of \$4.50
29. gain of $3\frac{1}{5}$ inches
30. loss of $4\frac{1}{2}$ pounds
31. loss of $32\frac{1}{8}$ ounces

Write a rational number for each description. Use a number line to help you.

32. number halfway between 2 and 3
33. number halfway between 0 and $-1\frac{1}{2}$
34. number four and one-half units less than 1
35. number two and one-quarter units greater than -3

EXTEND/ SOLVE PROBLEMS

MIXED REVIEW

Write in scientific notation.

1. 9,432
2. 0.000035
3. 0.00073
4. 25,038

Complete.

5. 72 c = ■ qt
6. 19 ft = ■ in.
7. 9 mL = ■ L
8. 200 m = ■ cm

Solve.

9. A garden has 3 rows of corn plants with the same number of plants in each row. Each row is 30 feet long. The plants are placed 3 feet apart starting at the corner of the garden. How many corn plants are in the garden?

THINK CRITICALLY/ SOLVE PROBLEMS

Write *true* or *false* for each statement. If you think the statement is false, give a counterexample using rational numbers that proves the statement false.

Suppose that p and q are both positive rational numbers.

36. $p + q$ is positive
37. $q - p$ is positive

Suppose that p and q are both negative rational numbers.

38. $p - q$ is negative
39. $p(p + q)$ is positive

Suppose that p is positive and q is negative.

40. pq is less than p
41. $p + q$ is greater than p

6-8

- READ
- PLAN
- SOLVE
- ANSWER
- CHECK

Problem Solving/ Decision Making:

CHOOSE A STRATEGY

PROBLEM SOLVING TIP

Here is a checklist of the problem solving strategies that you have studied so far in this book.

Find a pattern
Logical reasoning
Solve a simpler problem
Draw a picture

In this book you have been studying a variety of problem solving strategies. Experience in applying these strategies will help you decide which will be most appropriate for solving a particular problem. Sometimes only one strategy will work. In other cases, any one of several strategies will offer a solution. There may be times when you will want to use two different approaches to a problem in order to be sure that the solution you found is correct. For certain problems, you will need to use more than one strategy in order to find the solution.

PROBLEMS

Solve. Name the strategy you used to solve the problems.

1. The members of the geology club went on a one-week trip. They traveled 16 mi the first day, then traveled $1\frac{1}{2}$ times farther each succeeding day. How far would they travel by the end of the third day? the fifth day? the seventh day?

COMPUTER TIP

For Problem 3, you may want to RUN this program several times and make a chart of the results to help discover the pattern.

```
10. INPUT "HOW MUCH
    MONEY? ";M: PRINT
20  FOR X = 3 TO M STEP 2
30  W = (M − 1) / 2
40  NEXT X
50  PRINT "$";M;"
    DEPOSITED IN WEEK
    ";W;"."
```

2. Suppose five athletes enter a swimming competition. They will compete in pairs until each athlete has competed against every other athlete one time. How many competitions will there be?

3. Stefanie plans to deposit \$3 into a savings account the first week, \$5 the second week, \$7 the third week, and so on until she reaches her goal of depositing \$25 in a week. In which week will she make a \$25 deposit?

4. Peter, José, Dwayne, Yolanda, Dolores, and Alma are three brother-and-sister pairs who play doubles tennis. Brother and sister cannot play in the same pair. Peter and Yolanda play Dwayne and Dolores on one day. Dwayne and Yolanda play José and Alma on another day. Name each brother-and-sister pair.

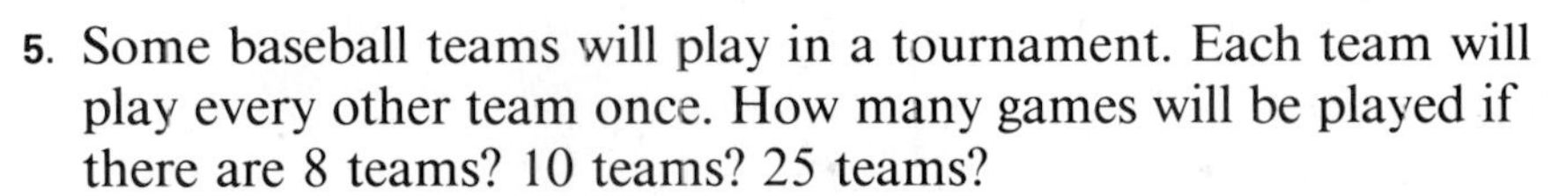

5. Some baseball teams will play in a tournament. Each team will play every other team once. How many games will be played if there are 8 teams? 10 teams? 25 teams?

6. Some players enter an elimination tournament. Once beaten by another player, they cannot play another round. How many matches will be played if there are 4 players? 8 players? 16 players?

7. The members of the Explorer's Club traveled 26 mi on the first day of their hike. On each of the following days, they hiked only half as far as they had the day before. How many miles would they cover on the second day? On the third day? On the fourth day?

8. Find the next row of the pattern. $\frac{L}{M}$ $\frac{N}{O}$ $\frac{L}{O}$ $\frac{N}{M}$

$\frac{P}{Q}$ $\frac{R}{S}$ $\frac{P}{S}$ $\frac{R}{Q}$

9. The producers of a TV game show award money to contestants who answer questions correctly. If a question is answered correctly by the first contestant, that person will receive \$300. If the first contestant misses and the second contestant answers correctly, the amount is doubled, and so on. How much money could someone win who answered correctly and was the third contestant? the fifth contestant? the twelfth contestant?

10. Geologists examined rock formations at a road-cut site in Ridgeway. Ridgeway is 8 mi due west of Mountain Pass. Dalton is 6 mi due west of Haber, which is 9 mi south and 4 mi west of Mountain Pass. How many miles and in which direction is Dalton from the rock formations in Ridgeway?

COMPUTER TIP

For Problem 5, you may want to RUN this program several times and make a chart of the results to help discover the pattern.

```
10 INPUT "HOW MANY
   TEAMS? ";T: PRINT
20 FOR X = 1 TO T - 1
30 S = S + X
40 NEXT X
50 PRINT S;" GAMES WILL
   BE PLAYED."
```

CHAPTER 6 ● REVIEW

1. A number greater than zero is a(an) ___?___ number.
2. A number less than zero is a(an) ___?___ number
3. Two numbers are ___?___ when they are the same distance from 0, but in different directions.
4. The distance a number is from zero is called its ___?___.
5. A number that can be expressed as $\frac{a}{b}$, where a is any integer and b is any integer except 0, is a(an) ___?___ number.

a. absolute value
b. rational
c. negative
d. opposites
e. positive

SECTION 6–1 EXPLORING INTEGERS (pages 186–189)

► Opposite integers have the same absolute value.
► If two numbers are graphed on a horizontal number line, the number that is to the right is the greater number.

Find each absolute value.

6. $|-23|$ 7. $|-17|$ 8. $|17|$ 9. $|45|$ 10. $|38|$

Replace each ● with <, >, or =.

11. 23 ● −23 12. −8 ● −7 13. −2 ● 0

14. *USING DATA* Use the table on page 185 to help you solve. Express the distance from the crust to the center of the earth as a negative number.

SECTIONS 6–2 and 6–3 ADDING and SUBTRACTING INTEGERS (pages 190–197)

► To add integers with the same signs, add the absolute values. Give the sum the sign of the addends.
► To add integers with different signs, subtract the absolute values. Give the sum the sign of the addend with the greater absolute value.
► To subtract an integer, add its opposite.

Add.

15. $18 + 15$ 16. $18 + (-15)$ 17. $-18 + 15$ 18. $-18 + (-15)$

Subtract.

19. $-34 - 18$ 20. $36 - 38$ 21. $-23 - (-4)$ 22. $-10 - (-4)$

SECTIONS 6–4 and 6–5 MULTIPLYING and DIVIDING INTEGERS (pages 198–205)

► The product or quotient of two integers with the *same* sign is positive.
► The product or quotient of two integers with *different* signs is negative.

Multiply.

23. 7×9 **24.** -8×4 **25.** $-6 \times (-8)$ **26.** $9 \times (-3)$

27. $0 \times (-14)$ **28.** -6×7 **29.** $-7 \times (-8)$ **30.** $4 \times (-6)$

Divide.

31. $81 \div 9$ **32.** $-49 \div 7$ **33.** $18 \div (-3)$ **34.** $-72 \div (-8)$

35. $-36 \div 4$ **36.** $-16 \div (-8)$ **37.** $56 \div (-7)$ **38.** $-48 \div (-6)$

SECTION 6–7 RATIONAL NUMBERS (pages 208–211)

► A **rational number** is any number that can be expressed in the form $\frac{a}{b}$, where a is any integer and b is any integer except 0.
► A rational number can be graphed as a point on a number line.

Write each rational number as a ratio of two integers.

39. 6 **40.** 0.6 **41.** $5\frac{2}{3}$ **42.** -3.3

Name the point that corresponds to each rational number.

43. -0.6 **44.** $\frac{2}{5}$ **45.** -1.5 **46.** 3.5

SECTIONS 6–6 and 6–8 PROBLEM SOLVING (pages 206–207 and 212–213)

► Sometimes only one strategy will work in solving a problem. In other cases, you can use any one of several strategies.

► The relationship between the numbers in a problem is often the key to the operation needed to solve the problem.

Choose the operation. Then solve the problem.

47. A number is nine more than negative three. What is the number?

Choose the strategy to help you solve.

48. Twelve baseball teams will play in a tournament. Each team will play every other team once. How many games will be played?

CHAPTER 6 ● TEST

Find each absolute value.

1. $|-6|$
2. $|12|$
3. $|-3|$
4. $|-18|$

Replace each ● with <, >, = .

5. 12 ● -12
6. -13 ● -18
7. -14 ● 2
8. -5 ● 0
9. -9 ● -5
10. -12 ● 7

Add.

11. $-8 + 3$
12. $9 + (-11)$
13. $-7 + (-13)$
14. $2 + (-4)$
15. $-12 + (-4)$
16. $20 + (-20)$

Subtract.

17. $5 - (-9)$
18. $-3 - 7$
19. $-8 - (-10)$
20. $-6 - 8$
21. $7 - 13$
22. $6 - (-2)$

Multiply.

23. 5×7
24. -4×8
25. $-6 \times (-9)$
26. $8 \times (-8)$
27. $-2 \times (-4)$
28. -3×9

Divide.

29. $25 \div 5$
30. $-36 \div 9$
31. $-42 \div (-7)$
32. $-54 \div 6$
33. $-48 \div (-8)$
34. $-27 \div 3$

Write each rational number as a ratio of two integers.

35. 5
36. 0.4
37. $-4\frac{3}{4}$
38. -5.1
39. -7
40. $2\frac{9}{10}$
41. -0.24
42. 0.014

Solve.

43. A number is fourteen more than negative three. What is the number?
44. A number is six less than negative fourteen. What is the number?
45. Ten players enter an elimination tournament. Once a player is beaten by another player, he or she cannot play another round. How many matches will be played?

Find the mean and median for each set of data.

1. 4 7 9 9 11
2. 17 29 17 12 50 17 17 25
3. 6 7 9 19 19 35 35 67
4. 150 123 128 149 123 19

5. The school newspaper wants to know what its readers think of its new movie review feature. They question the first ten students to enter the school. What kind of sampling does this situation represent?

Complete.

6. $4(8 + 3) = (4 \times 8) + (\blacksquare \times 3)$
7. $8(\blacksquare - 2) = (8 \times 3) - (8 \times 2)$
8. $\blacksquare(4 + 3) = (6 \times 4) + (6 \times 3)$

---

9. Determine whether the following argument is *valid* or *invalid*.

   If a figure is a rectangle, then it has four sides.
   Figure *ABCD* has four sides.
   Therefore, figure *ABCD* is a rectangle.

10. Pick a number. Add 1. Multiply by 2. Add 8. Divide by 2. Subtract the original number. What number results? Repeat the process with several other numbers. Make a conjecture about the result.

---

11. Write 0.16 as a fraction in lowest terms.
12. Write $\frac{3}{24}$ as a terminating decimal.

Add or subtract. Write your answer in lowest terms.

13. $\frac{5}{8} + \frac{2}{3}$
14. $8 - 2\frac{1}{3}$
15. $\frac{7}{9} + \frac{1}{3}$
16. $\frac{7}{12} - \frac{3}{8}$

Solve.

17. The length of one side of a regular octagon measures 12.7 cm. Find the perimeter.
18. Use the formula to find the perimeter of a rectangular figure with these dimensions: length: 5.2 mi, width: 2.5 mi

---

Find each circumference. Round your answer to the nearest tenth or whole number.

19.

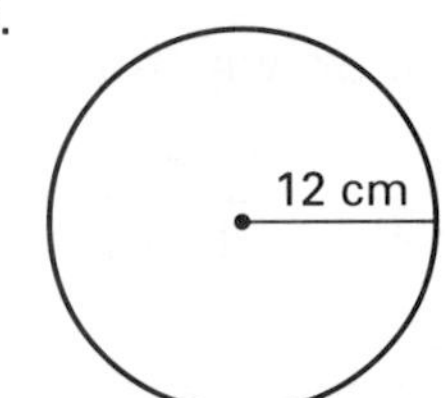

20.

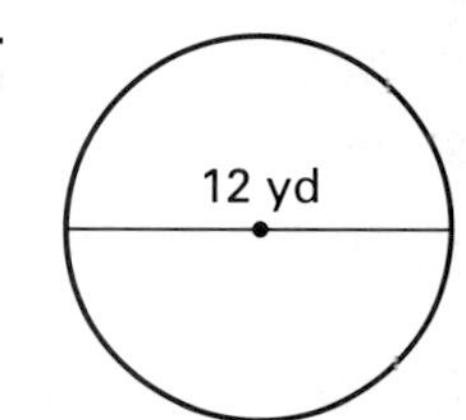

21.

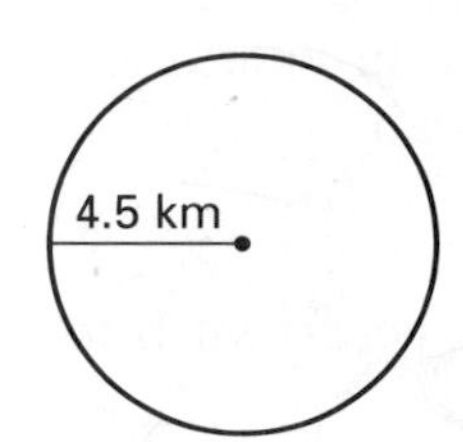

22.

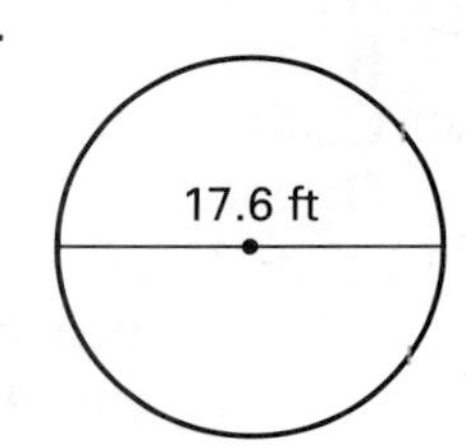

Find each answer.

23. $-64 \div 8$
24. $-7 + (-6)$
25. $-6 - (-8)$
26. $-63 \div 9$
27. negative twenty-four divided by negative eight
28. the product of fourteen and negative seven

CHAPTER 6 ● CUMULATIVE TEST

1. To decide which shirt to buy in a store, Will asked the opinions of the other shoppers in the clothing department. What type of sampling did he use?

   A. systematic B. cluster
   C. convenience D. random

2. What is the mean number of days in a month for a year that is not a leap year?

   A. 30 B. 29 C. 31 D. 30.4

---

3. Complete.
   $6(\blacksquare - 3) = (6 \times 5) - (6 \times 3)$

   A. 12 B. 3 C. 5 D. 6

4. Simplify. $4 \times 8 + 10 \div 2$

   A. 37 B. 21 C. 40 D. 80

---

5. Which conclusion can be drawn from these statements?

   If people live in Puerto Rico, then they live on an island.
   Marie lives in Puerto Rico.

   A. Marie lives on an island.
   B. Marie does not live in Puerto Rico.
   C. Marie lives in Hawaii.
   D. none of these

6. Which conclusion can be drawn from these statements?

   If the animal is a hare, then it hops.
   The animal hops.

   A. The animal is not a hare.
   B. The animal is a hare.
   C. The animal is a kangaroo.
   D. none of these

---

7. Express 0.312 as a fraction in lowest terms.

   A. $\frac{8}{25}$ B. $\frac{39}{125}$ C. $\frac{156}{500}$ D. $\frac{312}{1000}$

8. Which is the decimal for $\frac{7}{8}$

   A. 1.142 B. 7.8
   C. 0.625 D. 0.875

9. Which is the decimal for $\frac{3}{20}$?

   A. 6.6 B. 3.2
   C. 0.15 D. 1.05

10. Express 2.125 as a fraction in lowest terms.

    A. $2\frac{1}{8}$ B. $2\frac{1}{125}$
    C. $\frac{125}{1000}$ D. $2\frac{1}{4}$

---

11. Which is the circumference of a circle with a diameter of 39 yd? Round your answer to the nearest tenth or whole number.

    A. 122.5 yd B. 244.9 yd
    C. 61.2 yd D. 192.2 yd

12. Which is the circumference of a circle with a radius of 10.8 in.? Round your answer to the nearest tenth or whole number.

    A. 53.4 in. B. 21.6 in.
    C. 17 in. D. 67.8 in.

---

13. Add. $-8 + (-13)$

    A. -5 B. -21 C. -27 D. 12

14. Subtract. $5 - (-4)$

    A. 9 B. 1 C. -9 D. -1

15. Divide. $63 \div (-9)$

    A. 54 B. -7 C. 9 D. -54

16. Multiply. $9 \times (^{-}3)$

    A. 6 B. -3 C. -27 D. 12

17. Subtract. $-10 - 6$

    A. -4 B. 4
    C. 16 D. none of these

CHAPTER 7 SKILLS PREVIEW

Tell whether the equation is *true, false,* or an *open* sentence.

1. $9 = 6 - w$
2. $-15 = 5(4 - 7)$

Solve each equation.

3. $-2k = -9$
4. $t - 3 = -8$
5. $6.3 = y + 8.5$
6. $-3 = \frac{n}{-7}$
7. $8c + 11 = -13$
8. $14 = 3z - 4$
9. $\frac{8}{9}n = -\frac{2}{3}$
10. $\frac{5}{4}x + 6 = 1$
11. $3(x + 3) = 15$
12. $8c - 9 = 6c + 5$
13. Find the height of a triangle with a base measuring 16 cm and an area of 360 cm^2. The formula for the area of a triangle is $A = \frac{1}{2}bh$ where A = area, b = length of base, and h = height.
14. Solve the formula $r = t - n$ for t.

Simplify.

15. $\frac{1}{3}y + \frac{1}{3}y + 4y + \frac{1}{3}y$
16. $4(x - 3) + 2(x + 5) - 3x$

Graph each open sentence on a number line.

17. $x \geq -3$
18. $-1 = 3y + 5$
19. $1.6 > 2.1x - 6.8$
20. $\frac{m}{-2} + 3 \leq 2$

Write an equation and solve.

21. Ms. Flynn bought a melon for \$0.79 and three grapefruit. The total cost of the fruit was \$1.84. How much did each grapefruit cost?
22. Lisa bought a package of postcards. She hung 4 on her bulletin board, then gave 3 to her friend José. Finally, she sent half of the remaining cards to pen pals. She had 7 cards left. How many postcards were in the package she bought?
23. Midori earned \$120.90 before taxes. Her hourly wage is \$7.80. How many hours did she work?
24. Lena paid $(a + 3b)$ cents for a pint of cream and $(2a + b)$ cents for 2 quarts of milk. Write and simplify an expression for the total cost of her purchases.

CHAPTER 7

EQUATIONS AND INEQUALITIES

THEME Recreation

Algebra is a branch of mathematics in which symbols are used to represent numbers. Unlike symbols in arithmetic, algebraic symbols can vary in value. Only when numbers are substituted for them do algebraic symbols take on specific values. Their variability gives them a wide range of applications in solving problems, particularly problems involving equations and inequalities.

Because formulas use letters to represent numbers, they are studied in algebra. Formulas are often used to compute statistics in sports. You might use a formula like the one that follows to determine a bowler's handicap.

formula:

$h = (0.8)(200 - a)$

meaning of letters:

h = bowler's handicap

a = average score

BOWLING RESULTS			
	Score		
Name	Game 1	Game 2	Game 3
Christy	85	121	145
Mark	114	102	126
Evan	96	97	95
Rox	156	177	186

DECISION MAKING

Using Data

Use the table to answer the following questions.

1. Calculate each bowler's handicap. Round to the nearest whole number.
2. A bowler's handicap was 40. Describe a method for finding the bowler's average. What was the average?
3. Why is it harder to use the formula to find an average than it is to use it to find a handicap?

Working Together

Your group's task is to find at least five formulas that are used to calculate sports statistics. Use your own knowledge, library reference materials, or interviews with people involved in sports. Write the formulas, clearly identifying the meaning of the letters in each. Then write two problems involving each formula. Exchange your problems with another group.

7-1 Introduction to Equations

EXPLORE

Tug-of-war teams are to be chosen so that the total weight on both sides is equal. Students participating are Mario, who weighs 135 lb; Julie, 120 lb; Gwen, 125 lb; Pao, 143 lb; Heather, 104 lb; Mike, 139 lb; Ed, 112 lb; Jill, 122 lb; William, 142 lb; and Pat, who weighs between 120 lb and 134 lb.

How would you make up the teams so that each side has the same total weight? What is Pat's weight in your arrangement? Check with other students. Did they make the same arrangements?

SKILLS DEVELOPMENT

An **equation** is a statement that two numbers or expressions are equal.

This equation is **true.**	This equation is **false.**	This equation is neither true nor false.
$12 - 7 = 5$	$13 - 7 = 5$	$x - 7 = 5$

An **open sentence** is a sentence that contains one or more variables. It can be true or false, depending upon what values are substituted for the variables. A value of the variable that makes an equation true is called a **solution of the equation.**

Open sentence: $x - 7 = 5$
Solution: $x = 12$

Example 1

Tell whether the equation is *true*, *false*, or an *open sentence*.

a. $3(8 - 2) = 20$ b. $2y - 6 = 10$ c. $-9 + 5 = 2(7 - 9)$

Solution

a. $3(8 - 2) = 20$
$3(6) \stackrel{?}{=} 20$
$18 \stackrel{?}{=} 20$
The equation is *false*.

b. $2y - 6 = 10$
The equation contains a variable, so it is an *open sentence*.

c. $-9 + 5 = 2(7 - 9)$
$-4 \stackrel{?}{=} 2(-2)$
$-4 = -4$
The equation is *true*. ◄

Example 2

Which of the values 5 and 6 is a solution of the equation $3x - 1 = 17$?

Sometimes the solution of an equation is said to satisfy the equation. An everyday meaning of the word satisfy is "to meet the needs or requirements of." So a solution of an equation like $3x - 1 = 17$ meets a certain requirement. What is that requirement?

Solution

Substitute 5 for x in the equation.

$3x - 1 = 3(5) - 1$ ← **Remember the order of operations: Multiply first, then subtract.**
$= 15 - 1$
$= 14$

So 5 is not a solution.

Substitute 6 for x in the equation.

$3x - 1 = 3(6) - 1$
$= 18 - 1$
$= 17$

So 6 is a solution. ◄

Example 3

Use mental math to solve the equation $x + 9 = 14$.

Solution

Think: What number added to 9 equals 14?
You know that $5 + 9 = 14$, so $x = 5$. ◄

Example 4

Phil, who weighs 191 lb, will have a tug of war against Andy, who weighs 98 lb, and another person. Andy's teammate will be either Anna (97 lb) or Tonio (93 lb). Use the equation $98 + x = 191$. Find the partner Andy should team with so that the weights will be equal on both sides.

Solution

$$98 + x = 191$$

Substitute Anna's weight for x.

$98 + 97 \stackrel{?}{=} 191$
$195 \neq 191$ ← **$\neq$ means "is not equal to."**

Substitute Tonio's weight for x.

$98 + 93 \stackrel{?}{=} 191$
$191 = 191$

Andy should team up with Tonio. ◄

TRY THESE

Tell whether each equation is *true, false,* or an *open sentence.*

1. $-3(5-7)=6$
2. $n-4=4$
3. $-6+3=\frac{12}{7-3}$

Which of the given values is a solution of the equation?

4. $3-x=-1$; 2, 4, -4
5. $2m-1=9$; 3, 4, 5

Use mental math to solve the following equations.

6. $t+18=25$
7. $h-3=-1$
8. $p+5=-2$

Solve.

9. On a seesaw, Art and Beth balance Craig. Art weighs 113 lb and Craig weighs 219 lb. Use the equation $113+x=219$ to find whether Beth weighs 96 lb or 106 lb.

EXERCISES

PRACTICE/ SOLVE PROBLEMS

Tell whether each equation is *true, false,* or an *open sentence.*

1. $8=9-y$
2. $-11=-15-4$
3. $24\div(-6)=-4$
4. $4(2+5)=30-(4-6)$
5. $\frac{-3+11}{2}=5-(4-3)$
6. $8+(7-5)=3-x+2$

Which of the given values is a solution of the equation?

7. $b-5=-1$; 4, 6
8. $9+c=9$; 0, 1
9. $d+7=-1$; -6, -8
10. $1-e=4$; -3, -5
11. $-6=6+h$; 0, -12
12. $8=f-4$; 4, 12
13. $2k=2$; 0, 1, -1
14. $8m-2=-10$; -8, 1, -1
15. $17=2n$; $8\frac{1}{2}$, $9\frac{1}{2}$, $\frac{1}{2}$
16. $19=10-2p$; 5.5, 4.5, -4.5
17. $\frac{p}{4}=12$; 3, 8, 48
18. $7+\frac{1}{2}k=9$; 0, 2, 4

Use mental math to solve each equation.

19. $x+5=7$
20. $n-4=3$
21. $2=10-k$
22. $3+h=0$
23. $-4m=20$
24. $-3=18\div m$
25. $-9+p=-5$
26. $6=12+y$

Solve.

27. Zoa scored 13 more points than Tad did. Zoa scored 41 points. Use the equation $41 - T = 13$ and these values for x: 22, 25, 28. Find T, the number of points that Tad scored.

28. Pete's earnings equaled the sum of Doug's and Sam's earnings. Pete earned \$82 and Sam earned \$39. Use the equation $39 + x = 82$ and these values for x: 37, 43, 45. Find x, the amount that Doug earned.

EXTEND/ SOLVE PROBLEMS

Which of the given values is a solution of the equation?

29. $4b = -3 + 5$; $1, \frac{1}{2}, -1$

30. $-\frac{1}{4} + c = -\frac{1}{2}$; $\frac{1}{2}, -\frac{1}{4}, \frac{1}{4}$

31. $2a + (-2) = 0$; $0, 1, -1$

32. $-9 + 3 = 3z - 15$; $0, 1, -1$

33. $(5 - 7)d = \frac{-12}{9-5}$; $-1.5, 1, 1.5$

34. $4.6f - 2.3 = 1 - 7.9$; $1, 2, -1$

Solve.

35. A stack of nickels is worth \$1.35. Use the equation $0.05n = 1.35$ and these values for n: 26, 27, 28. Find n, the number of nickels in the stack.

36. The sum of Mike's and Hedda's ages equals the sum of Nan's and Sol's ages. Hedda is 26, Nan is 33, and Sol is 41. Use the equation $M + 26 = 33 + 41$ and these values for M: 48, 39. Find M, Mike's age.

37. A radio was on sale for \$48, which was $\frac{2}{3}$ of its regular price. Use the equation $\frac{2}{3}r = 48$ and these values for r: 32, 72, 81. Find r, the regular price.

THINK CRITICALLY/ SOLVE PROBLEMS

Use mental math to find two solutions of the equation.

38. $x^2 = 25$

39. $h^2 - 4 = 5$

40. Verify that 5 is a solution of the equation $-2 + k = 3$. Do you think any other value of k is a solution? Explain.

USING DATA In baseball, a pitcher's earned run average (ERA) gives the average number of earned runs the pitcher gives up per game. Use the Data Index on page 546 to find ERA statistics, and solve.

41. Ed Plank's ERA is 0.46 greater than that of another pitcher. Use the values 1.82 and 1.88 and the equation $P - p = 0.46$ to find the other pitcher's name (P = Ed Plank's ERA; p = the other pitcher's ERA).

7-2 Using Addition or Subtraction to Solve an Equation

EXPLORE

A double-pan balance has a 16-g weight on the left side and weights of 11 g and 5 g on the right side.

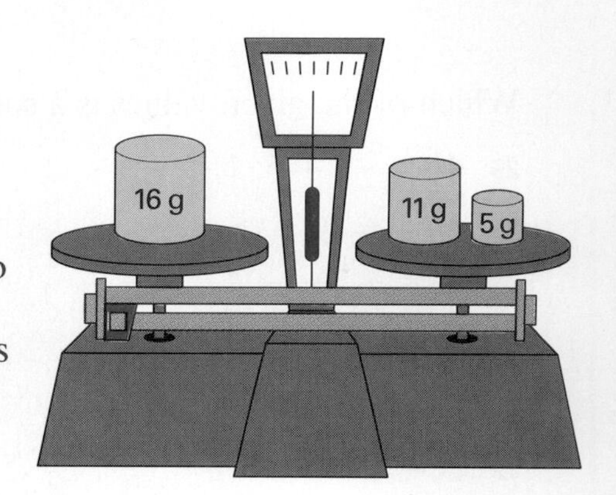

1. Is the scale in balance? How can you tell?
2. Suppose you could choose any whole-number gram weight up to 10 grams. Could you replace the 5-g weight with two other weights without losing balance? Give an example. Could you replace the 11-g weight with three other weights without losing balance? Give an example.
3. What will happen if you add a 4-g weight to the left side? What can you do on the right side to bring the scale into balance?
4. What will happen if you remove the 11-g weight from the right side? What can you do on the left side to bring the scale into balance?

SKILLS DEVELOPMENT

When you **solve an equation,** you find all the values of the variable that make the equation true. In solving an equation, it helps to think of the equation as a balance scale.

"Undo" each operation performed on the variable by using the inverse operation. Keep the equation in balance by performing the same operation on both sides.

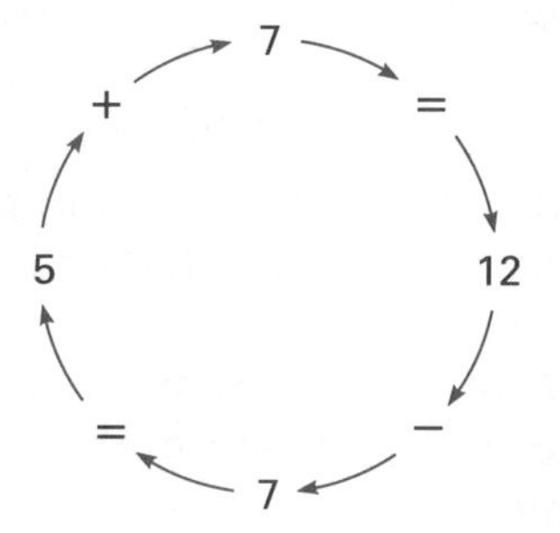

Addition and subtraction are inverse operations.

Continue until the variable is alone on one side of the equals sign.

Check your solution by substituting it in the original equation.

Example 1

Solve $x + 7 = 11$. Check the solution.

Solution

$$x + 7 = 11$$

Undo addition with subtraction.

$$x + 7 - 7 = 11 - 7$$

↑ ↑ **Keep the equation in balance.**

$$x + 0 = 4$$
$$x = 4$$

Check.

$$x + 7 = 11 \leftarrow \text{original equation}$$
$$(4) + 7 \stackrel{?}{=} 11$$
$$11 = 11 \checkmark \text{ The solution checks.}$$

The solution is 4. ◄

Example 2

Solve $-3 = n - 5$. Check the solution.

Solution

$$-3 = n - 5$$

Undo subtraction with addition.

$$-3 + 5 = n - 5 + 5$$

↑ ↑ **Keep the equation in balance.**

$$2 = n + 0$$
$$2 = n$$

Check.

$$-3 = n - 5$$
$$-3 \stackrel{?}{=} (2) - 5$$
$$-3 = -3 \checkmark$$

The solution is 2. ◄

CHECK UNDERSTANDING

Would you add or subtract to solve each equation? Explain.

1. $x - 7 = 3$
2. $t + 6 = -2$
3. $-5 = m + 1$

Example 3

Eight less than a certain number equals -2 times 3. To model this situation, Nancy let x = "a certain number." Then she wrote the equation $x - 8 = -2(3)$. Solve the equation.

Solution

$$x - 8 = -2(3)$$
$$x - 8 = -6$$
$$x - 8 + 8 = -6 + 8$$
$$x = 2$$

Check.

$$x - 8 = -2(3)$$
$$(2) - 8 \stackrel{?}{=} -6$$
$$-6 = -6$$

The solution is 2. ◄

TRY THESE

Solve each equation. Check the solution.

1. $y + 8 = -3$
2. $4 = h - 6$
3. $5 - z = -3$
4. $4 + t = -5$
5. $x - 7 = 2$
6. $2 + y = -8$
7. $35 = x - 15$
8. $t + 18 = 56$
9. $r + 7 = -16$
10. $k - 45 = 22$
11. $-15 + a = 19$
12. $-3 = w + 2$
13. Two more than a certain number equals negative ten divided by negative two. To model this situation, Lucy let r equal "a certain number." Then she wrote the equation $r + 2 = -10 \div (-2)$. Solve the equation.
14. Forty is 12 less than a certain number. To model this situation, Eileen let n equal "a certain number." Then she wrote the equation $40 = n - 12$. Solve the equation.
15. Five more than a certain number equals half of -10. To model this situation, Chi let m equal "a certain number." Then he wrote the equation $m + 5 = \frac{1}{2}(-10)$. Solve the equation.

EXERCISES

PRACTICE/ SOLVE PROBLEMS

Write the number that must be added or subtracted to solve the equation. Do not solve.

1. $n - 5 = 6$
2. $y + 9 = -3$
3. $-8 + k = 3$
4. $2.7 = m - 0.4$
5. $p - \frac{4}{5} = 1\frac{1}{5}$
6. $-4.6 = 35 + c$

Solve each equation. Check the solution.

7. $c - 6 = -3$
8. $d + 9 = -2$
9. $4 = e - 4$
10. $3 + f = -5$
11. $g - 1.2 = 2.8$
12. $0 = k + 7$
13. $\frac{3}{4} + m = -1$
14. $-4.5 = n - 2.7$
15. $-6 = y - 6$
16. $t + 1\frac{5}{8} = 2\frac{3}{4}$

Solve.

17. Tai has 87 rock samples and wants to collect a total of 100. To model the situation, he let x equal "the number of samples needed." Then he wrote the equation $87 + x = 100$. Solve the equation.

EXTEND/ SOLVE PROBLEMS

Solve each equation. Check the solution.

18. $x - 8 = -2(4 - 5)$
19. $2 + 3 \times 4 = k + 5$
20. $\frac{8(9)}{3(2)} = n + 17$
21. $6 + p + 4 = -2 + 8$
22. $2(9 - 5) = c - 36 + 14$
23. $x + 2.5 + 4.7 = 8.1 - 5.3$
24. $x + 21{,}677 = 13{,}291$
25. $m - 7.48 = 25.793$
26. $465.91 = 209.8 + y$
27. $53{,}809 = e - 84{,}277$

Solve.

28. After Mark scored -5 he had a score of 12. To model the situation, he let s equal "my score before this turn." Then he wrote the equation $s + (-5) = 12$. Solve the equation to find his score before this turn.

29. Today's average temperature of $-9°C$ is $8°C$ higher than yesterday's average. Ann let y equal "yesterday's average temperature." Then she wrote the equation $y + 8 = -9$. Solve the equation to find yesterday's average temperature.

30. Death Valley's elevation is 14,776 ft lower than the top of Mount Whitney, 14,494 ft above sea level. Let d equal Death Valley's elevation. Solve the equation $d + 14{,}776 = 14{,}494$ to find Death Valley's elevation.

THINK CRITICALLY/ SOLVE PROBLEMS

31. Write an equation involving addition that has 4 as its solution.

32. Write an equation involving subtraction that has -7 as its solution.

Solve each equation.

33. $x^2 - 2.9 = 6.1$
34. $62.8 = n^2 + 13.8$

7-3 Problem Solving Skills:

WRITING AN EQUATION

Often you can solve a problem by writing an equation and then finding the solution of the equation. Read the problem carefully. Think about how the numbers in the problem are related. Choose a variable to represent the unknown quantity. Then, write an equation representing the situation described in the problem.

PROBLEM

Members of the Chess Club are earning money to buy new chess sets. Their goal is to earn \$225. Today's accounting shows that they have earned \$157 so far. How much more must they earn in order to reach their goal?

SOLUTION

Given: The target amount is \$225.
Their current total is \$157.
Find: how much more is needed

Let a represent the amount needed to reach \$225.	← **The variable represents the unknown quantity.**
The \$157 earned so far plus the amount needed equals \$225.	← **This sentence describes how the numbers in the problem are related.**
$157 + a = 225$ $157 - 157 + a = 225 - 157$	← **The equation translates the problem situation into numbers and variables.**
$a = 68$	← **The solution of the equation is 68.**

Since a represents the amount needed, the club needs to earn \$68 more.

PROBLEMS

Choose the letter of the equation that represents the problem situation correctly. Do not solve the problem.

1. After Harriet spent \$21 to rent roller skates, she had \$23 left. How much did she have to begin with? Let x represent the amount she had to begin with.
   a. $x + 21 = 23$ b. $x - 23 = 21$ c. $x - 21 = 23$
2. Stock in the ABC Corporation rose $2\frac{3}{4}$ points to a value of $17\frac{1}{8}$ points. Find the value before the increase. Let v represent the original value.
   a. $v + 2\frac{3}{4} = 17\frac{1}{8}$ b. $v - 2\frac{3}{4} = 17\frac{1}{8}$ c. $v + 17\frac{1}{8} = 2\frac{3}{4}$

Solve each problem by writing an equation.

3. Sandra gave her brother 16 baseball cards, leaving her with 42 cards. How many cards did she have to begin with?
4. If a certain number is decreased by 56, the result is 36. Find the number.
5. A sweater was on sale for \$47. The sale price was \$16 lower than the regular price. Find the regular price.
6. Spike rode his bike for $3\frac{1}{2}$ h on Saturday. If he rode $1\frac{1}{4}$ h in the morning, how long did he ride the rest of the day?
7. On her first turn in a game Verna scored -18. After her second turn her total score was 2. How many points did she score on her second turn?
8. Janine has read 230 pages of a novel. That is 39 more pages than Harrison has read. How many pages has Harrison read?

WRITING ABOUT MATH

Write two different word problems that could be solved using the equation $8 + n = 14$.

7-4 Using Multiplication or Division to Solve an Equation

EXPLORE

a. Choose a number.

b. Multiply the number by 7.

c. Write the result on a slip of paper.

d. Exchange slips with a classmate.

e. Find your classmate's original number.

f. Choose a new number.

g. Divide the number by 3.

h. Write the result on a slip of paper.

i. Exchange slips with a classmate.

j. Find your classmate's original number.

SKILLS DEVELOPMENT

Like addition and subtraction, multiplication and division are inverse operations. You can solve equations involving multiplication or division by using the inverse operation.

2 × 9 = 18 ÷ 9 = 2

Multiplication and division are inverse operations.

Example 1

Solve $4m = -52$. Check the solution.

Solution

$$4m = -52$$

Undo multiplication with division.

$$\frac{4m}{4} = \frac{-52}{4}$$

Keep the equation in balance.

$$(1)m = -13$$

$$m = -13$$

Check.

$$4m = -52$$

$$4(-13) \stackrel{?}{=} -52$$

$$-52 = -52 \checkmark$$

The solution is -13. ◄

Example 2

Solve $\frac{k}{8} = -2$. Check the solution.

Solution

$$\frac{k}{8} = -2$$

Undo division with multiplication.

$$8\left(\frac{k}{8}\right) = 8(-2)$$
$$(1)k = -16$$
$$k = -16$$

Check.

$$\frac{k}{8} = -2$$
$$\frac{(-16)}{8} \stackrel{?}{=} -2$$
$$-2 = -2 \checkmark$$

The solution is −16. ◄

Example 3

Ted needs 60 sandwiches to serve at his party. He can fit 12 sandwiches on a serving platter. How many platters are needed?

Solution

Write and solve an equation that represents the situation.
Let p = number of platters needed

$$12p = 60$$
$$\frac{12p}{12} = \frac{60}{12}$$
$$p = 5$$

Ted needs 5 serving platters. ◄

MIXED REVIEW

Write in exponential form.

1. $11 \times 11 \times 11 \times 11$
2. 4.3×4.3

Add or subtract.

3. $-7 + (-17)$
4. $11 - 13$
5. $-5 - (-8)$

Change the following measurements.

6. 5 qt to cups
7. 1,500 m to kilometers
8. 3 kg to grams

Solve.

9. Of 42 members of the Roller Skate Club, 16 are at the rink today. Write in lowest terms the fraction of the members who are at the rink.

TRY THESE

Solve each equation. Check the solution.

1. $\frac{p}{-3} = 9$
2. $51 = 15e$
3. Kirk earned $83.75 working for an hourly wage of $6.70. How many hours did he work?

EXERCISES

PRACTICE/ SOLVE PROBLEMS

Write the number by which you must multiply or divide to solve the equation. Do not solve.

1. $-8y = -10$
2. $5.2h = -15.6$
3. $\frac{c}{2} = -2$
4. $19 = -3.5m$
5. $-3 = \frac{d}{16}$
6. $\frac{n}{-9} = -8$

Solve each equation. Check the solution.

7. $228 = 4x$
8. $\frac{h}{-7} = -5$
9. $28 = \frac{k}{14}$
10. $-12p = 3$
11. $-15 = \frac{y}{-2.2}$
12. $6k = 6$
13. $30r = 18$
14. $\frac{w}{8} = -21$
15. $\frac{m}{-3} = \frac{4}{9}$
16. $1 = \frac{h}{5}$
17. $3.7x = 8.51$
18. $-y = 6$

Solve.

19. A rectangular garden with an area of 105 ft^2 has a length of 12 ft. Find the width.
20. Board games are on sale for $8.95 each. During the sale, members of the Seniors Club spent $107.40 on board games. How many games did club members buy?

EXTEND/ SOLVE PROBLEMS

Solve and check.

21. $-4(-5) = 8n$
22. $\frac{K}{3} = 2(6 - 8 + 12)$
23. $\frac{p}{4 - 10} = \frac{4 + 5}{8 - 5}$
24. $\frac{64.8}{3} = -2.4x$
25. $6 + 18 \div 9 = (9 - 29)y$
26. $4(-2.1) = \frac{n}{12(-3.5)}$
27. $56.8g = 1{,}209.84$
28. $13.6 = \frac{e}{-9.7}$
29. $\frac{h}{123} = 0.88$
30. $2{,}142 = -2{,}520p$
31. $8.036 = \frac{y}{-0.45}$
32. $2.7x = 8.1(-9.3)$
33. $7.2 = 3(y - 2.3)$
34. $12\left(n + \frac{2}{9}\right) = -11.3$

PROBLEM SOLVING TIP

For Exercise 36, ask yourself how many *changes* in temperature the forecaster measured.

Solve.

35. Mary Beth bought 4 pens priced at $0.79 each and 2 pens priced at $0.97 each. Find the average price she paid per pen.
36. A weather forecaster made 7 temperature measurements on Sunday afternoon. The first reading was 11°F and the last was −1°F. What was the average temperature change between measurements?

THINK CRITICALLY/ SOLVE PROBLEMS

Describe all solutions of each equation.

37. $2n = 0$
38. $0n = 2$
39. $0n = 0$
40. What is the result if you add 0 to each side of an equation?
41. What is the result if you multiply each side of an equation by 0?

► READ
► PLAN
► SOLVE
► ANSWER
► CHECK

Problem Solving Applications:

STOCK MARKET REPORTS

Most stock market reports tell the value at which a stock opens (value of the stock in the morning), what change in value happens that day, and the value at which a stock closes (value at the end of the day).

Marilyn likes to keep track of what is happening in the stock market. Part of her hobby is to keep track of the stock prices and the changes that occur during a day. One day she spilled juice on part of a report. However, she could still read some information. From what she could read she was able to find the opening price, change for the day, and closing price of each stock listed.

Show how Marilyn could use an equation to represent each situation. Then solve the equation.

1. ABC stock closed at $14\frac{1}{2}$. It had opened at 12 that day. What was the change for the day?
2. Innercomp Communications Systems stock opened at 33. It closed at 28. What was the change for the day?
3. Big Company stock has a change of $+2$ in one day. The value of all Mr. Ruiz's shares in Big Company increased by $300 that day. How many shares of Big Company stock does Mr. Ruiz own?
4. Bruns Corporation had a change of -5. If its stock closed at 93, at what price did the stock open that day?
5. An investor sold 120 shares of Williams Wardrobes for $180 less than was paid for those shares. What was the change in price of a share between the time of purchase and the time of sale?

7-5 Using More Than One Operation to Solve an Equation

EXPLORE

You can use algebra tiles to model equations.

represents x. represents 1.

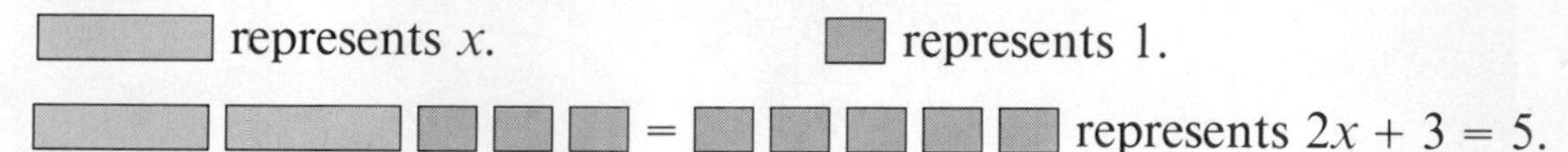
represents $2x + 3 = 5$.

Write the equation represented by each diagram.

1.

2.

3.

4.

Use tiles to model the equation.

5. $x + 1 = 5$
6. $x + 4 = 3$
7. $3x = 6$
8. $2x + 2 = 4$
9. $3x + 1 = 7$
10. $4x + 3 = 9$
11. Model the equation $x + 4 = 7$. Describe how you can use tiles to solve the equation. What is the solution?
12. Model the equation $3x = 9$. Describe how you can use tiles to solve the equation. What is the solution?

Model the equation $2x + 5 = 9$.

13. Show how to subtract 5 from both sides.
14. Show how to divide both sides into two equal groups.
15. What is the solution of the equation? How do you know?

Use tiles to solve each equation.

16. $2x + 1 = 7$
17. $3x + 2 = 8$
18. $4x + 1 = 9$
19. $2x + 5 = 7$

MATH: WHO, WHERE, WHEN

Methods of solving equations by performing the same operation on both sides had been developed by Arab and Hindu mathematicians before the ninth century. Around A.D. 830, the Arab mathematician Al-Khowarizmi wrote a book describing methods of solving equations, including equations of the types studied in this chapter. Al-Khowarizmi was a teacher who had earlier worked at the court of the Moslem ruler Harun-al-Rashid in Baghdad. The word *algebra* is derived from the word *al-jabr,* which appeared in the title of his book.

Until now you have been able to solve every equation by applying a single inverse operation. Such an equation is called a **one-step equation.**

SKILLS DEVELOPMENT

Some equations involve two operations. To solve these **two-step equations,** undo addition and subtraction first. Then undo multiplication and division.

Example 1

Solve $4x + 7 = 23$. Check the solution.

Solution

$$4x + 7 = 23$$

Undo the addition.

$$4x + 7 - 7 = 23 - 7$$

$$4x = 16$$

Undo the multiplication.

$$\frac{4x}{4} = \frac{16}{4}$$

$$x = 4$$

Check.

$$4x + 7 = 23$$
$$4(4) + 7 \stackrel{?}{=} 23$$
$$16 + 7 \stackrel{?}{=} 23$$
$$23 = 23 \checkmark$$

The solution is 4. ◄

PROBLEM SOLVING TIP

Double-check to be sure you are substituting your solution in the original equation.

Example 2

Solve $7(x + 4) = 21$. Check the solution.

Solution

$$7(x + 4) = 21$$

Multiply each term in parentheses by 7.

$$7x + 28 = 21$$

Undo addition with subtraction.

$$7x + 28 - 28 = 21 - 28$$

Simplify.

$$7x + 0 = -7$$
$$7x = -7$$

Undo multiplication with division.

$$\frac{7x}{7} = \frac{-7}{7}$$

$$x = -1$$

Check.

$$7(x + 4) = 21$$
$$7(-1 + 4) \stackrel{?}{=} 21$$
$$7(3) \stackrel{?}{=} 21$$
$$21 = 21 \checkmark$$

The solution is -1. ◄

CALCULATOR

You can solve equations using your calculator.

To solve $3.2x - 4.8 = -17.92$, use this key sequence.

17.92 [+/−] [+] 4.8 [÷] 3.2 [=]

On some calculators, you may have to press the [=] key after you enter the 4.8.

The solution is -4.1.

Example 3

Arnie, Megan, and Piero picked apples, which they shared equally. Piero ate 2 apples, leaving him with 7. How many apples did Arnie, Megan, and Piero pick?

Solution

Write and solve an equation that represents the situation.

Let n = total number of apples picked.

Piero's share minus 2 left Piero with 7 apples.

$$\frac{n}{3} - 2 = 7$$

Undo the subtraction. $$\frac{n}{3} - 2 + 2 = 7 + 2$$

$$\frac{n}{3} = 9$$

Undo the division. $$3\left(\frac{n}{3}\right) = 3(9)$$

$$n = 27$$

The three picked 27 apples. ◄

TRY THESE

Solve each equation. Check the solution.

1. $3x - 5 = 16$
2. $\frac{n}{-6} + 8 = 5$
3. $1 = -9y + 7$
4. $2(y - 6) = 18$
5. A bicycle can be rented for $4.50 per hour plus a $3 service fee. Pete paid $25.50 to rent a bike. For how many hours did he rent it?

EXERCISES

PRACTICE/ SOLVE PROBLEMS

Write the first step in solving the equation. Do not solve.

1. $\frac{y}{-7} - 3 = 4$
2. $\frac{1}{2} = 8x + 5$
3. $8 + 3n = -1$
4. $9 = -6 + \frac{k}{2}$

Solve each equation. Check the solution.

5. $6p - 7 = 5$
6. $11 = -2x + 5$
7. $\frac{k}{-2} + 4 = -8$
8. $3 = 3h + 6$
9. $7 + \frac{e}{6} = 0$
10. $34 = \frac{f}{-2} + 47$
11. $9n - 1.5 = 10.2$
12. $\frac{3}{5} + 3c = 3$
13. $\frac{x}{7} - 15 = -4$
14. $5(k + 3) = 30$
15. $-7(p - 9) = 14$
16. $26 = 8(m + 7)$

Greg's age of 15 is 6 years more than Marv's age divided by 4.

17. Let M = Marv's age. Write an equation relating Marv's age to Greg's.

18. Solve the equation you wrote in Exercise 17. Check your solution.

EXTEND/ SOLVE PROBLEMS

Solve each equation. Check the solution.

19. $\frac{-45}{-9} + (5p - 3p) = 4 + 7 \times 3$

20. $\frac{y}{15.6} + 319.7 = 334$

21. $47.3k - 8.77 = -131.75$

It costs \$24 per day plus \$0.08 per mile to rent a car. Jeanne's charge for a one-day rental was \$42.32.

22. Let m = number of miles driven. Write an equation representing Jeanne's charge.

23. Solve the equation you wrote in Exercise 22. Check the solution.

THINK CRITICALLY/ SOLVE PROBLEMS

Solve.

24. $9\left(x - \frac{1}{3}\right) = -21$

25. $-28 = -4(2y - 3)$

26. Write a two-step equation involving addition and division that has -12 as its solution.

27. Write a two-step equation involving subtraction and multiplication that has 3 as its solution.

7-6 Using a Reciprocal to Solve an Equation

EXPLORE/ WORKING TOGETHER

Jeff has many different hobbies. His favorite hobby is building model airplanes. The table belows gives the amount of time he spent working on some of his models. Study the table carefully.

Plane	Number of hours to build	Fraction of job completed per hour
B-29	12	$\frac{1}{12}$
Tigercat	5	■
Stinger	$\frac{15}{4}$	■
Shark	$\frac{11}{2}$	■

a. Work with a partner to complete the table. Find what fraction of the job Jeff completed per hour. Assume he worked at a steady rate.

b. What relationship do you notice between the numbers in the second and third columns?

c. For each model, find the product of the numbers in the second and third columns.

SKILLS DEVELOPMENT

When an equation involves a variable that is multiplied by a fraction, you can use the reciprocal of the fraction to solve the equation.

Example 1

Solve $\frac{3}{8}k = -27$. Check the solution.

Solution

$$\frac{3}{8}k = -27$$

Multiply both sides by the reciprocal of $\frac{3}{8}$.

$$\frac{8}{3}\left(\frac{3}{8}\right)k = \frac{8}{3}(-27)$$
$$(1)k = -72$$
$$k = -72$$

Check.

$$\frac{3}{8}k = -27$$
$$\frac{3}{8}(-72) \stackrel{?}{=} -27$$
$$-27 = -27 \checkmark$$ The solution is -72. ◄

Recall that the product of a fraction and its reciprocal is 1. You can find the reciprocal of a fraction by interchanging the numerator and denominator.

Example 2

Solve $-\frac{9}{2}n - 6 = 12$. Check the solution.

Solution

$$-\frac{9}{2}n - 6 = 12$$

Undo the subtraction.

$$-\frac{9}{2}n - 6 + 6 = 12 + 6$$
$$-\frac{9}{2}n = 18$$

Multiply both sides by the reciprocal of $-\frac{9}{2}$.

$$-\frac{2}{9}\left(-\frac{9}{2}n\right) = -\frac{2}{9}(18)$$
$$n = -4$$

Check.

$$-\frac{9}{2}n - 6 = 12$$
$$-\frac{9}{2}(-4) - 6 = 12$$
$$\frac{36}{2} - 6 = 12$$
$$18 - 6 = 12$$
$$12 = 12 \checkmark$$

The solution is -4. ◄

Example 3

Copies of a math book are $1\frac{1}{8}$ in. thick. How many can be stored on a shelf $20\frac{1}{4}$ in. wide?

Solution

Write and solve an equation that represents the situation.

Let $n =$ the number of copies.

$$1\frac{1}{8}n = 20\frac{1}{4}$$
$$\frac{9}{8}n = \frac{81}{4}$$ Write the mixed numbers as fractions.
$$\frac{8}{9}\left(\frac{9}{8}n\right) = \frac{8}{9}\left(\frac{81}{4}\right)$$
$$n = 18$$

Eighteen books can be stored on the shelf. ◄

CHECK UNDERSTANDING

How does multiplying by the reciprocal simplify the process of solving an equation?

TRY THESE

Solve each equation. Check the solution.

1. $\frac{5}{6}v = 35$
2. $\frac{10}{3}g + 5 = 8$
3. $\frac{4}{5}m = \frac{3}{4}$
4. Marci planted a seedling plum tree 28 in. tall. The tree grew at an average rate of $8\frac{1}{4}$ in. per year. After how many years was the tree 127 in. tall?

EXERCISES

PRACTICE/ SOLVE PROBLEMS

Solve each equation. Check the solution.

1. $\frac{3}{4}x = -6$
2. $9 = -\frac{2}{3}x$
3. $\frac{4}{5}x = 8$
4. $-\frac{3}{7}h = 9$
5. $1 = \frac{4}{5}c$
6. $\frac{4}{3}e = 16$
7. $\frac{2}{5}x - 3 = 9$
8. $8 = -\frac{3}{7}x + 4$
9. $\frac{5}{4}x - 6 = 14$
10. $\frac{1}{9}m = 1\frac{1}{3}$
11. $\frac{1}{3} = -\frac{1}{2}t$
12. $0.4 = \frac{2}{9}k$
13. A snail crawled at a rate of $8\frac{3}{4}$ inches per hour. How long did it take the snail to crawl $41\frac{7}{8}$ inches?

EXTEND/ SOLVE PROBLEMS

Solve each equation. Check the solution.

14. $3 = \frac{7}{8}p + \frac{1}{4}$
15. $\frac{8}{3}f - \frac{3}{4} = 7\frac{1}{4}$
16. $\frac{8}{5}y - 1\frac{1}{3} = -\frac{2}{3}$
17. $\frac{7}{9}k - 6\frac{1}{3} = 1\frac{2}{3}$
18. $0 = \frac{15}{4}n + 1\frac{1}{2}$
19. $\frac{9}{4}c + 2\frac{1}{4} = 3\frac{3}{4}$
20. $\frac{16}{15}x + \frac{1}{3} = -\frac{1}{3}$
21. $9\left(\frac{3}{4}m - 1\right) = -5\frac{5}{8}$
22. Bricks $2\frac{1}{4}$ in. thick are stacked atop each other to make a wall. A $9\frac{1}{2}$-in. wood border decorates the top of the wall. If the wall plus the border is $65\frac{3}{4}$ in. high, how many rows of bricks are there?

THINK CRITICALLY/ SOLVE PROBLEMS

23. If the product of two numbers is 1, the numbers are reciprocals of one another. What can you say about two numbers whose quotient is 1? Two numbers whose product is 0?

Problem Solving Applications:

BREAKING THE CODE

The young of various animals have been given special names.

In the table, each name is given a code number.

CODE	
Code number	**Name of young**
-4	leveret
-3	foal
-2	calf
-1	gosling
0	parr
1	squab
2	piglet
3	cygnet
4	chick
5	lamb
6	duckling

To find the names of the young of various animals in the questions that follow, you need to solve the equations, then use the codes given in the table. Here is an example.

What is a young goose called?

Solve the equation $-6 = 2g - 4$ and then use the code.

$$-6 = 2g - 4$$
$$-6 + 4 = 2g - 4 + 4$$
$$-2 = 2g$$
$$-1 = g$$

From the table, -1 is the code number of "gosling." So a young goose is called a gosling.

Solve each equation. Then use the code to determine the name of the young of each animal given.

1. What is a young donkey called?
   Solve the equation $7d + 6 = -15$ and use the code.
2. What is a young elephant called?
   Solve the equation $-10 = 2e - 6$ and use the code.
3. What is a young pigeon called?
   Solve the equation $3p + (-12) = -9$ and use the code.
4. What is a young swan called?
   Solve the equation $5s - 15 = 0$ and use the code.
5. What is a young turkey called?
   Solve the equation $1.5t - (-3) = 9$ and use the code.

7-7 Working with Formulas

EXPLORE

Several members of the Kite Klub ordered fancy kite tails from a catalog. All the kite tails have the same price, and the mail-order house charges a fixed amount for shipping any order.

Member	Number of Kite Tails	Total Cost of Order
Allie	4	\$18.50
Bob	2	\$10.50
Mi-Su	7	\$30.50
Greg	5	\$22.50

a. Use the information in the table. Notice that Allie ordered 2 more kite tails than Bob. How much do 2 more tails add to the cost of an order? How much does one kite tail cost?

b. How much does shipping cost per order?

c. Suppose you want to find the total cost of ordering a certain number of kite tails. You can begin by multiplying the number of tails by the cost per tail. What would you do next?

d. Write a formula for the total cost, T, of ordering n kite tails.

SKILLS DEVELOPMENT

A formula is an equation stating a relationship between two or more quantities. For example, the number of square units in the area (A) of a rectangle is equal to the number of units of length (l) multiplied by the number of units of width (w).

Here is the formula for the area of a rectangle: $A = lw$

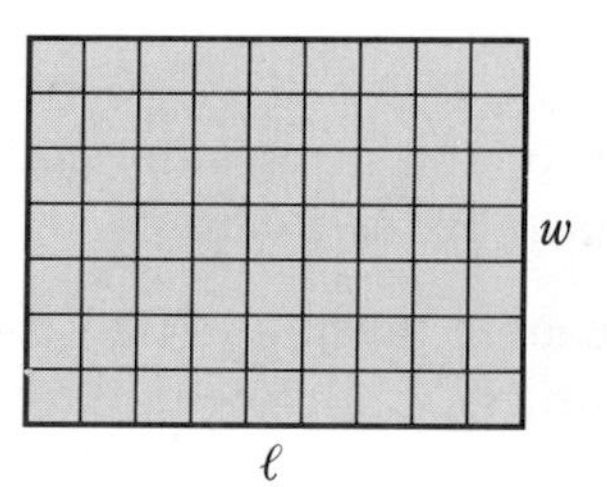

Sometimes you can evaluate a variable in a formula directly, using the given information.

$$A = lw$$
$$A = 9 \times 7$$
$$A = 63$$

The area is 63 cm^2.

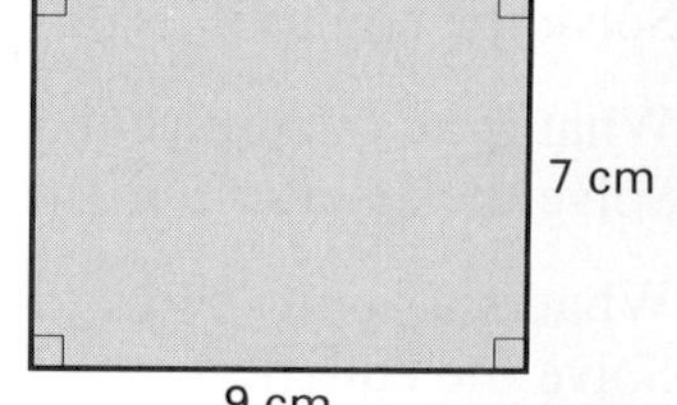

At other times, however, you must use your knowledge of equations to solve for a variable in a formula.

Example 1

A car traveled 120 mi in $2\frac{1}{2}$ h. Find its average rate.

Solution

Use the formula $d = rt$, where d = distance, r = rate, and t = time.

$$d = rt$$

$$120 = r\left(2\frac{1}{2}\right)$$

Substitute the known values of the variables into the formula.

$$120 = r\left(\frac{5}{2}\right)$$

$$120\left(\frac{2}{5}\right) = r\left(\frac{5}{2}\right)\left(\frac{2}{5}\right)$$

Solve the equation.

$$48 = r$$

The car's average rate was 48 mi/h. ◄

MIXED REVIEW

Write in standard form.

1. $7^3 + 7$
2. $(0.3)^2$

Round to the nearest hundredth.

3. 4.059
4. 0.2331
5. 1.111

Multiply or divide.

6. $\frac{5}{7} \times \frac{4}{9}$
7. $\frac{1}{5} \div \frac{11}{15}$
8. $0.08 \div 0.16$
9. 3.4×6.002

Solve.

10. A rectangular garden plot has a width of $8\frac{1}{2}$ feet and a length of $10\frac{1}{4}$ feet. What is the perimeter of the plot?

Example 2

The formula $F = \frac{9}{5}C + 32$ relates Fahrenheit and Celsius temperatures, where F = Fahrenheit temperature and C = Celsius temperature. Use the formula to convert 113°F to a Celsius temperature.

Solution

$$F = \frac{9}{5}C + 32$$

$$113 = \frac{9}{5}C + 32$$

$$113 - 32 = \frac{9}{5}C + 32 - 32$$

$$81 = \frac{9}{5}C$$

$$\left(\frac{5}{9}\right)81 = \left(\frac{5}{9}\right)\frac{9}{5}C$$

$$45 = C$$

The temperature is 45°C. ◄

Sometimes you must solve a formula for a variable that is *not* alone on one side of the equals sign.

Example 3

Solve the formula $d = rt$ for r.

Solution

To solve a formula for a given variable, rewrite the formula so that the variable stands alone on one side of the equals sign.

$$d = rt$$

$$\frac{d}{t} = \frac{rt}{t}$$ **Divide both sides by t to get r alone on one side.**

$$\frac{d}{t} = r$$ ◄

The new formula indicates that rate can be found by dividing distance by time.

TRY THESE

Solve.

1. Jolene jogged $8\frac{1}{8}$ mi at an average rate of $6\frac{1}{2}$ mi/h. How long did she jog?
2. A formula for the perimeter of a rectangle is $P = 2l + 2w$, where P = perimeter, l = length, and w = width. A rectangle has a width of 7 cm and a perimeter of 46 cm. Find the length of this rectangle.
3. The formula for the area of a rectangle is $A = lw$, where A = area, l = length, and w = width. Solve this formula for w.

EXERCISES

PRACTICE/ SOLVE PROBLEMS

Solve.

1. The formula for the area of a rectangle is $A = lw$, where l = length and w = width. A rectangle has an area of 336 cm^2 and a width of 14 cm. Find the length of this rectangle.
2. The formula for the area of a triangle is $A = \frac{1}{2}bh$, where b = length of base and h = height. A triangle has a base measuring 13 cm and an area of 91 cm^2. Find the height.
3. The formula for mass is $M = DV$, where D = density and V = volume. The largest gold nugget ever found had a mass of 216,160 g. The density of gold is 19.3 g/cm^3. Calculate the volume of the nugget.
4. The formula for the volume of a rectangular prism is $V = lwh$, where l = length, w = width, and h = height. The volume of a rectangular prism is 121.5 cm^3. The length of the prism is 12 cm and the height is 4.5 cm. Find the width.

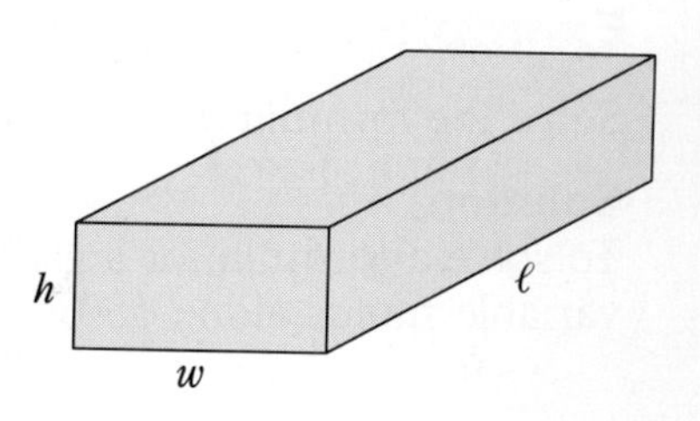

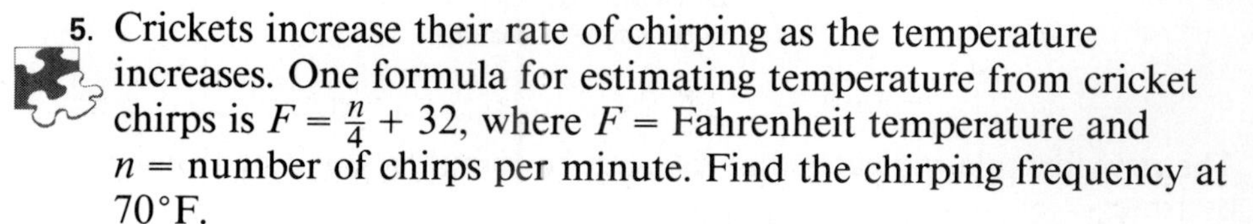

5. Crickets increase their rate of chirping as the temperature increases. One formula for estimating temperature from cricket chirps is $F = \frac{n}{4} + 32$, where F = Fahrenheit temperature and n = number of chirps per minute. Find the chirping frequency at 70°F.

6. Juanita sells magazine subscriptions. She earns a salary of $840 per month plus $3 for each subscription she sells. Her earnings can be computed using the formula $E = 840 + 3s$, where E = earnings and s = number of subscriptions sold. One month she earned $1,932. Find the number of subscriptions she sold.

USING DATA The earned run average (ERA) of a baseball pitcher can be computed using the formula $E = \frac{9R}{I}$, where E = ERA, R = number of runs given up, and I = number of innings pitched. Use the Data Index on page 546 to find the ERA statistics. Use them to solve Exercises 7 and 8.

7. Find the number of runs Rube Waddell gave up in 150 innings.

8. Jeff Tesreau pitched 1,800 innings. How many runs did he give up?

Solve each formula for the indicated variable.

9. $I = prt$, for t

10. $E = mc^2$, for m

11. $A = p + i$, for i

12. $P = 3s$, for s

13. $K = 273 + C$, for C

EXTEND/ SOLVE PROBLEMS

Solve each formula for the indicated variable.

14. $P = 2l + 2w$, for w

15. $A = \frac{1}{2}bh$, for h

16. $F = \frac{9}{5}C + 32$, for C

17. $y = mx + b$, for x

18. $T = p + prt$, for r

THINK CRITICALLY/ SOLVE PROBLEMS

19. A traffic court judge fined speeding offenders $75 plus $2 for every mile per hour they exceeded the speed limit. Calculate the fine for driving at the given rate.

    a. 45 mi/h; speed limit: 35 mi/h

    b. 70 mi/h; speed limit: 55 mi/h

    c. 92 mi/h; speed limit: 65 mi/h

20. A motorist drove r mi/h on a highway posting a speed limit of s mi/h.

    a. If $r > s$, by how much did the motorist exceed the speed limit?

    b. Refer to Exercise 19 for the method of calculating a fine. Write a formula for computing the fine, f.

7-8 Problem Solving Strategies:

WORK BACKWARD

► READ
► PLAN
► SOLVE
► ANSWER
► CHECK

Some problems tell you the outcome of a series of steps and ask you to find the fact that sets those steps in motion. To solve such a problem, start at the end and work backward to the beginning.

PROBLEM

Tanya purchased a $930 clarinet using her $5.20/h earnings as a cashier. While she was saving, her uncle donated $150 toward her goal. How many hours did Tanya work to earn the total?

SOLUTION

?	×	$5.20	+	$150	=	$930
number of hours		hourly wages		uncle's contribution		total
	↑ Multiply.		↑ Add.			

TALK IT OVER

How can you use a two-step equation to find the number of hours Tanya worked?

The diagram shows the steps and operations leading to the $930 total. To solve, work backward from $930.

Start with the total.	$930
Subtract the contribution.	− 150
	$780
Divide by the hourly wage.	$780 ÷ $5.20 = 150

Tanya worked 150 h.

Check: (150 × $5.20) + $150 = $930

PROBLEMS

1. During April the price of an apple tripled. In May the price dropped $0.18. In June it halved. In July the price rose $0.06 to $0.30 per apple. Find the price at the beginning of April.

2. A mountain climber ascended at a rate of 1,200 ft/h, fell and slid 600 ft down, got up and ascended for $2\frac{1}{3}$ h at a rate of 900 ft/h. The total gain in altitude was 4,500 ft. How long did the mountaineer climb at a rate of 1,200 ft/h?

3. Marcus spends 55 min each morning getting ready for work. He walks 7 min to a cafeteria where he spends 27 min eating breakfast. Then he walks 4 min to a subway station where he catches a subway for a 12-min ride to work. At what time must he get up in order to get to his job at 8:30 a.m.?

4. Lee wants to save $900 for his vacation. When he has saved four times as much as he already has saved, he'll need just $44 more to reach his goal. How much has he saved?

5. Julie multiplied her age by 2, subtracted 5, divided by 3, and added 9. The result was 20. How old is she?

6. Abbie, Bill, Colin, and Diane divided up a flat of petunia plants. Abbie took half. Bill took half of what was left. Colin took a third of what was left. Diane took the remaining 6 plants. How many petunia plants were there to begin with?

JUST FOR FUN

Try working backward to solve this problem, which the ninth-century Hindu teacher Mahavira posed for his students. "Of a basket of mangoes, the king took $\frac{1}{6}$, the queen $\frac{1}{5}$ of the remainder, and the three oldest princes $\frac{1}{4}$, $\frac{1}{3}$ and $\frac{1}{2}$ of the successive remainders. There were then only three mangoes left for the youngest member of the royal family. Tell me, all you who are clever in such problems, how many mangoes there were in the basket."

7. A new car carried a sticker price of $11,586. The price consisted of the base price plus $1,360 in options, multiplied by 1.05 for sales tax. To this amount a $180 dealer preparation fee and $66 license fee were added. Find the base price.

8. On Monday, the price of a share of Heavy Metal Industries fell $2\frac{1}{4}$ points. On Tuesday the price rose $5\frac{1}{2}$ points, but it fell $\frac{7}{8}$ on both Wednesday and Thursday. On Friday the stock gained $1\frac{3}{4}$ points to close the week at $28\frac{3}{8}$. Find the price of a share of Heavy Metal at the start of trading on Monday.

7-9 Combining Like Terms

EXPLORE

Model each expression using algebra tiles. Use your model to find a simpler way to write each expression. Recall that [long tile] represents x and [small tile] represents 1.

1. $3x + 2x$
2. $6x - x$

SKILLS DEVELOPMENT

The parts of a variable expression that are separated by addition or subtraction signs are called **terms.** The expression $7a + 2b + 5a - 3b^2$ contains four terms, $7a$, $2b$, $5a$, and $3b^2$. The terms $7a$ and $5a$ are called **like terms** because they have identical variable parts. The terms $2b$, $5a$, and $3b^2$ are called **unlike terms** because they have different variable parts.

Example 1

Identify the terms as like or unlike.

a. $3x, 3y$ b. $7k, -3k$ c. $2p, 2pt$

Solution

a. The terms are unlike because the variable parts, x and y, are different.

b. The terms are like because the variable parts, k and k, are identical.

c. The terms are unlike because the variable parts, p and pt, are different. ◄

You simplify an expression when you perform as many of the indicated operations as possible. You can use the distributive property to simplify a variable expression that contains like terms. This process is called **combining like terms.** Unlike terms cannot be combined.

Example 2

Simplify. a. $3m + 5m$ b. $9x - x + 5y + 2y^2$ c. $4k - 3h + 2k$

Solution

a. Use the distributive property.

$$3m + 5m = (3 + 5)m$$
$$= 8m$$

b. Use the distributive property.

Recall that $x = 1x$.

$$9x - x + 5y + 2y^2 = (9 - 1)x + 5y + 2y^2$$
$$= 8x + 5y + 2y^2$$

c. Rewrite using the commutative property.

$$4k - 3h + 2k = 4k + 2k - 3h$$

Use the distributive property.

$$= (4 + 2)k - 3h$$
$$= 6k - 3h$$ ◄

Example 3

Jesse worked $2p$ hours in the morning, $5r$ hours in the afternoon, and $4p$ hours in the evening. Write and simplify an expression for the total number of hours he worked.

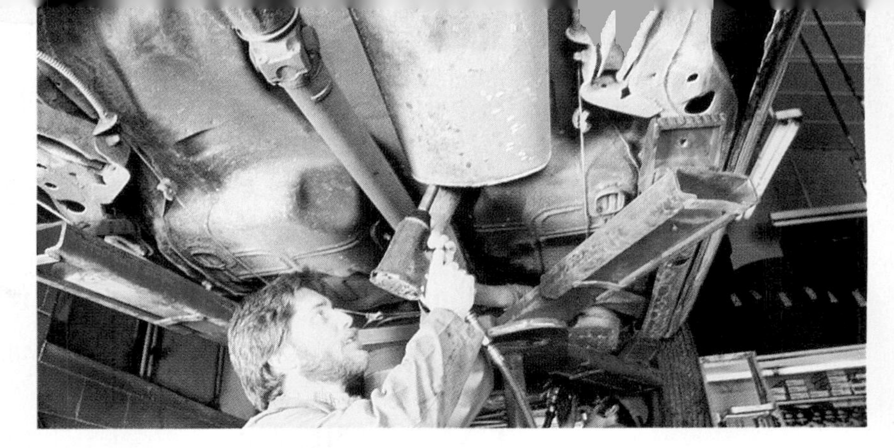

Solution

Total times worked = sum of morning, afternoon, and evening hours

$$= 2p + 5r + 4p$$

$$= 6p + 5r$$ ◄

Some equations have a variable term on both sides of the equals sign. Combining like terms is an important part of the process of solving these equations.

Example 4

Solve $5x - 7 = 3x + 13$. Check the solution.

Solution

$$5x - 7 = 3x + 13$$

Subtract $3x$ from both sides. $5x - 3x - 7 = 3x - 3x + 13$

Simplify. $2x - 7 = 13$

Add 7 to both sides. $2x - 7 + 7 = 13 + 7$

$$2x = 20$$

Divide both sides by 2. $\frac{2x}{2} = \frac{20}{2}$

$$x = 10$$

Check.

$$5x - 7 = 3x + 13$$
$$5(10) - 7 = 3(10) + 13$$
$$50 - 7 = 30 + 13$$
$$43 = 43 \checkmark$$

The solution is 10. ◄

TRY THESE

Identify the terms as like or unlike.

1. $-4a, -5a$
2. $3mn, 3mn^2$
3. $6x, -4x$
4. $7a, 7ab$

Simplify.

5. $2y + 13y$
6. $8x + 5z + 4 - 3z$
7. $7n + 3m - m + 2n$
8. $e + 5e + e^2 - 3e$

9. On four successive days, Meg's driving times were $18k$, $11h$, $13k$, and $9h$ hours. Write and simplify an expression for the total amount of time she drove.

Solve and check.

10. $4n + 7 = n - 2$
11. $9x = 6x - 15$
12. $7c - 3 = 3c + 5$
13. $8m = 2m + 18$

EXERCISES

PRACTICE/ SOLVE PROBLEMS

Identify the terms as like or unlike.

1. $9n, 3m$
2. $-2p, -2p^2$
3. $x, -2x$
4. $5y, 5$
5. $-3h, -3h$
6. xy, xz
7. $-8a, 8a$
8. $3s, 6s$
9. $-k, k$

Simplify.

10. $4m + 8m$
11. $8s + 5t - 3s$
12. $x + 11x$
13. $7y + 3y - 4$
14. $2 + 9p - 6p$
15. $a + 2a + 3a$
16. $3w + 2w + 5 - 7$
17. $14n - 6x + (-5n) + 8x$
18. $2c + 8c^2 + 11c + 9a$
19. $-3x - 3x + 3xy$

20. Emilio spent $5a$ min on his math homework, $3a$ min on history, $8b$ min on social studies, and $7b$ min on science. Write and simplify an expression for the total amount of time he spent on homework.

21. In four successive basketball games, Ruth scored $11s$, $9p$, $7s$, and $6k$ points. Write and simplify an expression for the total number of points she scored.

Solve each equation. Check the solution.

22. $5x - 7 = 2x + 2$
23. $3n - 4 = 2n + 3$
24. $4n + 9 = n - 3$
25. $8r - 11 = 6r - 6$
26. $3n - 10 = -2n + 5$
27. $n - 3 = 5n + 1$

EXTEND/ SOLVE PROBLEMS

Simplify.

28. $4.6x + 3.5y - 6.9x + y$
29. $\frac{1}{2}x + \frac{1}{2}x + 3x$
30. $\frac{2}{3}m + \frac{3}{4}n + \frac{5}{8}n$
31. $7 + 14.2a - 8 + 3.8a$

32. $z + z + z + z^2 + z$

33. $297k + 403k$

34. $3(x - 5) + 2(x + 2)$

35. $-4n + 5(3n) - 2(-2n)$

36. $6(-3y) + x + 2(2x + y)$

37. $8(a + b + c) + 2(a + 2b + 3c)$

38. $9.177x + 8.591y + 13.962x - 12.413y$

39. $173x + 23(14x - 27y) + 45(22x + 12y)$

40. Benito bought 3 quarts of milk costing $(a + 2b)$ cents per quart and 2 loaves of bread costing $(3a + b)$ cents per loaf. Write and simplify an expression for the total cost of his purchases.

41. Write and simplify an expression for the combined areas of the two rectangles.

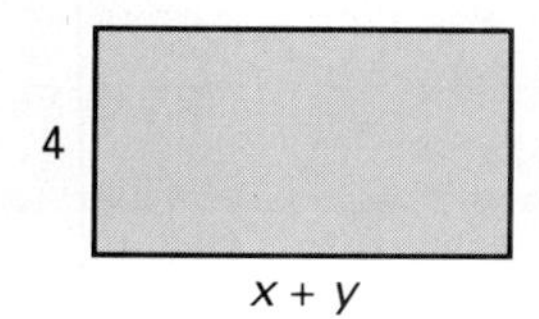

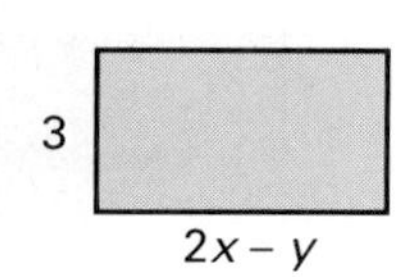

Solve each equation. Check the solution.

42. $7(x - 2) = 5(x + 4)$

43. $3n + 5(n - 2) = 3(n - 5)$

44. $6a + 4a + 5 = 3 + 4(a + 5)$

45. $\frac{3}{4}(x + 12) = \frac{5}{8}(x + 8)$

46. Choose three numbers. Copy and complete the table. Describe the unusual results you obtain.

	First number	Second number	Third number
Add 3.	■	■	■
Multiply by 4.	■	■	■
Subtract 6.	■	■	■
Divide by 2.	■	■	■
Subtract twice the original number.	■	■	■

47. Follow these steps to discover why you obtained the above result. Let $n =$ the chosen number. Write and simplify the expression that results when you apply the given operation.

a. Add 3 to n.

b. Multiply the result obtained in a by 4.

c. Subtract 6 from the result obtained in b.

d. Divide the result obtained in c by 2.

e. Subtract twice the original number from the result obtained in d.

f. Explain why you obtained the unusual results in the table above.

THINK CRITICALLY/ SOLVE PROBLEMS

COMPUTER

This program will give unusual results, too. Run it several times and compare the result with the number you entered. Can you "predict" the result for any number entered? Why is this true? Can you change lines 20 and 30 to create a different result?

```
10 INPUT "ENTER A
   NUMBER: ";N: PRINT
20 LET X =
   (3 * (N - 2) - 12) / 6
30 LET R = X + 3
40 PRINT "THE RESULT IS
   ";R: PRINT
50 INPUT "TRY IT AGAIN? Y
   OR N: ";X$: PRINT
60 IF X$ = "Y" THEN GO TO
   10
```

7-10 Graphing Open Sentences

EXPLORE

INEQUALITY SYMBOLS	
Symbol	**Meaning**
$>$	"is greater than"
$<$	"is less than"
$\geq$	"is greater than or equal to"
$\leq$	"is less than or equal to"

Match each statement with a symbolic representation.

1. The temperature is less than 17°F.
2. No more than 17 tickets remain.
3. No one under 17 admitted.
4. The blouse cost at least $17.
5. More than 17 students attended.
6. She finished in under 17 minutes.

a. $x > 17$
b. $x < 17$
c. $x \geq 17$
d. $x \leq 17$

SKILLS DEVELOPMENT

A mathematical sentence that contains one of the symbols $<$, $>$, $\leq$, or $\geq$ is called an **inequality.**

$5 < 7$ — 5 is less than 7.

$7 \geq y - 2$ — 7 is greater than or equal to $y - 2$.

In Section 7-1 you learned that an open sentence is a sentence that contains one or more variables. An open sentence may be either an equation or an inequality. By itself, an open sentence like $x + 7 = 11$ or $k > 3$ is neither true nor false. When you substitute a number for the variable, however, you can determine whether the result is true or false. Any value of the variable that makes the sentence true is called a **solution of the open sentence.**

equation: $x + 7 = 11$
solution: 4

inequality: $k > 3$
some solutions: 4
$7\frac{1}{3}$
π (3.141592...)

How many solutions are there for the inequality $k > 3$? Every integer greater than 3 is a solution. In fact, any rational number, such as $7\frac{1}{3}$, is a solution. The number π, which is approximately equal to 3.14, is also a solution.

Numbers such as π that are non-terminating, non-repeating decimals are called **irrational numbers.** There are many irrational numbers beside π that are greater than 3. Together the set of rational numbers and the set of irrational numbers make up the set of **real numbers.** So we say that the solutions of $k > 3$ are *all the real numbers that are greater than 3.*

Since solutions of open sentences are real numbers, you can graph them on a number line.

Example 1

Graph the equation $m + 9 = 14$.

Solution

$$m + 9 = 14$$
$$m + 9 - 9 = 14 - 9$$
$$m = 5$$

The equation has one solution. Graph the solution on a number line by drawing a solid dot at the point 5.

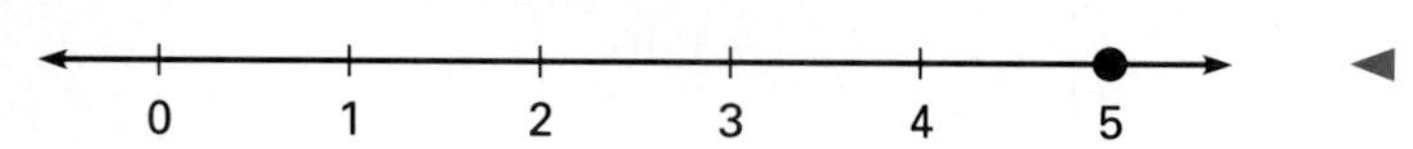

Example 2

Graph the inequality $p > -2$.

Solution

The solution consists of all real numbers greater than -2. Graph the solution by drawing a solid arrow beginning at -2 and pointing to the right. To indicate that -2 is not part of the solution, draw an open circle at the point -2.

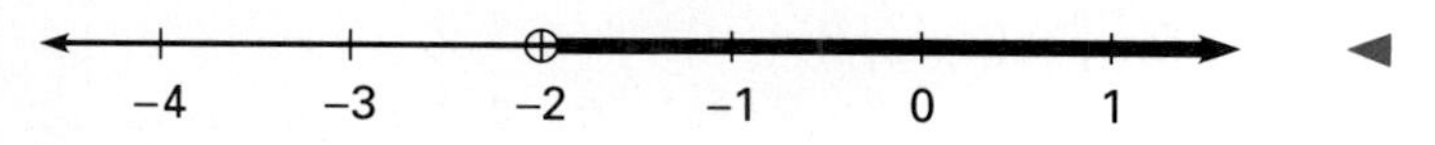

CHECK UNDERSTANDING

Is 4 a solution of each inequality?

1. $x < 4.1$
2. $x > 4$
3. $x \geq 4.1$
4. $x \geq 3.999$
5. $x \leq 4$

Example 3

Graph the inequality $k \leq 4$.

Solution

The solution consists of all real numbers less than or equal to 4. Graph the solution by drawing a solid arrow beginning at 4 and pointing to the left. To indicate that 4 is part of the solution, draw a solid dot at the point 4.

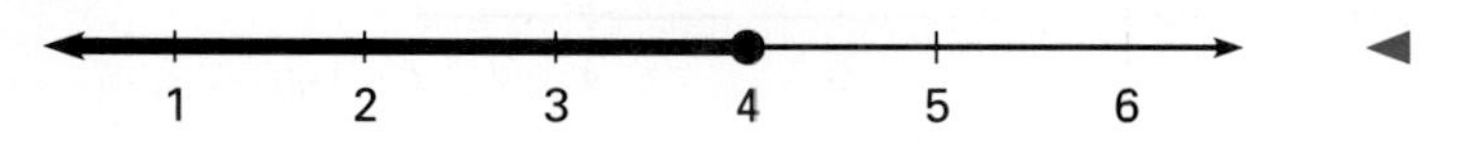

TRY THESE

Graph each open sentence on a number line.

1. $x - 7 = 3$
2. $h \geq -5$
3. $e < 1$

EXERCISES

PRACTICE/ SOLVE PROBLEMS

Graph each open sentence on a number line.

1. $x > 4$
2. $m < -1$
3. $h + 5 = 2$
4. $y \leq 0$
5. $p \geq 1$
6. $-5 = e - 6$
7. $x > -5$
8. $0 < w$

Tell whether the given number is a solution of the inequality.

9. $n < 4$: 0; 4; 5; -4
10. $y \geq 6$: 0; -5; -6; -7
11. $x \leq -1$: 1; 0; -1; -2
12. $m > -2$: -5; -1; -2; -4
13. $k \leq -\frac{1}{2}$: -1; $-\frac{3}{4}$; -0.5; 0
14. $e > -3.4$: -2; -4; -3.45; -3.40

Write an inequality to describe the situation.

15. Car repairs cost more than \$300. Let r equal the repair cost.
16. Joaquin's height is greater than or equal to 2 m. Let h equal Joaquin's height.
17. Every frame in the store was priced at under \$25. Let c equal the cost of a frame.

EXTEND/ SOLVE PROBLEMS

Write an open sentence for each graph.

18.

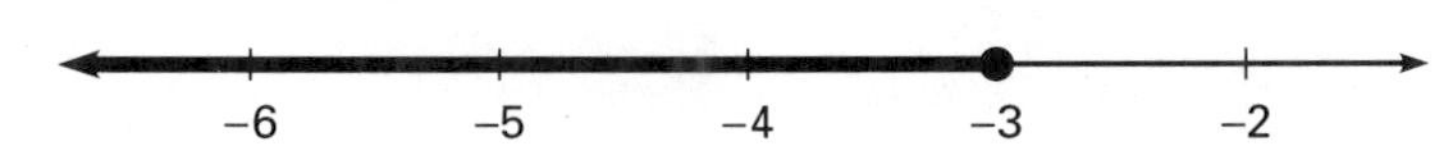

19.

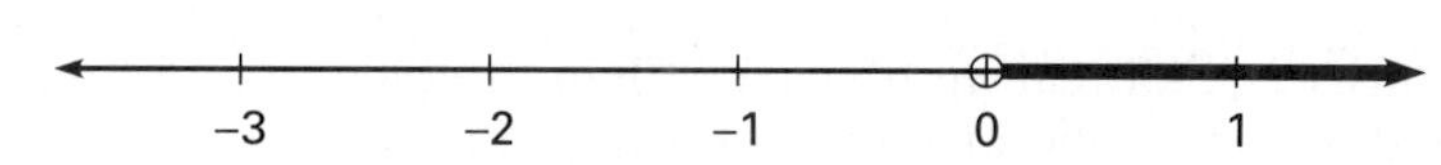

20.

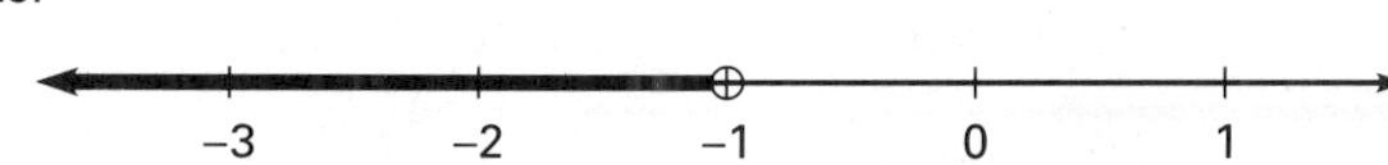

21.

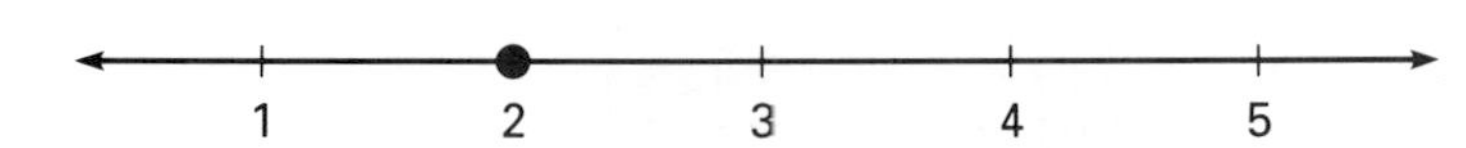

22.

Write an inequality to describe the situation.

23. Gregor hit at least 8 foul shots in every game. Let f equal the number of foul shots that he hit.

24. Sue planned to spend no more than \$2,000 on advertising. Let a equal the amount she planned to spend.

25. Police estimated the size of the crowd as not less than 50,000 people. Let p equal the number of people present.

THINK CRITICALLY/ SOLVE PROBLEMS

Graph on a number line.

26. all values of n that are greater than -3 and less than 3

27. all values of n that are greater than or equal to -2 and less than 2

28. all values of n that are greater than -3 and greater than 1

29. all values of n that are less than 0 and less than or equal to -2

USING DATA Use the table on page 221.
Let a = average of the four scores in Game 1
b = average of the four scores in Game 2
c = average of the four scores in Game 3

Write *true* or *false*.

30. $a \geq b$	31. $c > a$	32. $b < c$
33. $b < 125$	34. $c \leq 138$	35. $a > 113$

7-11 Solving Inequalities

EXPLORE

−8	−4	0	4	12

The integers above are ordered from least to greatest. Write the 5 integers that result when you perform the indicated operation on each of the above integers.

1. add 6
2. add −4
3. subtract 8
4. subtract −3
5. multiply by 5
6. multiply by −1
7. divide by 2
8. divide by −4
9. List the exercises in which the operations you performed resulted in a set of integers ordered least to greatest.
10. List the exercises in which the operations you performed resulted in a set of integers ordered greatest to least.

SKILLS DEVELOPMENT

Most of the equations you have solved have had only one solution. As you discovered in the last section, an inequality may have an infinite number of solutions. To find the solutions of an inequality containing several operations, you follow the rules for solving equations. The single exception to the rules is the following.

► When multiplying or dividing both sides of an inequality by a negative number, reverse the direction of the inequality.

Example 1

Solve and graph $k + 3 \geq -4$.

Solution

$$k + 3 \geq -4$$

Undo the addition. $$k + 3 - 3 \geq -4 - 3$$

Simplify. $$k \geq -7$$

Draw the graph.

The closed circle shows that −7 is a solution.

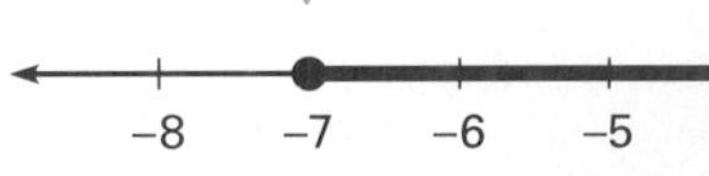

The solution is all numbers greater than or equal to −7. ◄

Example 2

Solve and graph $-3n - 5 < 1$.

Solution

$$-3n - 5 < 1$$

Undo the subtraction. $-3n - 5 + 5 < 1 + 5$

Simplify. $-3n < 6$

Undo the multiplication. $\frac{-3n}{-3} > \frac{6}{-3}$

Dividing by a negative number reverses the direction of the inequality.

Simplify. $n > -2$

Draw the graph.

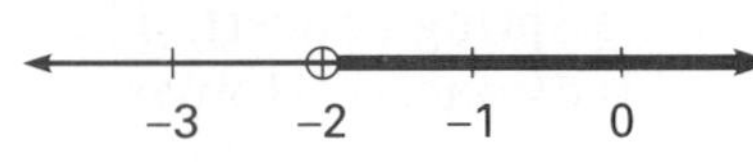

The open circle shows that −2 is not a solution. ◄

Example 3

The pilot of a small plane wanted to stay at least 1,400 ft beneath the storm clouds, which are at 11,000 ft. After takeoff, for how long could the pilot ascend at a rate of 640 ft/min?

Solution

Write and solve an inequality that represents the situation.

Let m = number of minutes ascending at 640 ft/min.

$$640m + 1{,}400 \leq 11{,}000$$

$$640m + 1{,}400 - 1{,}400 \leq 11{,}000 - 1{,}400$$

$$640m \leq 9{,}600$$

$$m \leq 15$$

The pilot can climb for up to 15 min. ◄

CALCULATOR

You can solve inequalities using your calculator.

Solve as though you were solving an equation, but remember to use the correct inequality symbol instead of an equals sign in your answer.

To solve $5.1x - 20.8 \leq -3.46$, use this key sequence.
3.46 [+/−] [+] 20.8 [÷] 5.1 [=] 3.4

On some calculators, you have to press the [=] key after you enter the 20.8.

The solution is $x \leq 3.4$.

TRY THESE

Solve and graph each inequality.

1. $x - 9 \geq -5$
2. $5y < -45$
3. $-2p \leq 10$
4. $-7 < 2k + 3$

Solve.

5. Members of the Music Club hope to raise at least $1,800 from the sale of tickets to the spring concert. They estimate that 450 people will attend the concert. How much should they charge for tickets?

EXERCISES

PRACTICE/ SOLVE PROBLEMS

Solve and graph each inequality.

1. $w - 6 < -2$
2. $3p \geq -12$
3. $x + 9 \leq 3$
4. $5n > 0$
5. $\frac{y}{-3} \geq 1$
6. $4 < \frac{2}{3}h$
7. $15 + t > 17$
8. $12 \leq -4e$
9. $\frac{1}{7}m > -1$
10. $13 \leq c + 8$
11. $-x < 2$
12. $k - (-3) > -2$
13. $6w + 2 > -4$
14. $4 + 5y < 29$
15. $7n + 4 \geq -10$

Solve.

16. Marcia needs to score at least 270 points on 3 tests to earn an A. On her first 2 tests she scored 88 and 87. What must she score on the final to receive an A?
17. Multiply a number by -2. Then subtract 5 from the product. The result is no more than 7. Find the number.

EXTEND/ SOLVE PROBLEMS

Solve.

18. $4\left(x - \frac{1}{2}\right) < 6$
19. $2y - 5y \geq 12$
20. $-2 + 5(2) > -2(x - 1)$
21. $\frac{2}{3}x - \frac{3}{4}x \leq -1$
22. $-1 < 4n - 5n$
23. $15(e + 4) \leq 5(12 - 3e)$
24. Ted and Tom worked more than 14 h painting a garage. Ted worked 2 h more than Tom. Find the least whole number of hours each of them might have worked.

25. On an ocean dive, Tracy wants to stay at least 7 m above her deepest previous depth, 40 m. How long can she descend at a rate of 2.2 m/min?

This program will check your answer for Exercise 28. What happens if the number is both < -4 and > -8? What would you change in line 20 to have it check your answers for Exercises 29–32?

```
10 FOR X = -10 TO 10
20 IF X < -4 AND X > -8
     THEN PRINT X
30 NEXT X
```

THINK CRITICALLY/ SOLVE PROBLEMS

26. Write a problem that could be solved using the inequality $t - 5 > 9$.

27. Write a problem that could be solved using the inequality $2x + 4 < 30$.

Find all integers, if any, that are solutions of both inequalities.

28. $x < -4$ and $x > -8$

29. $x \geq 3$ and $x < 9$

30. $x < -2$ and $x > -5$

31. $x > 0$ and $x < -6$

32. $x \geq 4$ and $x \leq 4$

CHAPTER 7 ● REVIEW

Choose a phrase from the list to complete each statement.

1. An equation that is neither true nor false is ___?___.
2. Multiplication is ___?___ of division.
3. The product of a fraction and ___?___ equals 1.

a. its reciprocal
b. an open sentence
c. the inverse

SECTION 7–1 INTRODUCTION TO EQUATIONS (pages 222–225)

► An equation is a statement that two numbers or expressions are equal.
► An open sentence is a sentence that contains one or more variables.

State whether the equation is *true, false,* or an *open sentence.*

4. $k - 7 = -1$
5. $-4 + 6 = -6 + 4$
6. $-2(3 + 5) = -16$

SECTIONS 7–2 AND 7–4 SOLVE EQUATIONS (pages 226–229 and 232–235)

► Undo an operation by using the inverse operation.

Solve.

7. $f + 22 = -16$
8. $c + 1\frac{1}{2} = 2$
9. $1.4 = e - 1.4$
10. $36 = -18k$
11. $\frac{m}{7} = 4$
12. $3.42 = 2p$

SECTIONS 7–3 AND 7–8 PROBLEM SOLVING (pages 230–231 and 248–249)

► To solve a problem by writing an equation, decide how the numbers in the problem are related. Choose a variable to represent the unknown quantity. Write an equation representing the problem situation.

USING DATA Solve using the table on page 221.

13. The sum of Val's, Mark's, and Evan's scores in Game 1 equaled 307. Write an equation and solve to find Val's score.

SECTION 7–5 EQUATIONS: MORE THAN ONE OPERATION (pages 236–239)

► Undo addition and subtraction first, then multiplication and division.

Solve.

14. $7r + 12 = 5$
15. $6(s - 3) = 12$
16. $3.2x - 2.1 = 4.3$

SECTION 7–6 USING A RECIPROCAL TO SOLVE AN EQUATION (pages 240–243)

► To solve an equation in which the variable is multiplied by a fraction, multiply both sides of the equation by the reciprocal of the fraction.

Solve.

17. $\frac{3}{8}m = 21$

18. $\frac{5}{6}k = 10$

SECTION 7–7 WORKING WITH FORMULAS (pages 244–247)

► When working with a formula, substitute for known values of the variables. Then solve the resulting equation.

19. Find the time it takes to walk $19\frac{1}{8}$ mi at $4\frac{1}{4}$ mi/h.

SECTION 7–9 COMBINING LIKE TERMS (pages 250–253)

► You combine like terms as part of the process of solving an equation with a variable term on both sides.

Simplify.

20. $2n + 5 = n - 4$

21. $2n = 4n + 10$

22. $3y - 6 = -4y + 7$

SECTION 7–10 GRAPHING SOLUTIONS (pages 254–257)

► To indicate that a number is a solution of an equality or inequality, draw a solid dot at the corresponding point on a number line. To indicate that the number is not a solution, draw an open circle at the point.

Graph each open sentence on a number line.

23. $m < -1$

24. $-7 = p - 2$

25. $m \geq 2$

SECTION 7–11 SOLVING INEQUALITIES (pages 260–261)

► Follow all steps for solving equations except for the following situation: When multiplying or dividing both sides of an inequality by a negative number, reverse the direction of the inequality.

Solve and graph each inequality.

26. $x - 4 < -3$

27. $-3n \geq 12$

28. $\frac{h}{-6} + 5 > -2$

CHAPTER 7 ● TEST

Tell whether the equation is *true, false,* or an *open sentence.*

1. $v + 8 = -6$
2. $3(4 - 5) = 3$

Solve each equation.

3. $e - 5 = 2$
4. $-6d = 420$
5. $m + 4.4 = 12.2$
6. $\frac{x}{3} = -5$
7. $5m - 3 = 17$
8. $19 = 2c + 3$
9. $-\frac{5}{6}y = \frac{1}{5}$
10. $\frac{3}{2}x + 4 = 2$
11. $2(y - 3) = 14$
12. $5d + 5 = 2d - 4$

Solve.

13. Convert a temperature of 167°F to the Celsius scale. Use the formula $F = \frac{9}{5}C + 32$ (F = Fahrenheit, C = Celsius).
14. Solve the formula $P = 24 + d$ for d.

Simplify.

15. $x + 2x + 3x^2 + 5x$
16. $7(a + b - c) - 3(a + 2b - 4c)$

Graph each open sentence on a number line.

17. $f \leq -2$
18. $2x - 5 = -7$
19. $2 < 0.5x + 1.5$
20. $-4p + 9 \leq -3$

Write an equation and solve.

21. For a bike-a-thon, Wendy had pledges that totaled \$34.35 for each kilometer she rode. She also collected \$64.55 in other donations. If she collected \$751.55 in all, how far did she ride?
22. Mr. Cintron spent \$5.60 in the delicatessen. Then he spent $\frac{1}{2}$ of the money he had left in the bakery. Finally, he spent \$10.80 in the hardware store. He had \$7.50 left. How much did he have before making his purchases?
23. Marcia earns \$6.90 per hour. Her weekly salary before taxes is \$241.50. How many hours does she work per week?
24. On four trips to the supermarket, Lou drove $3s$ miles, $4p$ miles, $5s$ miles, and $2p$ miles. Write and simplify an expression for the number of miles he drove.

CHAPTER 7 • CUMULATIVE REVIEW

1. If the average of three chemistry test scores was 80, could all three test scores be above 80? Would they all have to be equal to 80?

Simplify.

2. $24 \div (3 + 5) - 1$
3. $64 \div 4 - 5 \times 3$
4. $5 + (7 - 2)^2$
5. $3 \times 5 - (5 + 1)$

6. Determine whether the following argument is *valid* or *invalid*.

   If the fiber is wool, then it is a natural fiber.
   The fiber is a natural fiber.
   Therefore, the fiber is wool.

7. Write two conditional statements from the following two sentences:

   The sun has set.
   The sky will grow dark.

8. Teams A, B, and C played three other teams, X, Y, and Z, but not necessarily in that order. Team A did not play team X. Team A did not play teams X or Y. Which teams played each other?

Find the prime factorization of each number.

9. 40
10. 84
11. 64
12. 120

Write each in exponential form.

13. $\frac{1}{12 \times 12 \times 12}$
14. $\frac{1}{7 \times 7 \times 7 \times 7 \times 7}$

Find each circumference. Round your answer to the nearest tenth or whole number.

15.
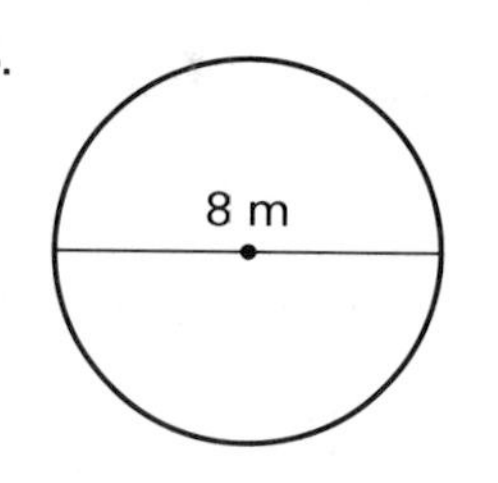

16.
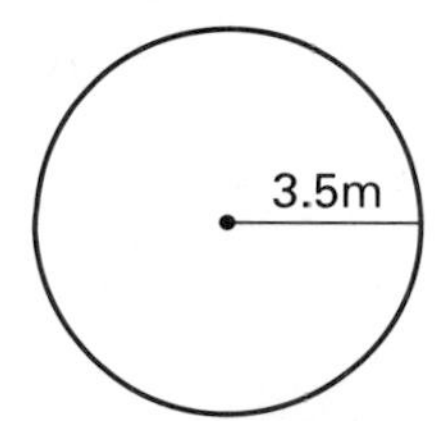

17.
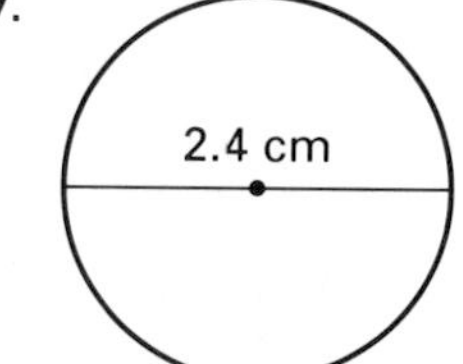

18.
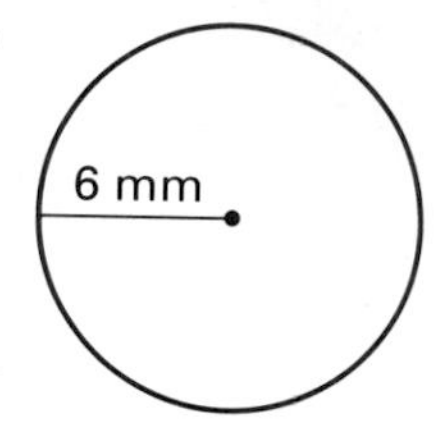

19. Find the perimeter of the figure.

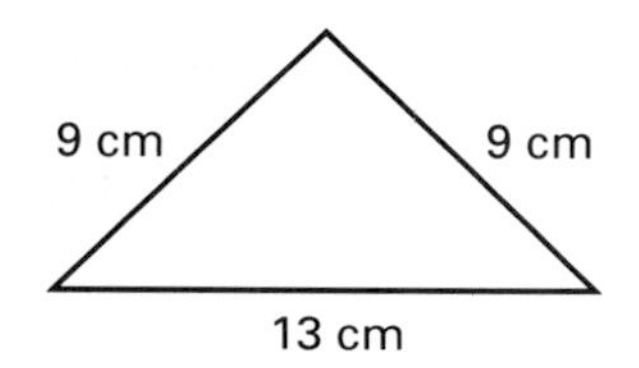

20. Use the formula to find the perimeter of a rectangular figure with these dimensions.
    length: 4.5 m
    width: 10.1 m

Find each answer.

21. $-3 \times (-4)$
22. $-72 \div 6$
23. $-36 \div 12$
24. $6 \times (-25)$
25. $-6 + -2$
26. $7 - (-3)$

Solve each equation or inequality.

27. $-18 > -3t$
28. $24 = 2m - 26$
29. $15 \leq 2d + 10$
30. $7x = -14$

CHAPTER 7 ● CUMULATIVE TEST

1. The mean of three test scores was 85. Two of the scores were 78 and 86. What was the third score?

   A. 91 B. 85 C. 88 D. 100

2. Complete.
   ■$(2 + 9) = (5 \times 2) + (5 \times 9)$

   A. 2 B. 5 C. 9 D. 24

3. Complete.
   $3(6 - 1) = (3 \times$ ■$) - (3 \times 1)$

   A. 15 B. 1 C. 3 D. 6

4. Which is a counterexample for the following statement?

   If a number is divisible by 5, then it is divisible by 50.

   A. 100 B. 75 C. 150 D. 2000

5. Express 3.2×10^{-2} in standard form.

   A. 3,200 B. 0.0032
   C. 3.200 D. 0.032

6. Express 5^{-4} as a fraction.

   A. $\frac{1}{20}$ B. $\frac{1}{625}$ C. $\frac{1}{256}$ D. $-\frac{1}{20}$

7. Compare. 0.6 ■ $\frac{3}{7}$

   A. < B. >
   C. = D. none of these

8. Express $\frac{1}{2 \times 2 \times 2 \times 2 \times 2 \times 2 \times 2}$ in exponential form.

   A. 2^{-7} B. 2^{7} C. 7^{2} D. 7^{-2}

9. Express 0.000045 in scientific notation.

   A. 45×10^{5} B. 4.5×10^{-5}
   C. $4^{5} \times 10$ D. 4.5×10^{-6}

10. Complete. 36 ft = ■ yd

    A. 12 yd B. 3 yd
    C. 48 yd D. 432 yd

11. Complete. 6 yd = ■ in.

    A. 216 in. B. 72 in.
    C. 3 in. D. 0.5 in.

12. Use the formula to find the perimeter of a rectangular figure whose length is 340 mi and whose width is 216 mi.

    A. 1,112 mi B. 556 mi
    C. 73,440 mi D. 1,090 mi

13. Which group of rational numbers is in order from least to greatest.

    A. 0.23, −0.21, −0.7
    B. −0.7, −0.21, 0.23
    C. −0.7, 0.23, −0.21
    D. −0.21, 0.23, −0.7

14. Which group of rational numbers is in order from least to greatest?

    A. $\frac{11}{16}, \frac{5}{8}, -\frac{13}{24}$ B. $\frac{5}{8}, \frac{11}{16}, -\frac{13}{24}$
    C. $-\frac{13}{24}, \frac{11}{16}, \frac{5}{8}$ D. $-\frac{13}{24}, \frac{5}{8}, \frac{11}{16}$

15. Solve. $x + 18 = 15$

    A. $x = 33$ B. $x = 3$
    C. $x = -3$ D. none of these

16. Solve. $-6z = 24$

    A. $z = 4$ B. $z = -4$
    C. $z = 144$ D. none of these

17. Solve. $8 = \frac{x}{2}$

    A. $x = 4$ B. $x = 2$
    C. $x = 16$ D. none of these

18. Solve. $5x - 3 = 12$

    A. $x = 1\frac{4}{5}$ B. $x = 5$
    C. $x = -3$ D. none of these

CHAPTER 8 SKILLS PREVIEW

Write each ratio in two other ways.

1. 2 to 7 **2.** $\frac{3}{5}$ **3.** 6:3 **4.** 5 to 6

5. $\frac{1}{4}$ **6.** 3:4 **7.** 4:2 **8.** $\frac{6}{1}$

9. $\frac{1}{7}$ **10.** 8:4 **11.** 9 to 10 **12.** 5:15

Find the unit rate.

13. \$42 to 7 h **14.** 630 mi:15 h **15.** 304 m to 9.5 s

16. 150 words:5 min **17.** $\frac{\text{392 revolutions}}{\text{7 min}}$ **18.** 200 students to 4 buses

19. 144 cans:12 cartons **20.** $\frac{\text{1,512 m}}{\text{18 s}}$ **21.** 273 mi to 4.2 h

Write three ratios equivalent to the given ratios.

22. $\frac{1}{5}$ **23.** 4:12 **24.** $\frac{50}{60}$

25. $\frac{1}{6}$ **26.** 15:45 **27.** $\frac{3}{5}$

28. 12 to 18 **29.** 28:42 **30.** $\frac{30}{80}$

Solve each proportion.

31. $\frac{7}{8} = \frac{x}{24}$ **32.** $\frac{9}{x} = \frac{3}{2}$ **33.** $\frac{x}{10} = \frac{12}{15}$

34. $\frac{5}{3} = \frac{15}{x}$ **35.** $\frac{26}{65} = \frac{2}{x}$ **36.** $\frac{21}{x} = \frac{7}{2}$

37. $\frac{8}{9} = \frac{32}{x}$ **38.** $\frac{3}{x} = \frac{42}{56}$ **39.** $\frac{x}{5} = \frac{90}{150}$

40. The length of a living-room floor in a scale drawing is 8 in. The actual length of the floor is 20 ft. What is the scale of the drawing?

41. The actual width of a door is 3 feet. If the scale of a drawing of the room is 1 in.:12 ft, what is the drawing length of the width of the door?

42. The scale on a map is 1 in.:30 mi. If the map distance from Acton to Parker is 3 in., what is the actual distance between these two towns?

43. On a scale drawing, the width of a parking lot is 6 in. The actual width of the lot is 108 ft. What is the scale of the drawing?

CHAPTER

8 EXPLORING RATIO AND PROPORTION

THEME Hobbies

Many of our hobbies and sports pastimes involve comparing two numbers or quantities. Batting averages, football turnover ratios, and model building all incorporate the concept of ratios. In this chapter you will solve problems that involve ratios, rates, proportions, and scale drawings.

Statisticians calculate a baseball player's batting average by using this ratio.

$$\text{batting average} = \frac{\text{number of hits}}{\text{number of times at bat}}$$

The ratio is expressed as a decimal rounded to the nearest thousandth.

This table gives statistics for some leading American League batters at the end of a recent season.

PLAYER	TEAM	HITS	AT BATS
Boggs	Boston	187	619
Brett	Kansas City	179	544
Thomas	Chicago	63	191
Polonia	California	135	403
Mattingly	New York	101	394
Sierra	Texas	170	608
Gallego	Oakland	80	389

DECISION MAKING

Using Data

Use the information in the table to answer the following questions.

1. Who had the greatest number of hits?
2. Who had the highest batting average?
3. What was Boggs's batting average for the baseball season?
4. Who had the lower batting average, Mattingly or Sierra? How much lower was it?
5. At the end of the season, who had the higher batting average, Brett or Thomas?

Working Together

Your group's task is to find the players with the top batting averages for a recent baseball season. For your research, use almanacs, sports magazines, or newspapers. Organize your results in a chart. Compare your group's chart with those made by other groups. Did you include both National League and American League players? Were some of the players selected by more than one group?

In 1901, batter Napoleon Lajoie was the American League batting champion with a .422 batting average. Since then, most batting champions have not batted over .366. Why do you think this is so? What possible changes could influence the drop in averages over time?

8-1 Ratios

EXPLORE/ WORKING TOGETHER

Work with a partner. Supply the missing word in this analogy:

Hot is to cold as warm is to ___?___ .

Each partner should make up four analogies. Exchange papers with your partner and find the missing words in the analogies. Discuss the following questions in your group.

1. What is meant by an analogy?
2. How does the relationship of the words in each of your analogies show comparison?
3. The word *analogy* comes from the Greek *analogos,* which means "in due ratio." How is a ratio of one number to another like the relationship between one word and another in an analogy?

SKILLS DEVELOPMENT

A **ratio** is the quotient of two numbers and is used to compare one number to the other. The order of the numbers in a ratio is important.

Example 1

Write each ratio in two other ways.

a. 1 to 3 **b.** $\frac{3}{5}$ **c.** 9:10

Solution

a. 1 to 3 can be written as $\frac{1}{3}$ or as 1:3. ← Read: "one to three"

b. $\frac{3}{5}$ can be written as 3 to 5 or as 3:5. ← Read: "three to five"

c. 9:10 can be written as 9 to 10 or as $\frac{9}{10}$. ← Read: "nine to ten" ◄

Example 2

Write each ratio as a fraction in lowest terms.

a. 2 cm to 6 cm **b.** 50 cm to 2 m

Solution

a. When a ratio is written as a fraction, write it in lowest terms.

$$\frac{2}{6} = \frac{2 \div 2}{6 \div 2} = \frac{1}{3}$$

b. To compare two measurements, use the same unit for both measures. If necessary, rename the larger unit. Then write the ratio.

1 m = 100 cm, so 2 m = 200 cm.

$$\frac{50}{200} = \frac{50 \div 50}{200 \div 50} = \frac{1}{4}$$ ◄

Example 3

Carlene grows geraniums. She feeds the plants once a week with a mixture of liquid plant food and water. This mixture calls for three parts plant food to eight parts water. How much food should Carlene mix with 40 fl oz of water? What is the ratio of plant food to water?

Solution

For 8 fl oz of water, Carlene would need 3 fl oz of plant food. Then, for each additional 8 fl oz of water, she would need an additional 3 fl oz of plant food. Use this pattern to make a table. Extend the table until you arrive at 40 fl oz of water.

	+3	+3	+3	+3	
fluid ounces of plant food	3	6	9	12	15
fluid ounces of water	8	16	24	32	40
	+8	+8	+8	+8	

Carlene should mix 15 fl oz of plant food with 40 fl oz of water. The ratio of plant food to water is $\frac{15}{40}$, or $\frac{3}{8}$. ◄

CHECK UNDERSTANDING

In Example 3, how much plant food should Carlene mix with 56 fl oz of water? What would be the ratio of plant food to water?

TRY THESE

Write each ratio in two other ways.

1. 1 to 2
2. $\frac{4}{8}$
3. 10 to 9
4. 9:13
5. $\frac{6}{7}$
6. 7:10
7. $\frac{8}{9}$
8. 4 to 1

Write the ratio of the first quantity to the second as a fraction in lowest terms.

9. 8 cm to 3 cm
10. 5 kg to 2 kg
11. 25 qt to 5 qt
12. 2 h to 6 h
13. $1.10: 55¢
14. 7 qt to 4 gal
15. 65 min to 3 h
16. 5 m: 2.5 cm

Solve.

17. Andres mixes 2 parts peat moss with 5 parts potting soil for planting his hybrid petunias. How much peat moss should be mixed with 30 parts of potting soil? What is the ratio of peat moss to potting soil?

EXERCISES

PRACTICE/ SOLVE PROBLEMS

Write each ratio in two other ways.

1. 6 to 9
2. 10:3
3. $\frac{8}{1}$
4. 5:4
5. $\frac{1}{3}$
6. 2 to 3

Write each ratio as a fraction in lowest terms.

7. 5 cm to 2 cm
8. 4 min to 20 min
9. 1.5 m to 30 cm
10. 16 gal to 12 qt

Solve. Write the ratio as a fraction in lowest terms.

11. Molly planted marigold seedlings in a window box. There were 21 yellow and 12 orange marigold plants. What is the ratio of yellow to orange plants?
12. A recipe calls for 4 parts cornstarch to 7 parts water. How much cornstarch should be mixed with 28 parts water? What is the ratio of cornstarch to water?

EXTEND/ SOLVE PROBLEMS

Copy and complete each of the following *ratio tables*.

13. The ratio of part A to part B is 5 to 6.

part A	5	10	?	?	?
part B	?	?	18	24	30

14. The ratio of part A to part B is 3 to 4.
The ratio of part B to part C is 1 to 3.

part A	?	6	?	12	?
part B	4	?	12	?	20
part C	?	?	?	48	?

What is the ratio of the width to length in each rectangle? Express the ratio as a fraction in lowest terms.

15.
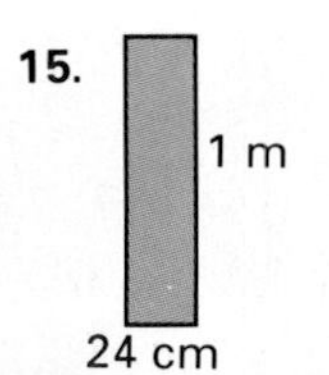

16.

17.
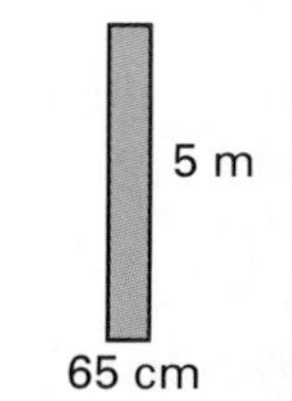

THINK CRITICALLY/ SOLVE PROBLEMS

Compare the ratios using < or >. Describe the method you used to compare.

18. 1 to 3 and 2 to 9
19. 6 to 11 and 5 to 13
20. 3 to 4 and 10 to 12
21. 4 to 9 and 7 to 12

- ► READ
- ► PLAN
- ► SOLVE
- ► ANSWER
- ► CHECK

Problem Solving Applications:

RATIOS IN SPORTS

Football and baseball are two sports in which ratios are used to make comparisons.

For football, statisticians keep track of the percent of successful field goals made by an individual kicker. The kicking ratio is the ratio of successful field goals to the total number of field goals attempted.

$$\text{kicking ratio} = \frac{\text{successful field goal kicks}}{\text{attempted field goals}}$$

For example, if a player had 17 successful kicks out of 21 attempted field goals, that player's *kicking ratio* is $\frac{17}{21}$, or .810. Note that the ratio is expressed as a three-place decimal rounded to the nearest thousandth.

In baseball, a player's *batting average* is calculated using the following ratio.

$$\text{batting average} = \frac{\text{number of hits}}{\text{number of times at bat}}$$

A batting average also is expressed as a decimal rounded to the nearest thousandth.

CHECK UNDERSTANDING

The ratio of 4 to 2 is a whole-number ratio.

$\frac{4}{2} = \frac{2}{1}$, and $\frac{2}{1} = 2$

The first number is divisible by the second.

Rewrite each ratio as a fraction with a denominator of 1.

1. 20 to 4
2. 50 to 2
3. 48 to 8
4. 21 to 3

USING DATA Use the information in the chart on page 268 to answer Exercises 1–2.

1. Brett and Thomas had almost the same batting average for the season. For each, compare the number of hits to the number of times at bat. Tell how their averages can be very close even though Thomas's number of at-bats is much lower than Brett's.
2. What was Polonia's batting average for the season?
3. If a player successfully kicks 15 field goals in 20 attempts, then his kicking ratio is ___?___ .

8-2 Rates

EXPLORE/ WORKING TOGETHER

Work in small groups. Look through newspapers or magazines and cut out articles or advertisements that give examples of rates. Look for advertisements that specify two different units, such as 10 ears of corn for $1.00 or 35 miles per gallon of gasoline.

1. Display the group's advertisements and explain how they may relate to each other.
2. Make a conjecture about why retailers may offer customers a discount, such as 5 cans of motor oil for $9, when the price of a single can is $1.99.
3. Look at the advertisements and discuss how rates are related to ratios.

SKILLS DEVELOPMENT

A **rate** is a ratio that compares two different kinds of quantities. For example, suppose a typist can type 275 words in 5 minutes. The typist's rate is a ratio comparing the total number of words typed (275) to the amount of time it takes to type that number of words (5 minutes).

Example 1

Simplify the rate *275 words in 5 minutes.*

Solution

$$\begin{matrix}\text{words} \rightarrow \\ \text{minutes} \rightarrow\end{matrix} \frac{275}{5} = \frac{275 \div 5}{5 \div 5} = \frac{55}{1}$$

The rate is 55 words in 1 minute. ◄

A rate that has a denominator of 1 unit is called a **unit rate.** The rate *55 words in 1 minute* is an example of a unit rate. You can write this unit rate as 55 words per minute, or 55 words/min.

READING MATH

How is a rate related to ratio? What makes a rate different from a ratio?

Example 2

The Hortons drove to a sports car show in another town. Their car consumed 12 gallons of gasoline on the 390-mile trip. How many miles per gallon did they average?

Solution

$$\begin{matrix}\text{miles} \rightarrow \\ \text{gallons} \rightarrow\end{matrix} \frac{390}{12} = \frac{390 \div 12}{12 \div 12} = \frac{32.5}{1}$$

The Hortons' car averaged 32.5 miles per gallon of gasoline. You can write this as 32.5 mi/gal. ◄

TRY THESE

Find the unit rate.

1. $\frac{\$63}{7\text{ h}}$
2. $\frac{702\text{ mi}}{13\text{ h}}$
3. $\frac{299\text{ meters}}{11.5\text{ seconds}}$
4. $\frac{140\text{ students}}{5\text{ teachers}}$
5. $\frac{450\text{ words}}{9\text{ min}}$
6. $\frac{190\text{ revolutions}}{5\text{ min}}$
7. 80 km in 2 h
8. $120 for 15 days
9. 65 mi in 2 h
10. 125 words in 5 min
11. 608 mi in 8 h
12. 360 km on 6 gal

Solve.

13. Darlene works in a bakery. She makes 520 muffins in an 8-hour shift. How many muffins does she make in one hour?
14. Chuck read 45 pages in 60 minutes. What is his reading rate in pages per minute?

EXERCISES

PRACTICE/ SOLVE PROBLEMS

Find the unit rate.

1. $\frac{300 \text{ mi}}{6 \text{ h}}$
2. $\frac{\$42}{5 \text{ h}}$
3. $\frac{440 \text{ words}}{8 \text{ min}}$
4. $\frac{\$12.15}{9 \text{ gal}}$
5. $\frac{\$20}{4 \text{ lb}}$
6. $\frac{264 \text{ mi}}{22 \text{ gal}}$
7. $260 for 4 rooms
8. $272 for 4 tires
9. 6,000 km in 12 days
10. 420 ft in 14 min

Solve.

11. Fiona drove 550 mi in 10 h. What was her average speed in miles per hour?
12. Elaine bought 7 yd of fabric for $69.65. What did the fabric cost per yard?

EXTEND/ SOLVE PROBLEMS

Which is the lower rate?

13. 2,133 km in 6 h or 1,498 km in 7 h
14. 495 km on 45 L or 175 km on 25 L
15. 144 students in 3 buses or 208 students in 4 buses
16. 88 h in 4 weeks or 108 h in 5 weeks
17. $3.96 for 36 oz or $6.30 for 42 oz

Solve.

18. Will Raymond earn more money working 10 h per week at $4.75/h or 12 h per week at $4.25/h?

THINK CRITICALLY/ SOLVE PROBLEMS

19. Jerome's car uses 18.75 gal of gasoline to travel 900 mi. What is the rate of fuel consumption in miles per gallon? What is the rate in gallons per mile?

20. The answers to Exercise 19 are each unit rates. Explain how they differ from one another.

► READ
► PLAN
► SOLVE
► ANSWER
► CHECK

Problem Solving Applications:

UNIT PRICE

A rate that identifies the cost of an item per unit is called the **unit price.** Many retailers and grocers post the unit price of an item as well as the total price. Sometimes, the retailer will offer a lower unit price to consumers when they buy a larger quantity.

Price-conscious consumers often calculate the unit price to determine the better buy. To find the unit price:

► Write a ratio of the price to the number of units.
► Rewrite the ratio so that the denominator represents 1 unit.
► Round the numerator to the nearest cent, if necessary.

Find the unit price. Round to the nearest cent, if necessary.

1. As the office manager, Megan buys supplies for a doctor's office. One brand of copy paper sells for $11.98 for 500 sheets. What is the unit price?
2. A large pizza costs $14.98 and serves 6 people. What is the price per serving?
3. The manager at the Fruit Stand is selling a 3-lb bag of apples for $2.19. What is the unit price?
4. The Deli King is selling 1 dozen bagels for $4.20. What does one bagel cost?
5. A one-gallon container of skim milk sells for $1.99. A half gallon sells for $.98. Which is the better buy?
6. Pasta Palace is selling a 3-lb package of spaghetti for $2.45. The cost of a 32-oz package is $1.70. There is a 25-cents-off coupon on the 32-oz package which is redeemable at the checkout stand. Which package is the better buy?

COMPUTER

This computer program asks you to enter a price and number of units. It then computes, rounds to the nearest cent, and prints the unit price.

```
10 INPUT "ENTER THE PRICE:
   ";P: PRINT
20 INPUT "ENTER THE
   NUMBER OF UNITS: ";N:
   PRINT
30 U = INT (10 ^ 2 * (P / N) +
   0.5) / 10 ^ 2
40 PRINT "UNIT PRICE = ";U
```

TALK IT OVER

What factors, other than price, might influence a shopper to buy one item over another?

8-3 Equivalent Ratios

EXPLORE/ WORKING TOGETHER

Work with a partner.

Carmine is making fruit punch for a party. The recipe calls for 2 parts cranberry juice to 7 parts pineapple juice.

1. Copy and complete the table to show the ratios of cranberry juice to pineapple juice, in fluid ounces.

parts of cranberry juice	2	4	?	?	?	?	14	?
parts of pineapple juice	?	?	21	28	35	42	?	56

2. How much cranberry juice is needed for 35 fl oz of pineapple juice?
3. How many fluid ounces of pineapple juice are needed for 12 fl oz of cranberry juice?
4. Look at your table. What patterns do you notice?
5. How could you have found the answers to questions 2 and 3 without using a table?

SKILLS DEVELOPMENT

Two ratios that represent the same comparison are called **equivalent ratios.** One way to find a ratio equivalent to a given ratio is to multiply both terms of the given ratio by the same nonzero number.

Example 1

Find three ratios equivalent to 2:3.

Solution

Write the ratio as a fraction. Multiply the numerator and the denominator by the same number.

$$\frac{2}{3} = \frac{2 \times 2}{3 \times 2} = \frac{4}{6}$$

$$\frac{2}{3} = \frac{2 \times 3}{3 \times 3} = \frac{6}{9}$$

$$\frac{2}{3} = \frac{2 \times 4}{3 \times 4} = \frac{8}{12}$$

The ratios 4:6, 6:9, and 8:12 are equivalent to 2:3. ◄

Another way to find a ratio equivalent to another ratio is to *divide* both terms of the given ratio by the same nonzero number.

Example 2

Find three ratios equivalent to 20 to 30.

Solution

Write the ratio as a fraction.

$$\frac{20}{30} = \frac{20 \div 2}{30 \div 2} = \frac{10}{15} \qquad \frac{20}{30} = \frac{20 \div 5}{30 \div 5} = \frac{4}{6} \qquad \frac{20}{30} = \frac{20 \div 10}{30 \div 10} = \frac{2}{3}$$

The ratios 10 to 15, 4 to 6, and 2 to 3 are equivalent to 20 to 30. ◄

To determine if two given ratios are equivalent, you write each ratio as a fraction in lowest terms.

Example 3

Is each pair of ratios equivalent?

a. 10:12, 15:18

b. 8 to 6, 12 to 8

Solution

a. $\frac{10}{12} = \frac{10 \div 2}{12 \div 2} = \frac{5}{6}$

$\frac{15}{18} = \frac{15 \div 3}{18 \div 3} = \frac{5}{6}$

Yes, the ratios 10:12 and 15:18 are equivalent.

b. $\frac{8}{6} = \frac{8 \div 2}{6 \div 2} = \frac{4}{3}$

$\frac{12}{8} = \frac{12 \div 4}{8 \div 4} = \frac{3}{2}$

No, the ratios 8 to 6 and 12 to 8 are not equivalent. ◄

Example 4

Jane answered 5 out of 6 questions correctly on the lesson quiz and 25 out of 30 correctly on the chapter test. Was the ratio of the number of correct answers to the number of questions the same for both the quiz and the test?

Solution

Write the ratios as fractions in simplest form.

5 out of 6 questions → $\frac{5}{6}$

25 out of 30 questions → $\frac{25}{30} = \frac{25 \div 5}{30 \div 5} = \frac{5}{6}$

Yes, the ratios of Jane's scores—5 out of 6 correct on the lesson quiz and 25 out of 30 correct on the chapter test—were the same. ◄

TRY THESE

Use multiplication to find three equivalent ratios.

1. $\frac{1}{4}$
2. 4:6
3. 2 out of 7

Use division to find three equivalent ratios.

4. $\frac{30}{50}$
5. 18 to 24
6. 15:75

Are the ratios equivalent? Write *yes* or *no*.

7. 1 to 3, 25 to 75
8. 3:2, 15:5
9. $\frac{7}{3}, \frac{42}{18}$
10. 30:98, 10:15
11. $\frac{4}{5}, \frac{12}{18}$
12. 3 to 4, 6 to 8
13. Julio answered 51 out of 60 questions correctly on a history test. He answered 17 out of 20 questions correctly on a math quiz. Was the ratio of correct answers to questions the same for both the test and the quiz?

EXERCISES

PRACTICE/ SOLVE PROBLEMS

Write three equivalent ratios for each given ratio.

1. $\frac{6}{7}$
2. 15 to 18
3. 3:4
4. 16:20
5. 5 to 3
6. $\frac{7}{8}$
7. $\frac{8}{9}$
8. 12:14
9. 3:10

Are the ratios equivalent? Write *yes* or *no*.

10. 4 to 5, 12 to 15
11. 8:2, 16:9
12. $\frac{14}{20}, \frac{170}{100}$
13. 24 to 40, 144 to 80
14. $\frac{2}{3}, \frac{12}{18}$
15. 5:8, 15:32
16. $\frac{8}{14}, \frac{24}{44}$
17. $\frac{6}{7}, \frac{36}{42}$
18. $2:5, \frac{100}{250}$
19. Janna appeared on the show *Trivia Bowl.* She answered 7 out of 10 questions correctly during the first round and 28 out of 40 questions correctly in the second round. Was the ratio of the number of correct answers to the number of trivia questions the same for both rounds?
20. The school reading club took a survey. They discovered that in the winter 17 out of 21 readers regularly read mystery novels, but in the spring 68 out of 84 readers chose adventure novels. Was the ratio of the number of readers of mystery novels the same as the ratio for readers of adventure novels?

MIXED REVIEW

1. Solve : $\frac{2}{5}x = 6$
2. Solve : $3x + 5 = 14$
3. Estimate the quotient: $4{,}507 \div 23$
4. Find the product: $\frac{3}{4} \times \frac{5}{8}$
5. Find the sum: $\frac{2}{9} + \frac{7}{12}$

21. The local car dealership was running a clearance sale in November. They sold 6 out of 9 midsize cars and 24 out of 45 economy cars. Was the ratio for midsize cars sold the same as the ratio for economy cars sold? Write a number statement to justify your answer.

EXTEND/ SOLVE PROBLEMS

22. Write a ratio equivalent to $a{:}b$, if $a = 5$ and $b = 4$.

23. Write a ratio equivalent to x *to* y, if $x = 45$ and $y = 50$.

24. Write a ratio equivalent to $\frac{d}{e}$, if $d = 17$ and $e = 25$.

Jewelry that contains gold is measured in karats (k), with pure gold being 24 k. If you have a 10-k gold ring, the ratio of the mass of pure gold in the ring to the mass of the ring is 10:24.

25. Express the gold content of a gold ring marked 18 k as a ratio in lowest terms.

26. Express the gold content of a gold earring marked 14 k as a ratio in lowest terms.

THINK CRITICALLY/ SOLVE PROBLEMS

It is possible to write ratios with three numbers. For example, 30:85:60 means 30 compared to 85 compared to 60.

27. Write the ratio 30:85:60 in lowest terms.

28. Write two 2-number ratios using the ratio 30:85:60.

29. Write two equivalent 3-number ratios for 30:85:60.

8-4 Solving Proportions

EXPLORE/ WORKING TOGETHER

Work with a partner.

Twenty-eight boys were accepted on the football team. Two out of every seven boys who tried out were accepted on the team. How many boys tried out for the football team?

1. Write a ratio to represent how many boys were accepted on the football team.
2. Explain how you might solve this problem using an equation.
3. Tell how to find an equivalent ratio for $\frac{2}{7}$ using the number 28 as the numerator of the ratio.

SKILLS DEVELOPMENT

A **proportion** is an equation that states that two ratios are equivalent.

You can write a proportion two ways.

$\frac{2}{3} = \frac{6}{9}$ $\qquad$ $2:3 = 6:9$

Read: "Two is to three as six is to nine."

COMPUTER

Identify what A, B, C, and D represent in this program. What happens if the cross-products are not equal?

```
10 INPUT "ENTER A,B,C,D:
     ";A,B,C,D
20 IF A * D = B * C THEN
     GOTO 40
30 PRINT "NOT A
     PROPORTION"
35 END
40 PRINT "PROPORTION"
```

The numbers 2, 3, 6, and 9 are called the **terms** of the proportion. If a statement is a proportion, the **cross-products** of the terms are equal.

$\frac{2}{3} = \frac{6}{9}$ $\qquad$ $3 \times 6 = 18$
$2 \times 9 = 18$ ← cross-products

Example 1

Tell whether the statement is a proportion.

$$\frac{3}{4} \stackrel{?}{=} \frac{21}{28}$$

Solution

Find the cross-products.

$3 \times 28 = 84$
$4 \times 21 = 84$

$84 = 84$, so $\frac{3}{4} = \frac{21}{28}$

Yes, the statement is a proportion. ◄

Example 2

Tell whether the statement is a proportion.

$$2:7 \stackrel{?}{=} 6:10$$

Solution

Write each ratio as a fraction. Find the cross products.

$$\frac{2}{7} \stackrel{?}{=} \frac{6}{10} \qquad \begin{aligned} 2 \times 10 &= 20 \\ 7 \times 6 &= 42 \end{aligned}$$

$$20 \neq 42, \text{ so } 2:7 \neq 6:10$$

No, the statement is not a proportion. ◀

Sometimes one term of a proportion is unknown.
You can use cross-products to *solve* this type of proportion.

Example 3

Solve the proportion.

$$\frac{n}{16} = \frac{15}{24}$$

Solution

Write the cross-products.

$$n \times 24 = 15 \times 16$$
$$24n = 240$$
$$\frac{24n}{24} = \frac{240}{24} \leftarrow \text{Divide both sides of the equation by 24.}$$
$$n = 10$$

Check by substituting the value for n. Then determine if the two ratios are equivalent.

$$\frac{10}{16} = \frac{10 \div 2}{16 \div 2} = \frac{5}{8}$$
$$\frac{15}{24} = \frac{15 \div 3}{24 \div 3} = \frac{5}{8} \qquad \text{So } n = 10.$$ ◀

Write the cross-products of the terms.

1. $\frac{24}{6} = \frac{12}{3}$
2. $\frac{25}{40} = \frac{5}{8}$
3. $\frac{x}{2} = \frac{148}{12}$

Example 4

A baseball pitcher pitched 12 losing games. The ratio of wins to losses was 3:2. How many games did the pitcher win?

Solution

Let n represent the number of games the pitcher won.
Write and solve a proportion.

$$\text{wins} \rightarrow \frac{3}{2} \leftarrow \text{losses} \quad = \quad \frac{n}{12} \begin{array}{l}\leftarrow \text{wins} \\ \leftarrow \text{losses}\end{array}$$

$$3 \times 12 = 2 \times n$$
$$36 = 2n$$
$$\frac{36}{2} = \frac{2n}{2}$$
$$18 = n$$

Check: $\frac{3}{2} = \frac{3}{2} \qquad \frac{18}{12} = \frac{18 \div 6}{12 \div 6} = \frac{3}{2}$

The pitcher won 18 games. ◀

TRY THESE

Tell whether each statement is a proportion. Write *yes* or *no*.

1. $\frac{3}{9} \stackrel{?}{=} \frac{9}{27}$
2. $\frac{2}{5} \stackrel{?}{=} \frac{12}{11}$
3. $4:12 \stackrel{?}{=} 7:36$
4. $7:9 \stackrel{?}{=} 49:63$

Solve each proportion.

5. $\frac{x}{6} = \frac{30}{36}$
6. $\frac{4}{3} = \frac{a}{45}$
7. $n:5 = 16:40$
8. $12:3 = y:18$
9. $\frac{x}{8} = \frac{21}{56}$
10. $3:15 = b:45$
11. If a cyclist travels 10 mi in 2 h, how far will the cyclist travel in 5 h?

EXERCISES

PRACTICE/ SOLVE PROBLEMS

Tell whether each statement is a proportion. Write *yes* or *no*.

1. $\frac{2}{3} \stackrel{?}{=} \frac{4}{6}$
2. $\frac{4}{8} \stackrel{?}{=} \frac{1}{4}$
3. $6:8 \stackrel{?}{=} 3:4$
4. $5:6 \stackrel{?}{=} 40:48$
5. $\frac{8}{5} \stackrel{?}{=} \frac{15}{24}$
6. $2:7 \stackrel{?}{=} 6:21$

CALCULATOR

You can use a calculator to solve a proportion.

Solve: $9:m = 15:10$

Write the ratios as fractions.

$\frac{9}{m} = \frac{15}{10}$

Then use cross-products to find the answer.

Enter: 9 [×] 10 [÷] 15 [=]

The solution is 6.
9:6 = 15:10

Write the key sequence for each proportion.

1. $\frac{2.4}{3.6} = \frac{x}{1.8}$
2. $m:22.5 = 7:9$
3. $\frac{4.4}{z} = \frac{11}{2.5}$

Solve each proportion. Check your answers.

7. $\frac{1}{2} = \frac{n}{16}$
8. $\frac{h}{15} = \frac{20}{75}$
9. $\frac{3}{10} = \frac{x}{60}$
10. $r:28 = 7:4$
11. $b:12 = 2:24$
12. $\frac{6}{15} = \frac{z}{40}$
13. $\frac{n}{8} = \frac{1}{32}$
14. $3:2 = s:50$
15. $8:y = 2:4$

Solve.

16. An artist makes green paint by mixing blue paint and yellow paint in the ratio 1:2. How many parts of yellow paint would be mixed with 6 parts of blue paint to make green paint?
17. Robert swims 30 laps in 20 minutes. How many laps can he swim in 40 minutes?

EXTEND/ SOLVE PROBLEMS

Solve each proportion.

18. $\frac{13}{10} = \frac{52}{x}$ **19.** $\frac{78}{m} = \frac{18}{12}$ **20.** $56:32 = x:4$

21. $9:21 = 15:m$ **22.** $z:39 = 4:12$ **23.** $\frac{x}{81} = \frac{5}{9}$

24. $\frac{11}{44} = \frac{12}{n}$ **25.** $6:x = 42:77$ **26.** $\frac{0.004}{0.12} = \frac{x}{1.8}$

27. $\frac{4.5}{6} = \frac{a}{20}$ **28.** $\frac{0.9}{3.6} = \frac{1.2}{b}$ **29.** $\frac{2.4}{32} = \frac{y}{16}$

30. The basketball team scored 38 points in 30 minutes. At that rate, how many points are they likely to score in 1 hour?

31. Jan can run 525 yards in 7.5 minutes. Express her average speed in yards per minute.

32. ***USING DATA*** Refer to the Data Index on page 546 to answer this question: The ratio of the length of a 2-penny nail to the length of a 20-penny nail is 1:4. How long is a 20-penny nail?

THINK CRITICALLY/ SOLVE PROBLEMS

33. Write a proportion using the numbers 9, 27, 35, and 105.

34. The track-team star runs 0.5 miles in 4 minutes. What is the star's rate in feet per minute?

35. The width and length of a rectangular plot are in the ratio 3:8. The perimeter of the lot is 440 ft. What are the dimensions of the width and length?

36. The cost of a tool shed and the cost of a lawn mower are in the ratio 5:2. The total cost of the shed and the mower is $2,100. What is the cost of each?

8-5 Problem Solving Skills:

USING PROPORTIONS TO SOLVE PROBLEMS

- ► READ
- ► PLAN
- ► SOLVE
- ► ANSWER
- ► CHECK

Before you begin to solve a problem, you need to decide which mathematical ideas or skills will help you find the solution. How would you solve this problem?

PROBLEM

Sue's hobby is making ceramic masks. Out of every 20 masks she makes, 3 are clown masks. If Sue makes 240 masks, how many of them will be clown masks?

SOLUTION

You know that 3 out of every 20 masks are clown masks. This tells you that the ratio of clown masks to all the masks is 3:20.

If Sue makes 240 masks, the ratio of clown masks to all masks will still be 3:20. So, you can solve the problem by finding a ratio equivalent to 3:20 in which the second term is 240.

To find an equivalent ratio, you can write and solve a proportion. Remember that the two sides of the proportion must compare the quantities in the same order. Let a variable such as x represent the number of clown masks.

$$\frac{x}{240} = \frac{3}{20}$$

(number of clown masks → x; total number of masks → 240; 3 ← number of clown masks; 20 ← total number of masks)

$$\frac{3}{20} = \frac{3 \times 12}{20 \times 12} = \frac{36}{240}$$ ← Write a ratio equivalent to $\frac{3}{20}$.

$$x = 36$$ ← number of clown masks in 240

Notice that an alternative way to solve the proportion would be to use cross-products. Therefore, whenever Sue makes 240 masks, 36 of them will be clown masks.

The words *ratio* and *proportion* did not appear in the problem, yet you found that you could use a proportion to solve it.

PROBLEMS

MENTAL MATH TIP

You may be able to solve certain problems mentally, rather than by writing a proportion. For Problem 1, ask yourself how many groups of 25 statuettes there are in 200. Then how many groups of 8 birds would there be?

1. Evan's hobby is carving wooden statuettes. Out of every 25 statuettes he makes, 8 of them are birds. If Evan carves 200 statuettes, how many of them will be birds?
2. Cindy makes model airplanes. She paints 8 out of every 20 models she makes. If Cindy makes 65 model planes, how many of them will she paint?
3. Marta uses color film for 48 out of every 72 photographs she takes. If she takes 156 photographs, how many will be on color film?

4. The students in the Art Club used 3 L of paint for 36 pictures. How many liters of paint would they need for 60 pictures?
5. Out of every 30 students, 23 belong to an after-school club. How many students out of every 120 students belong to an after-school club?
6. Ben makes ceramic pots. He decorates 11 out of every 15 pots he makes and leaves the rest undecorated. If Ben makes 60 pots, how many of them will be undecorated?
7. The Crafts Club used 4 lb of clay to make 20 sculptures. If 5 lb more clay had been used, how many sculptures could have been made?

8-6 Scale Drawings

EXPLORE/ WORKING TOGETHER

Work with a partner. Each partner should create a simple design on a section of graph paper. (Be sure that your design is drawn only on the grid lines of the graph paper.) Exchange designs with your partner. Copy your partner's design, but make the following change. For every small square that your partner's design covers, your design should cover 4 small squares that are arranged in the shape of a large square. Discuss the results with your group.

SKILLS DEVELOPMENT

A **scale drawing** is a drawing that represents a real object. All lengths in the drawing are proportional to the actual lengths in the object. The ratio of the size of the drawing to the size of the actual object is called the **scale** of the drawing.

To find the actual length or the drawing length, set up and solve a proportion.

Example 1

The scale of this drawing is 1 in.:2 ft. Find the actual length of the mural.

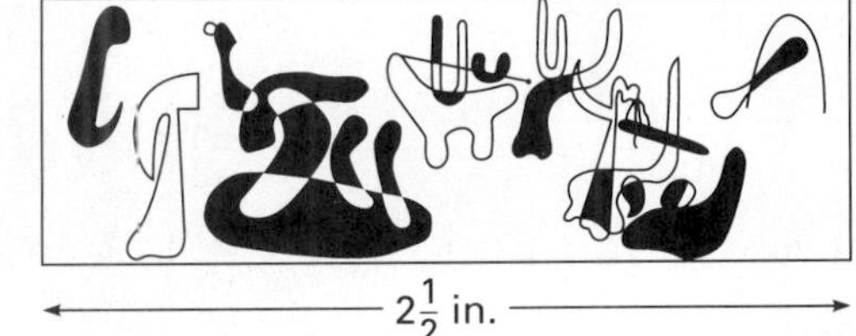

Solution

Write and solve a proportion. Let n represent the actual length.

$$\begin{array}{r} \text{drawing length (inches)} \rightarrow \\ \text{actual length (feet)} \rightarrow \end{array} \frac{1}{2} = \frac{2\frac{1}{2}}{n}$$

$$1 \times n = 2 \times 2\frac{1}{2} \quad \leftarrow \text{Write the cross-products.}$$

$$n = 5$$

The actual length of the mural is 5 ft. ◄

Example 2

The scale of a drawing is 1 in.:3 ft. Find the drawing length that would be used to represent an actual length of 12 ft.

Solution

Write and solve a proportion.

Let n represent the length in the drawing.

$$\begin{array}{r} \text{drawing length (inches)} \rightarrow \\ \text{actual length (feet)} \rightarrow \end{array} \frac{1}{3} = \frac{n}{12}$$

$$1 \times 12 = 3n \quad \leftarrow \text{Write the cross-products.}$$

$$\frac{12}{3} = \frac{3n}{3}$$

$$4 = n$$

The drawing length would be 4 in. ◄

Example 3

The length of the bedroom in this scale drawing is 3 in. The actual length is 15 ft.

a. What is the scale of the drawing?

b. What is the actual width of the bedroom?

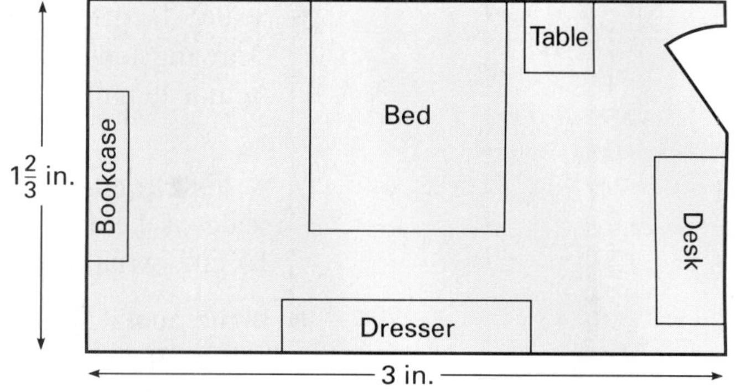

Solution

a. Use the information given to write a ratio. Then simplify the ratio.

drawing length (inches) → / actual length (feet) → $\frac{3}{15} = \frac{3 \div 3}{15 \div 3} = \frac{1}{5}$

The scale of the drawing is 1 in.:5 ft.

b. Use the scale to write and solve a proportion. Let w represent the actual width of the bedroom.

drawing length (inches) → / actual length (feet) → $\frac{1}{5} = \frac{1\frac{2}{3}}{w}$

$1 \times w = 5 \times 1\frac{2}{3}$ ← Write the cross-products.

$w = 8\frac{1}{3}$ The actual width of the bedroom is $8\frac{1}{3}$ ft. ◄

TRY THESE

Find the actual length for each drawing length given.

1. Drawing length: 8 in.
   Scale: 1 in.:2 ft
2. Drawing length: 4 cm
   Scale: 1 cm:0.5 m
3. Drawing length: 12 in.
   Scale: 1 in.:6 yd
4. Drawing length: 5 in.
   Scale: 1 in.:7 mi

Find the drawing length for each actual length given.

5. 35 ft
   Scale: 1 in.:7 ft
6. 32 mi
   Scale: 1 in.:8 mi
7. 4.5 km
   Scale: 1 cm:0.5 km
8. 30 yd
   Scale: 1 in.:$2\frac{1}{2}$ yd
9. Jane made a scale drawing of her living room. The length of the living room in her drawing is 4 in. The actual length is 20 ft. What scale did Jane use for her drawing?

MIXED REVIEW

1. Express 3.8 as the quotient of two integers.
2. Write $\frac{14}{54}$ in lowest terms.
3. Estimate the sum: 54 + 39 + 21 + 15
4. Write an equation that represents the sentence: A number, *b*, divided by five is twenty.

EXERCISES

PRACTICE/ SOLVE PROBLEMS

Find the actual or drawing length.

1. Scale: 1 cm:6 m
   Drawing length: 3 cm
   Actual length: ?
2. Scale: 1 cm:12 km
   Drawing length: ?
   Actual length: 120 km
3. Scale: 1 in.:22 ft
   Drawing length: ?
   Actual length: 330 ft
4. Scale: 1 in.:7 mi
   Drawing length: 9 in.
   Actual length: ?

Find the actual or drawing length.

5. Scale: 1 cm:5 m
   Drawing length: ?
   Actual length: 55 m

6. Scale: 1 cm:8.5 m
   Drawing length: 6 cm
   Actual length: ?

7. A scale model of a square-based statue is constructed using the scale of 1 cm: 0.1 m. If the length of the base of the model is 14 cm, what is the length of the base of the statue?

8. If the actual height of the statue in Exercise 7 is 3.5 m, find the height of the scale model.

EXTEND/ SOLVE PROBLEMS

9. A movie company made a large model of a domino. The actual width of the domino is 15 mm. If the width on the scale model is 6 m, what is the scale of the model?

Actual and drawing lengths are given in the table below. Express the scale in the form 1 cm:_?_ km or 1 in.:_?_ mi.

	ACTUAL LENGTH	DRAWING LENGTH
10.	8 km	2 cm
11.	50 mi	5 in.
12.	100 km	25 cm
13.	$416\frac{1}{2}$ mi	$8\frac{1}{2}$ in.

14. The scale 1 in.:3 ft was used to make a scale drawing of Anne's bedroom. In the drawing, the bedroom is 5 in. long and 2 in. wide. Could a bed 72 in. long and 42 in. wide fit along one of the walls?

15. The actual dimensions of a deck are 10 ft by 15 ft. If the dimensions of a scale drawing of the deck are 2 in. by 3 in., what is the scale of the drawing?

THINK CRITICALLY/ SOLVE PROBLEMS

16. Make a scale drawing of a room in your home. Measure the length and width of the room to the nearest inch. Choose a scale for your drawing. Make and cut out a scale model for each piece of furniture. Position the cutouts on the scale drawing. Be sure to write the scale on your drawing.

17. Make a scale drawing of your classroom. Measure the room to the nearest inch. Decide on a scale to use. Show where windows, desks, and tables are located. Be sure to include the scale.

Problem Solving Applications:

MAPS

Cartography is the art of representing a geographical area graphically, usually on a flat surface such as a map. It is an ancient discipline that dates from the prehistoric depiction of hunting and fishing territories.

The discovery of the New World led to the need for new techniques in map making, and by the seventeenth and eighteenth centuries printed maps had increased in accuracy. Today, cartography involves the use of aerial photographs as a base for some maps.

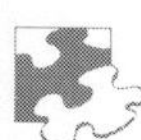

Use a customary ruler to measure each map distance. Then use the scale 1 in. : 15 mi to find the actual distance.

1. school to library
2. town pool to post office
3. fire station to general store
4. hospital to school
5. Which two locations are farthest apart?
6. Which two locations are closest together?

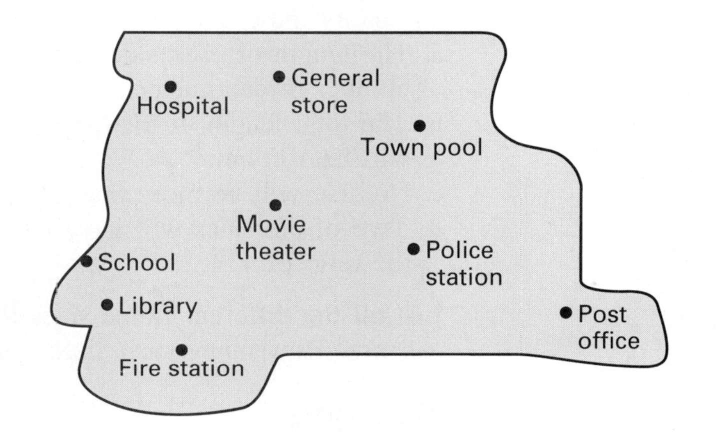

Copy and complete the table.

	Scale	Distance on a Map	Actual Distance
7.	1 cm represents 10 km	22 cm	?
8.	1 : 1500	?	15,000 m
9.	1 in. : 200 mi	?	250 mi
10.	1 cm : 0.4 km	20 cm	?

11. On a map, 1 cm represents 50 km. If the actual distance between two cities is 230 km, what is the distance on the map?
12. Create a map of your neighborhood. Choose an appropriate scale and include your favorite places to visit.
13. How is a map like a scale drawing?

8-7 Problem Solving Strategies:

MAKE AN ORGANIZED LIST

► READ
► PLAN
► SOLVE
► ANSWER
► CHECK

Often a problem calls for finding the number of ways in which something can be done. You may decide to list all the possible ways and then count them. Depending on how you choose to organize your list, though, the process could be either long and confusing or clear and systematic.

PROBLEM

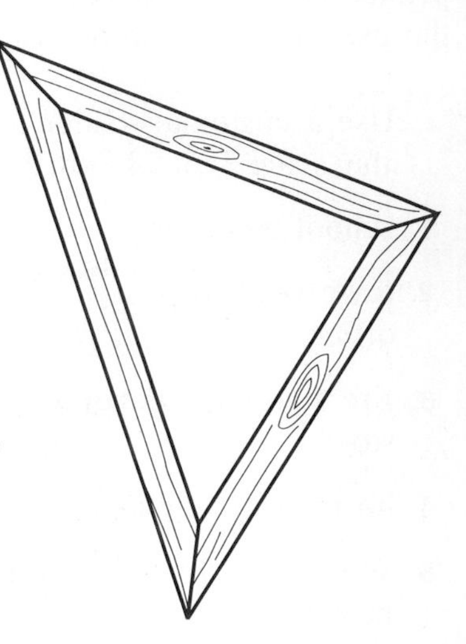

Ramón makes triangular frames for collages. He has decided that the frames he makes this month will all satisfy these conditions:

a. The length of each side is a whole number of centimeters.
b. The total length of the three sides will be 100 cm.
c. No side will be more than 50 cm long.
d. Two of the sides will have lengths in the ratio of 4:3.

List all the different frame sizes Ramón can use. How many such sizes are there?

SOLUTION

You could try every possible grouping of three numbers that add up to 100 and then check whether each grouping satisfies all of Ramón's conditions. However, this would take a very long time and you would probably overlook some groupings. Instead, you can organize your list of frame sizes in a way that saves you from considering most of the impossible groupings.

From conditions **a** and **d** you can reason that one side of any frame must have a length that is a multiple of 4. Since no side can be longer than 50 cm, begin by listing a frame with a side 48 cm long.

Lengths of sides in ratio of 4:3		Length of third side
48 cm	36 cm	16 cm
44 cm	33 cm	23 cm
40 cm	30 cm	30 cm
36 cm	27 cm	37 cm
32 cm	24 cm	44 cm
28 cm	21 cm	51 cm

$\frac{48}{36} = \frac{4}{3}$;

16 cm → $100 - (48 + 36) = 16$

51 cm → This frame would not satisfy condition c.

If you continue listing frames whose first two sides are shorter than 28 cm, the third side will always violate condition **c.** Thus, by making your list in a logical and organized way, you have quickly found the only frame sizes Ramon could make.

PROBLEMS

1. Suppose Ramón changed only condition **d** to have two sides of the frame in the ratio of 3:2. Complete the following chart to find how many different frames he could then make.

Lengths of sides in ratio of 3:2		Length of third side
■ cm	32 cm	■ cm
■ cm	■ cm	25 cm
42 cm	■ cm	■ cm
■ cm	■ cm	■ cm
■ cm	■ cm	■ cm
■ cm	■ cm	■ cm
■ cm	■ cm	■ cm

Make an organized list to find all the possible answers for each problem.

2. How many different 4-digit numbers can be made with the digits 2, 3, 4, and 5? No digit may be used more than once in any number.
3. In how many different ways can you make 50 cents using nickels, dimes, or nickels and dimes?
4. In how many different ways can you make a one-dollar-and-fifteen-cent bus fare with nickels, dimes, and at least two quarters?
5. How many different 4-digit odd numbers can be made with the digits 5, 6, 7, and 0? No digit may be used more than once in any number.
6. How many 4-digit even numbers can be made with the digits 4, 5, 6, and 8? No digit may be used more than once in any number.
7. In a student-council election, there are 3 candidates for president: Mack, Billie, and Sue; 3 candidates for vice-president: Marge, Lu, and Dana; and 2 candidates for secretary-treasurer: Steve and Mel. In how many different combinations could these candidates be elected?
8. How many different 3-digit numbers can be made from the digits 6, 7, 8, and 9? No digit may be used more than once in any number.

COMPUTER

This program will let you determine how many numbers can be made from a given number of digits when no digits repeat. Can you use the program to figure out how you could do it on a calculator?

```
10 LET P = 1
20 INPUT "HOW MANY
   DIGITS? ";N: PRINT
30 FOR X = N TO 1 STEP
   - 1
40 LET P = P * X
50 NEXT X
60 PRINT "THERE ARE ";P;"
   NUMBERS."
```

CHAPTER 8 • REVIEW

1. A statement that two ratios are equal is called a ___?___ .
2. A ___?___ is a ratio of two quantities with different units of measure.
3. A drawing of an object in which the lengths in the drawing are proportional to the actual lengths is called a ___?___ drawing.
4. A quotient of two numbers that is used to compare the numbers is called a ___?___ .
5. The cost per unit of a given item is the ___?___ .

a. ratio
b. rate
c. proportion
d. unit price
e. scale

SECTION 8–1 RATIO (pages 270–273)

► A ratio compares two numbers by division.
► A ratio can be written in three ways:

1:2   1 to 2 $\frac{1}{2}$

Write each ratio in two other ways.

6. 9:10
7. $\frac{1}{3}$
8. 4 to 5

SECTION 8–2 RATES (pages 274–277)

► A rate is a ratio that is used to compare two different kinds of quantities.
► A unit rate is the ratio of a number to 1.

Find the unit rate.

9. $\frac{\$150}{5\text{ h}}$
10. $\frac{48\text{ revolutions}}{6\text{ s}}$
11. $\frac{112.5\text{ m}}{7.5\text{ s}}$

SECTION 8–3 EQUIVALENT RATIOS (pages 278–281)

You can find equivalent ratios in either of these ways.

► Multiply both terms of the ratio by the same nonzero number.
► Divide both terms of the ratio by the same nonzero number.

Write two equivalent ratios for each given ratio.

12. 3:9
13. 20:200
14. $\frac{5}{50}$
15. 16 to 20

SECTION 8–4 SOLVING PROPORTIONS (pages 282–285)

- A proportion is an equation that states that two ratios are equivalent.
- To solve a proportion, use the cross-products to write an equation.

Solve each proportion.

16. $\frac{6}{4} = \frac{y}{1.4}$

17. $\frac{16}{8} = \frac{x}{3}$

18. $\frac{x}{15} = \frac{2.2}{3.3}$

SECTION 8–5 PROBLEM SOLVING SKILLS (pages 286–287)

- You can solve problems by writing and solving a proportion.

19. Stan collects postage stamps. For every 3 United States stamps in his collection, he has 7 foreign stamps. If Stan has 420 foreign stamps, how many United States stamps does he have?

SECTION 8–6 SCALE DRAWINGS (pages 288–291)

- A scale drawing represents a real object. All lengths in the drawing are proportional to the actual lengths in the object.

Find the actual or scale length.

20. Scale: ?
Drawing length: 4 in.
Actual length: 80 mi

21. Scale: 1 cm:16 m
Drawing length: 21 cm
Actual length: ?

22. Scale: 1 cm:4 km
Drawing length: 6 mm
Actual length: ?

SECTION 8–7 PROBLEM SOLVING STRATEGIES (pages 292–293)

- An organized list helps you solve problems by showing information in a way that accounts for all possibilities and avoids repetition.

23. In how many different ways can you make $1.00 using quarters and nickels or nickels and at least five dimes?

24. ***USING DATA*** Refer to the data index on page 546 to find the chart for baseball at-bats. Write a ratio of Gallego's hits to his at-bats.

CHAPTER 8 • TEST

Write each ratio in two other ways.

1. 5 to 4
2. $\frac{6}{7}$
3. 7:11
4. 8 to 10
5. $\frac{1}{3}$
6. 2:3
7. 9:7
8. $\frac{5}{2}$
9. $\frac{4}{7}$
10. 8:4
11. 7 to 21
12. 4:28

Find the unit rate.

13. \$72 to 8 h
14. 636 mi:12 h
15. 342 m to 9.5 s
16. 220 words:5 min
17. $\frac{252 \text{ revolutions}}{6 \text{ min}}$
18. 120 students to 4 teachers
19. 72 boxes:6 cases
20. 228 m/12 h
21. 852 mi to 12 h

Write three ratios equivalent to each given ratio.

22. $\frac{1}{6}$
23. 2:10
24. $\frac{100}{200}$
25. $\frac{1}{8}$
26. 16:24
27. $\frac{2}{5}$
28. 11 to 22
29. 18:90
30. $\frac{30}{70}$

Solve each proportion.

31. $\frac{22}{10} = \frac{x}{5}$
32. $\frac{x}{25} = \frac{4}{10}$
33. $\frac{1.5}{x} = \frac{10}{20}$
34. $\frac{28}{32} = \frac{56}{x}$
35. $x:1.5 = 8:3$
36. $\frac{15}{21} = \frac{40}{x}$
37. $\frac{52}{39} = \frac{x}{3}$
38. $\frac{5}{8} = \frac{x}{56}$
39. $\frac{x}{2.8} = \frac{5}{20}$

40. The length of a bedroom floor in a scale drawing is 6 in. The actual length of the floor is 15 ft. What is the scale of the floor plan?
41. The actual height of a door on a train is 6 ft. If the scale of a model train is 1 in.:4 ft, what is the height of the door in the model?
42. The scale on a map is 1 in.:50 mi. If the map distance from Morton to Curryville is 3.5 in., what is the actual distance between these two towns?
43. On a scale drawing, the width of a corner lot is 8 in. The actual width of the lot is 200 ft. What is the scale of the drawing?

The following data show the temperature readings, in degrees Celsius, at ten locations inside three kilns.

Kiln A: 240 250 230 260 250 250 220 240 250 210
Kiln B: 260 250 240 270 250 250 230 260 250 240
Kiln C: 220 240 280 275 260 210 255 240 230 285

1. Find the mean, median, mode, and range for each.

Write in standard form.

2. 356×10^4
3. 6.324×10^5
4. 82.64×10^5
5. 2.0001×10^4

Write each answer in exponential form.

6. $4^5 \times 4^2$
7. $b^{14} \div b^7$
8. $25^{10} \div 25^2$
9. $p^6 \times p^3$

---

10. Sonia, Theo, Jay, and Kim are not in the same first period class. English, math, French, and computer lab are taught in first period. Theo has English first period. Sonia does not have French. Jay and Sonia have computer lab during second period. Which class is each student in during first period?

11. Determine whether the following argument is *valid* or *invalid*.

    If the water is too cold, then you will not be able to swim.
    The water is too cold.
    Therefore, you will not be able to swim.

Replace ● with $<$, $>$, or $=$.

12. 0.47 ● $\frac{7}{13}$
13. $\frac{5}{9}$ ● 0.54
14. 0.6 ● $\frac{3}{7}$
15. $\frac{1}{8}$ ● 0.125

Complete.

16. 0.6 m = ■ mm
17. 62 g = ■ kg
18. 64 in. = ■ ft ■ in.
19. 100 oz = ■ lb ■ oz

20. Find the perimeter.

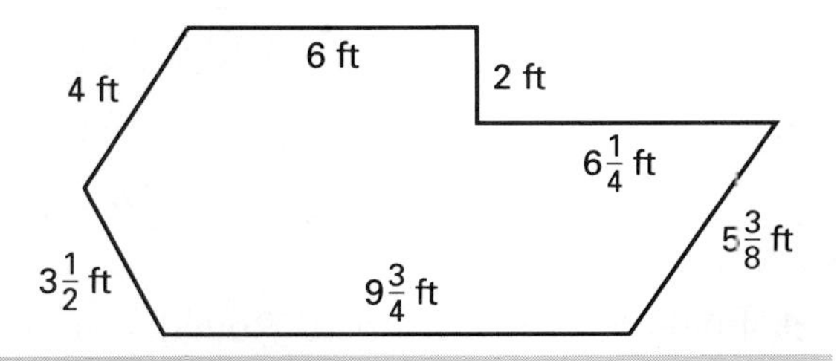

Find each answer.

21. $-4 + (-21)$
22. $9 - (-13)$
23. $5 \times (-8)$
24. $27 \div (-3)$

Solve each inequality. Graph the solution set on a number line.

25. $-6 + x \geq 2$
26. $-e < 5$
27. $t + 2 \leq -1$
28. $x - 2 < -6$

Write each ratio in two other ways.

29. 16:20
30. 2 to 5

Solve each proportion.

31. $\frac{3}{6} = \frac{9}{x}$
32. $5:8 = x:24$
33. $\frac{5}{x} = \frac{30}{42}$
34. $\frac{5}{6} = \frac{x}{48}$

Are the ratios equivalent? Write *yes* or *no*.

35. 10:15, 12:18
36. $\frac{15}{10}, \frac{10}{15}$

CHAPTER 8 ● CUMULATIVE TEST

1. If you had 250 total points for three tests, what score would you need on the next test to receive an 85 average?
   A. 50 B. 95 C. 130 D. 90

2. Which conclusion can be drawn from these statements?
   If the plant is a violet, its flowers are small.
   The flowers are small.
   A. The plant is a violet.
   B. The plant is not a violet.
   C. The plant is sick.
   D. none of these

3. Express 3^{-5} as a fraction.
   A. $\frac{1}{243}$ B. $\frac{1}{125}$ C. $\frac{3}{3{,}125}$ D. $\frac{3}{243}$

4. Find the circumference. Round your answer to the nearest tenth.
   A. 75.6 m
   B. 37.7 m
   C. 80.4 m
   D. 40.1 m

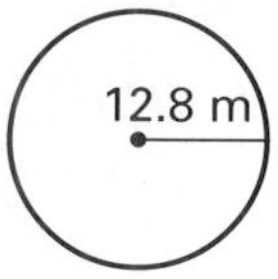

5. Find the perimeter of a rectangular figure whose length is 7.2 km and whose width is 5.4 km.
   A. 25.2 km B. 12.6 km
   C. 18 km D. 38.88 km

6. Add. $-16 + (-8)$
   A. -24 B. -8
   C. -8 D. 128

7. Subtract. $24 - (-6)$
   A. -4 B. -18
   C. 30 D. none of these

8. Multiply. $-6 \times (-2)$
   A. 3 B. -8 C. -4 D. 12

9. Divide. $-15 \div 3$
   A. -12 B. -5 C. -18 D. 45

10. Which graph shows the solution set?
    $-180 \leq 60x$

    A. 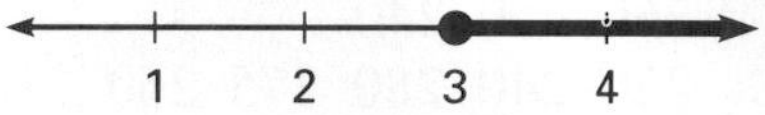

    B. 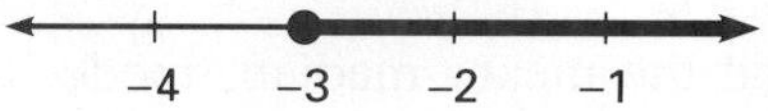

    C. 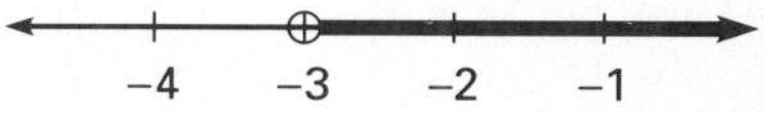

    D. none of these

11. Solve. $4 + r = 4$
    A. $r = 0$ B. $r = -1$
    C. $r = 1$ D. none of these

12. Solve. $40 = 6x - 8$
    A. $x = \frac{1}{3}$ B. $x = 6\frac{2}{3}$
    C. $x = 8$ D. none of these

13. What is the value of y in the proportion $\frac{32}{8} = \frac{16}{y}$?
    A. 3 B. 5 C. 4 D. 6

14. Which of the following ratios is equivalent to 12:9?
    A. 9:12 B. 4:2
    C. 6:9 D. 4:3

15. Which of the following ratios is equivalent to 26:130?
    A. 5:1 B. 1:5
    C. 1:6 D. none of these

16. What is the unit rate for $\frac{\$54}{9\text{ h}}$?
    A. $\frac{\$54}{9}$ B. $\frac{\$6}{1\text{ h}}$
    C. $\frac{1}{6}$ D. none of these

CHAPTER 9 SKILLS PREVIEW

Write each as a percent.

1. 0.3
2. $\frac{1}{4}$
3. 0.007

Find the percent of each number.

4. 18% of 30
5. 40% of 500
6. $12\frac{1}{2}$% of 70
7. 3.2% of 45
8. 15% of 120
9. 33% of 512

Find each percent.

10. What percent of 50 is 30?
11. What percent of 56 is 39.76?
12. 63 is what percent of 450?
13. 4.38 is what percent of 6?
14. What percent of 20 is 7?
15. 12 is what percent of 15?

Find the percent of increase or decrease. Round to the nearest tenth.

16. Original amount: $535
    New amount: $642
17. Original amount: $145
    New amount: $174
18. Original amount: 125
    New amount: 120
19. Original amount: 80
    New amount: 70
20. Original amount: $64
    New amount: $40
21. Original amount: $30
    New amount: $39

Find the unknown number.

22. 17 is 50% of what number?
23. $37\frac{1}{2}$% of what number is 72?
24. 12 is 75% of what number?
25. 150% of what number is 36?
26. 12 is 40% of what number?
27. 5% of what number is 12.8?

Find the discount and the sale price. Round to the nearest cent.

28. Regular price: $179.89
    Percent of discount: 20%
29. Regular price: $45.23
    Percent of discount: 15%

Find the interest and the amount.

30. Principal: $320
    Rate: 5%/year
    Time: 1 year
31. Principal: $7,200
    Rate: 8.5%/year
    Time: 24 months
32. Principal: $819
    Rate: 13%/year
    Time: 6 months

CHAPTER 9 EXPLORING PERCENT

THEME Shopping

Buying a car involves some of the most important—and most difficult—purchasing decisions most people ever make. Often the first decision is whether to buy a new or a used car. There are other factors to consider as well. What make and model should be purchased? What options should be chosen? How will the car be paid for? If a car loan is needed, what is the best financing arrangement?

In this chapter, you will solve problems that involve percents, discount, and simple interest.

This chart shows what happens to the value of a new car as it gets older. The car's value drops dramatically during the first three years of ownership.

NEW AUTOMOBILE DEPRECIATION SCHEDULE

Year	1	2	3	4	5–7	8+
% Decrease*	30–32	24–26	18–20	7–8	3–5	1–2

*The percent given in each column represents a decrease in the value of the car from the year before.

DECISION MAKING

Using Data

Use the information in the chart to answer the following questions.

1. Why do you think the percent of decrease in a car's value drops so dramatically after the first three years?
2. Suppose you wanted to buy a new car. You already own a car that is about three years old. Would you sell the car independently or would you try to trade it in at the dealership? What factors would you consider before making the decision? Does the depreciation schedule assist you?

Working Together

Your group's task will be to research and "buy" a new car. One person will be the car buyer, another will be the car dealer, and the third will be the banker. Look through newspapers, car magazines, and consumer magazines and decide which make and model car to buy. The car dealer should look for ads with the different prices and options for the model the car buyer has chosen. The banker should meet with both the dealer and the buyer and decide whether or not the bank will loan the buyer some of the money to buy the car. The banker should explain the terms and conditions of the bank loan. The buyer should analyze all the information and decide if this is the car to buy.

Make a list of all of the factors to consider before making your decision. Compare your lists with those of other groups. What is the most reasonable thing to do?

9-1 Exploring Percent

EXPLORE/ WORKING TOGETHER

Work with a partner. Outline a ten-unit-by-ten-unit square on grid paper. Shade 25 one-unit squares.

a. Write a ratio that compares the number of shaded squares to the total number of squares.

b. Write the ratio as a fraction and a decimal.

c. Make a similar model to show the ratio 32:100.

d. What does the ratio 32:100 represent?

SKILLS DEVELOPMENT

A ratio that compares a number to 100 is called a **percent.** Percent means *per one hundred.* The symbol % is used to show percent.

This grid has 35 out of the 100 squares shaded blue.

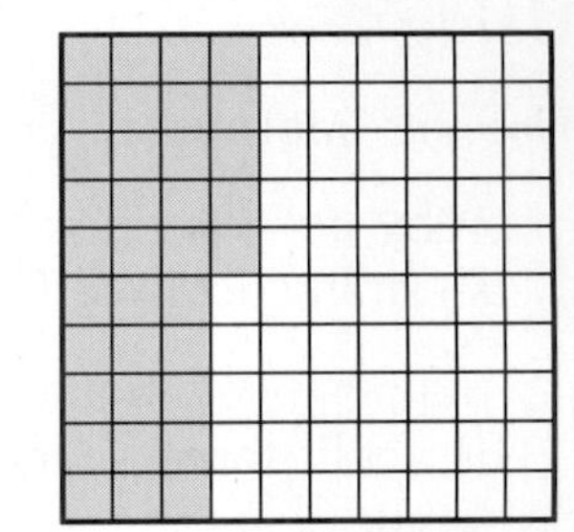

The ratio $\frac{35}{100}$ means 35 per 100, so 35% of the grid is shaded blue.

Example 1

Write each percent as a decimal.

a. 45% **b.** 5.9% **c.** $7\frac{1}{2}\%$

Solution

a. $45\% = \frac{45}{100} = 0.45$

b. $5.9\% = \frac{5.9}{100} = \frac{5.9 \times 10}{100 \times 10} = \frac{59}{1,000} = 0.059$

c. $7\frac{1}{2}\% = \frac{7\frac{1}{2}}{100} = \frac{7.5}{100} = \frac{7.5 \times 10}{100 \times 10} = \frac{75}{1,000} = 0.075$ ◄

Example 2

Write each percent as a fraction.

a. 35% **b.** 0.2% **c.** $\frac{3}{4}\%$

CHECK UNDERSTANDING

How is 0.2 different from 0.2%?

Solution

Write as a fraction with a denominator of 100 and simplify.

a. $35\% = \frac{35}{100} = \frac{7}{20}$

b. $0.2\% = \frac{0.2}{100} = \frac{0.2 \times 10}{100 \times 10} = \frac{2}{1000} = \frac{1}{500}$

c. $\frac{3}{4}\% = \frac{\frac{3}{4}}{100} = \frac{3}{4} \div 100 = \frac{3}{4} \times \frac{1}{100} = \frac{3}{400}$ ◄

Example 3

Write each decimal as a percent.

a. 0.6 **b.** 0.025 **c.** $0.66\frac{2}{3}$

Solution

a. $0.6 = 0.60 = \frac{60}{100} = 60\%$

b. $0.025 = \frac{25}{1{,}000} = \frac{25 \div 10}{1{,}000 \div 10} = \frac{2.5}{100} = 2.5\%$

c. $0.66\frac{2}{3} = \frac{66\frac{2}{3}}{100} = 66\frac{2}{3}\%$ ◄

Example 4

Write each fraction as a percent.

a. $\frac{3}{5}$ **b.** $\frac{5}{8}$ **c.** $\frac{1}{12}$

Solution

a. $\frac{3}{5} = \frac{3 \times 20}{5 \times 20} = \frac{60}{100} = 60\%$

b. $\frac{5}{8} = 0.625 = \frac{625}{1{,}000} = \frac{625 \div 10}{1{,}000 \div 10} = \frac{62.5}{100} = 62.5\%$

c. $\frac{1}{12} = 0.08\frac{1}{3} = \frac{8\frac{1}{3}}{100} = 8\frac{1}{3}\%$ ◄

MENTAL MATH TIP

Here is a shortcut for changing a percent to a decimal or a decimal to a percent.

Since percent means per hundred, 45% means 45 per hundred, or 45/100, or 45 ÷ 100.

So a quick way to change 45% to a decimal is to divide 45 by 100 by moving the decimal point 2 places to the left.

45% = 0.45

A quick way to change a decimal such as 0.68 to a percent is to multiply by 100 by moving the decimal point 2 places to the right.

0.68 = 68%

Example 5

A basketball team won 15 out of 25 games. What percent of the total games played did the team win?

Solution

Write the ratio as a fraction.

$\frac{15}{25}$ ← games won / ← total games played

Write a fraction that has a denominator of 100, and then write the percent.

$$\frac{15}{25} = \frac{15 \times 4}{25 \times 4} = \frac{60}{100} = 60\%$$

The basketball team won 60% of the total games played. ◄

TRY THESE

Write each percent as a decimal.

1. 38%
2. 6.3%
3. $8\frac{1}{4}\%$

Write each percent as a fraction.

4. 32%
5. 15%
6. $\frac{1}{2}\%$

Write each decimal as a percent.

7. 0.4 8. 0.032 9. $0.56\frac{1}{4}$

Write each fraction as a percent.

10. $\frac{7}{10}$ 11. $\frac{3}{8}$ 12. $\frac{5}{12}$

Solve.

13. John's batting average is 0.325. What percent of the times he has been at bat has he had a hit?

EXERCISES

PRACTICE/ SOLVE PROBLEMS

Write each percent as a decimal.

1. 50% 2. 42% 3. 17% 4. 95%
5. 8.5% 6. 16.4% 7. 88.1% 8. 7.2%
9. $13\frac{1}{2}\%$ 10. $3\frac{1}{4}\%$ 11. $5\frac{3}{4}\%$ 12. $2\frac{1}{3}\%$

Write each percent as a fraction.

13. 55% 14. 19% 15. 20% 16. 64%
17. 0.4% 18. 2.4% 19. 0.7% 20. 0.5%
21. $\frac{3}{8}\%$ 22. $5\frac{1}{2}\%$ 23. $16\frac{1}{4}\%$ 24. $1\frac{3}{4}\%$

Write each decimal as a percent.

25. 0.2 26. 0.55 27. 0.79 28. 0.09
29. 0.015 30. 0.113 31. 0.005 32. 0.061
33. $0.12\frac{1}{2}$ 34. $0.09\frac{1}{4}$ 35. $0.16\frac{2}{3}$ 36. $0.03\frac{1}{2}$

Write each fraction as a percent.

37. $\frac{12}{100}$ 38. $\frac{4}{5}$ 39. $\frac{3}{4}$ 40. $\frac{1}{2}$
41. $\frac{7}{8}$ 42. $\frac{5}{6}$ 43. $\frac{5}{12}$ 44. $\frac{1}{3}$

Solve.

45. Michelle attended three out of every 20 away football games. What percent of the away football games did she attend?

46. Jonathan received a score of 80% on a history quiz. What fraction of the questions did he answer correctly?

CONNECTIONS

Only about a quarter of the earth's surface is covered by land. The rest is covered by water. The table below shows what fraction of the total land is occupied by each continent. Write each fraction as a percent.

Continent	Fraction of Earth's Land
Africa	$\frac{1}{5}$
Antarctica	$\frac{19}{200}$
Asia	$\frac{3}{10}$
Australia	$\frac{1}{20}$
Europe	$\frac{13}{200}$
North America	$\frac{4}{25}$
South America	$\frac{3}{25}$

EXTEND/ SOLVE PROBLEMS

Copy and complete the chart.

	Fraction	Decimal	Percent
47.	$\frac{1}{20}$	■	■
48.	■	■	30%
49.	$\frac{6}{15}$	■	■
50.	■	■	15%

Solve.

51. Sally received \$60 for her birthday. She bought a game for \$45. What percent of her money did she spend on the game?

52. In a referendum, 18 people out of 200 voted against adding a slide to the town pool. What percent voted against the slide?

53. Over the summer, Peter read 9 out of 20 library books. What percent of the books did he read?

THINK CRITICALLY/ SOLVE PROBLEMS

In Exercises 54–57, decide whether it makes sense to rewrite the fraction or the decimal as a percent. Write *yes* or *no*. If your answer is *yes*, then write the percent.

54. Renee hiked a distance of 3.8 kilometers.

55. Two thirds of the people surveyed did their grocery shopping after work or on the weekend.

56. By the end of the day, Sharon had finished only 0.7 of the job.

57. Harold added one quarter pound of ground meat to the recipe.

Solve.

58. What percent of 1 meter is 1 centimeter?

59. What percent of 1 yard is 1 foot?

60. What percent of 1 year is 1 month?

9-2 Finding the Percent of a Number

EXPLORE

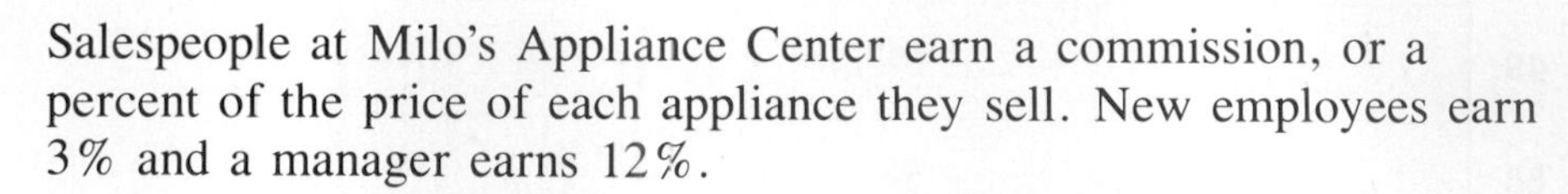

Salespeople at Milo's Appliance Center earn a commission, or a percent of the price of each appliance they sell. New employees earn 3% and a manager earns 12%.

This 10-by-10 grid represents the price of a large-screen television. The shaded part represents the manager's commission.

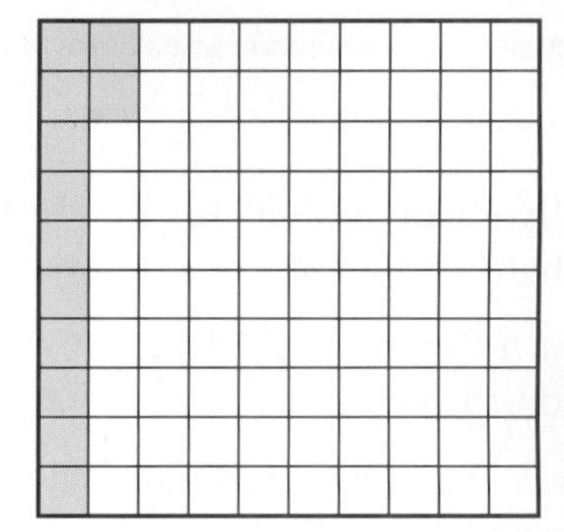

a. What is the ratio of the number of shaded parts to the total number of squares?

b. What fraction of the squares is shaded?

c. Suppose the price of the television was $2,000. How much money would the manager receive as a commission?

SKILLS DEVELOPMENT

CALCULATOR TIP

% KEY

You can use the percent key (%) on your calculator to find the percent of a number.

To find 40% of 360, you can use this key sequence:

360 [×] 40 [%]

On some calculators, you have to press the [=] key to get the answer.

To find a percent of a number, you can write and solve an equation.

Example 1

What number is 60% of 45?

Solution 1

Find the percent of a number by writing an equation using a decimal.

What number is 60% of 45?

$$x = 0.60 \times 45$$
$$x = 27$$

Let x represent the unknown number

Solution 2

Find the percent of a number by writing an equation using a fraction.

What number is 60% of 45?

$$x = \frac{3}{5} \times 45$$
$$x = 27$$

So 60% of 45 is 27. ◄

Another way to find the percent of a number is to write and solve a proportion. You write the proportion using this relationship.

$$\frac{\text{part}}{\text{whole}} = \frac{\text{part}}{\text{whole}}$$

One of the ratios in this proportion represents the percent. The "whole" referred to in this ratio is 100.

COMPUTER

If you enter the correct "part" of the percent ratio and the correct "whole" of the number ratio when asked, this program will find a given percent of a given number for you.

```
10 INPUT "ENTER % PART:
   ";P: PRINT
20 INPUT "ENTER NUMBER
   WHOLE: ";W: PRINT
30 PRINT P;"*";W;" = 100X"
35 PRINT P * W;" = 100X"
40 X = (P * W) / 100
50 PRINT X;" = X"; PRINT
60 INPUT "RUN AGAIN?
   Y OR N: ";X$: PRINT
70 IF X$ = "Y" THEN
   GOTO 10
```

Example 2

Find $5\frac{1}{2}\%$ of 30.

Solution

Write $5\frac{1}{2}\%$ as the ratio $\frac{5\frac{1}{2}}{100}$. Use this ratio to write a proportion.

$$\begin{matrix}\text{part} \rightarrow \\ \text{whole} \rightarrow\end{matrix} \frac{5\frac{1}{2}}{100} = \frac{x}{30} \begin{matrix}\leftarrow \text{part} \\ \leftarrow \text{whole}\end{matrix}$$

Then solve the proportion.

$$\frac{5\frac{1}{2}}{100} = \frac{x}{30}$$

$$5\tfrac{1}{2} \times 30 = 100x$$

$$165 = 100x$$

$$\frac{165}{100} = \frac{100x}{100}$$

$$1.65 = x$$

So $5\frac{1}{2}\%$ of 30 is 1.65. ◄

Example 3

Keiko bought a pocket camera for 20% off the original price of \$45. How much money did she save?

Solution

Write and solve an equation.

What amount is 20% of \$45?

$$x = \frac{1}{5} \times 45$$

$$x = 9$$

Keiko saved \$9. ◄

TRY THESE

Find the percent of each number using an equation.

1. 40% of 75
2. 22% of 50
3. 25% of 48
4. 90% of 30

Find the percent of each number using a proportion.

5. 16% of 125
6. 30% of 70
7. $7\frac{1}{2}\%$ of 60
8. 15% of 85

Solve.

9. What number is 35% of 25?
10. Larry bought stereo speakers on sale for 32% off the original price of \$735. How much money did he save?

EXERCISES

PRACTICE/ SOLVE PROBLEMS

Find the percent of each number using an equation.

1. 87% of 80
2. 25% of 600
3. 20% of 93
4. 72% of 90
5. 60% of 55
6. 25% of 10
7. $33\frac{1}{3}$% of 189
8. $66\frac{2}{3}$% of 210
9. $87\frac{1}{2}$% of 560

Find the percent of each number using a proportion.

10. 30% of 210
11. 80% of 25
12. 40% of 125
13. 60% of 80
14. 45% of 60
15. 75% of 36
16. $12\frac{1}{2}$% of 92
17. $8\frac{1}{2}$% of 100
18. $66\frac{2}{3}$% of 60

Solve.

19. What number is 83% of 20?
20. Mario plans to make a 15% down payment on a new mountain bike that sells for $355. What will his down payment be?
21. The regular price of a compact disc player is $210. It is now on sale for 22% off its original price. What is the amount of the savings?
22. A gallon of gasoline costs $1.50. Of that amount, 32% goes to state, local, and highway taxes. How much money per gallon goes to taxes?

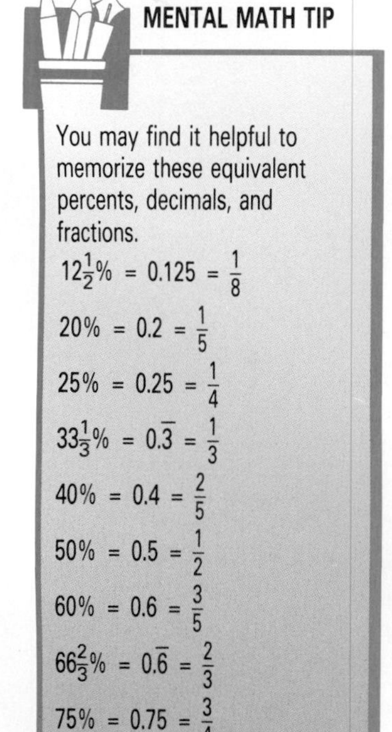

MENTAL MATH TIP

You may find it helpful to memorize these equivalent percents, decimals, and fractions.

$12\frac{1}{2}\% = 0.125 = \frac{1}{8}$

$20\% = 0.2 = \frac{1}{5}$

$25\% = 0.25 = \frac{1}{4}$

$33\frac{1}{3}\% = 0.\overline{3} = \frac{1}{3}$

$40\% = 0.4 = \frac{2}{5}$

$50\% = 0.5 = \frac{1}{2}$

$60\% = 0.6 = \frac{3}{5}$

$66\frac{2}{3}\% = 0.\overline{6} = \frac{2}{3}$

$75\% = 0.75 = \frac{3}{4}$

$80\% = 0.8 = \frac{4}{5}$

Estimate.

23. 42% of 495
24. 70% of 189
25. 21% of 204
26. 32% of 63
27. 11% of 375
28. 49% of 93

Find the percent of each number.

29. 20% of 4,225
30. 2% of 1,550
31. 6% of 16,450
32. 9.5% of 60
33. 54.3% of 300
34. 12.5% of 200
35. 0.1% of 150
36. $\frac{3}{4}$% of 80
37. $0.1\frac{3}{4}$% of 12

38. Mavis bought a $90 blazer on sale for 54% off. How much did she spend on the blazer?

39. The manager at Milo's Appliance Center marked up the price of a $129 vacuum cleaner by 43%. How much was the final cost of the vacuum cleaner?

40. Seven percent of Paul's monthly salary is deducted from his payroll check each month for the company's savings plan. He saves $315 per month. What is Paul's monthly salary?

EXTEND/ SOLVE PROBLEMS

ESTIMATION TIP

One way to estimate a percent of a number is to use compatible numbers. For example, this is how you might estimate 34% of 1,098.

Think:
34% is close to $33\frac{1}{3}$%, or $\frac{1}{3}$.
1,098 is close to 1,200.
$\frac{1}{3} \times 1{,}200 = 400$
So 34% of 1,098 is about 400.

THINK CRITICALLY/ SOLVE PROBLEMS

USING DATA Use the Automobile Depreciation Schedule on page 301 for Exercises 41–45. Some of your answers should be given as a range of numbers.

Consider a new car that is worth $10,000.

41. How much does the car depreciate, in dollars, in the first year?

42. What is the trade-in value of the $10,000 car after the first year?

43. Find the maximum dollar amount of depreciation in the second year.

44. Find the minimum dollar amount of depreciation in the second year.

45. Find the trade-in value of the car at the end of the second year. Write an explanation of how you found the trade-in value of the car.

9-3 Finding What Percent One Number Is of Another

EXPLORE/ WORKING TOGETHER

Work with a partner. Write the digits 1 to 8 on 8 different cards and mix up the cards in a paper bag. One partner should select a card from the bag 20 times while the other partner creates a tally chart for the number of times each card is taken from the bag. Remember to put each card back in the bag before taking another.

Copy and complete the tally chart by writing a ratio for the number of times out of 20 that each card was selected. Then calculate the percents from the ratios.

	1	2	3	4	5	6	7	8
Tally								
Ratio								
Percent								

Explain how you might find the tally count if you knew only the percent and the total number of times cards were drawn from the bag.

SKILLS DEVELOPMENT

To find what percent one number is of another, you can write and solve an equation or a proportion.

Example 1

What percent of 65 is 13?

Solution

Write an equation. Let b = the percent.

$$\underbrace{\text{What percent}}_{b} \times 65 = 13?$$

$$b \times 65 = 13$$

$$65b = 13$$

$$\frac{65b}{65} = \frac{13}{65}$$

$$b = \frac{13}{65}$$

$$b = 0.2$$

So 13 is 20% of 65. ◄

50% of a number is half of that number.

100% of a number is that number.

200% of a number is twice that number.

What is 200% of 5?

What is 300% of 5?

Example 2

20 is what percent of 10?

Solution

Let the unknown percent be $x\%$, and write this as $\frac{x}{100}$. Use this as the basis for writing a proportion.

$$\begin{array}{l}\text{part} \rightarrow \\ \text{whole} \rightarrow\end{array} \frac{x}{100} = \frac{20}{10} \begin{array}{l}\leftarrow \text{part} \\ \leftarrow \text{whole}\end{array}$$

$$10x = 2{,}000$$

$$\frac{10x}{10} = \frac{2{,}000}{10}$$

$$x = 200$$ So 20 is 200% of 10. ◄

You can change the program on page 307 to find what percent one number is of another. Below is the adapted program.

```
10 INPUT "ENTER NUMBER
   PART: ";P: PRINT
20 INPUT "ENTER NUMBER
   WHOLE: ";W: PRINT
30 PRINT W:"X = 100*";P
35 PRINT W;"X = ";100 * P
40 X = (100 * P) / W
50 PRINT "X = ";X: PRINT
60 INPUT "RUN AGAIN? Y
   OR N: "; X$: PRINT
70 IF X$ = "Y" THEN
   GOTO 10
```

Example 3

The sale price of the VCR Alicia bought was \$360. The sales tax she paid was \$23.40. What was the sales tax rate?

Solution

Let the sales tax rate be $s\%$. Write and solve a proportion.

$$\begin{array}{l}\text{part} \rightarrow \\ \text{whole} \rightarrow\end{array} \frac{s}{100} = \frac{23.40}{360} \begin{array}{l}\leftarrow \text{part} \\ \leftarrow \text{whole}\end{array}$$

$$360s = 2{,}340$$

$$\frac{360s}{360} = \frac{2{,}340}{360}$$

$$s = 6.5$$ So the sales tax rate was 6.5% ◄

TRY THESE

Find each percent.

1. What percent of 25 is 15?
2. What percent of 64 is 32?
3. What percent of 56 is 14?
4. What percent of 25 is 45?
5. What percent of 3 is 12?
6. What percent of 125 is 87.5?
7. 48 is what percent of 32?
8. 3 is what percent of $4\frac{1}{2}$
9. 12 is what percent of 72?
10. 34 is what percent of 80?
11. 250 is what percent of 625?
12. 36 is what percent of 10?

Solve.

13. A 6-volume CD anthology was marked down from \$120 to \$96. What percent of the original price was the sale price?
14. Four concert tickets cost \$110. Ellison paid \$4.73 in sales tax. What was the sales tax rate?

EXERCISES

PRACTICE/ SOLVE PROBLEMS

Find each percent.

1. What percent of 16 is 4?
2. What percent of 25 is 24?
3. What percent of 200 is 68?
4. What percent of 500 is 450?
5. What percent of 770 is 77?
6. What percent of 120 is 18?
7. 56 is what percent of 160?
8. 1 is what percent of 40?
9. 15 is what percent of 15?
10. 2.4 is what percent of 200?

Solve.

11. Last April it rained 9 out of 30 days. What percent of the days did it not rain last April?
12. Ellen purchased a snowboard for $625. The sales tax she paid was $50. What was the sales tax rate?

EXTEND/ SOLVE PROBLEMS

13. One weekend at a mountain resort, 7 out of 10 guests skied on Saturday and 69 out of 100 guests skied on Sunday. What percent of guests skied on Saturday? on Sunday?
14. Refer to Exercise 13. On which day did the greater percent of guests ski?
15. Out of a total annual income of $24,000, the Davis family saves $1,500. To the nearest tenth of a percent, what percent of their total income do they save? What percent do they spend? Write and solve an equation to answer each question.

TALK IT OVER

Why is the sum of the answers to Exercise 15 greater than 100%?

THINK CRITICALLY/ SOLVE PROBLEMS

Refer to Exercise 15 to complete Exercises 16–17.

16. Assume that the amount of money saved by the Davis family doubles but that their income remains the same. Does the percent saved also double?
17. Assume that the Davises' income doubles but the amount they save remains the same. What happens to the percent saved?

Problem Solving Applications:

SPORTS AND STANDINGS

The performances of players or teams can be compared by using percents. This chart gives information about field goals for five top-scoring basketball players at Ashton High School.

Player	Field Goals Attempted	Field Goals Made
Gilson	993	560
Johnson	1,307	790
McNance	1,062	585
Parker	937	557
Williams	1,057	588

1. For each player, find the percent of field goals made out of field goals attempted. Round each answer to the nearest tenth of a percent.

2. List the players in order, from the one with the highest percent to the one with the lowest.

This chart shows the won–lost records of the baseball teams in the Western Division of the American League partway through a recent season.

Team	Games Won	Games Lost
California	55	58
Chicago	66	48
Kansas City	60	52
Minnesota	68	47
Oakland	64	50
Seattle	60	53
Texas	57	54

3. Write as a percent each of the seven ratios of games won to games played.

4. Which team is in first place? Which team is in last place?

The chart below gives passing records for five of the top quarterbacks in the National Football League during a recent season.

Quarterback	Passes Attempted	Passes Completed
Eason	448	276
Esiason	469	273
Krieg	375	225
Marino	623	378
O'Brien	482	300

5. Rank the quarterbacks in order of percent of passes completed, from least to greatest.

9-4 Percent of Increase and Decrease

EXPLORE/ WORKING TOGETHER

Work in your group to complete these sentences. For each of the following sentences, first simplify the fraction and then write the equivalent percent.

a. When 5 is added to 25, the whole, 25, has been increased by $\frac{5}{25}$, or __?__. So 5 added to 25 represents an increase of __?__ %.

b. When 5 is subtracted from 30, the whole, 30, has been decreased by $\frac{5}{30}$, or __?__. So 5 subtracted from 30 represents a decrease of __?__ %.

Discuss the following with your group.

c. It is true that $25 + 5 = 30$ and $30 - 5 = 25$.
It is false that 5 added to 25 and 5 subtracted from 30 represent the same percent of increase and decrease.

SKILLS DEVELOPMENT

The **percent of increase** tells what percent the amount of increase is of the original amount.

The **percent of decrease** tells what percent the amount of decrease is of the original amount.

To find the percent of increase or decrease, express the ratio of the amount of increase or decrease to the original amount as a percent.

Example 1

Find the percent of increase.

Original number: 70
New number: 84

Solution

Find the amount of increase.

$$84 - 70 = 14$$

Write a ratio.

$$\frac{\text{amount of increase}}{\text{original number}} = \frac{14}{70}$$

Find the percent.

$$\frac{14}{70} = 0.2 = 20\%$$

So the percent of increase is 20%. ◄

Example 2

Find the percent of decrease. \$200 to \$175

Original amount: \$200
New amount: \$175

Solution
Find the amount of decrease.

Write a ratio.

$$\frac{\text{amount of decrease}}{\text{original number}} = \frac{\$25}{\$200} = \frac{1}{8}$$

Find the percent.

$$\frac{1}{8} = 12\tfrac{1}{2}\%$$

So the percent of decrease is $12\frac{1}{2}\%$. ◄

Example 3

At the beginning of Barbara's exercise program she walked 40 minutes a day. After six weeks she walked 55 minutes a day. Find the percent of increase.

Solution
Find the amount of increase.

$$55 - 40 = 15$$

Write a ratio.

$$\frac{15}{40} = \frac{3}{8} = 0.375$$

Find the percent.

$$0.375 = 37.5\%$$

So the percent of increase from 40 to 55 is 37.5%. ◄

Example 4

In May, the round-trip airfare from Denver to Newark was \$440. In January, the airlines reduced the fare to \$330 round trip. Find the percent of decrease.

Solution
Find the amount of decrease.

$$\$440 - \$330 = \$110$$

Write a ratio.

$$\frac{\$110}{\$440} = \frac{1}{4}$$

Find the percent.

$$\frac{1}{4} = 25\%$$

So the percent of decrease is 25%. ◄

TRY THESE

Find the percent of increase.

1. Original price of refrigerator: $600
   New price of refrigerator: $630
2. Original number of vacation days: 14
   New number of vacation days: 21
3. Original weight: 140 lb
   New weight: 161 lb
4. Original salary: $450
   New salary: $513
5. Original fare: $75
   New fare: $99
6. Original number of employees: 375
   New number of employees: 600

Find the percent of decrease.

7. Original population: 840
   New population: 504
8. Original price: $250
   New price: $175
9. Original beats per minute: 80
   New beats per minute: 72
10. Original airfare: $680
    New airfare: $646
11. Original score: 185
    New score: 148
12. Original weight: 200 lb
    New weight: 144 lb

Solve.

13. The Jacksons bought a house for $145,000 four years ago. They sold the house this year for $174,000. What is the percent of increase?
14. The original price of a gold bracelet is $325. The sale price is $260. What is the percent of decrease?

EXERCISES

PRACTICE/ SOLVE PROBLEMS

Find the percent of increase.

1. Original rent: $500
   New rent: $550
2. Original price: $0.25
   New price: $1.50
3. Original price: $120
   New price: $132
4. Original weight: 125 lb
   New weight: 135 lb
5. Original number: 75
   New number: 105
6. Original number: 32
   New number: 128

Find the percent of decrease.

7. Original number of players: 80
   New number of players: 60
8. Original population: 320
   New population: 208
9. Original price: $500
   New price: $410
10. Original fare: $800
    New fare: $750
11. Original salary: $900
    New salary: $765
12. Original price: $680
    New price: $578

Solve.

13. The regular price of a handheld video game is $40. The sale price is $34. What is the percent of decrease?
14. Molly bought a used car for $5,500. She sold it two years later for $2,475. What is the percent of decrease?
15. After a workout, Bill's pulse rate rose from 70 beats per minute to 91 beats per minute. What is the percent of increase?

EXTEND/ SOLVE PROBLEMS

Find the percent of increase or decrease.

16. $1500 to $300
17. 16,000 to 19,840
18. 0.45 to 0.27
19. $360 to $378

Solve.

20. Daniel is shopping for a new computer monitor that sells for $750 retail. The monitor is on sale at Computer Discount for $479. Byteland is having a 40% off sale on the same monitor. Which is the better buy? How much could Daniel save off the original price?

21. A leather jacket was priced at $550 in the store. The jacket cost the owner of the store $250. What is the owner's percent of increase over the owner's cost?

THINK CRITICALLY/ SOLVE PROBLEMS

22. An $8 item is marked up 50%. Then the new price is marked down 50%. Is the final price $8? Explain.
23. ***USING DATA*** Use the Data Index on page 546 to find the table of United States Resident Population Change by Region. Use a calculator. Determine which regions of the United States had the greatest and least percent of increase in population from 1980 to 1990.

9-5 Finding a Number When a Percent of It Is Known

EXPLORE/ WORKING TOGETHER

The manager of a discount clothing outlet estimated that 75% of the number of items sold were sold to people under 20 years old. Last year the outlet sold about 52,500 items. About how many items were sold to people under 20 years old?

You can draw a triangle diagram to model the parts of any percent problem.

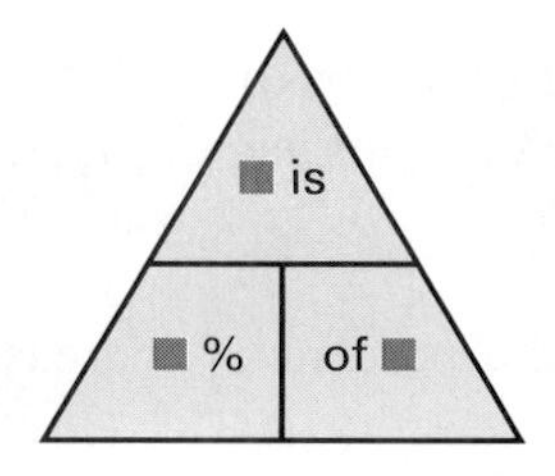

a. Draw and complete a triangle diagram to represent the situation above.

b. Use the diagram to write an equation.

c. Solve the equation.

d. Answer the question.

SKILLS DEVELOPMENT

To find a number when a percent of it is known, you can write and solve an equation or a proportion.

Example 1

4.2% of what number is 2.31?

Solution

Write an equation. Let n represent the whole amount.

4.2% of what number is 2.31?

$$4.2\% \times n = 2.31$$

$$0.042n = 2.31$$

$$\frac{0.042n}{0.042} = \frac{2.31}{0.042}$$

$$n = 55$$

So 4.2% of 55 is 2.31. ◄

Example 2

$\frac{3}{8}\%$ of what number is 30?

Solution

Write and solve an equation. Let n represent the whole amount.

$\frac{3}{8}\%$ of $\underbrace{\text{what number}}$ is 30?

$$\frac{3}{8}\% \times n = 30$$

$$\frac{3}{800}n = 30$$

$$\frac{800}{3}\left(\frac{3}{300}n\right) = \frac{800}{3}(30)$$

$$n = 8{,}000$$

So $\frac{3}{8}\%$ of 8,000 is 30. ◄

COMPUTER TIP

If you want to solve these problems using a proportion, use the adaptation of the program on p. 307 to find a number when a percent of it is known.

```
10 INPUT "ENTER % PART:
   ";P: PRINT
20 INPUT "ENTER NUMBER
   PART: ";N: PRINT
30 PRINT P;"X = 100*";N
35 PRINT P;"X";" = ";100
   * N
40 X = (100 * N) / P
50 PRINT "X = ";X: PRINT
60 INPUT "RUN AGAIN? Y
   OR N: ";X$: PRINT
70 IF X$ = "Y" THEN
   GOTO 10
```

Example 3

20 is 25% of what number?

Solution

Write 25% as $\frac{25}{100}$. Use this as the basis for writing a proportion. Let x represent the whole amount.

$$\frac{\text{part} \rightarrow 25}{\text{whole} \rightarrow 100} = \frac{20 \leftarrow \text{part}}{x \leftarrow \text{whole}}$$

$$25x = 2{,}000$$

$$\frac{25x}{25} = \frac{2{,}000}{25}$$

$$x = 80$$

So 20 is 25% of 80. ◄

Example 4

Joanne took Annemarie to a restaurant for lunch. Joanne left a tip of \$5.25, 15% of the bill. How much was the bill for lunch?

Solution

Write and solve an equation. Let n represent the whole amount.

\$5.25 is 15% of $\underbrace{\text{what number}}$?

$$\$5.25 = 15\% \times n$$

$$5.25 = 0.15n$$

$$\frac{5.25}{0.15} = \frac{0.15n}{0.15}$$

$$35 = n$$

So the bill for lunch was \$35. ◄

TRY THESE

Find each answer.

1. 25% of what number is 6?
2. 42 is 60% of what number?
3. 31.5 is 75% of what number?
4. 29% of what number is 145?
5. $26\frac{2}{3}$% of what number is 21.6?
6. 80 is 20% of what number?
7. A real estate agent received a $9,900 commission, which was 6% of the selling price. At what price did the agent sell the house?

EXERCISES

PRACTICE/ SOLVE PROBLEMS

Find each answer.

1. 35% of what number is 54.6?
2. 4.68 is 18% of what number?
3. $33\frac{1}{3}$% of what number is $29\frac{1}{3}$?
4. 2 is 0.8% of what number?
5. 8.3 is 10% of what number?
6. 88 is 22% of what number?
7. 68% of what number is 17?
8. 15 is 18% of what number?
9. 20% of what number is 65?
10. 3 is 0.1% of what number?
11. $\frac{1}{2}$% of what number is 3?
12. 18 is $2\frac{1}{2}$% of what number?
13. Suppose that 50,400 people attended a football game at the Cotton Bowl Stadium in Dallas, Texas. If the people who attended the game filled 70% of the seats in the stadium, what is the seating capacity of the stadium?
14. The manager of a construction company totals the costs of labor and materials and then adds 16%. The company has just been awarded a bid for a recreational center. If the added cost is $32,000, what is the cost of labor and materials?

MIXED REVIEW

1. Write 19.5% as a fraction and a decimal.
2. Find 5% of 15.6.
3. What percent of 84 is 7?
4. 11 is what percent of 88?

Multiply or divide. Be sure the answer is in lowest terms.

5. $6\frac{2}{3} \times 2\frac{1}{3}$
6. $2\frac{2}{3} \div 8$
7. $6\frac{3}{10} \times 1\frac{2}{3}$
8. Find the perimeter of a square patio that measures $22\frac{1}{3}$ yd on a side.

EXTEND/ SOLVE PROBLEMS

Do you know why ice floats? Any object will float if it is less dense, or less concentrated, than the fluid in which it is immersed. The **density** of a material is found by dividing its mass by its volume.

When water freezes, its volume increases. Its mass remains the same, so its density decreases. Therefore, ice is less dense than water and so ice will float.

Assume that, when water freezes, the volume of ice that is formed is 112% of the volume of the water. Find the volume of water that formed each amount of ice.

15. 112 cm^3 **16.** 168 cm^3 **17.** 672 cm^3

18. 95.2 cm^3 **19.** 336 cm^3 **20.** 476 cm^3

THINK CRITICALLY/ SOLVE PROBLEMS

21. ***USING DATA*** Use the Data Index on page 546 to find state sales tax rates.
During a recent vacation, Sharon bought these souvenirs.

Flag $3.69 Postcard set $6.25
Slides $16.25 Wood carvings $36.95

If Sharon paid $66.77 for the souvenirs, including sales tax, in which state did she make the purchase?

The computer program is written in BASIC.

22. Describe what the computer program does.

23. How would you use the program for a sales tax rate of 8.5%?

24. Explain why .5 is added to the product of P and R in Line 50. (Hint: The INT function on a computer, INT(X), always gives the greatest integer that is less than or equal to x.)

```
10 PRINT "SALES TAX AND
   TOTAL AMOUNTS"
20 INPUT "ENTER THE
   PERCENT TAX RATE";R
30 PRINT "PRICE",
   "PERCENT TAX RATE",
   "TAX", "TOTAL PRICE"
40 READ P
50 TAX=INT(P*R+.5)/100:
   TOTAL=P+TAX
60 PRINT "$"P, R"%", "$"TAX,
   "$"TOTAL
70 IF P=45.89 THEN END
80 GOTO 40
90 DATA 28.79, 56.87, 112.13,
   67.34, 81.08, 45.89
```

9-6 Problem Solving Skills:
CHOOSE A COMPUTATION METHOD

► READ
► PLAN
► SOLVE
► ANSWER
► CHECK

Many problems can be solved using different computation methods such as mental math, estimation, paper and pencil, or a calculator.

When you encounter a problem, first read it carefully. Then, develop a plan. Ask:

► Do you need an exact answer or an estimate?
► Which computation method is appropriate?

Try to solve the problem using mental math. If this method is not appropriate, try pencil and paper or a calculator.

PROBLEM

Stacey works as a clerk. She earns $6.73 per hour. One week she worked 10.5 hours. How much money did she earn?

SOLUTION

Which computation method is appropriate?

Pencil and paper

```
  $6.73
×  10.5
   3365
   000
  673
$70.665 ≈ $70.67
```

Can the computation be done quickly with pencil and paper?

Estimation

```
 $6.70
×   10
$67.00
```

Is an exact answer needed?

Calculator

6.73 [×] 10.5 [=] [70.665] ↓ $70.67

Can the computation be done quickly with a calculator?

Using a calculator is probably the best method to solve this problem.

PROBLEMS

Indicate which computation method you would use to solve each problem.

1. Dawn bought mascara for \$7.00, nail polish for \$2.25, and shampoo for \$6.00. How much did she spend?
2. Justin is hanging a wallpaper border in his rectangular-shaped bathroom. The bathroom is 6 feet 3 inches long and 4 feet 6 inches wide. What is the perimeter of his bathroom?
3. Penny stopped at the fruit stand on her way home from school. She spent \$5.27 for watermelon, \$2.39 for pears, \$6.19 for green peppers, and \$3.05 for apples. How much money did she spend?

4. Justin wants to buy 4 new custom rims for his car tires. Each rim sells for \$129.99. He has \$500. Does he have enough money?
5. Bonnie is dividing her collection of 335 baseball cards evenly among herself and her 4 brothers. How many baseball cards will each person receive?
6. Mel wants to buy 3 compact discs. Each disc sells for \$14.98, including tax. If Mel has \$45, does he have enough money?
7. The bowling team has decided to buy new shoes for the upcoming tournament. There are 6 people on the team. Bowling shoes cost \$58.95 per pair plus 8% sales tax. How much money does the team need to buy the shoes?

9-7 Discount

EXPLORE

Drucilla is shopping for a new TV set. She finds this ad in Sunday's paper. As a Video Vibes club member, Drucilla gets 12% off the regular price of store merchandise on Tuesdays. If she wants the TV set advertised, should she buy it at the sale price, or should she wait until Tuesday to buy it?

a. What is 12% of $459.99?

b. How much would Drucilla save by waiting until Tuesday to make the purchase?

SKILLS DEVELOPMENT

If an item is on sale, you can save money by buying it at less than the regular price.

The **discount** is the amount that the regular price is reduced.

The **sale price** is the regular price less the discount.

WRITING ABOUT MATH

Write a paragraph in your journal explaining the difference between the terms *regular price, discount,* and *sale price.*

Example 1

Find the discount and sale price of an item that sells for $298.95 with an 18% discount.

Solution

Find the discount.

discount = percent of discount × regular price

discount = 18% × $298.95
= 0.18 × $298.95
= $53.811, or $53.81

The discount is $53.81.

Find the sale price.

sale price = regular price − discount
sale price = $298.95 − $53.81
= $245.14

The sale price is $245.14 ◄

Example 2

Use a calculator to find the sale price if it is $15\frac{1}{2}\%$ off the regular price of $359.62.

Solution

Since discount = percent of discount × regular price, you can use this key sequence.

359.62 [×] 15.5 [%] | 55.74111 |

[+/−] [+] 359.62 [=] | 303.88789 |

The sale price is $303.88. ◄

READING MATH

Explain the difference between 25% *of* $90 and 25% *off* $90.

Example 3

Kayla went to a 25%-off sale at Lacy's Department Store. The regular price of a leather belt is $48. Find the sale price.

Solution

Think: $25\% = \frac{1}{4}$

$\frac{1}{4}$ of $48 = 12$

$48 − $12 = $36

The sale price of the belt is $36. ◄

TRY THESE

Find the discount and the sale price. Round your answers to the nearest cent.

1. regular price: $415.38
   percent of discount: 15%
2. regular price: $63.45
   percent of discount: 25%
3. regular price: $89.95
   percent of discount: 50%
4. regular price: $369.99
   percent of discount: 20%
5. regular price: $115.18
   percent of discount: 3%
6. regular price: $3,880
   percent of discount: 12.5%
7. regular price: $1,385
   percent of discount: 5%
8. regular price: $8,175
   percent of discount: 16%

Solve.

9. Francie bought a $49.95 sweater on sale for 20% off. What was the sale price?
10. Bruce went to a 50% end-of-ski-season sale. How much did he pay for $1,580 worth of regular-priced merchandise?

EXERCISES

PRACTICE/ SOLVE PROBLEMS

Find the discount and the sale price. Round your answers to the nearest cent.

1. regular price: $150.50
   percent of discount: 10%
2. regular price: $389.55
   percent of discount: 20%
3. regular price: $689.99
   percent of discount: 18.5%
4. regular price: $1,450.39
   percent of discount: 25%
5. regular price: $511.09
   percent of discount: 13%
6. regular price: $48.88
   percent of discount: 3%

PROBLEM SOLVING TIP

Since the sale price in Exercise 7 is the original price (100%) minus the discount (75%), the sale price is 100% − 75%, or 25%, of the original price. Multiply.
$410 × 0.25 = ?

Solve.

7. A $410 ski jacket is on sale for 75% off. What is the sale price?
8. The regular price of a Caribbean cruise is $895. For a limited time, the price is being reduced by 28%. How much would you save if you traveled at the reduced rate?

EXTEND/ SOLVE PROBLEMS

The Pants Palace discounts all merchandise that has been on the racks for six weeks. Any merchandise remaining another four weeks is discounted again by taking a percent of the first discount price. Complete the table to show the final sale price.

	Regular Price	First Discount	Second Discount	Sale Price
9.	$89.00	10%	15%	■
10.	$384.50	5%	20%	■
11.	$25.75	25%	10%	■
12.	$2,853.27	30%	5%	■

THINK CRITICALLY/ SOLVE PROBLEMS

13. During a special sale, a guitar was marked 20% off and sold for $85. A banjo was marked down 25% and sold for $72. Which instrument had the greater amount of discount?
14. James received a 25% discount on a suit during a sale. He also received a 5% discount on the sale price for paying cash. If he paid $120.50 for the suit, what was the regular price of the suit?
15. A blouse that is reduced by 20% one week is marked down an additional 15% after two weeks. Does this blouse cost the same as if it had been reduced by 35% at the start of the sale? Explain.

► READ
► PLAN
► SOLVE
► ANSWER
► CHECK

Problem Solving Applications:

COMMISSIONS ON SALES

Many salespeople work on commission. This means that they earn an amount of money that is a percent of their total sales. The percent is called the **commission rate,** and the amount of money is called the **commission.** Often, their income is made up of a combination of salary and commission.

Solve. Express dollar answers to the nearest cent.

1. Corretta is a real estate agent. Her commission rate is 3%. Find her commission on a house that sold for $85,000.

2. Laura's commission rate for selling cosmetics is 15%. If her sales totaled $225 one week, how much commission did she receive?

3. One week, Alberto sold $42,000 worth of computers. His commission was $1,260. Find his commission rate.

4. Shirley receives a base salary of $135 a week and a commission rate of 2% of sales. What did she earn in a week in which she sold $1,859 worth of cosmetics?

5. Rosemarie receives commissions at the rate of 12% for the first $15,000 of sales and 30% for sales in excess of $15,000. What is her commission on $27,000 of sales?

6. Jackson sells two cars. Each car is priced at $9,245.80, and his commission rate is $2\frac{1}{2}$%. How much commission does he receive?

7. Robin sells cars on commission. She receives 5% for new cars and 4% for used cars. Today she sold two used cars totaling $7,120 and one new car selling for $6,895. How much commission did she earn?

8. Lemuel sells furniture. He receives a base salary of $150 a week and a commission on sales. One week he sold $6,800 worth of furniture. His combined salary and commission was $320. What is his commission rate?

9. Juanita receives a commission of 25% on sales of merchandise in excess of $2,000. One week her sales were $7,500. What was the amount of her commission for that week?

MIXED REVIEW

Find each answer. Round decimal quotients to the nearest thousandth.

1. 7 ÷ 12
2. 0.03 × 650
3. 1.45 × 62
4. 745 ÷ 640
5. −18 + 7
6. 42 − (−8)
7. 9 × (−12)
8. −250 ÷ (−10)
9. −25 + (−16)
10. If the cost of fencing is $4.60 per meter, how much would it cost to enclose a rectangular garden that measures 9.2 m by 4.6 m?

9-8 Simple Interest

EXPLORE

a. Follow the pattern. Copy and complete the table.

Number of Years	Amount
1	$100 + 1 × 5% of $100 = ■
2	$100 + 2 × 5% of $100 = ■
3	$100 + 3 × 5% of $100 = ■
4	$100 + 4 × 5% of $100 = ■
5	$100 + 5 × 5% of $100 = ■

b. What would be the entry in the Amount column for 8 years?

c. What do you think the amount would be for 6 months?

SKILLS DEVELOPMENT

Interest is money that is paid for the use of money over a period of time. For example, if you put money into a savings account, a bank will pay you interest for the use of that money over a given period of time. If you borrow money from a bank, the bank will charge you interest for the use of their money for a given period of time.

The amount of money that is earning interest or that you are borrowing at interest is called the **principal. Simple interest** is paid only on the principal.

COMPUTER TIP

The spreadsheet is a type of computer application software that is widely used today. A spreadsheet is made up of rows and columns into which you can enter information. The computer will perform calculations on the numerical information in a spreadsheet according to formulas that you enter. If you have spreadsheet software available, you can use its power to investigate various "what if" situations.

The **rate** is the percent of the principal charged for the use of the money over a given period of time, usually a year.

The **time** is the number of time periods during which the principal remains in the bank account or has not been paid back. Time is usually expressed in years or parts of a year.

The interest (I) earned or paid on a given principal (p) at a given rate (r) over a time period (t) is given by a formula.

$$I = prt$$

Example 1

Find the interest on $3,000 for 2 years at a rate of 8% per year.

Solution

Use the formula $I = prt$ to solve.
Substitute the values into the formula.

$I = \$3{,}000 \times 8\% \times 2$
$I = \$3{,}000 \times 0.08 \times 2$
$I = \$480$ The interest is $480. ◄

Example 2

Find the interest and the total amount in the account for a \$6,300 deposit for 4 years at a rate of 5.4% per year.

Solution

Find the interest.

$I = prt$
$I = \$6{,}300 \times 5.4\% \times 4$
$I = \$6{,}300 \times 0.054 \times 4$
$I = \$1{,}360.80$ ← interest

Find the total amount by adding the interest to the principal.

$\$6{,}300 + \$1{,}360.80 = \$7{,}660.80$

So the total amount in the account after 4 years is \$7,660.80. ◄

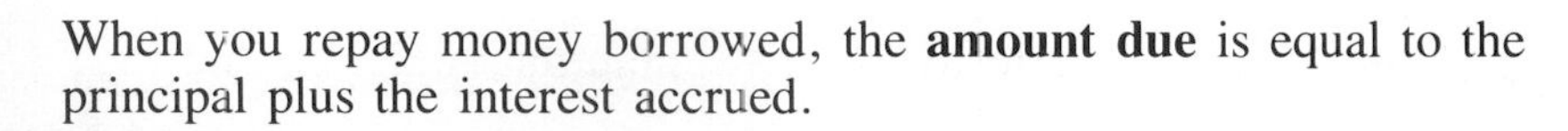

When you repay money borrowed, the **amount due** is equal to the principal plus the interest accrued.

Example 3

Kendra borrowed \$4,125 from a bank at an annual interest rate of 18.5%. The term of the loan was 36 months. Find the simple interest and the total amount due on the loan.

Solution

Find the interest.

$I = prt$
$I = \$4{,}125 \times 0.185 \times 3$ ← Change 36 months to 3 years.
$I = \$2{,}289.375 \approx \$2{,}289.38$

Find the total amount due.

$\$4{,}125 + \$2{,}289.38 = \$6{,}414.38$

Kendra paid \$2,289.38 in simple interest and \$6,414.38 over the life of the loan. ◄

Find the interest.

1. Principal: \$1,000
   Rate: 8%/year
   Time: 2 years
2. Principal: \$895
   Rate: 17.5%/year
   Time: 3 years
3. Principal: \$1,300
   Rate: 5.4%/year
   Time: 1 year
4. Principal: \$475
   Rate: 6%/year
   Time: 6 months
5. Principal: \$900
   Rate: 7.5%/year
   Time: 24 months
6. Principal: \$3,450
   Rate: 13%/year
   Time: 12 months

Find the interest and the amount due.

7. Principal: \$650
   Rate: 4%/year
   Time: 5 years

8. Principal: \$980
   Rate: 7.5%/year
   Time: 3 years

9. Principal: \$1,500
   Rate: 15%/year
   Time: 36 months

Solve.

10. Clint borrowed \$6,400 from a bank at 14% interest per year. How much simple interest will he owe after 4 years?

11. Gina opened a savings account with a deposit of \$1,100. What was the total amount of money in her account after 2 years if she earned simple interest at 6% per year and she made no other deposits or withdrawals?

EXERCISES

PRACTICE/ SOLVE PROBLEMS

Find the interest and the amount due.

	Principal	Annual Rate	Time	Interest	Amount Due
1.	\$410	5%	6 months	■	■
2.	\$35	7%	2 years	■	■
3.	\$1,575	12.5%	3 years	■	■
4.	\$4,853	8%	48 months	■	■
5.	\$3,000	5.4%	5 years	■	■
6.	\$8,957	4%	12 months	■	■
7.	\$505.50	10.5%	1 year	■	■
8.	\$2,200.75	11%	8 years	■	■
9.	\$775	15%	36 months	■	■
10.	\$8,000	16.5%	5 years	■	■
11.	\$10,100	6.5%	24 months	■	■
12.	\$69.75	8.5%	4 years	■	■

Solve.

13. Bruce purchased a certificate of deposit for \$2,000. The certificate paid 6% simple interest per year. How much money will Bruce have in his account after 2 years?

14. Erica borrows \$10,000 for 4 years at 11.5% simple interest per year. How much money does she repay the bank after 4 years?

15. Rowland opened a money market account with a deposit of $3,000. The account pays 8.2% annual interest. How much interest will be credited to his account after 3 years if he makes no other deposits or withdrawals?

EXTEND/ SOLVE PROBLEMS

Solve.

16. Teresa invested $2,280 at 9% per year for 3 months. How much interest did she earn?

17. Rob loaned $625 to his niece. She agreed to repay the loan in 9 months, at a rate of 13% per year. How much interest will Rob receive?

18. Patsy borrowed $6,800 at $14\frac{1}{8}$% simple interest per year for 2 years. How much money will she have to repay the bank?

19. Chuck borrowed $18,000 to expand his courier business. He took out two different loans. One loan was for $10,000 at 12% per year and the other loan was for $8,000 at $15\frac{1}{4}$% per year. Both loans were for 5 years. How much money must he repay at the end of 5 years?

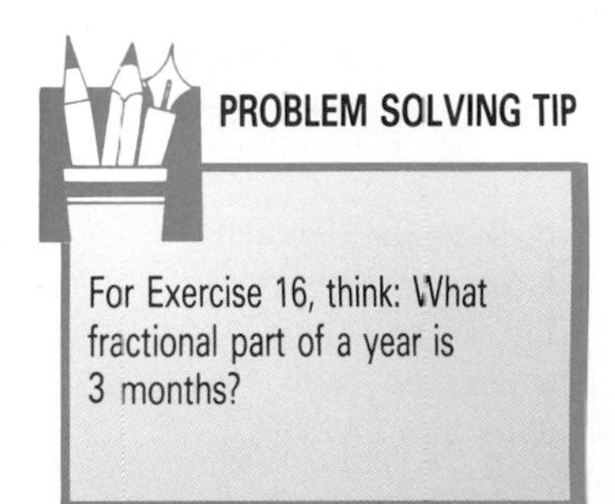

For Exercise 16, think: What fractional part of a year is 3 months?

THINK CRITICALLY/ SOLVE PROBLEMS

Find the principal, rate, time, or interest:

	Principal	Annual Rate	Time	Interest
20.	$1,300	■	$2\frac{1}{2}$ years	$455
21.	■	12.5%	4 years	$450
22.	$1,500	12%	■	$90
23.	$4,500	10%	■	$675
24.	■	11%	26 weeks	$371.25
25.	$3,000	■	6 months	$180
26.	$3,500	11%	18 months	■

Solve.

27. Janice opened a savings account with a deposit of $900. She made no other deposits or withdrawals. She received $90 in interest after one year. What was the annual interest rate?

28. Norm purchased a certificate of deposit for $4,800 at $6\frac{1}{4}$% interest per year. The term is for 2 years and 5 months. How much interest did he receive on his deposit at the end of the term?

29. Suppose you earn an annual salary of $32,000. You are offered a 10.5% raise for a 4-year period, or you can have a 3% raise each year for 4 years. Which is the better offer?

9-9

- READ
- PLAN
- SOLVE
- ANSWER
- CHECK

Problem Solving/ Decision Making:

CHOOSE A STRATEGY

PROBLEM SOLVING TIP

Here is a checklist of the problem solving strategies that you have studied so far in this book.

Solve a simpler problem
Find a pattern
Guess and check
Work backward
Make an organized list
Make a table
Use logical reasoning
Act it out

In this book you have been studying a variety of problem solving strategies. Experience in applying these strategies will help you decide which will be most appropriate for solving a particular problem. Sometimes, only one strategy will work. In other cases, any one of several strategies will offer a solution. There may be times when you will want to use two different approaches to a problem in order to be sure that the solution you found is correct. For certain problems, you will need to use more than one strategy in order to find the solution.

PROBLEMS

Solve. Name the strategy you used to solve the problem.

1. Six months ago, Joe received an 8% pay raise. His salary was then \$256. Two months ago, Joe took an 8% pay cut. What was Joe's salary after the pay cut? How does that amount compare with his salary six months ago? (Always round Joe's salary to the nearest dollar.)

2. The Walters are having an outdoor barbecue. They plan to set up square card tables joined in a row to seat 36 people. How many card tables will be needed to seat 36 people if just one person sits on each side of a table?

3. The annual membership fee of the Movie-Goers Club is \$15. Admission to each movie is \$3.75 for members and \$5.50 for nonmembers. How many movies would someone have to see before it would be worth becoming a member?

4. A game is played by drawing three of these number cards from a bag. Each player's score is the sum of the digits on the cards picked. How many possible scores are there for each pick?

6 2 4 0 7

Use this information for Problems 5 and 6.

These are pentominoes.

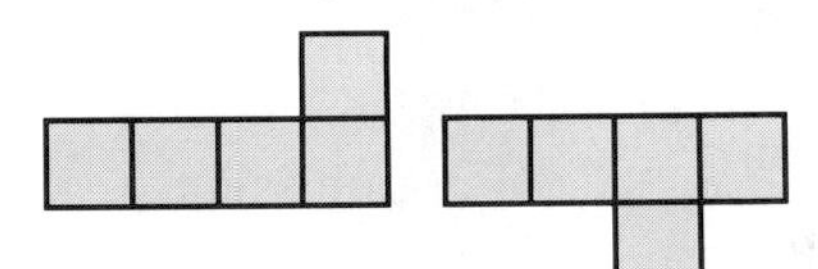

These are *not* pentominoes.

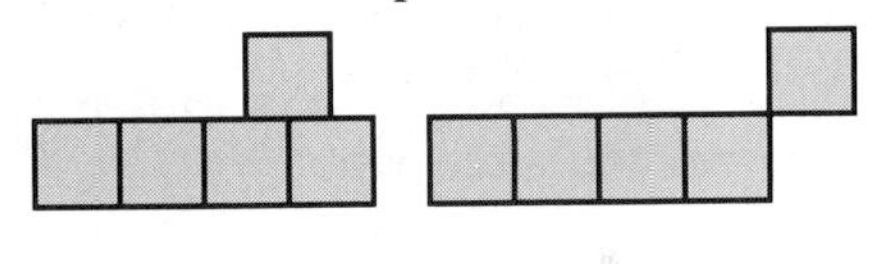

5. Find how many unique pentominoes are possible.

6. How many pentominoes can be folded to form open-top boxes?

7. A construction foreman determined that 3 workers can do half a job in 20 days. Working at the same rate, how long would it take 12 workers to do the whole job?

8. Complete the pattern.

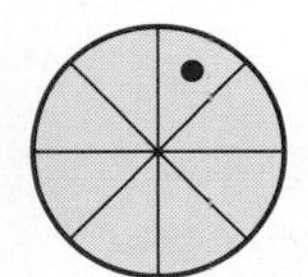 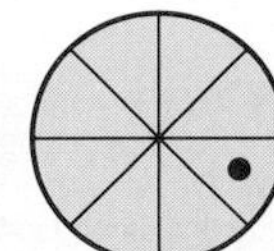 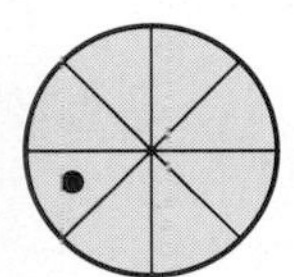 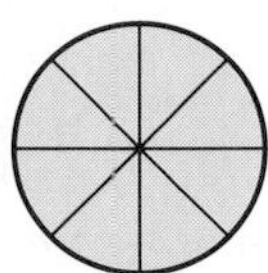 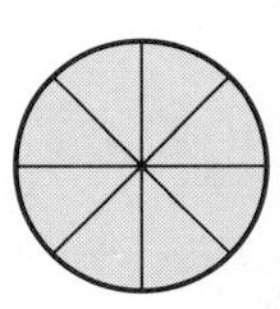 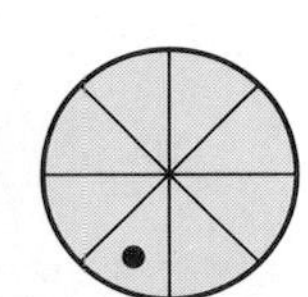

9. A grandmother divided her figurine collection among her four grandchildren. She gave the oldest grandchild half of her collection, the second oldest one fourth of her collection, the third oldest one fifth of her collection, and the youngest 48 figurines. How many figurines were there in her collection altogether?

10. Six cubes make a 3-step staircase.

    How many cubes are needed to make a 9-step staircase?

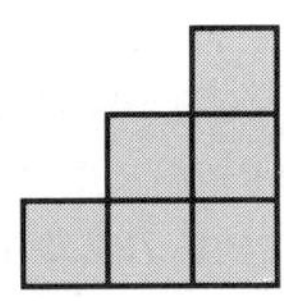

In each of these puzzles a letter represents a digit. Different letters represent different digits. Find the numbers that solve each puzzle.

11.	12.	13.	14.
A	A B	T W O	H O C U S
× A	+ B A	+ T W O	+ P O C U S
BA	C A C	F O U R	P R E S T O

CHAPTER 9 • REVIEW

1. A __?__ is a ratio that compares a number to 100.
2. The percent of the regular price you save when you buy an item on sale is the __?__.
3. The __?__ is the amount you save when you buy an item on sale.
4. The percent of sales received as income is __?__.
5. The __?__ is the amount saved or the amount borrowed.

a. amount of discount
b. commission rate
c. discount rate
d. percent
e. principal

SECTION 9–1 EXPLORING PERCENTS (pages 302–305)

► A ratio that compares a number to 100 is called a percent.
► Percent means *per one hundred.*

Write each as a percent.

6. $\frac{7}{8}$
7. $\frac{3}{5}$
8. $\frac{23}{100}$

SECTION 9–2 FINDING THE PERCENT OF A NUMBER (pages 306–309)

► To find a percent of a number, you can write and solve an equation or a proportion.

What number is 75% of 80?

$$x = 75\% \times 80$$

4% of 15 is what number?

$$\text{part} \rightarrow \frac{4}{100} = \frac{n}{15} \leftarrow \text{part} \quad (\text{whole} \rightarrow 100,\ 15 \leftarrow \text{whole})$$

Find the percent of each number.

9. 75% of 64
10. 7% of 2,000
11. 64.3% of 200

SECTION 9–3 FINDING WHAT PERCENT ONE NUMBER IS OF ANOTHER (pages 310–313)

► To find what percent one number is of another, you can write and solve an equation or a proportion.

12. What percent of 80 is 12?
13. What percent of 86 is 215?
14. 5 is what percent of 30?
15. 1 is what percent of 40?

SECTION 9–4 PERCENT OF INCREASE AND DECREASE (pages 314–317)

► The percent of increase tells what percent the amount of increase is of the original amount.
► The percent of decrease tells what percent the amount of decrease is of the original amount.

Find the percent of increase or decrease.

16. 96 to 108
17. 950 to 836
18. 800 to 716

SECTION 9–5 FINDING A NUMBER WHEN A PERCENT OF IT IS KNOWN (pages 318–321)

► To find a number when a percent of it is known, use an equation or a proportion.

30 is 50% of what number?

$$30 = 50\% \times n$$

25% of what number is 12?

$$\frac{25}{100} = \frac{12}{n} \quad \begin{matrix}\leftarrow \text{part} \\ \leftarrow \text{whole}\end{matrix}$$

Find the unknown number.

19. 60 is 40% of what number?

20. 70% of what number is 147?

SECTION 9–7 DISCOUNT (pages 324–327)

► The discount is the amount that the regular price is reduced.
► The sale price is the regular price minus the discount.

Find the discount and the sale price. Round to the nearest cent.

21. Regular price: $899.98
Discount: 25%

22. Regular price: $510.20
Discount: 10%

SECTION 9–8 SIMPLE INTEREST (pages 328–331)

► The principal is the amount of money in a bank account or the amount of money to be borrowed from the bank.
► The rate is the percent of principal earned or paid per year.
► The time is the number of months or years the money remains in the bank account or the length of the loan.

Find the interest.

23. Principal: $600
Rate: 7%/year
Time: 2 years

24. Principal: $598
Rate: 3.4%/year
Time: 36 months

25. Principal: $9,000
Rate: 14%/year
Time: 4 years

SECTIONS 9–6 and 9–9 PROBLEM SOLVING (pages 322–0323, 332–333)

► Sometimes more than one strategy will be needed in order to find the solution to a problem.

Choose a strategy. Solve the problem.

26. The sum of two numbers is 212. If the smaller number is subtracted from the greater number, the result is 14. What are the numbers?

USING DATA Use the Automobile Depreciation Schedule on page 301.

27. Find the minimum dollar amount of depreciation on a 2-year-old car that cost $12,000 when it was new.

CHAPTER 9 ● TEST

Write each as a percent.

1. $\frac{16}{100}$
2. $\frac{13}{20}$
3. 0.05

Find the percent of each number.

4. 17% of 50
5. 35% of 800
6. $13\frac{1}{2}$% of 80
7. 5.8% of 70
8. 20% of 140
9. 62% of 410

Find each percent.

10. What percent of 65 is 29.9?
11. What percent of 95 is 49.4?
12. 84 is what percent of 350?
13. 2.61 is what percent of 3?
14. What percent of 50 is 10?
15. 27 is what percent of 9?

Find the percent of increase or decrease.

16. Original amount: $90
    New amount: $144
17. Original amount: $300
    New amount: $405
18. Original amount: 500
    New amount: 320
19. Original amount: 75
    New amount: 60
20. Original amount: $280
    New amount: $126
21. Original amount: $72
    New amount: $99

Find the unknown number.

22. 441 is 9% of what number?
23. 60 is 40% of what number?
24. 2% of what number is 3?
25. 25% of what number is 36?
26. 8 is 4% of what number?
27. 25% of what number is 48?

Find the discount and the sale price. Round to the nearest cent.

28. Regular price: $419
    Discount percent: 12%
29. Regular price: $53.69
    Discount percent: 25%

Find the interest and the amount.

30. Principal: $170
    Rate: 6%/year
    Time: 2 years
31. Principal: $8,100
    Rate: 7.5%/year
    Time: 36 months
32. Principal: $517
    Rate: 12%/year
    Time: 6 months

Use the pictograph for Exercises 1–5.

STAR VIDEO RENTALS
(for June 16)

Drama

Comedy

Foreign

Sci Fi

Sports

Key: represents 10 rentals

1. How many sports videos were rented?
2. Which group had the most rentals?
3. Suppose 35 Do-It-Yourself videos were rented. How many video symbols would be needed to represent this number
4. What was the total number of videos rented for June 16?
5. The number of drama rentals was how many times greater than the number of foreign rentals?

Write each answer in exponential form.

6. $a^3 \times a^5$
7. $n^{12} \div n^4$

8. Write two conditional statements using the following two sentences:

   Clara can read French.
   She can translate this letter.

9. Write 0.000025 in scientific notation.

10. A fence that encloses a square garden has 10 fence posts on each side of the garden. There is a post in each corner. How many fence posts are there in all?

Replace ● with <, >, or =.

11. -16 ● -13
12. -15 ● -11
13. 3 ● -8
14. 0 ● -3

Solve.

15. $\frac{x}{3} = 12$
16. $4d - 18 = -18$
17. $-y + 3 = 19$
18. $\frac{5}{3}k = 25$

Write each ratio in two other ways.

19. 4 to 1
20. 8:2
21. $\frac{19}{20}$
22. Write the unit rate for $\frac{125 \text{ words}}{5 \text{ min}}$.

23. All merchandise in Lucy's Gift Shop is marked 20% off. Write this reduction as a fraction.
24. Find 75% of 48.
25. 16 is what percent of 25?
26. A \$450 television set is on sale at 10% off. Find the discount and the sale price.
27. Find the interest for one year on a \$2,500 deposit if the interest rate is $8\frac{1}{2}\%$ per year.

CHAPTER 9 • CUMULATIVE TEST

1. If the mean score on a math test was 80, what would a student who was absent have to score on a makeup test to keep the same mean?

A. 85　　B. 90
C. 80　　D. none of these

2. Multiply. $x^8 \times x^2$

A. x^6　　B. x^{16}　　C. x^4　　D. x^{10}

3. Which conclusion can be drawn from these statements?

If Willa is Scandinavian, then she is not Italian.
Willa is Scandinavian.

A. Willa is not Italian.
B. Willa is from Rome.
C. Willa is Scandinavian.
D. none of these

4. Which is the decimal for $2\frac{3}{8}$?

A. 2.6　　B. 2.375
C. 0.375　　D. 2.38

5. Complete. 716 mg = ■ kg.

A. 7.16 kg　　B. 0.716 kg
C. 0.0716 kg　　D. 0.000716 kg

6. Find the circumference. Round your answer to the nearest tenth or whole number.

A. 34.5 cm
B. 8.6 cm
C. 27 cm
D. 84.8 cm

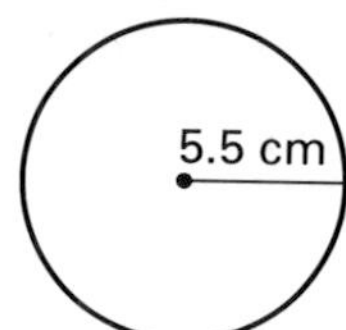

7. Add. $25 + (-24)$

A. 49　　B. 1　　C. -49　　D. -1

8. Solve. $3 = \frac{t}{4}$

A. $t = 16$　　B. $t = 15$
C. $t = 11$　　D. none of these

9. Solve. $5x + 11 > -4$

A. $x > -3$　　B. $x < -3$
C. $x > 1\frac{2}{5}$　　D. none of these

10. Solve. $3:4 = b:16$

A. 7　　B. 3　　C. 12　　D. 6

11. Which of the following ratios is equivalent to 12:8?

A. 12:2　　B. 36:24
C. 6:2　　D. none of these

12. Which is the unit rate for 87 mi on 6 gal?

A. $\frac{13 \text{ mi}}{\text{gal}}$　　B. $\frac{15.4 \text{ mi}}{\text{gal}}$
C. $\frac{14.5}{\text{gal}}$　　D. none of these

13. The scale of a map is 1 in.: 10 mi. What is the actual distance between two cities if the distance on the map is $7\frac{1}{2}$ in.?

A. 70 mi　　B. $7\frac{1}{2}$
C. 75 mi　　D. none of these

14. Express $\frac{2}{5}$ as a percent.

A. 20%　　B. 40%
C. 25%　　D. 50%

15. What is $33\frac{1}{3}$% of 58.5?

A. 175.5　　B. 195
C. 1.95　　D. none of these

16. 8 is what percent of 12?

A. $66\frac{2}{3}$%　　B. 80%
C. 75%　　D. $33\frac{1}{3}$%

17. An $800 refrigerator is on sale at 15% off. What is the discount?

A. $785　　B. $120
C. $680　　D. none of these

CHAPTER 10 SKILLS PREVIEW

Give all the names that apply to each polygon or polyhedron.

1. 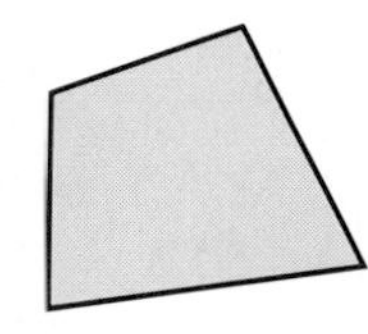

2. 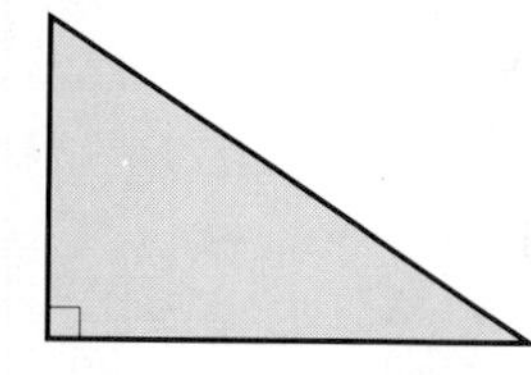

3. 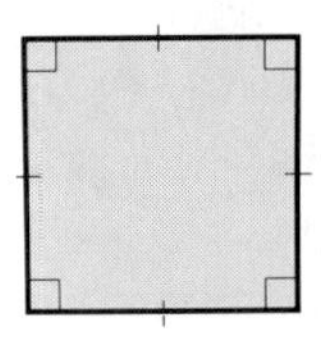

4. 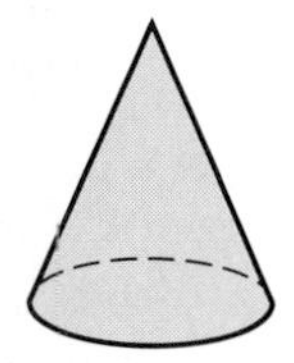

5. 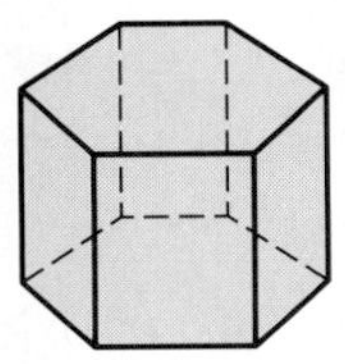

6. 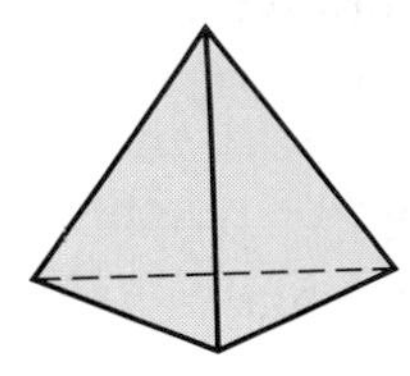

Classify each angle as *acute, right, straight,* or *obtuse.*

7. 15°
8. 180°
9. 90°
10. 98°

11. In the figure below, lines *RS* and *TU* are parallel. Identify the numbered angles that have equal measures.

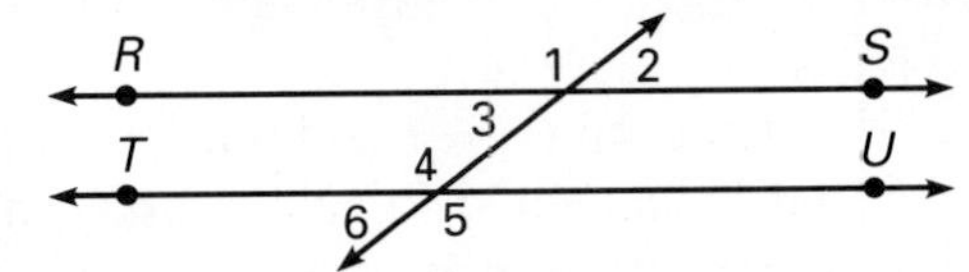

Classify each triangle as *right, acute,* or *obtuse.*

12. 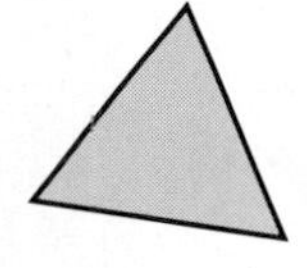

13. 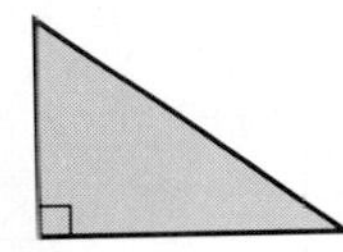

14. 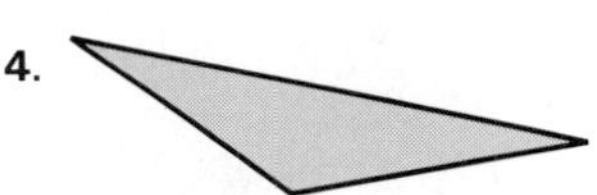

15. 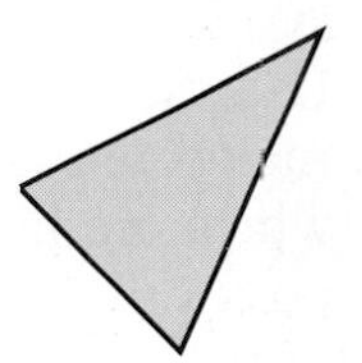

16. The measure of $\angle ABC$ is 98°. Ray *BD* bisects $\angle ABC$. What is the measure of $\angle DBC$?

17. Use a protractor to draw an angle that measures 118°. Construct its bisector.

Find the unknown angle measure.

18. 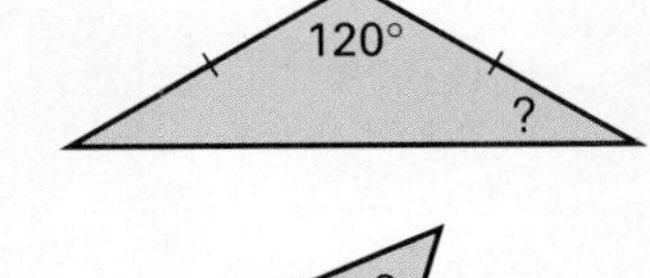

19. 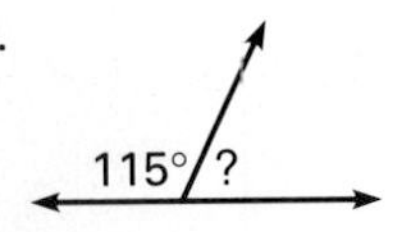

20. 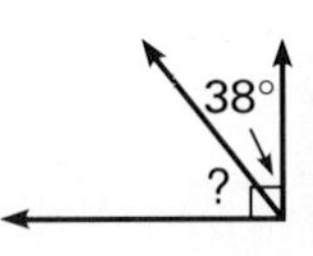

21. 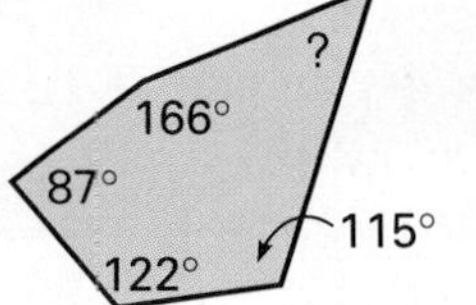

22. 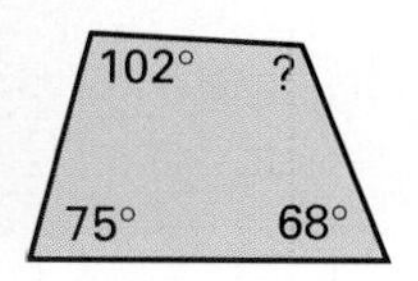

23. 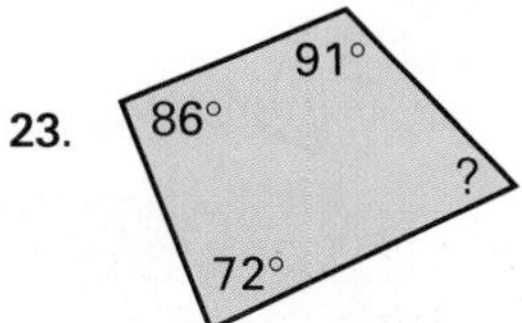

CHAPTER 10

GEOMETRY OF SIZE AND SHAPE

THEME Road Signs and Other Signs

Since ancient times, every culture has communicated with the aid of signs and symbols. In the past, piles of stones or notches in tree trunks marked the trails of wilderness travelers. Today, signs of different geometric shapes tell drivers to stop, yield, or slow down for a curve. You will learn the meanings of many symbols as you study the geometry of size and shape.

Some signals form a code. Semaphore, for example, is a system of signals used in the U.S. Navy. Semaphore code makes use of two flags, which are held in different positions to signal the letters of the alphabet. When preceded by the signal for *numeral,* the positions in which the flags are held to represent each of the first ten letters represent the numbers from 1 through 9 and 0.

DECISION MAKING

Using Data

Use the semaphore chart to answer the questions.

1. What is the shape of each flag?
2. What is the shape of the sections of each flag?
3. For which letters are the signaler's arms at a 90° angle?
4. For which letters are the signaler's arms at a 45° angle?
5. For which letters are the signaler's arms at a 180° angle?

Working Together

In small groups, look at the semaphore code to discover ways of learning it easily. Are there patterns that would allow you to learn the signals for several letters at a time instead of individually? What pattern do you detect in the sequence for the letters from A through D? What other patterns can you find?

Make a set of semaphore flags for your group, using light-colored and dark-colored triangles. Practice signaling and reading the words *add, subtract, multiply,* and *divide* and the numbers from 1 through 99. Then each group can signal math problems to other groups, which can signal the answers in return. Begin with problems like $3 + 5$ and go on to problems like 53×49.

10-1 Points, Lines, and Planes

EXPLORE

Geometric shapes appear in many places. You can see many of them in the superstructures of buildings and bridges. What geometric shapes do you see in the photograph?

SKILLS DEVELOPMENT

Here is a chart of some basic geometric figures. Note how each is drawn, named, and symbolized.

Figure	Name	Description
• A	point A	A **point** is a location in space. Although a point has no dimension, it is usually represented by a dot.
B C	line BC ($\overleftrightarrow{BC}$) or line CB ($\overleftrightarrow{CB}$)	A **line** is a set of points that extends without end in two opposite directions. Two points determine a line. Points that are on the same line are said to be **collinear points.**
D E	line segment DE ($\overline{DE}$) or line segment ED ($\overline{ED}$)	A **line segment** is a part of a line that consists of two **endpoints** and all points between them.
F G	ray FG ($\overrightarrow{FG}$)	A **ray** is a part of a line that starts at one endpoint and extends without end in one direction.
A B C	angle ABC ($\angle ABC$) or angle CBA ($\angle CBA$) or angle B ($\angle B$).	An **angle** is the figure formed by two rays that share a common endpoint. The endpoint is called the **vertex** of the angle. The rays are called the **sides** of the angle.
4	angle 4 or $\angle 4$	If there is a number between the rays that form an angle, the angle can be named by that number.
m X Y Z	plane XYZ or plane m	A **plane** is a flat surface that extends without end in all directions. It is determined by three noncollinear points. Points that are on the same plane are said to be **coplanar points.**

CHECK UNDERSTANDING

When you use three letters to name an angle, which letter must be in the middle?

Why would it be incorrect to also refer to the angle at the right as $\angle A$?

Example 1

Write the symbol for each figure.

a.

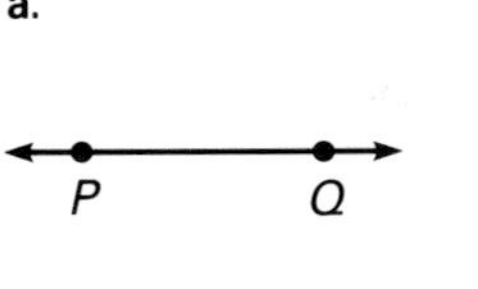

b. **c.**

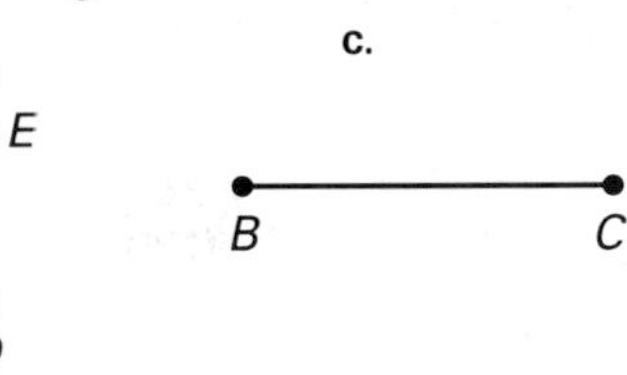

d.

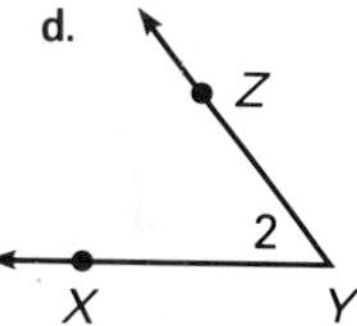

Solution

a. $\overleftrightarrow{PQ}$ or $\overleftrightarrow{QP}$

b. $\overrightarrow{DE}$

c. $\overline{BC}$ or $\overline{CB}$

d. $\angle XYZ$ or $\angle ZYX$ or $\angle Y$ or $\angle 2$ ◄

A **polygon** is a closed plane figure that is formed by joining three or more line segments at their endpoints. Each line segment joins exactly two others and is called a **side of the polygon.** The point at which two sides meet is a **vertex of the polygon.** Some polygons have special names.

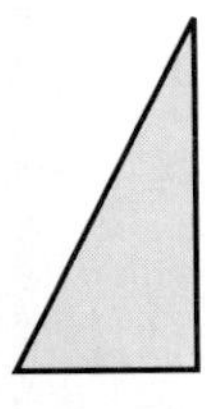

triangle (3 sides)

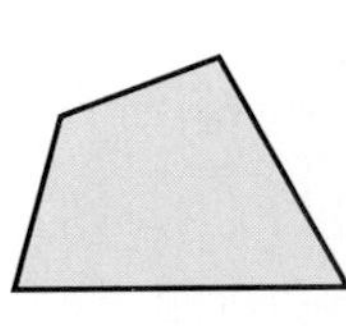

quadrilateral (4 sides)

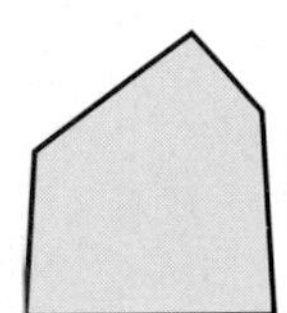

pentagon (5 sides)

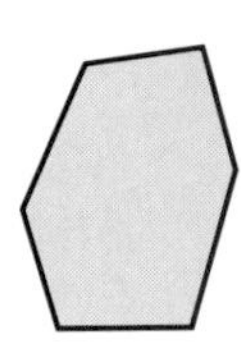

hexagon (6 sides)

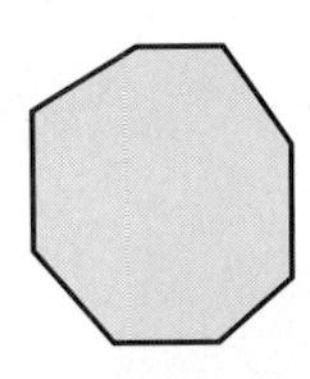

octagon (8 sides)

A polygon is named by the letters at its vertices. You list the vertices in order.

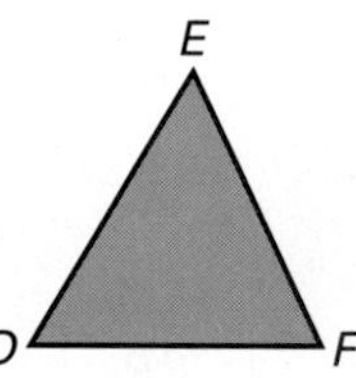

triangle DEF or $\triangle DEF$

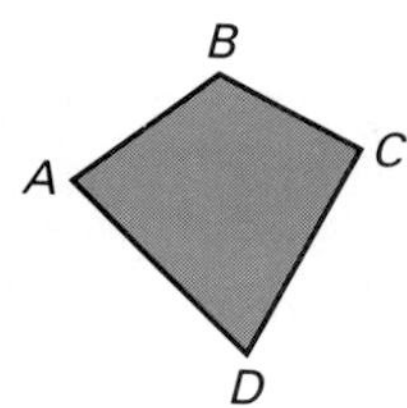

quadrilateral $ABCD$

Example 2

Identify the type of polygon.

a.

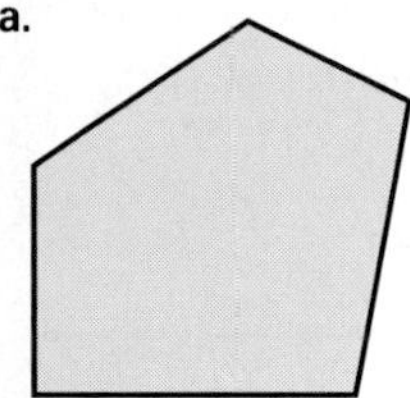

b.

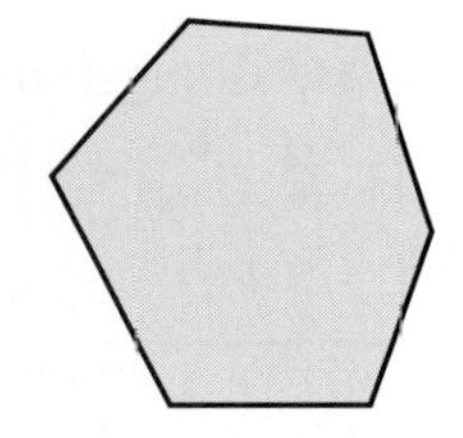

c.

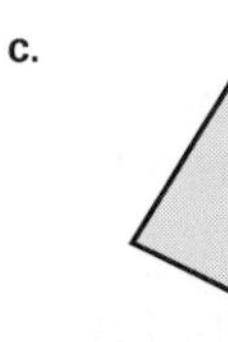

Solution

a. There are 5 sides. It is a pentagon.

b. There are 6 sides. It is a hexagon.

c. There are 3 sides. It is a triangle. ◄

READING MATH

The prefixes in the names of polygons tell you how many sides and angles the polygon has. Use a dictionary to find out what number these prefixes signify: *hepta-, nona-, deca-, dodeca-, icosa-.* What do you think is the definition of a *nonagon*? What do you think is the name of a polygon with twelve sides?

COMPUTER

Complete this program by writing lines 40–60. With these lines, the program can be RUN for all polygons from 3-sided to 6 -sided.

```
10  INPUT "ENTER THE
    NUMBER OF SIDES IN
    THE POLYGON: ";S:
    PRINT
20  INPUT "ENTER THE
    NAME OF THE POLYGON:
    ";N$: PRINT
30  IF S = 3 AND N$ =
    "TRIANGLE" THEN
    GOTO 80
70  PRINT "SORRY, NOT
    QUITE RIGHT.":PRINT :
    GOTO 90
80  PRINT "GREAT JOB":
    PRINT
90  INPUT "RUN AGAIN? Y
    OR N:";X$: PRINT
100 IF X$ = "Y" THEN
    GOTO 10
```

TRY THESE

Write the symbol for each figure.

1. 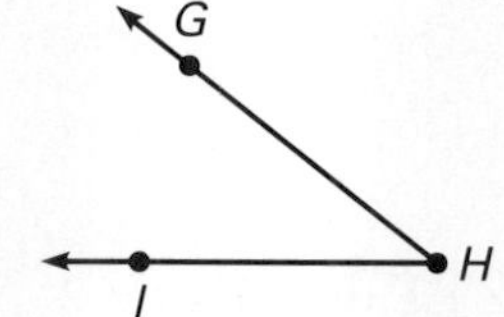

2. 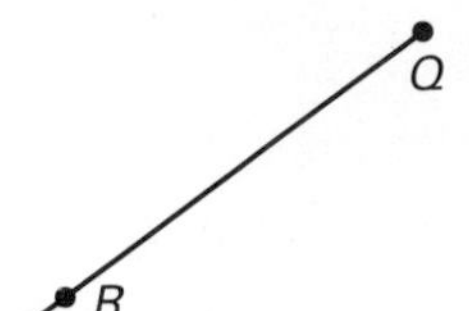

3. 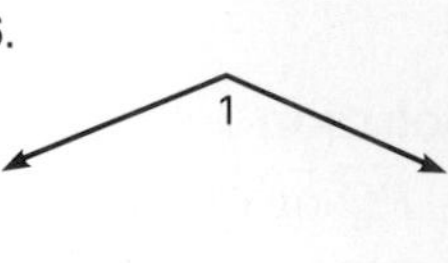

4. 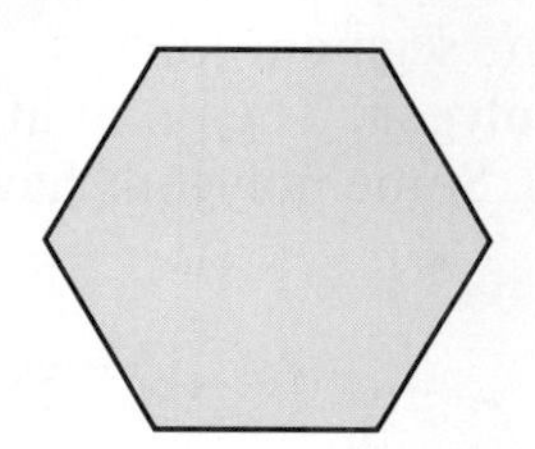

5. 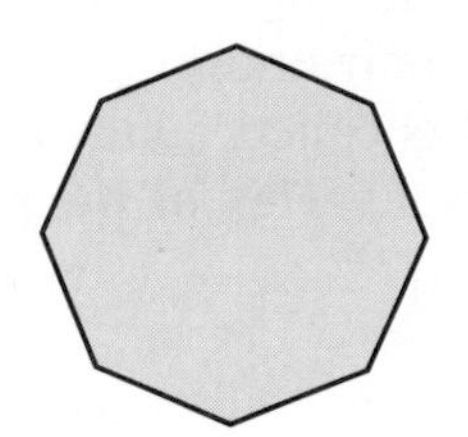

6. 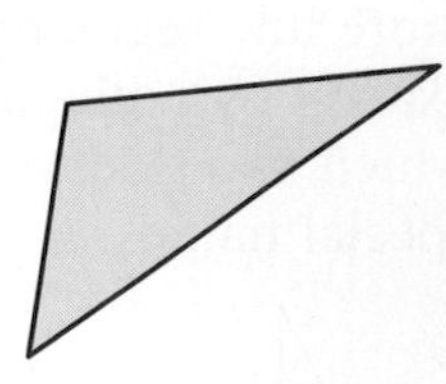

Identify the type of polygon.

7.

8.

9.

EXERCISES

PRACTICE/ SOLVE PROBLEMS

Write the symbol for each figure.

1. 

2. E F

3. 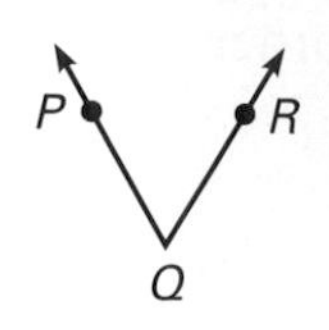

Draw a representation of each figure.

4. $\overleftrightarrow{MN}$ 5. $\overrightarrow{PQ}$ 6. plane m 7. $\overline{PR}$ 8. $\angle XYZ$

9. pentagon 10. hexagon 11. quadrilateral 12. octagon

USING DATA Refer to the Data Index on page 546 to answer this question.

13. What two-dimensional figure is suggested by the shape of the Mosque of Omar?

EXTEND/ SOLVE PROBLEMS

Use this figure for Exercises 14–15.

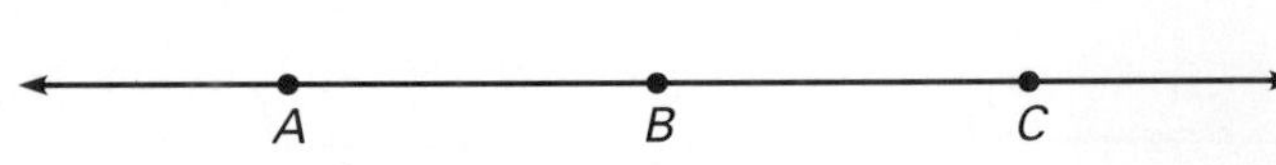

14. Use symbols to name the line in at least four ways.

15. Use symbols to name two rays with B as an endpoint.

Use the terms in the box at the right. Write the letters of all the terms that match each description.

a. ray
b. line
c. polygon
d. line segment
e. point
f. plane
g. angle

16. It is a location in space.
17. It extends without end in two opposite directions.
18. It has one vertex.
19. It extends without end in only one direction.
20. It can be named with two endpoints.
21. It is determined by three points.
22. It can be named by two points.
23. It has three or more vertices.

Refer to the figure at the right, then write *true* or *false* for each.

24. Point M is on $\overleftrightarrow{NO}$.
25. Point M is on $\overleftrightarrow{PQ}$ but not on $\overleftrightarrow{NO}$.
26. $\overrightarrow{MO}$ can also be named $\overrightarrow{OM}$.
27. $\overleftrightarrow{PQ}$ and $\overleftrightarrow{PM}$ name the same figure.
28. $\overleftrightarrow{MN}$ has two endpoints.
29. $\overleftrightarrow{MQ}$ has no endpoints.

P O M N Q

THINK CRITICALLY/ SOLVE PROBLEMS

Suppose that you have eight straws that you can use to represent eight line segments.

30. What polygon and what combinations of polygons could you model using all eight straws each time? Name all the possibilities.
31. Refer to your answers to Exercise 30. What is the total number of vertices in each polygon or combination of polygons?

MATH: WHO, WHERE, WHEN

Archie Alexander (1867–1958) is among the most innovative of United States bridge builders. One of the first African Americans to graduate from the University of Iowa with a degree in engineering, Alexander went on to continue his studies in bridge design at the University of London.

While he implemented the construction of many other civil engineering (construction) projects, Alexander's greatest achievement was the River Bridge at Mt. Pleasant, Iowa. When it was built, this bridge was known to be the longest and most efficient in the state of Iowa.

10-2 Angles and Angle Measures

EXPLORE

Cut out two paper strips. Attach them at one end with a paper fastener. Notice that the two strips in the diagram are positioned to make a "square angle." Make a matching angle with your strips.

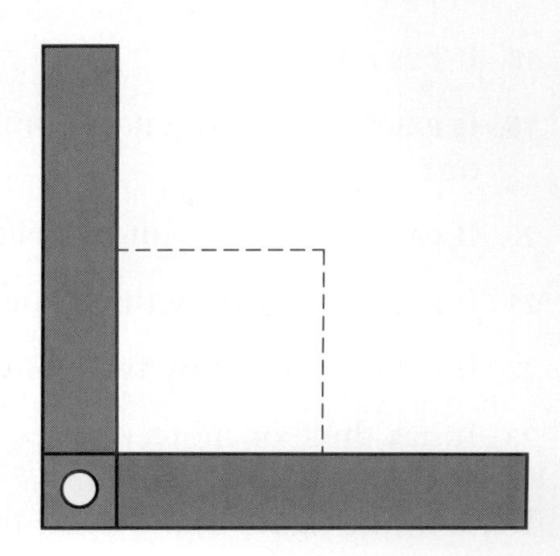

Now use the strips to make an angle larger than the square angle. Then make an angle that is smaller than the square angle. Describe all the differences you observe between the larger and the smaller angles.

SKILLS DEVELOPMENT

The size of an angle is commonly measured in the unit called the **degree.** The number of degrees in the angle's measure indicates the amount of openness between the sides of the angle. Angles are measured with the instrument called a **protractor.** On the protractor, one scale shows degree measures from 0° to 180° in a clockwise direction; the other scale shows these degree measures in a counterclockwise direction.

Example 1

Use a protractor to measure $\angle PQR$.

Solution

Place the center of the protractor on vertex Q. Place the 0° line of the protractor on $\overrightarrow{QR}$. Locate the position of $\overrightarrow{QP}$ on the counterclockwise scale.

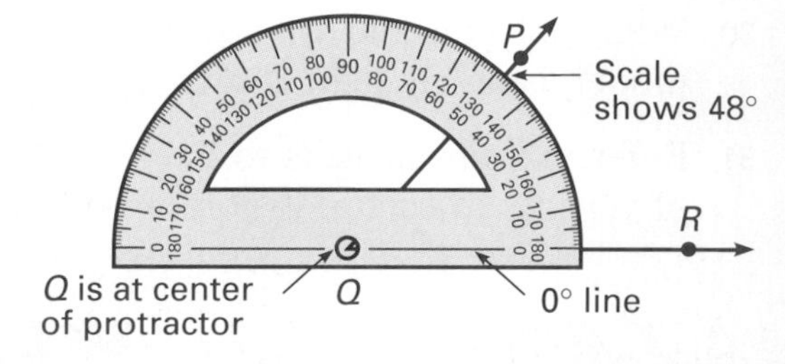

The measure of $\angle PQR$ is 48°.
This is written $m\angle PQR = 48°$. ◄

A protractor can also be used to draw angles.

Example 2

Use a protractor to draw $\angle JKL$ so that $m\angle JKL = 120°$.

Solution

Step 1
Draw a ray from point K through point L.

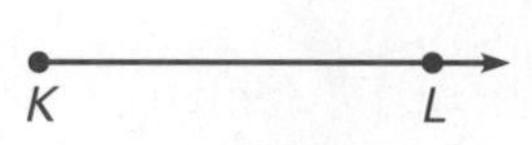

Step 2
Place the center of the protractor on the vertex, K. Place the $0°$ line of the protractor on $\overrightarrow{KL}$. Locate the $120°$ point on the protractor. Mark point J at $120°$.

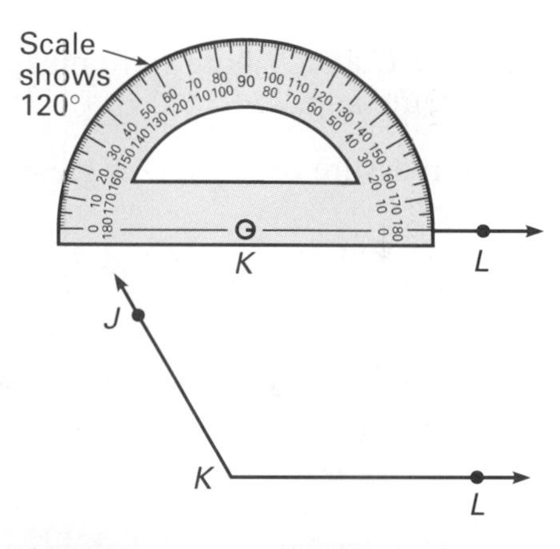

Step 3
Remove the protractor. Draw $\overrightarrow{KJ}$.

$m\angle JKL = 120°$ ◄

Angles are classified according to their measures, as listed in the chart.

Name of Angle	Measure of Angle
acute	between 0° and 90°
right	90°
obtuse	between 90° and 180°
straight	180°

Certain pairs of angles are related in important ways.

Two angles are called **complementary angles** if the sum of their measures is $90°$. In the figure at the right, $\angle ABC$ is the *complement* of $\angle CBD$.

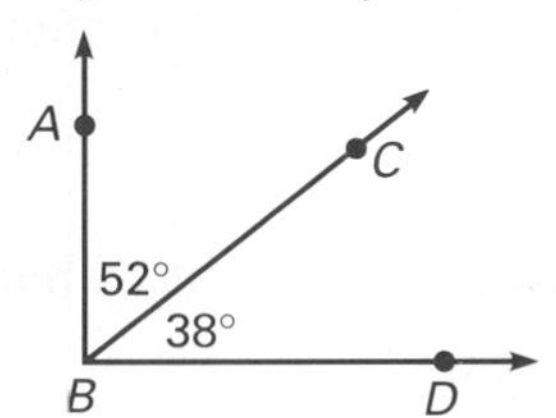

Two angles are called **supplementary angles** if the sum of their measures is $180°$. In the figure at the right, $\angle LOM$ is the *supplement* of $\angle MON$.

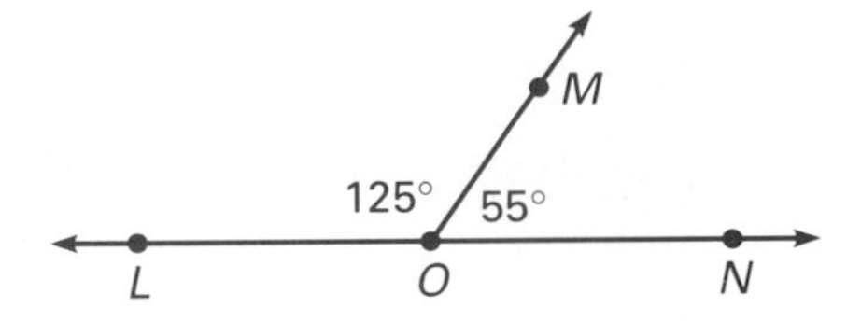

Adjacent angles are angles that have a common vertex and a common side, but have no interior points in common. In the figure at the right, $\angle QPR$ is adjacent to $\angle RPS$.

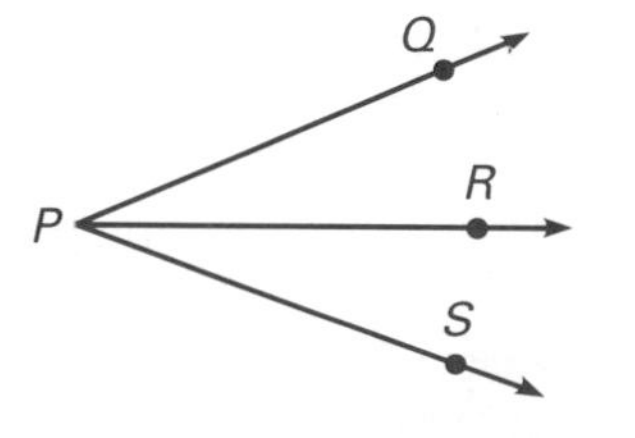

Example 3

Use the figures at the right to name the following.

a. all acute angles
b. all obtuse angles
c. a pair of complementary angles
d. a pair of supplementary angles

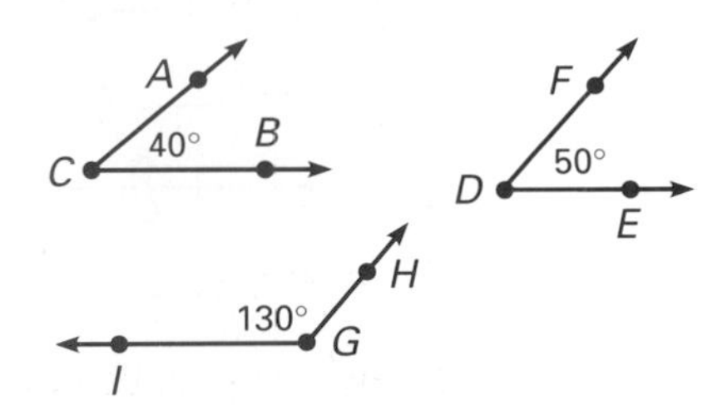

Solution

a. Angles C and D both have measures less than 90°. So, $\angle C$ and $\angle D$ are acute angles.

b. The measure of angle G is greater than 90°. So $\angle G$ is an obtuse angle.

c. $m\angle C + m\angle D = 40° + 50° = 90°$, so $\angle C$ and $\angle D$ are a pair of complementary angles.

d. $m\angle D + m\angle G = 50° + 130° = 180°$, so $\angle D$ and $\angle G$ are a pair of supplementary angles. ◄

TRY THESE

Use a protractor to measure each angle.

1.

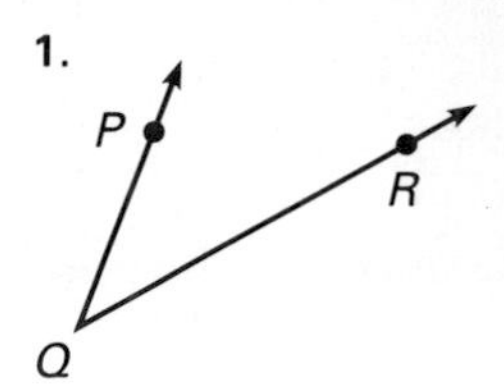

2.

A
B
C

3.

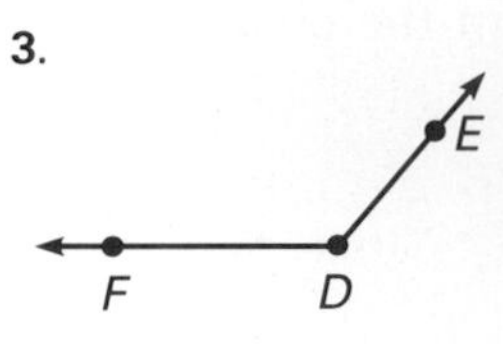

Use a protractor to draw an angle of the given measure.

4. 40°
5. 110°
6. 44°
7. 135°
8. 19°

Use the figures at the right to name the following.

9. an acute angle
10. a pair of complementary angles
11. an obtuse angle
12. a pair of supplementary angles

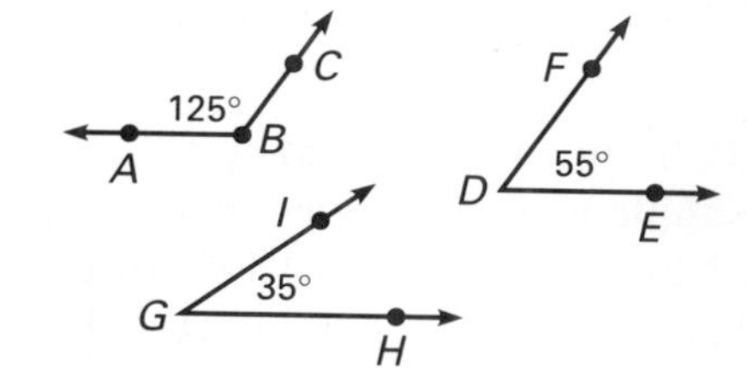

EXERCISES

PRACTICE/ SOLVE PROBLEMS

Find the measure of each angle. Classify the angle.

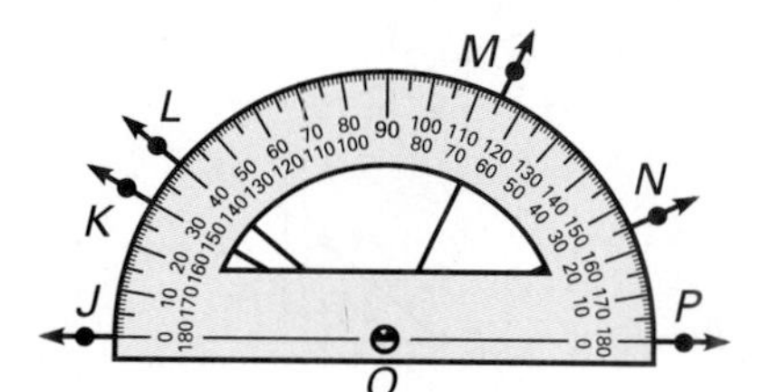

1. $\angle NOP$
2. $\angle KOP$
3. $\angle LOP$
4. $\angle MOP$
5. $\angle JOM$
6. $\angle JOL$
7. $\angle JOK$
8. $\angle JON$

Use a protractor to draw an angle of the given measure.

9. 10°
10. 140°
11. 65°
12. 175°
13. 28°

Use the figures at the right to name the following.

14. a right angle
15. a pair of adjacent angles
16. a pair of supplementary angles
17. a pair of complementary angles

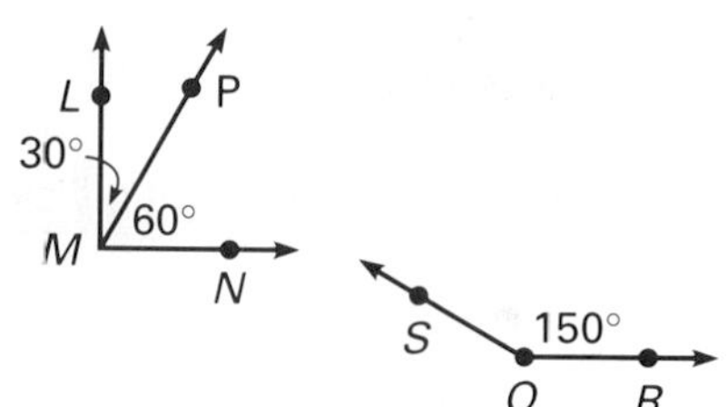

EXTEND/ SOLVE PROBLEMS

For each of Exercises 18–20, classify the angle, estimate its size, and then measure the angle.

18.

19.

20.

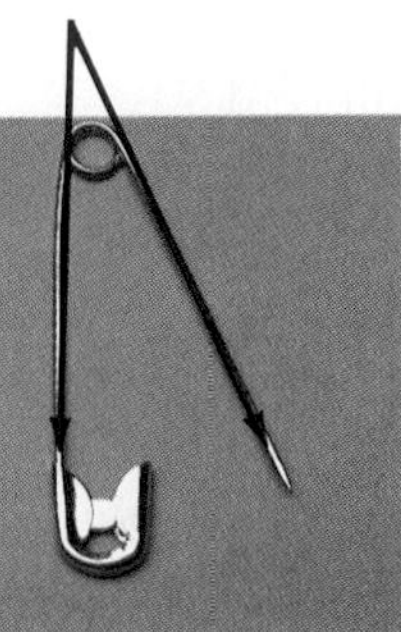

ESTIMATION TIP

To estimate the measure of an angle, think visually using 90°, 45°, and 180° as benchmarks. Knowing that a right angle measures 90°, you can estimate the size of an angle that is half as big, or 45°. In the same way, you can estimate the measure of a larger angle by deciding if it is just a little larger than a right angle, or if it is closer to a *straight angle,* which measures 180°.

Use the figure at the right for Exercises 21–23. The ⌜ symbol is used to show a right angle.

21. Name all the angles with a measure of 40°.
22. Name all the angles with a measure of 50°.
23. Name all the right angles.

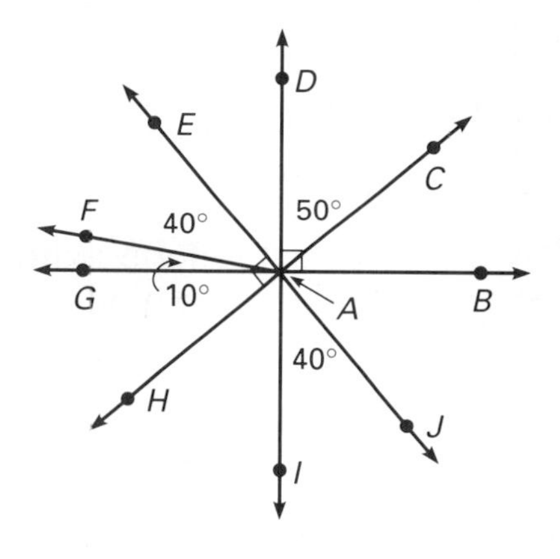

THINK CRITICALLY/ SOLVE PROBLEMS

Determine whether each statement is *true* or *false.*

24. The supplement of an obtuse angle is an acute angle.
25. The complement of an acute angle is a right angle.

Solve.

26. The measure of an angle is 50° more than its supplement. Find the measure of both angles.
27. Suppose that $\angle A$ and $\angle B$ are complementary angles and that $\angle A$ and $\angle C$ are supplementary angles. What is the relationship between the measure of $\angle B$ and the measure of $\angle C$?

10-3 Parallel and Perpendicular Lines

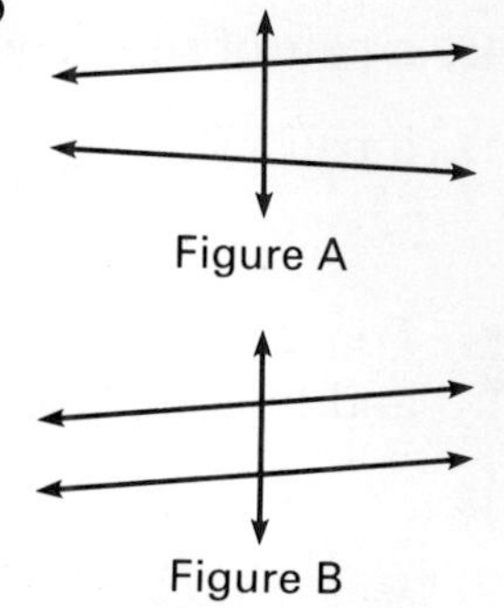
Figure A

Figure B

EXPLORE/ WORKING TOGETHER

Look at Figures A and B. In each diagram, one line crosses two other lines. Measure the angles formed by the lines in each diagram. What is the same in both diagrams? What is different?

SKILLS DEVELOPMENT

Any two lines in a plane are related in exactly one of these two ways.

The two lines might **intersect** at a single point.

$\overleftrightarrow{AB}$ and $\overleftrightarrow{CD}$ intersect at point E.

The two lines might be **parallel** and have no points in common.

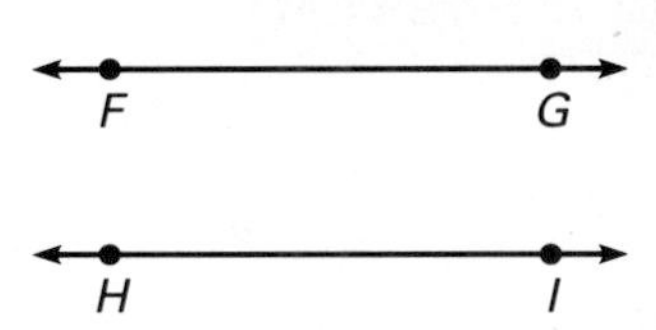

$\overleftrightarrow{FG}$ is parallel to $\overleftrightarrow{HI}$.
Write this as $\overleftrightarrow{FG} \parallel \overleftrightarrow{HI}$.

If two lines intersect to form right angles, then the lines are said to be **perpendicular** to each other.

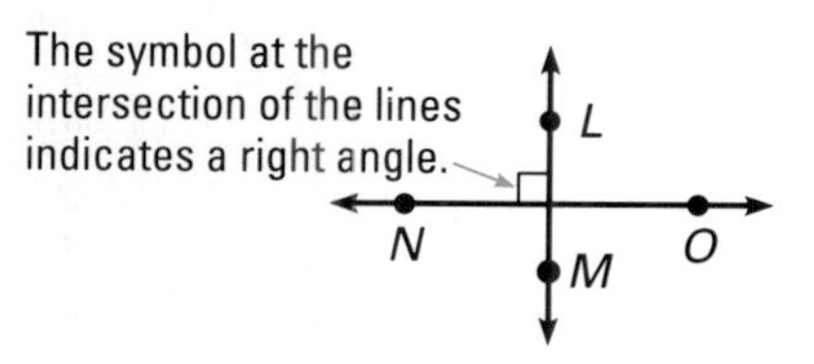

$\overleftrightarrow{LM}$ is perpendicular to $\overleftrightarrow{NO}$. Write this as $\overleftrightarrow{LM} \perp \overleftrightarrow{NO}$.

When two lines intersect, the angles that are not adjacent to each other are called **vertical angles.** In the figure at the right, $\angle 1$ and $\angle 3$ form a pair of vertical angles, and $\angle 2$ and $\angle 4$ form another pair.

The double marks indicate that angles 2 and 4 are equal in measure.

The single marks indicate that angles 1 and 3 are equal in measure.

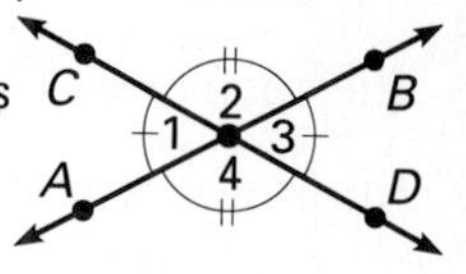

Vertical angles have the same measure. So, in this figure, $m\angle 1 = m\angle 3$ and $m\angle 2 = m\angle 4$.

WRITING ABOUT MATH

Use what you know about straight angles and supplementary angles to write a convincing argument explaining why vertical angles must be equal in measure. Use a figure like the one below to illustrate your reasoning.

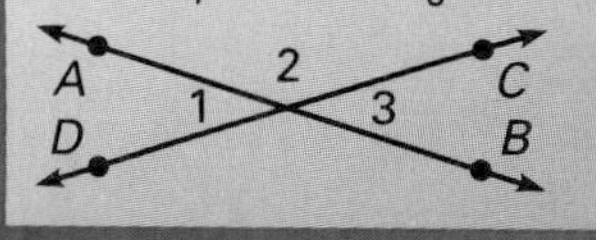

Example 1

Use the figure at the right.

a. Find $m\angle 3$. b. Find $m\angle 2$.

Solution

a. $\angle 1$ and $\angle 3$ are not adjacent to each other, so they are vertical angles. Vertical angles have the same measure.

$m\angle 1 = m\angle 3$ $\quad$ $m\angle 3 = 145°$

b. $\angle 2$ and $\angle 3$ are supplementary angles, so the sum of their measures is 180°.

$m\angle 2 + 145° = 180°$
$m\angle 2 = 180° - 145°$
$m\angle 2 = 35°$ ◄

A **transversal** is a line that intersects two or more lines in a plane at different points. In the figure at the right, $\overleftrightarrow{AB}$ is a transversal intersecting $\overleftrightarrow{CD}$ and $\overleftrightarrow{EF}$.

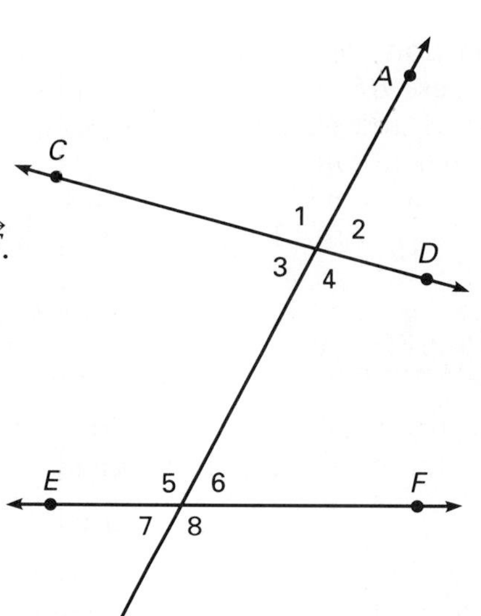

TALK IT OVER

In the definition of a transversal, why do you think it is important to include the phrase *at different points?*

A transversal such as $\overleftrightarrow{AB}$ forms several special pairs of angles with the lines it intersects.

Angles that are in the same position relative to the transversal and the lines are called **corresponding angles.** In this figure, there are four pairs of corresponding angles.

$\angle 1$ and $\angle 5$ $\quad$ $\angle 2$ and $\angle 6$
$\angle 3$ and $\angle 7$ $\quad$ $\angle 4$ and $\angle 8$

Alternate interior angles are pairs of angles that are *interior* to the lines but on *alternate* sides of the transversal. In the figure above, there are two pairs of alternate interior angles.

$\angle 3$ and $\angle 6$ $\quad$ $\angle 4$ and $\angle 5$

Alternate exterior angles are pairs of angles that are *exterior* to the lines and on *alternate* sides of the transversal. In the figure above, there are two pairs of alternate exterior angles.

$\angle 1$ and $\angle 8$ $\quad$ $\angle 2$ and $\angle 7$

When a transversal intersects a pair of parallel lines, there are special relationships among these pairs of angles.

- ► Corresponding angles have the same measure.
- ► Alternate interior angles have the same measure.
- ► Alternate exterior angles have the same measure.

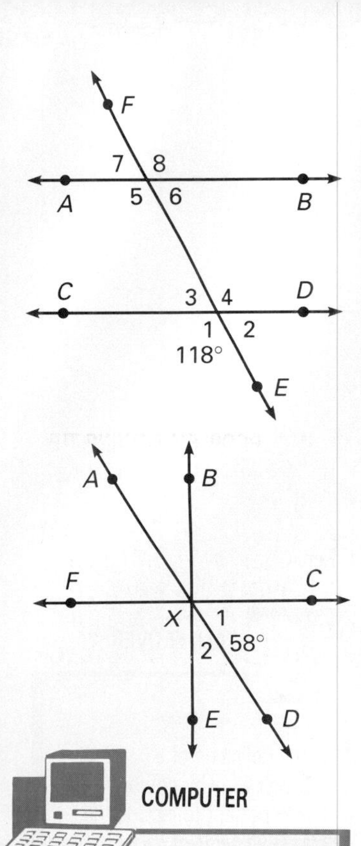

Example 2

In the figure at the left, $\overleftrightarrow{AB} \parallel \overleftrightarrow{CD}$. Find the measure of each angle.

a. $m\angle 8$ b. $m\angle 7$

Solution

a. $\angle 8$ and $\angle 1$ are a pair of alternate exterior angles, so $m\angle 8 = m\angle 1$.
$m\angle 8 = 118°$

b. $\angle 7$ and $\angle 8$ are a pair of supplementary angles, so the sum of their measures is $180°$.
$m\angle 7 + m\angle 8 = 180°$
$m\angle 7 + 118° = 180°$
$m\angle 7 = 180° - 118° = 62°$ ◄

Example 3

In the figure at the left, $\overleftrightarrow{BE} \perp \overleftrightarrow{CF}$. Find $m\angle 2$.

Solution

Because $\overleftrightarrow{BE} \perp \overleftrightarrow{CF}$, $\angle CXE$ is a right angle, with a measure of $90°$. So, $\angle 1$ and $\angle 2$ are a pair of complementary angles.
$m\angle 1 + m\angle 2 = 90°$
$58° + m\angle 2 = 90°$
$m\angle 2 = 90° - 58° = 32°$ ◄

COMPUTER

This program will print the measure of the complement or the supplement of an angle based on your input.

```
10 INPUT "WHAT IS THE
   MEASURE OF THE ANGLE?
   ";M: PRINT
20 INPUT "ARE THE ANGLES
   COMPLEMENTARY OR
   SUPPLEMENTARY?
   C OR S: ";A$: PRINT
30 IF A$ = "C" THEN GOTO
   50
40 PRINT "THE MEASURE OF
   THE SUPPLEMENT IS
   ";180 - M: END
50 PRINT "THE MEASURE OF
   THE COMPLEMENT IS";
   90 - M
```

TRY THESE

Use the figure at the right to find the measure of each angle.

120°
3
4 2
1

1. $\angle 1$ 2. $\angle 4$

In the figure at the right, $\overleftrightarrow{LM} \parallel \overleftrightarrow{NO}$. Find the measure of each angle.

3. $\angle 5$ 4. $\angle 6$

In the figure at the right, $\overleftrightarrow{RS} \perp \overleftrightarrow{TU}$. Find the measure of each angle.

5. $\angle 2$ 6. $\angle 3$

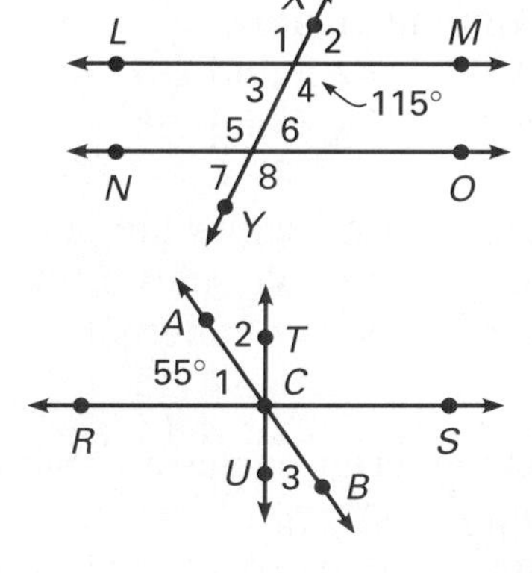

EXERCISES

PRACTICE/ SOLVE PROBLEMS

Use the figure at the right to find the measure of each angle.

1. $\angle 3$ 2. $\angle 1$ 3. $\angle 4$

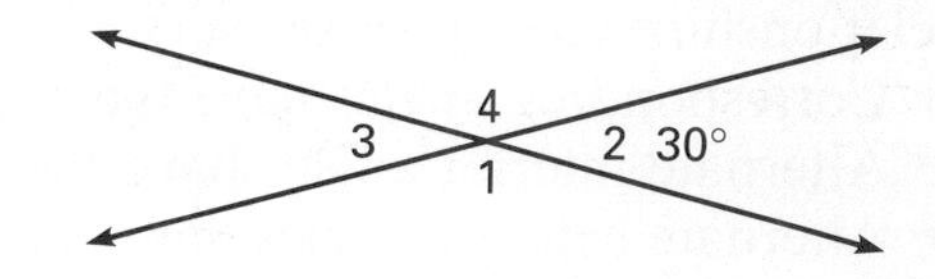

In the figure at the right, $\overleftrightarrow{AB} \parallel \overleftrightarrow{CD}$. Find the measure of each angle.

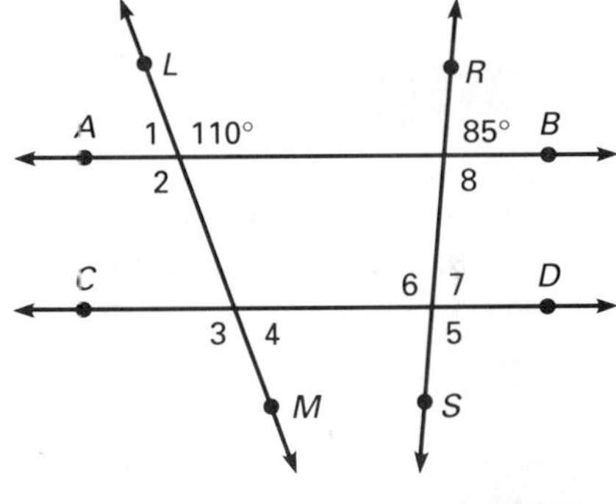

4. $\angle 1$
5. $\angle 2$
6. $\angle 3$
7. $\angle 4$
8. $\angle 5$
9. $\angle 6$
10. $\angle 7$
11. $\angle 8$

12. Assume that, over a small area, the surface of the earth can be considered a plane. If a freighter going north crosses the equator at a 59° angle and continues on in the same direction, what will be the angle of the freighter's path as it crosses the 5°N latitude line?

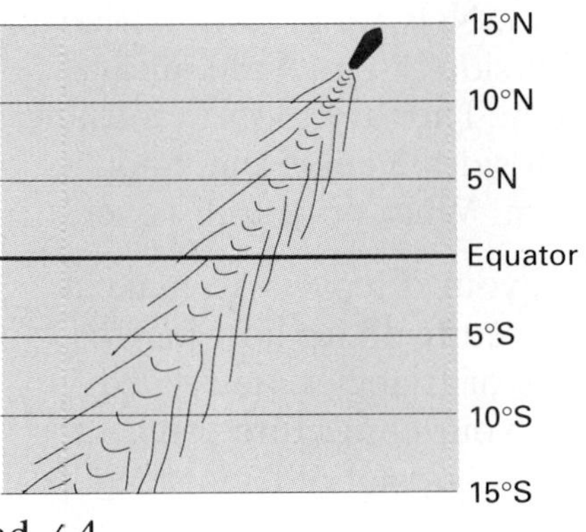

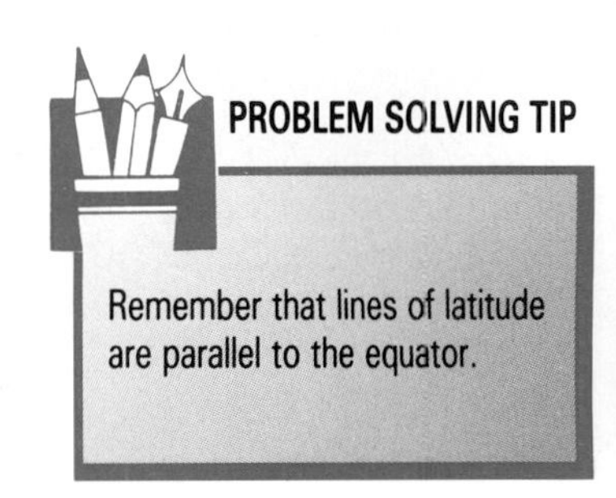

EXTEND/ SOLVE PROBLEMS

Find the measure of $\angle 1$, $\angle 2$, $\angle 3$, and $\angle 4$.

13.

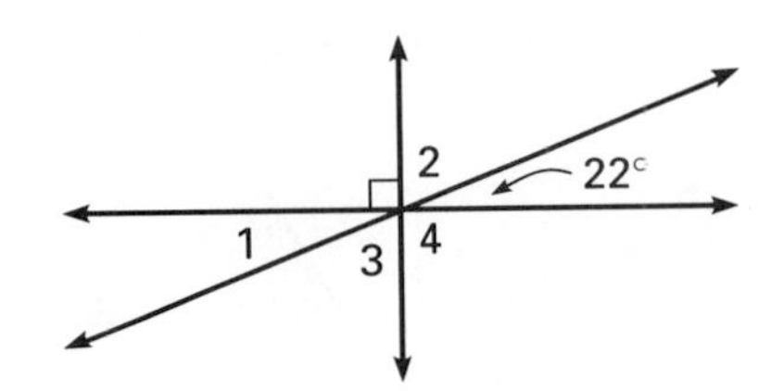

PROBLEM SOLVING TIP

In Exercises 14 and 15, write an equation that represents the relationship between the angles whose measures are labeled.

14. Find $m\angle ATS$ and $m\angle STB$ in the figure below.

15. Find $m\angle QRN$ and $m\angle PRN$ in the figure below.

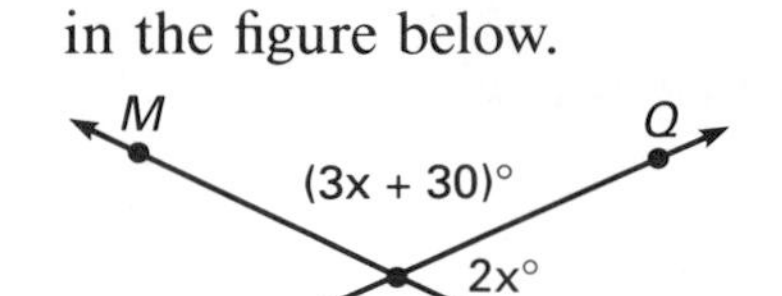

16. Do you think that $\overleftrightarrow{AB}$ and $\overleftrightarrow{CD}$ in the figure below are parallel? Explain.

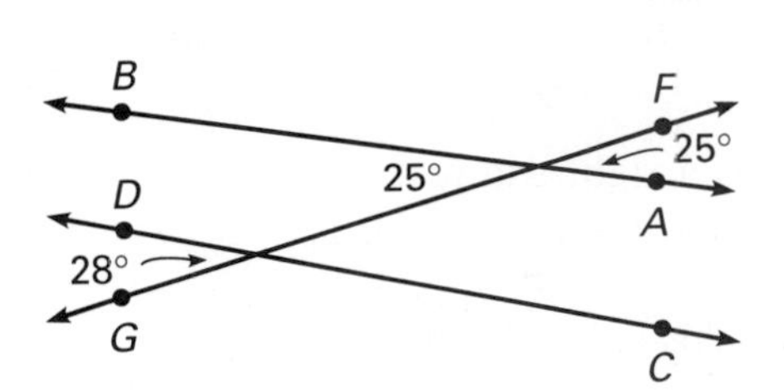

17. In the figure below, $m\angle BAC = m\angle DCE$. Which line segments or lines do you think are parallel? Explain.

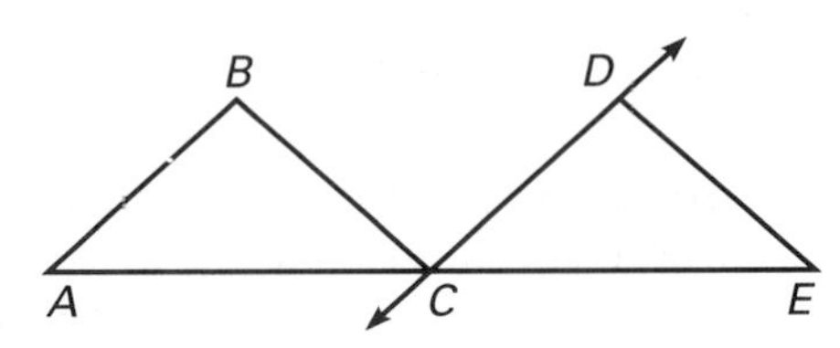

THINK CRITICALLY/ SOLVE PROBLEMS

10-4 Triangles

EXPLORE

TALK IT OVER

Suppose that the strips of paper had lengths of 3 in., 5 in., 8 in., and 9 in. Can you predict which combination of strips could be used to form a triangle?

Cut four thin strips of paper that have lengths 2 cm, 5 cm, 6 cm, and 9 cm. Using the strips, try to form a triangle with sides of 5 cm, 6 cm, and 9 cm. Now try to form a triangle with sides 2 cm, 5 cm, and 9 cm. Then try to form a triangle with sides 2 cm, 5 cm, and 6 cm. What do you discover?

Use your discovery to make a conjecture about how lengths of sides of triangles are related. Test your conjecture using paper strips of different lengths.

SKILLS DEVELOPMENT

Triangles can be classified by the measures of their angles.

right triangle: A triangle that has one right angle.

acute triangle: A triangle that has three acute angles.

obtuse triangle: A triangle that has one obtuse angle.

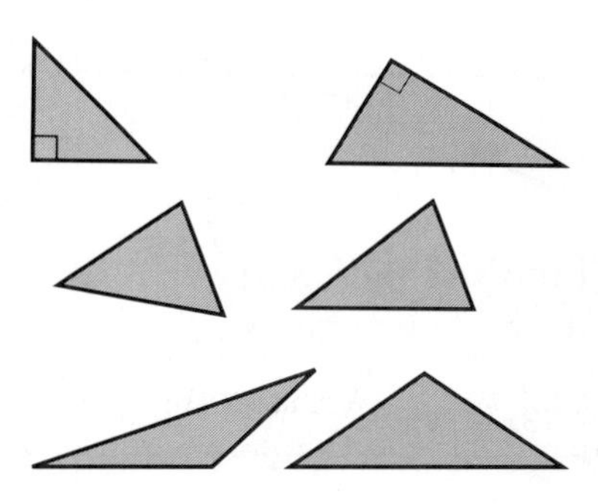

Example 1

Classify each triangle as *acute, obtuse,* or *right.*

a.

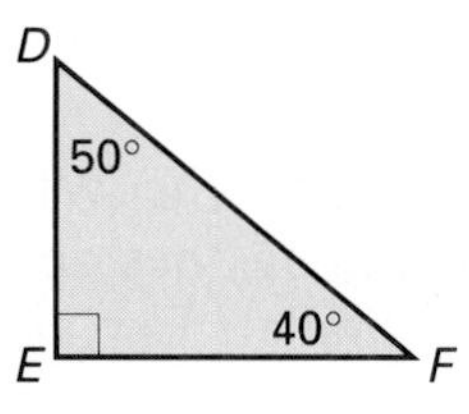

b.

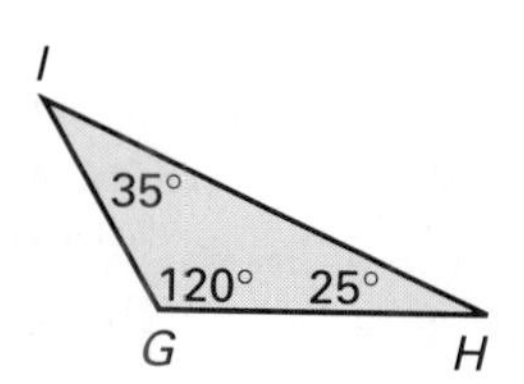

c.

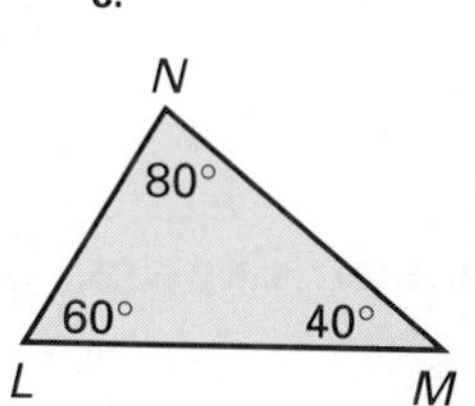

Solution

a. Angle E is a right angle. So, $\triangle DEF$ is a right triangle.

b. The measure of angle G is greater than $90°$. So, $\triangle GHI$ is an obtuse triangle.

c. The measure of each of the three angles is less than $90°$. So, $\triangle LMN$ is an acute triangle. ◄

Triangles can also be classified by the lengths of their sides.

isosceles triangle: A triangle that has two sides of equal length.

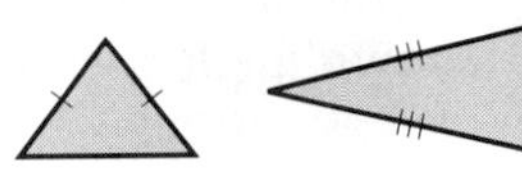

equilateral triangle: A triangle that has all three sides of equal length.

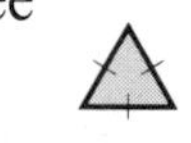

scalene triangle: A triangle that has no sides of equal length.

The marks on the figures indicate which sides are of equal length.

WRITING ABOUT MATH

How would you describe the different classifications of triangles? In your journal, describe each type of classification.

Example 2

Classify each triangle as *isosceles, equilateral,* or *scalene.*

a.

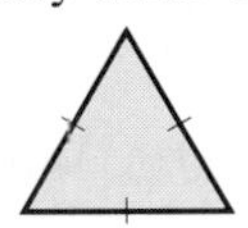

b.

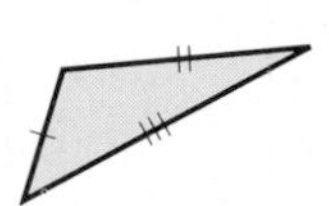

c.

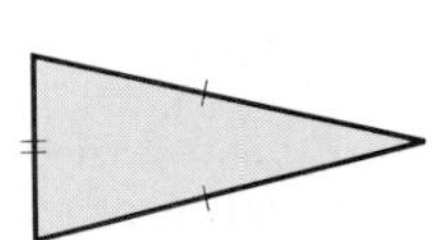

Solution

a. All three sides are of equal length. The figure is an equilateral triangle.
b. No sides are of equal length. The figure is a scalene triangle.
c. Two sides are of equal length. The figure is an isosceles triangle. ◄

The sum of the measures of the angles of a triangle is 180°.

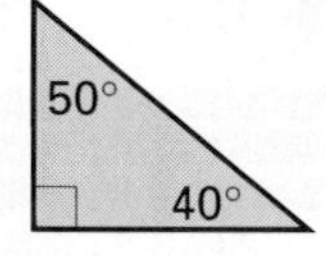

$90° + 50° + 40° = 180°$

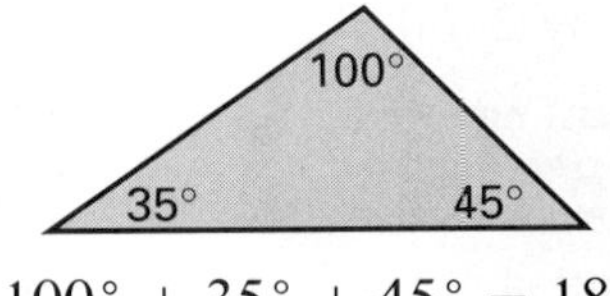

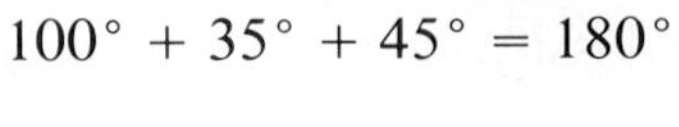

$100° + 35° + 45° = 180°$

CHECK UNDERSTANDING

It is possible for a triangle to have two right angles? Explain.

The three angles of an equilateral triangle have equal measures. That is, an equilateral triangle is also **equiangular.** Each angle of an equilateral triangle measures 60°.

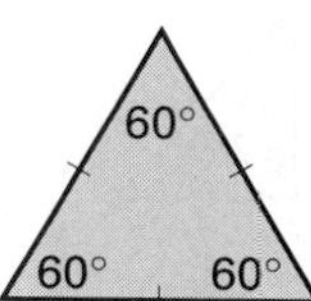

Example 3

a. The angles of a scalene triangle have the measures shown in the figure at the right. What is the value of x?
b. The measure of one acute angle of a right triangle is 30°. What is the measure of the other acute angle?

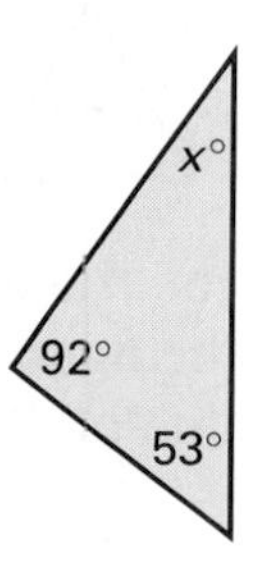

Solution

a. The sum of the angles of a triangle → 180
Known angle measures 92° + 53° → −145

$$\begin{array}{r} 180 \\ -145 \\ \hline 35 \end{array}$$

The measure of the third angle is 35°, so $x = 35$.

b. The sum of the angles of a triangle $\rightarrow$ 180
The measure of the right angle is 90°.
Known angle measures $90° + 30°$ $\rightarrow -120$
60
The measure of the other acute angle is 60°. ◄

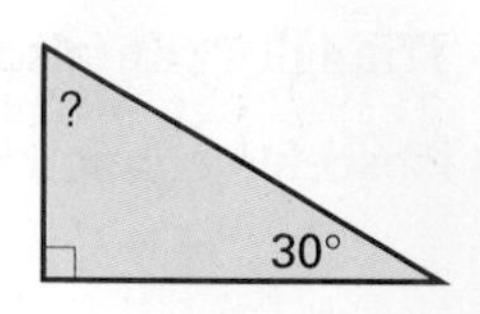

In an isosceles triangle the sides and angles have special names.

Notice that the legs of an isosceles triangle are of equal length. The base angles, which are opposite the legs, also have equal measures.

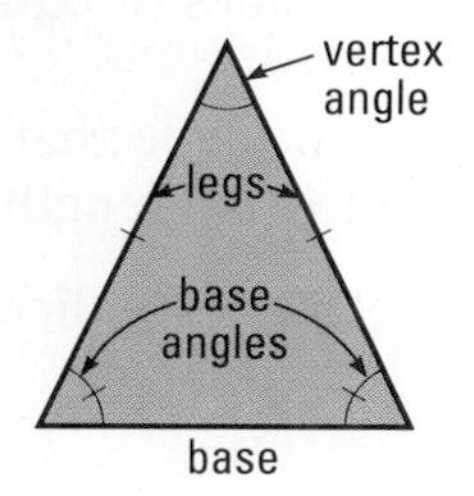

Example 4

The measure of the vertex angle of an isosceles triangle is 120°. What is the measure of each of the base angles?

Solution

The sum of the angles of a triangle $\rightarrow$ 180
Known angle measure $\rightarrow -120$
60

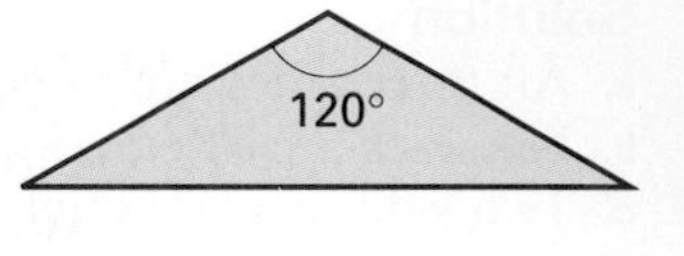

Since the base angles of an isosceles triangle are equal in measure, divide 60 by 2: $60 \div 2 = 30$. The measure of each base angle is 30°. ◄

TRY THESE

Classify each triangle as *acute, obtuse,* or *right.*

1. 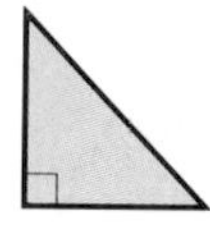
2. 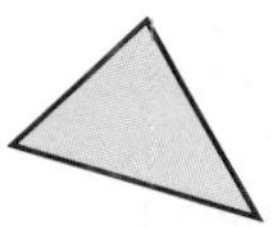
3. 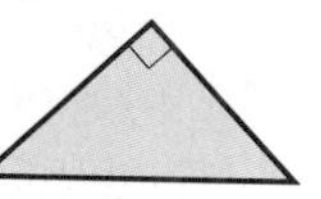
4. 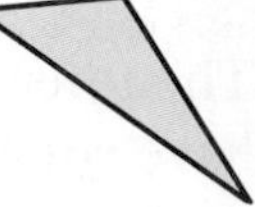

Classify each triangle as *isosceles, equilateral,* or *scalene.*

5. 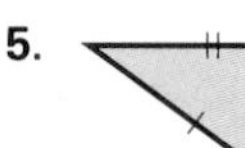
6. 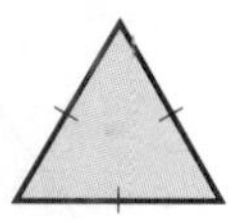
7. 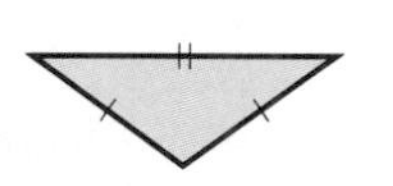
8. 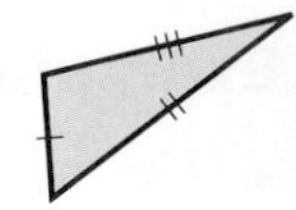

9. The angles of a scalene triangle have the measures shown in the figure at the right. What is the value of x?

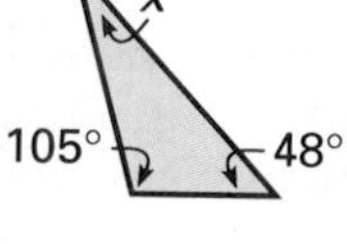

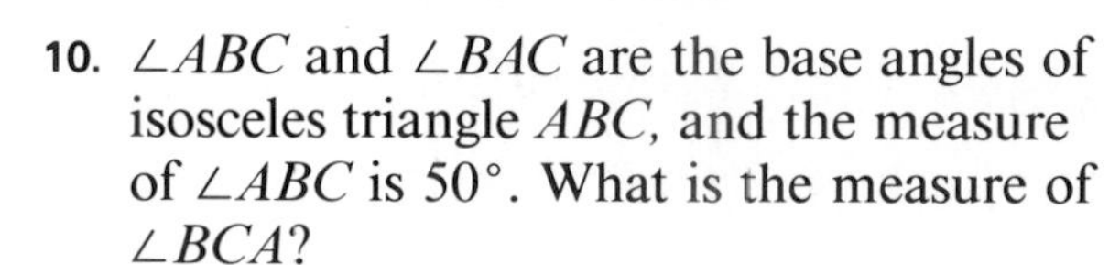

10. $\angle ABC$ and $\angle BAC$ are the base angles of isosceles triangle ABC, and the measure of $\angle ABC$ is 50°. What is the measure of $\angle BCA$?

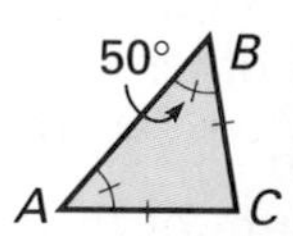

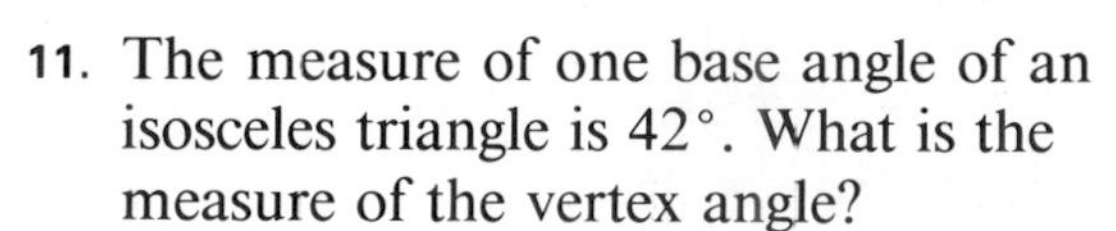

11. The measure of one base angle of an isosceles triangle is 42°. What is the measure of the vertex angle?

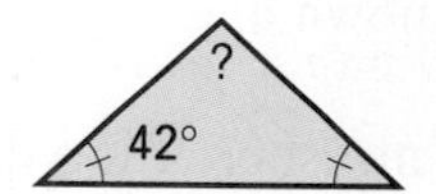

EXERCISES

PRACTICE/ SOLVE PROBLEMS

Classify each triangle as *acute, obtuse,* or *right.*

1. $\triangle ABC$ with $m\angle A = 35°$, $m\angle B = 70°$, and $m\angle C = 75°$.
2. $\triangle LMN$ with $m\angle M = 40°$, $m\angle L = 90°$, and $m\angle N = 50°$
3. $\triangle HIJ$ with $m\angle I = 104°$, $m\angle H = 46°$, and $m\angle J = 30°$

Classify each triangle as *isosceles, equilateral,* or *scalene.*

4. 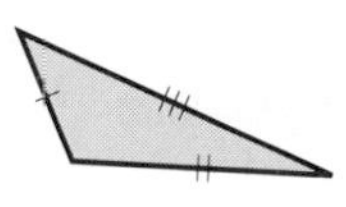

5. 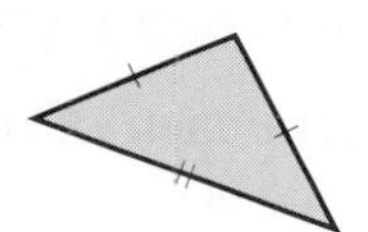

6. 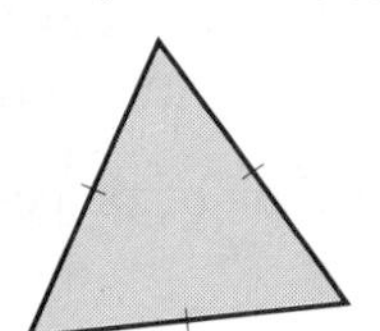

7. 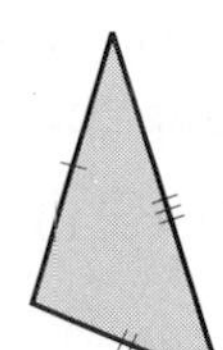

Use the figures at the right for Exercises 8 and 9.

8. The angles of a scalene triangle have the measures shown. What is the value of x?
9. The measure of the vertex angle of an isosceles triangle is 66°. What is the measure of each of the base angles?

EXTEND/ SOLVE PROBLEMS

10. Name all the triangles in the figure at the right. Classify each triangle by its angles.

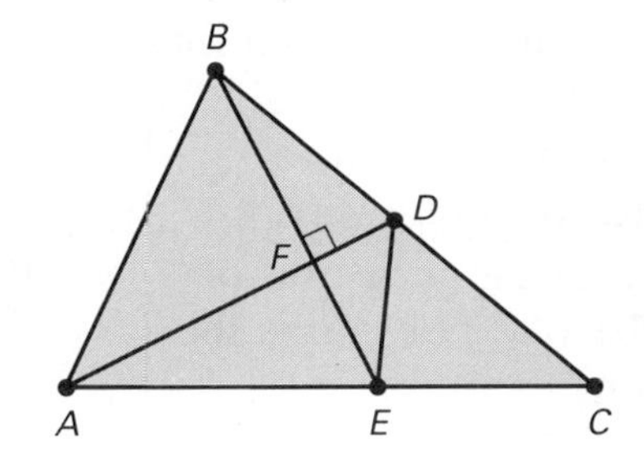

THINK CRITICALLY/ SOLVE PROBLEMS

Determine whether each statement is *true* or *false.*

11. A scalene triangle can also be an obtuse triangle.
12. An equilateral triangle can also be a right triangle.
13. An isosceles triangle can also be an obtuse triangle.
14. An equilateral triangle must also be an acute triangle.
15. Some right triangles are also isosceles.
16. All equilateral triangles are also equiangular.

10-5 Polygons

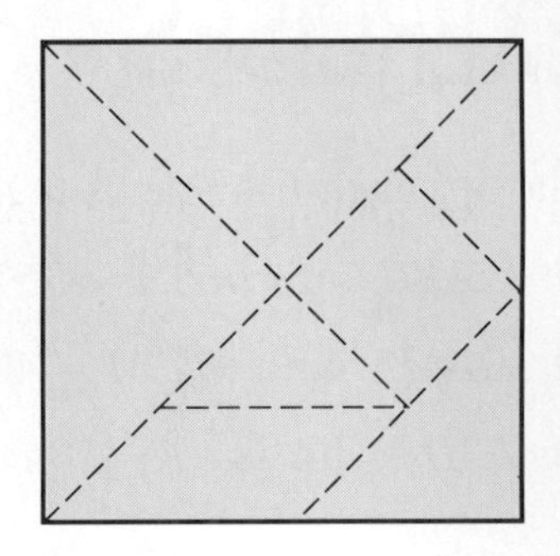

EXPLORE

A tangram is an ancient Chinese puzzle made up of seven pieces. The pieces can be combined to form different figures. Use a set of tangram pieces or trace the figure to make your own.

(If you trace the figure, cut apart your tracing along the lines to form the seven tangram pieces.

a. Use any number of pieces to form each of these figures.

   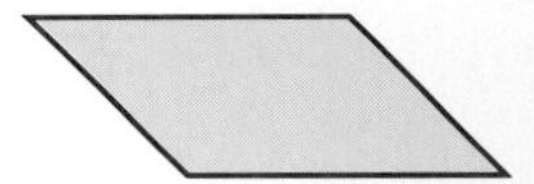

b. Record which pieces were used to make each figure.

c. Compare the pieces you used to make each figure with a classmate. Is there more than one way to make each figure? Discuss.

SKILLS DEVELOPMENT

A **diagonal** of a polygon is a line segment that joins two vertices and is not a side. Polygons can be separated into nonoverlapping triangular regions by drawing all the diagonals from one vertex.

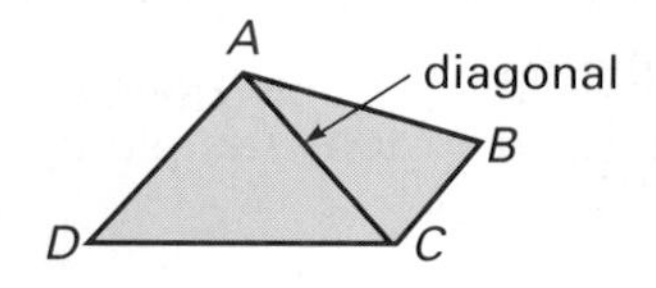

The sum of the angles of $\triangle ABC$ equals 180°.
The sum of the angles of $\triangle ACD$ equals 180°.

The sum of the measures of the angles of a polygon is the product of the number of triangles formed and 180°.

CHECK UNDERSTANDING

What pattern do you see that relates the number of sides of a polygon and the number of triangles formed by diagonals from one vertex?

Example 1

Find the sum of the angle measures for this polygon.

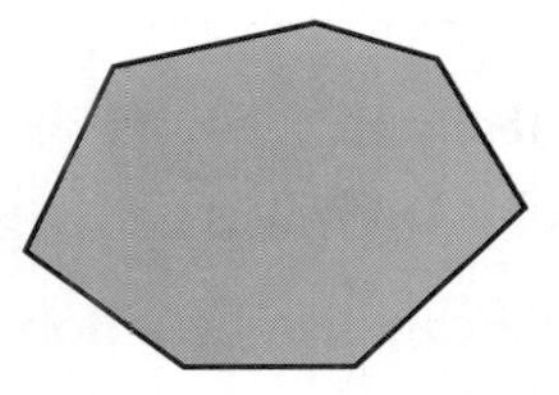

Solution

The polygon has 7 sides. When all the diagonals are drawn from one vertex, there are 5 triangular regions. Since $5 \times 180° = 900°$, the sum of the angle measures is 900°. ◄

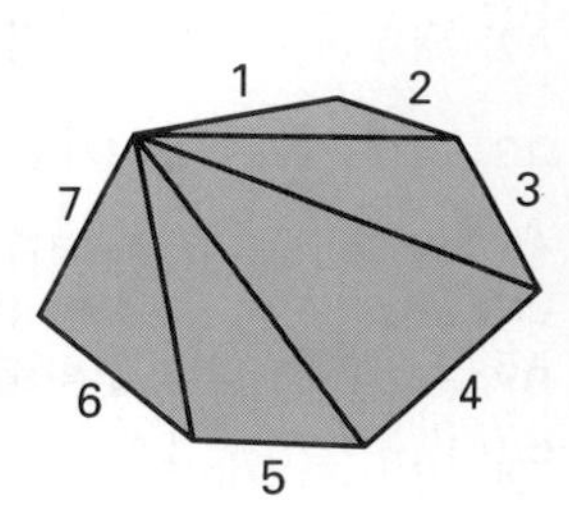

Example 2

Find $m\angle D$ in the polygon at the right.

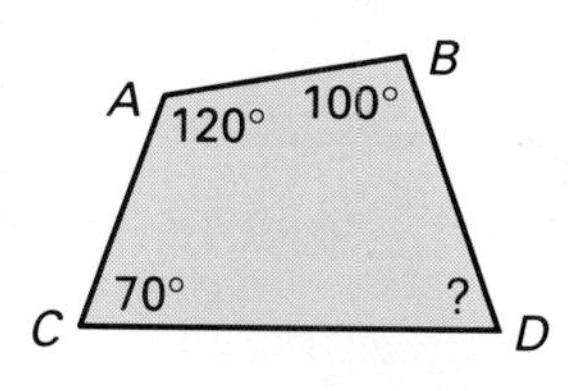

Solution

Since the polygon can be separated into two triangles, the sum of the angle measures is 360°.

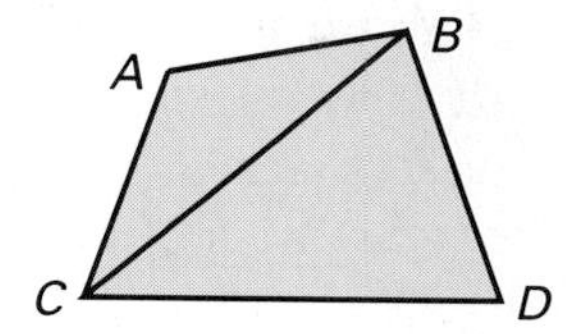

$$m\angle A + m\angle B + m\angle C + m\angle D = 360°$$
$$120° + 100° + 70° + m\angle D = 360°$$
$$290° + m\angle D = 360°$$
$$m\angle D = 70°$$ ◄

Recall that a four-sided polygon is called a *quadrilateral.* Some quadrilaterals have special names.

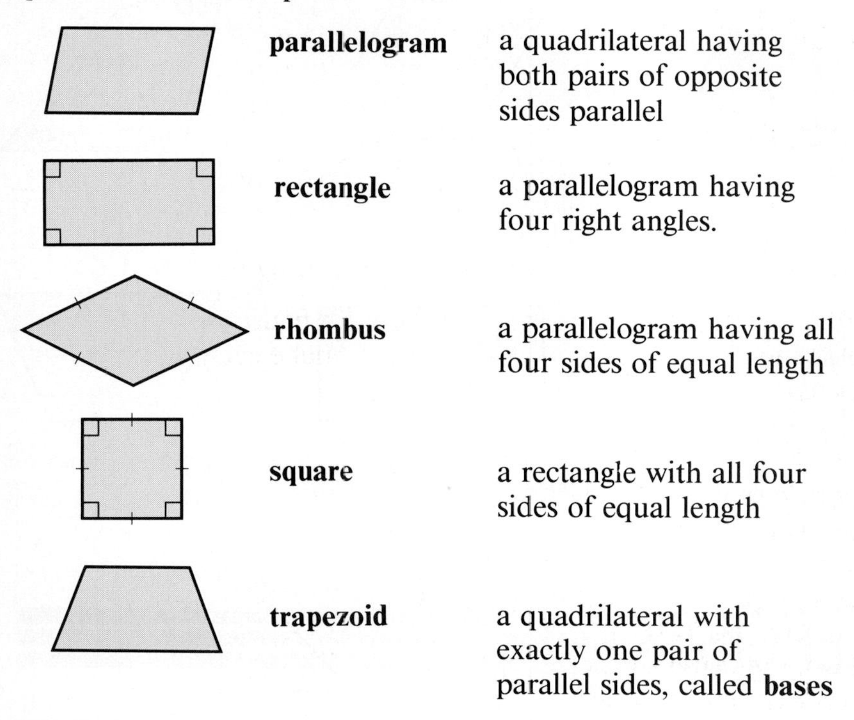

parallelogram a quadrilateral having both pairs of opposite sides parallel

rectangle a parallelogram having four right angles.

rhombus a parallelogram having all four sides of equal length

square a rectangle with all four sides of equal length

trapezoid a quadrilateral with exactly one pair of parallel sides, called **bases**

Example 3

Give all the names that apply to the figure at the right.

Solution

It is a four-sided figure that has both pairs of opposite sides parallel. So, the figure is a polygon, a quadrilateral, and a parallelogram. ◄

A polygon is **regular** if all its sides are of equal length and all its angles are of equal measure. That is, a regular polygon is equilateral and equiangular.

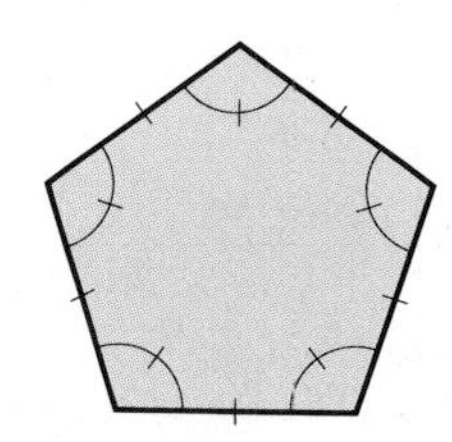

TRY THESE

Find the sum of the angle measures for each polygon.

1.
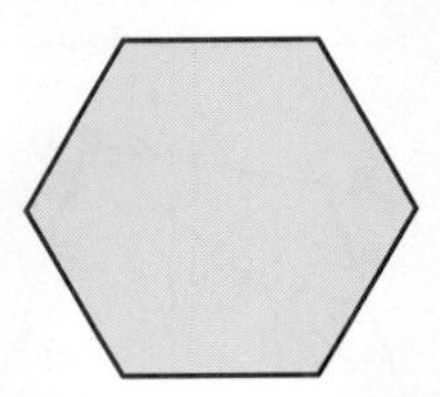

2.
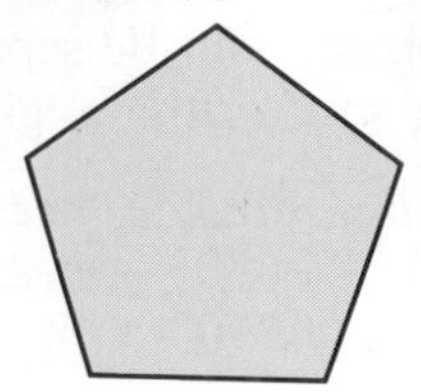

Find the unknown angle measure for each polygon.

3.
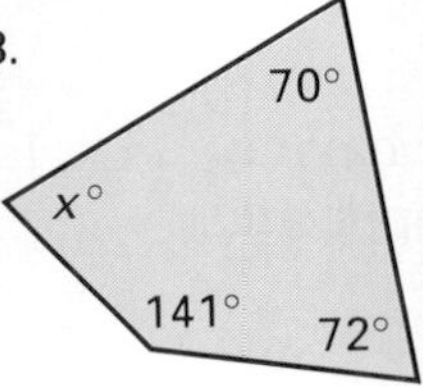

4.
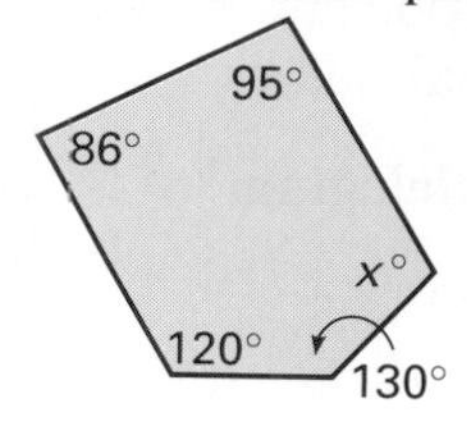

5.
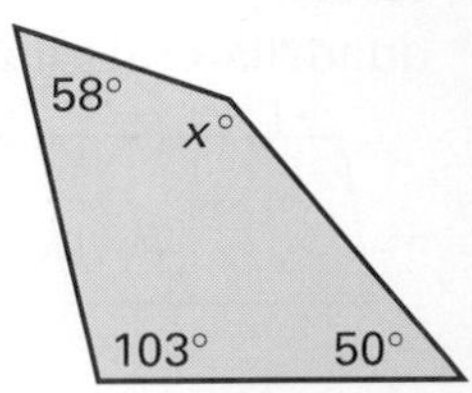

Give all the names that apply to the figure.

6.
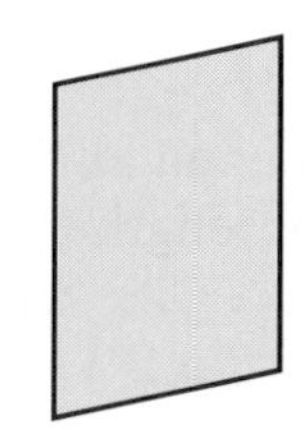

7.
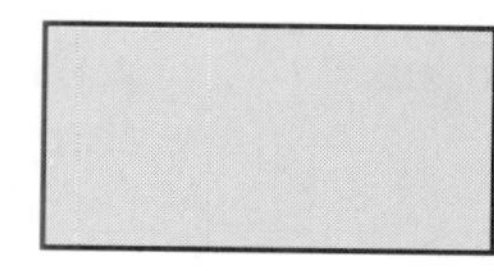

8.
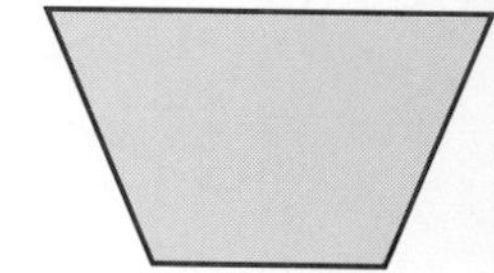

COMPUTER

This program tells you the sum of the angle measures of a polygon when you enter the number of sides. Can you explain how the formula in line 20 was derived?

```
10 INPUT "HOW MANY
   SIDES? ";S: PRINT
20 LET D = (S – 2) * 180
30 PRINT "THE SUM OF THE
   MEASURES OF THE
   ANGLES IS ";D;"."
```

EXERCISES

PRACTICE/ SOLVE PROBLEMS

Copy and complete the chart.

	Polygon	Number of sides	Number of triangles formed by drawing all possible diagonals from one vertex	Sum of the angle measures
	triangle	3	1	180
	quadrilateral	4	2	$2 \times 180 = 360°$
1.	pentagon	■	■	■
2.	hexagon	■	■	■
3.	heptagon	7	■	■
4.	octagon	■	■	■
5.	nonagon	9	■	■
6.	decagon	■	■	■

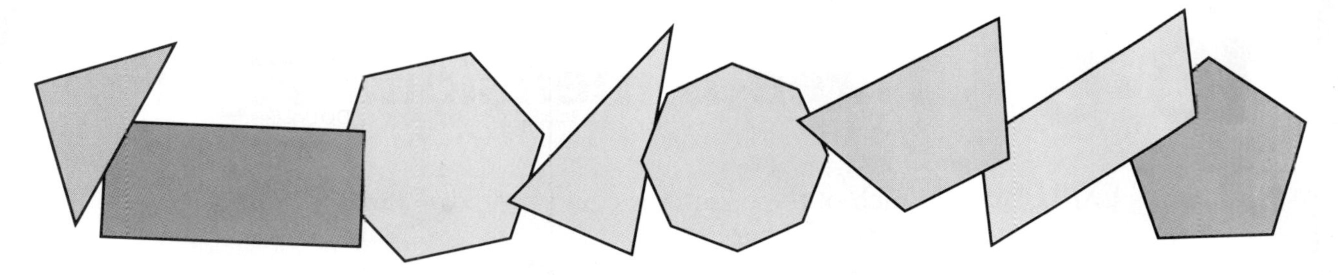

EXTEND/ SOLVE PROBLEMS

7. The perimeter of a regular octagon is 29.92 cm. What is the length of each side?

8. What is the measure of each angle of a regular hexagon?

Trace each polygon. Then draw all its diagonals.

9. 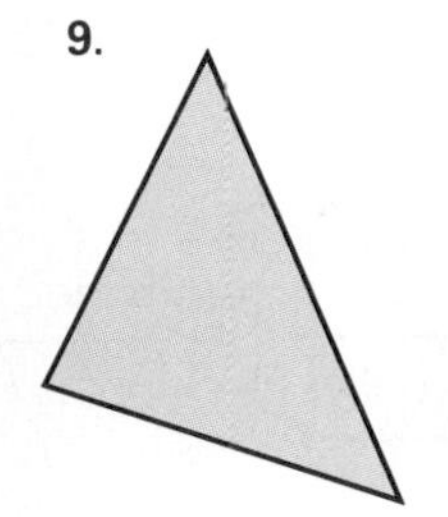

10.  

11. 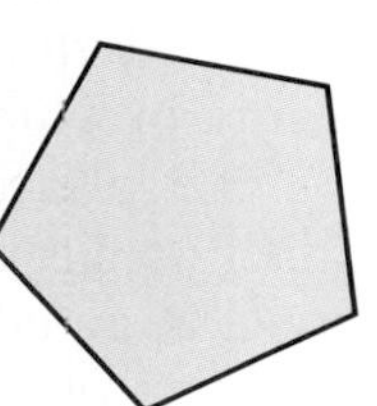

12. 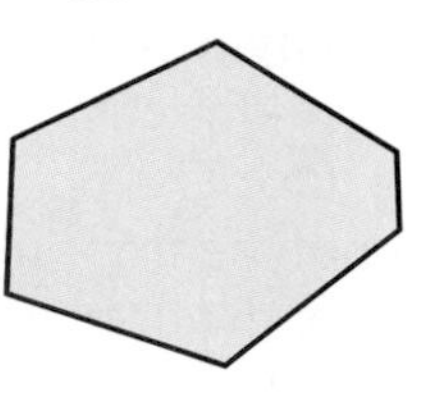

13. Copy and complete using your results from Exercises 9–12.

number of sides	3	4	5	6	7	8	9
number of diagonals	■	■	■	■	14	20	27

14. Refer to the table you completed in Exercise 13. What is the pattern that shows the relationship between the number of sides of a figure and the number of diagonals? Use the pattern to predict the number of diagonals of a decagon.

COMPUTER

Can you complete and use the chart to compute the number of diagonals in a 10-sided or 11-sided polygon quicker than the computer? Have a friend use this program while you challenge the computer!

```
10 D = 0
20 INPUT "HOW MANY
   SIDES? ";N: PRINT
30 FOR X = 0 TO N − 1
40 D = D + X − 1
50 NEXT X
60 PRINT "THERE ARE ";D;"
   DIAGONALS."
```

THINK CRITICALLY/ SOLVE PROBLEMS

Determine whether each statement is *true* or *false.*

15. Some quadrilaterals are trapezoids.

16. A rhombus is an equilateral parallelogram.

17. Not every square is a parallelogram.

18. Every parallelogram is also a rectangle.

19. Is it possible for a hexagon to have six sides of equal length, but not to be a regular hexagon? Explain.

20. Is it possible for a hexagon to have six angles of equal measure, but not to be a regular hexagon?

10-6 Three-Dimensional Figures

EXPLORE

Which of these patterns could be folded along the dotted lines to form a cube?

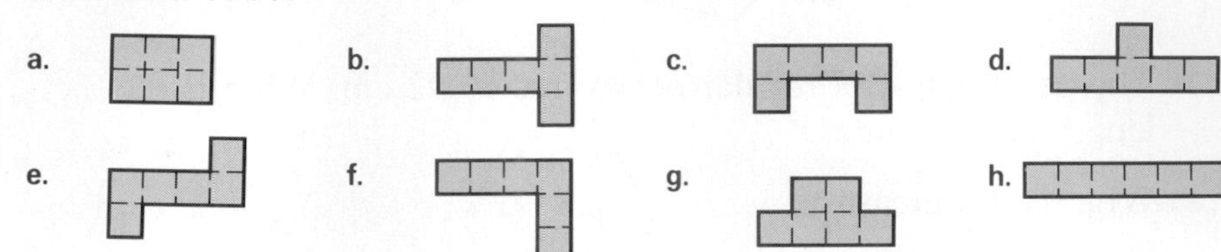

There are eleven different patterns of six squares that can be folded to form a cube. Sketch the eleven patterns on a piece of graph paper.

SKILLS DEVELOPMENT

A **polyhedron** (plural: *polyhedra*) is a three-dimensional figure in which each surface is a polygon. The surfaces of a polyhedron are called its **faces.** Two faces meet, or intersect, at an **edge.** Three or more edges intersect at a **vertex** (plural: *vertices*).

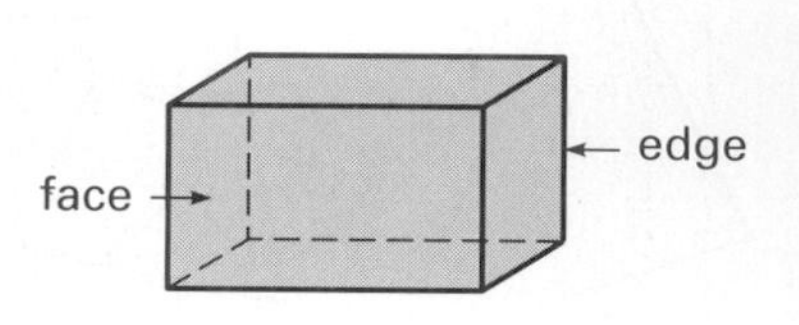

Each of these polyhedra is a prism. A **prism** has two identical, parallel faces, which are called **bases.** The other faces are parallelograms. A prism is named by the shape of its base.

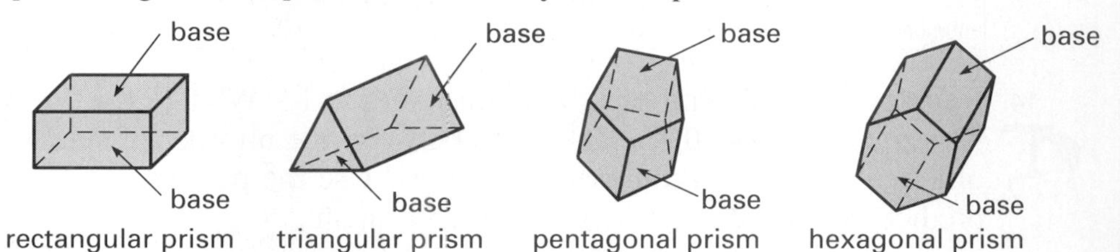

A rectangular prism with all edges of the same length is called a **cube.**

Example 1

Identify the polyhedron.

a.

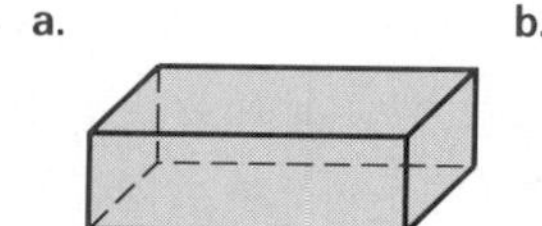

b.

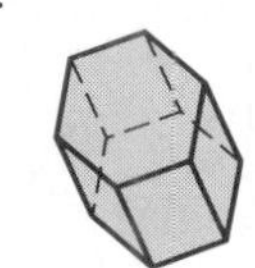

c.

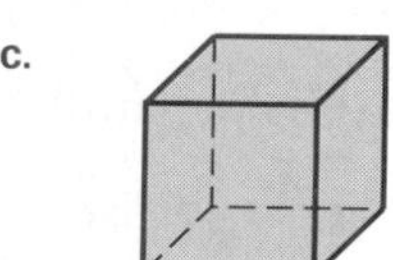

d.

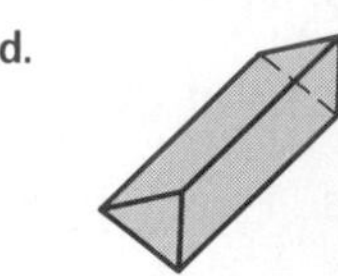

Solution

a. The bases are identical rectangles. It is a rectangular prism.
b. The bases are identical hexagons. It is a hexagonal prism.
c. It is a rectangular prism with all edges of the same length. It is a cube.
d. The bases are identical triangles. It is a triangular prism. ◄

A **pyramid** is a polyhedron with only one base. The other faces are triangles. A pyramid is named by the shape of its base.

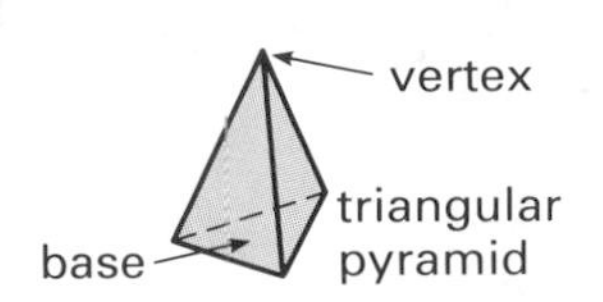

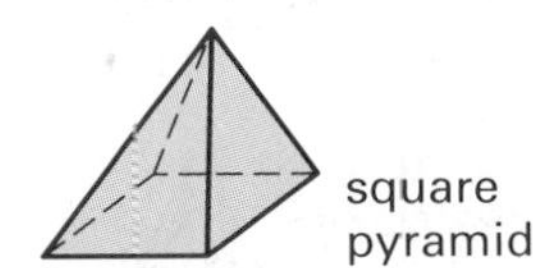

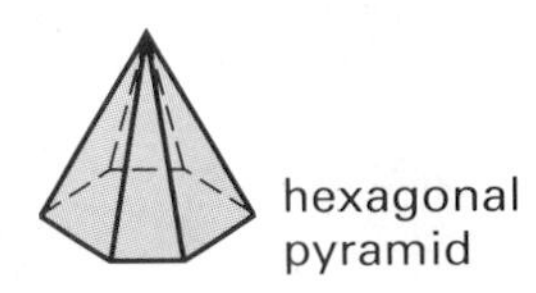

Example 2

Identify the pyramid.

a. 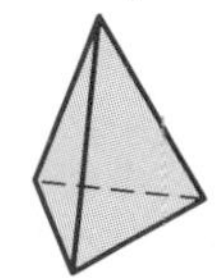

b. 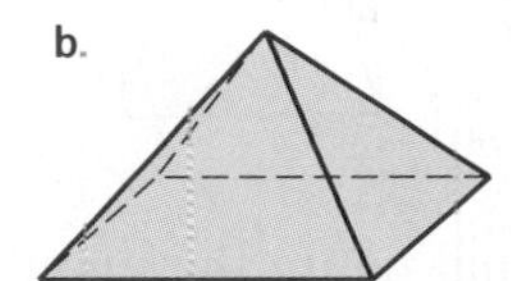

c. 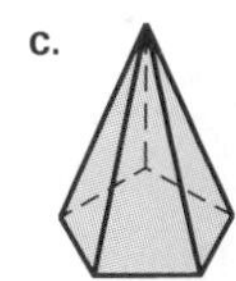

a. The base is a triangle. It is a triangular pyramid.
b. The base is a rectangle. It is a rectangular pyramid.
c. The base is a pentagon. It is a pentagonal pyramid. ◄

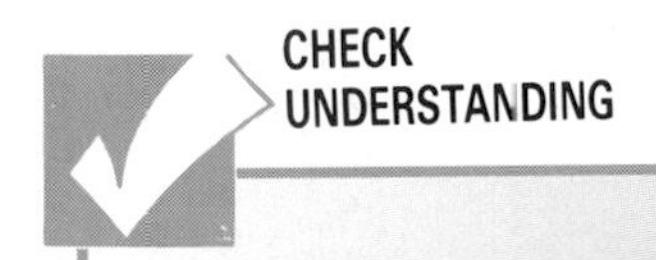

Why are cylinders, cones, and spheres *not* called polyhedra?

Some three-dimensional figures have a curved surface.

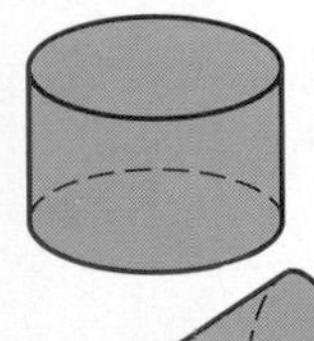

A **cylinder** has two identical, parallel, circular bases.

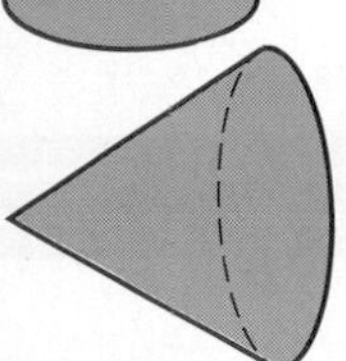

A **cone** has one circular base and one vertex.

A **sphere** is the set of all points in space that are the same distance from a given point, called the center of the sphere.

Example 3

Identify the figure.

a. 

b. 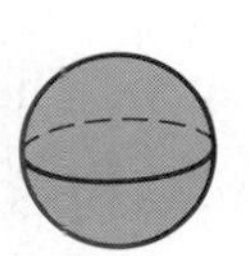

c. 

a. It has a curved surface and one circular base. It is a cone.
b. It has a curved surface with no bases, and all points are the same distance from the center. It is a sphere.
c. It has a curved surface with two circular bases. It is a cylinder. ◄

Example 4

Identify the three-dimensional figure that would be formed if the pattern at the right were folded along the dotted lines.

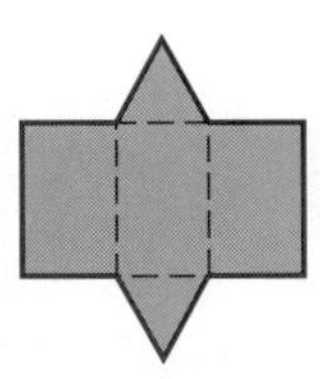

Solution

The figure would have two triangular bases and three other faces that are rectangles. It would be a triangular prism. ◄

TRY THESE

Identify each figure.

1. 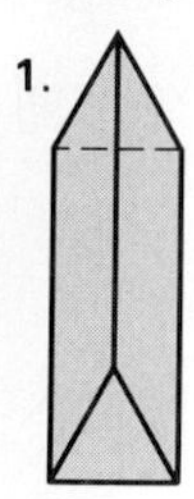

2. 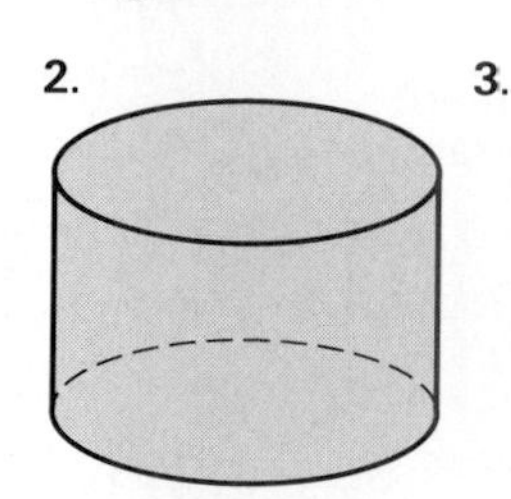

3. 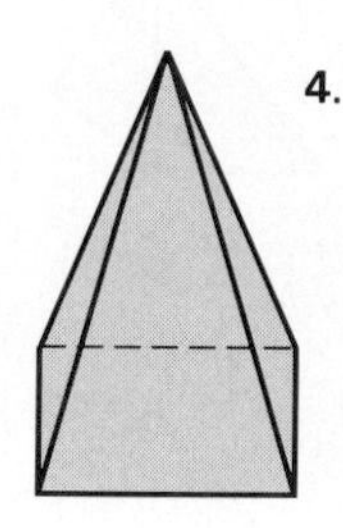

4. 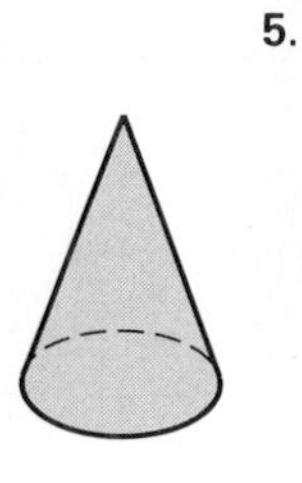

5. 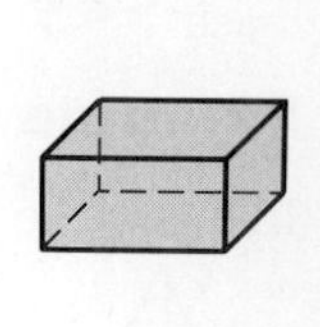

Identify the three-dimensional figure that would be formed by folding the pattern.

6 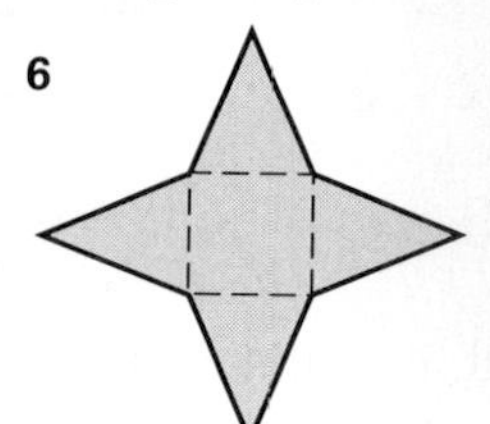

7. 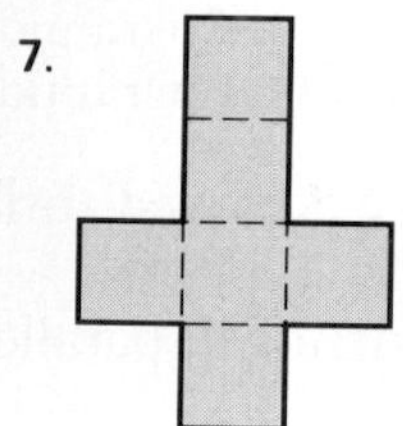

8. 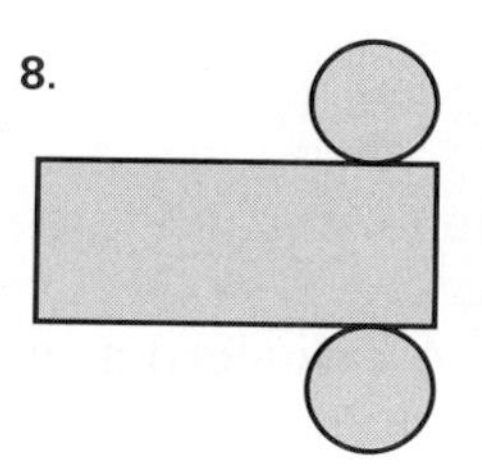

9. 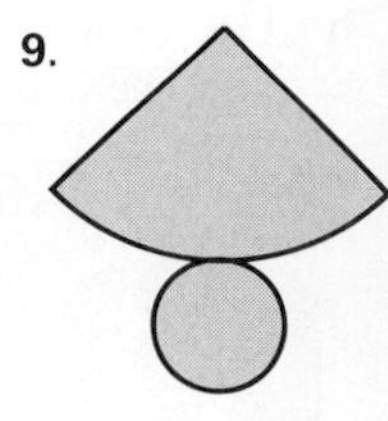

EXERCISES

PRACTICE/ SOLVE PROBLEMS

Make a drawing of each three-dimensional figure.

1. triangular prism
2. rectangular pyramid
3. cube
4. cone

Identify each three-dimensional figure.

5. It has one base that is an octagon. The other faces are triangles.
6. Its two bases are identical, parallel pentagons. The other faces are parallelograms.

Identify the three-dimensional figure that would be formed by folding each pattern along the dotted lines.

7. 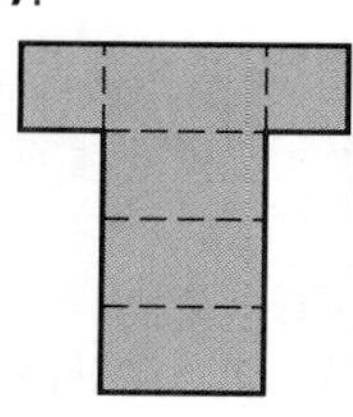

8.

9. 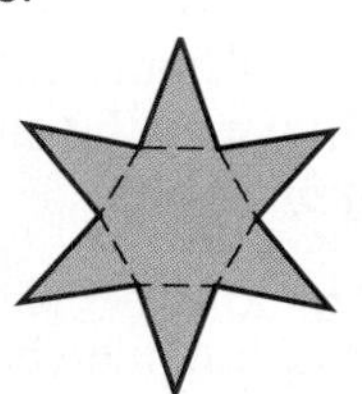

10.

EXTEND/ SOLVE PROBLEMS

Name three everyday objects that have the given shape.

11. rectangular prism
12. cylinder
13. sphere

Copy and complete this table.

	Polyhedron	Number of Faces (F)	Number of Vertices (V)	Number of Edges (E)	F + V − E
14.	triangular prism	■	■	■	■
15.	rectangular prism	■	■	■	■
16.	pentagonal prism	■	■	■	■
17.	triangular pyramid	■	■	■	■
18.	rectangular pyramid	■	■	■	■
19.	pentagonal pyramid	■	■	■	■

20. Use your results from Exercises 14–19. Make a generalization about the relationship between the faces, edges, and vertices of a polyhedron. Write a brief paragraph to explain your generalization.

21. The offices of the U.S. Department of Defense are located in Arlington, Virginia, in a building called the Pentagon. Why do you think the building was given this name?

THINK CRITICALLY/ SOLVE PROBLEMS

22. Each base of a prism is a polygon with n sides. Write a variable expression that represents the total number of faces of the prism.

23. The base of a pyramid is a polygon with t sides. Write a variable expression that represents the total number of edges of the pyramid.

24. Suppose that a cylinder is cut in half along a plane that is perpendicular to the bases. What is the shape of the flat face of each half of the cylinder?

25. Suppose that a cube is cut in half along a plane that passes through the diagonal of one of its faces. What is the shape of each half of the cube?

10-7 Problem Solving Skills:

MAKE A CONJECTURE

- READ
- PLAN
- SOLVE
- ANSWER
- CHECK

Sometimes you can solve a problem by making a conjecture. A **conjecture** is a guess that seems reasonable based on knowledge that you have. After making a conjecture, test it to find out whether it is the correct solution to the problem.

PROBLEM

In triangle ABC, point D is the midpoint of $\overline{AC}$ and point E is the midpoint of $\overline{BC}$. How would the line segment joining points D and E relate to side AB of the triangle?

SOLUTION

Make a conjecture. From the figure, it looks as though $\overline{DE}$ would be parallel to $\overline{AB}$. Also, it looks as though the length of $\overline{DE}$ would be about two thirds the length of $\overline{AB}$.

Test both parts of your conjecture. Trace the triangle onto a sheet of paper and draw line segment DE. Then use a protractor to measure $\angle CDE$ and $\angle BAD$, which are corresponding angles. If $\overline{DE}$ is parallel to $\overline{AB}$, $\angle CDE$ and $\angle BAD$ should be equal in measure.

Use a ruler to measure the length of $\overline{DE}$ and the length of $\overline{AB}$. Compare the measurements.

When the conjecture is tested, you find that $\overline{DE} \parallel \overline{AB}$. The length of $\overline{DE}$ is one half the length of $\overline{AB}$.

PROBLEMS

For each problem, use the question in part **a** to make a conjecture. Then use a ruler or protractor to answer the question in part **b.**

1. For parallelogram $ABCD$, find $m\angle C$.
   a. Conjecture: What is the relationship between the *opposite angles* of a parallelogram?
   b. What is $m\angle C$?

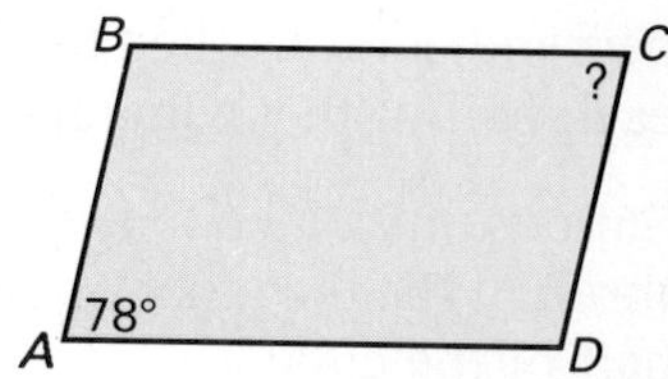

2. For parallelogram $PQRS$, find the length of $\overline{RS}$.
   a. Conjecture: What is the relationship between the *opposite sides* of a parallelogram?
   b. What is the length of $\overline{RS}$?

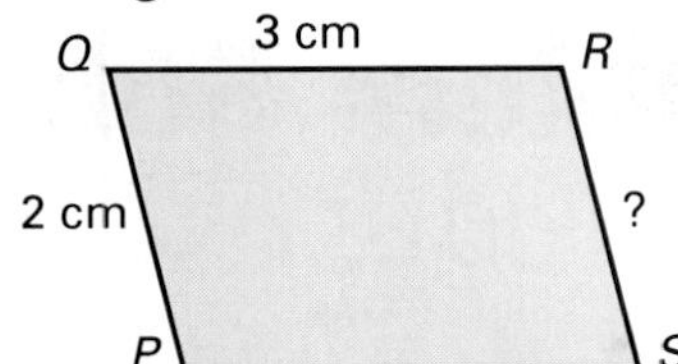

3. For rectangle $WXYZ$, where $\overline{WY}$ measures $1\frac{3}{4}$ in., find the length of $\overline{XZ}$.
   a. Conjecture: What is the relationship between the diagonals of a rectangle?
   b. What is the length of $\overline{XZ}$?

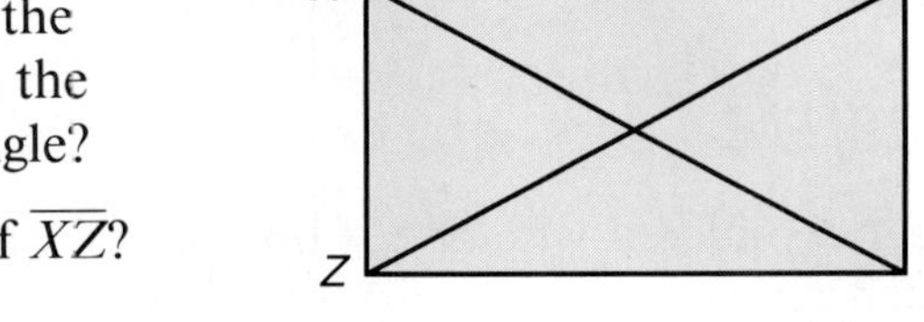

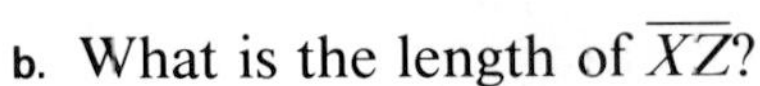

4. For rhombus $JKLM$, find $m\angle 1$.
   a. Conjecture: What is the relationship between the diagonals of a rhombus?
   b. What is $m\angle 1$?

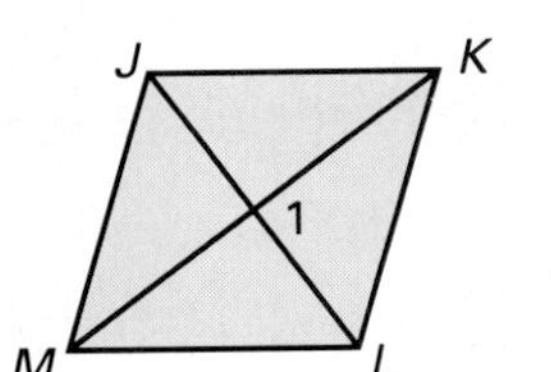

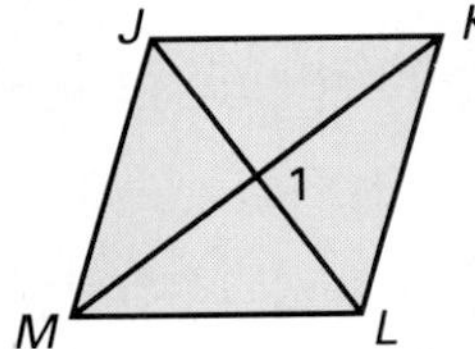

COMPUTER TIP

Geometric drawing software can help you to draw geometric figures quickly and easily. With such software a simple command will usually give you a quick reading of the length of a line segment or the measure of an angle. If you have this kind of software available, you might find it helpful in testing conjectures about geometric figures.

5. For isosceles trapezoid $EFGH$, find $m\angle H$. (In an isosceles trapezoid the nonparallel sides are equal in length.)
   a. Conjecture: What is the relationship between the base angles of an isosceles trapezoid?
   b. What is $m\angle H$?

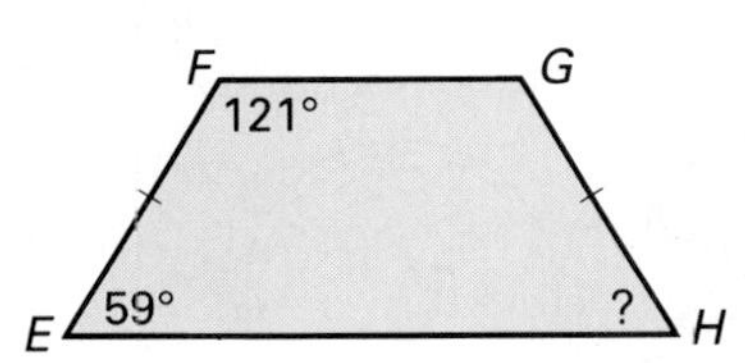

6. For isosceles trapezoid $MNOP$, where $\overline{MO}$ measures 3.5 cm, find the length of $\overline{NP}$.
   a. Conjecture: What is the relationship between the diagonals of an isosceles trapezoid?
   b. What is the length of $\overline{NP}$?

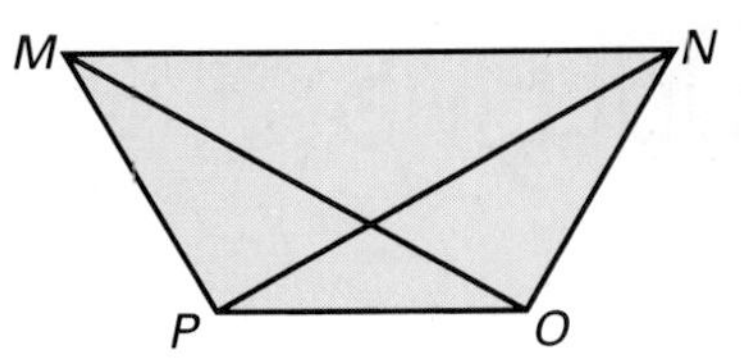

Test each of these conjectures to determine whether it is *true* or *false*.

7. The diagonals of a parallelogram are equal in length.
8. Opposite angles of a trapezoid are equal in measure.
9. Each diagonal of a rhombus separates the other diagonal into two segments of equal length.
10. The diagonals of a parallelogram are perpendicular to each other.

10-8 Problem Solving Strategies:
ACT IT OUT

The most efficient way to solve some problems is by acting out the situation described in the problem. This strategy can be especially helpful in solving problems that involve geometric shapes.

PROBLEM

Eileen wondered whether, if one penny was placed next to another and then was rolled halfway around the other, Lincoln's head would be right-side-up or upside down.

SOLUTION

Act it out. Take two pennies, place them as shown, and then roll the right penny halfway around the left penny. Lincoln's head is right-side-up.

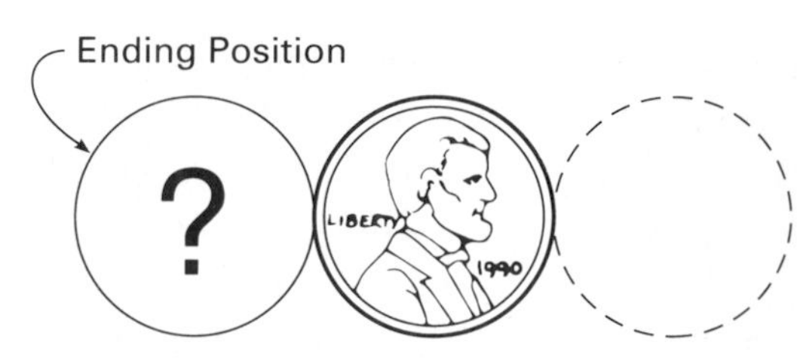

PROBLEMS

Solve.

1. Susan and Kathy were making geometric shapes with toothpicks. Susan challenged Kathy to make three squares by moving exactly three toothpicks in the pattern shown here. Act it out. Draw a picture of the three squares.

2. Starting with the original pattern in Exercise 1, Kathy challenged Susan to make three squares by moving exactly four toothpicks. Act it out. Draw a picture of the three squares.

Trace and cut out square *ABCD*. Label the corners as shown. What figure is formed as each of the following moves is made? Check by acting it out.

3. *A* folded onto *B*
4. *A* folded onto *D*
5. *A* folded onto *B*, then *B* folded onto *C*
6. *A* folded onto *D*, then *B* folded onto *C*

Use your cutout of square *ABCD* to solve.

7. Fold *A* onto *B* and then *B* onto *C*. The perimeter of the resulting figure is what fractional part of the perimeter of the original square?
8. Fold *A* onto the midpoint of side *AB*. If the area of the resulting figure is 12 cm^2, what was the area of the original square?

Use coins or cardboard cutouts of coins to act out each problem. In how many ways can you make change for each amount of money?

9. 25¢
10. 28¢
11. $1.00, using 7 coins
12. 65¢, using 8 coins

13. A heavy log is being moved by rolling it on cylinders. If the circumference of each of the cylinders is 6 ft, how far will the log move for each revolution of the cylinders? Use cardboard or paper to make a model.

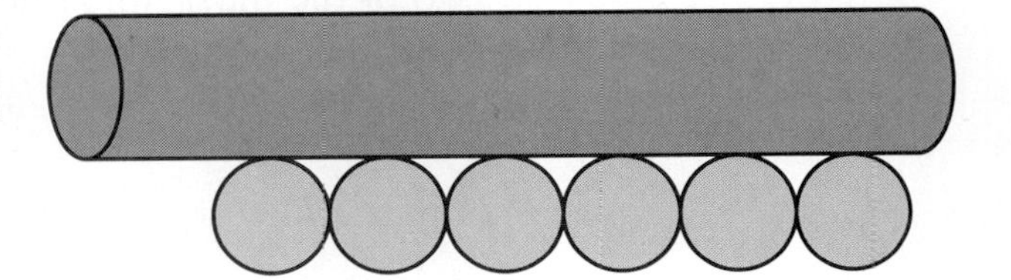

10-9 Constructing and Bisecting Angles

EXPLORE

a. Trace $\angle ABC$ onto a piece of paper.

b. Fold the paper so that $\overrightarrow{BA}$ lies directly on top of $\overrightarrow{BC}$.

c. Unfold the paper. Along the fold line, draw $\overrightarrow{BD}$.

d. If $m\angle ABC = 120°$, what is $m\angle ABD$? Explain.

e. Devise a method for locating $\overrightarrow{BE}$ so that $m\angle ABE = 30°$.

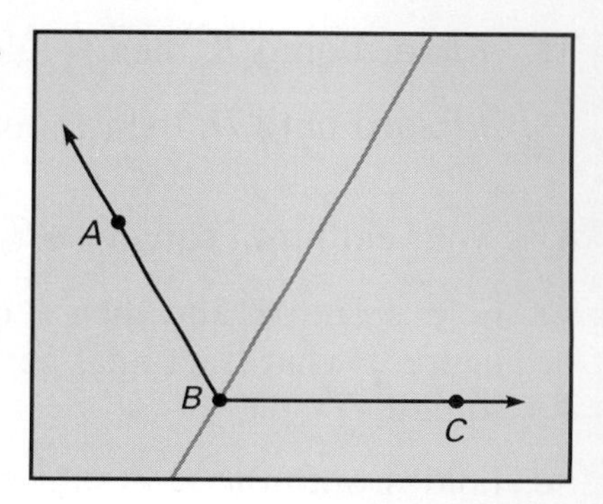

SKILLS DEVELOPMENT

A **construction** is a drawing of a geometric figure that is made using only two tools: a **compass** and an unmarked **straightedge.** (You can use the edge of a ruler as a straightedge, but you must ignore the markings on the ruler.)

Two of the most fundamental constructions are copying an angle and bisecting an angle.

Example 1

Copy $\angle ABC$ using a compass and straightedge.

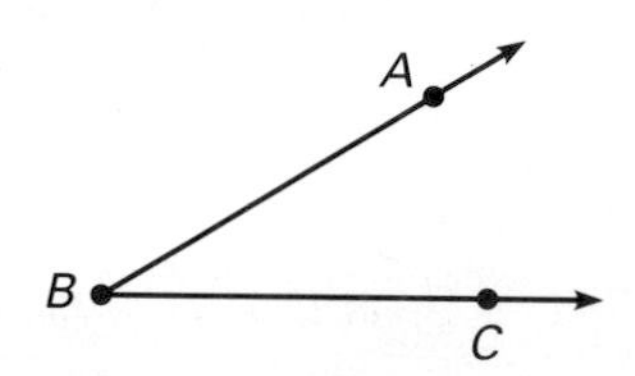

Solution

Step 1

Place the point of the compass at B and draw an arc that intersects both $\overrightarrow{BA}$ and $\overrightarrow{BC}$. Label the intersection points D and E.

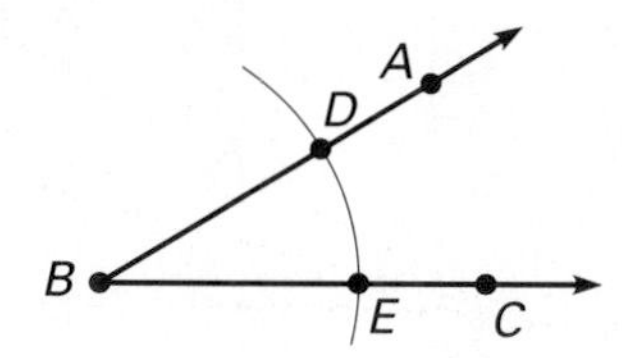

Step 2

Use a straightedge to draw any ray $\overrightarrow{FG}$. Place the point of the compass at F. With the same setting used in Step 1, draw an arc that intersects $\overrightarrow{FG}$. Label the intersection point H.

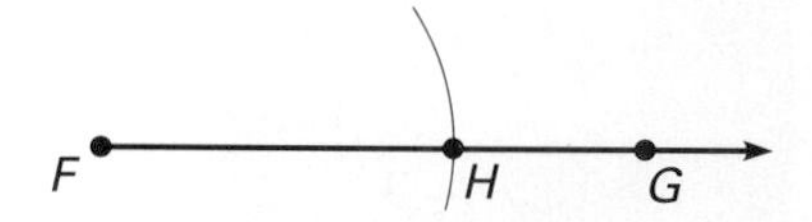

Step 3
Place the point of the compass at D and the point of the pencil at E. Keep that setting and place the point of the compass at H. Draw an arc that crosses the first arc. Label the intersection point I.

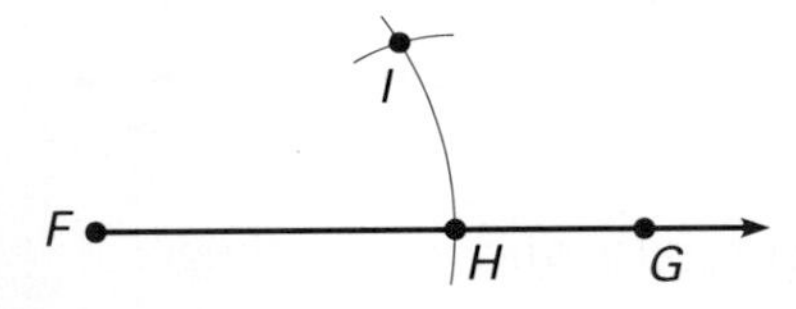

Step 4
Use a straightedge to draw $\overrightarrow{FI}$.

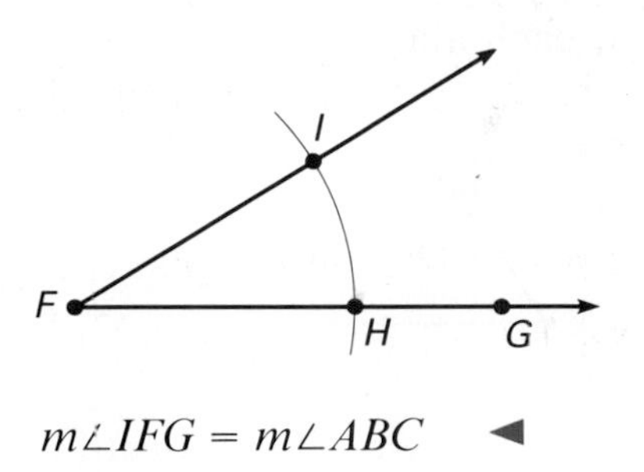

$m\angle IFG = m\angle ABC$ ◄

To *bisect* means "to divide into two equal parts." The **bisector of an angle** is the ray that separates the angle into two adjacent angles of equal measure.

Example 2

Bisect $\angle ABC$ using a compass and a straightedge.

Solution

Step 1
With the compass point at B, draw an arc that intersects $\overrightarrow{BA}$ and $\overrightarrow{BC}$. Label the intersection points P and Q.

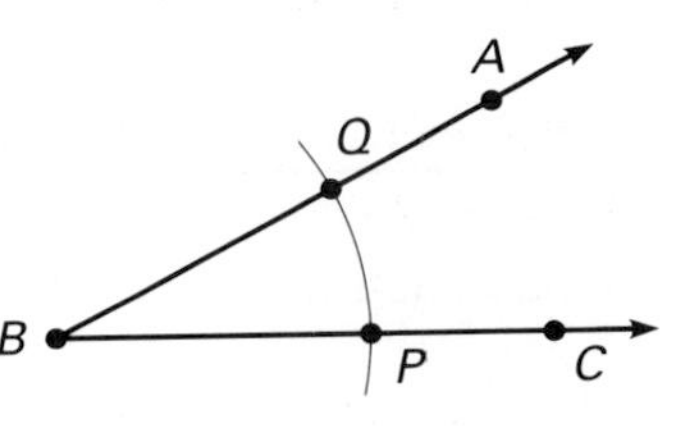

Step 2
With the compass point at P, adjust the compass so that the opening is a little greater than half of arc PQ. Draw an arc inside the angle, as shown.

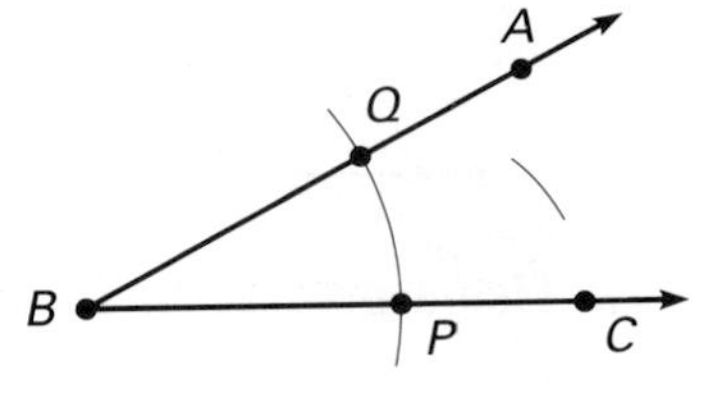

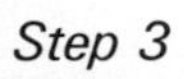

Step 3
Place the point of the compass at Q. With the same compass opening you used in Step 2, draw an arc that intersects the first arc at T.

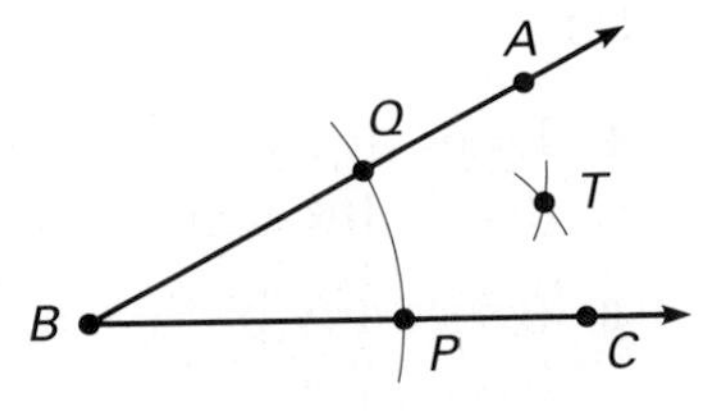

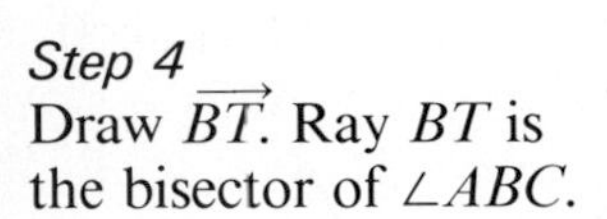

Step 4
Draw $\overrightarrow{BT}$. Ray BT is the bisector of $\angle ABC$.

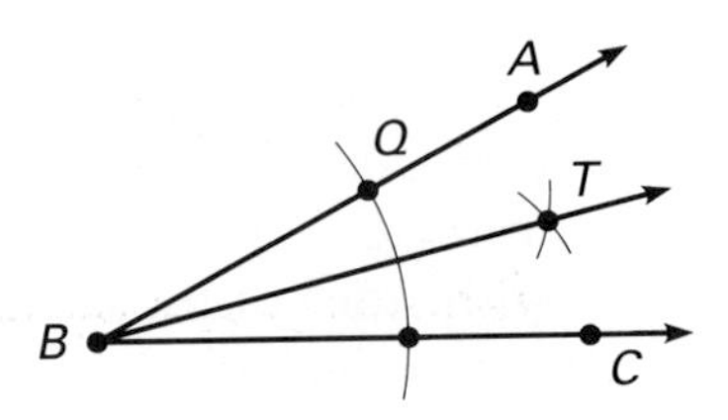

$m\angle ABT = m\angle TBC$ ◄

TRY THESE

Trace each angle. Then copy it using a compass and straightedge.

1.

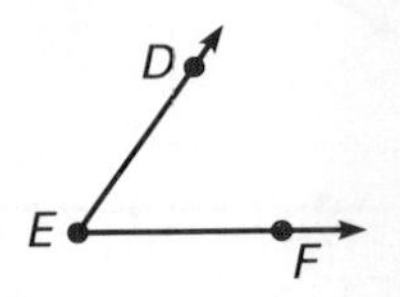

2.

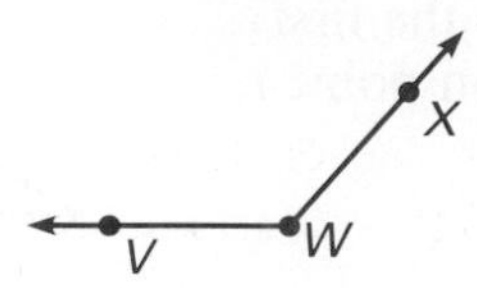

3.

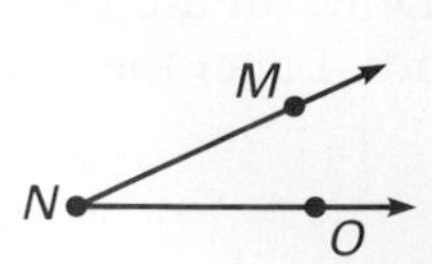

Trace each angle. Then bisect it using a compass and straightedge.

4.

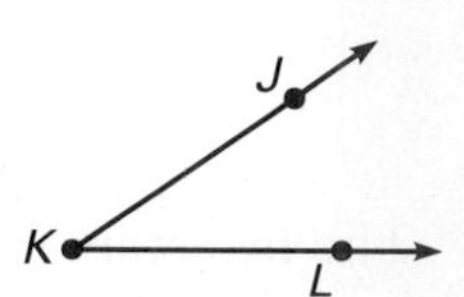

5.

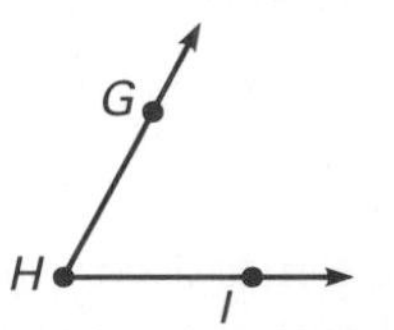

6.

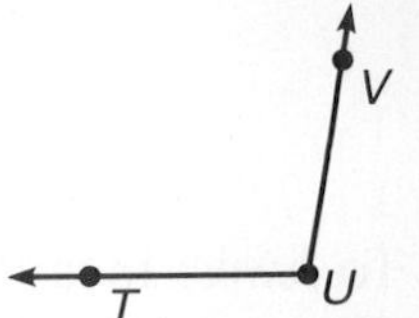

EXERCISES

PRACTICE/ SOLVE PROBLEMS

Trace each angle. Then copy it using a compass and straightedge.

1.

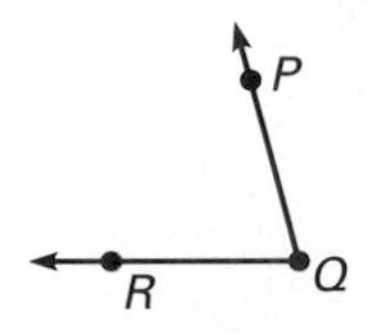

2.

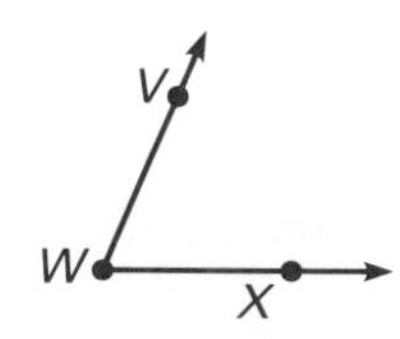

3.

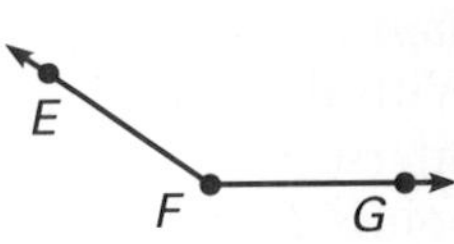

Trace each angle. Then bisect it using a compass and straightedge.

4.

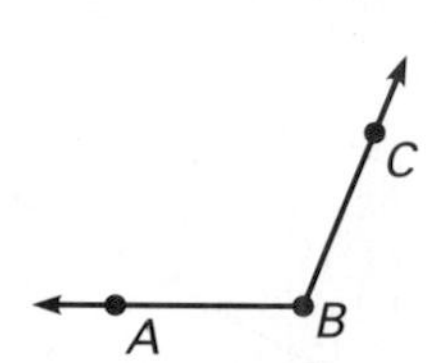

5.

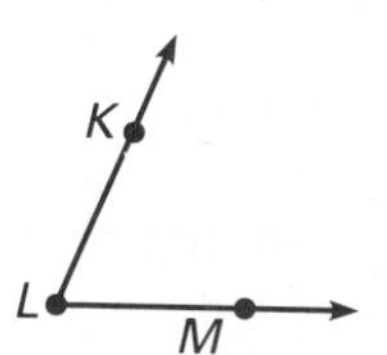

6.

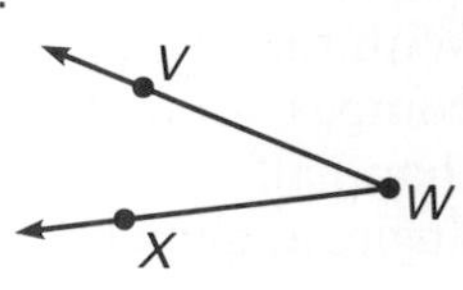

7. Use a protractor to draw an acute angle that measures 54°. Construct the bisector of this angle.

EXTEND/ SOLVE PROBLEMS

8. Draw any obtuse angle. Then use a compass and straightedge to divide the angle into four equal parts.

9. Bisect any of the angles that resulted from your construction in Exercise 8.

THINK CRITICALLY/ SOLVE PROBLEMS

10. Use a compass and protractor to construct a triangle with the same angle measures and with sides of the same length as the triangle at the right.

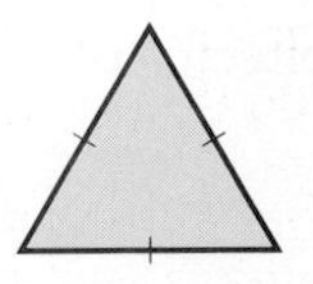

► READ
► PLAN
► SOLVE
► ANSWER
► CHECK

Problem Solving Applications:

CONSTRUCTING CIRCLE GRAPHS

When data are expressed as parts of a whole, you can display the data in a **circle graph.** The entire circle represents 100% of the data. The different parts are represented as **sectors** of the circle. Since there are 360° in a circle, the sum of the **central angles** formed by the sectors must be 360°.

Example

Draw a circle graph to represent the following data. In a survey, 350 people named their favorite movie: *Star Wars,* 119; *Indiana Jones and the Temple of Doom,* 98; *Gone With the Wind,* 77; *It's a Wonderful Life,* 56.

Solution

Step 1 Find the decimal for each part of the whole.

Star Wars:
$119 \div 350 = 0.34$
Indiana Jones:
$98 \div 350 = 0.28$
Gone With the Wind:
$77 \div 350 = 0.22$
Wonderful Life:
$56 \div 350 = 0.16$

Step 2 *Star Wars:* $0.34 \times 360° = 122.4° \approx 122°$
Indiana Jones: $0.28 \times 360° = 100.8° \approx 101°$

Step 3 *Gone With the Wind:* $0.22 \times 360° = 79.2° \approx 79°$
It's a Wonderful Life: $0.16 \times 360° = 57.6° \approx 58°$

Step 4 Draw a circle. Use a protractor to construct the central angle for each sector.
Label each sector and give the graph a title. ◄

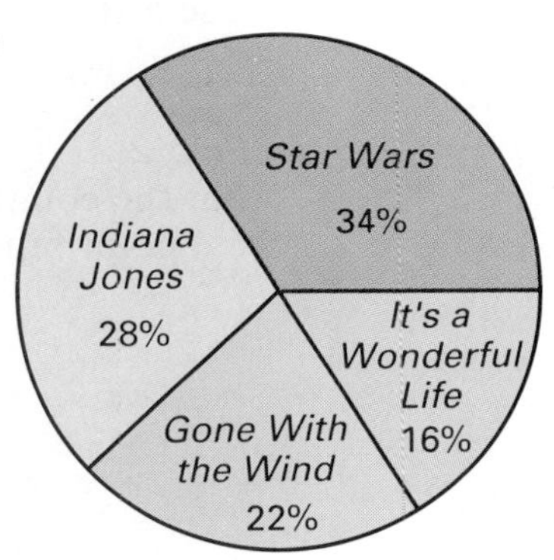

Construct a circle graph to represent each set of data.

1. The content of cheddar cheese is as follows: protein, 26%; water, 40%; fat, 34%. (*Hint:* Write each percent as a decimal.)
2. A concrete mix consists of 1.5 L of water, 3.8 L of sand, 1 L of cement, and 4.2 L of gravel.
3. A recipe for chili con carne contains the following ingredients. hamburger, 520 g; kidney beans, 280 g; tomato sauce, 150 g; tomato paste, 150 g; seasoning, 60 g; water 440 g.
4. During a sale, an appliance store sold 70 refrigerators, 35 stoves, 60 dishwashers, and 85 color TVs.

10-10 Constructing Line Segments and Perpendiculars

EXPLORE

Trace $\overline{AB}$ on a sheet of paper. Fold the paper so that point A of the line segment meets point B. Open the folded paper. Label the crease $\overleftrightarrow{CD}$. Label the point where $\overleftrightarrow{CD}$ intersects $\overline{AB}$ point M. What can you say about $\overline{AM}$ and $\overline{MB}$?

SKILLS DEVELOPMENT

Copying a line segment and bisecting a line segment are two more important constructions.

Example 1

Construct a line segment that is equal in length to $\overline{AB}$. Use a compass and straightedge.

Solution

Step 1

Use a straightedge to draw $\overrightarrow{CD}$.

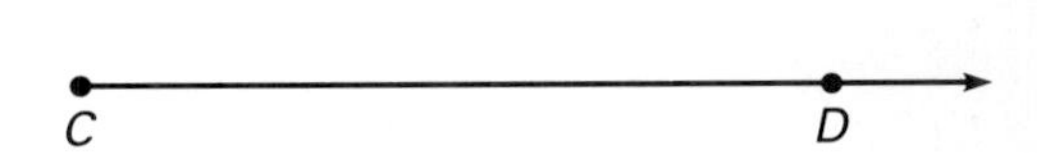

TALK IT OVER

Tell how geometric constructions were used to make this perspective design. Make a copy of the design.

Step 2

Put the compass point at A and the point of the pencil at B.

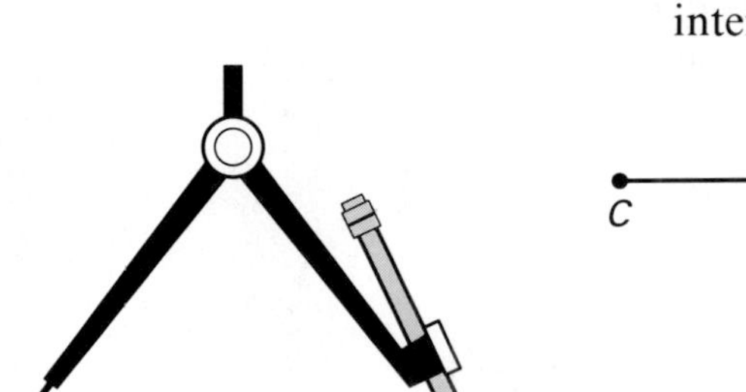

Step 3

Place the point of the compass at C. With the same setting used in Step 2, draw an arc. Label the intersection point E.

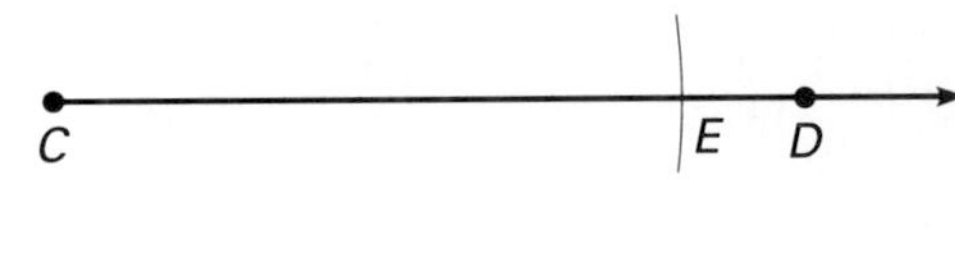

$CE = AB$ ◄

The **midpoint** of a line segment is the point that separates it into two line segments of equal length. The **perpendicular bisector** of a line segment is a line, ray, or line segment that is perpendicular to a line segment at its midpoint.

Example 2

Construct the perpendicular bisector of $\overline{JK}$.

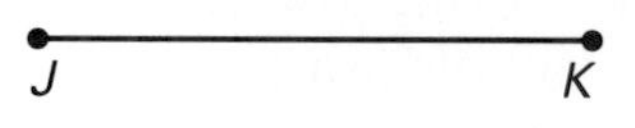

Solution

Step 1

Place the point of the compass at J. Open the compass a little more than half the length of $\overline{JK}$. Draw one arc above $\overline{JK}$ and another below $\overline{JK}$.

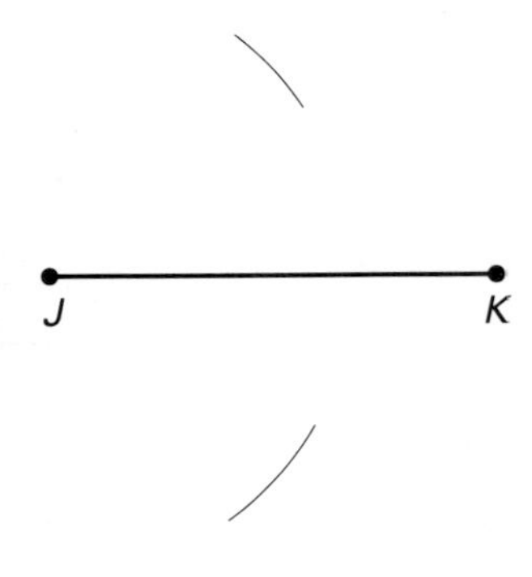

Step 2

Place the point of the compass at K. With the same setting used in Step 1, draw arcs above and below $\overline{JK}$. Label the points of intersection L and N. Use a straightedge to draw $\overleftrightarrow{LN}$. Label the midpoint of $\overline{JK}$ point M.

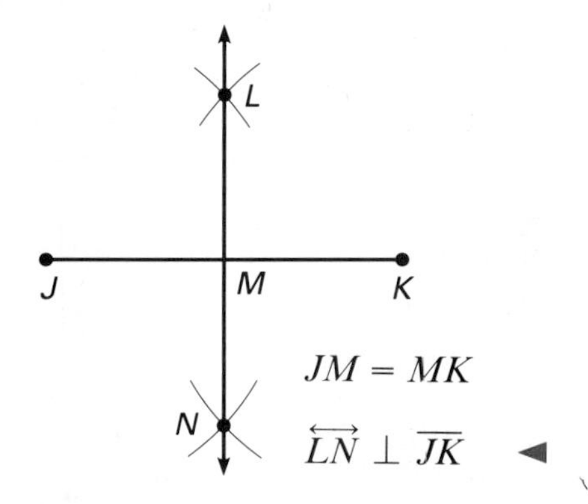

$JM = MK$

$\overleftrightarrow{LN} \perp \overline{JK}$ ◀

Example 3

Construct a line perpendicular to $\overleftrightarrow{RS}$ from point A.

Solution

Step 1

With the point of the compass at A, draw an arc that intersects $\overleftrightarrow{RS}$ at B and C.

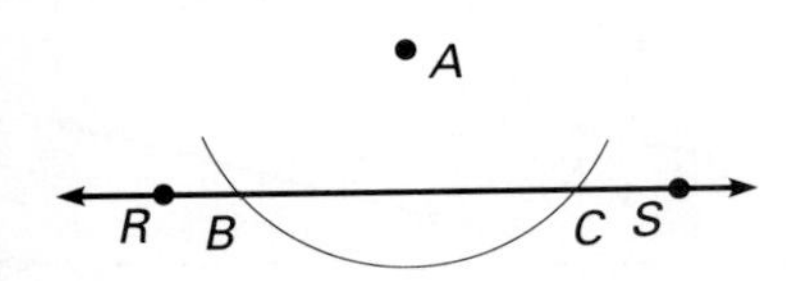

Step 2

With the point of the compass at B, open the compass a little more than half the length of $\overline{BC}$. Draw an arc below $\overleftrightarrow{RS}$.

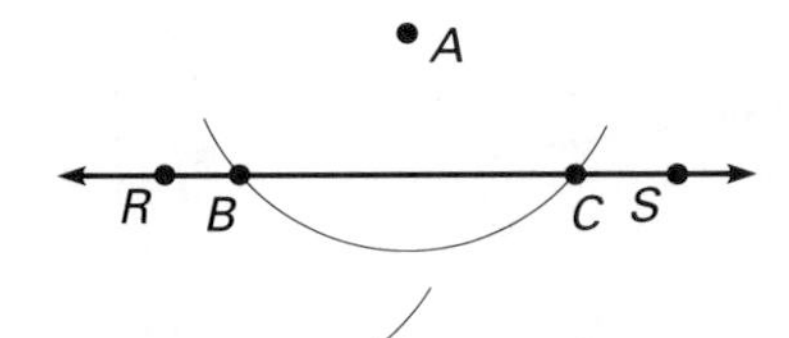

Step 3

Place the point of the compass at C. With the same setting used in Step 2, draw an arc below $\overleftrightarrow{RS}$ that intersects the arc drawn in Step 2. Label the point of intersection D.

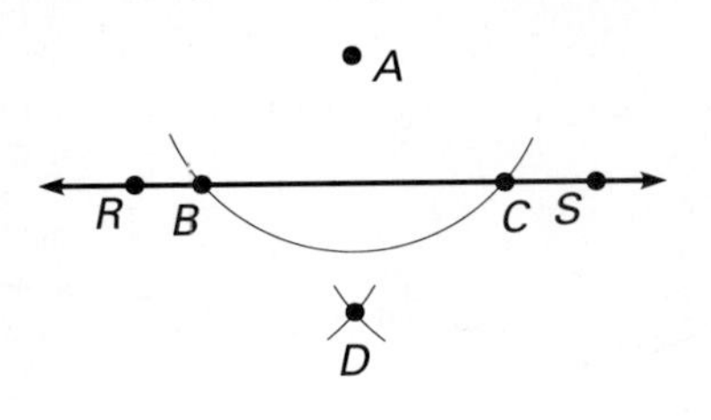

Step 4

Draw $\overleftrightarrow{AD}$.

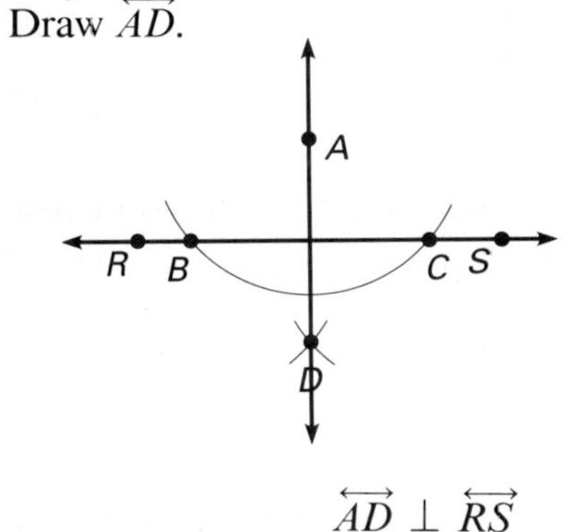

$\overleftrightarrow{AD} \perp \overleftrightarrow{RS}$ ◀

TRY THESE

Trace each line segment. Then construct a line segment equal in length.

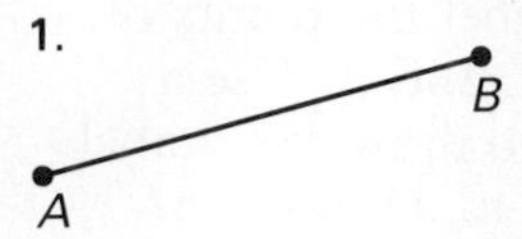

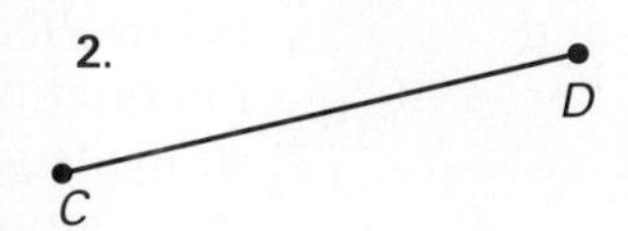

Trace each line segment. Then construct its perpendicular bisector.

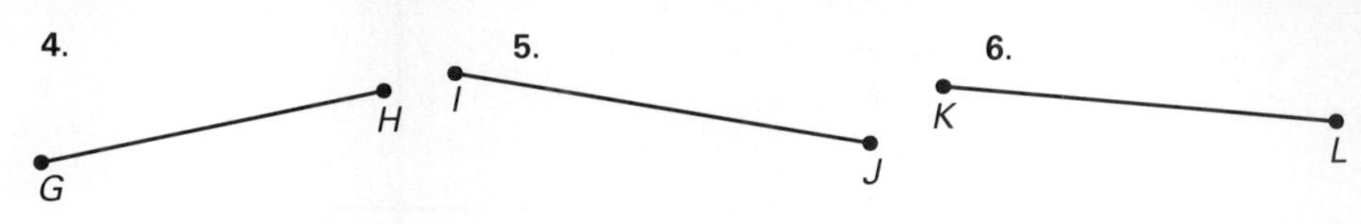

Trace each figure. Then construct a line perpendicular to line *AB* from point *P*.

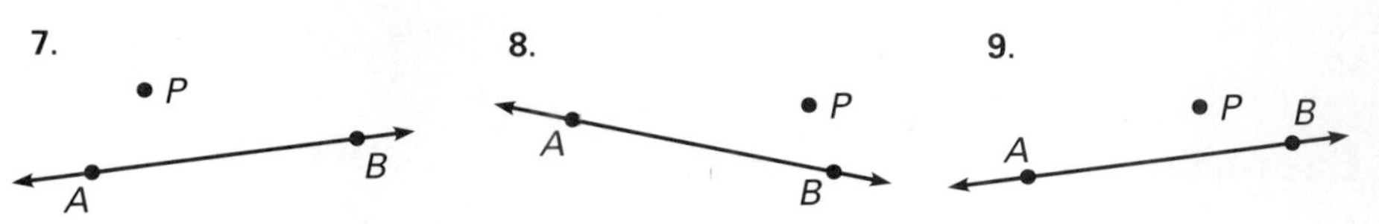

EXERCISES

PRACTICE/ SOLVE PROBLEMS

Trace each line segment. Then construct a line segment equal in length.

Trace each line segment. Then construct its perpendicular bisector.

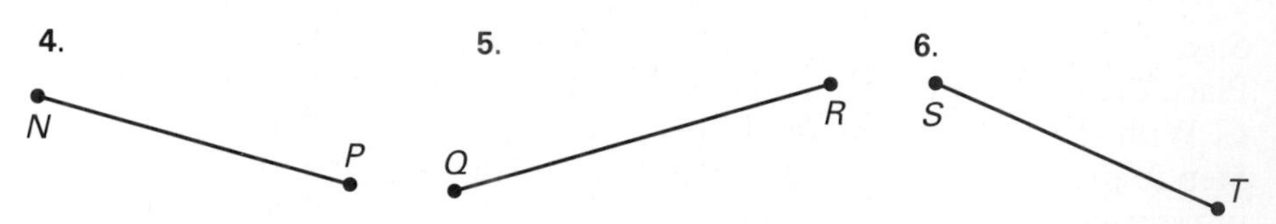

Trace each figure. Then construct a line perpendicular to line *CD* from point *K*.

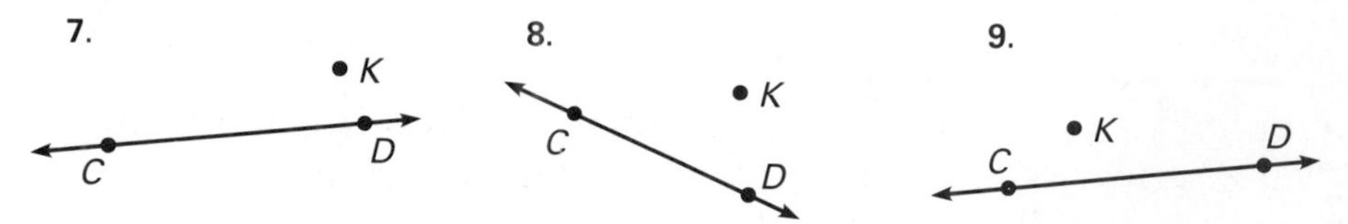

EXTEND/ SOLVE PROBLEMS

An **altitude of a triangle** is a line segment from a vertex of the triangle perpendicular to the opposite side or to a line containing that side. You can use the construction described in Example 3 to construct any altitude of a triangle.

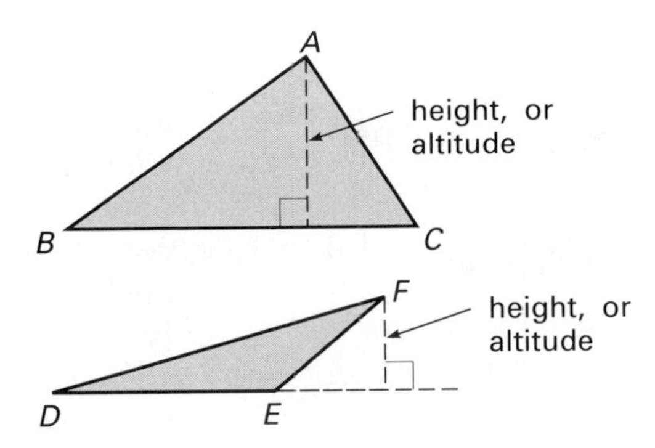

Trace each triangle. Construct an altitude of the triangle from point R to $\overline{PS}$.

10.

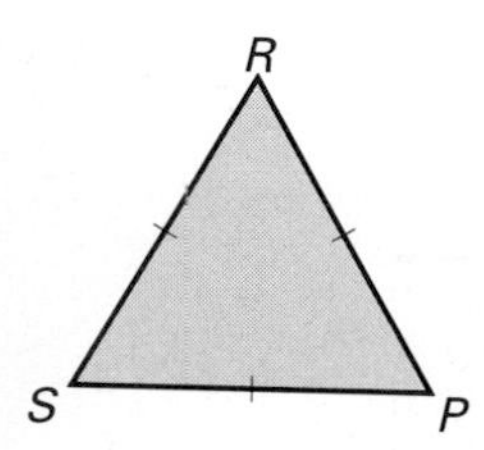

11.

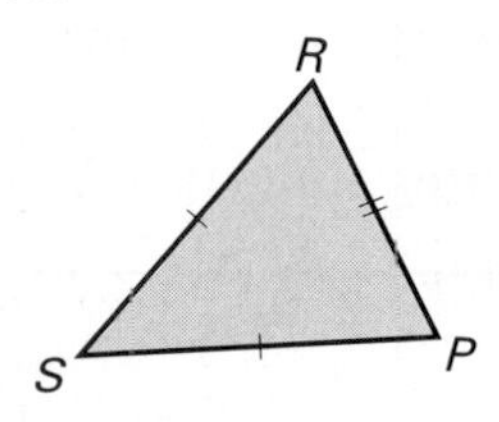

12.

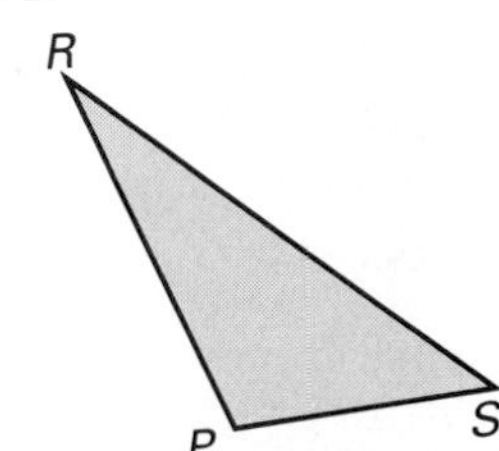

THINK CRITICALLY/ SOLVE PROBLEMS

13. Trace $\triangle ABC$. Bisect each side of $\triangle ABC$. The bisectors will meet at a point. Label the point P. Measure the distance from point P to each vertex. What do you observe?

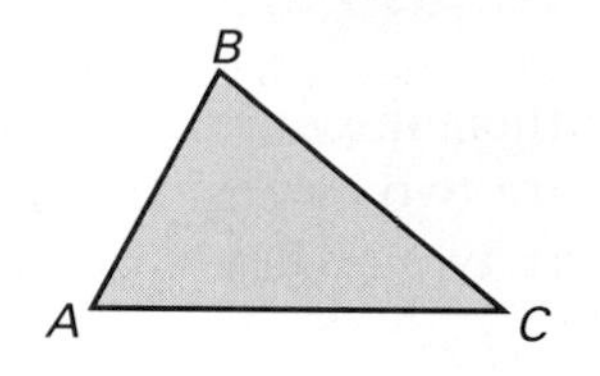

14. Trace $\triangle DEF$. Bisect each side of the triangle and label the point where the bisectors intersect point Q. Measure the distance from point Q to each vertex. What do you observe?

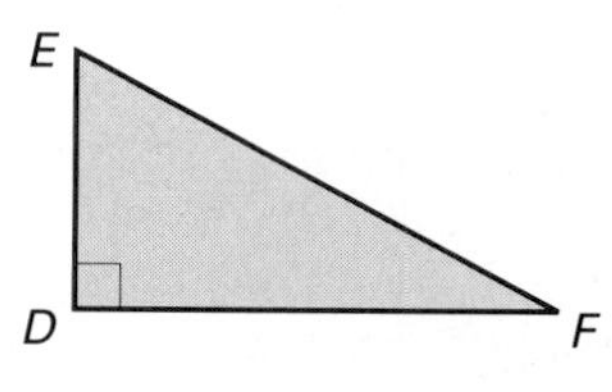

15. Trace $\triangle GHI$. Bisect each side of the triangle and label the point where the bisectors intersect point R. Measure the distance from point R to each vertex. What do you observe?

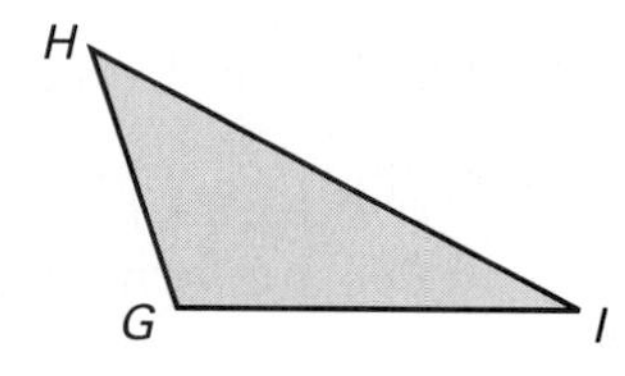

16. Compare your answers to Exercises 13–15. How is the position of the intersection point of the perpendicular bisectors of the sides related to the type of triangle?

CHAPTER 10 ● REVIEW

Match the letters of the words at the right with the descriptions at the left.

1. has two identical, parallel bases that are triangles and three other faces that are parallelograms
2. the sum of their measure is 180°
3. has two sides of equal length
4. flat surface that extends without end in all directions
5. parallelogram with four sides of equal length
6. the sum of their measures is 90°
7. has one base and three other faces that are triangles

a. plane
b. supplementary angles
c. triangular pyramid
d. complementary angles
e. isosceles triangle
f. rhombus
g. triangular prism

SECTION 10–1 POINTS, LINES, AND PLANES (pages 342–345)

► Basic geometric figures include the **point**, the **line**, the **line segment**, the **ray**, the **angle**, and the **plane**.

► A **polygon** is a closed plane figure formed by joining three or more line segments at their endpoints.

8. Name the line in four ways.

P Q R

SECTION 10–2 ANGLES AND ANGLE MEASURES (pages 346–349)

► Angles are classified by their measure as **acute, right, obtuse,** or **straight.**

► **Complementary angles** are two angles whose measures add to 90°.

► **Supplementary angles** are two angles whose measures add to 180°.

9. Measure and classify $\angle ABC$.
10. Find $m\angle DBA$.

SECTION 10–3 PARALLEL AND PERPENDICULAR LINES (pages 350–353)

► Two lines in a plane either intersect at a single point or are parallel and have no points in common.

► When a transversal intersects two parallel lines, **vertical angles** along with several other pairs of angles are formed: **corresponding angles, alternate interior angles,** and **alternate exterior angles.**

11. Name a pair of alternate interior angles.
12. Determine $m\angle 2$ in the figure.

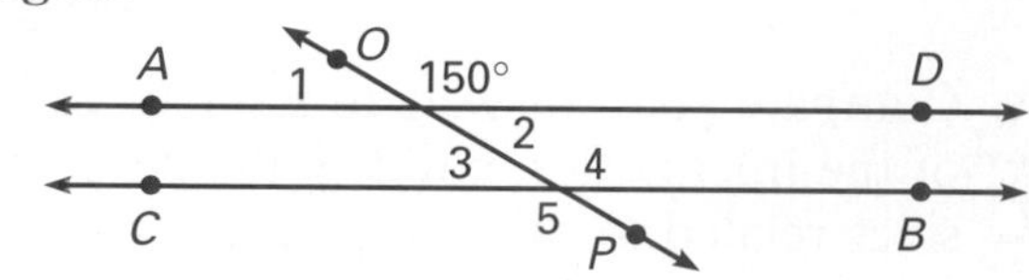

SECTIONS 10–4 AND 10–5 TRIANGLES AND POLYGONS (pages 354–361)

► Triangles are classified by the measures of their angles and sides.
► Polygons are named by the number of their sides.

Classify each triangle as *acute, obtuse,* or *right.*

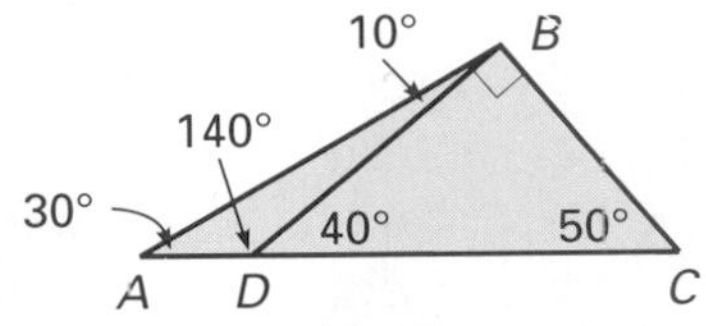

13. $\triangle CBD$
14. $\triangle CBA$
15. $\triangle ADB$

Name each polygon and give the sum of the angle measures.

16. 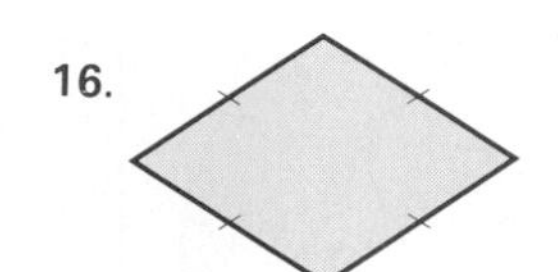

17. 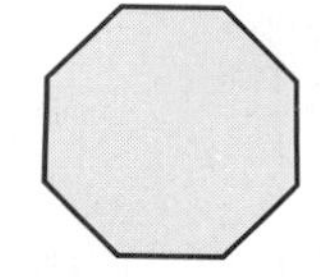

SECTION 10–6 THREE-DIMENSIONAL FIGURES (pages 362–365)

► Some three-dimensional figures are **polyhedra** —figures in which each surface is a polygon. Others have curved surfaces.

Draw the figure.

18. rectangular prism
19. rectangular pyramid
20. cylinder
21. cone

SECTIONS 10–7 AND 10–8 PROBLEM SOLVING (pages 366–369)

► Acting out a problem is a useful problem solving strategy.

22. In how many ways can you make change for 70¢ with 4 coins?

SECTIONS 10–9 AND 10–10 CONSTRUCTIONS (pages 370–377)

► You can copy and bisect any angle using a compass and straightedge.
► With a straightedge and compass, you can construct a line segment equal in length to a given line segment, a perpendicular bisector of a given line segment, and a perpendicular to a line from a given point not on a line.

23. Trace $\angle LMN$. Then bisect it.

24. Construct a line segment congruent to $\overline{JK}$ and bisect it.

USING DATA Refer to the chart of semaphore code signals on page 340.

25. For which letters are the flags held at an obtuse angle?

CHAPTER 10 ● TEST

Give all the names that apply to each polygon or polyhedron.

1. 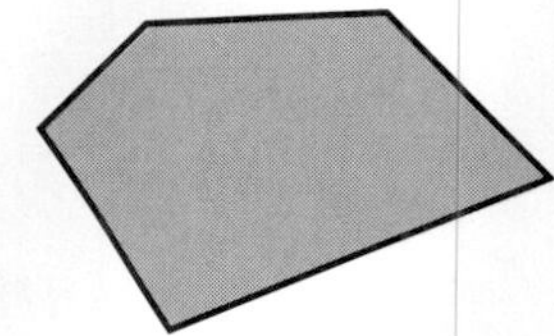

2. 

3. 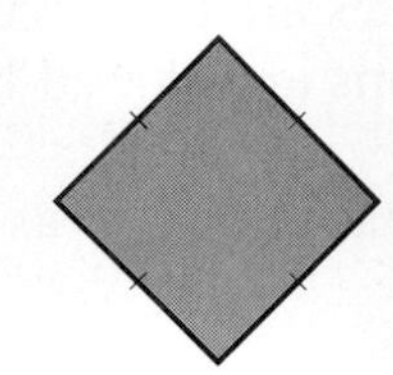

4. 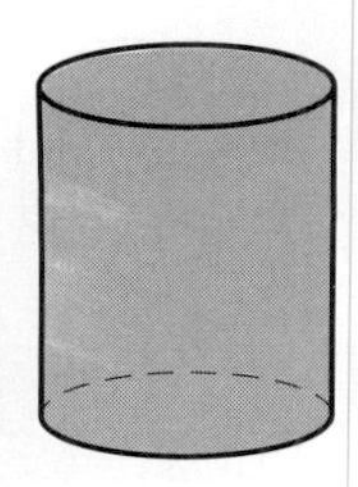

5. 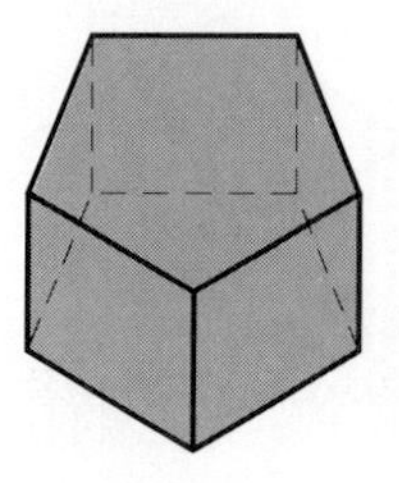

6. 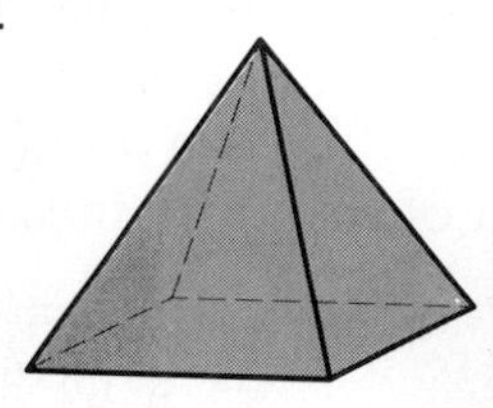

Classify each angle as *acute, right,* or *obtuse.*

7. 150°
8. 95°
9. 32°
10. 90°

Use the figure at the right for Exercises 11 and 12.

11. Name a pair of vertical angles.
12. Name a pair of adjacent angles.

Classify each triangle as *isosceles, equilateral,* or *scalene.*

13. 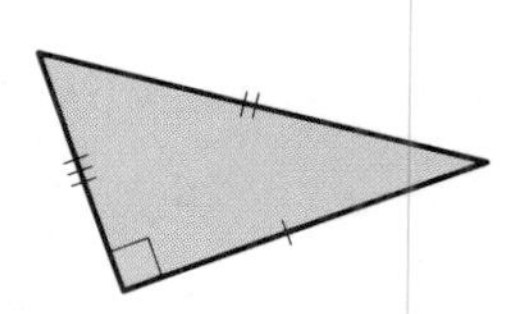

14. 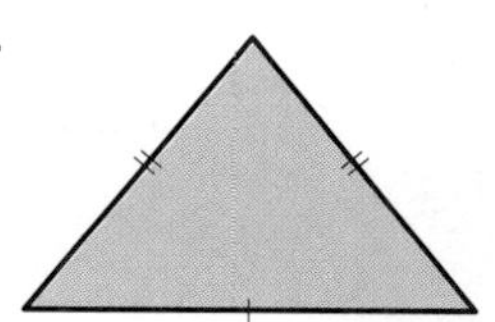

15 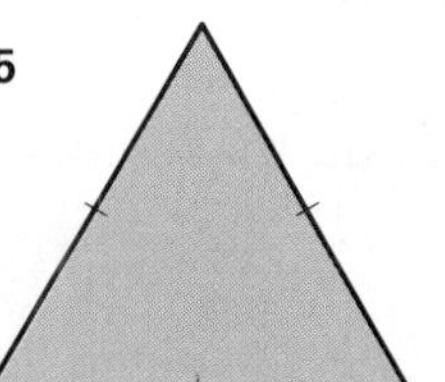

16. 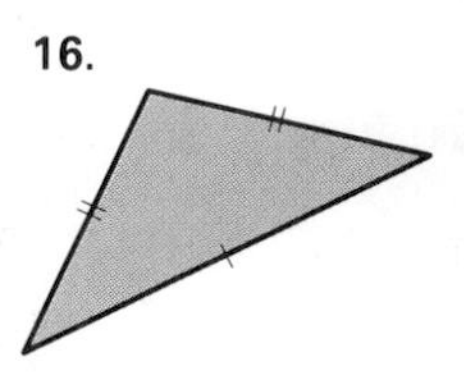

17. Trace $\overleftrightarrow{JK}$ with point L. Construct a line perpendicular to $\overleftrightarrow{JK}$ from point L.

Find the unknown angle measure.

18. 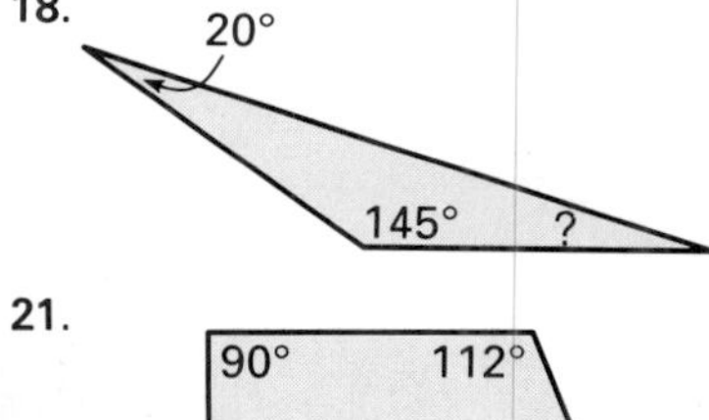

19. 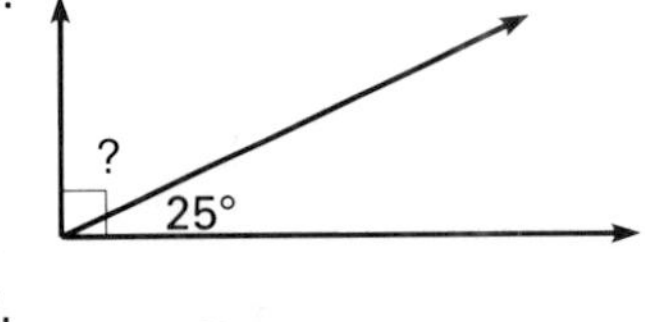

20. 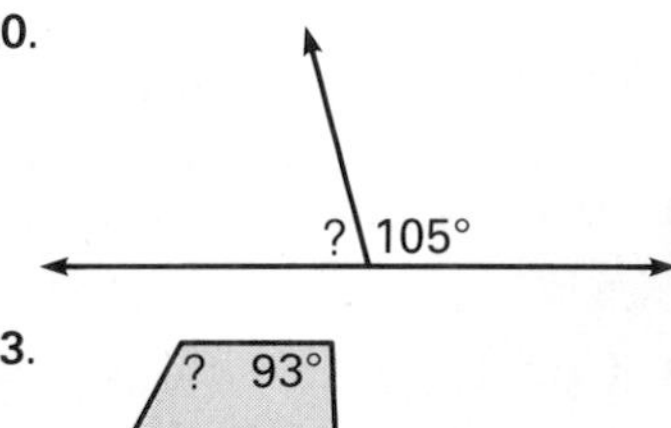

21. 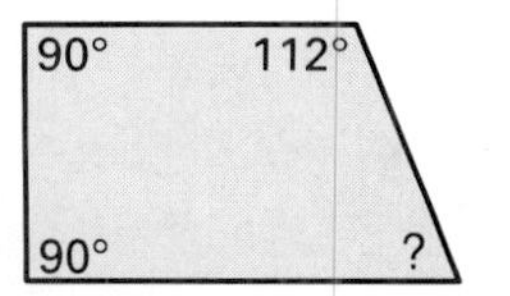

22. 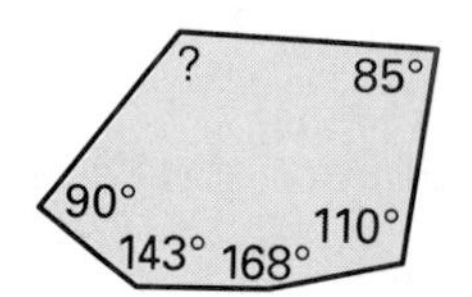

23. 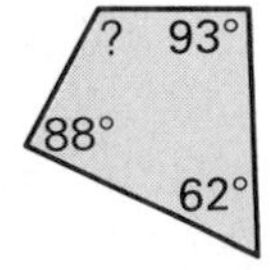

CHAPTER 10 ● CUMULATIVE REVIEW

1. A quality control manager inspects every seventh knapsack sewn. Which term describes this sample?

Write in scientific notation.

2. 8,794
3. 15,096

4. Write a statement about the width and the height of the figure below. Then check to see if your statement is true or false.

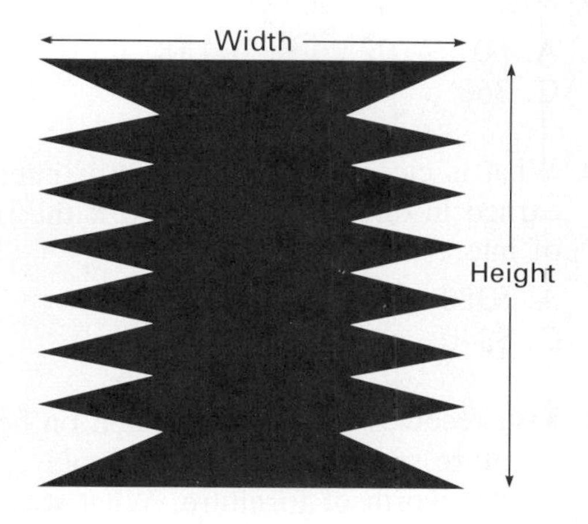

5. Write two conditional statements using the following two sentences:

   Rhea helped us study.
   We passed the test.

Write each fraction as a decimal.

6. $\frac{7}{8}$
7. $\frac{7}{9}$
8. $\frac{5}{16}$

9. Find the perimeter of a rectangular figure whose length is 10.5 cm and whose width is 40.2 cm.

Replace ● with $<$, $>$, or $=$.

10. $5 - 16$ ● $10 - (-1)$
11. $-4 + (16)$ ● $8 + 4$
12. $13 - (-5)$ ● $45 - 64$

Solve each equation.

13. $2\frac{1}{2}x = 5$
14. $r - 18 = 23$
15. $150 = 10x + 30$
16. $\frac{b}{7} = -15$

Solve each proportion.

17. $\frac{4}{5} = \frac{12}{x}$
18. $8:x = 6:15$

19. Write $\frac{16}{20}$ and $\frac{45}{60}$ as percents. Which percent is greater?

20. Write a percent equivalent to $\frac{12}{8}$.

21. What percent of 50 is 35?

22. A jacket that regularly sells for $80 is on sale for $60. What is the discount rate?

23. Jeff receives a 5% commission on sales of sports equipment. He sold $2,500 worth of equipment last week. What was the amount of his commission?

Classify each triangle as *acute, obtuse,* or *right.*

24.

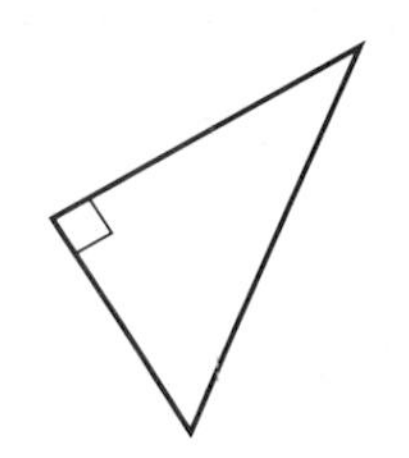

25.

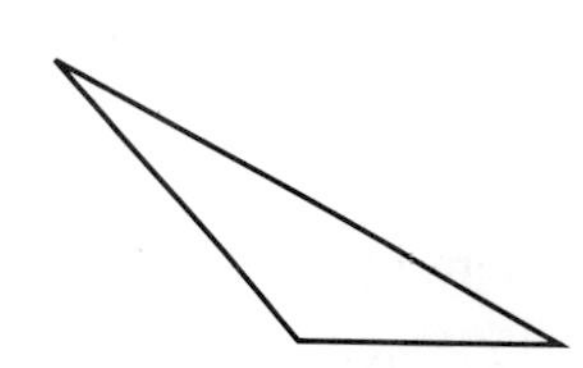

26.

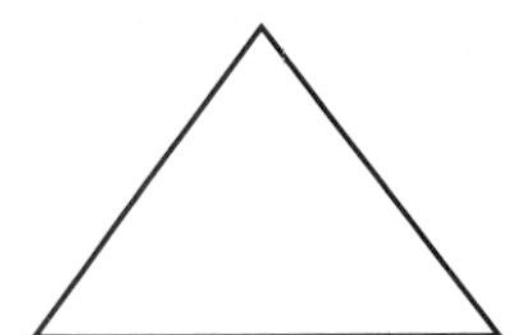

27.

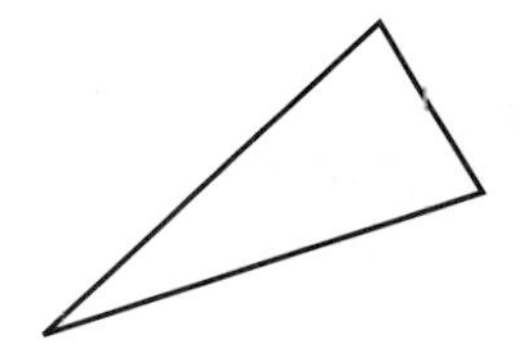

1. If you had a total of 288 test points and an average of 72, how many tests did you take?

   A. 7 B. 8 C. 4 D. 5

2. Divide. $1.836 \div 1.2$

   A. 153 B. 15.3
   C. 1.53 D. 0.153

3. Which is a counterexample for the following statement?

   If a number is a multiple of 4, then it is a multiple of 8.

   A. 8 B. 16 C. 64 D. 20

4. Which is the decimal for $\frac{5}{6}$?

   A. $0.8\overline{3}$ B. 1.2
   C. 0.8 D. $0.\overline{83}$

5. Complete. 1.73 kg = ■ g

   A. 173 g B. 17.3 g
   C. 1,730 g D. 17,300 g

6. Complete. 2 lb 8 oz × 4

   A. 8 lb 10 oz B. 10 lb 8 oz
   C. 10 lb D. 9 lb 4 oz

7. Subtract. $25 - (-15)$

   A. -40 B. -10
   C. 10 D. 40

8. Stan biked 44 km in $3\frac{2}{3}$ h. What was his average rate of speed? Use the formula $d = rt$.

   A. 12 km/h B. $161\frac{1}{3}$ km/h
   C. 15 km/h D. $\frac{2}{3}$ km/h

9. Which of the following ratios is equivalent to 4:5?

   A. 8:10 B. 2:3
   C. 10:20 D. none of these

10. The scale of a drawing is 2 cm:5 m. What is the actual width of a room if the width in the scale drawing is 4 cm?

    A. 2 m B. 1 cm
    C. 10 m D. none of these

11. What is 30% of 650?

    A. 195 B. 19.5
    C. 620 D. 1,950

12. 80% of what number is 20?

    A. 60 B. 18
    C. 360 D. 25

13. What is closest to the amount of interest earned in one year on $4,500 if the rate of interest is 8.5% per year?

    A. $380 B. $34
    C. $340 D. $400

14. Risa receives a 4% commission on her furniture sales. Last week she sold $2,800 worth of furniture. What was the amount of her commission?

    A. $400 B. $2,912
    C. $112 D. none of these

15. $\overrightarrow{DB}$ bisects $\angle ADC$. $\angle BDC = 25°$. Find $m\angle ADC$.

    A. 25°
    B. 50°
    C. 1.5°
    D. none of these

16. Classify $\angle GHI$.

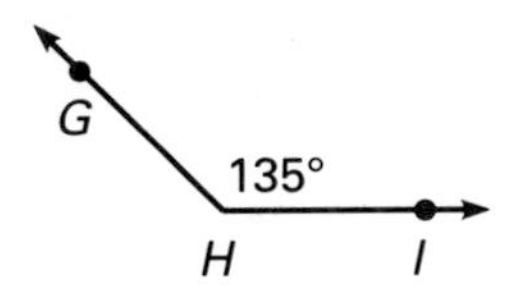

    A. right B. straight
    C. acute D. none of these

CHAPTER 11 SKILLS PREVIEW

Find the area of each figure.

1. 5 cm 9 cm

2. 7 in. 15 in.

3.

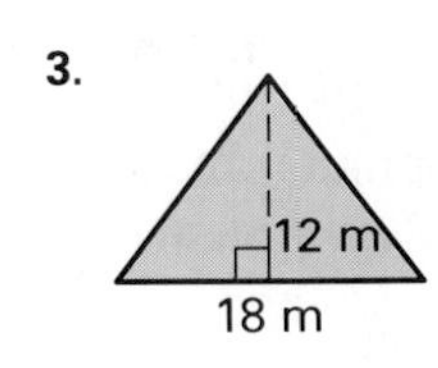

4. 4 ft 11 ft 5 ft 8 ft

Find the square root to the nearest tenth.

5. $\sqrt{49}$ **6.** $\sqrt{4000}$ **7.** $\sqrt{38}$ **8.** $\sqrt{110}$

Find the unknown length.

9.

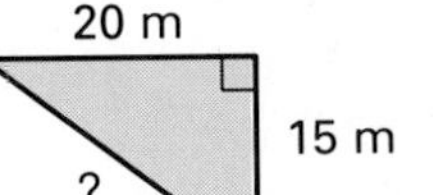

10.

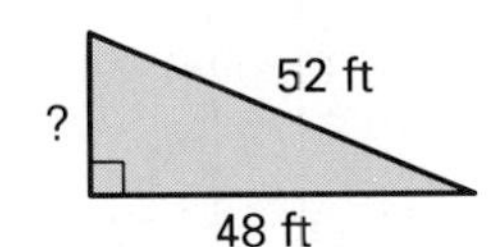

11.

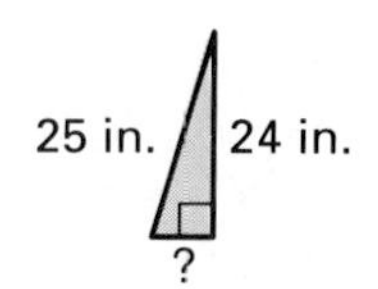

Find the volume of each figure. Round to the nearest whole number.

12.

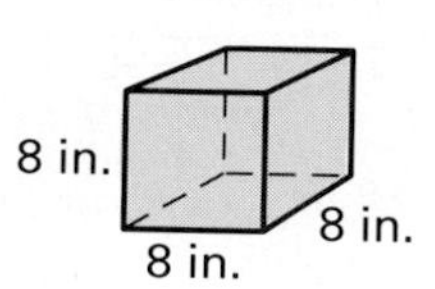

13.

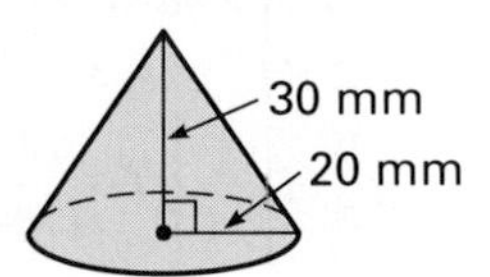

14.

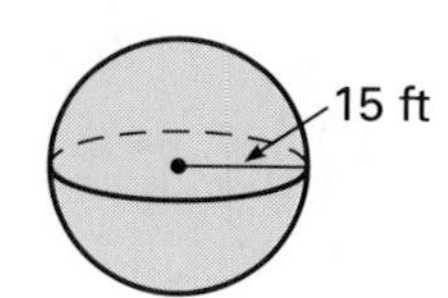

15.

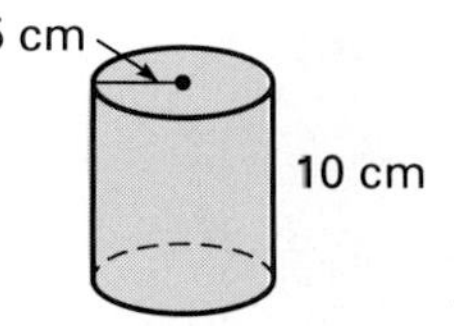

16. The diameter of a can of tomatoes is 9 cm and the height is 16 cm. Find the volume. Use $\pi \approx 3.14$. Round to the nearest whole number.

Copy and complete the table.

	radius (r)	diameter (d)	area (A)
17.	■	30 cm	■
18.	8 cm	■	■

Find the surface area of each figure.

19.

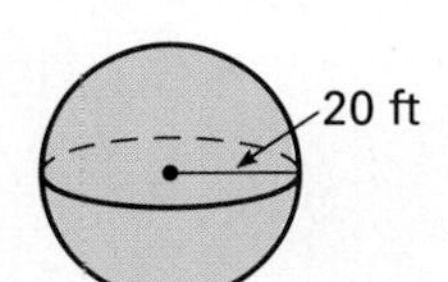

20.

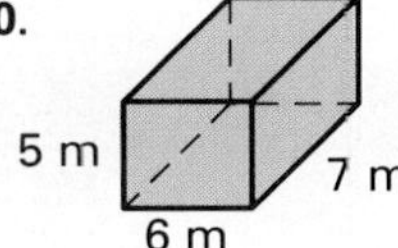

21.

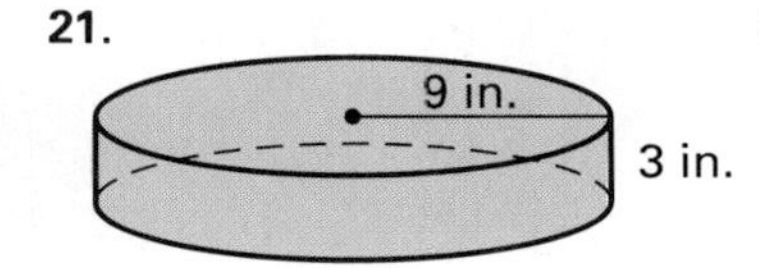

22.

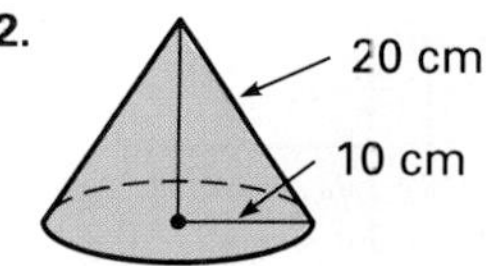

Solve.

23. The area of a rectangle is 32 in.^{2} The length is twice the width. Find the dimensions of the rectangle.

CHAPTER 11

AREA AND VOLUME

THEME Sports

You can find a lot of geometry in sports. For example, the equipment and the playing area suggest geometric shapes. In this chapter, you will solve problems that involve finding the area, volume, and surface area of geometric shapes.

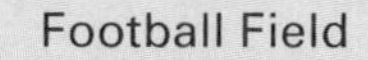

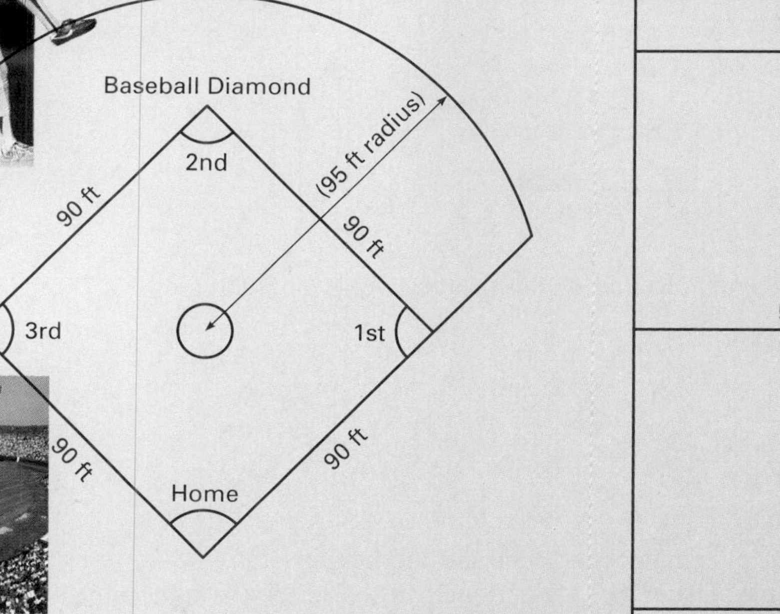

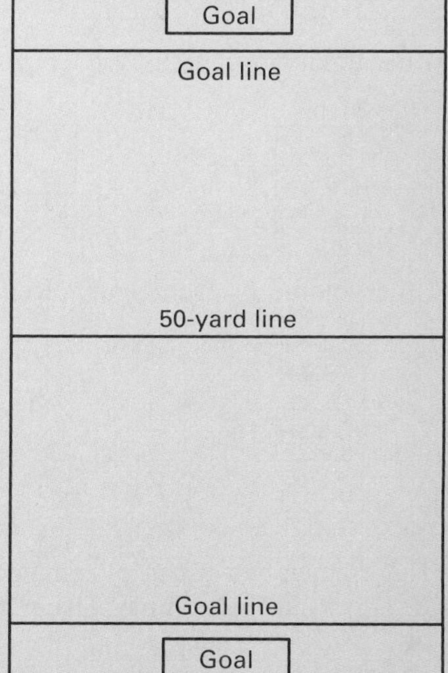

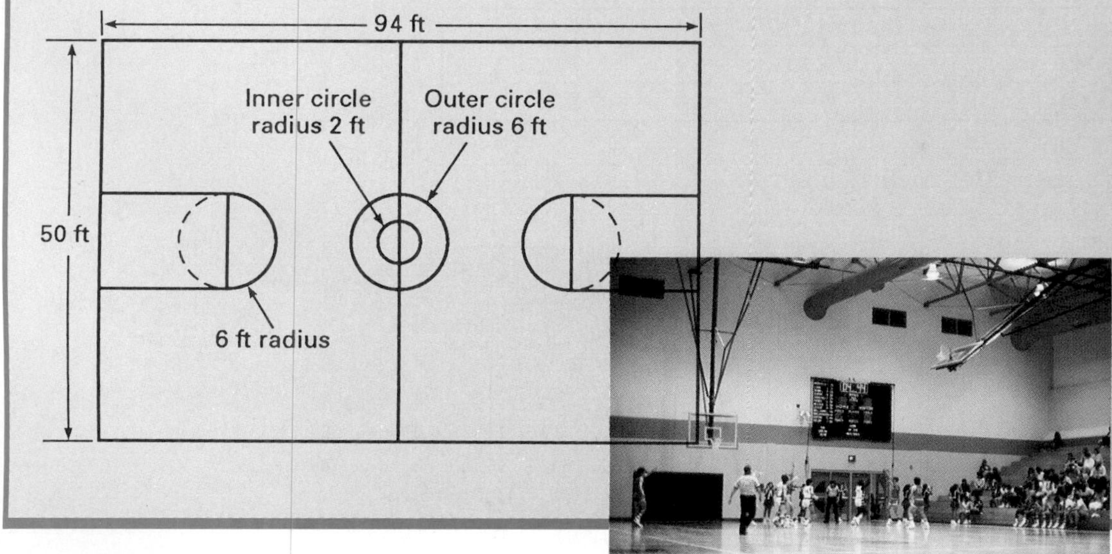

DECISION MAKING

Using Data

Use the information in the diagrams to answer the following questions.

1. What is the perimeter of the baseball "diamond"?
2. How great an area would you have to cover to wax the floor of the basketball court?
3. How much clay would you need to cover the tennis court with a 1.5-in. layer?

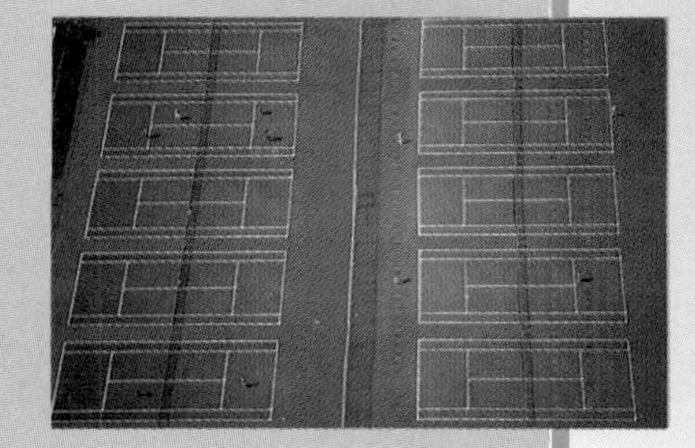

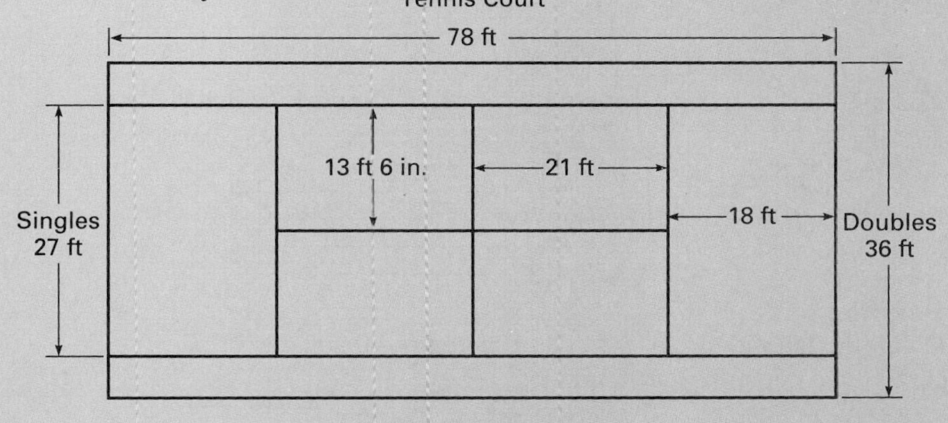

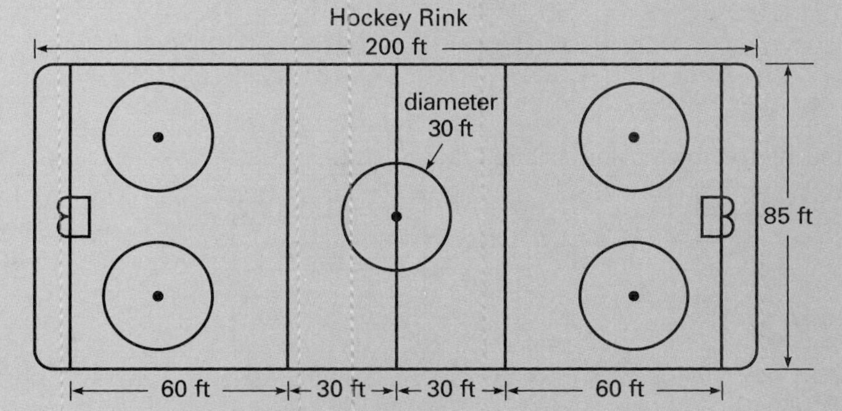

Working Together

Your group's task is to choose a sport other than basketball, baseball, tennis, or hockey. Gather information about the field or court area used for that sport. You might choose football or soccer or even swimming. (An Olympic-sized swimming pool can also be thought of as a "field.") Prepare a table of facts about the field and draw a diagram illustrating it and the critical distances and lines on it. Prepare an oral report on your sport to present to the class.

11-1 Area and Volume

EXPLORE

This large square, made up of 9 small squares, has a perimeter of 12 units.

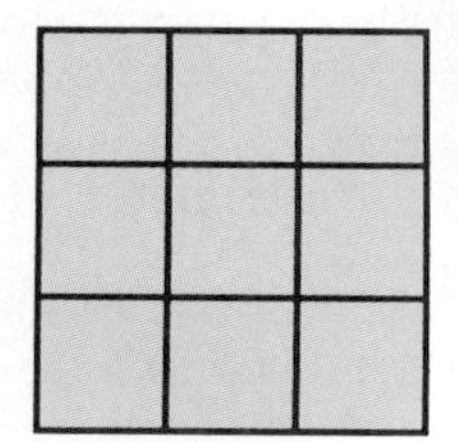

Use grid paper to draw other closed figures, each with a perimeter of 12 units. Try figures that are made up of 8, 7, 6, and 5 squares.

SKILLS DEVELOPMENT

The amount of surface covered by a plane figure is called its **area.**

Example 1

Find the area of this rectangle.

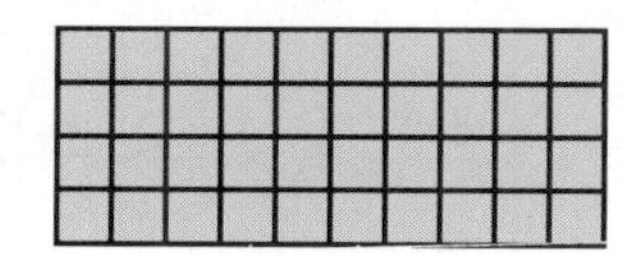

Solution

Count the number of square units. There are 40 squares in all.

The area of the rectangle is 40 square units. ◄

Plane figures with the same area may have different perimeters.

Example 2

Find the perimeter and area of each figure.

a.

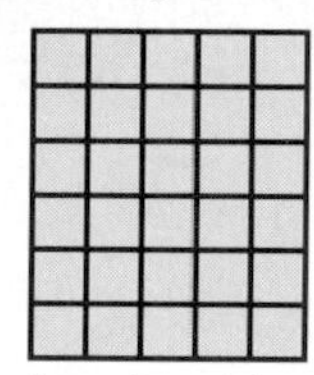

1 unit = 1 in.

b.

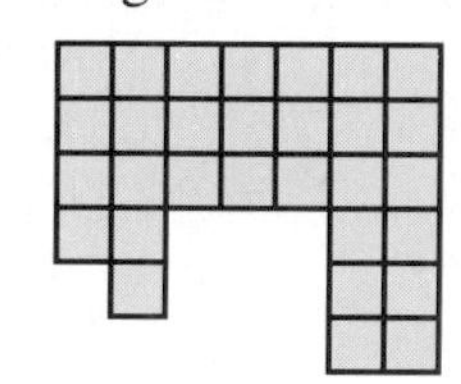

1 unit = 1 in.

Solution

a. To find the perimeter, count the number of units along the sides of the figure. There are 22 units, so the is 22 in.

To find the area, count the number of *square units* covered by the figure. There are 30 square units, so the area is 30 in.^{2}

b. Count the number of units along the sides of the figure. There are 30, so the perimeter is 30 in.

Count the number of *square units*. There are 30, so the area is 30 in.^{2} ◄

The amount of space a three-dimensional figure occupies is called its **volume.** Volume can be measured by counting the number of cubic units that fill a space.

Example 3

Find the volume of this rectangular prism.

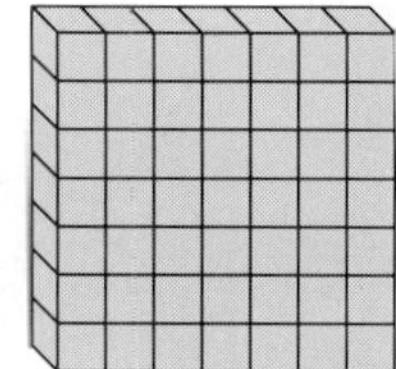

1 unit = 1 cm

Solution

The number of cubic centimeters that fill the space is 49.

The volume of the rectangular prism is 49 cm^3. ◄

Sometimes, not all of the cubic units can be seen in the drawing of a three-dimensional figure.

Example 4

Find the volume of the rectangular prism.

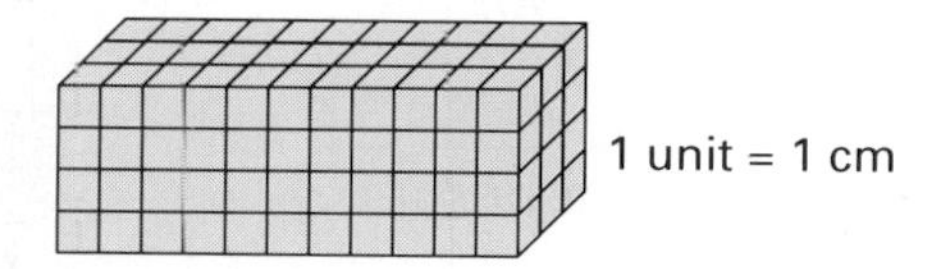

Solution

Count the number of cubic units in the top layer.

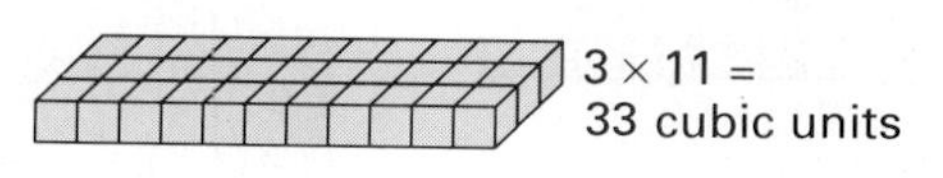

Count the number of layers.

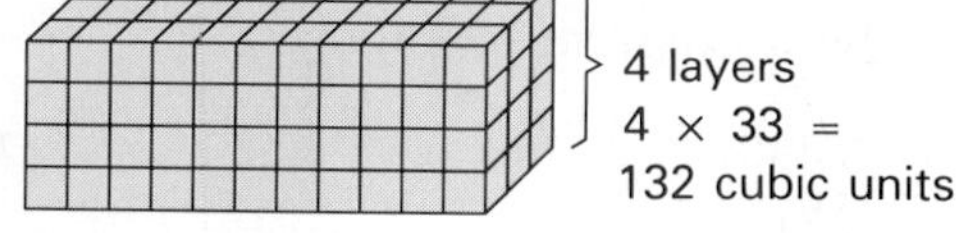

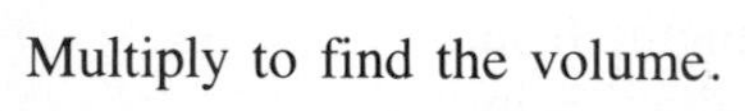

Multiply to find the volume.

The volume of the figure is 132 cubic centimeters, or 132 cm^3. ◄

TRY THESE

In each of Exercises 1–4, 1 unit = 1 cm.

1. Find the area of the figure.

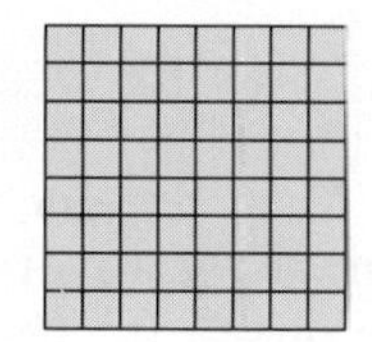

2. Find the perimeter and the area of the figure.

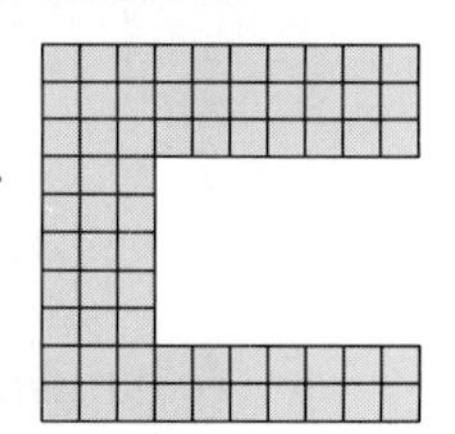

3. Find the volume of the rectangular prism.

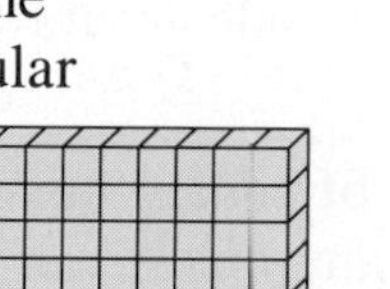

4. Find the volume of the rectangular prism.

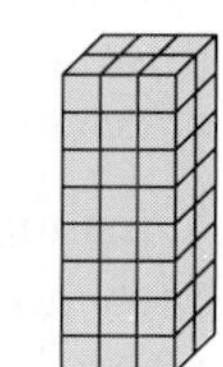

EXERCISES

PRACTICE/ SOLVE PROBLEMS

For each of Exercises 1–9, 1 unit = 1 cm.

Find the area of each rectangle.

1.

2. 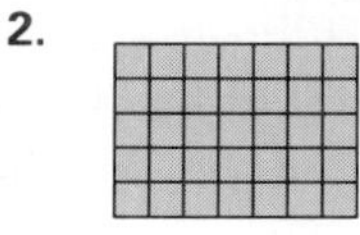

3. 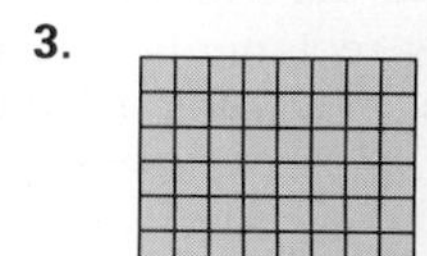

Find the perimeter and the area of each figure.

4. 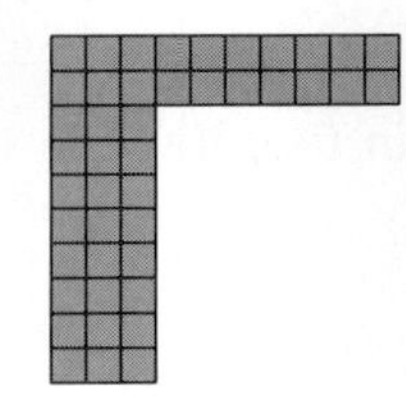

5. 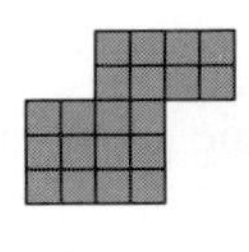

6. 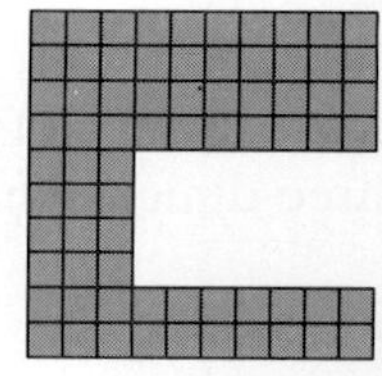

Find the volume of each prism.

7. 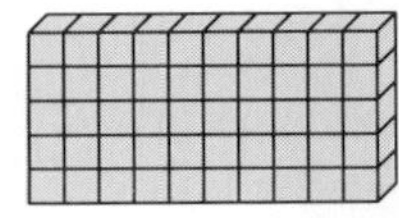

8. 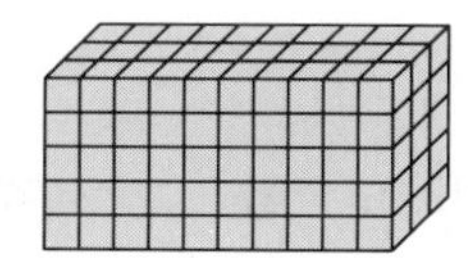

9. 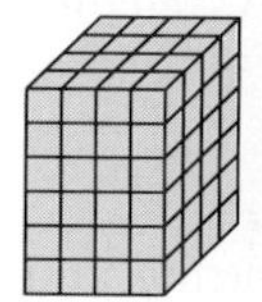

Find the volume of each rectangular prism. Visualize the space or draw a diagram to solve.

10. Length: 5 in.
    Width: 1 in.
    Height: 2 in.

11. Length: 6 in.
    Width: 3 in.
    Height: 1 in.

12. Length: 4 cm
    Width: 3 cm
    Height: 2 cm

13. Length: 7 cm
    Width: 2 cm
    Height: 1 cm

Solve.

14. A builder wanted to cover the floor of the foyer of a house with square tiles. The width of the foyer was 6 ft. The length was 8 ft. How many tiles measuring 1 ft on a side would the builder need to cover the floor?

15. A manufacturer can pack 3 layers of square boxes in one carton. Each layer holds 4 boxes. How many boxes can be packed in each carton? Draw a rectangular prism to help you solve.

MIXED REVIEW

Using Data. Use the Data Index on page 546 to find the statistics you need for each answer.

1. Draw a stem-and-leaf plot showing the number of stories in the tallest buildings in the world as of 1990.
2. Draw a bar graph that represents the length, in meters, of the spans of the longest bridges as of 1988.
3. Decide whether a line graph is an appropriate way to express the data in the famous-waterfalls table. Explain your decision.

Find the area of each shaded figure. Use 1 unit = 1 ft.

16. 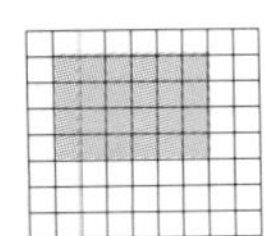

17. 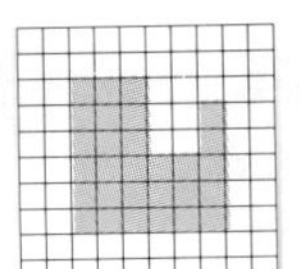

18. 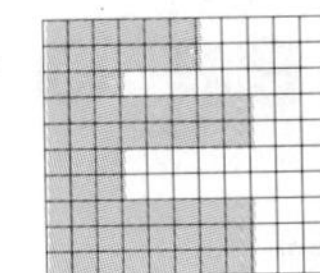

EXTEND/ SOLVE PROBLEMS

COMPUTER

Use this program to strengthen your understanding about the relationship between the area and perimeter of a figure. Run the program several times, each time using a different set of measurements for length and width.

```
10 INPUT "ENTER AREA: "
     ;A: PRINT
20 PRINT "LENGTH"; TAB(10 );
     "WIDTH","PERIMETER"
30 FOR L = 1 TO A
40 LET W = A / L
50 IF W <> INT (A / L)
     THEN GOTO 80
60 P = 2 * L + 2 * W
70 PRINT L; TAB( 10);W,P
80 NEXT L
```

Find the area of each shaded figure. Use 1 unit = 1 cm. Hint: two halves of a square unit can be added to make one square unit.

19. 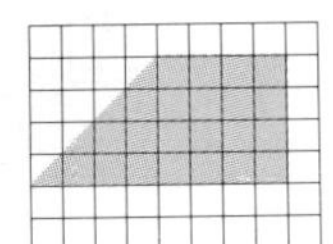

20. 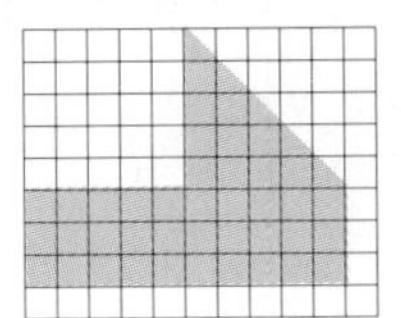

21. 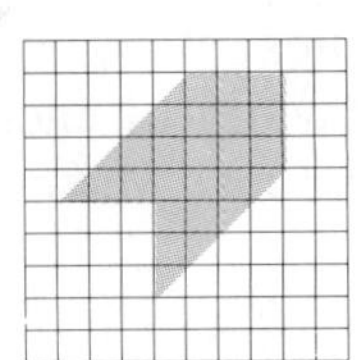

Find the volume of each figure. Use 1 unit = 1 in.

22. 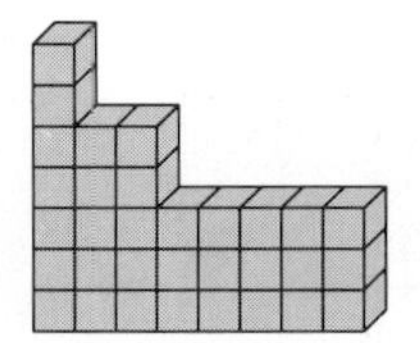

23. 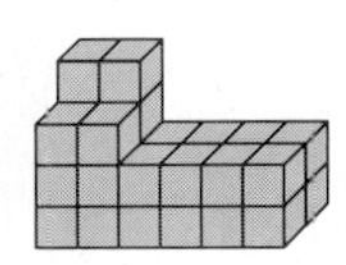

24. 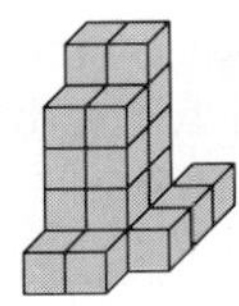

Find the volume of each figure. Use 1 unit = 1 m.

25. 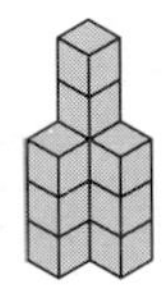

26. 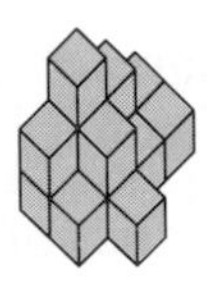

27. A landscape architect plans to use stones to cover a rectangular area that measures 2 ft by $2\frac{1}{2}$ ft. Each bag of stones will cover 2 ft^2. How many bags of stones will be needed? Explain.

THINK CRITICALLY/ SOLVE PROBLEMS

28. Draw as many rectangles as you can that have whole-number dimensions and an area of 16 square units. Then find the perimeter of each. Write a short paragraph stating your conclusions about the relationship between area and perimeter.

29. How many differently shaped three-dimensional figures can be made using four cubic units?

30. The volume of a carton is 132 cubic units. If there are 4 layers, how many units are in each layer? Visualize the problem or draw a rectangular prism to solve.

31. Find the volume of this figure.

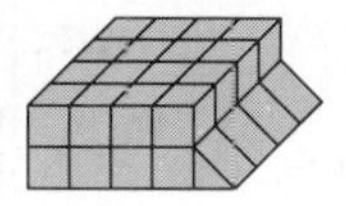

11-2 Area of Rectangles

EXPLORE

How many different rectangles can you construct using 24 tiles measuring 1 in. by 1 in.? In each rectangle, you must use all 24 tiles. What are the length and width of each rectangle you construct? Make a list.

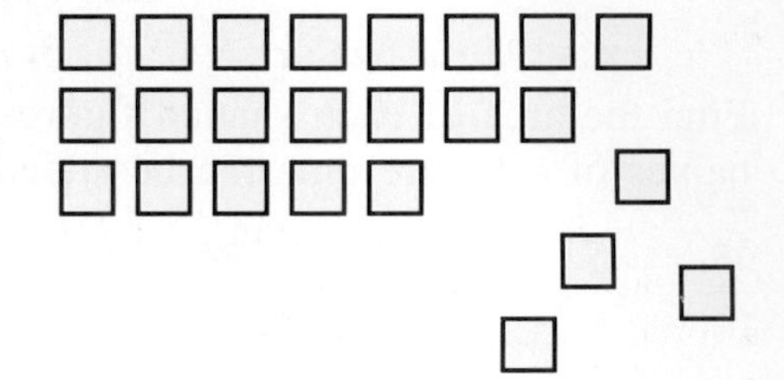

SKILLS DEVELOPMENT

You have found the area of a rectangle by counting the number of square units. The rectangle below has 24 one-inch square units. Its area is 24 in.^{2}

The area of the rectangle can also be found by using this formula.

$$A = l \times w$$

Area ↑ length ↑ width ↑

The length of the rectangle measures 6 inches. The width measures 4 inches.

6 in.

4 in.

$A = 6 \times 4 = 24$ The area of the rectangle is 24 in.^{2}

TALK IT OVER

A special formula for the area of a square is $A = s^2$, where s is the length of a side. Explain how this formula is related to the formula for the area of a rectangle.

Example 1

Use the formula $A = l \times w$ to find the area of a rectangle having a length of 16 in. and a width of 7 in.

Solution

$A = l \times w$ ← Substitute 16 for l and 7 for w.

$A = 16 \times 7$

$A = 112$ The area of the rectangle is 112 in.^{2} ◄

To find the area of an irregular figure, separate it into smaller, rectangular regions.

Example 2

Find the area of the irregular figure at the right. All angles are right angles.

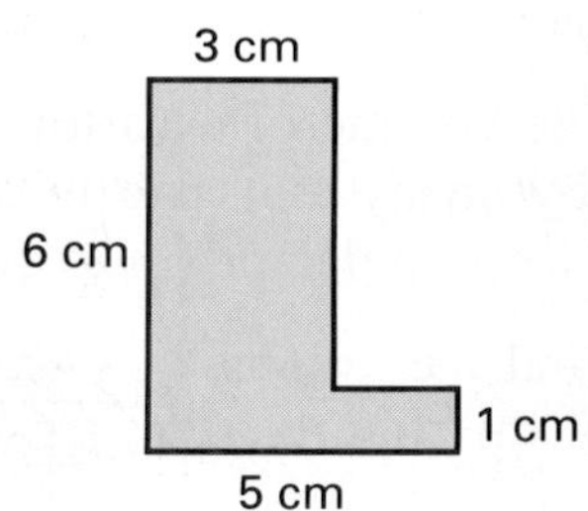

Solution

Total area	=	Area of region 1	+	Area of region 2
A	=	5×3	+	5×1
A	=	15	+	5
A	=		20	

The area of the irregular figure is 20 cm^2. ◄

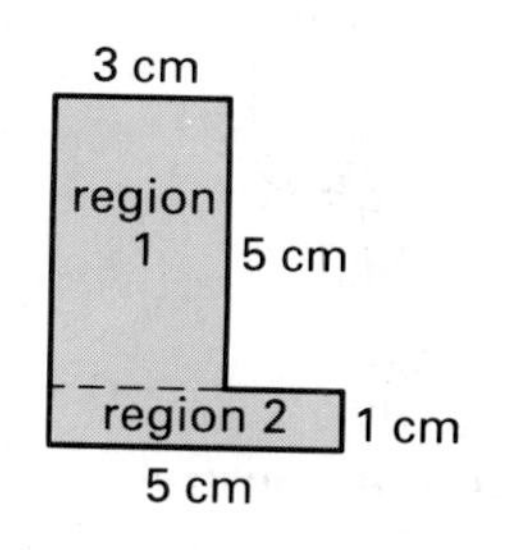

Example 3

The dimensions of a rectangular field on a sod farm are shown.

a. Find the area of the field.

b. At 7.5¢/ft^2, find the total cost of plowing the field.

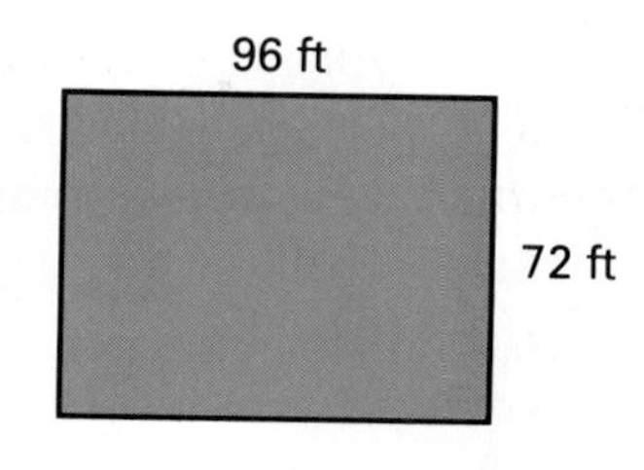

Solution

a. The field is shaped like a rectangle. Use the formula for the area of rectangles.

$A = l \times w$

$A = 96 \times 72$

$A = 6{,}912$ The area is 6,912 ft^2

b. It costs 7.5¢ to plow 1 ft^2. Write 7.5¢ as \$0.075, then multiply.

$6{,}912 \times 0.075 = 518.4$

The cost of plowing 6,912 ft^2 is \$518.40. ◄

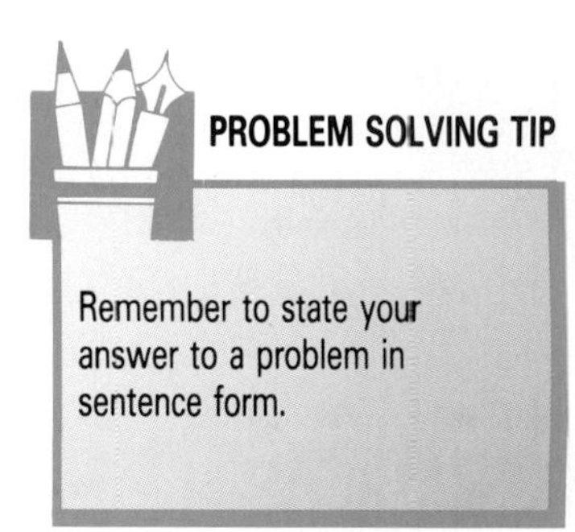

TRY THESE

Use the formula $A = l \times w$ to find the area of each rectangle.

1.

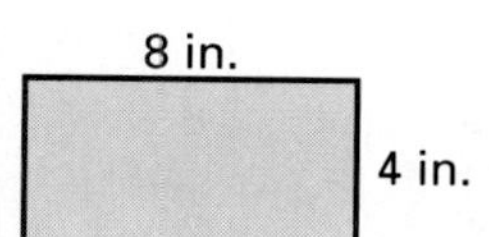

2.

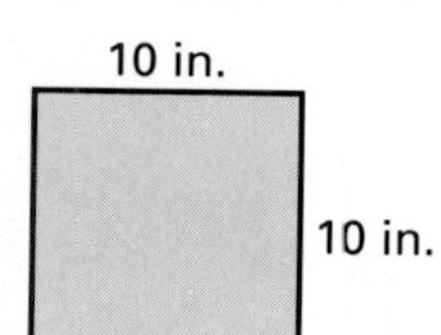

3.

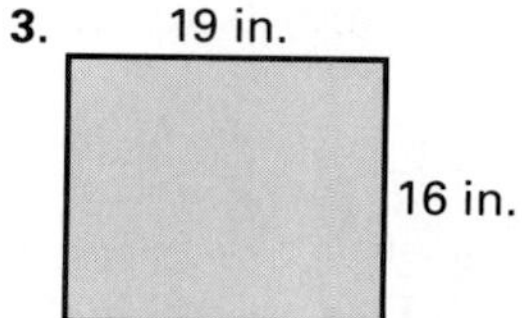

Find the area of each irregular figure. All angles are right angles.

4.

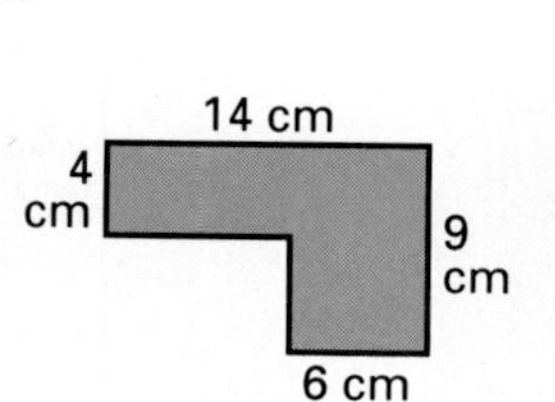

5.

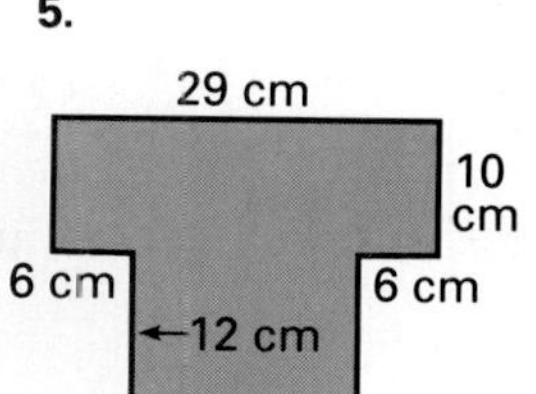

6.

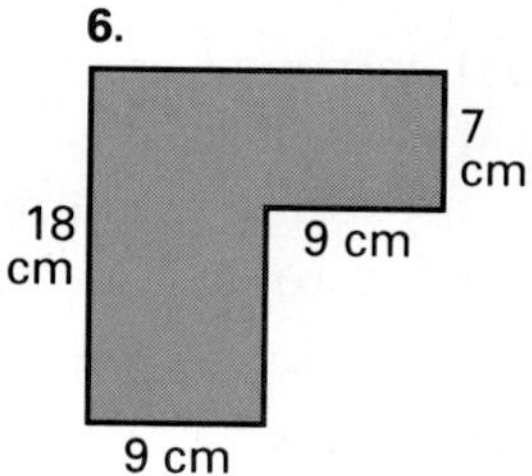

A rectangular floor measures 8 yd by 10 yd.

7. Find the area of the floor.

8. If carpeting costs \$12.75/yd^2, what would be the cost of carpeting the floor?

EXERCISES

PRACTICE/ SOLVE PROBLEMS

Here is the way to change the program on page 389 so that it can be used to find the area of rectangles of all sizes:

In line 20, delete the following.
"PERIMETER"
Delete line 60.
In line 70, delete the following.
,P

Find the area of each rectangle.

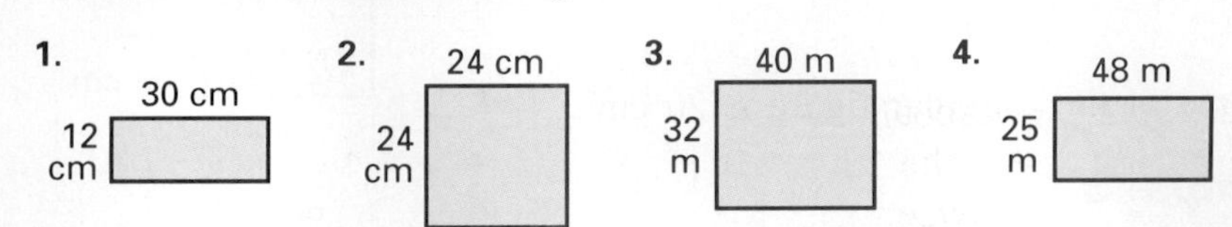

Find the area of each plot of land. Round to the nearest square foot.

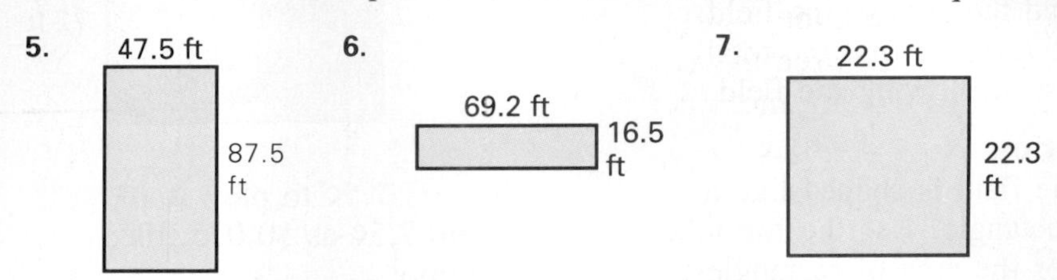

Find the area of each figure. All angles are right angles.

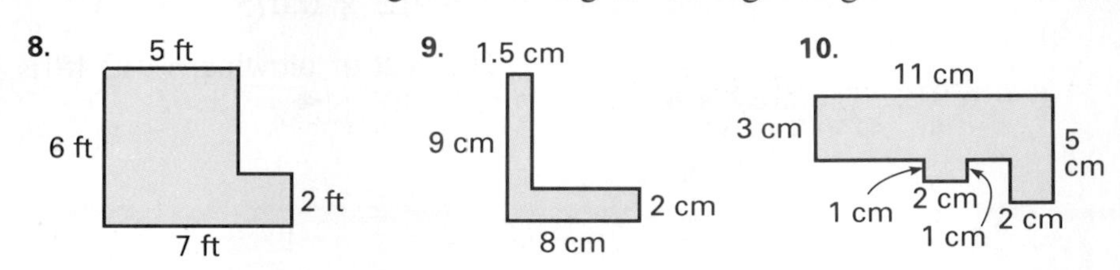

11. Rose's vegetable garden is square. It measures 24.5 ft on a side. Find its area.

EXTEND/ SOLVE PROBLEMS

In Exercise 13, don't forget to change square feet to square yards.

 Use this sketch of Ann's living room for Exercises 12 and 13.

12. What is the area of the living room?

13. If carpeting costs $8.25/yd², what will be the cost of carpeting the entire living room?

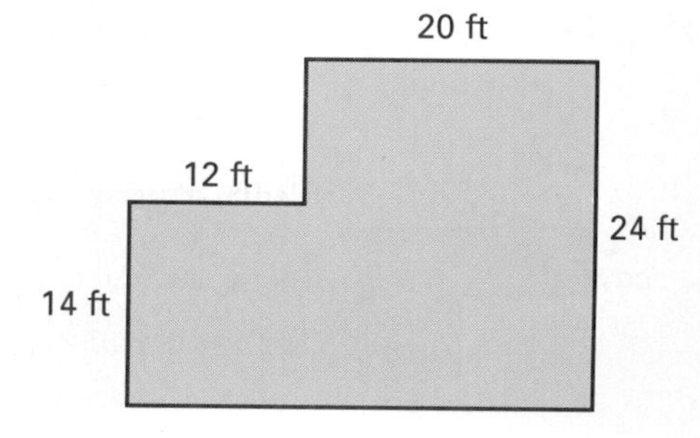

THINK CRITICALLY/ SOLVE PROBLEMS

14. A rectangular lot 50 ft by 100 ft is surrounded by a sidewalk 5 ft wide. What is the area of the sidewalk?

15. The width of a rectangle is four-fifths its length. If its perimeter is 72 in., what is its area?

- READ
- PLAN
- SOLVE
- ANSWER
- CHECK

Problem Solving Applications:

BASEBALL STRIKE ZONE

In major league baseball, a pitcher must throw the ball over home plate and between the batter's armpits and knees. The shaded region *ABCD* is called the **strike zone.** You can find the area of the strike zone by using the area formula for a rectangle.

A B 31 in. D 12 in. C

$A = l \times w$
$A = 31 \times 12$
$A = 372$ The area of the strike zone is 372 in.^{2}

Assume that the width of home plate is 12 in. Find the area of each strike zone.

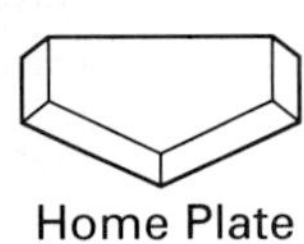
Home Plate

1. 

2. 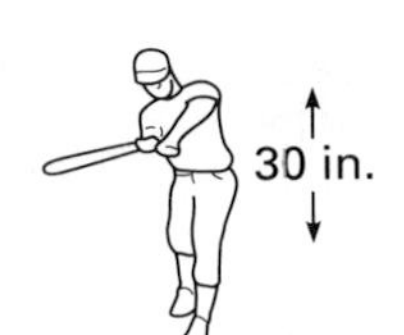

3. 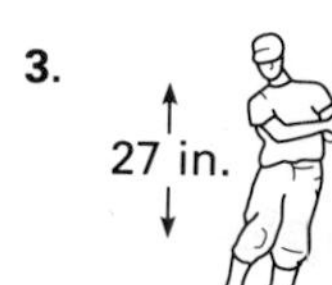

4. 

The dimensions of the strike zone for each player are listed in the table. Find the area of each strike zone.

	Player	Dimensions	
		l	*w*
5.	Jose Canseco	27 in.	12 in.
6.	Carlton Fisk	32 in.	12 in.
7.	Pete Incaviglia	37 in.	12 in.
8.	Rickey Henderson	34 in.	12 in.
9.	Lance Parrish	28 in.	12 in.

10. Two batting positions are shown.

By how much does the area of the strike zone decrease if this batter crouches?

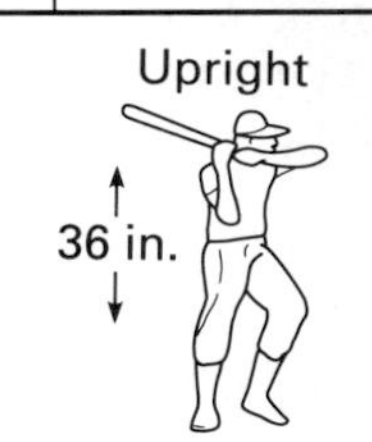

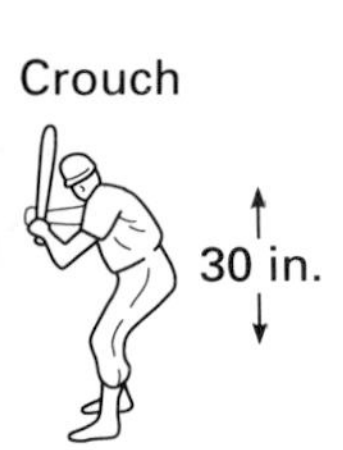

Find the decrease in the area of the strike zone if each batter crouches as shown.

11. 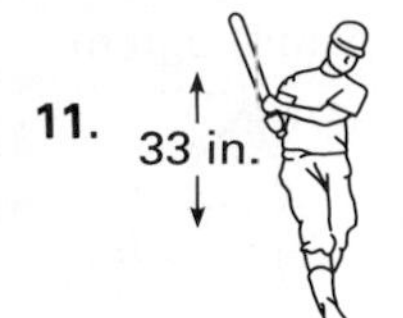

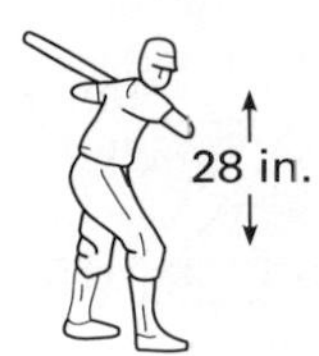

12. 

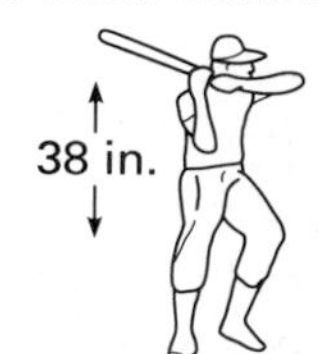

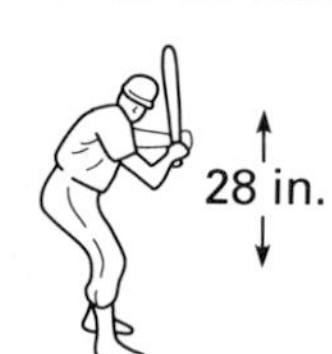

11-3 Area of Parallelograms and Triangles

EXPLORE

Count the square units to find the area of the rectangle and the parallelogram. Compare.
Now copy each figure onto grid paper. Cut out the parallelogram. How can you cut the parallelogram into two pieces that can be rearranged to form the rectangle? Try it!

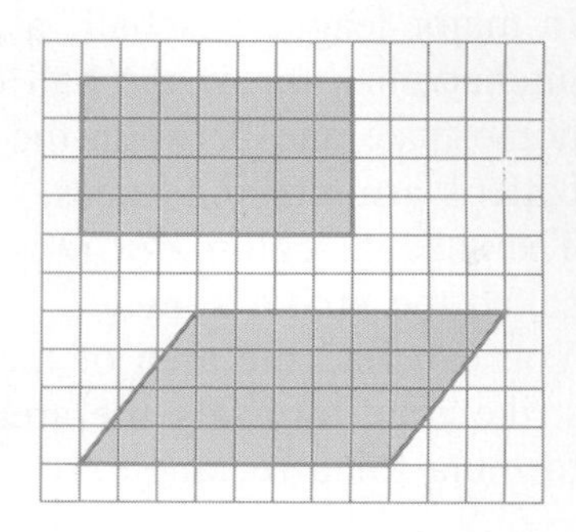

SKILLS DEVELOPMENT

In the Explore activity, you discovered how a parallelogram can be "rearranged" to form a rectangle. Because of this relationship, the area formula for a parallelogram is very much like the area formula for a rectangle.

$$A = b \times h$$

A ↑ Area; b ↑ length of base; h ↑ height

Parallelogram

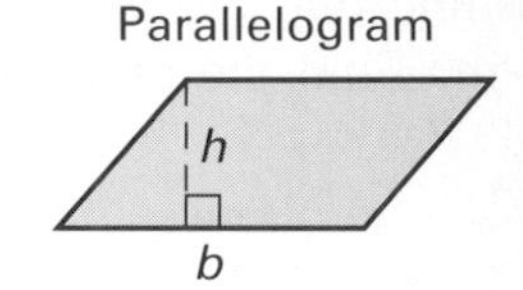

Example 1

Find the area of each parallelogram.

a.

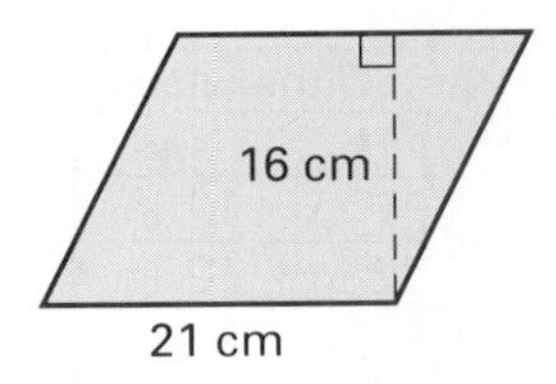

b.

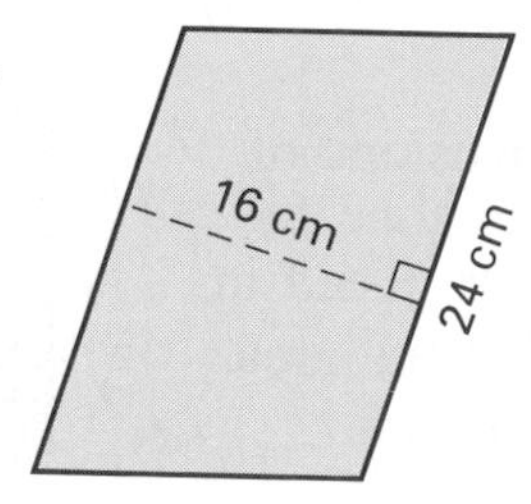

Solution

a. $A = b \times h$ ← Substitute 21 for b and 16 for h.
$A = 21 \times 16$
$A = 336$

The area is 336 cm^2.

b. $A = b \times h$ ← Substitute 24 for b and 16 for h.
$A = 24 \times 16$
$A = 384$

The area is 384 cm^2. ◄

The area of a triangle is one half the area of a parallelogram with the same base and height.

Area of parallelogram

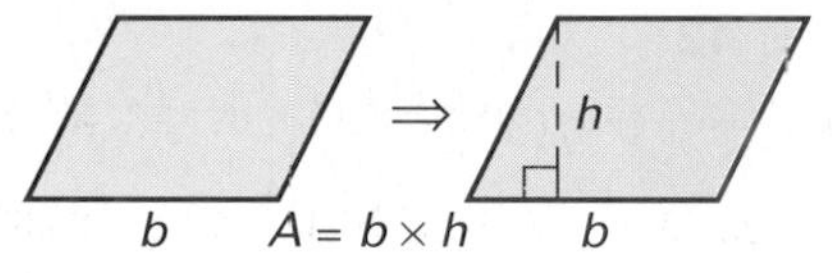

Area of triangle = $\frac{1}{2}$ area of parallelogram

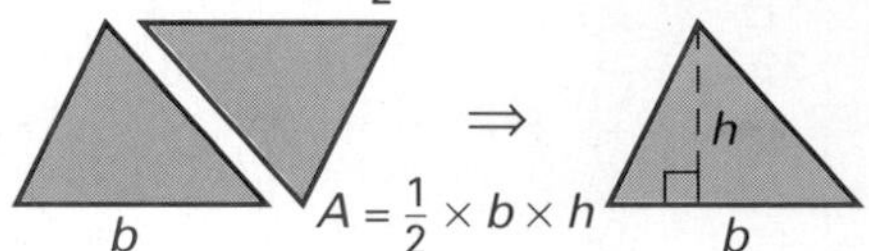

Example 2

Find the area of each triangle.

a.

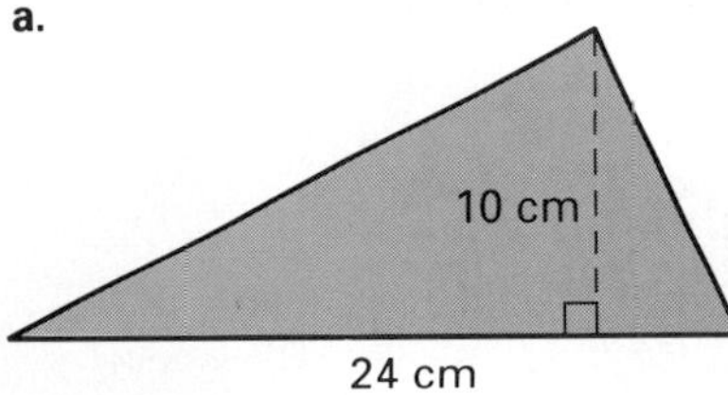

b.

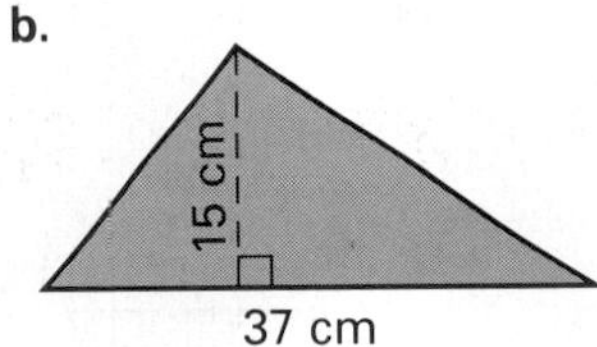

Solution

a. $A = \frac{1}{2} \times b \times h$ ← Substitute 24 for *b* and 10 for *h*.

$A = \frac{1}{2} \times 24 \times 10$

$A = 120$

The area is 120 cm^2.

b. $A = \frac{1}{2} \times b \times h$ ← Substitute 37 for *b* and 15 for *h*.

$A = \frac{1}{2} \times 37 \times 15$

$A = 18.5 \times 15$

$A = 277.5$

The area is 277.5 cm^2. ◄

Example 3

The area of a triangle is 450 cm^2. The length of the base is 36 cm. Find its height.

Solution

$A = \frac{1}{2} \times b \times h$

$450 = \frac{1}{2} \times 36 \times h$ ← Substitute 450 for *A* and 36 for *b*.

$450 = 18 \times h$

$\frac{450}{18} = \frac{18h}{18}$

$25 = h$ The height of the triangle is 25 cm. ◄

TRY THESE

Find the area of each parallelogram.

1.

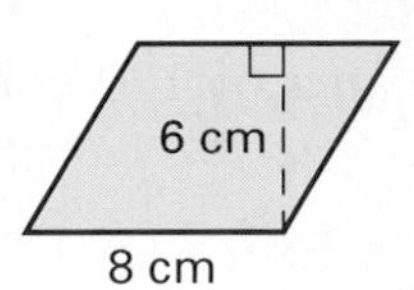

2.

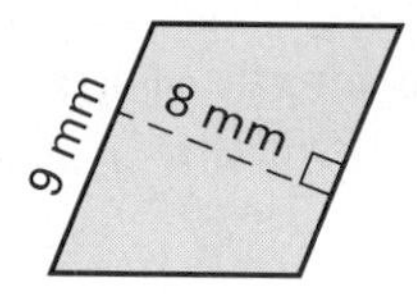

Find the area of each triangle.

3.

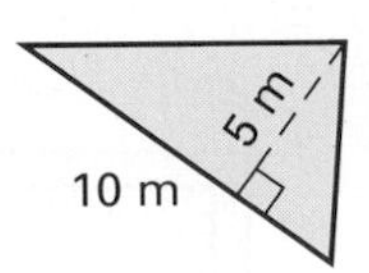

4.

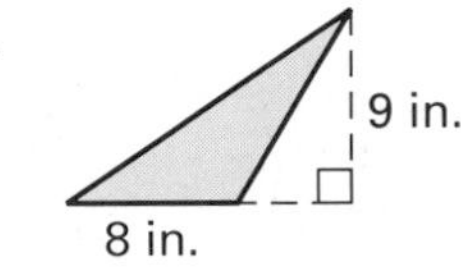

5. The area of a parallelogram is 72 $in.^2$ Its height is 9 in. Find its base.
6. The base of a triangle is 40 cm. Its area is 1,200 cm^2. Find its height.

EXERCISES

PRACTICE/ SOLVE PROBLEMS

Find the area of each figure.

1.

25 cm

36 cm

2.

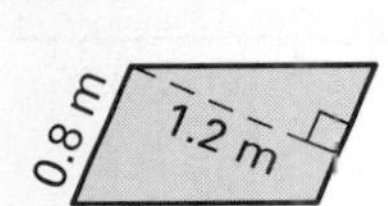

3.

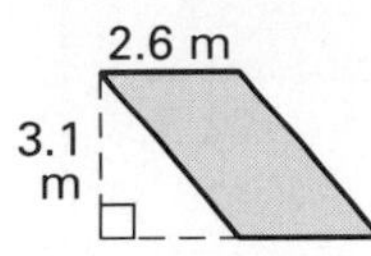

4.

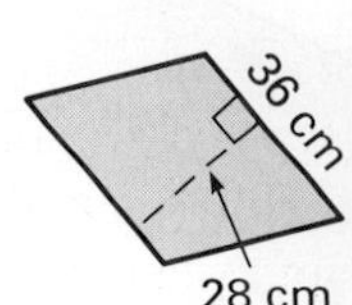

5.

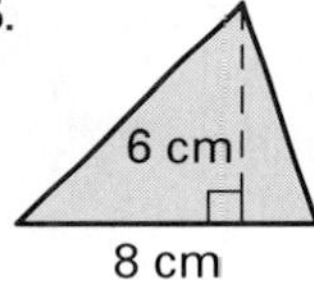

6.

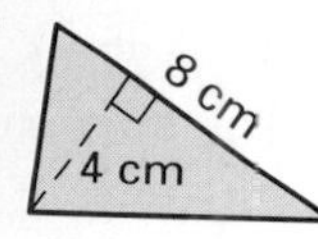

7.

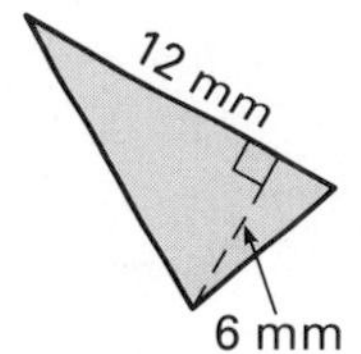

8.

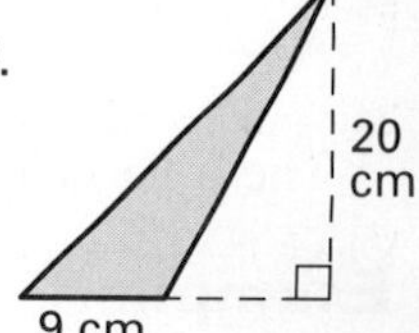

Find the area of a parallelogram with the given dimensions.

9. $b = 10$ in., $h = 6$ in.

10. $b = 20$ ft, $h = 20$ ft

11. $b = 25$ cm, $h = 20$ cm

12. $b = 60$ mm, $h = 40$ mm

Find the area of a triangle with the given dimensions.

13. $b = 14$ yd, $h = 8$ yd

14. $b = 20$ in., $h = 10$ in.

15. $b = 25$ ft, $h = 20$ ft

16. $b = 28$ cm, $h = 12$ cm

Solve.

17. A parallelogram has an area of 180 cm^2 and a height of 10 cm. Find its base.

18. Find the length of a rectangle that has an area of 180 cm^2 and a width of 10 cm.

19. Draw a diagram to illustrate the answers in Exercises 17 and 18.

EXTEND/ SOLVE PROBLEMS

Find the area of each shaded region.

20.

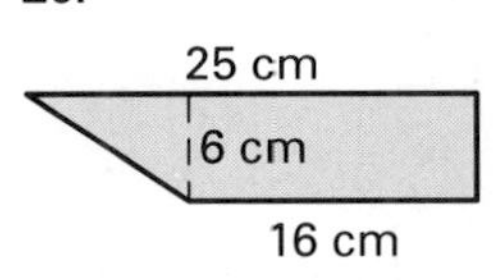

21.

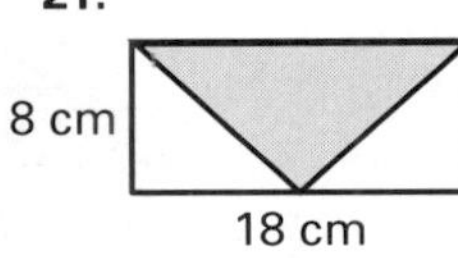

22.

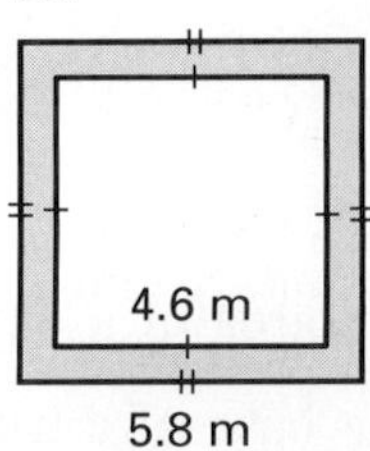

23. Find the amount of aluminum siding needed to cover the side of the house.

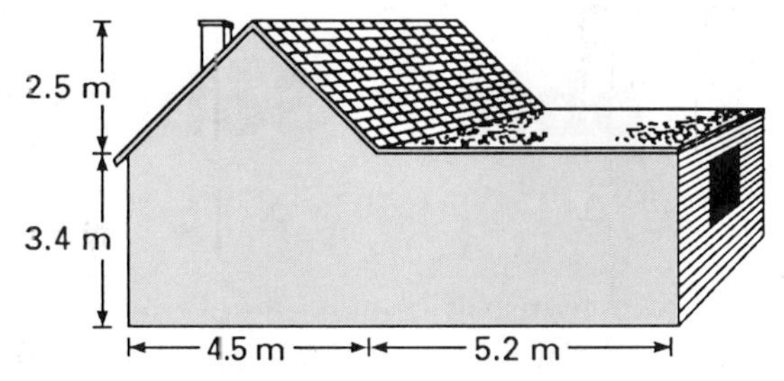

Most sailboats have two basic sails, the jib and the mainsail. Assume that the sails are triangles. Using the data in the diagram, find the area of each sail, to the nearest tenth.

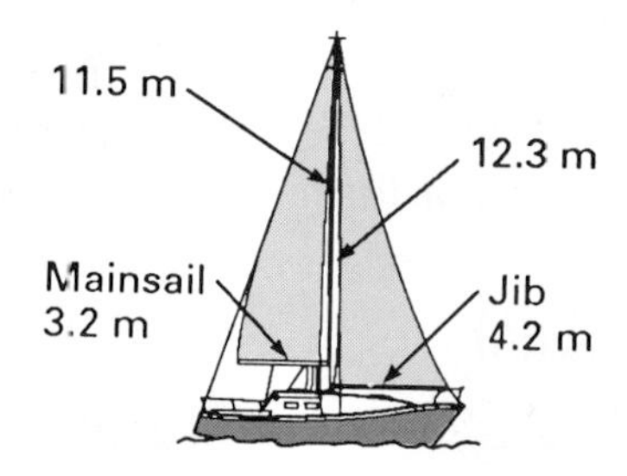

24. mainsail

25. jib

Find the total area of the sails on each boat, to the nearest tenth.

26.

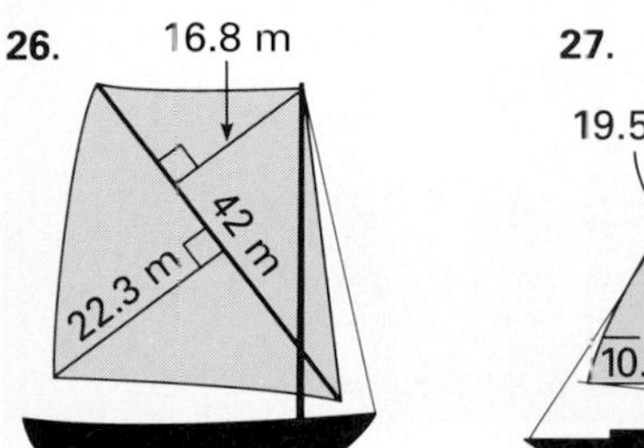

27.

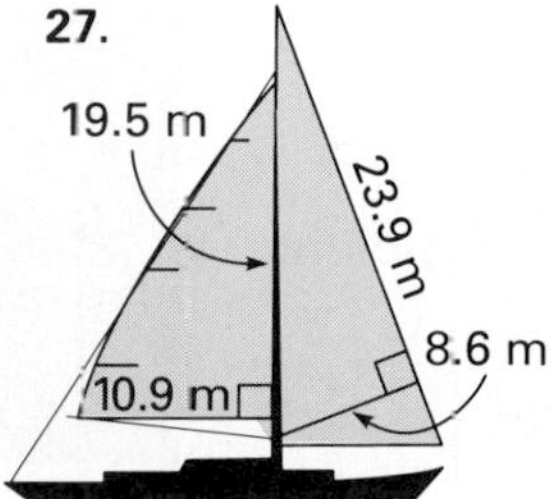

28.

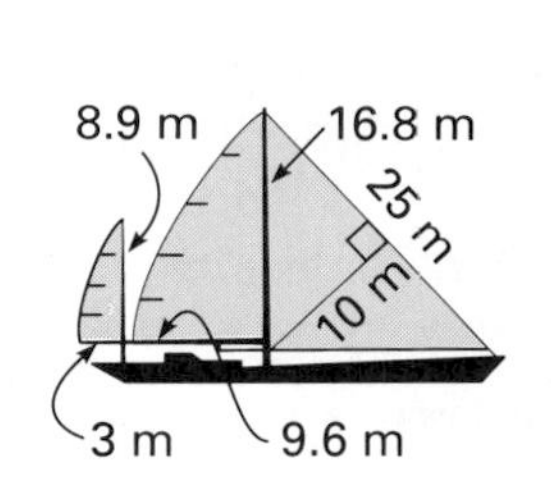

THINK CRITICALLY/ SOLVE PROBLEMS

Use the triangle at the right for Exercises 29 and 30.

29. What measurements are needed to find the area of this triangle?

30. Use a metric ruler. Make the measurements needed and find the area.

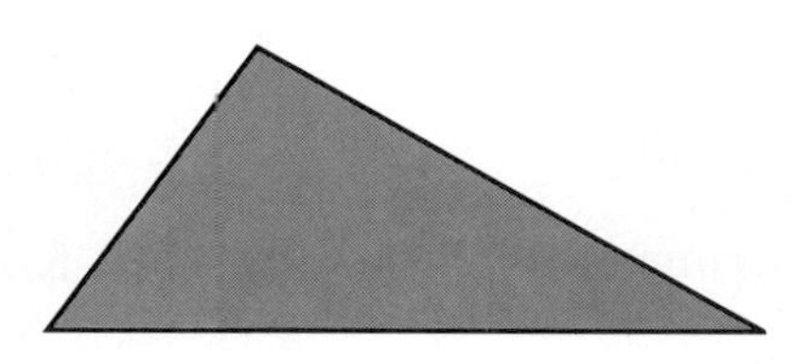

A box lid is constructed from the pattern shown.

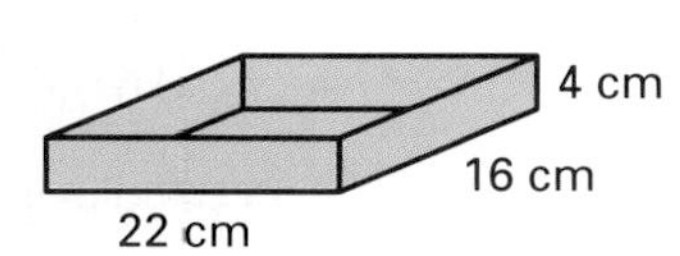

31. Find the total area of the triangular tabs.

32. Find the total amount of material used to make a box lid.

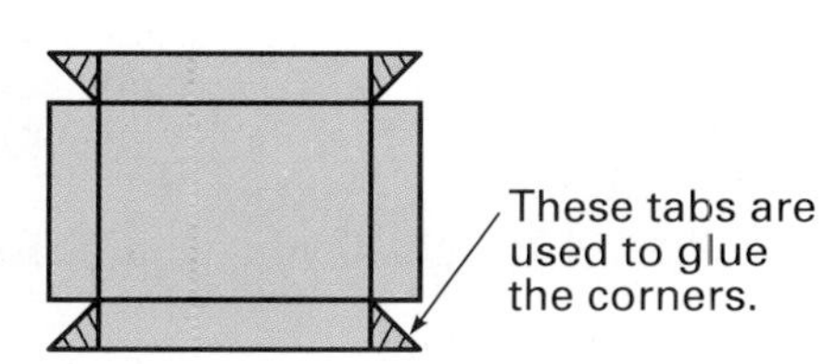

11-4 Problem Solving Skills:

USE A FORMULA FOR PERIMETER OR AREA

► READ
► PLAN
► SOLVE
► ANSWER
► CHECK

In order to use formulas in solving problems, you must decide which formula is the one you need. For example, if you want to find the amount of wallpaper that will be needed to cover the walls of a room, you must use a formula for area. If you want to know the amount of fencing that is needed to enclose a field, you will need to use a formula for perimeter.

PROBLEM

Molding is to be installed along the base of each wall of a rectangular room that measures 36 ft by 24 ft. On one of the longer walls, there is a 4-ft-wide door. No molding is needed along the floor at the doorway. Find the length of molding needed for the room.

SOLUTION

The words "along the base of each wall" and "length of molding" suggest that the goal is to find the distance around the room.

Use the formula for perimeter.

$$\begin{aligned} P = 2l + 2w &= 2(36) + 2(24) \\ &= 72 + 48 \\ &= 120 \end{aligned}$$

The perimeter of the room is 120 ft.

No molding will be used along the floor at the doorway. Therefore, you subtract the width of the door from the perimeter.

$$120 - 4 = 116$$

The amount of molding needed for the room is 116 ft.

PROBLEMS

Write P or A to tell if you would use a *perimeter* or an *area* formula.

1. amount of fencing
2. amount of floor covering

3. amount of material in a bedspread

4. amount of fringe needed to decorate the edge of a scarf

Choose a formula that can be used to solve the problem. Then solve.

$$P = 2l + 2w \qquad A = l \times w \qquad A = s^2$$

$$A = \frac{1}{2} \times b \times h \qquad P = 4 \times s$$

5. A garden is 7.9 m wide and 14 m long. How much fencing will be needed to enclose it?

6. Carpeting costs \$25/yd². How much will it cost to carpet a room that is 9 ft on each side?

7. A square quilt measures 425 cm on each side. How much binding is needed to bind the edges of the quilt?

8. Find the area of the canvas covering the rectangular sides of the tent shown.

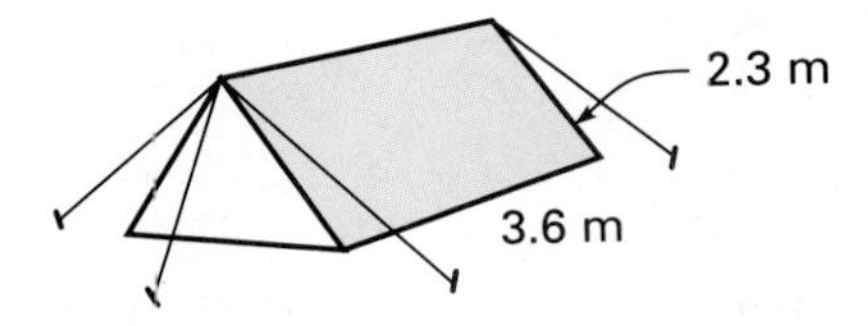

9. A new rectangular park measures 88 yd by 72 yd. Sod costs \$1.10/yd². Find the total cost of covering the entire park with sod.

10. A new garden plot has the dimensions shown. If plastic covering costs \$0.89/m², find the cost of covering the entire plot.

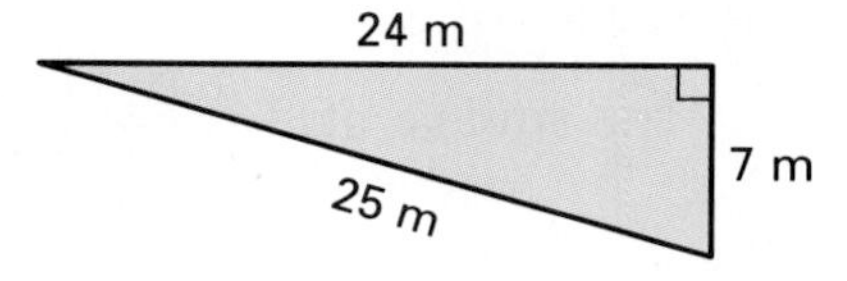

Use the diagram showing the layout of a farm for Exercises 11–15.

11. What is the area of land used to grow hay?

12. It costs 63¢/ft² to fertilize the vegetable garden. What will be the total cost?

13. A fence is to enclose the property around the house. How many feet of fencing are needed?

14. Each apple tree requires 25 ft² of space. How many apple trees can there be in the orchard?

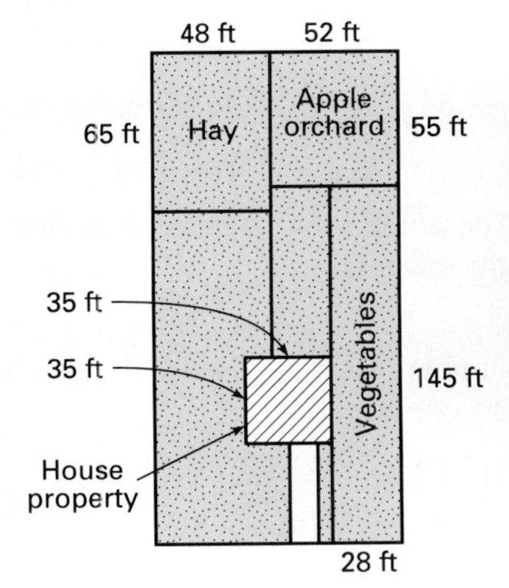

15. How many square feet of the farm are used neither for the house nor for planting?

11-5 Squares and Square Roots

EXPLORE

A square field has an area of 100 m². What do you think is the measure of its sides?

$A = 100 \text{ m}^2$

Suppose another square field has an area of 81 m². What do you think is the measure of its sides?

$A = 81 \text{ m}^2$

Another square field has an area of 90 m². What do you think is the measure of its sides?

SKILLS DEVELOPMENT

The area of the square at the right is 64 square units. The length of each side of the square is 8 units. From the figure you can see why the expression 8^2 is read "8 squared" or "the square of 8."

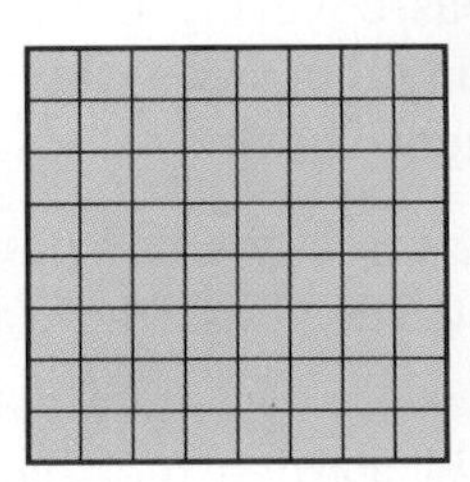

The **square** of 8 is 64.
$8^2 = 64$

The **square root** of 64 is 8.
$\sqrt{64} = 8$

A number a is a square root of another number b if $a^2 = b$. The number 64 has two square roots.

$64 = (8)^2$ 8 is the positive square root of 64.
You write $\sqrt{64} = 8$.

$64 = (-8)^2$ -8 is the negative square root of 64.
You write $-\sqrt{64} = -8$.

The number 64 is called a **perfect square** because its square roots are integers.

COMPUTER TIP

The square root function, SQR, will give you the positive square root of any non-negative integer, N. Type:
? SQR(N) then press ENTER.

Numbers such as $\frac{4}{9}$ and 0.49 also have two square roots, one positive and one negative.

$\sqrt{\frac{4}{9}} = \frac{2}{3}$ $-\sqrt{\frac{4}{9}} = -\frac{2}{3}$

$\sqrt{0.49} = 0.7$ $-\sqrt{0.49} = -0.7$

The number 0 has only one square root: $\sqrt{0} = 0$.

Example 1

Find each square.

a. 5^2 b. $(-3)^2$ c. $\left(-\frac{1}{4}\right)^2$ d. $(.09)^2$

MENTAL MATH TIP

Memorize the squares of integers from 1 to 12. You'll find that you can recall their square roots and write them faster than you can compute them on a calculator or look them up in a table.

Solution

a. $5^2 = 5 \times 5 = 25$

b. $(-3)^2 = (-3) \times (-3) = 9$

c. $\left(-\frac{1}{4}\right)^2 = \left(-\frac{1}{4}\right) \times \left(-\frac{1}{4}\right) = \frac{1}{16}$

d. $(0.09)^2 = 0.09 \times 0.09 = 0.0081$ ◀

Example 2

Find each square root.

a. $\sqrt{\frac{9}{16}}$ b. $-\sqrt{0.36}$

Solution

a. $\frac{3}{4} \times \frac{3}{4} = \frac{9}{16}$, so $\sqrt{\frac{9}{16}} = \frac{3}{4}$

b. $-0.6 \times (-0.6) = 0.36$,

so $-\sqrt{0.36} = -0.6$. ◀

The square root of any rational number that is not a perfect square is a nonterminating, nonrepeating decimal. In Chapter 5 you learned that this type of decimal is called an irrational number.

CHECK UNDERSTANDING

Tell whether each square root is rational or irrational.

1. $\sqrt{169}$

2. $\sqrt{27}$

So, numbers like $\sqrt{2}$ and $\sqrt{3}$ cannot be expressed as common fractions or as terminating or repeating decimals. For numbers such as these you can use a table of square roots or a calculator to find an *approximate* value. For those numbers that are not perfect squares, the values in the table are approximations of square roots, rounded to the nearest thousandth.

TABLE OF SQUARE ROOTS

Number	Square root	Number	Square root	Number	Square root	Number	Square root
1	1.000	**26**	5.099	**51**	7.141	**76**	8.718
2	1.414	**27**	5.196	**52**	7.211	**77**	8.775
3	1.732	**28**	5.292	**53**	7.280	**78**	8.832
4	2.000	**29**	5.385	**54**	7.348	**79**	8.888
5	2.236	**30**	5.477	**55**	7.416	**80**	8.944
6	2.449	**31**	5.568	**56**	7.483	**81**	9.000
7	2.646	**32**	5.657	**57**	7.550	**82**	9.055
8	2.828	**33**	5.745	**58**	7.616	**83**	9.110
9	3.000	**34**	5.831	**59**	7.681	**84**	9.165
10	3.162	**35**	5.916	**60**	6.746	**85**	9.220
11	3.317	**36**	6.000	**61**	7.810	**86**	9.274
12	3.464	**37**	6.083	**62**	7.874	**87**	9.327
13	3.606	**38**	6.164	**63**	7.937	**88**	9.381
14	3.742	**39**	6.245	**64**	8.000	**89**	9.434
15	3.873	**40**	6.325	**65**	8.062	**90**	9.487
16	4.000	**41**	6.403	**66**	8.124	**91**	9.539
17	4.123	**42**	6.481	**67**	8.185	**92**	9.592
18	4.243	**43**	6.557	**68**	8.246	**93**	9.644
19	4.359	**44**	6.633	**69**	8.307	**94**	9.695
20	4.472	**45**	6.708	**70**	8.367	**95**	9.747
21	4.583	**46**	6.782	**71**	8.426	**96**	9.798
22	4.690	**47**	6.856	**72**	8.485	**97**	9.849
23	4.796	**48**	6.928	**73**	8.544	**98**	9.899
24	4.899	**49**	7.000	**74**	8.602	**99**	9.950
25	5.000	**50**	7.071	**75**	8.660	**100**	10.000

Number — Square root of the number

Example 3

Find $\sqrt{23}$.

Solution 1

Use the table on page 401. Find 23 in the number column. Read to the right to locate the square root, rounded to the nearest thousandth.

$\sqrt{23} \approx 4.796$ ◄ Because the value is an approximation, this symbol, meaning "approximately equal to," is used.

Solution 2

Use a calculator. Enter this key sequence.

23 [√]

Display: 4.7958315

Round the number in the display to the nearest thousandth. $\sqrt{23} \approx 4.796$ ◄

MATH: WHO, WHERE, WHEN

One way of computing square roots is a method developed by the English mathematician Sir Isaac Newton (1642–1727). Here are the steps in this process, known as Newton's Method.

1. Choose a number: 13
2. Estimate $\sqrt{13}$.
   $\sqrt{9} < \sqrt{13} < \sqrt{16}$
   $3 < ? < 4$
3. Try $\sqrt{13} \rightarrow 3.8$.
4. Divide: $13 \div 3.8 \approx 3.42$
5. Find the average of your estimate in Step 3 and the quotient from Step 4.
   $\frac{3.8 + 3.42}{2} = 3.61 \approx 3.6$
6. Divide the number 13 by the result in Step 5.
   $13 \div 3.6 = 3.6\overline{1} \approx 3.6$
7. Check by multiplying.
   $(3.6)^2 = 12.96$
   $(3.61)^2 \approx 13.03$
   Therefore, $\sqrt{13} \approx 3.6$, rounded to the nearest tenth.

Use Newton's method to approximate $\sqrt{27}$ to the nearest tenth.

TRY THESE

Find each square.

1. 14^2 **2.** 16^2 **3.** $(-4)^2$ **4.** $\left(\frac{5}{8}\right)^2$ **5.** $(0.6)^2$

Find each square root.

6. $\sqrt{\frac{81}{169}}$ **7.** $\sqrt{\frac{144}{289}}$ **8.** $\sqrt{\frac{16}{121}}$ **9.** $\sqrt{\frac{289}{361}}$

10. $-\sqrt{0.25}$ **11.** $\sqrt{0.09}$ **12.** $\sqrt{0.16}$ **13.** $-\sqrt{1.44}$

Use the table of square roots on page 401 to find each square root.

14. $\sqrt{87}$ **15.** $\sqrt{19}$ **16.** $\sqrt{52}$ **17.** $\sqrt{7}$

Use a calculator to find each square root. Round to the nearest thousandth.

18. $\sqrt{22}$ **19.** $\sqrt{53}$ **20.** $\sqrt{58}$ **21.** $\sqrt{85}$

EXERCISES

PRACTICE/ SOLVE PROBLEMS

Find each square.

1. 18^2 **2.** 17^2 **3.** 21^2 **4.** 15^2

5. $(1.8)^2$ **6.** $(1.7)^2$ **7.** $(2.1)^2$ **8.** $(1.5)^2$

9. $\left(\frac{1}{5}\right)^2$ **10.** $\left(\frac{3}{7}\right)^2$ **11.** $\left(-\frac{8}{9}\right)^2$ **12.** $\left(-\frac{18}{23}\right)^2$

Find each square root.

13. $\sqrt{256}$ 14. $\sqrt{144}$ 15. $-\sqrt{100}$ 16. $-\sqrt{0.36}$

17. $-\sqrt{0.49}$ 18. $\sqrt{\frac{1}{49}}$ 19. $\sqrt{\frac{4}{81}}$ 20. $-\sqrt{\frac{16}{121}}$

Use the table of square roots on page 401 to find the square root. Round to the nearest tenth.

21. $\sqrt{12}$ 22. $\sqrt{90}$ 23. $\sqrt{84}$ 24. $\sqrt{72}$

25. $\sqrt{42}$ 26. $\sqrt{94}$ 27. $\sqrt{6}$ 28. $\sqrt{38}$

Use a calculator to find each square root. Round to the nearest thousandth.

29. $\sqrt{153}$ 30. $\sqrt{527}$ 31. $\sqrt{976}$ 32. $\sqrt{369}$

33. $\sqrt{101}$ 34. $\sqrt{482}$ 35. $\sqrt{815}$ 36. $\sqrt{239}$

37. The area of a square stamp is 23 cm^2. Find the measure of a side. Round your answer to the nearest tenth.

COMPUTER

```
10 INPUT "ENTER A
     NUMBER: ";N: PRINT
20 INPUT "ESTIMATE ITS
     SQUARE ROOT: ";E:
     PRINT
30 Q = N / E
40 IF Q = N THEN GOTO 90
50 A = (E + Q) / 2
60 Q = N / A
62 IF INT (10 * A + 0.5) /
     10 = INT (10 * Q +
     0.5) / 10 THEN GOTO 70
65 GOTO 50
70 IF INT (10 * A + 0.5) /
     10 = INT (10 * Q +
     0.5) / 10 THEN GOTO 90
80 GOTO 50
90 PRINT "TO NEAREST
     TENTH, APPROXIMATE
     SQUARE ROOT OF ";N;
     " IS "; INT (10 * Q +
     0.5) / 10
```

EXTEND/ SOLVE PROBLEMS

Find each square.

38. 29^2 39. 34^2 40. $(-37)^2$ 41. $(0.2)^2$

42. $(0.16)^2$ 43. $(0.8)^2$ 44. $(0.08)^2$ 45. $(0.008)^2$

46. $(0.0008)^2$ 47. $(\sqrt{8})^2$ 48. $(\sqrt{19})^2$ 49. $(-\sqrt{24})^2$

50. The area of a square plot is 72 ft^2. Find the perimeter of the plot to the nearest tenth.

51. A square lot has an area of 500 ft^2. If fencing costs $10.25 per foot, how much would it cost to enclose the lot?

THINK CRITICALLY/ SOLVE PROBLEMS

The amount of time (t) in seconds it takes an object to fall a distance (d) in meters is expressed in this formula. $t = \sqrt{\frac{d}{4.9}}$

Use this formula for Exercises 52 and 53. Round each answer to the nearest tenth.

52. An object fell 40 m. How long did it take the object to hit the ground?

53. A rock falls over a cliff 75 m in height. How long will it take the rock to hit the water at the bottom of the cliff?

11-6 The Pythagorean Theorem

EXPLORE

Trace this right triangle. Draw a square on each side of the triangle, so that each side forms one side of a square. Measure a side of each square using a centimeter ruler. Find the area of each square. Look for a relationship between the numbers representing the areas of the squares.

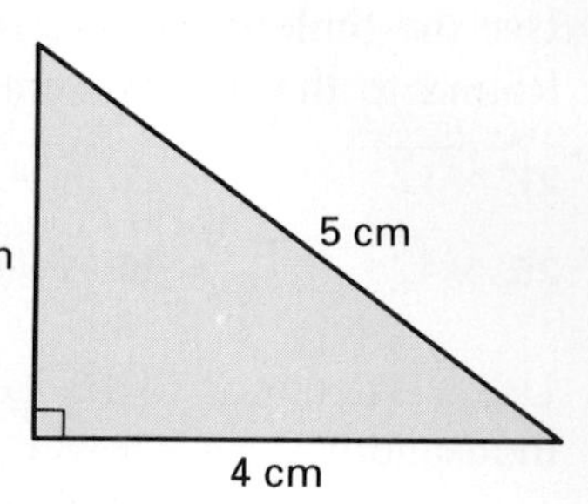

SKILLS DEVELOPMENT

In a right triangle, the side opposite the right angle is called the **hypotenuse.** The other sides are called the **legs.** There is a special relationship between the legs of a right triangle and the hypotenuse.

> In any right triangle, the square of the length of the hypotenuse is equal to the sum of the squares of the lengths of the legs.

In right triangle ABC, where c is the length of the hypotenuse, and a and b are the lengths of the legs, you can state the relationship as follows.

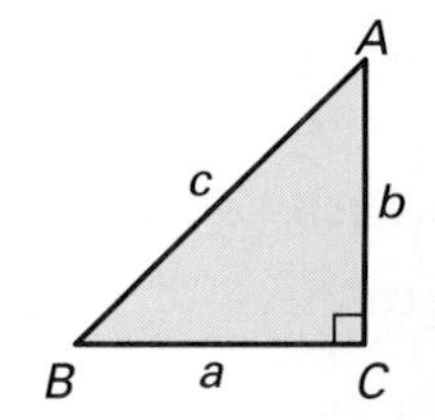

$$c^2 = a^2 + b^2$$

This property is called the **Pythagorean Theorem,** after the Greek mathematician Pythagoras. The Pythagorean Theorem can be used to find the measure of one side of a right triangle if the measures of the other two sides are known.

COMPUTER

You can use this program to check your work when finding the hypotenuse of a right triangle.

```
10 INPUT "ENTER LENGTHS
   OF LEGS: ";A,B:
   PRINT
20 LET C = SQR (A ^ 2 +
   B ^ 2)
30 PRINT "HYPOTENUSE
   IS ";C
```

Example 1

Find the length of the hypotenuse of triangle ABC.

Solution

Use the Pythagorean Theorem.

$$c^2 = a^2 + b^2$$
$$c^2 = 5^2 + 12^2$$
$$c^2 = 25 + 144$$
$$c^2 = 169$$
$$c = \sqrt{169}$$
$$c = 13$$

The length of the hypotenuse is 13 cm. ◄

Example 2

In triangle ABC, find the unknown length b to the nearest tenth.

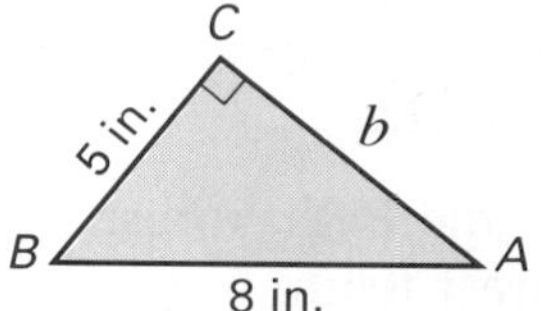
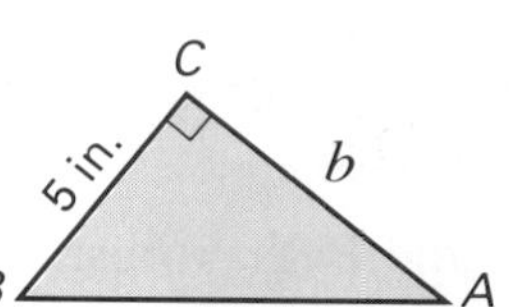

Solution

In triangle ABC, you know that $c^2 = a^2 + b^2$.
If you subtract a^2 from each side of the equation, the relationship becomes $b^2 = c^2 - a^2$.

$$b^2 = c^2 - a^2$$
$$b^2 = 8^2 - 5^2$$
$$b^2 = 64 - 25$$
$$b^2 = 39$$
$$b = \sqrt{39}$$
$$b \approx 6.245$$
$$b \approx 6.2 \text{ (nearest tenth)}$$

Use the table of square roots (page 401) or a calculator to find an approximate value of $\sqrt{39}$.

The length b is about 6.2 in. ◄

Example 3

A ladder 6.1 m in length is placed against a wall. If the foot of the ladder is a distance of 2.5 m from the wall, how far up the wall does the ladder reach? Express your answer as an approximation to the nearest tenth.

Solution

Draw a diagram of the problem. Since the wall meets the ground at a right angle, the ladder serves as the hypotenuse of a right triangle. Use the Pythagorean Theorem.

$$c^2 = a^2 + b^2$$
$$b^2 = c^2 - a^2$$
$$b^2 = (6.1)^2 - (2.5)^2$$
$$b^2 = 37.31 - 6.25$$
$$b^2 = 30.96$$
$$b = \sqrt{30.96} \approx 5.6$$

A
6.1 m
b = ?
B 2.5 m C

The ladder reaches about 5.6 m up the wall. ◄

TRY THESE

Use the Pythagorean Theorem to find the unknown length, to the nearest tenth.

1.
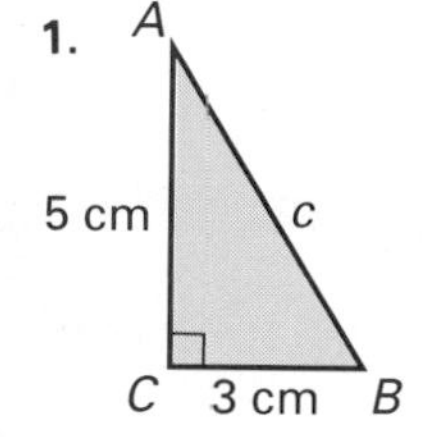

2.
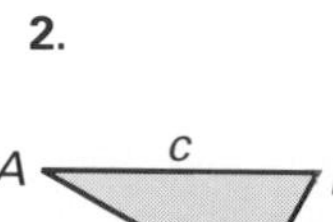

3.
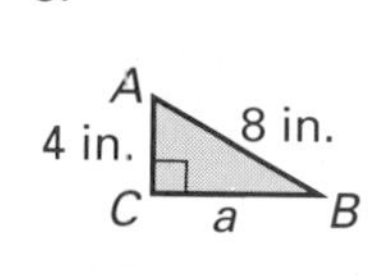

4.
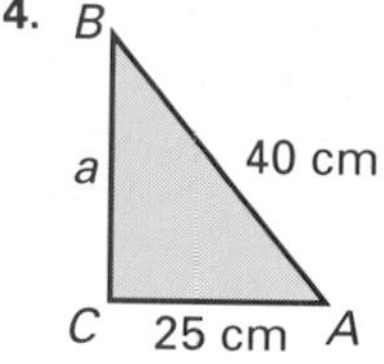

5. A 15-ft-long wooden brace is built against a wall at a height of 12 ft. How far from the wall is the foot of the brace?

COMPUTER TIP

You can use this program to find a missing leg of a right triangle. Compare this program with the one used to find the hypotenuse on page 404.

```
10 INPUT "ENTER LEG, THEN
   HYPOTENUSE ";A,C
20 LET B = SQR (C ^ 2 -
   A ^ 2)
30 PRINT "OTHER LEG
   IS ";B
```

MATH IN THE WORKPLACE

Here is how carpenters use math to make right angles.

Mark off 3 ft (or 6 ft)
Right angle
BOARD
BOARD
Mark off 4 ft (or 8 ft)
Tape measure
Adjust boards to get a reading of 5 ft (or 10 ft) on the tape measure.

EXERCISES

PRACTICE/ SOLVE PROBLEMS

Use the Pythagorean Theorem to find the unknown length to the nearest tenth.

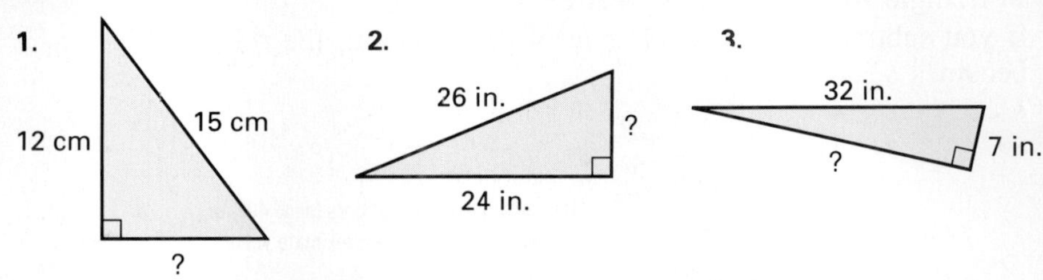

Find the length of the hypotenuse in each right triangle. Round to the nearest tenth.

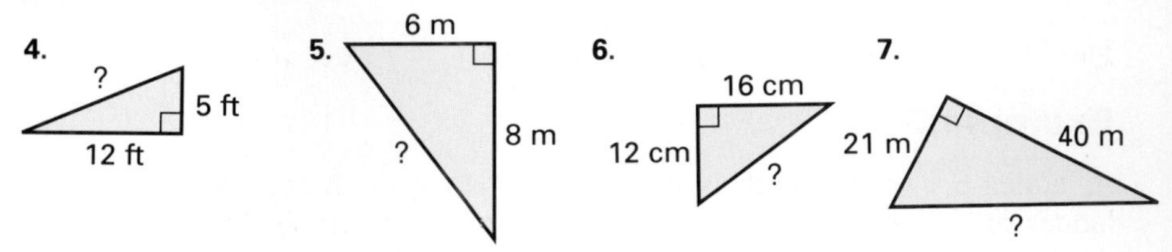

Find the unknown length in each triangle. Round to the nearest tenth.

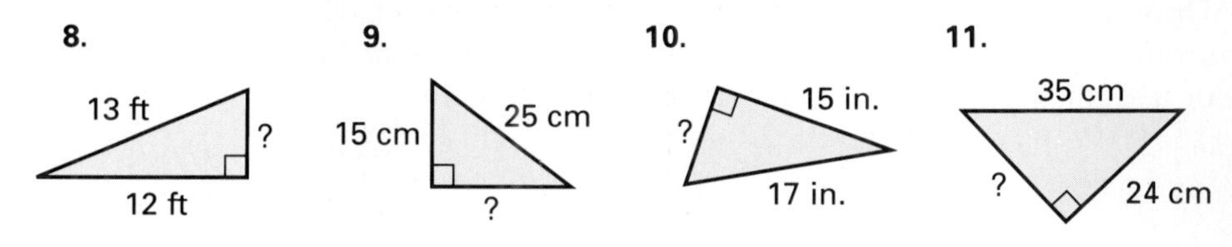

Use the Pythagorean Theorem to find each unknown measurement. Round to the nearest tenth.

12. length of the ladder

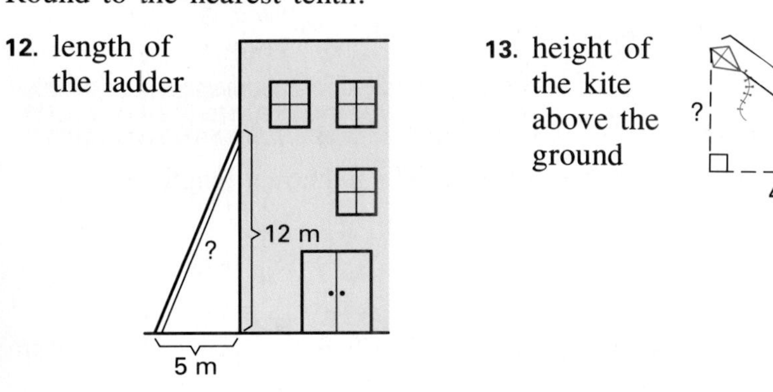

13. height of the kite above the ground

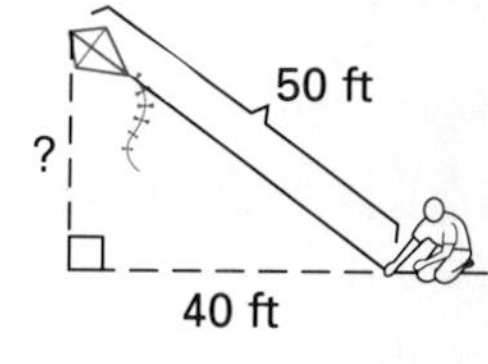

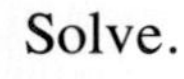

Solve.

14. A ladder 7 m long is placed against a wall so that the foot of the ladder is 2.5 m from the wall. Find how high up the wall the ladder reaches. Round to the nearest tenth.

EXTEND/ SOLVE PROBLEMS

Calculate the value of the variable. Estimate the square root to the nearest tenth.

15. $c^2 = (1.8)^2 + (3.2)^2$

16. $b^2 = (8.6)^2 - (5.3)^2$

17. $a^2 = (7.3)^2 - (4.5)^2$

18. $(5.8)^2 + (3.0)^2 = c^2$

Find the unknown length for each. Round to the nearest hundredth.

19.

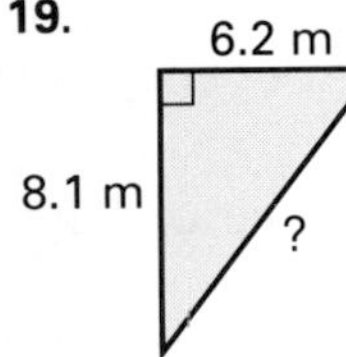

20.

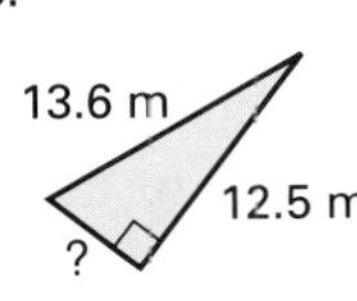

21.

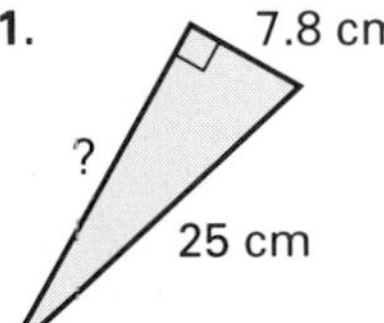

22.

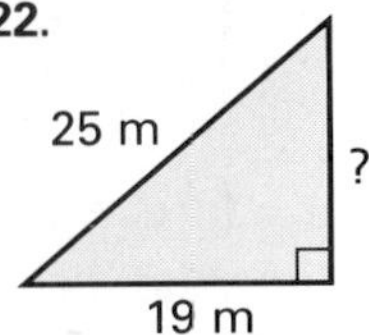

Find the length of each diagonal to the nearest tenth.

23.

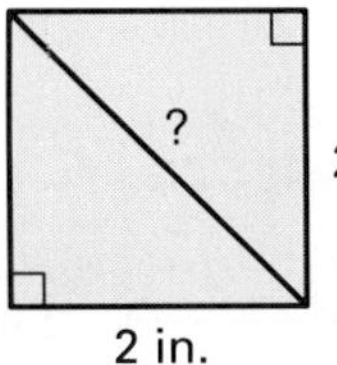

24.

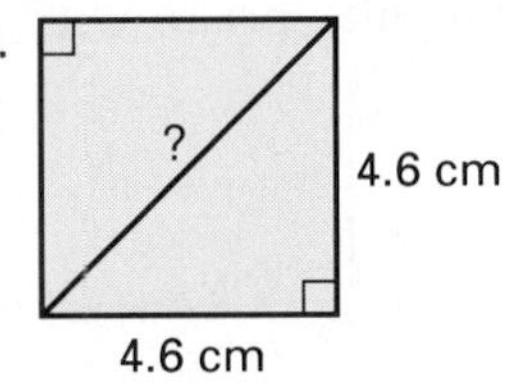

25.

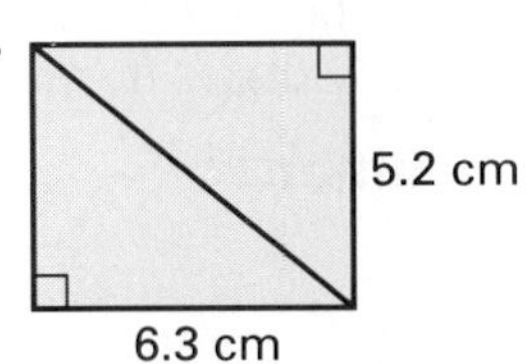

26. The figure is a diagram of a roof. Find the length of the span (SP) of the roof.

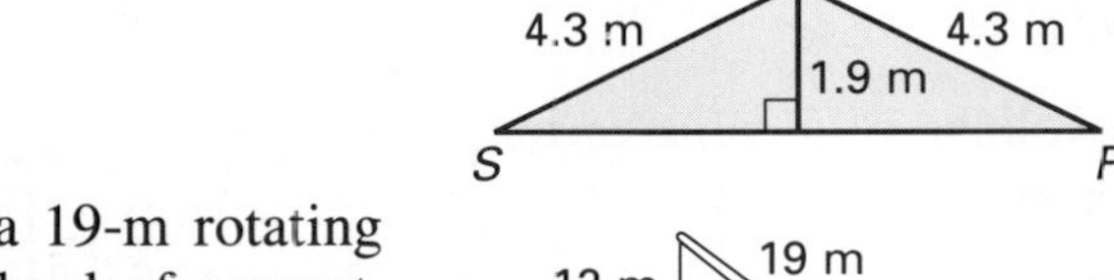

27. A crane with a 19-m rotating lever drops a load of cement at the end of a 12-m cable so that the load is level with the base of the rotating arm. How far is the load from the edge of the building?

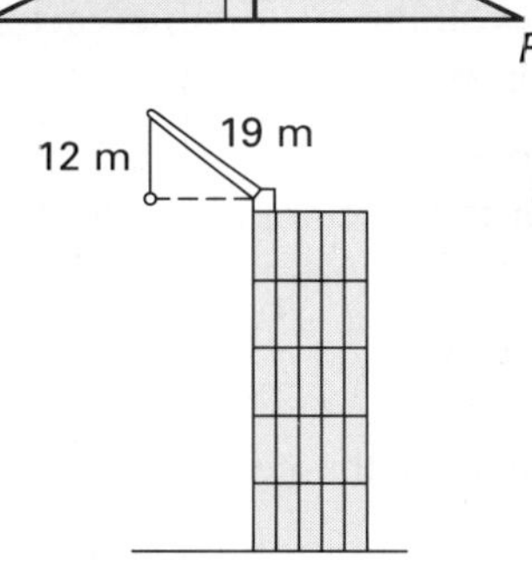

THINK CRITICALLY/ SOLVE PROBLEMS

In a square room, the distance between opposite corners measures 12.8 m. For Exercises 28–30, find these measurements to the nearest tenth.

28. length of one side

29. perimeter

30. area

31. Find the length of $\overline{AB}$, the diagonal of this box.

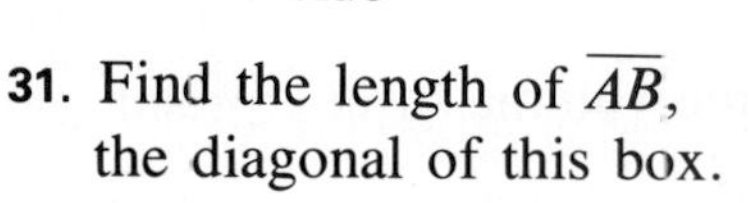

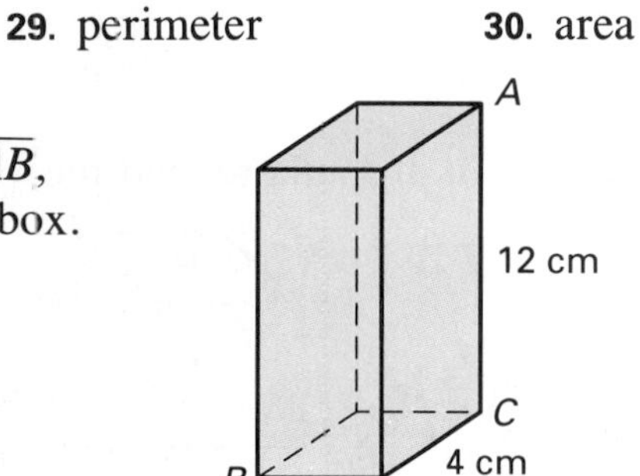

11-7 Volume of Prisms

EXPLORE

How many one-inch cubes would be needed to build an L-shaped structure that is 3 inches tall and has a floor plan as shown?

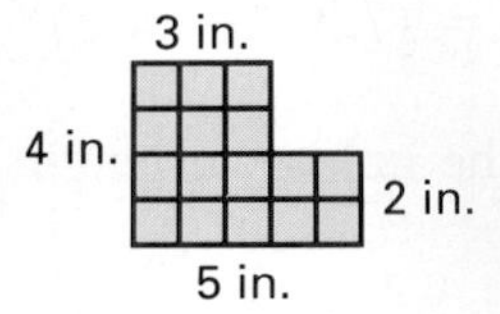

SKILLS DEVELOPMENT

You can use the formula $V = l \times w \times h$ to find the volume of any rectangular prism.

Example 1

Find the volume of a rectangular prism with length 3 ft, width 2 ft, and height 2 ft.

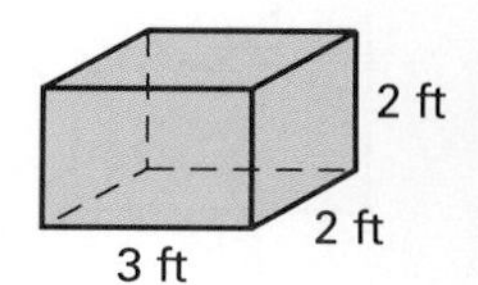

Solution

$V = l \times w \times h$

$V = 3 \times 2 \times 2$ ← Substitute 3 for l, 2 for w, and 2 for h.

$V = 12$

The volume of the rectangular prism is 12 ft^3. ◄

Example 2

Find the volume of a cubic carton with edges measuring 6 in.

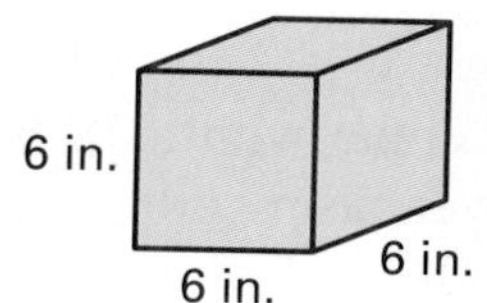

Solution

The cube is a rectangular prism in which each dimension is the same. $V = l \times w \times h$

$V = 6 \times 6 \times 6$ ← Substitute 6 for l, w, and h.

$V = 216$

The volume of the cube is 216 $in.^3$ ◄

The area of any rectangular prism is found by multiplying the area of the **base,** $l \times w$, by the height of the prism, h.

$$V = B \times h$$

V ↑ Volume; B ↑ area of base; h ↑ height of prism

You can generalize this formula to find the volume of any prism.

Triangular prism

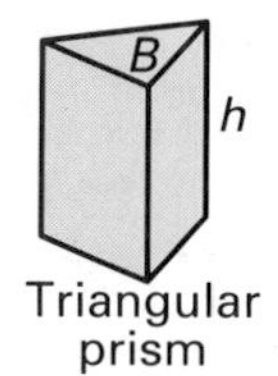

Rectangular prism

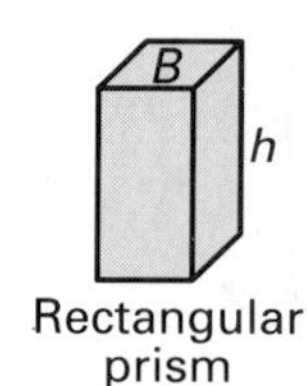

Cube

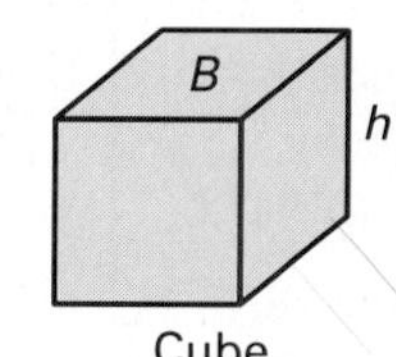

Pentagonal prism

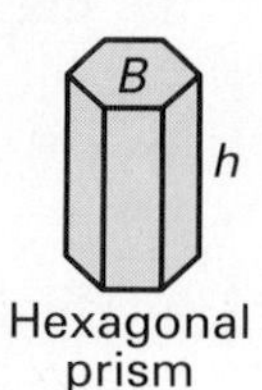

Hexagonal prism

Example 3

Find the volume of a triangular prism with the dimensions shown.

Solution

First find the area of the base. The base is a triangle with a base of 12 cm and a height of 6 cm.

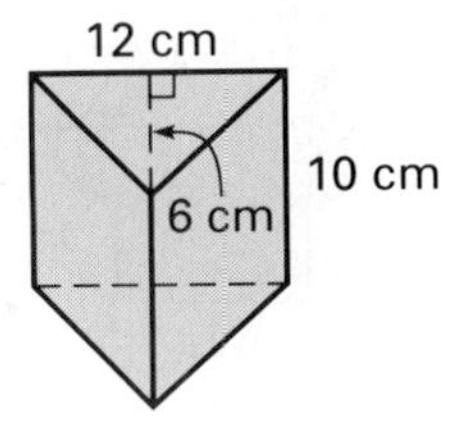

$B = \frac{1}{2} \times b \times h$ — Use the dimensions of the triangle.

$B = \frac{1}{2} \times 12 \times 6$ ← Substitute 12 for *b* and 6 for *h*.

$B = 36$

The area of the base is 36 cm^2.

To find the volume of the prism, use the formula $V = B \times h$, where h is the height of the prism.

$V = B \times h$

$V = 36 \times 10$ ← Use the dimensions of the prism. Substitute 36 for *B* and 10 for *h*.

$V = 360$

The volume of the triangular prism is 360 cm^3. ◄

Example 4

A trench for a pipe is dug with dimensions as shown in the diagram. Find the volume of earth removed in digging the trench.

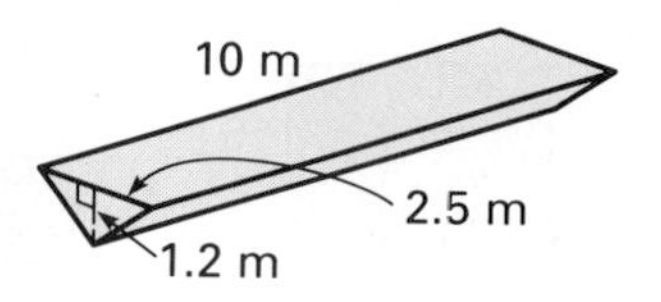

Solution

The trench is in the shape of a triangular prism. Use the volume formula for prisms. First, calculate the area of one triangular base (B).

$B = \frac{1}{2} \times b \times h$ — Use the dimensions of the triangular end of the prism.

$B = \frac{1}{2} \times 2.5 \times 1.2$ ← Substitute 2.5 for *b* and 1.2 for *h*.

$B = 1.5$

The area of the end (base) of the trench is 1.5 m^2.
Now find the volume of the prism.

$V = B \times h$

$V = 1.5 \times 10$ ← Use the dimensions of the entire prism. Substitute 1.5 for *B* and 10 for *h*.

$V = 15$

The volume of earth removed is 15 m^3. ◄

COMPUTER

You may want to use this program to check your work when finding the volume of a rectangular or triangular prism.

```
10 INPUT "ENTER BASE
   AND HEIGHT OF BASE
   OF PRISM: ";B,H: PRINT
20 INPUT "ENTER HEIGHT
   OF PRISM: ";HP: PRINT
30 INPUT "ENTER THE TYPE
   PRISM (TRIANGULAR,
   RECTANGULAR OR
   CUBE) T,R OR C: ";P$:
   PRINT
40 IF P$ = "T" THEN AB =
   1 / 2 * B * H
50 IF P$ = "R" OR P$ =
   "C" THEN AB = B * H
60 V = AB * HP
70 PRINT "AREA OF BASE
   = ";AB:
80 PRINT "VOLUME = ";V
```

TRY THESE

1. Find the volume of a rectangular prism that measures 12 ft by 6 ft by 4 ft.
2. Find the volume of a cube with edges measuring 7 in.
3. Find the volume of a triangular prism with dimensions shown.

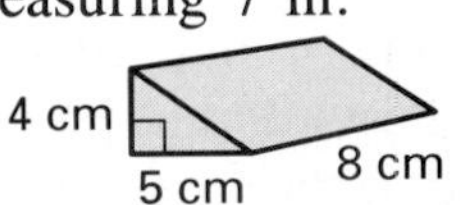

4. A triangular trough has the dimensions shown. Find the volume of the trough.

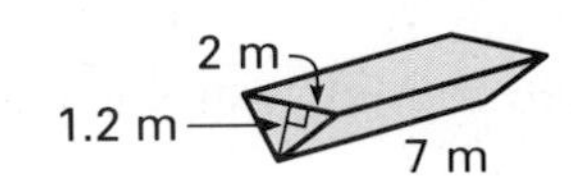

EXERCISES

PRACTICE/ SOLVE PROBLEMS

Find the volume of each prism. If necessary, round to the nearest tenth.

1. 2 in. 3 in. 4 in.

2. 4 in. 2 in. 5 in.

3. 6 cm 9 cm 12 cm

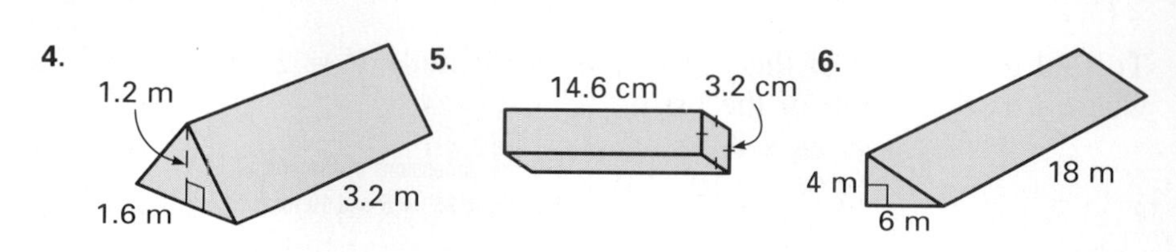

Which prism in each pair has the greater volume?

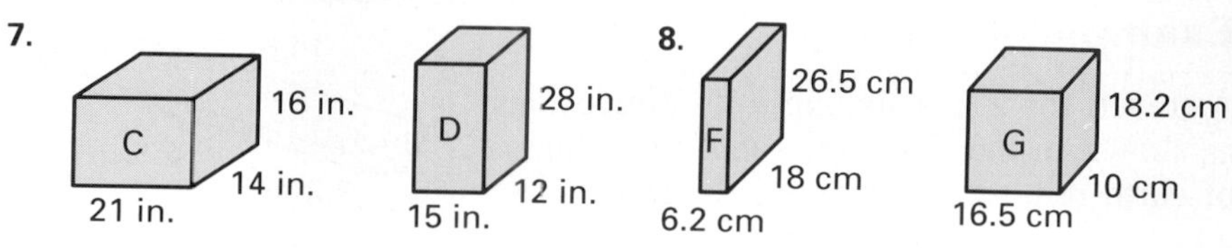

9. Find the volume of the cheese.

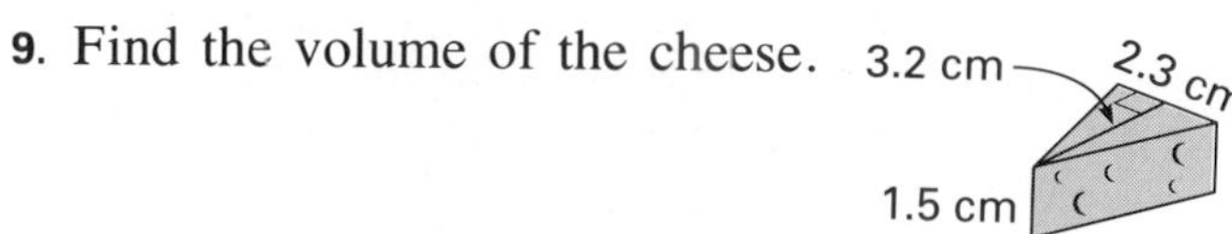

10. A bale of hay is 1.5 m long, 0.8 m wide, and 1.4 m high. What is the volume of the bale?

EXTEND/ SOLVE PROBLEMS

Find the volume. If necessary, round to the nearest tenth.

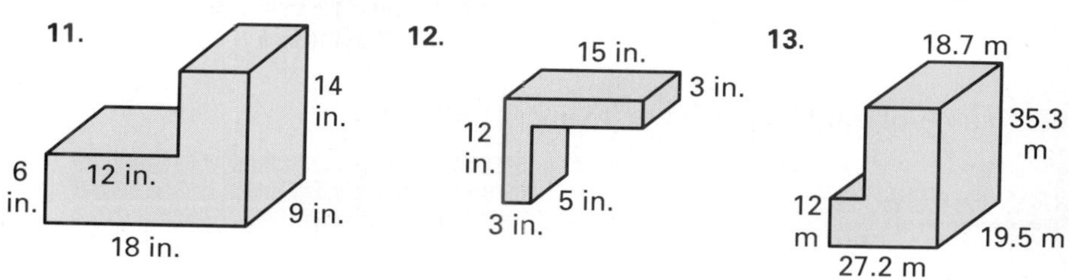

14. Find the volume of water in the swimming pool if the surface of the water is 0.3 m below the top edge of the pool.

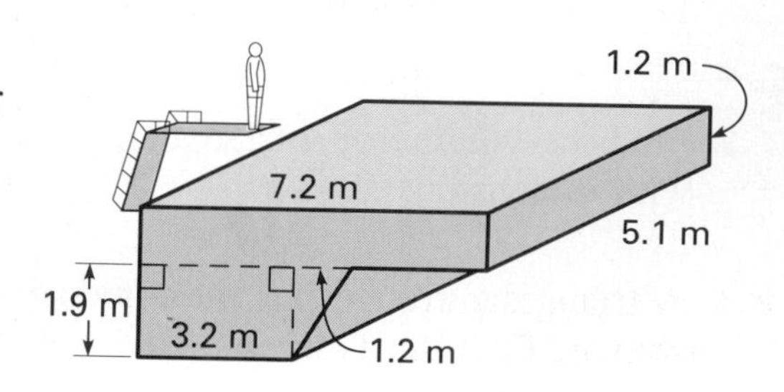

15. The cost of heating the air in a room measuring 3.2 m by 4.8 m by 2.9 m is 1.5¢/m³. What is the cost of heating the room?

16. To construct the basement of a house, a hole with dimensions 10 m by 5 m by 2 m must be dug. A dump truck can haul 25 m³ of earth on each trip. How many trips must the truck make to remove all the earth?

17. There are 10 garbage containers at an apartment site. Each container holds 4.5 yd³ of garbage. If the truck that picks up the garbage has a closed trailer with dimensions 4.5 yd by 2.5 yd by 2 yd, how many trips will it take to haul away the garbage?

18. Before the surface of an ice rink is safe for skating, an area 75.1 m long and 74.8 m wide must be frozen to a depth of 0.5 m. What is the volume of ice that will result?

19. A cubical tank for storing syrup has sides measuring 1.2 m. What volume of syrup is in the tank when the tank is half full?

20. The concrete pillar at the right has a hexagonal base. All the triangular areas are identical. Find the volume of the pillar.

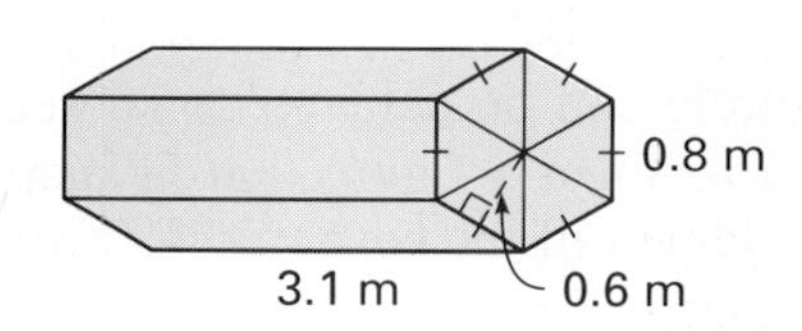

COMPUTER TIP

For Exercise 20, you can use the program on page 409. Change line 50 to V = 6 * AB * HP to find the volume of the pillar.

THINK CRITICALLY/ SOLVE PROBLEMS

For Exercises 21–25, use these relationships between volume and liquid capacity. 1 L = 1,000 cm³ and 1 mL = 1 cm³

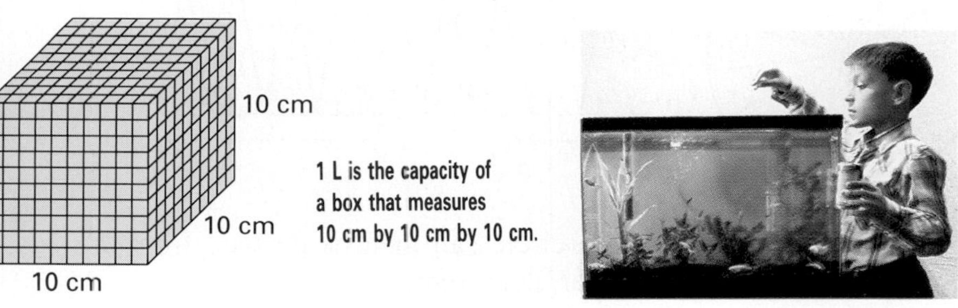

1 L is the capacity of a box that measures 10 cm by 10 cm by 10 cm.

A fish tank is 180 cm long, 150 cm wide, and 70 cm high.

21. What is its volume?

22. In liters of water, what is the capacity of the tank?

23. A goldfish requires about 1,000 cm³ of tank space to survive. How many goldfish can this tank support?

TALK IT OVER

Which two plans could you use to solve Exercise 25?

A fish tank is 42.8 cm long, 19.5 cm wide, and 15.2 cm deep.

24. Find, to the nearest thousandth, the number of liters of water the tank will hold when full.

25. The tank is filled so that the water is 4.7 cm from the top. How much water is in the tank?

11-8 Area of Circles

EXPLORE/ WORKING TOGETHER

With a partner, draw several circles of various sizes on a sheet of grid paper. Count the number of square units within each circular region. Estimate the area of each circle in square units. Compare your estimates with those of other classmates.

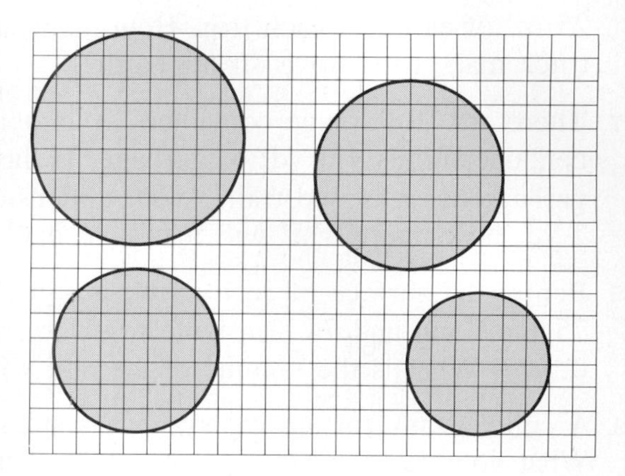

SKILLS DEVELOPMENT

A circle can be cut into equal sections. These sections can then be used to form a figure that is shaped very much like a parallelogram. The length of the base of this "parallelogram" will approximate one half the circumference of the circle. The height of the "parallelogram" will equal the radius of the circle.

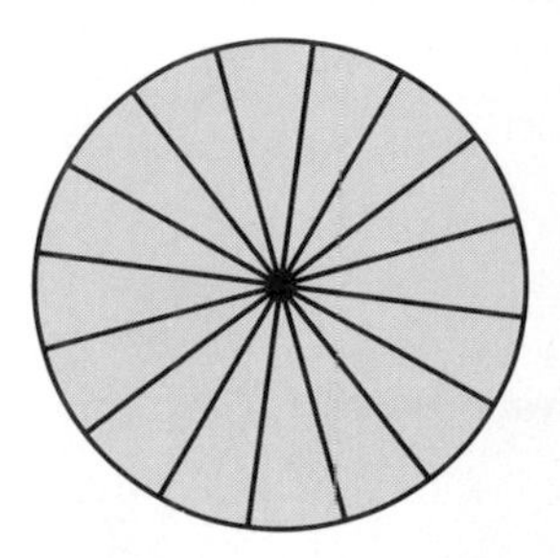

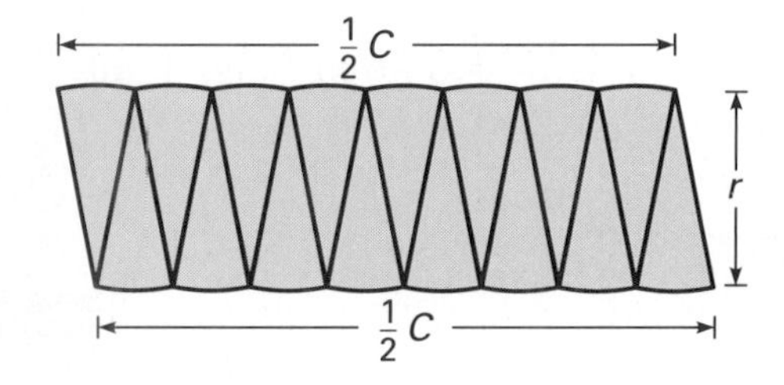

A formula for the area of a circle can then be derived from the formula for the area of a parallelogram.

$$A = b \times h$$
$$A = \tfrac{1}{2}C \times r$$
$$A = \tfrac{1}{2} \times (2\pi r) \times r$$
$$A = \pi r \times r$$
$$A = \pi r^2$$

You can use the formula $A = \pi r^2$ to solve problems involving the area of a circle.

Example 1

Find the area of circle O to the nearest square centimeter.

Use $\pi \approx 3.14$.

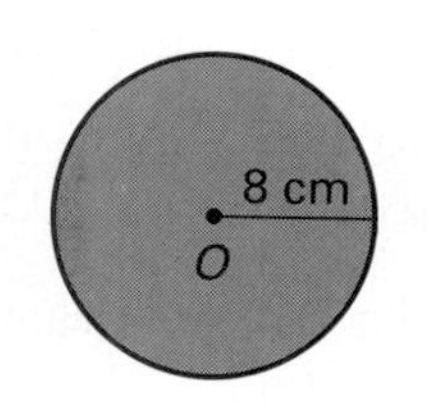

A circle is named by its center point.

Solution

$$A = \pi r^2$$
$$A \approx 3.14 \times 8^2 \leftarrow \text{Substitute 8 for } r.$$
$$A \approx 3.14 \times 64$$
$$A \approx 200.96$$
$$A \approx 201 \leftarrow \text{Round to the nearest whole number.}$$

The area of the circle is approximately 201 cm^2. ◄

COMPUTER TIP

You can use the computer to check your calculation of the area of a circle with radius R by typing
? 3.14 * R ^ 2 or
? 3.14 * R * R .

Example 2

Find the area of a circular garden that has radius 2.6 m. Use $\pi \approx 3.14$. Round your answer to the nearest tenth.

Solution

The garden is shaped like a circle. Use the area formula for circles.

$$A = \pi r^2$$
$$A = \pi \times (2.6)^2 \leftarrow \text{Substitute 2.6 for } r.$$
$$A \approx 3.14 \times 6.76$$
$$A \approx 21.2264$$
$$A \approx 21.2$$

The area of the garden is approximately 21.2 m^2. ◄

Use $\pi \approx 3.14$. Round answers to the nearest tenth.

Find the area of each circle.

1. 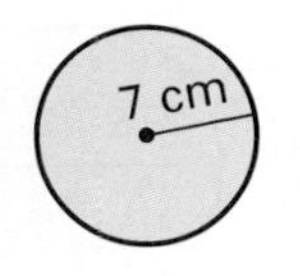

2. 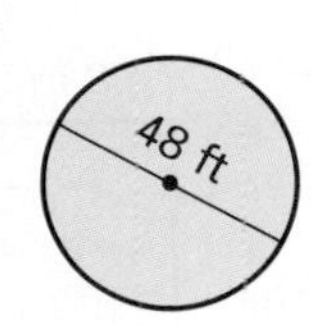

3. 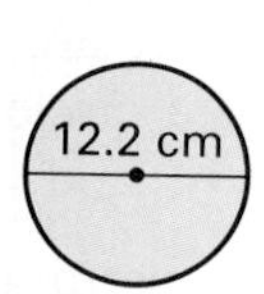

4. 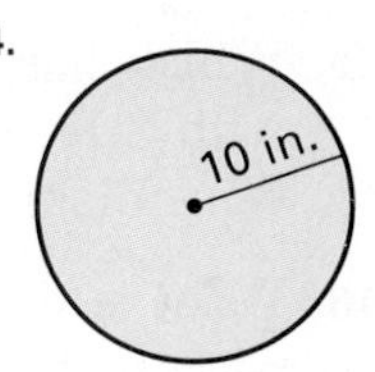

CHECK UNDERSTANDING

If you know the diameter of a circle, how would you find its radius?

5. The diameter of a circle is 12.6 cm. Find the area of the circle.

EXERCISES

PRACTICE/ SOLVE PROBLEMS

Use $\pi \approx 3.14$. Round answers to the nearest tenth.

Find the area of each circle.

1. 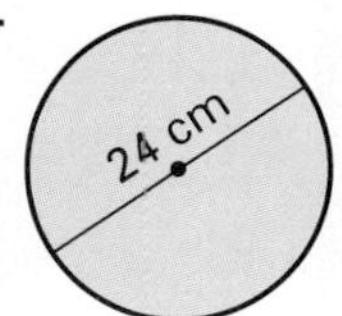

2. 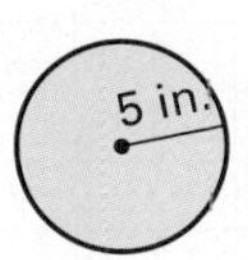

3. 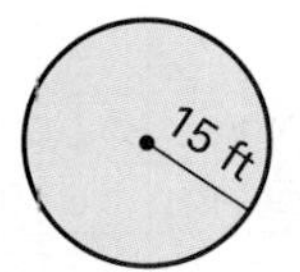

4. 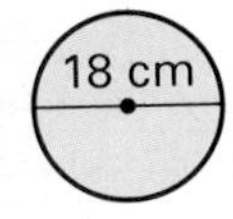

Find the area of a circle with the given radius.

5. 8 ft 6. 36 m 7. 9.8 in. 8. 10.6 m

Write the letter of the corresponding circle. The measures are in centimeters.

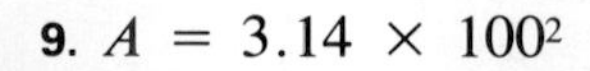

9. $A = 3.14 \times 100^2$

10. $A = 3.14 \times 50^2$

11. $A = 3.14 \times 100$

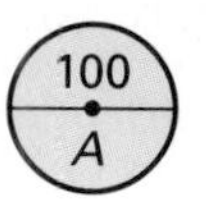

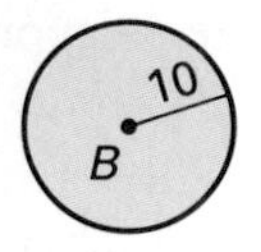

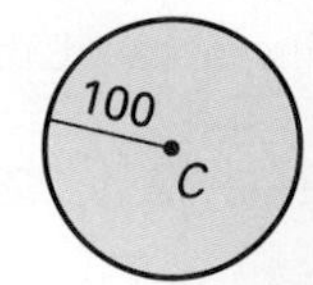

Solve.

12. By how much does the area of circle O exceed the area of circle P?

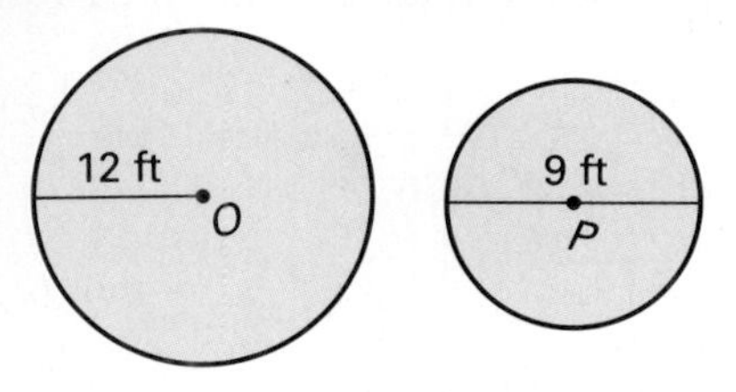

13. Throwing the discus is one of the Olympic field events. The circular discus has a radius of 11 cm. Find its area.

14. The discus is thrown by an athlete standing within a circle with a diameter of about 2.8 m. Find the area of the circle.

EXTEND/ SOLVE PROBLEMS

15. Which has the greater area: a square with 8-in. sides or a circle with a 9-in. diameter?

16. What is the difference in area, to the nearest thousandth, of the figures used for Exercise 15?

17. The radius of a circle is 1.3 yd. Find the area of one quadrant (one-fourth) of the circle.

Use this table for Exercises 18–28.

PIZZA MENU

	Small 10 in.	Medium 14 in.	Large 16 in.	Extra Large 18 in.
BASIC TOMATO SAUCE AND CHEESE	$ 5.00	$ 7.00	$ 8.75	$ 9.25
One choice of topping	5.50	7.75	9.25	10.00
Two choices of toppings	6.00	8.50	9.85	10.45
Three choices of toppings	6.50	8.75	10.00	10.75
Four choices of toppings	7.25	9.25	10.50	12.00

18. What is the cost of 3 small pizzas with 3 choices of toppings?

Find the area of the pizza of each size.

19. small **20.** medium **21.** large

Find the cost per square inch of each of the following pizzas. Express your answers in cents per square inch.

22. small with 2 choices

23. medium with 2 choices

24. large with 2 choices

25. Which of the pizzas described in Exercises 22–24 costs least per square inch?

26. How much greater in size is a large pizza than a small pizza?

27. Find the area of 4 small pizzas.

28. Which is the better buy, 4 small pizzas with 2 choices or 2 large pizzas with 2 choices?

THINK CRITICALLY/ SOLVE PROBLEMS

Find the area of each shaded region to the nearest tenth.

29.

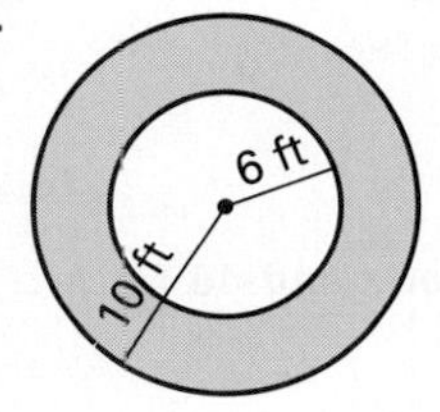

30.

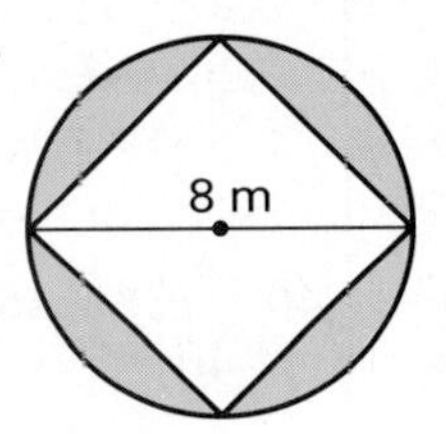

31.

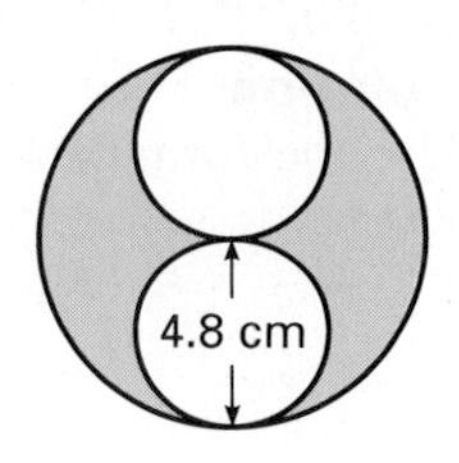

Find the area of each figure to the nearest tenth.

32.

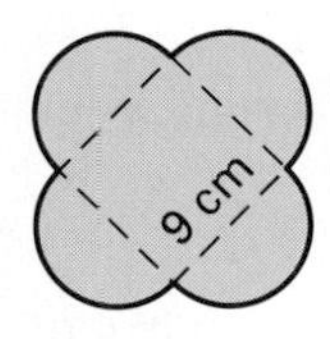

33.

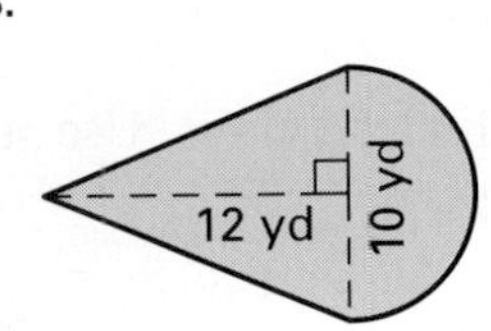

34.

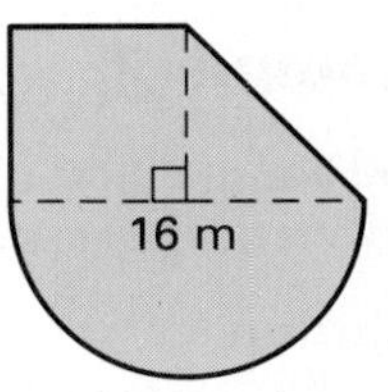

35. Find the area of the window at the right.

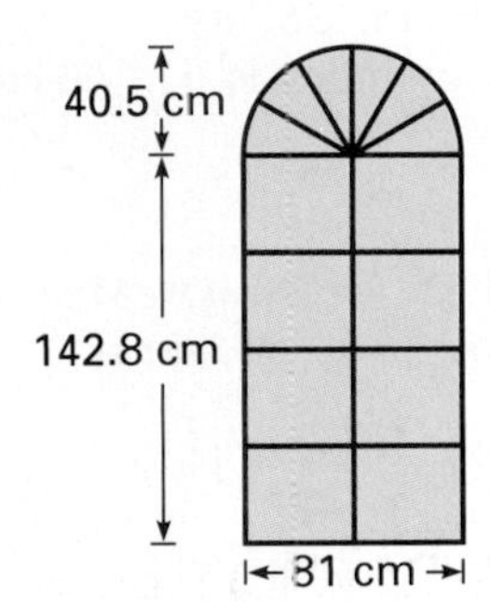

11-9 Volume of Cylinders

EXPLORE

The three-dimensional figure in the diagram below is a cylinder. How is a cylinder different from a prism? How is it like a prism?

SKILLS DEVELOPMENT

Recall the general formula for finding the volume of a prism is $V = B \times h$, where B is the area of the base. Since the base of a cylinder is a circle, replace B in this formula with πr^2.

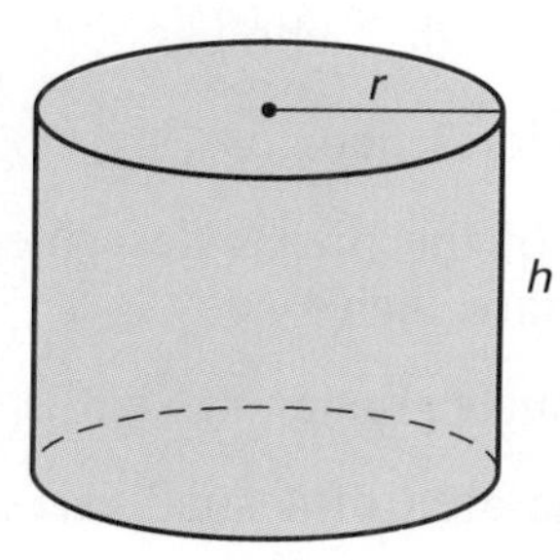

$$V = B \times h$$
$$V = \pi r^2 \times h, \text{ or } V = \pi r^2 h$$

COMPUTER

This program will find the volume of a cylinder when you INPUT the radius and the height. Remember to label your answer in cubic units.

```
10 INPUT "ENTER RADIUS
   AND HEIGHT: ";R,H:
   PRINT
20 LET V = 3.14 * R ^ 2 * H
30 LET A = INT (V * 10 ^ 2
   + 0.5) / 10 ^ 2
40 PRINT "THE VOLUME IS
   ABOUT ";A
```

Example 1

Find the volume of a cylinder with a radius of 4 in. and a height of 40 in. Use $\pi \approx 3.14$.

Solution

Use the formula $V = \pi r^2 \times h$.

$$V = \pi r^2 \times h$$
$$V = \pi \times 4^2 \times 40 \quad \leftarrow \textbf{Substitute 4 for } r \textbf{ and 40 for } h.$$
$$V \approx 3.14 \times 16 \times 40$$
$$V \approx 2{,}009.6$$

The volume of the cylinder is approximately 2,009.6 in.^{3} ◀

Example 2

Find the volume of the cylinder below. Use $\pi \approx 3.14$. Round your answer to the nearest tenth.

Solution

Use the formula $V = \pi r^2 \times h$.

16 cm

36 cm

The diameter of the cylinder is 16 cm. So, its radius is 8 cm. The height of the cylinder is 36 cm.

$$V = \pi r^2 \times h$$
$$V = \pi \times 8^2 \times 36 \quad \leftarrow \textbf{Substitute 8 for } r \textbf{ and 36 for } h.$$
$$V \approx 3.14 \times 64 \times 36$$
$$V \approx 7{,}234.56$$
$$V \approx 7{,}234.6$$

The volume of the cylinder is about 7,234.6 cm^3. ◀

Example 3

Find the volume of the can. Use $\pi \approx 3.14$.

Solution

The can has a radius of 5 cm. Use the volume formula for a cylinder.

$V = \pi r^2 \times h$
$V = \pi \times 5^2 \times 14.5$ ← **Substitute 5 for *r* and 14.5 for *h*.**
$V \approx 3.14 \times 25 \times 14.5$
$V \approx 1{,}138.25$
$V \approx 1{,}138.3$

The volume of the can is about 1,138.3 cm^3. ◄

TRY THESE

Use $\pi \approx 3.14$. Round answers to the nearest tenth.

1. Find the volume of a cylinder with a radius of 3 in. and a height of 28 in.

Use the dimensions given to find the volume of each cylinder.

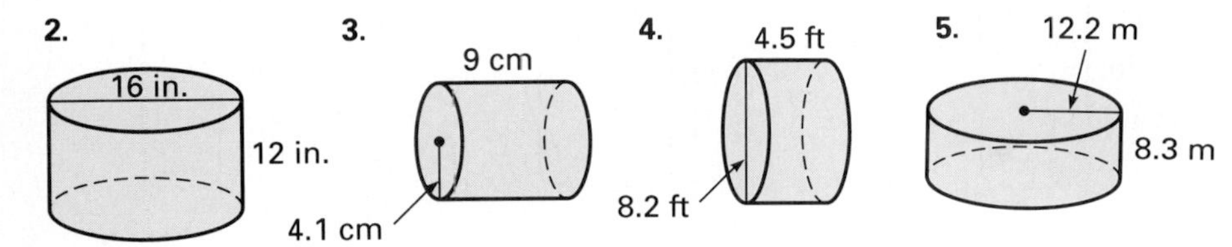

Solve.

6. Find the volume of a cylindrical water drum with a height of 4 ft and a radius of 2.5 ft.

EXERCISES

PRACTICE/ SOLVE PROBLEMS

Use $\pi \approx 3.14$. Round answers to the nearest whole number.

Find the volume of each cylinder.

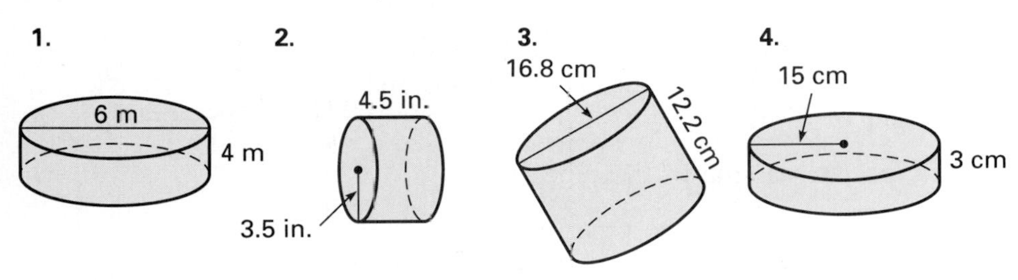

Copy and complete the table.

	diameter of base (d)	radius of base (r)	height (h)	Volume (V)
5.	6 ft	■	8 ft	■
6.	18 cm	■	23 cm	■
7.	■	6.9 in.	4.2 in.	■
8.	■	2.5 m	4.8 m	■

9. The diameter of a cylinder is 24 in. The height is 16 in. Find its volume.

10. An oil drum is 94 cm high. The radius of the base is 35 cm. What is the volume?

EXTEND/ SOLVE PROBLEMS

Use $\pi \approx 3.14$. Round answers to the nearest whole number.

Use these diagrams for Exercises 11 and 12.

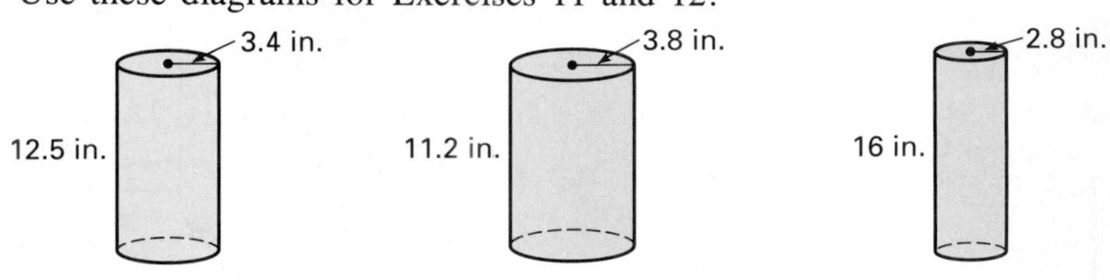

11. Estimate which can holds the most.

12. Find the volume of each can. Was your estimate correct?

13. A car engine has 8 cylinders, each with a diameter of 7.2 cm and a height of 8.4 cm. The total volume of the cylinders is called the *capacity* of the engine. Find the engine capacity.

THINK CRITICALLY/ SOLVE PROBLEMS

Use $\pi \approx 3.14$. Round answers to the nearest whole number.

14. Suppose that the mass of wheat is 120 kg/m^3. A silo with a radius of 3.4 m and a height of 7.3 m is filled with wheat. Find the total mass of the wheat in the silo.

A section of water pipe has an inner radius of 14.6 cm, an outer radius of 18.2 cm, and a height of 6 m.

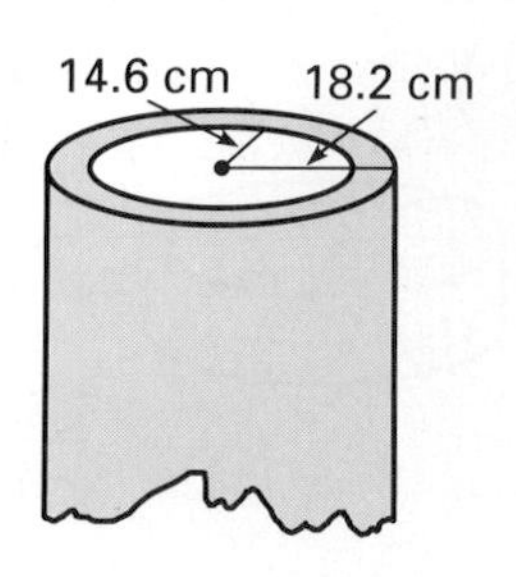

15. Find the volume of the pipe.

16. Find the mass of the pipe (without the water) if the piping has a mass of 10 g/cm^3.

READ
PLAN
SOLVE
ANSWER
CHECK

Problem Solving Applications:

THUNDERSTORMS AND TORNADOS

Thunderstorms and tornados are two of nature's most destructive phenomena. After a tornado, the average total property loss is about $24,300,000.

On average, a tornado lasts for less than 30 s. A thunderstorm lasts an average of 2 h.

The atmospheric disturbance created by a thunderstorm is almost cylindrical, as shown at the right.

Use $\pi \approx 3.14$. Round your answer to the nearest tenth.

1. Find the volume of the atmospheric disturbance shown above.
2. The smallest thunderstorm on record measured 2.7 km in diameter and its cloud formation had a height of 8.4 km. Find the total volume of the disturbance in the atmosphere.
3. The largest thunderstorm on record measured 43.9 km in diameter. It had a height of 20.3 km. Find the total volume this storm occupied in the atmosphere.
4. A tornado is a narrow funnel-shaped cloud that extends downwards from cumulonimbus clouds. Find the zone of destruction of a tornado that destroys a tract of land 8 mi wide and 6.9 mi long.

Specially equipped aircraft and ground crews record data about thunderstorms. These measurements were taken of two thunderstorms.

Storm A	
diameter:	9.6 km
height:	5.9 km

Storm B	
diameter:	8.5 km
height:	6.3 km

5. Which storm, A or B, occupied the greater space?
6. A tornado in Massachusetts covered a path 0.256 km wide and 27.5 km long. Find the zone of destruction.
7. A small tornado swept across a tract of land 0.04 km wide and 7.8 km long. Find the land area that was destroyed.

11-10 Volume of Pyramids, Cones, and Spheres

EXPLORE

The containers shown are a pyramid and a prism that have identical bases. The containers are equal in height.

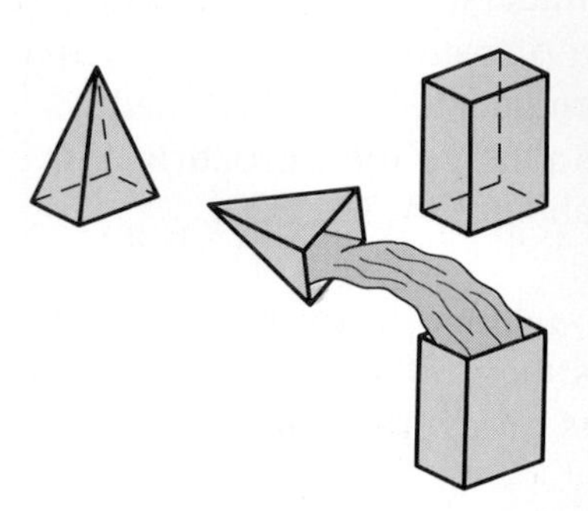

Suppose that the pyramid were filled with water, and the water was then poured into the prism. What part of the prism would be filled?

SKILLS DEVELOPMENT

The formula for the volume of a pyramid is related to the formula for the volume of a prism.

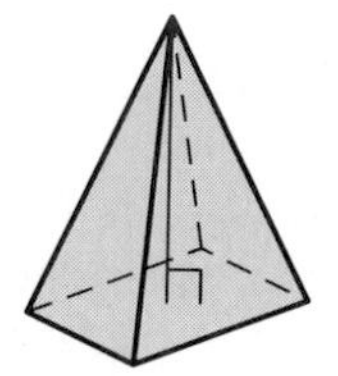

prism: $V = B \times h$

pyramid: $V = \frac{1}{3} \times B \times h$

That is, the volume of a pyramid with a given base is one third the volume of a prism with a base of the same size.

Example 1

Find the volume of a rectangular pyramid with a base of length 6 in. and width 4 in. and with height 8 in.

Solution

Use the volume formula for a pyramid.

$$V = \frac{1}{3} \times B \times h$$

$$V = \frac{1}{3} \times 6 \times 4 \times 8$$ ← Substitute 6×4 for B and 8 for h.

$$V = 64$$

The volume of the pyramid is 64 in.^{3} ◄

In a similar way, the formula for the volume of a cone is related to the formula for the volume of a cylinder.

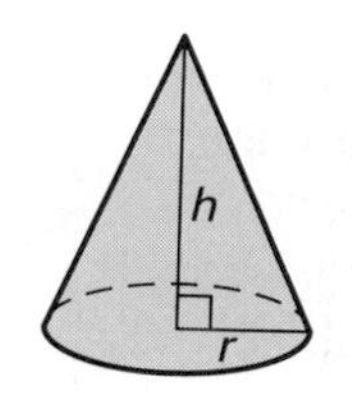

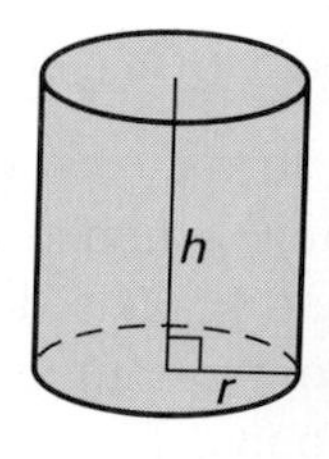

cylinder: $V = \pi r^2 \times h$

cone: $V = \frac{1}{3} \times \pi r^2 \times h$

That is, the volume of a cone with a given radius and height is one third the volume of a cylinder with the same radius and height.

Example 2

Find the volume of the cone at the right.
Use $\pi \approx 3.14$.

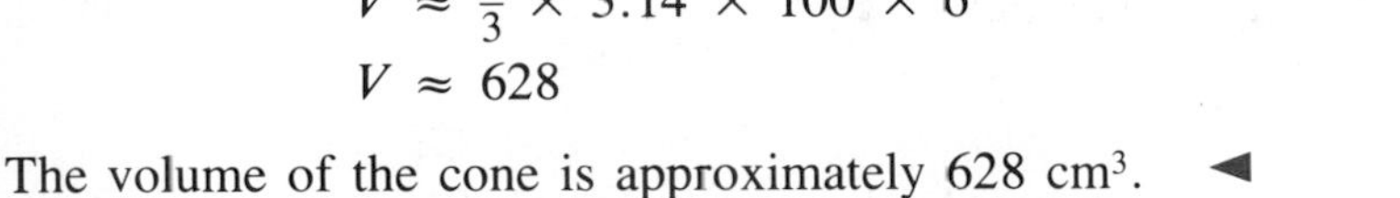

Solution

$$V = \frac{1}{3} \times \pi r^2 \times h$$

$$V = \frac{1}{3} \times \pi(10)^2 \times 6 \quad \leftarrow \text{Substitute 10 for } r \text{ and 6 for } h.$$

$$V \approx \frac{1}{3} \times 3.14 \times 100 \times 6$$

$$V \approx 628$$

The volume of the cone is approximately 628 cm^3. ◄

The formula for the volume of a sphere is as follows.

$$\text{sphere:} \quad V = \frac{4}{3} \times \pi r^3$$

Example 3

Find the volume of a sphere with radius 6 in. Use $\pi \approx 3.14$. Round your answer to the nearest whole number.

Solution

$$V = \frac{4}{3} \times \pi r^3$$

$$V = \frac{4}{3} \times \pi \times 6^3 \quad \leftarrow \text{Substitute 6 for } r.$$

$$V \approx \frac{4}{3} \times 3.14 \times 216$$

$$V \approx 904.32$$

$$V \approx 904$$

The volume of the sphere is approximately 904 in.^3 ◄

Example 4

To the nearest whole number, find the volume of a cone-shaped storage bin with a height of 27 m and a radius of 15 m.

Solution

The storage bin is a cone with radius 15 m and height 27 m. Use the volume formula for a cone.

$$V = \frac{1}{3} \times \pi r^2 \times h$$

$$V = \frac{1}{3} \times \pi(15)^2 \times 27 \quad \leftarrow \text{Substitute 15 for } r \text{ and 27 for } h.$$

$$V \approx \frac{1}{3} \times 3.14 \times 225 \times 27$$

$$V \approx 6{,}358.5$$

$$V \approx 6.359 \quad ◄$$

The volume of the bin is approximately 6.359 m^3.

TRY THESE

Use $\pi \approx 3.14$. Round your answer to the nearest tenth.

Find the volume of each figure.

1.
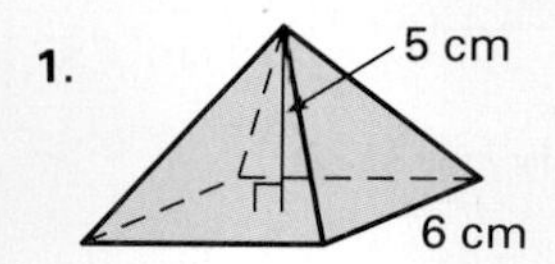

2.
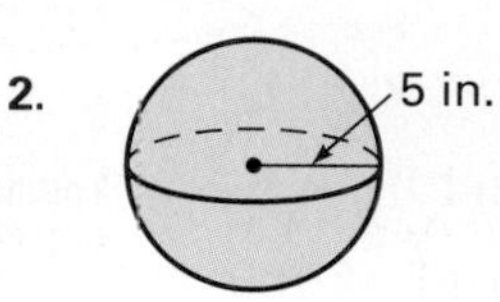

3.
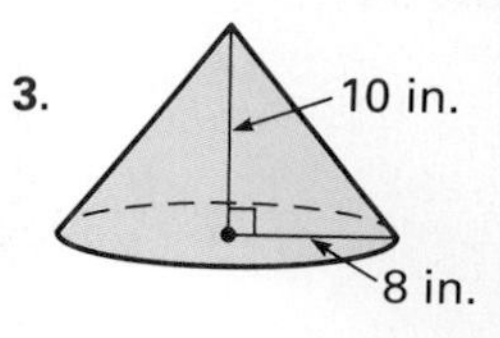

Solve.

4. A pyramid has a rectangular base with length 8 ft and width 3 ft. Its height is 9 ft. Find the volume of the pyramid.

EXERCISES

PRACTICE/ SOLVE PROBLEMS

Use $\pi \approx 3.14$. Round your answer to the nearest tenth.

Find the volume of each figure.

1.
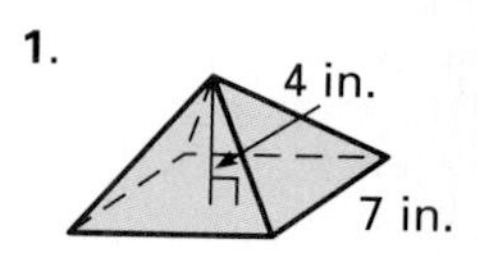

2.
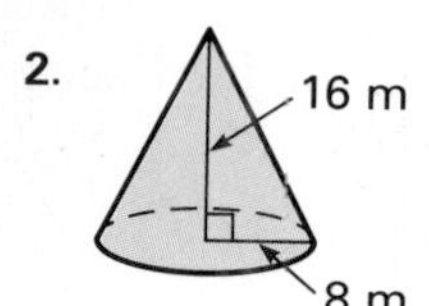

3.
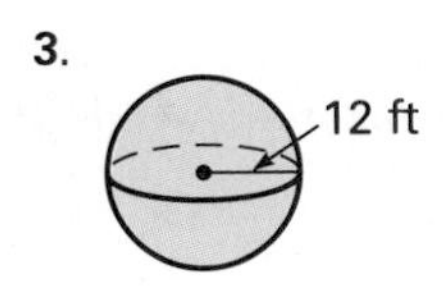

4.
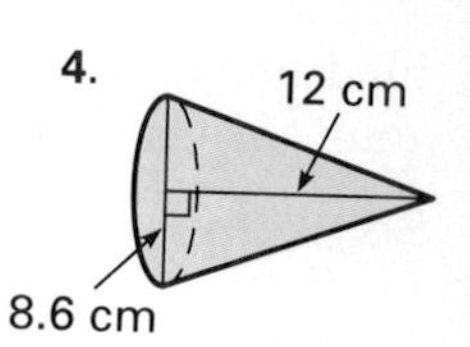

5. Find the volume of a cone 3.8 ft in diameter and 5.1 ft high.

6. Find the volume of a pyramid that is 10.2 m high and has a rectangular base measuring 7 m by 3 m.

Find the volume of a sphere with the given radius.

7. 8 in.
8. 25 cm
9. 6.8 ft
10. 10.4 cm

Find the volume of a sphere with the given diameter.

11. 18 m
12. 9.2 ft
13. 13.6 mm
14. 24.8 cm

15. A baseball has a radius of 3.6 cm. What is its volume?

16. Find the volume of a cone-shaped funnel with radius 15 cm and height 22 cm.

17. Phosphate is stored in a conical pile. Find the volume of the pile if the height is 7.8 m and the radius is 16.2 m.

18. A tent is in the form of a pyramid with a base 2 yd wide and 3 yd long, and it has a height of 2.4 yd. Find the volume of the tent.

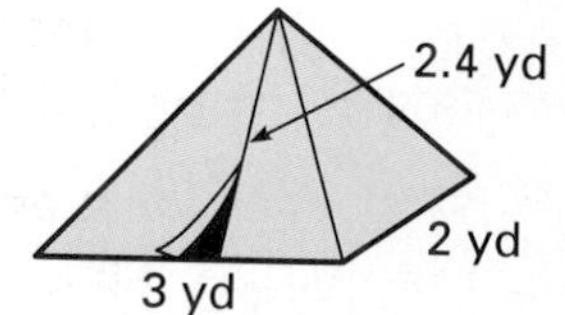

USING DATA Use the Data Index on page 546 to locate information on ball sizes for international competition sports. Find the volume of the ball specified in each of Exercises 19 and 20. Round your answer to the nearest tenth.

19. soccer ball

20. tennis ball

EXTEND/ SOLVE PROBLEMS

Use $\pi \approx 3.14$. Unless otherwise stated, round to the nearest tenth.

21. A half sphere is called a *hemisphere.* A hemispherical tank and a conical tank have the dimensions shown. How much greater is the volume of the hemisphere?

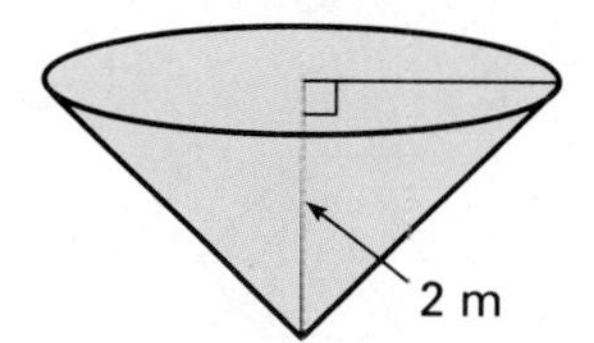

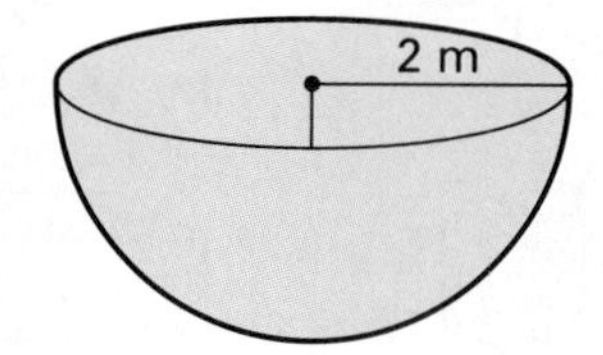

Find the volume of each figure.

22.

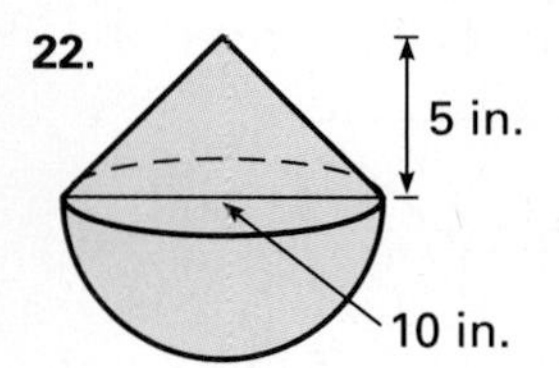

23.

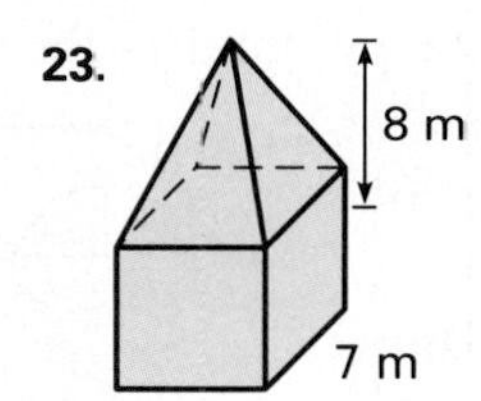

24.

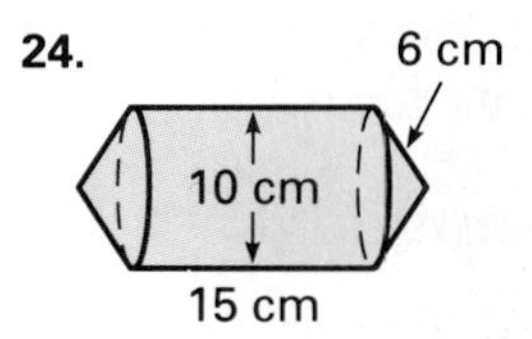

25. Find the volume of the storage bin.

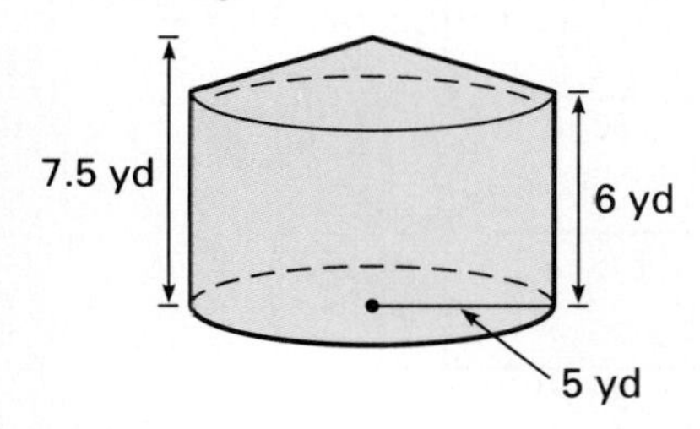

26. Find the volume of the barn to the nearest whole number.

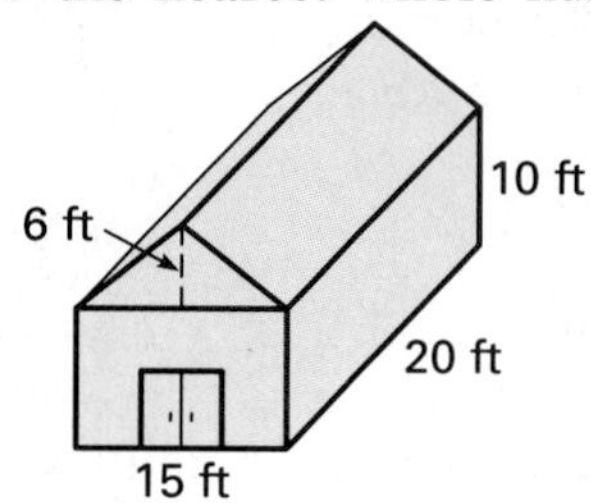

27. The thickness of a grapefruit peel is 0.5 cm. If the diameter of an unpeeled grapefruit is 12.8 cm, what part of the grapefruit is peel? Express your answer as a percent.

THINK CRITICALLY/ SOLVE PROBLEMS

28. How would doubling the height of a cone affect the volume, if the radius of the base is unchanged?

29. How would doubling the radius of the base of a cone affect the volume if the height is unchanged?

30. How would the volume of a cone be affected if both the height and the radius were doubled?

11-11 Surface Area

EXPLORE

Each of these two-dimensional patterns can be folded to form a polyhedron. Match each pattern to the polyhedron it would form.

1.

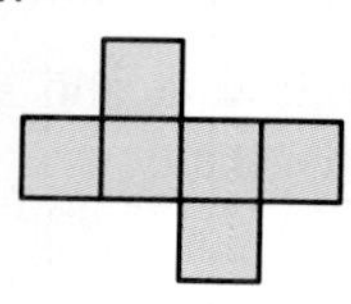

2.

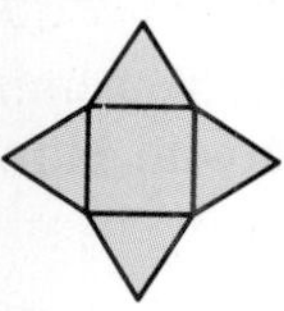

3.

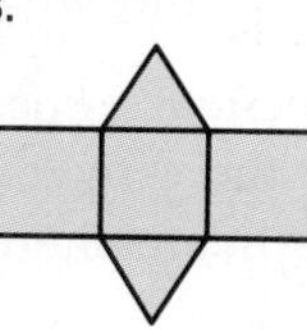

4.

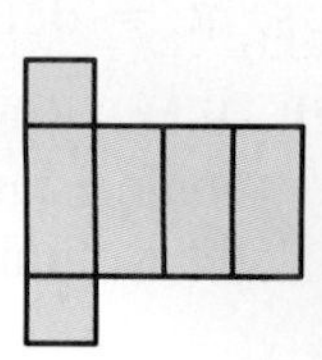

A.

B.

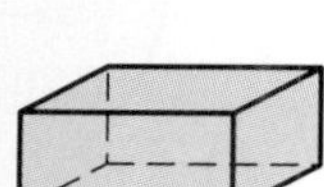

C.

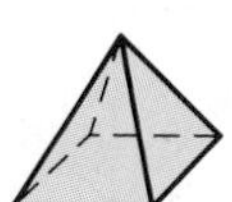

D.

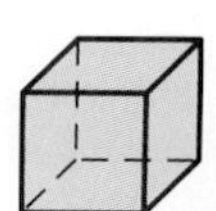

SKILLS DEVELOPMENT

The **surface area** of a prism or pyramid is the sum of the areas of all its faces.

Example 1

Find the surface area of the prism shown at the right.

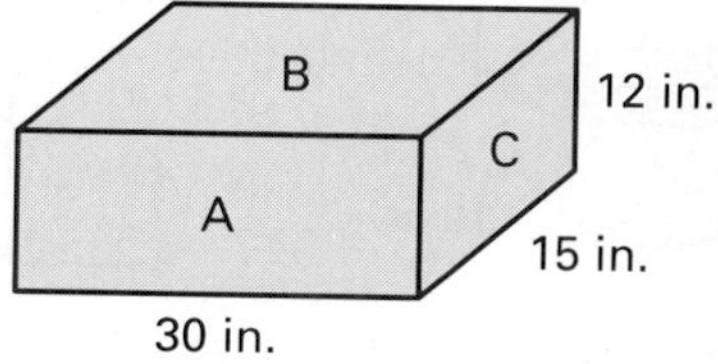

CHECK UNDERSTANDING

In example 1, why do you multiply Area A, Area B, and Area C by 2 to find the total surface area?

Solution

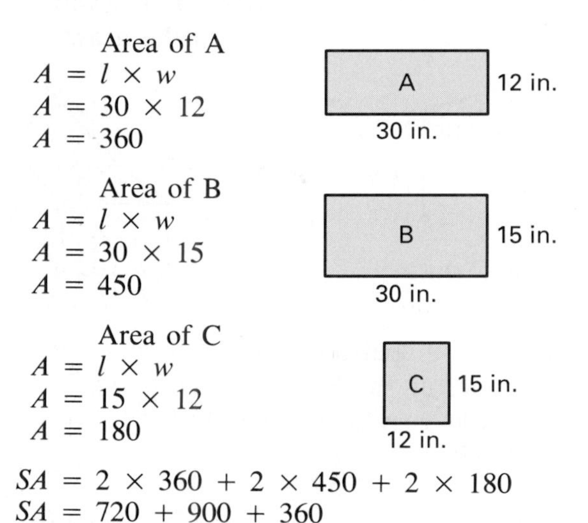

Area of A
$A = l \times w$
$A = 30 \times 12$
$A = 360$

Area of B
$A = l \times w$
$A = 30 \times 15$
$A = 450$

Area of C
$A = l \times w$
$A = 15 \times 12$
$A = 180$

$SA = 2 \times 360 + 2 \times 450 + 2 \times 180$
$SA = 720 + 900 + 360$
$SA = 1{,}980$

The surface area is 1,980 in.^{2} ◄

To find the surface area of a cylinder, add the area of the curved surface to the sum of the areas of the two bases.

$$SA = 2\pi rh + 2\pi r^2$$

Example 2

Find the total surface area of the cylinder. Round your answer to the nearest whole number.

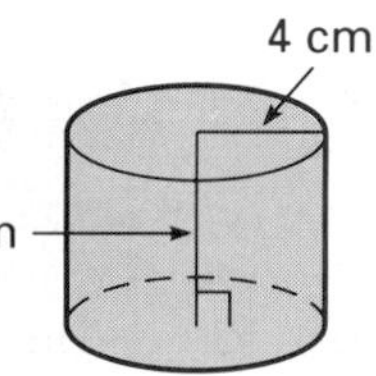

Solution

First, find the area of the curved surface of the cylinder. When "unrolled," this surface is a rectangle with length equal to the circumference of the circle. The width of the rectangle is equal to the height of the cylinder.

$$A = 2\pi r \times h$$
$$A = 2\pi \times 4 \times 9$$
$$A \approx 2 \times 3.14 \times 4 \times 9$$
$$A \approx 226.08$$

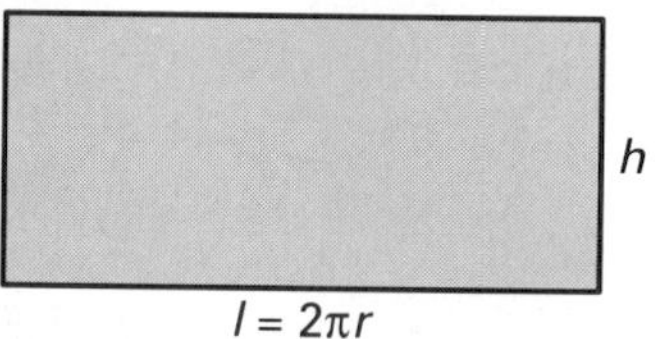

Next, find the area of a base and multiply it by 2.

$$A = \pi r^2 \qquad 2 \times A \approx 2 \times 50.24$$
$$A = \pi(4)^2 \qquad 2 \times A \approx 100.48$$
$$A \approx 3.14 \times 16$$
$$A \approx 50.24$$

Add these areas to find the total surface area of the cylinder.

$$SA \approx 226.08 + 100.48$$
$$SA \approx 326.56$$
$$SA \approx 327$$

The total surface area of the cylinder is approximately 327 cm^2. ◄

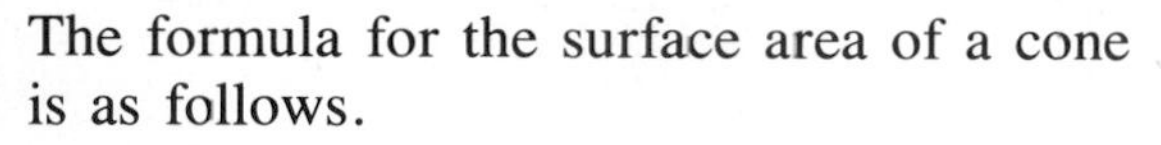

The formula for the surface area of a cone is as follows.

$$SA = \pi rs + \pi r^2$$

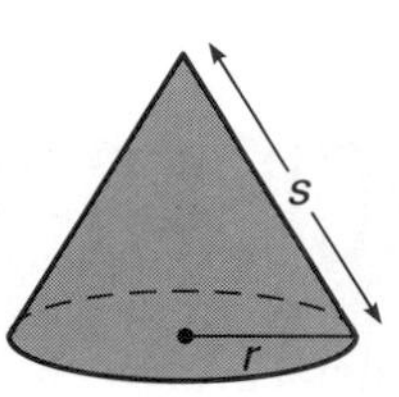

In the formula s represents the *slant height* of the cone.

Example 3

Find the total surface area of the cone at the right. Round your answer to the nearest whole number.

Solution

$$SA = \pi rs + \pi r^2$$
$$SA = \pi \times 4 \times 10 + \pi \times (4)^2$$
$$SA \approx 3.14 \times 4 \times 10 + 3.14 \times 16$$
$$SA \approx 125.6 + 50.24$$
$$SA \approx 175.84$$
$$SA \approx 176$$

The surface area of the cone is approximately 176 cm^2. ◄

Tell what kind of graph would best represent the data.

1. average monthly temperature change in one year
2. world's deepest bodies of water
3. team's budget for the year
4. Find the median.
   39, 52, 25, 16

Find the perimeter.

5. 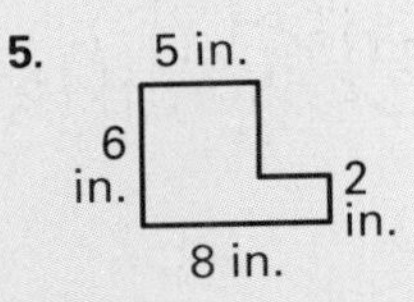

Simplify.

6. $2 + 9 \cdot 6 - 24 \div 3 + 5$
7. 2^3
8. $9^2 - 4^3$

The formula for the surface area of a sphere is as follows.

$$SA = 4\pi r^2$$

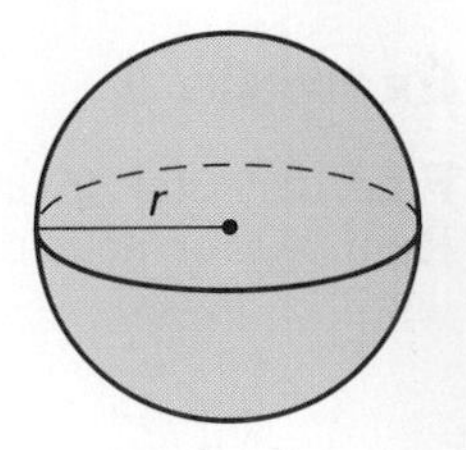

In the formula, r represents the radius of the sphere.

Example 4

Find the surface area of the sphere at the right. Round your answer to the nearest whole number.

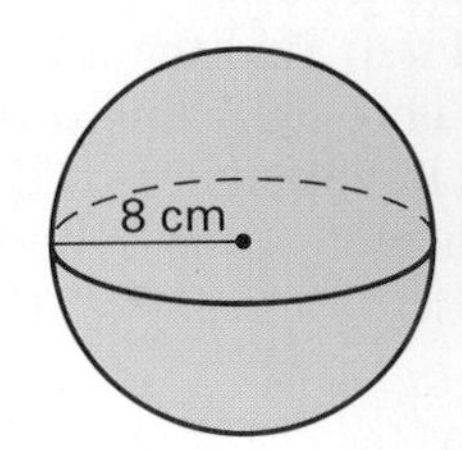

Solution

$$SA = 4\pi r^2$$
$$SA = 4 \times \pi \times (8)^2$$
$$SA \approx 4 \times 3.14 \times 64$$
$$SA \approx 803.84$$
$$SA \approx 804$$

The surface area of the sphere is approximately 804 cm^2. ◄

TRY THESE

Find the surface area of each figure. Use $\pi \approx 3.14$. Round your answer to the nearest whole number.

1. 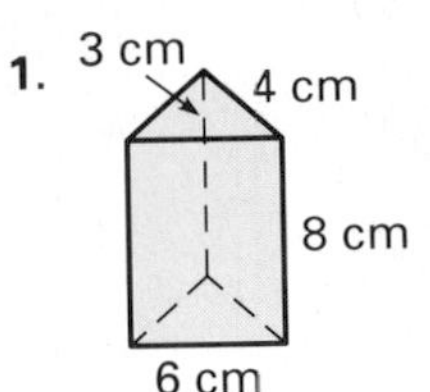

2. 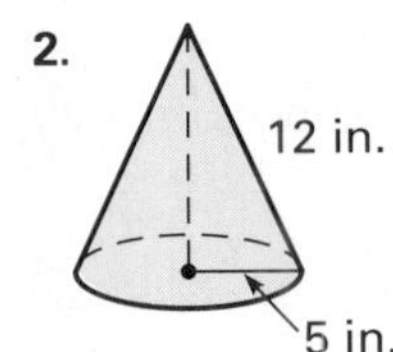

3. 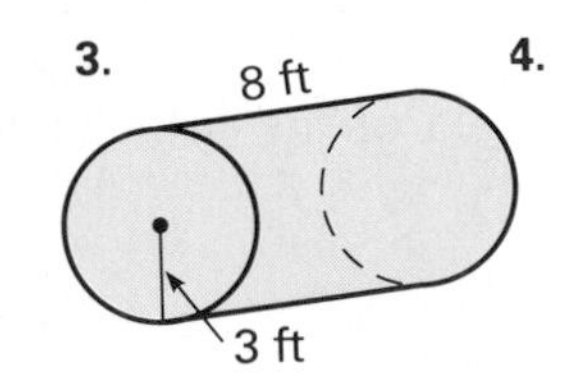

4. 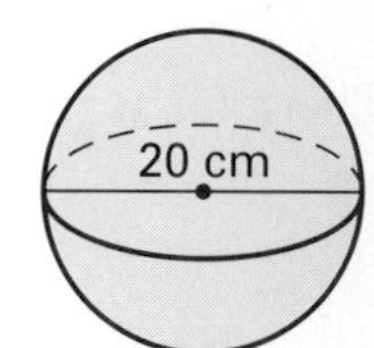

EXERCISES

PRACTICE/ SOLVE PROBLEMS

Find the surface area of each figure. Use $\pi \approx 3.14$. Round your answer to the nearest whole number.

1. 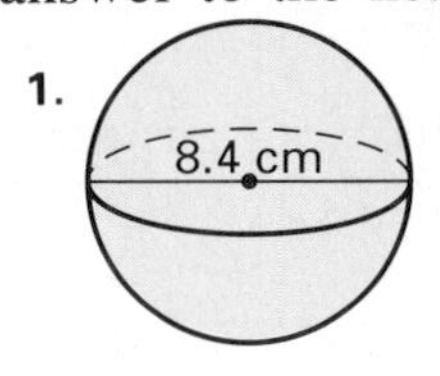

2. 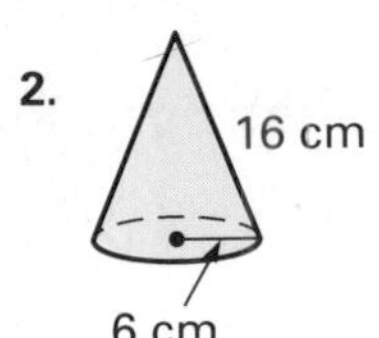

3. 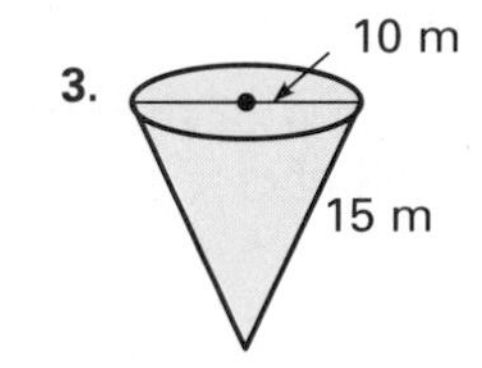

4. 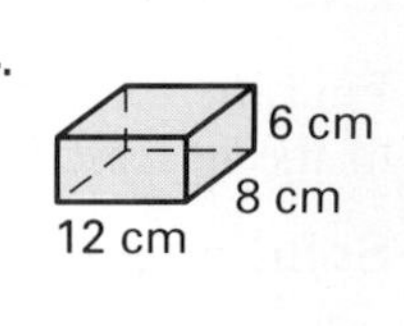

5. 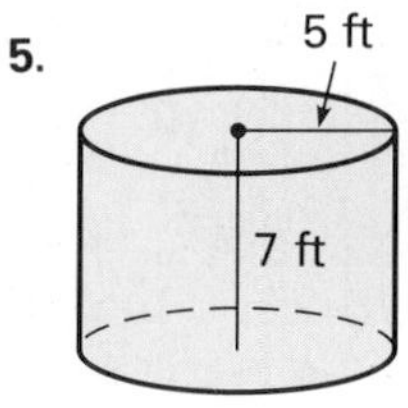

6. 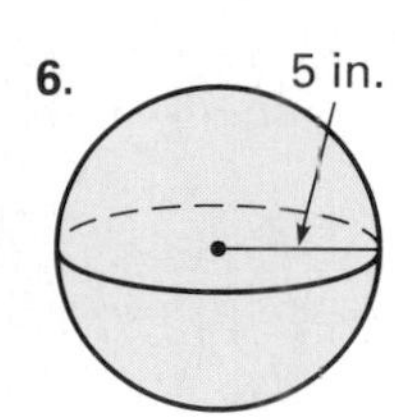

7. 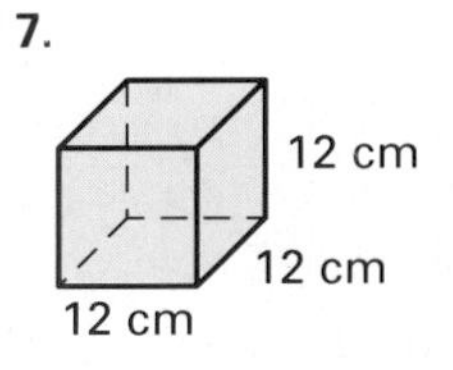

8. 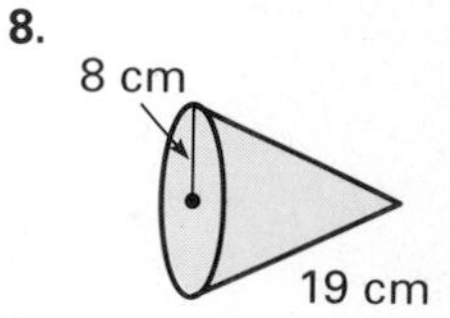

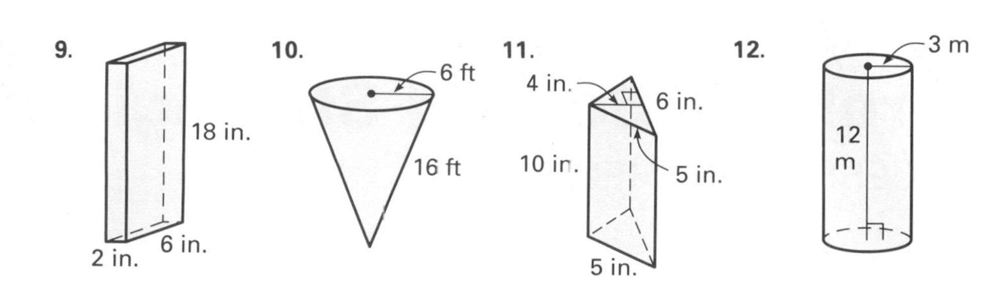

Find the surface area of a sphere with the given diameter. Round your answer to the nearest tenth. Use $\pi \approx 3.14$.

13. 16 cm **14.** 58 cm **15.** 9.6 cm

Solve.

16. The dimensions of a gift box are 8 in. by 12 in. by 3 in. How much gift wrap is needed to cover the box?

17. A major league baseball has a radius of 3.8 cm. What is the surface area of the ball? Round your answer to the nearest hundredth.

EXTEND/ SOLVE PROBLEMS

Round your answer to the nearest hundredth. Use $\pi \approx 3.14$.

Use the diagrams below for Exercises 18–20.

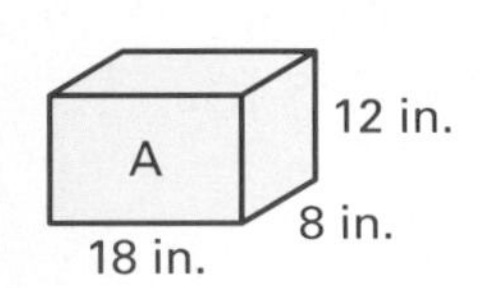

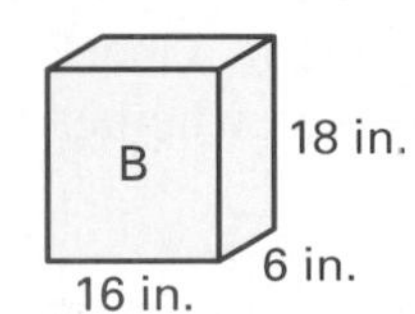

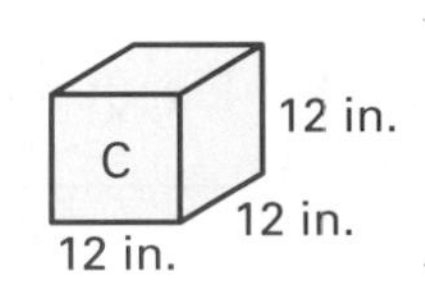

18. Find the volume of each box. What do you notice?

19. Find the surface area of each box.

20. Which box uses the least amount of material?

21. The radius of a sphere is 15 cm. By how much does the surface area of the sphere increase if the radius increases by 2 cm?

THINK CRITICALLY/ SOLVE PROBLEMS

22. The base of a rectangular prism is l units long and w units wide. Its height is h units. Write a formula that can be used to find the surface area of the prism.

11-12 Problem Solving Strategies:

GUESS AND CHECK

► READ
► PLAN
► SOLVE
► ANSWER
► CHECK

To solve some area problems, you may not be able to apply a geometric formula directly. For example, you may know the area of a figure and may need to find the dimensions. An effective approach to such a problem is the use of the strategy, *guess and check.* Make guesses for the length and width of the figure, each time checking to see if the product of the pair of numbers you guessed equals the known area.

PROBLEM

The area of a rectangular rug is 40.5 ft^2. The length is twice the width. What are the dimensions of the rug?

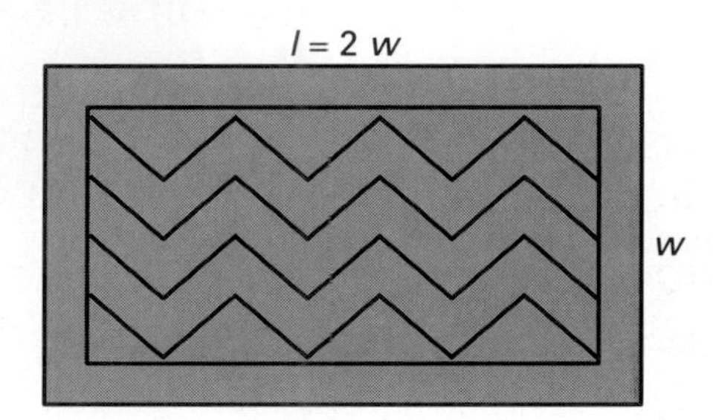

SOLUTION

Use a guess-and-check approach. After you make each guess for the width, find the corresponding length. Check each pair of your guesses for length and width to see whether the area is 40.5 ft^2.

	width	length	area	
	w	$l = 2w$	$A = l \times w$	
guess →	10 ft	20 ft	200 ft^2	(too large)
guess →	5 ft	10 ft	50 ft^2	(too large)
	4 ft	8 ft	32 ft^2	(too small)

Notice that 40.5 is between 50 and 32.

The width must be between 4 ft and 5 ft. Since the last digit of 40.5 is 5, try 4.5 ft for the width.

$w = 4.5$
$l = 2 \times 4.5 = 9$
$A = 9 \times 4.5 = 40.5$

The rug is 4.5 ft wide and 9 ft long.

PROBLEMS

1. The area of a triangular tile is 60 in.^{2} The height is 2 in. more than the base. Copy and complete the table to find the base and height of the tile.

base	height	area
b	$h = b + 2$	$\frac{1}{2} \times b \times h$
6	■	■
12	■	■

The number 60 ends in zero, so try ___?___.

2. The volume of a rectangular box is 500 cm^3. The length is twice the width. The height is 5 cm more than the width. Find the dimensions of the box.

3. The volume of a pyramid is 98 cm^3. The area of the base is 49 cm^2. Find the height.

4. The area of a rectangular garden is 105 m^2. The width is 8 m less than the length. Find the dimensions of the garden.

5. The area of a triangular park is 2,400 yd^2. The height of the triangle is three times the base. Find the dimensions of the park.

6. A park in the shape of a parallelogram has a perimeter of 160 m. The longer sides of the park are 10 m longer than the shorter sides. How long are the sides of the park?

7. The volume of a cube is 1,728 cm^3. Find the length of an edge of the cube.

8. The distance around the track at the right is 714 yd. The length of the rectangle is twice the diameter of each semicircle. Find the dimensions of the rectangle.

CHAPTER 11 ● REVIEW

Match the expression at the left with the name of the figure in the box. Tell whether the expression is used in the formula for *area, volume,* or *surface area.* Some words are used more than once.

1. $B \times h$
2. $\frac{4}{3} \times \pi r^3$
3. $\frac{1}{2} \times b \times h$
4. $\frac{1}{3} \times B \times h$
5. $\frac{1}{3} \times \pi r^2 \times h$
6. $4\,\pi r^2$

a. triangle	d. sphere
b. rectangular prism	e. pyramid
c. cone	f. cylinder
	g. circle

SECTION 11–1 AREA AND VOLUME (pages 384–387)

► **Area** is the amount of surface enclosed by a plane figure. Area is measured in *square units.*

► **Volume** is the amount of space enclosed by a three-dimensional figure. Volume is measured in *cubic units.*

Find the area.

7. 

8. 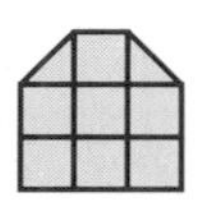

Find the volume.

9. 

10. 

SECTIONS 11–2, 11–3 AREA OF POLYGONS (pages 388–395)

► You can use these formulas to find area:
rectangle: $A = l \times w$ square: $A = s^2$
parallelogram: $A = b \times h$ triangle: $A = \frac{1}{2} \times b \times h$

Find the area of each figure.

11. 

12. 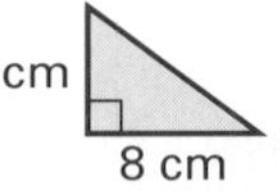

13. 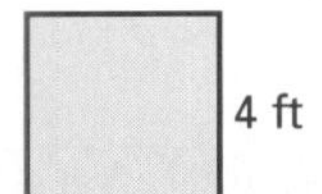

14. 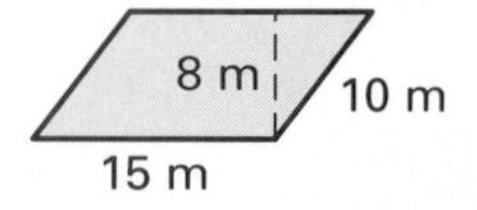

SECTIONS 11–5, 11–6 THE PYTHAGOREAN THEOREM (pages 398–405)

► Since $9^2 = 81$, $\sqrt{81} = 9$. Since $-9^2 = 81$, $-\sqrt{81} = -9$.

► For all right triangles, $c^2 = a^2 + b^2$, where c is the length of the hypotenuse and a and b are the lengths of the legs.

15. Write the positive square roots of 121 and 78 rounded to the nearest tenth.

16. Find the measure of the hypotenuse of right triangle ABC with legs that measure 9 in. and 12 in.

SECTION 11–8 AREA OF CIRCLES (pages 410–413)

► Find the area of a circle with radius r by using the formula $A = \pi r^2$

17. Find the area of a circle with a radius of 7 in. Round to the nearest whole number.

SECTIONS 11–7, 11–9, 11–10 VOLUME (pages 406–409, 414–421)

► The **volume** of any prism is given by the formula $V = B \times h$.
► You can use these special formulas to find volume:

pyramid: $V = \frac{1}{3} \times B \times h$ cone: $V = \frac{1}{3} \times \pi r^2 \times h$ sphere: $V = \frac{4}{3} \times \pi r^3$

18. Find the volume of a prism with a height of 7 in. and a base that measures 6 in. by 8 in. Round to the nearest whole number.
19. Find the volume of a sphere with a radius of 7 in. Round to the nearest whole number.

► To find the volume of a cylinder, use the formula $V = \pi r^2 \times h$.

20. Find the volume of a cylinder whose base has a radius of 7 ft and whose height is 10 ft. Round to the nearest tenth. Use $\pi \approx 3.14$.

SECTION 11–11 SURFACE AREA (pages 422–425)

► The **surface area** of a prism or pyramid is the sum of all its faces. These figures have special formulas for surface area:

sphere: $SA = 4\pi r^2$ cone: $SA = \pi rs + \pi r^2$

Find the surface area of each figure.

21.

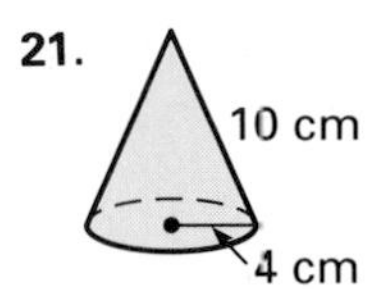

22.

5 in.
8 in.
4 in.

23.

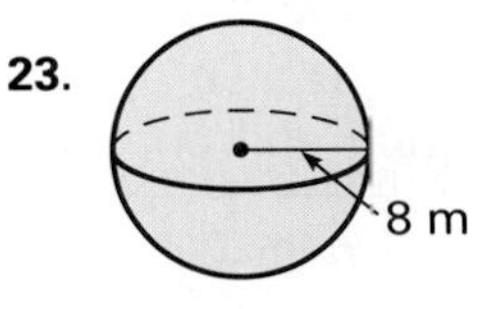

SECTIONS 11–4, 11–12 PROBLEM SOLVING (pages 396–397, 426–427)

► You need to recognize whether the formula for perimeter or for area is the formula you need to solve word problems.
► **Guess and check** is a strategy for solving certain kinds of problems.

24. The area of a triangular arm patch is 20 cm. The base is 3 cm longer than the height. Find the base and height of the patch.

USING DATA Use the diagram on page 385 to answer the following.

25. What is the total playing surface on a hockey rink?

CHAPTER 11 ● TEST

Find the area of each figure.

1. 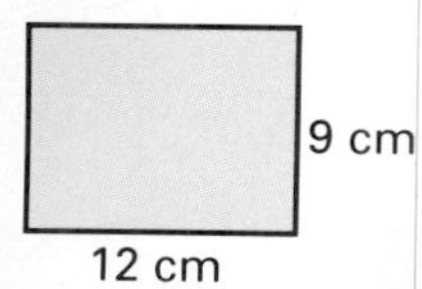

2. 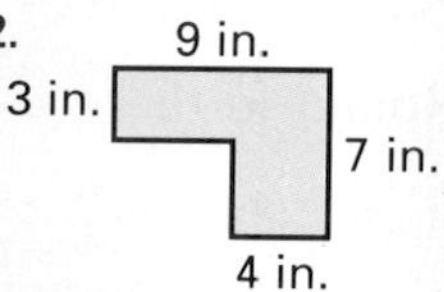

3. 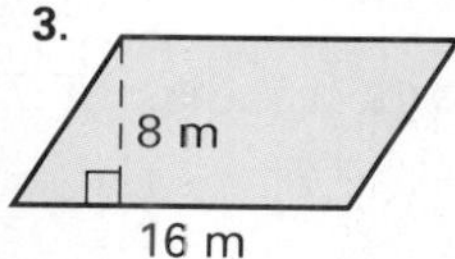

4. 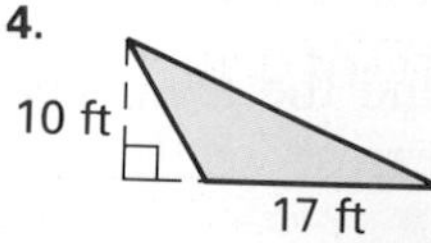

Find the square root to the nearest tenth.

5. $\sqrt{169}$
6. $\sqrt{1600}$
7. $\sqrt{91}$
8. $\sqrt{159}$

Find the unknown length.

9. 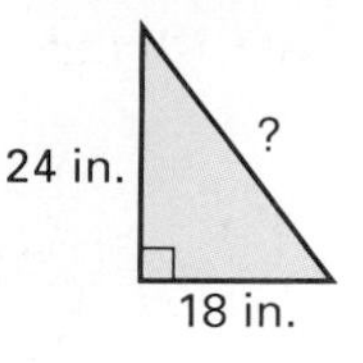

10. 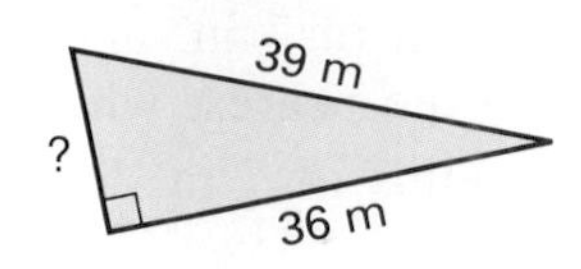

11. 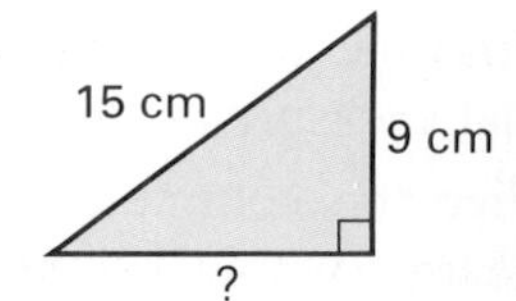

Find the volume of each figure. Round to the nearest whole number.

12. 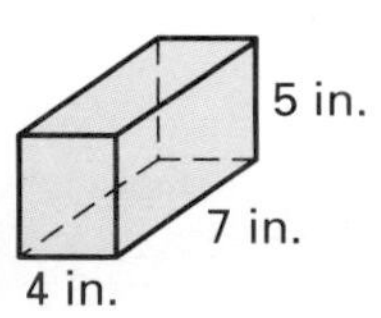

13. 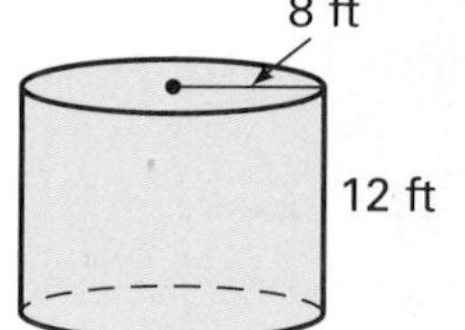

14. 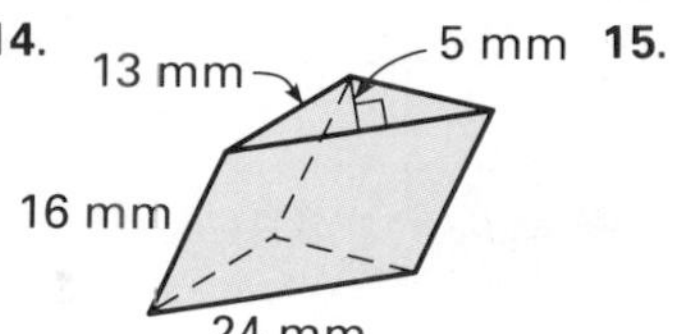

15. 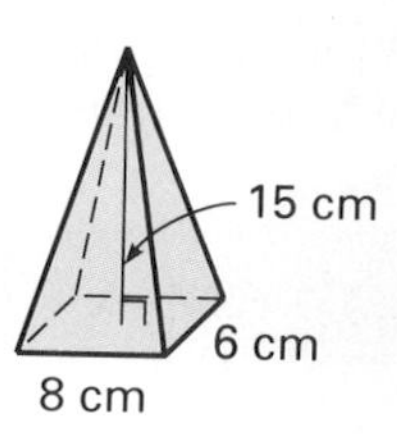

16. A large mound of sand is in the shape of a cone. The diameter is 42 ft and the height is 9 ft. Find the volume. Use $\pi \approx 3.14$. Round to the nearest whole number.

Copy and complete the table.

	radius (r)	diameter (d)	area (A)
17.	20 cm	■	■
18.	■	10 cm	■

Find the surface area of each figure.

19. 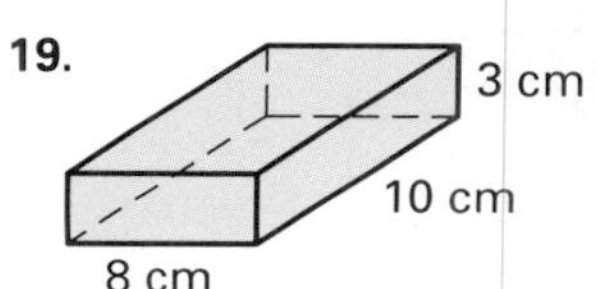

20. 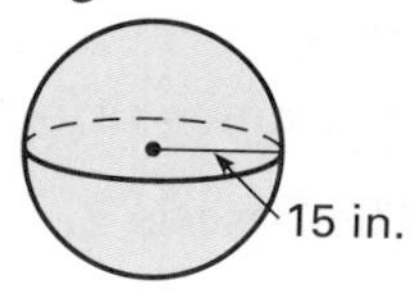

21. 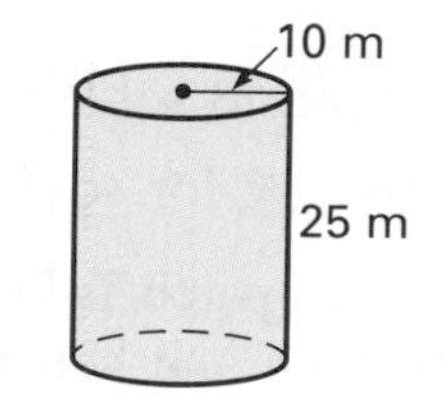

22. 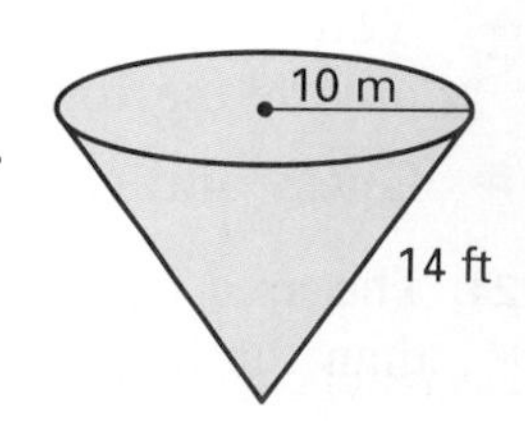

23. The area of a rectangle is 84 cm². The length is 5 cm more than the width. Find the dimensions of the rectangle.

CHAPTER 11 ● CUMULATIVE REVIEW

1. Draw a line graph to show the temperature data.

AVERAGE MONTHLY TEMPERATURE (BOISE, IDAHO)												
Month	J	F	M	A	M	J	J	A	S	O	N	D
Boise	30	36	41	49	57	66	75	72	63	52	40	32

Complete.

2. 1,468 + ■ = 2,193 + 1,468
3. 8,762 × 25 = 25 × ■
4. (6 + 8) + 4 = ■ + (8 + 4)

5. Write two conditional statements using the following two sentences:

   Melba is thirteen.
   She cannot be twelve.

Replace ● with <, >, or =.

6. $\frac{1}{8}$ ● $\frac{1}{15}$ 7. $\frac{3}{7}$ ● $\frac{3}{8}$ 8. $\frac{5}{8}$ ● $\frac{5}{6}$

Complete.

9. 17 ft = ■ yd ■ ft
10. 7 mm = ■ dm
11. 8,025 g = ■ kg
12. 19 pt = ■ qt ■ pt

Divide.

13. $56 \div (-7)$ 14. $-70 \div (-10)$
15. $-72 \div 8$ 16. $24 \div 3$

Solve each equation.

17. $77 = 10a + 7$ 18. $10.2 = 6t - 1.2$
19. $3m - 2 = 25$ 20. $\frac{x}{-3} + 1 = -1$

Solve each proportion.

21. $\frac{2.4}{4.8} = \frac{x}{4}$ 22. $2.5:10 = x:40$

23. What is 40% of 256?

24. What is the amount of interest earned in one year on $5,000 if the rate of interest is 7.5% per year?

Find the unknown angle in each triangle.

25.

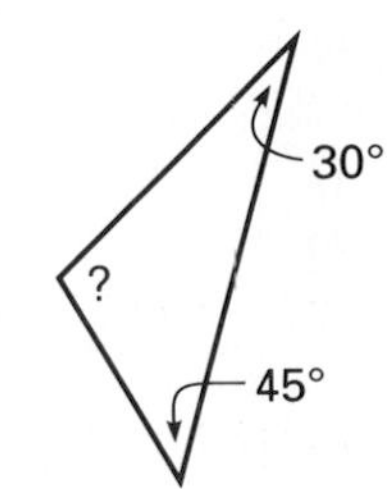

26.

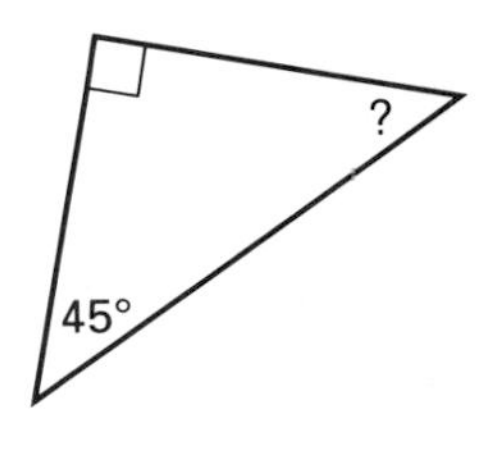

27.

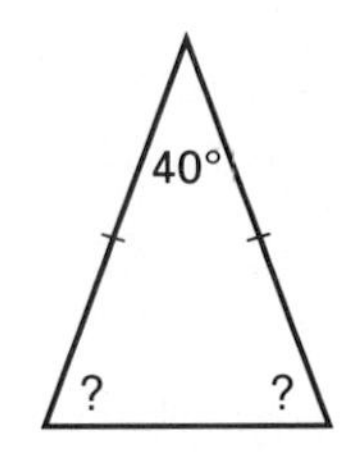

28.

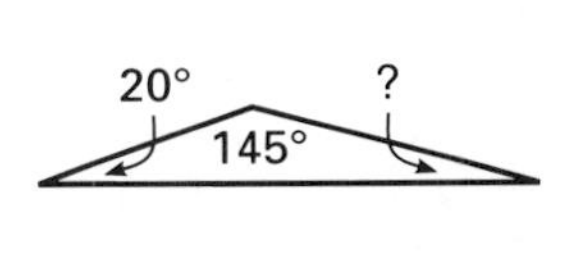

Find the area of each figure.

29.

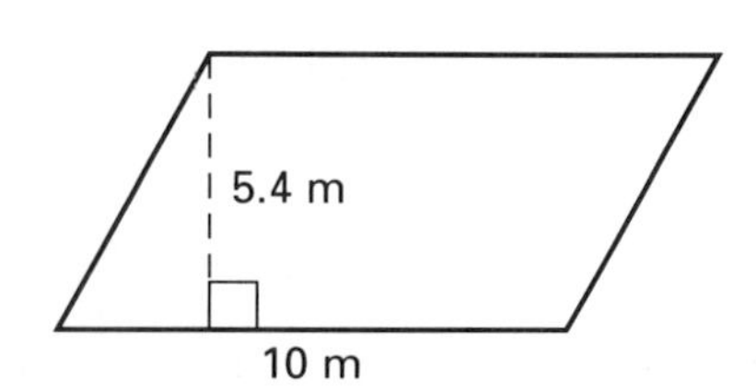

30.

20 cm
13 cm

CHAPTER 11 ● CUMULATIVE TEST

1. What is the mean weekly salary of someone who earns $20,000 per year?
   A. $1,667.67 B. $384.62
   C. $9.61 D. none of these

2. Complete.
   2,934 + ■ = 1,846 + 2,934
   A. 4,780 B. 1,088
   C. 2,934 D. 1,846

3. Which is a counterexample for the following statement?
   If a number is divisible by 6, then it is divisible by 12.
   A. 18 B. 36
   C. 48 D. 72

4. Which is the fraction for 7.16?
   A. $7\frac{1}{6}$ B. $7\frac{4}{25}$
   C. $\frac{7}{16}$ D. $7\frac{1}{4}$

5. Complete.
   76 oz = ■ lb ■ oz
   A. 4 lb 12 oz B. 6 lb 4 oz
   C. 9 lb 4 oz D. 19 lb 0 oz

6. Multiply. $10 \times (-3)$
   A. -7 B. 13
   C. -30 D. -13

7. Multiply. $-20 \times (-4)$
   A. 80 B. -16
   C. 16 D. -80

8. Solve. $\frac{2}{3}t = -18$
   A. $t = 12$ B. $t = 18$
   C. $t = -27$ D. $t = 6$

9. Solve. $4k - 42 = 18$
   A. $k = 24$ B. $k = 15$
   C. $k = -20$ D. $k = -18$

10. The scale of a drawing is 1 in.:3 ft. A room is drawn $4\frac{1}{4}$ in. wide. What is the width of the actual room?
    A. 12 ft B. $12\frac{3}{4}$ ft
    C. $7\frac{1}{4}$ ft D. none of these

11. A power tool sells for $125. James paid $5.75 in sales tax. What was the sales tax rate?
    A. 0.4% B. 6.6%
    C. 21.7% D. 4.6%

12. What is the interest for one year on a $5,000 deposit if the rate of interest is 9% per year?
    A. $4,500 B. $450
    C. $5,450 D. $4,550

13. $\angle JKL$ is 75°. Find the measure of its complement.
    A. 15° B. 105°
    C. 75° D. none of these

14. $\angle GHI$ is 35°. Find the measure of its supplement.
    A. 45° B. 55°
    C. 145° D. none of these

15. Find the area of a circle with a radius of 2.6 cm.
    A. 21.2 cm² B. 16.3 cm²
    C. 5.3 cm² D. 5.2 cm²

16. Find the volume of a cylinder with a radius of 1.5 in. and a height of 4 in.
    A. 11.3 in.³ B. 18.8 in.³
    C. 28.3 in.³ D. 7.1 in.³

CHAPTER 12 SKILLS PREVIEW

Simplify.

1. $x + 3x + (-2x)$

2. $4n^2 + (-3n^2) + 2n^2$

3. Three stars are located at the points of a triangular region, DEF. Write a simplified expression for the perimeter of the region.

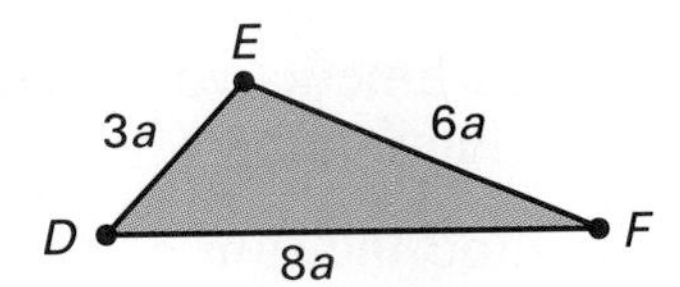

4. $(4n^2 + 2n - 8) + (2n^2 + n - 6)$

5. $(7r^2 + 3r - 9) - (4r^2 + 2r - 10)$

6. Write a simplified expression for the perimeter of the rectangle.

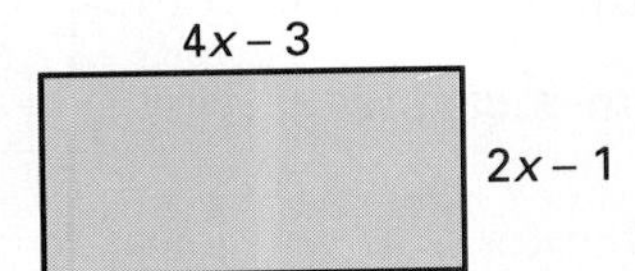

7. $(-3h)(2k)$

8. $(-4m)(-5n)$

9. $(4t)(2t^3)$

10. $(-3c^4)(-2c^2)$

11. $\frac{14xy}{7x}$

12. $\frac{-5h^3}{h^3}$

13. $\frac{12y^3z^4}{-2yz}$

14. $5(p + 1)$

15. $-2x(x + 2)$

16. $n^2(1 - n - n^2)$

Factor each polynomial.

17. $12n + 8$

18. $6a^2b - 18a$

19. $5mn + 10m^2n^2 - 20mnp$

Simplify.

20. $\frac{-16xy}{4y}$

21. $\frac{-24a^2b^3c}{-6ab}$

22. $\frac{35b^4 + 14b^2 - 21b^3}{7b}$

Solve.

23. The area of the rectangle shown is $14x^3y$. Write an expression for the unknown dimension.

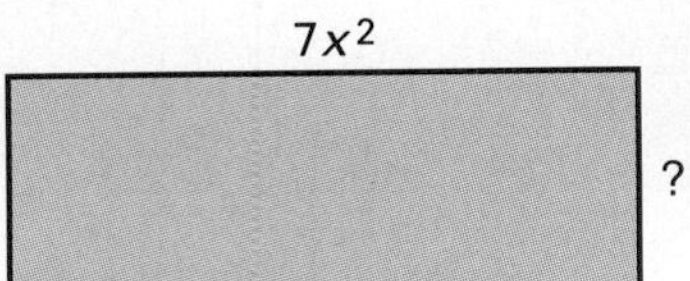

CHAPTER 12

POLYNOMIALS

THEME Astronomy

In arithmetic, you operate on quantities that are represented by specific numbers.

► Find the sum of 8 and 11.

► A room measures 10 ft by 10 ft. Find its area.

In algebra, you operate on quantities that vary in value and so must be represented by variable expressions.

► Find the sum of $3x$ and $6x$.

► A room measures y ft by $(y + 5)$ ft. Find its area.

In this chapter you will learn to perform operations on algebraic expressions called *polynomials.*

The measure of how bright a star appears from Earth is called its *apparent magnitude.* The brighter the star, the less its magnitude. Like the quantities in algebraic expressions, the magnitudes of some stars vary. The magnitude of the star Mira, for example, varies as the star expands and contracts over a long period of time. Mira is called a *long-period* variable star.

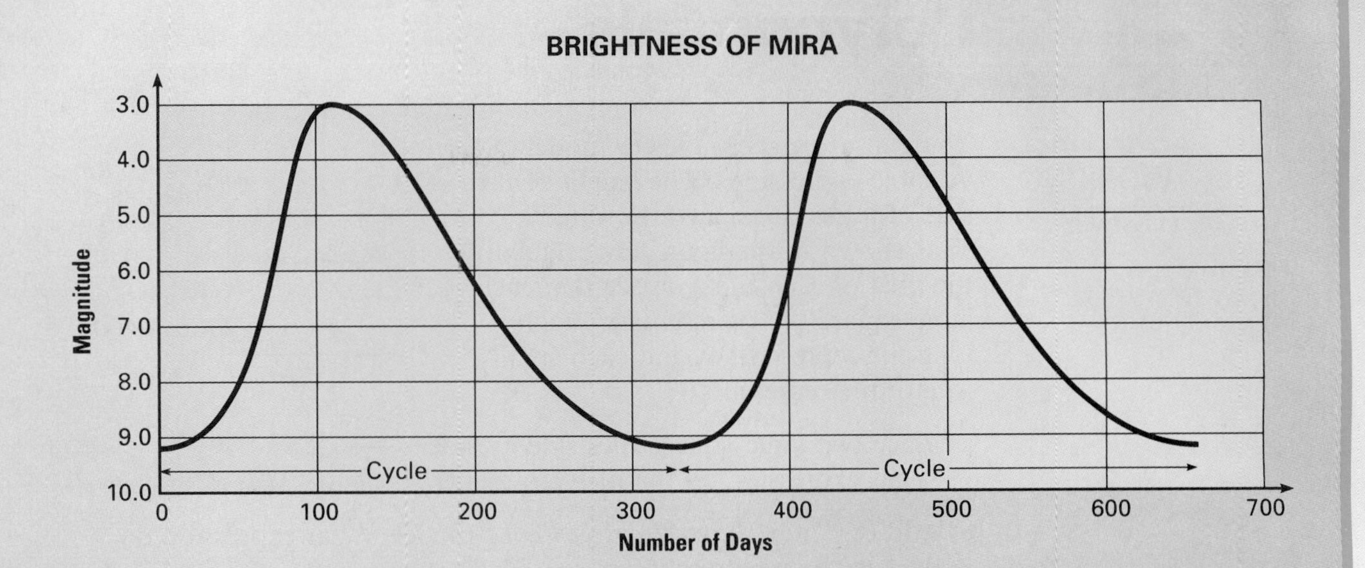

DECISION MAKING

Using Data

The graph above shows two of Mira's cycles. Use the graph to answer the following questions.

1. What is the maximum magnitude of Mira during a cycle?
2. About how many days long is each of Mira's cycles?
3. On day d Mira had a magnitude of m. Write an expression to represent the next day on which a magnitude of m occurred.
4. The star Sirius has a magnitude of -1.4. Is Sirius brighter or dimmer than Mira?

Working Together

Use earth-science books, astronomy books, or encyclopedias to find out about different kinds of variable stars. First, find out how *absolute magnitude* is related to apparent magnitude. Then, look for a Hertzsprung–Russell (H–R) diagram of the stars or some other indicator of the comparative temperature and brightness of the stars. Use this information to make lists of two kinds of stars—those with magnitudes greater than Mira's and those with magnitudes less than Mira's.

For what purpose do you think an astronomer might need the information about a star's variable period and its apparent magnitude?

12-1 Polynomials

EXPLORE/ WORKING TOGETHER

You can use algebra tiles to model variable expressions. The length of each side of a tile is assigned the value x or 1, as shown in the figure. By calculating the area of each tile, you see that each large square tile represents x^2, each long tile represents x, and each small square tile represents 1.

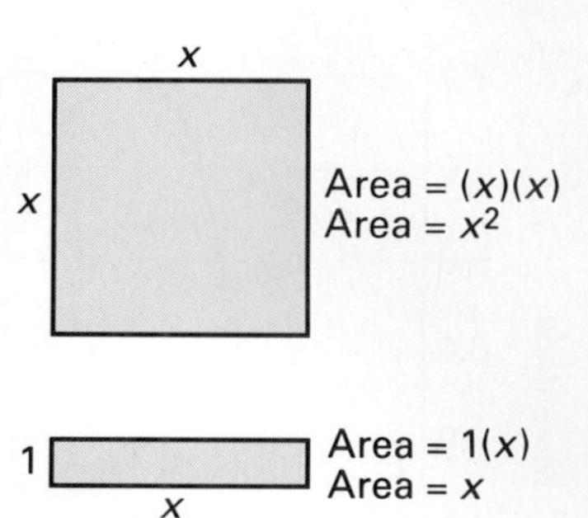

a. Place two large square tiles side by side. What expression do the tiles represent?

b. Add one more large square to your model. What expression do the tiles now represent?

c. Use long tiles to model the expression $4x$.

d. Add three more long tiles to your second model. What expression do the tiles now represent?

e. Combine the square tiles from the first model with the long tiles from the second model. What expression do the tiles now represent?

SKILLS DEVELOPMENT

Each of these expressions is called a monomial.

$$4 \qquad x \qquad 3y \qquad 2y^2 \qquad 5xy \qquad 6x^2y$$

A **monomial** is an expression that is either a single number, a variable, or the product of a number and one or more variables.

In a monomial such as $3y$, the number 3 is called the **numerical coefficient** or, more simply, the **coefficient.** In monomials such as x, xy, and x^2y, the coefficient is 1. A monomial such as -4, which contains no variable, is called a **constant monomial,** or a **constant.**

The *sum* of two or more monomials is an expression called a **polynomial.** Each monomial is called a **term** of the polynomial. A polynomial with two terms is called a **binomial.** A polynomial with three terms is called a **trinomial.**

binomials: $2x + 3$ (term, term) $\qquad x^2 + x^2y \qquad 4xy + (-5y)$

trinomials: $x^2 + y + 7$ (term, term, term) $\qquad 2xy^2 + (-5xy) + (-2)$

You can think of a monomial as a polynomial of one term.

READING MATH

The prefixes *mono–* and *poly–* are from Greek words meaning "one" and "many." The prefixes *bi–* and *tri–* are Latin prefixes meaning "two" and "three." Write these four prefixes across the top of a page in your journal. Under each word, list as many everyday words you can think of that begin with that prefix.

Sometimes a polynomial is written with subtractions rather than additions. Think of this kind of polynomial as a sum of monomials, some of which have negative coefficients.

$$7z^3 - 2z^2 - z + 6 = 7z^3 + (-2z^2) + (-1z) + 6$$

The coefficients are 7, −2, and −1.

A polynomial is written in **standard form** when its terms are arranged in order from greatest to least powers of one of the variables.

Example 1

Write each polynomial in standard form.

a. $8n^2 + 5n + 6n^3 + 9n^4$ **b.** $7y + 11 + y^2$ **c.** $5z - 2z^3 - 6 - 4z^2$

Solution

a. Write the terms of the polynomial in order from the greatest power of n to the least power of n.

$$8n^2 + 5n + 6n^3 + 9n^4 = 9n^4 + 6n^3 + 8n^2 + 5n$$

b. Order the terms from greatest to least power of y. Write the constant last.

$$7y + 11 + y^2 = y^2 + 7y + 11$$

c. First, write the polynomial as a sum of monomials with negative coefficients.

$$5z - 2z^3 - 6 - 4z^2 = 5z + (-2z^3) + (-6) + (-4z^2)$$

Then write the standard form.

$$-2z^3 + (-4z^2) + 5z + (-6), \text{ or } -2z^3 - 4z^2 + 5z - 6$$ ◄

In a polynomial, terms that are exactly alike, or that are alike *except* for their numerical coefficients, are called *like terms.*

These are examples of like terms: x^2y, $5x^2y$, $4x^2y$

These are not like terms: xy, x^2y, $2xy^2$

TALK IT OVER

Explain why x^2y and $2xy^2$ are not like terms.

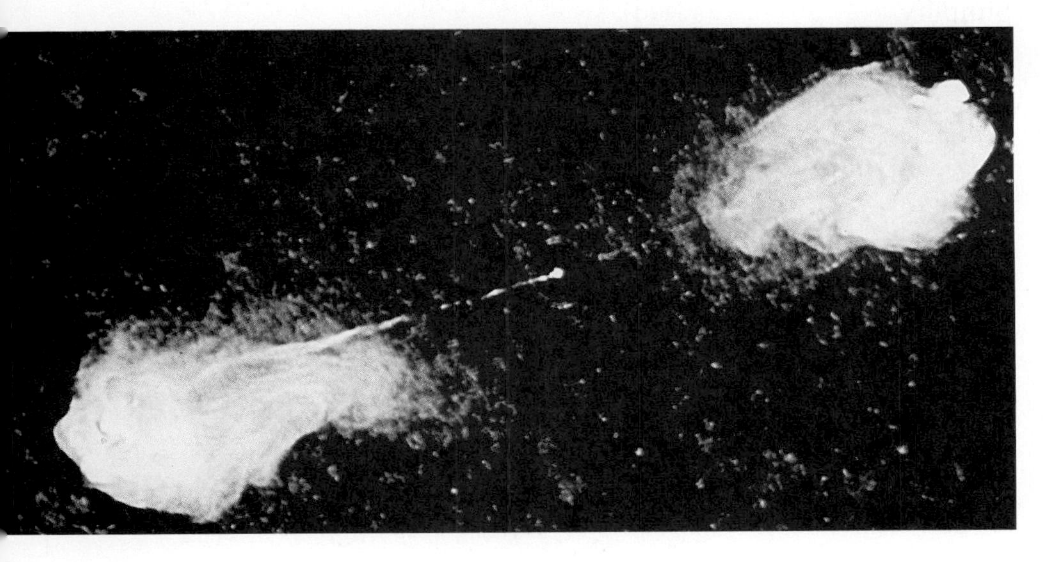

Tell whether the terms in each pair are *like* or *unlike* terms.

1. $2s, -5s$
2. $4ab, -3ab$
3. $7yz, 7xz^2$
4. $d^2t, -d^2t$

When a polynomial contains two or more like terms, you *simplify* the polynomial by adding these terms. Remember that this process is called *combining like terms.* A polynomial is simplified when all like terms have been combined and only unlike terms remain.

Example 2

Simplify.

a. $7x^2 + 2 + 9x^2$ **b.** $4d^2 - 6d - 3d^2 - 5d$ **c.** $-15m + 7m^2 + 9 - 3m$

Solution

a. $7x^2 + 2 + 9x^2 = 7x^2 + 9x^2 + 2$ ← **Rearrange, or collect, like terms.**

$= (7 + 9)x^2 + 2$ ← **Combine like terms by applying the distributive property.**

$= 16x^2 + 2$

b. $4d^2 - 6d - 3d^2 - 5d = 4d^2 + (-3d^2) + (-6d) + (-5d)$

$= [4 + (-3)]d^2 + [-6 + (-5)]d$ ← **Combine like terms and apply the distributive property.**

$= 1d^2 + (-11d)$

$= d^2 - 11d$

c. $-15m + 7m^2 + 9 - 3m = 7m^2 + (-15m) + (-3m) + 9$

$= 7m^2 + [-15 + (-3)]m + 9$

$= 7m^2 + (-18m) + 9$

$= 7m^2 - 18m + 9$ ◄

TRY THESE

Write each polynomial in standard form.

1. $5m + m^3 + 4m^2 + 2m^4$
2. $5d + d^2 + 14$
3. $-3x - 4x^2 + 2 + 5x^3$

Simplify.

4. $4x^2 + 7x^3 - 3x^2 - 8$
5. $5c^2 - 4c - 2c^2 - 3c$
6. $-4y + 2y^2 + 5 + 3y$

EXERCISES

PRACTICE/ SOLVE PROBLEMS

Write each polynomial in standard form.

1. $n^3 + 2n^2 + 7n + 4n^4$
2. $4b - 2b^2 + 7$
3. $-4z - 5z^2 + 4 + 5z^3$
4. $m^2 + 2m - 4m^3 + m^4$
5. $5n^5 - 8n + 2n^3$
6. $p^2 + 2p - 3p^3 - 9$

Simplify. Be sure your answer is in standard form.

7. $-6a - 3 + 9a$

8. $-4b + 7 + 8b$

9. $-2 + 4c - 9c$

10. $n^2 + 2n - 4n^2$

11. $7r^2 + 4r^2 + r^2 + r$

12. $14x^2 + 3x^2 + 5x^3 - 6x$

13. $7s^2 - s + 6s^2 - s^2 - s$

14. $3a^2 + 4a^3 + 2a^2 - 2a^3$

15. $-2 + 18h^2 - 9h^3 + 4h^2$

16. $-7r + 4r^2 + 8r - 12r^2 - 12$

17. $12s^3 + 4s - 2s^2 - 9s^3 + 7 - 9s$

EXTEND/ SOLVE PROBLEMS

So far in this lesson, you have been working with polynomials *in one variable*. Each term of the polynomial was either a constant term or contained a power of a single variable. You can extend what you have learned to the simplification of polynomials in two or more variables. Remember that two terms are like only if they contain the same variables raised to the same powers.

Simplify each expression. Be sure to write your answer in standard form.

18. $yz + 8yz$

19. $9x^2y - 4x^2y$

20. $xz^3 + 4xz^3 - 3yz^3$

21. $2mn - 3mn + 4mn$

22. $5g^2h + 4g^2h - 11g^2h$

23. $-12x^2y^2 + x^2y^2 - z^2 + 4z^2$

24. $-k - 2km - 9km + 15$

If possible, simplify each expression. Otherwise, write *not possible.*

25. $2xy + 3xy + 3x^2 - x^2 - x$

26. $r^3s - 4s + rs^2$

27. $a^3b + a^2b - ab + 4ab^2 + 8ab^3 - 3a^2b + 4ab$

28. $17x^3z + 18\,x^3y + 2xy^3 + xy - 4xz^3$

THINK CRITICALLY/ SOLVE PROBLEMS

The *degree of a polynomial* in one variable is the highest power of the variable that appears in the polynomial after it has been simplified. What is the degree of each polynomial?

29. $x^2 + 3x - 5$

30. $5a - 4 + a^3 - 6a^2$

31. $m + 4$

32. $-3x^2 + 2x^3 - x^4 - 7$

33. Tell whether the statement is *true* or *false:*
If the degree of a polynomial is 4, then the polynomial will have four terms.

34. Explain why the expression $\frac{3x}{2y}$ is not a polynomial.

35. Explain why $3x^4 - 3x^3 + 2x^2 - 5x$ cannot be simplified.

12-2 Adding and Subtracting Polynomials

EXPLORE

Find the term that makes the statement true.

a. $-9d + \underline{\quad ? \quad} = 5d$

b. $3m + (-7m) + \underline{\quad ? \quad} = m$

c. $-x + (-7x) + \underline{\quad ? \quad} = -x$

d. $7h^2 + \underline{\quad ? \quad} + (-h^2) + 4h^2 = 4h^2$

SKILLS DEVELOPMENT

Just as you can add, subtract, multiply, or divide real numbers, you can perform each of these basic operations with polynomials.

To add two polynomials, write the sum and simplify by combining like terms.

Example 1

Simplify.

a. $3m + (2m - 6)$

b. $(2a + 3) + (4a - 1)$

c. $(x^2 + 4x - 2) + (3x^2 + 7)$

Solution

a. $3m + (2m - 6) = (3m + 2m) - 6 \leftarrow$ **$3m$ and $2m$ are like terms. Use the associative property.**

$= (3 + 2)m - 6 \leftarrow$ **Use the distributive property.**

$= 5m - 6$

b. $(2a + 3) + (4a - 1) = (2a + 4a) + (3 - 1) \leftarrow$ **$2a$ and $4a$ are like terms. 3 and -1 are like terms.**

$= (2 + 4)a + 2$

$= 6a + 2$

c. $(x^2 + 4x - 2) + (3x^2 + 7) = (x^2 + 3x^2) + 4x + (-2 + 7)$

$= (1 + 3)x^2 + 4x + 5$

$= 4x^2 + 4x + 5$ ◄

You may also add polynomials by lining up like terms vertically. For example, at the right you see an alternative way to add the polynomials in part **c** of Example 1.

$$\begin{array}{r} x^2 + 4x - 2 \\ +3x^2 \qquad + 7 \\ \hline 4x^2 + 4x + 5 \end{array}$$

Subtraction of polynomials is like subtraction of real numbers. To subtract one polynomial from another, add its opposite and simplify. To find the opposite of a polynomial, find the opposite of each term of the polynomial.

Polynomial	*Opposite*
$x + 4$	$-x + (-4)$ or $-x - 4$
$2 + n$	$-2 + (-n)$ or $-2 - n$
$4f^2 - 2f - 7$	$-4f^2 + 2f + 7$
$-3x^2 + 5x + 2$	$3x^2 + (-5x) + (-2)$ or $3x^2 - 5x - 2$

Example 2

Simplify.

a. $9y - (2y - 4)$ b. $(-x + 2) - (-8x - 3)$

c. $(5x^2 - 11) - (3x^2 - 3x + 2)$

Solution

a. $9y - (2y - 4) = 9y + (-2y + 4)$ ← **Add the opposite of (2y − 4).**

$= [9y + (-2y)] + 4$ ← **9y and −2y are like terms.**

$= (9y - 2y) + 4$

$= (9 - 2)y + 4$ ← **Use the distributive property.**

$= 7y + 4$

b. $(-x + 2) - (-8x - 3) = (-x + 2) + (8x + 3)$ ← **Add the opposite of (−8x − 3).**

$= (-x + 8x) + (2 + 3)$

$= (-1 + 8)x + 5$ ← **Write the coefficient of −x as −1.**

$= 7x + 5$

c. $(5x^2 - 11) - (3x^2 - 3x + 2) = (5x^2 - 11) + [-3x^2 + 3x + (-2)]$

$= [5x^2 + (-3x^2)] + 3x + [-11 + (-2)]$

$= (5 - 3)x^2 + 3x + (-11 - 2)$

$= 2x^2 + 3x - 13$ ◄

As with addition, you may subtract polynomials by lining up like terms vertically. This is how you could subtract the polynomials in part **c** of Example 2.

$$\begin{array}{r} 5x^2 - 11 \\ -3x^2 + 3x - 2 \\ \hline 2x^2 + 3x - 13 \end{array}$$
← **Add the opposite of each term.**

MIXED REVIEW

Add or subtract.

1. $4 + 9$
2. $-3 + 4$
3. $19 + (-3)$
4. $18 - (-7)$
5. $-4 - (-8)$
6. $-180 - (-45)$

Multiply or divide.

7. $8 \times (-4)$
8. $-3 \times (-9)$
9. -18×4
10. $32 \div (-8)$
11. $-45 \div (-5)$
12. $-12 \div 2$

TRY THESE

Add.

1. $\begin{array}{r} 4x^2 + 3x - 2 \\ +\ x^2 \qquad + 6 \\ \hline \end{array}$

2. $\begin{array}{r} 2m^2 + 2m - 5 \\ + \qquad 3m + 4 \\ \hline \end{array}$

3. $\begin{array}{r} 5y^2 + 2y + 3 \\ 2y^2 \qquad + 4 \\ \hline \end{array}$

Simplify.

4. $4x + (3x - 5)$
5. $(4m + 2) + (5m - 3)$
6. $(2r^2 + 3r - 4) + (2r^2 + 3)$
7. $6y - (4y - 3)$
8. $(-z + 3) - (-4z - 2)$
9. $(3c^2 - 8) - (2c^2 - 2c + 6)$

EXERCISES

PRACTICE/SOLVE PROBLEMS

Simplify.

1. $2y + (4y - 3)$
2. $3k + (2k + 4)$
3. $9m + (-2m - 7)$
4. $(2x^2 - 1) + (3x^2 + 7)$
5. $(5a^2 + 3) + (-7a^2 - 4)$
6. $(2x^2 - 3x + 5) + (4x^2 + 6x - 8)$
7. $(a^2 - 2a + 1) + (a^2 - 4)$
8. $(2y - 3) + (4y + 5)$
9. $(3x) - (-2x)$
10. $2k - (2k + 4)$
11. $6z - (2z + 7)$
12. $(7x - 5) - (5x + 3)$
13. $(10 - 2k) - (7 - 3k)$
14. $(z^2 - 3z + 10) - (z^2 + 5z - 6)$
15. $(8p^2 + 5) - (3p^2 + 2p - 9)$
16. $(2a^2 - 3a + 5) - (a^2 - 2a - 4)$
17. $(11m^3 + 2m^2 - m) - (-6m^3 + 3m^2 + 4m)$
18. $(3d^3 - 10d^2 + d + 1) + (5d^3 + 2d^2 - 3d - 6)$

EXTEND/SOLVE PROBLEMS

Simplify.

19. $(2x - 4y) + (x + 3y)$
20. $(8m - 7n) + (-5m - 4n)$
21. $(a + b) + (a + b) + (a + b)$
22. $(7r - 3s) + (4r + 2s)$
23. $(8a - 3b) + (5a + 4b)$
24. $(2a + 6b - 1) + (5a + b - 3)$
25. $(3x - 4xy + y) + (-4x - 2xy - y)$
26. $(-4m^2n - 3mn^2) + (6m^2n - 3mn^2)$
27. $(3r^2t - 2rt^2) + (5r^2t - 4rt)$
28. $(-6x^2y - 4x^2) + (xy^2 - 2x^2y)$

29. $(4a - 3b) - (a - 4b)$

30. $(7m - 3n) - (-9m + 4n)$

31. $(2z - 3y) - (5z - 5y)$

32. $(7x - 2y) - (3x + 4y)$

33. $(8f - 7g - 1) - (-f + 4g + 3)$

34. $(-3k^2g - 4kg^2) - (6k^2g - 2kg^2)$

35. $(2r^2t^2 + 3rt^2) - (-4r^2t^2 - 7rt^2)$

36. $(4xy^2 - 4y^2) - (5xy^2 - x^2y)$

37. $(4a^2bc - 3abc^2 + 2abc) + (8abc^2 - 5a^2bc - ab^2c)$

TALK IT OVER

Is it possible for the sum of two trinomials to be a binomial?

Is it possible for the difference of two trinomials to be a monomial?

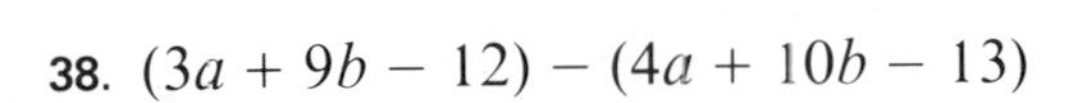

38. $(3a + 9b - 12) - (4a + 10b - 13)$

39. $(4x - 4xy + 3y) - (4x - 5xy + 3y)$

40. $(12abc - 3ab^2c + 4a^2bc) - (-3ab^2c - 4abc + 4a^2bc)$

Use the diagrams for Exercises 41–43.

41. Write an expression in simplest form for the perimeter of triangle *RST*.

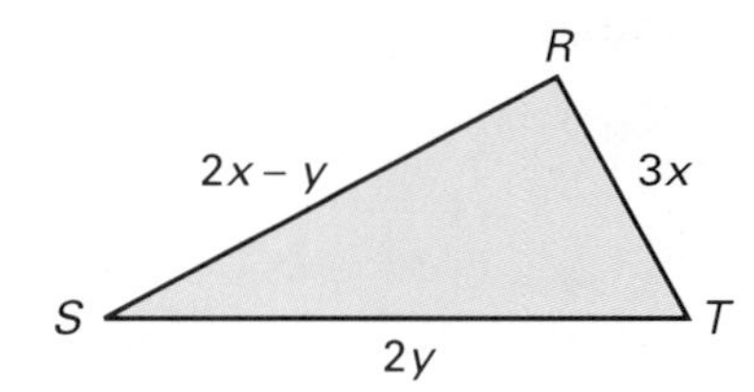

42. Write an expression in simplest form for the perimeter of quadrilateral *ABCD*.

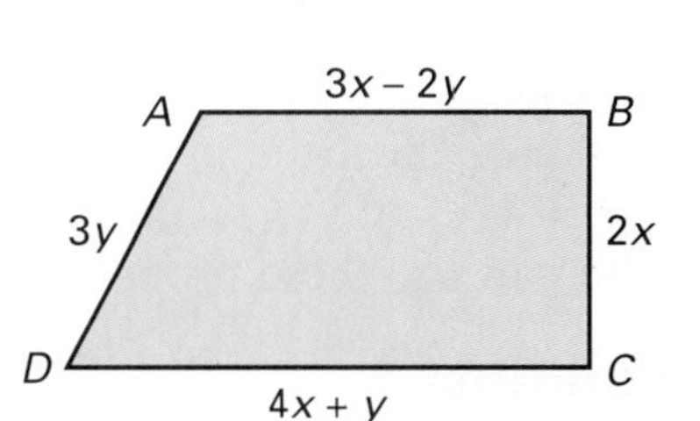

43. Find the difference when the perimeter of triangle *RST* is subtracted from that of quadrilateral *ABCD*.

THINK CRITICALLY/ SOLVE PROBLEMS

44. Find each difference.

   a. $(4n^2 - 2n + 1) - (n^2 + 6n - 5)$
   b. $(n^2 + 6n - 5) - (4n^2 - 2n + 1)$

45. Compare your answers to parts **a** and **b** of Exercise 44. How are they related?

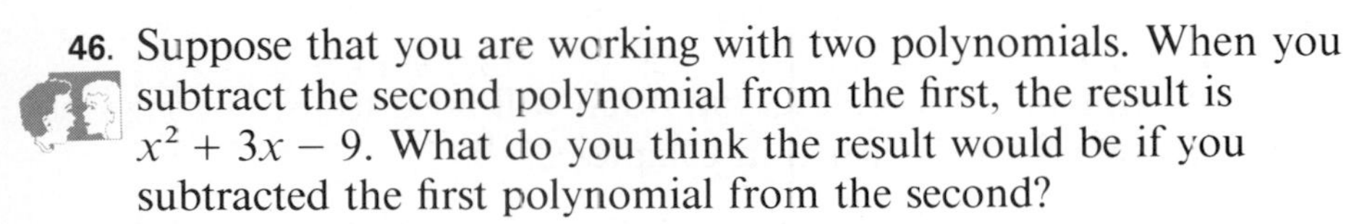

46. Suppose that you are working with two polynomials. When you subtract the second polynomial from the first, the result is $x^2 + 3x - 9$. What do you think the result would be if you subtracted the first polynomial from the second?

47. Analyze your answers to Exercises 44–46. Make a generalization about what happens when you reverse a subtraction of polynomials.

12-3 Multiplying Monomials

EXPLORE

Use algebra tiles. Assign the value a to the length of a long tile, and assign the value b to its width.

a. What is the area of each long tile?

b. Use long tiles to model a rectangle with length $3a$ and width $2b$. What is the total area of all the long tiles you used?

c. Using the area formula $A = l \times w$, give the area of the modeled rectangle as a product of its dimensions.

d. Use your model to give the product of $3a$ and $2b$ as a monomial.

e. Model the following products and give each result as a monomial. What pattern do you see?

$4a \times 3b$ $\quad a \times 3b$ $\quad 2a \times 2b$ $\quad 3a \times 5b$

SKILLS DEVELOPMENT

Recall that, by the commutative property of multiplication, the order in which factors are multiplied does not affect the product: $a \times b = b \times a$.

By the associative property of multiplication, the grouping of factors does not affect the product: $(a \times b) \times c = a \times (b \times c)$.

You can use these properties to find a product of two monomials.

Example 1

Simplify.

a. $(4a)(5b)$ b. $(-3k)(2p)$ c. $\left(-\frac{1}{2}x\right)(-4y)$

Solution

a. $(4a)(5b) = (4)(5)(a)(b)$ ← **Use the commutative and associative properties to regroup the coefficients and variables.**

$= 20ab$

b. $(-3k)(2p) = (-3)(2)(k)(p) = -6kp$

c. $\left(-\frac{1}{2}x\right)(-4y) = \left(-\frac{1}{2}\right)(-4)(x)(y) = 2xy$ ◄

READING MATH

Remember that parentheses are often used to show multiplication.

When multiplying monomials, you often will need to use the laws of exponents, which you learned in Chapter 2.

Recall the *product rule for exponents:* To multiply two powers having the same base, you add the exponents.

$$a^m \times a^n = a^{m+n}$$

You can use the product rule for exponents, together with the commutative and associative properties of multiplication, to find a product of two monomials having powers of the same base.

Example 2

Simplify.

a. $(-3k)(-2k^3)$

b. $(ab^3)(a^3b^4)$

Solution

a. $(-3k)(-2k^3) = (-3)(-2)(k \times k^3)$ ← **k means k^1**

$= 6k^{1+3}$

$= 6k^4$

b. $(ab^3)(a^3b^4) = (a \times a^3)(b^3 \times b^4)$

$= a^{1+3} \times b^{3+4}$ ← **Use the product rule for each base, a and b.**

$= a^4b^7$ ← **Because a and b are unlike bases, you cannot add exponents.** ◄

Recall the *power rule for exponents:* To find the power of a monomial that is itself a power, you multiply exponents.

$$(a^m)^n = a^{mn}$$

You can use the power rule, together with the commutative and associative properties for multiplication, to simplify a power of a product such as $(2c^3)^2$.

$$\begin{aligned}(2c^3)^2 &= 2c^3 \times 2c^3\\ &= (2 \times 2) \times (c^3 \times c^3)\\ &= 2^2 \times (c^3)^2\\ &= 4c^6\end{aligned}$$

Notice that, when the product $2c^3$ is squared, both 2 and c^3 are squared. This suggests the following rule, called the **power of a product rule:**

► To find the power of a product, find the power of each factor, and multiply.

$$(ab)^m = a^mb^m$$

Example 3

Simplify.

a. $(3z)^2$

b. $(4y^2)^2$

c. $(-2c^2)^3$

Solution

a. $(3z)^2 = (3^2)(z^2)$

$= 9z^2$

b. $(4y^2)^2 = (4)^2\,(y^2)^2$

$= 16y^{2 \times 2}$

$= 16y^4$

c. $(-2c^2)^3 = (-2)^3\,(c^2)^3$

$= -8c^{2 \times 3}$

$= -8c^6$ ◄

TRY THESE

Simplify.

1. $(3x)(3y)$
2. $(-b)(5d)$
3. $(4g)(-2h)$
4. $\left(-\frac{1}{3}a\right)(-3b)$
5. $(-2y)(y^2)$
6. $(2p^2r^3)(5pr^3)$
7. $(2y)^3$
8. $(3z^3)^2$
9. $(-3c^2)^2$

EXERCISES

PRACTICE/ SOLVE PROBLEMS

Simplify.

1. $(2a)(3b)$
2. $(-4m)(-3n)$
3. $(6k)(-2m)$
4. $(3x)(-3z)$
5. $(-2k)(-3p)$
6. $(-6a)\left(\frac{1}{2}b\right)$
7. $(2c)(3.5d)$
8. $(-3p)(-4.1r)$
9. $(2a)^2$
10. $(3m^2)^3$
11. $(7y)(y^4)$
12. $(5b^3)(-2b)$
13. $(4p^2)(-6p)$
14. $(-3a^2)(-2a^3)$
15. $(5k^4)(-3k^2)$
16. $(2p^3)(-4p^5)$
17. $(3x)^2$
18. $(3y^2)^3$
19. $(-2d^3)^3$
20. $(-3x^2)^3$
21. $(9y)(3y)^2$
22. $(3x^2)(2x)^3$
23. $(2c)^2(3c)^2$
24. $(2a^2)^2$
25. $(-4b^4)^3$
26. $(-y^2)^2(-5y^3)^2$

Write an expression for the area of each figure.

27.

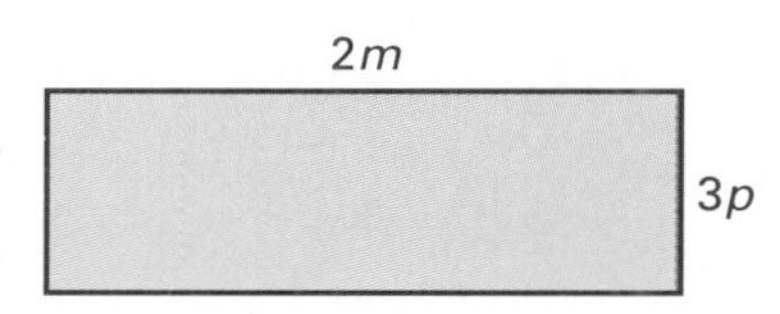

28.

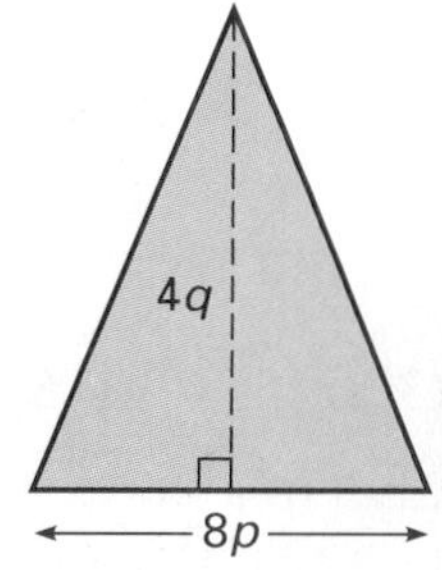

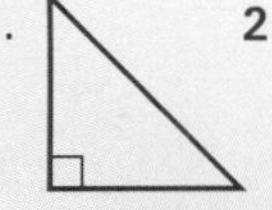

MIXED REVIEW

Identify each polygon.

1. 2.

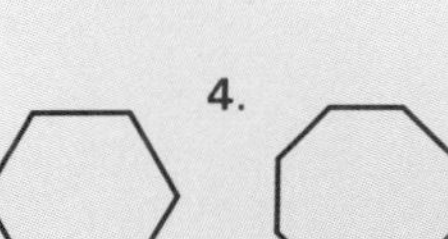

3. 4.

5. Find the area of a triangle whose base is 8 cm and whose height is 4.5 cm.

EXTEND/ SOLVE PROBLEMS

Simplify.

29. $(2a^3b^2)(6a^3b)$
30. $(-3x^3y^3)(-4xy^2)$
31. $\left(\frac{-1}{3}p^3r\right)(9p^2r^4)$
32. $(-2a^2b)^4\left(\frac{1}{2}ab^3\right)^2$
33. $\left(\frac{1}{3}x^2y\right)^3(3x^5y)^3$
34. $(2d)^3(4d^2)^3(-d)^2$
35. $(2x^3y^6)^4\left(\frac{1}{2}xy^3\right)^2$
36. $(4z)^2(2z^2)^3\left(\frac{1}{4}z\right)^3$
37. $(2p^2r^3)^2(3p^3r)^2$

Write an expression for the area of each figure.

38.

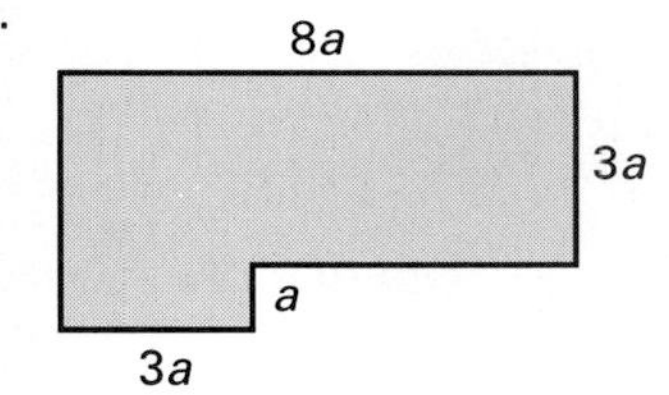

39.

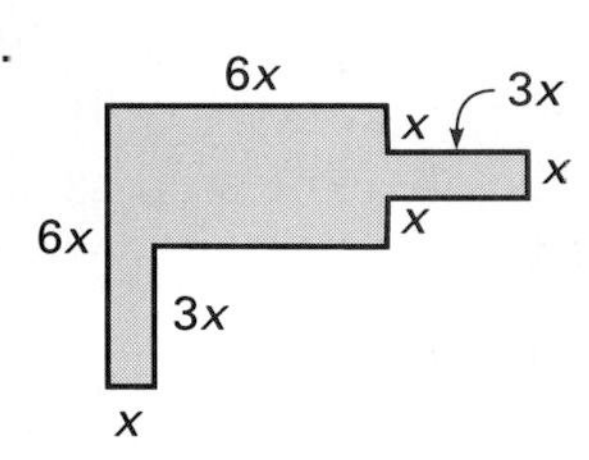

Write an expression for the volume of each cube.

40.

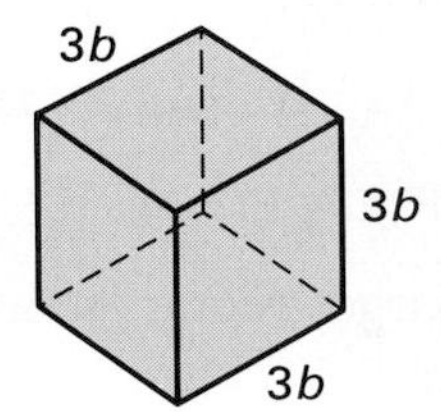

41.

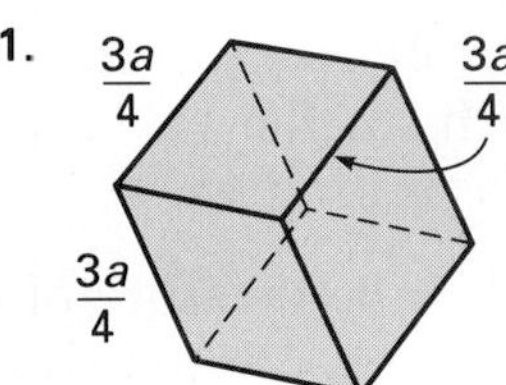

PROBLEM SOLVING TIP

For Exercise 45, ask yourself: How can the result of multiplying a number by 30 be less than the result of multiplying the same number by 28?

Write an expression for the volume of each prism.

42.

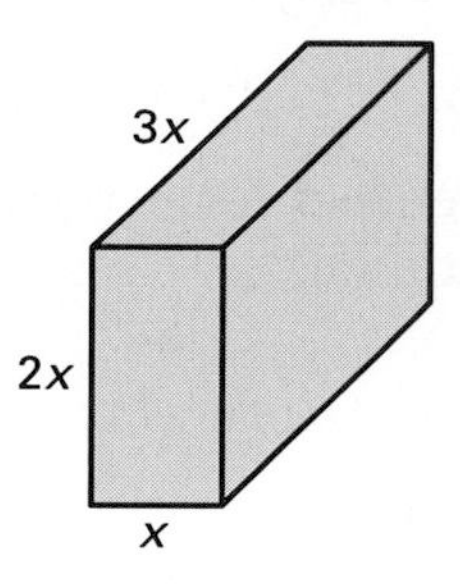

43.

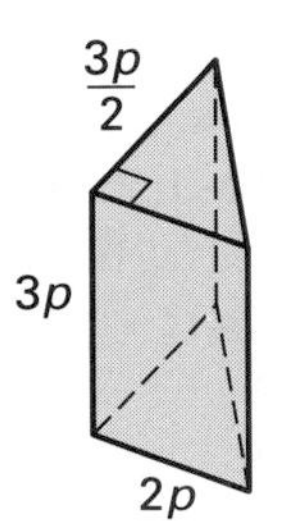

THINK CRITICALLY/ SOLVE PROBLEMS

44. Which triangle has the greater area?

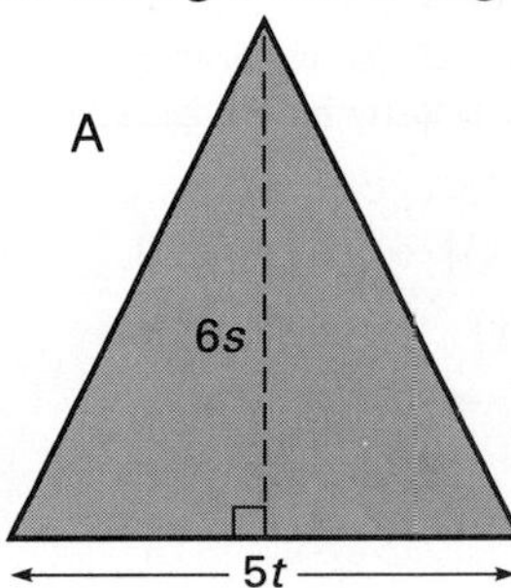

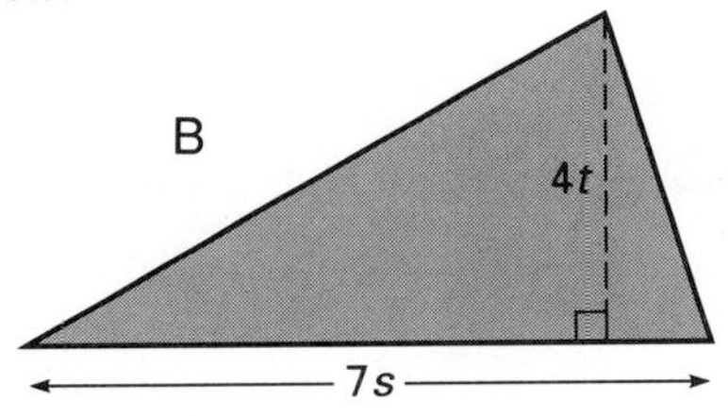

45. Give values of s and t for which $30st$ is less than $28st$.

46. As a result of answering Exercise 45, did you want to change your answer to Exercise 44 in any way? Explain why or why not.

12-4 Multiplying a Polynomial by a Monomial

EXPLORE

a. Figure A shows a rectangle with a length of $2x + 5$ and a width of $2x$. What is the measure of the area of the rectangle expressed as a product?

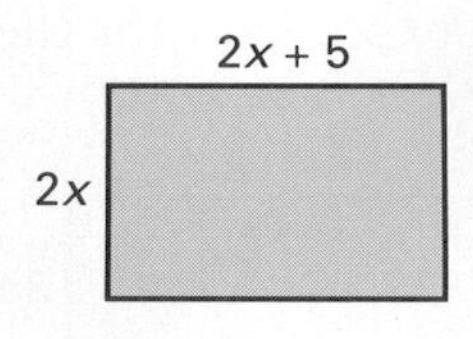

Figure A

b. Figure B shows the same rectangle divided into two smaller rectangles with dimensions as marked. Express the measure of the area of each smaller rectangle as a product.

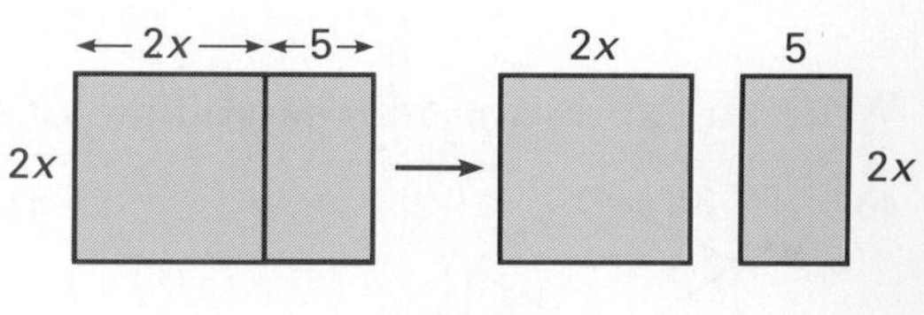

Figure B

c. Now express the measure of the area of the larger rectangle as the sum of the areas of the two smaller rectangles.

d. Compare your answer to *a* with your answer to *c*. What property is illustrated?

SKILLS DEVELOPMENT

You can use the distributive property and the rules for exponents to find the product of a monomial and a polynomial.

Example 1

Simplify.

a. $3a(a + 4)$

b. $-2x(x^2 - 3x + 11)$

Solution

a. $3a(a + 4) = 3a(a) + 3a(4)$ ← **Use the distributive property.**

$= 3a^2 + 12a$ ← **Apply the product rule for exponents and the commutative property for multiplication.**

b. $-2x(x^2 - 3x + 11) = -2x[x^2 + (-3x) + 11]$

$= -2x(x^2) + [-2x(-3x)] + [-2x(11)]$

$= -2x^3 + 6x^2 + (-22x)$

$= -2x^3 + 6x^2 - 22x$ ◄

Example 2

For a science exhibit, Gail uses a photograph of Saturn's rings that is 12 in. wide and 16 in. long. When she adds a caption below the photograph, the total width will be $(12 + c)$ in. Find an expression for the total area occupied by the photograph and the caption, then simplify the expression.

Solution

Use the formula for the area of a rectangle.

$A = l \times w$

$A = 16(12 + c)$ ← **Replace *l* with 16 and *w* with (12 + *c*).**

$A = 16(12) + 16(c)$

$A = 192 + 16c$

The total area occupied by the photograph and the caption is $(192 + 16c)$ in.^{2} ◄

TRY THESE

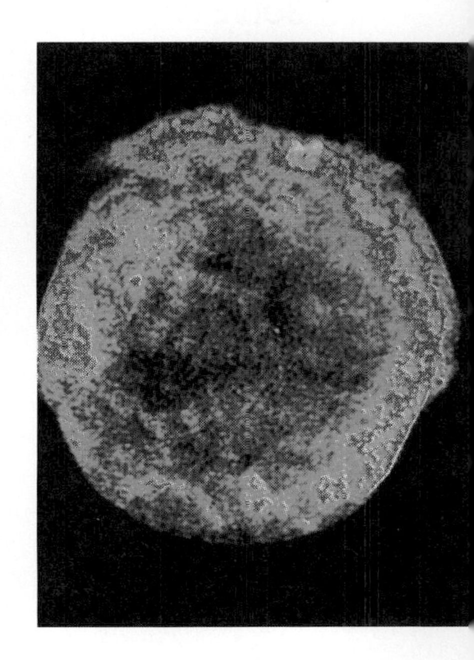

Simplify.

1. $4x^2(x + 7)$
2. $2b(b^2 + 12)$
3. $3g^2(g^2 - 7)$
4. $2z(3z^2 + z - 8)$
5. $6k(2k^2 + k)$
6. $-x(3x^2 - x - 6)$

Solve.

7. A photograph of a supernova was 12 in. long and 8 in. wide. After Pablo trimmed one side of the photograph, its width was $(8 - t)$ in. Find an expression for the area of the trimmed photograph and then simplify it.

EXERCISES

PRACTICE/ SOLVE PROBLEMS

Simplify.

1. $2x(x + 3x^2)$
2. $-4(-y + 2y^2)$
3. $2a(3a + 1)$
4. $10w(-3w + 2)$
5. $2(a^2 + 5a - 1)$
6. $-2m(m^2 - 3m - 4)$
7. $-3(f^2 - 2f - 8)$
8. $4p(p^2 - 3p - 2)$
9. $-r(r^3 - 9r + 6)$
10. $a(4 - a - 5a^2)$
11. $2x(2x^2 - 4x + 3)$
12. $4a^2(-a^2 - 3a + 9)$

Solve.

13. Write an expression for the area of the rectangle. Then simplify the expression.

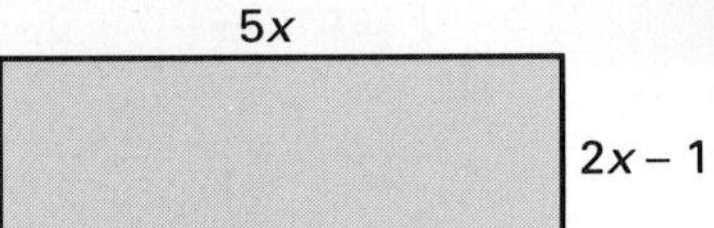

14. Let r represent an even integer. Then $r + 2$ is the next even integer. Write an expression for the product of the two integers. Then simplify the expression.

EXTEND/ SOLVE PROBLEMS

Simplify.

15. $4x + 2(x + 1)$
16. $4(a - 3) - 3a$
17. $-2p - 2(p - 3) + 6$
18. $4(m - 1) + 3(m + 5)$
19. $4(2a - b) - 3(a + b)$
20. $4a(2b - a) - 3a(2b + a)$
21. $x(2x^2 - x + 2) - x(x^3 + 4x^2 - 3x)$
22. $\frac{1}{2}x(2x^2 + 6xy - 14y^2)$
23. $\frac{1}{3}c^2(3c^2 - 9cd - 18d^2)$
24. $(4x^2y - 3xy^2 + 7y^3)(4xy)$
25. $(r^3 - 3r^2s + 4s^2)(-2rs)$

Exercises 26–28 refer to this diagram. The diagram indicates the number of seconds it took a beam of sunlight to travel first to the earth, then on to the moon.

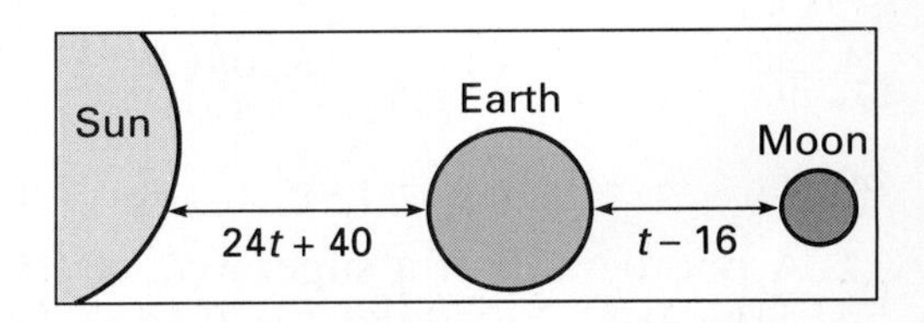

26. Write an expression for the total time it took the light to reach the moon.

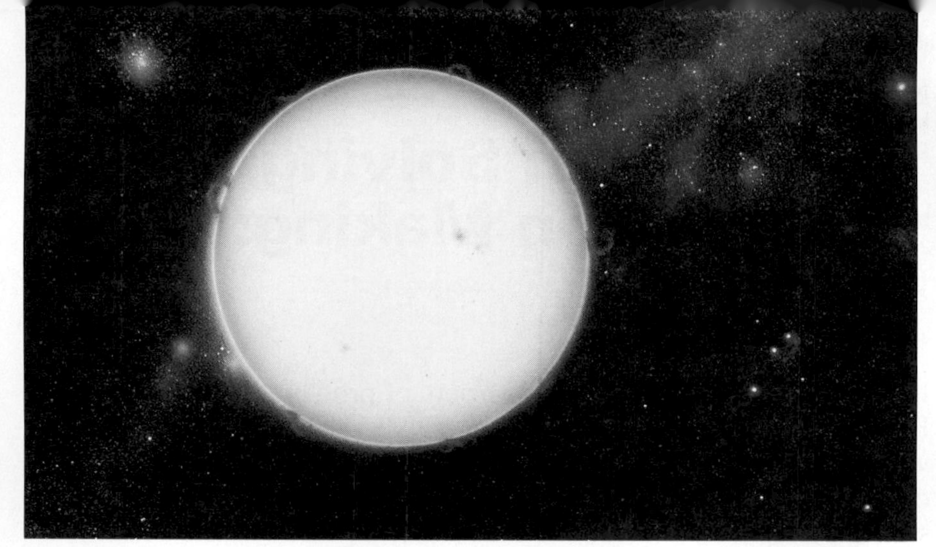

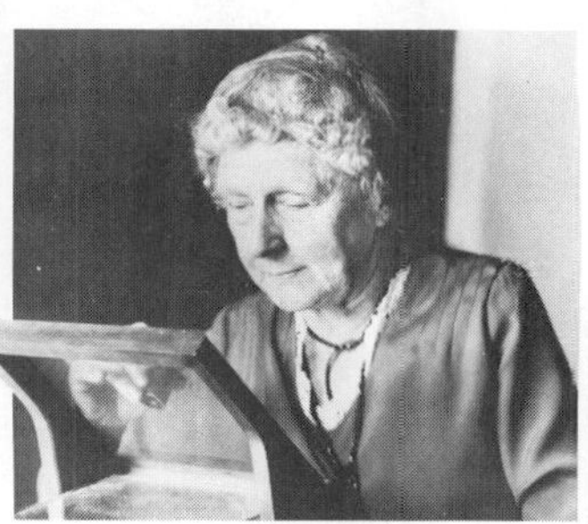

MATH:
WHO, WHERE, WHEN

Annie Jump Cannon (1863–1941) was one of the most well-known and honored astronomers of her day. She contributed much to the study of *variable* stars, stars whose magnitude is not constant, but changes. Ms. Cannon published a 10-volume catalog grouping over 1,000 stars according to their color spectra. Her careful and accurate work in compiling this information has benefited astronomers everywhere.

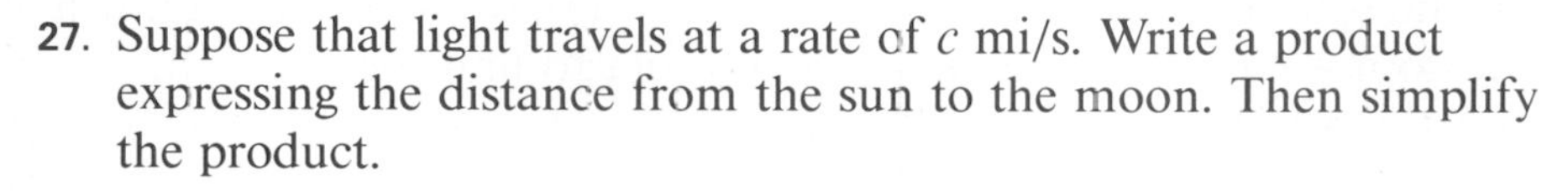

27. Suppose that light travels at a rate of c mi/s. Write a product expressing the distance from the sun to the moon. Then simplify the product.

28. Use $c = 186{,}000$ mi/s and $t = 19$ s to approximate the distance from the sun to the moon.

USING DATA Refer to the Data Index on page 546 to find the table listing the largest moons in the solar system.

29. One moon in our solar system has a diameter of $(w^2v)(w^5v)$ miles, where $w = 2$ and $v = 5$. Which moon is it?

THINK CRITICALLY/ SOLVE PROBLEMS

Write an expression for the area of the shaded region. Then simplify the expression.

30.

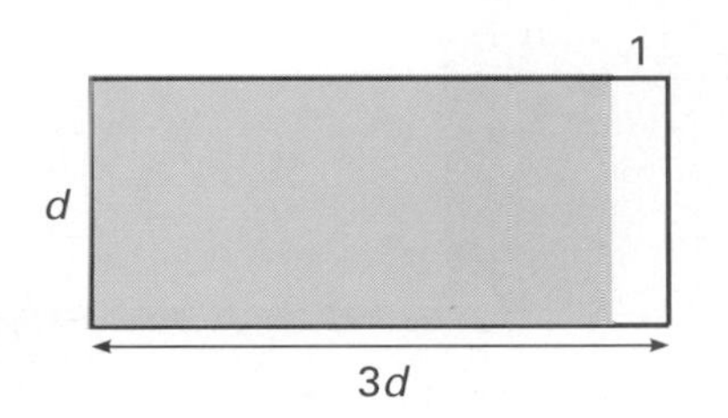

31.

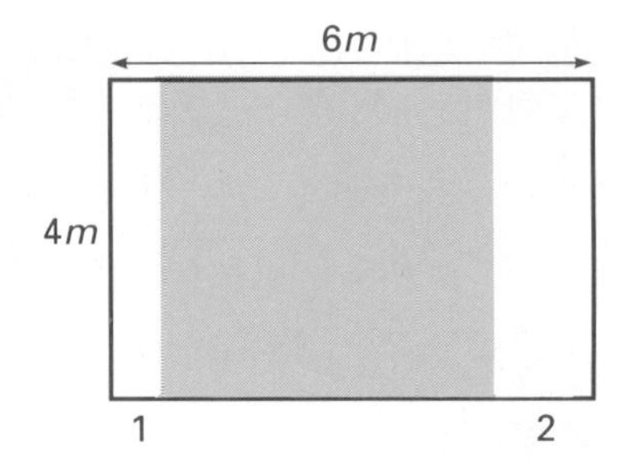

Simplify.

32. $3(x - y) - 2(x - y)$

33. $7(x - y) - 6(x - y)$

34. $13(x - y) - 12(x - y)$

35. $20(x - y) - 19(x - y)$

Examine your results for Exercises 32–35. Then, without multiplying, write a simplified expression for each.

36. $112(x - y) - 111(x - y)$

37. $50(w + z) - 49(w + z)$

12-5

Problem Solving/ Decision Making:

CHOOSE A STRATEGY

PROBLEM SOLVING TIP

Here is a checklist of the problem solving strategies that you have studied so far in this book.

Solve a simpler problem
Find a pattern
Guess and check
Work backward
Make an organized list
Make a table
Use logical reasoning
Act it out
Draw a picture

In this book you have studied a variety of problem solving strategies. Experience in applying these strategies will help you decide which will be most appropriate for solving a particular problem. Sometimes only one strategy will work. In other cases, any one of several strategies will offer a solution. There may be times when you will want to use two different approaches to a problem in order to be sure that the solution you found is correct. For certain problems, you will need to use more than one strategy in order to find the solution.

PROBLEMS

Solve. Name the strategy you used to solve each problem.

1. The sum of two monomials is $11n$. The product of the monomials is $30n^2$. Find the two monomials.

2. In 1609, Galileo discovered Jupiter's four largest moons—Callisto, Ganymede, Io, and Europa. In how many different orders might Galileo have sighted these moons?

3. The product of two monomials is x^6. How many different sums might the monomials have? What are the sums?

4. The height of a lighthouse is 25 yd plus half its height. How high is the lighthouse?

5. What is the sum of the first ten odd numbers? What is the sum of the first 75 odd numbers?

6. Take a certain monomial. Add $7x^3$ to it. Subtract $10x$. Add $4x^3 + 5x$. The result is $20x^3 - 5x$. Find the original monomial.

7. Akira knows he set his watch to the correct time at 6:00 a.m. When he hears a radio announcer say that it is exactly 10:30 a.m., Akira notices that his watch is 18 min fast. At that rate, how many minutes fast will his watch be by 5:00 p.m.?

8. Jana is thinking of a number. If she multiplies this number by itself and then adds the number to the result, she gets 420. What is Jana's number?

COMPUTER TIP

RUN this program to obtain some data that can help you in solving Problem 5.

```
10 PRINT "ODD NUMBERS",
   "SUM"
20 LET X = -1 : LET S = 0
30 FOR N = 1 TO 7
40 LET X = X + 2; LET
   S = S + X
50 PRINT N,S
60 NEXT N
```

9. A rectangle has a perimeter of $22x$. One side of the rectangle measures $3x$. Write an expression for the area of the rectangle.

10. Study this set of statements.

$$(x^1)^1 = x^1$$
$$(x^{11})^{11} = x^{121}$$
$$(x^{111})^{111} = x^{12,321}$$

Find $(x^{1,111})^{1,111}$ and $(x^{111,111})^{111,111}$.

11. On planet S there are four kinds of creatures: Lums, Mads, Nogs, and Ools. Each kind of creature has a standard number of eyes, 6, 8, 14, or 16. Nogs have more than twice as many eyes as Lums. The number of eyes of a Nog is not a perfect square. Mads have more eyes than Ools. How many eyes does each type of creature have?

12. A town is placing two rows of benches in a rectangular park, one row along the north end and one along the south end. Those ends of the park are 200 ft long. Each bench is 8 ft long, and they are to be 4 ft apart. If the rows of benches are to begin and end at a corner of the park, how many benches will be needed?

13. A checkerboard is made up of 8 rows of 8 squares. How many squares of all sizes are there in a checkerboard?

12-6 Factoring

EXPLORE

Suppose that a rectangle has an area of 36 ft^2. What are all the possible whole-number dimensions that the rectangle might have? Two possibilities are shown on the grid below.

What are all the factors of 36?

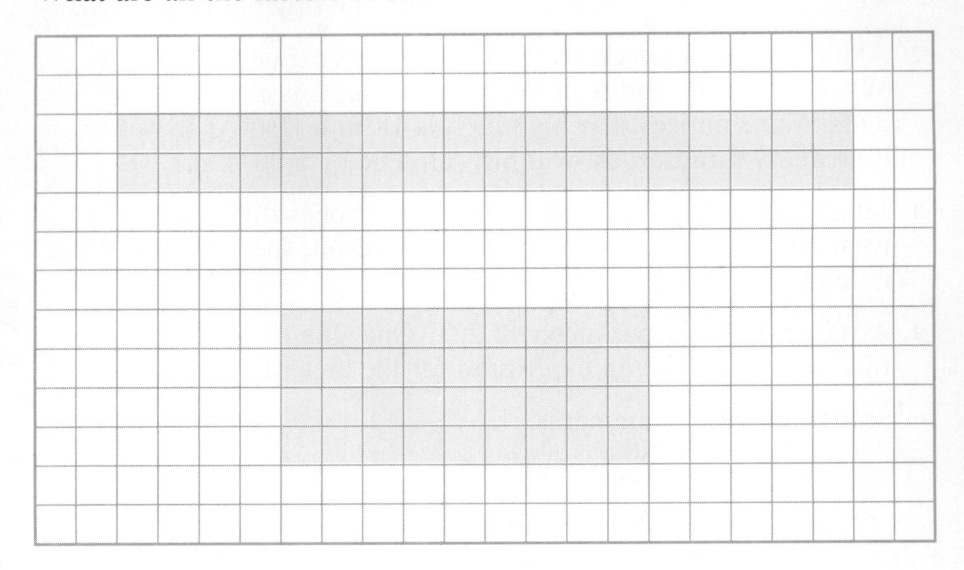

COMPUTER TIP

You may want to RUN this program to check your answers.

```
10 FOR L = 1 TO 6
20 FOR W = 1 TO 36
30 IF L * W = 36 THEN
   PRINT L; " FT BY ";W;"
   FT"
40 NEXT W
50 NEXT L
```

SKILLS DEVELOPMENT

Recall that, to simplify expressions, you can apply the distributive property.

$$2(3x + 4) = 2(3x) + 2(4) = 6x + 8$$

You began with the factors and multiplied them to find the product. You can reverse this process to find what factors were multiplied to obtain the product. The reverse process is called **factoring.**

The **greatest common factor** of two or more monomials is the *common factor* having the greatest numerical factor and the variable or variables of greatest degree.

Example 1

Find the greatest common factor of the monomials.

a. $4x$ and $16xy$ b. $-6a^2b$ and $12ab$

Solution

a. Write the factors of $4x$: $(2^2)(x)$

Write the factors of $16xy$: $(2^4)(x)(y)$

The GCF of $4x$ and $16xy$ is $(2^2)(x)$, or $4x$.

b. Write the factors of $-6a^2b$: $(2 \times 3)(a^2)(b)$

Write the factors of $12ab$: $(2^2 \times 3)(a)(b)$

The GCF of $-6a^2b$ and $12ab$ is $(2 \times 3)(a)(b)$, or $6ab$. ◄

To factor a polynomial, look for the greatest monomial factor common to all the terms. Then write the polynomial as the product of the greatest monomial factor and another polynomial.

Example 2

Factor each polynomial.

a. $3a + 6$ b. $4x^2y - 18x$

Solution

a. Find the GCF of each term of $3a + 6$.

factors of $3a$: $(3)(a)$
factors of 6: (2×3)

The GCF is 3.

Use the GCF to rewrite the polynomial.

$$3a + 6 = 3 \times a + 3 \times 2$$
$$= 3(a + 2) \leftarrow \text{Use the distributive property.}$$

So, $3a + 6 = 3(a + 2)$.

b. Find the GCF of each term.

factors of $4x^2y$: $(2^2)(x^2)(y)$
factors of $18x$: $(2 \times 3^2)(x)$

The GCF is $2x$.

Use the GCF to rewrite the polynomial.

$$4x^2y - 18x = (2x)(2xy) - (2x)(9)$$
$$= 2x(2xy - 9)$$

So, $4x^2y - 18x = 2x(2xy - 9)$. ◄

Example 3

The formula for the surface area of a rectangular prism is $SA = 2lw + 2wh + 2lh$, where l represents the length, w represents the width, and h represents the height. Use factoring to rewrite the formula.

Solution

Rewrite the formula with the right side as the polynomial in factored form. The GCF of the terms is 2.

$SA = 2lw + 2wh + 2lh$
$SA = 2(lw + wh + lh)$ ◄

TRY THESE

Find the greatest common factor of the monomials.

1. $3y$ and $15yz$
2. $4cd^2$ and $20cd$

Factor each polynomial.

3. $4x + 12$
4. $6y^2z - 9y$

5. The mean of four numbers a, b, c, and d can be found using the formula $M = \frac{1}{4}a + \frac{1}{4}b + \frac{1}{4}c + \frac{1}{4}d$. Rewrite the formula by factoring.

EXERCISES

PRACTICE/ SOLVE PROBLEMS

Find the greatest common factor of the monomials.

1. $8y$ and $4xy^2$
2. $3c^2$ and $18cd$
3. $5ab^2$ and $15a^2b$
4. $7r^2s$ and $9rs$
5. $10s^2t$ and $25s^3t^2$
6. $14x^2y^2$ and $42x^2y$

Match each set of monomials with their greatest common factor.

7. $2x^2, 2x^2y, 2x^2z$
8. $3y^3z, 3yz, 6y^2z$
9. $6ab^2, -4a^2b, 12a^2b^2$
10. $-2x^3y, -x^2y, -2xy^2$
11. $ax^2, 2bx^2, 3cx^2, 5dx^2$

a. $2ab$
b. $-xy$
c. x^2
d. $2x^2$
e. $3yz$

Factor each polynomial.

12. $35a - 7ab$
13. $9 - 18x$
14. $5x^2 - 4x$
15. $12ab - 3b^2$
16. $35mn - 15m^2n^2$
17. $24x^2 - 6xy$
18. $3x^2 + 6x^2y - 18xy$
19. $xy + xy^2 - y^2$
20. $8ab - 16a^2b + 32ab^2$
21. $7xy - 14x^2y^2 + 28xyz$

EXTEND/ SOLVE PROBLEMS

Find the missing factor.

22. $8x^5 = (2x^2)(\underline{\quad ? \quad})$
23. $-6a^3b^2 = (2ab^2)(\underline{\quad ? \quad})$
24. $21c^2d = (3c^2)(\underline{\quad ? \quad})$
25. $12xy^2 = (3x)(\underline{\quad ? \quad})$
26. $-15x^3y^5 = (-5x^2y^2)(\underline{\quad ? \quad})$

Factor each polynomial.

27. $\frac{1}{3}xyz - \frac{1}{3}xy^2z - \frac{1}{3}x^2yz^2$
28. $14f^2g - 42fg^2 - 56g^2h$
29. $4rs + 4st - 4tu + 4ru$
30. $12c^3d - 6c^2d^2 - 14c^2d + 4c^2$

THINK CRITICALLY/ SOLVE PROBLEMS

31. Explain the order of operations to follow in simplifying this expression. Simplify the expression, then factor.
$6x^2 - x[3x - 4x(2x - 3)]$

► READ
► PLAN
► SOLVE
► ANSWER
► CHECK

Problem Solving Applications:

USING POLYNOMIALS IN GEOMETRY

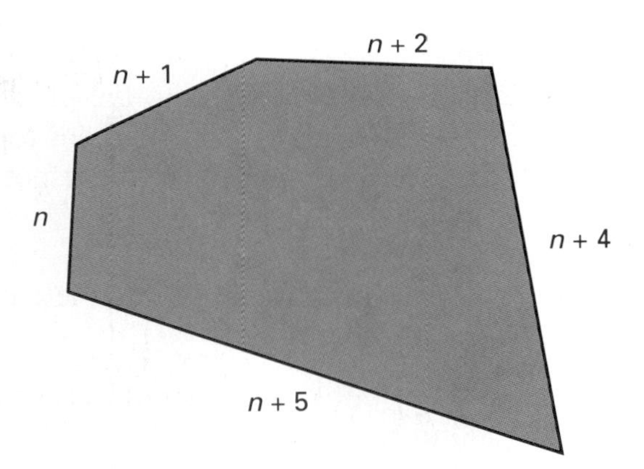

As you continue your study of geometry, you will find that the measures of the sides of a figure are sometimes given as variable expressions. Since the perimeter of a polygon is the sum of the measures of its sides, you can create a polynomial to represent the perimeter.

For instance, the perimeter of the figure above is represented by the polynomial $5n + 12$.

Write an expression for the perimeter of each figure. Then simplify the expression.

1. 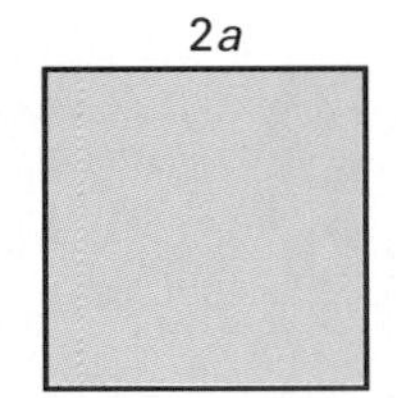

2. 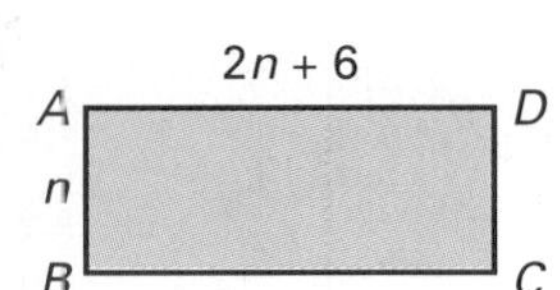

3. 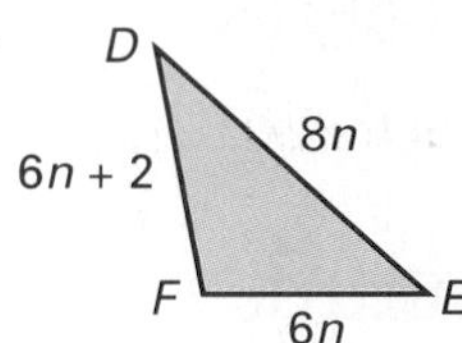

4. 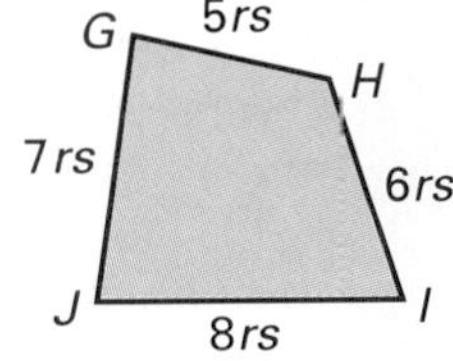

5. Write an expression for the area of the rectangle in Exercise 2.

6. The formula for the area of a trapezoid is $A = \frac{1}{2}hb_1 + \frac{1}{2}hb_2$, where h is the height, b_1 is the length of one base, and b_2 is the length of the other base. Rewrite the formula with the right side as a polynomial in factored form.

7. The formula for the surface area of a cylinder with radius r and height h is $SA = 2\pi rh + 2\pi r^2$. Rewrite the formula with the right side as a polynomial in factored form.

Write an expression in factored form for the area of the shaded region.

8. 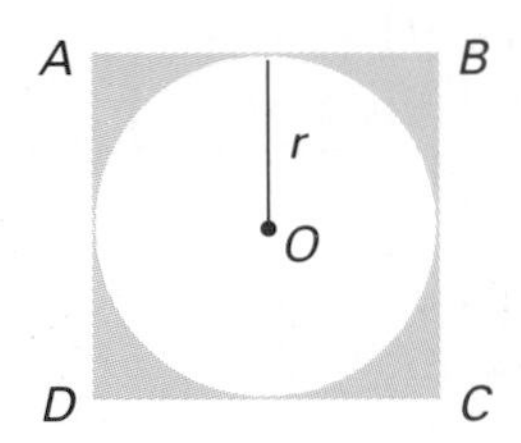

9. 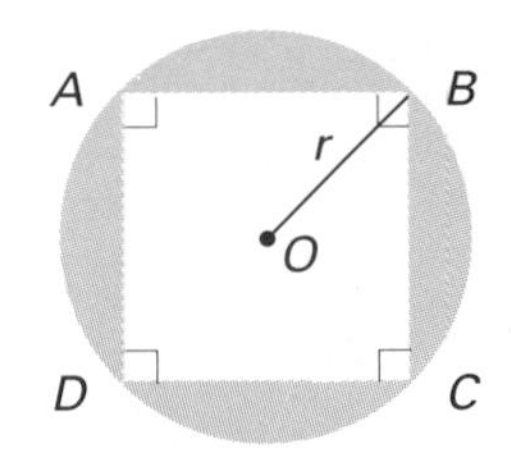

12-7 Problem Solving Skills:

CHUNKING

- ► READ
- ► PLAN
- ► SOLVE
- ► ANSWER
- ► CHECK

The term **chunking** refers to the process of collecting several pieces of information and grouping them together as a single piece of information. Although you might not think much about it, you use chunking every day.

> You chunk the individual letters *d*, *o*, and *g* and think of the word *dog*.
>
> You chunk your parents, sisters, brothers, and other relatives and talk about your *family*.

In algebra, you are chunking when you think of a polynomial, not as a group of individual terms, but rather as a single variable expression. Sometimes, chunking can help you solve a problem that is different from any problem you have encountered before.

PROBLEM

Factor $n(2 + n) + 5(2 + n)$.

SOLUTION

$$n(2 + n) + 5(2 + n) = n\boxed{(2 + n)} + 5\boxed{(2 + n)}$$ ← **Think of $(2 + n)$ as a chunk.**

$$= (n + 5)\boxed{(2 + n)}$$ ← **Use the distributive property to factor out the chunk.**

So, $n(2 + n) + 5(2 + n) = (n + 5)(2 + n)$.

You can use a diagram like the one below to prove that this answer is correct.

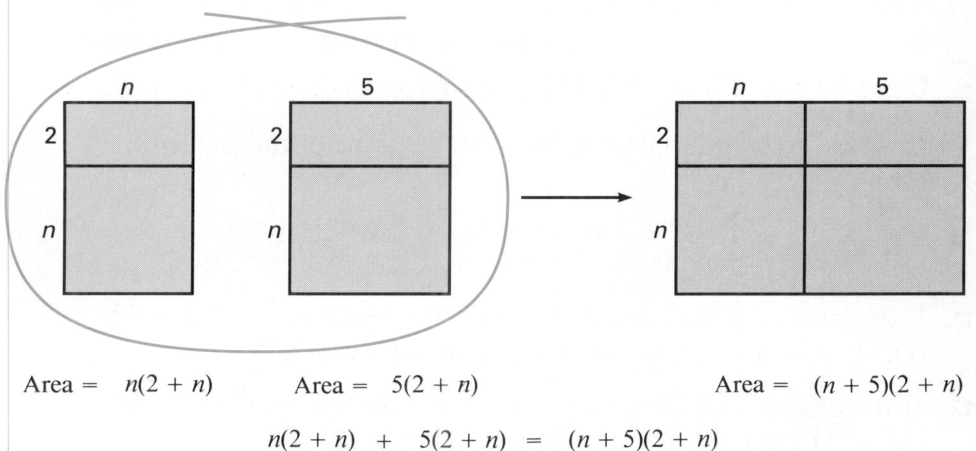

Area = $n(2 + n)$ Area = $5(2 + n)$ Area = $(n + 5)(2 + n)$

$$n(2 + n) + 5(2 + n) = (n + 5)(2 + n)$$

PROBLEMS

Write the mathematical sentence that each diagram represents.

1.

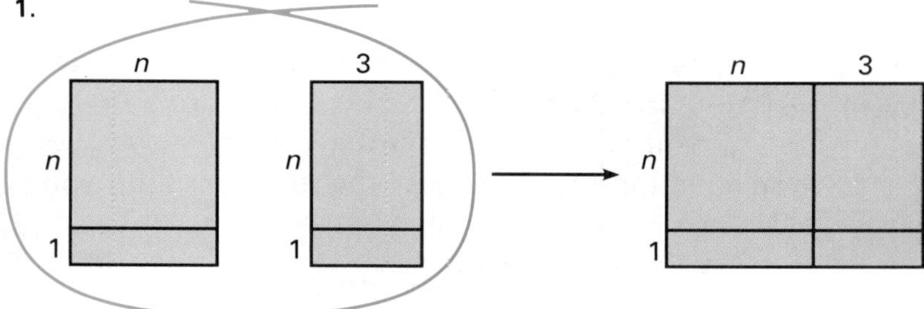

2.

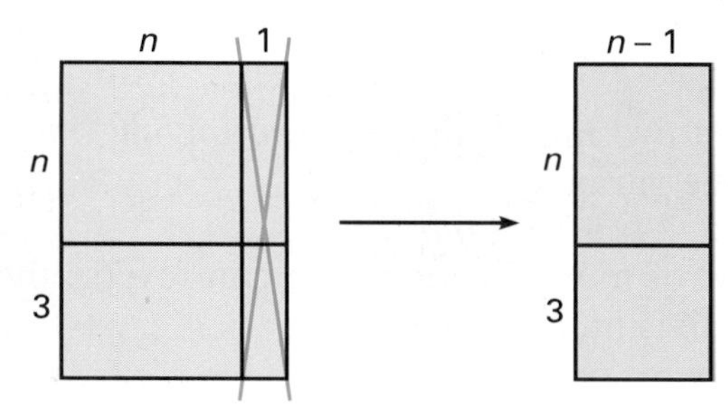

COMPUTER

After you complete Exercise 11, change 25 to 1, then to 4, then to 9, and then to 16. After each change, use chunking to solve. You can check your results by RUNning this program.

```
10 FOR X = 1 TO 4
20 LET A = X − 2
30 LET B = −X − 2
40 PRINT "IF (Y + 2) ^ 2 =
   ";X ^ 2;" THEN
   Y = ";A;" OR ";B
50 NEXT X
```

Use chunking to factor each expression.

3. $a(a - 3) + 2(a - 3)$

4. $(t + 5)t - (t + 5)7$

5. $r(r + 6) + 2(6 + r)$

6. $(k - 1)3 + k(k - 1)$

7. $c^2(c + 1) + c(c + 1) + 5(c + 1)$

8. $(z - 5)z^2 - (z - 5)z + (z - 5)2$

Describe how chunking can be used for each of Exercises 9–12.

9. How can you use chunking to solve the equation $2^{m+1} = 32$? (*Hint:* $32 = 2^?$)

10. How can you use chunking to solve the equation $|d + 1| = 3$? (*Hint:* Which two integers have an absolute value of 3?)

11. How can you use chunking to solve the equation $(y + 2)^2 = 25$? (*Hint:* Which two integers squared are equal to 25?)

12. How can you use chunking to solve the equation $\sqrt{b + 4} = 3$? (*Hint:* The square root of which number is 3?)

12-8 Dividing by a Monomial

EXPLORE

You know that division "undoes" multiplication.

$$-4 \times 9 = -36, \quad \text{so} \quad -36 \div 9 = -4.$$

1. Simplify the product $3g \times 5h$.
2. Write a division equation that "undoes" the multiplication you just performed in a.
3. Write a division equation that "undoes" each of these.
   a. $2a \times 5a = 10a^2$
   b. $6y \times 4y^2 = 24y^3$
   c. $2m \times 7mn = 14m^2n$

SKILLS DEVELOPMENT

You have found the product of two monomials by multiplying the coefficients and multiplying the variables.

$$3m \times 2n = (3 \times 2)(mn) = 6mn$$

Because division is the inverse of multiplication, you can reverse the process to obtain the related division sentence.

$$3m \times 2n = 6mn, \text{ so } 6mn \div 3m = 2n.$$

Now let's look at a different way to find the quotient $6mn \div 3m$.

$$\frac{6mn}{3m} = \frac{6}{3} \times \frac{mn}{m}$$

$$= \frac{\overset{2}{\cancel{6}}}{\underset{1}{\cancel{3}}} \times \frac{mn}{m} \leftarrow \text{Divide both coefficients by their GCF, which is 3.}$$

$$= \frac{\overset{2}{\cancel{6}}}{\underset{1}{\cancel{3}}} \times \frac{\overset{1}{\cancel{m}}n}{\underset{1}{\cancel{m}}} \leftarrow \text{Divide the variable parts by their GCF, which is } m.$$

$$= 2n$$

This division illustrates the following rule for dividing monomials:

► To simplify the quotient of two monomials, find the quotient of any numerical coefficients, then find the quotient of the variables.

CHECK UNDERSTANDING

What will be the coefficient of each quotient?

1. $\frac{-16ab}{8b}$
2. $\frac{-25pq}{-5p}$
3. $\frac{36st}{-6s}$

Example 1

Simplify.

a. $\frac{-4mn}{2m}$ b. $\frac{wxz}{wz}$

Solution

a. $\frac{-4mn}{2m} = \frac{-4}{2} \times \frac{mn}{m} = \frac{\overset{-2}{\cancel{-4}}}{\underset{1}{\cancel{2}}} \times \frac{\overset{1}{\cancel{m}}n}{\underset{1}{\cancel{m}}} = -2 \times n = -2n$

b. $\frac{wxz}{wz} = \frac{\overset{1}{\cancel{w}}x\overset{1}{\cancel{z}}}{\underset{1}{\cancel{w}}\underset{1}{\cancel{z}}} = \frac{x}{1} = x \leftarrow$ **Both coefficients are 1.** ◄

Recall the *quotient rule for exponents:* To divide two powers having the same base, you subtract the exponent of the denominator from that of the numerator.

$$\frac{a^m}{a^n} = a^{m-n}$$

You can use the quotient rule for exponents to find the quotients of monomials having the same base.

TALK IT OVER

Refer to Example 2b. Explain why the exponent 2 in the denominator is subtracted from the exponent 3 in the numerator but not from the exponent 5.

Example 2

Simplify.

a. $\frac{-16x^8}{4x^4}$ **b.** $\frac{-24x^3y^5}{-3x^2y}$

Solution

a. $\frac{-16x^8}{4x^4} = \left(\frac{-16}{4}\right)x^{8-4} = -4x^4$

b. $\frac{-24x^3y^5}{-3x^2y} = \left(\frac{-24}{-3}\right)x^{3-2}y^{5-1}$

$= 8xy^4$ ◄

Each of the statements is true.

$$\frac{18 + 24 + 36}{6} = \frac{78}{6} = 13 \quad \text{and} \quad \frac{18}{6} + \frac{24}{6} + \frac{36}{6} = 3 + 4 + 6 = 13$$

So you can arrive at this conclusion.

$$\frac{18 + 24 + 36}{6} = \frac{18}{6} + \frac{24}{6} + \frac{36}{6}$$

This example suggests the following general rule.

► When a, b, and c are real numbers, and c is not equal to 0,

$$\frac{a + b}{c} = \frac{a}{c} + \frac{b}{c}$$

CHECK UNDERSTANDING

Why is the quotient $\frac{12x^2}{2x^2}$ equal to 6?

You can use this rule to divide a polynomial by a monomial.

Example 3

Simplify.

a. $\frac{6a + 9}{3}$ **b.** $\frac{2x^4 + 8x^3 + 12x^2}{2x^2}$

Solution

a. $\frac{6a + 9}{3} = \frac{6a}{3} + \frac{9}{3}$ ← **Divide each term of the polynomial by the divisor.**

$= 2a + 3$

b. $\frac{2x^4 + 8x^3 + 12x^2}{2x^2} = \frac{2x^4}{2x^2} + \frac{8x^3}{2x^2} + \frac{12x^2}{2x^2} = x^2 + 4x + 6$ ◄

TRY THESE

Simplify.

1. $\frac{-6xy}{2y}$
2. $\frac{abc}{-ac}$
3. $\frac{8cd}{-4cd}$
4. $\frac{12wz}{3z}$
5. $\frac{24z^6}{6z^5}$
6. $\frac{27x^7y}{3x^2}$
7. $\frac{18x^6}{-9x^3}$
8. $\frac{-21a^7b^8}{-3ab^5}$
9. $\frac{4x + 8}{2}$
10. $\frac{6y - 12}{3}$
11. $\frac{-14b + 21}{7}$
12. $\frac{3y^4 + 6y^3 + 12y^2}{3y}$
13. $\frac{4x^5 - 8x^3 - 12x^2}{4x^2}$

EXERCISES

PRACTICE/ SOLVE PROBLEMS

Simplify.

1. $\frac{20cd}{4d}$
2. $\frac{18gh}{2g}$
3. $\frac{40ry}{10r}$
4. $\frac{-12ab}{2a}$
5. $\frac{-12pq}{3q}$
6. $\frac{-12st}{-4t}$
7. $\frac{y^3z}{y^2z}$
8. $\frac{x^3y^2z}{x^2y}$
9. $\frac{cd^6}{d^5}$
10. $\frac{4w^2x^2}{-2wx}$
11. $\frac{ab^5}{b^3}$
12. $\frac{-x^5y^5}{x^3y^4}$
13. $\frac{2a + 12}{2}$
14. $\frac{18y + 6}{2}$
15. $\frac{9a - 18b}{3}$
16. $\frac{32c - 8b - 4a}{4}$
17. $\frac{6x^3 - 9x^2 + 3x}{3x}$
18. $\frac{28x^2y^2 - 21xy^2 + 14xy}{7xy}$
19. $\frac{9a^2b^2c^2 - 15abc^2}{3abc}$

EXTEND/ SOLVE PROBLEMS

Simplify.

20. $\frac{28a^8}{7a^2}$
21. $\frac{18b^9}{-3b}$
22. $\frac{bm^3}{-m^2}$
23. $\frac{x^3}{x^2y}$
24. $\frac{x^2y^2}{y^2z}$
25. $\frac{5z^3}{15z}$
26. $\frac{-18r^6s^2t^3}{-3r^2st^2}$
27. $\frac{-42q^5r^2s^3}{7qrs^2}$
28. $\frac{-27a^3b^2c^3}{9abc}$
29. $\frac{24a^4b - 12a^2b^2 + 18a^3b^3}{3a^2b}$
30. $\frac{35c^3d^2e - 50c^4d^3e^2}{5cde}$
31. $\frac{8z^2xy - 12x^2zy + 16zx^2y^2 - 24z^2x^3y^3}{4zxy}$
32. $\frac{56a^4b^4c^3 - 49a^3b^3c^2 + 35ab^2c^2 - 28a^2bc^4}{-7abc^2}$

Write an expression for the unknown dimension of each rectangle.

33. Area: $25pq$

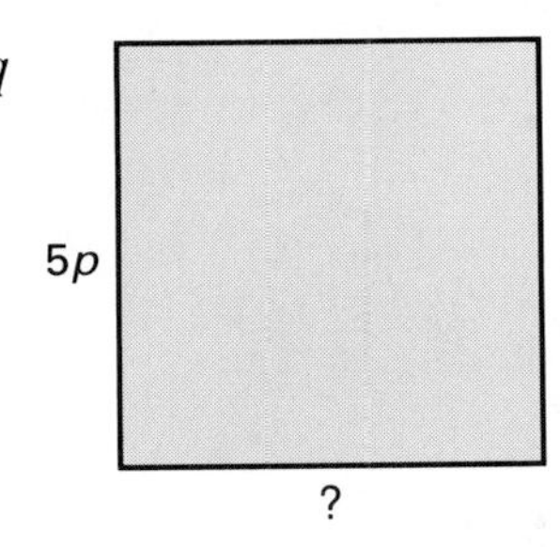

34. Area: $36abc$

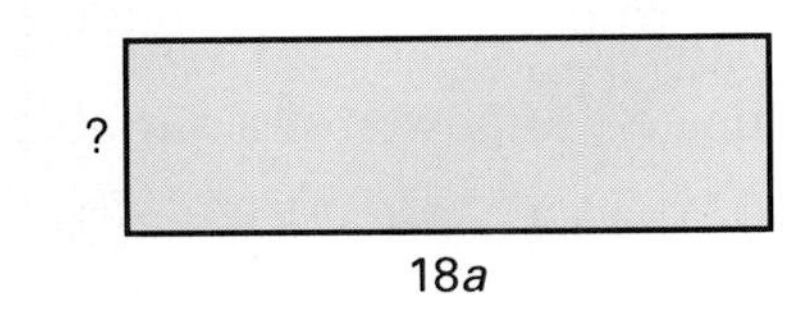

35. A rectangle has area $36ab$ square units. The width is $4b$ units. Write an expression for the length of the rectangle.

36. The area of the rectangle at the right is $28x^3y^2$. Write an expression for the unknown dimension.

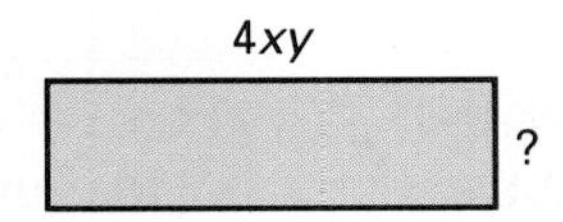

37. The area of the square at the right is $25m^4n^6$. Write an expression for the unknown dimension.

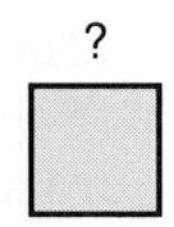

38. The product of $3x^2y^5$ and a certain monomial is $21x^7y^9z^2$. Find the missing monomial factor.

THINK CRITICALLY/ SOLVE PROBLEMS

39. Write an expression for the volume of a rectangular prism with a length of $6xz$, a width of $3y^2$, and a height of $5z$.

40. A prism with a volume equal to the one in Exercise 39 has a length of $15y$ and a width of $2z$. Write an expression for its height.

How could you use the distributive property to simplify these expressions?

41. $\dfrac{x^2(x+1) - 4(x+1)}{x+1}$

42. $\dfrac{y^2(y+5) - 9(y+5)}{y^2 - 9}$

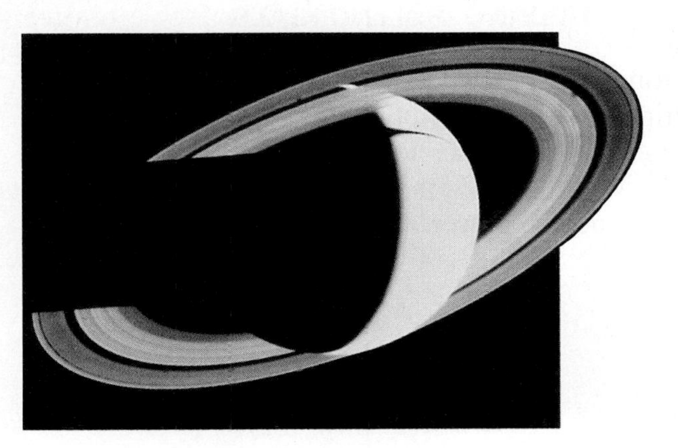

CHAPTER 12 ● REVIEW

Choose a word from the list to complete each statement.

1. Use a(n) __?__ to indicate the power of a variable.
2. The __?__ property is used to rewrite $a(b + c)$ as $ab + ac$.
3. $x^3 + x - 3$ is an example of a(n) __?__.
4. In $5m$, 5 is the numerical __?__.
5. $4x^2y^2z$ is an example of a(n) __?__.

a. coefficient
b. distributive
c. exponent
d. monomial
e. trinomial

SECTION 12–1 POLYNOMIALS (pages 436–439)

► A **monomial** is an expression that is either a single number, a variable such as x or y, or the product of a numerical **coefficient** and one or more variables. A monomial that contains no variable is called a **constant.**

► The sum of two or more monomials is a polynomial. **Like terms** are terms in a polynomial that are exactly alike, or are alike except for their numerical coefficients. One way to simplify a polynomial is to combine like terms.

Simplify.

6. $9n - n$
7. $4m + 5m$
8. $4x^2 - 3x + 2x^2$

SECTION 12–2 ADDING AND SUBTRACTING POLYNOMIALS (pages 440–443)

► To add polynomials, write the sum and then combine like terms.

► To subtract a polynomial, add its opposite and simplify.

Simplify.

9. $(8h + 4) + (5 - 5h)$
10. $(13a^2 + 7a) - (4a^2 - a)$

SECTIONS 12–3 AND 12–4 MULTIPLYING POLYNOMIALS (pages 444–451)

► To find the product of two monomials, use the commutative and associative properties of multiplication and simplify.

► To multiply two powers having the same base, add the exponents.

► To find the power of a monomial that is itself a power, multiply the exponents.

► To find the power of a product, find the power of each factor and multiply.

► To multiply a polynomial by a monomial, use the distributive property and the rules for exponents.

Simplify.

11. $(2a)(-4b)$
12. $(x^2y^2)^3$
13. $6w(3 - 4w + 2w^2)$
14. $-k(k^3 - 7k - 5)$

SECTIONS 12–5 AND 12-7 PROBLEM SOLVING (pages 452–453, 458–459)

- ▶ Experience in applying problem solving strategies will help you decide which one to use to solve a particular problem.
- ▶ You can group, or "chunk," by grouping pieces of data together.

Solve. Tell what strategy you used.

15. Rhoda is thinking of a number. If she multiplies the number by itself and then subtracts half the number from the result, she gets 248. What number is Rhoda thinking of?

Use chunking to factor each expression.

16. $x(x + 4) + 5(x + 4)$

17. $(m + 1)4 - m(m + 1)$

SECTION 12–6 FACTORING (pages 454–457)

- ▶ To **factor** an expression, write the expression as a product of factors.
- ▶ The greatest common factor of two or more monomials is the common factor having the greatest numerical factor (coefficient) and the variable or variables of greatest degree.

Factor each polynomial.

18. $6x^2y - 24x$

19. $ab - a^2b + b^2$

SECTION 12–8 DIVIDING BY A MONOMIAL (pages 460–464)

- ▶ To simplify the quotient of two monomials, find the quotient of any numerical coefficients and then find the quotient of the variables.
- ▶ To divide a polynomial by a monomial, divide each monomial term of the polynomial by the monomial.

Simplify.

20. $\frac{24mn}{-3m}$

21. $\frac{-15a^3b^2}{5ab}$

22. $\frac{12c + 15}{3}$

23. $\frac{4x^4 + 6x^3 + 14x^2}{2x^2}$

USING DATA For Exercises 24 and 25, use the graph on page 437.

Solve.

24. The magnitude (s) of star X is related to the magnitude (m) of the star Mira by the formula $s = 3m - 6$. Rewrite this formula by factoring.

25. What was the magnitude of star X on the 440th day from the beginning of Mira's first cycle?

CHAPTER 12 ● TEST

Simplify.

1. $4m + (-3m) + 2m$

2. $3x^2 + (-2x^2) + 4x^2$

3. Three meteorite fragments are located at the points of a triangular region, *ABC*. Write a simplified expression for the perimeter of the region.

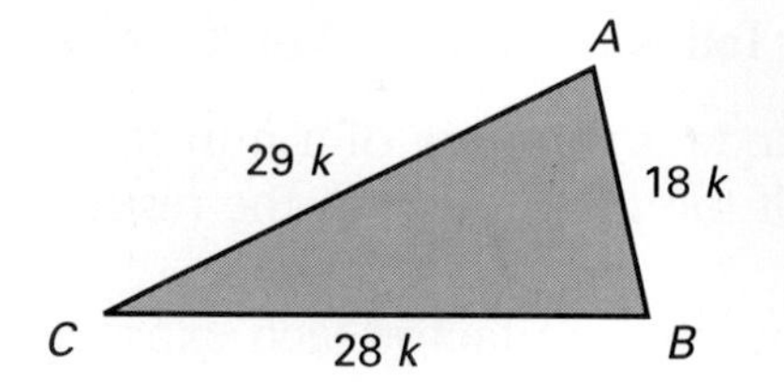

4. $(6z^2 + 4z - 5) + (3z^2 - z + 2)$

5. $(5p^2 + 4p - 8) - (3p^2 + 2p - 12)$

6. Write a simplified expression for the perimeter of the triangle.

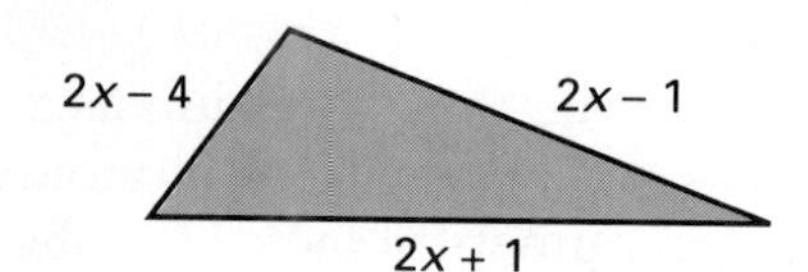

7. $(-2x)(3xy)$

8. $(-2m^2p^2)(pq)$

9. $(3x)(2x)$

10. $(4k^3)(-2k^4)$

11. $\dfrac{12ab}{4a}$

12. $\dfrac{-6p^2}{2p^2}$

13. $\dfrac{15t^2m^7}{-3tm^3}$

14. $3(y - 2)$

15. $(-3m)(2m - 1)$

16. $a^2(a^3 - 3a + c)$

Factor each polynomial.

17. $14p - 8$

18. $8m^2n + 24m$

19. $2rs - 12r^2s^2 + 10rst$

Simplify.

20. $\dfrac{-15ab}{3a}$

21. $\dfrac{-36x^3y^2z}{-9xz}$

22. $\dfrac{18q^5 + 21q - 9q^3}{3q}$

Solve.

23. A rectangular garden has an area of $8p^3t^5$ square units. The width is $4pt$ units. Write an expression for the length.

CHAPTER 12 ● CUMULATIVE REVIEW

1. Organize these numbers into a stem-and-leaf plot.

   72 68 64 49 51 58 54 62 82
   64 53 57 65 76 79 53 71 73

Complete.

2. $(21 + 45) + 16 = 21 + (■ + 16)$
3. $5(3 + 8) = (5 \times 3) + (■ \times 8)$

4. Is the argument *valid* or *invalid?*

   If the mattress is too soft, then you will sleep uncomfortably.
   You slept uncomfortably.
   Therefore, the mattress is too soft.

5. Write a statement about the size of each circle below. Then check to see if your statement is true or false.

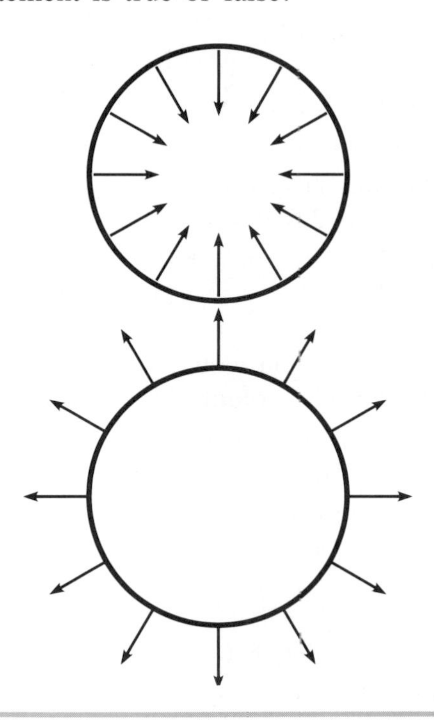

Replace ● with <, >, or =.

6. 0.18 ● $\frac{3}{16}$
7. $\frac{5}{8}$ ● 0.621
8. $\frac{3}{4}$ ● $\frac{3}{10}$
9. 0.02 ● $\frac{1}{5}$

10. Find the perimeter of a rectangle whose length is 7 in. and whose width is 6.3 in.

Subtract.

11. $-27 - (-15)$
12. $-21 - 4$
13. $4 - (-19)$
14. $-11 - (-11)$

Solve each equation.

15. $27 = -4y + 3$
16. $\frac{k}{-5} = 7$
17. $38 = 5b$
18. $3n - 4.2 = 1.2$

Write three equivalent ratios for each given ratio.

19. 7:3
20. 16 to 24

21. The scale on a map is 1 cm:8 m. If a distance on the map is 9.25 cm, what is the actual distance?

22. 78 is 20% of what number?

23. Find the discount and the sale price. Round to the nearest cent.
    regular price: \$265
    percent of discount: 25%

24. Draw an angle with a measure of 68°. Then construct its bisector.

Use a calculator to find each square root to the nearest thousandth.

25. $\sqrt{58}$
26. $\sqrt{546}$
27. $\sqrt{292}$

Simplify.

28. $5x(-3x^2 + 2x)$
29. $2x^2(y + 3)$

1. What is the mode for this set of data?

11.4 9.1 3.1 6.8 7.2 7.0 6.8 7.5 6.8

A. 7.3 B. 7.0
C. 6.9 D. none of these

2. Complete.

$9(7 + 4) = (9 \times 7) + (■ \times 4)$

A. 9 B. 7 C. 4 D. 11

3. Which is a counterexample for the following statement?

Any number divisible by 5 is divisible by 10.

A. 10 B. 20 C. 15 D. 40

4. Compare. $\frac{11}{15}$ ● 0.7

A. < B. >
C. = D. ≈

5. Find the perimeter.

A. 925 m
B. 121.3 m
C. 78.1 cm
D. 99.7 m

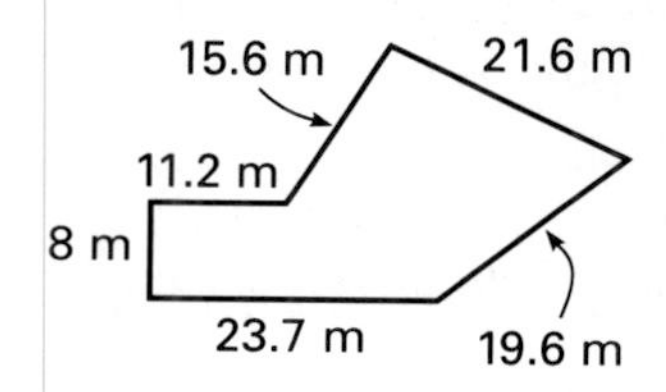

6. Compare. $4 \times (-6)$ ● $6 \times (-4)$

A. = B. > C. < D. ≈

7. Solve. $1\frac{3}{4}n = 17\frac{1}{2}$

A. $n = 30$ B. $n = 10$
C. $n = 1$ D. $n = 20$

8. Solve. $\frac{3}{4} = \frac{31.5}{x}$

A. $x = 4.2$ B. $x = 41.4$
C. $x = 21$ D. $x = 42$

9. 18 is $33\frac{1}{3}\%$ of what number?

A. 6 B. 54
C. 56 D. none of these

10. What is the sale price of an $800 copier if the discount rate is 20%?

A. $160 B. $600
C. $960 D. none of these

11. Find $m \angle 1$.

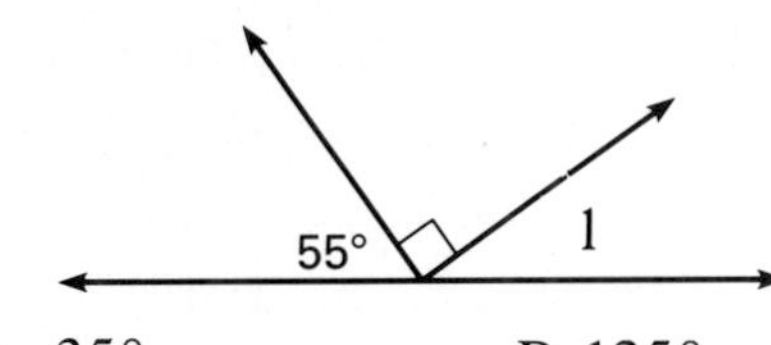

A. 35° B. 125°
C. 145° D. none of these

12. Find the missing angle measure.

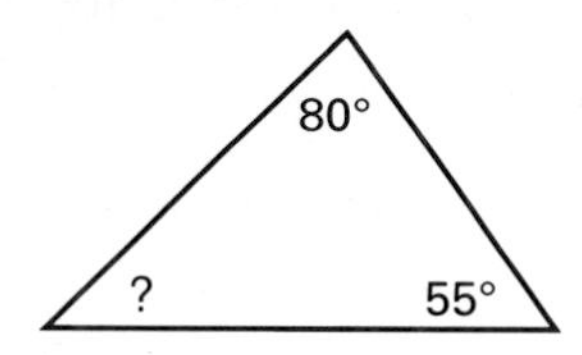

A. 55° B. 10°
C. 45° D. none of these

13. Find the surface area of a sphere with a radius of 2.5 ft.

A. 25 ft^2 B. 19.6 ft^2
C. 78.5 ft^2 D. 31.4 ft^2

14. Which formula would you use to calculate the volume of a cone?

A. $v = l \times w \times h$
B. $v = \frac{1}{3}Bh$
C. $v = \pi r^2 h$
D. $v = Bh$

15. Simplify. $2m + (m - 3)$

A. $2m^2 - 3$ B. $3m - 3$
C. $2m - 3$ D. none of these

16. Simplify. $(3x - 2) - (4x + 3)$

A. $x - 1$ B. $7x + 1$
C. $-x - 5$ D. $x + 5$

CHAPTER 13 SKILLS PREVIEW

Find each probability. Use this spinner.

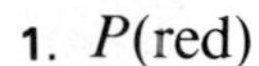

1. P(red)
2. P(not white)
3. P(2 or 3)
4. P(10)
5. P(red, white, or blue)

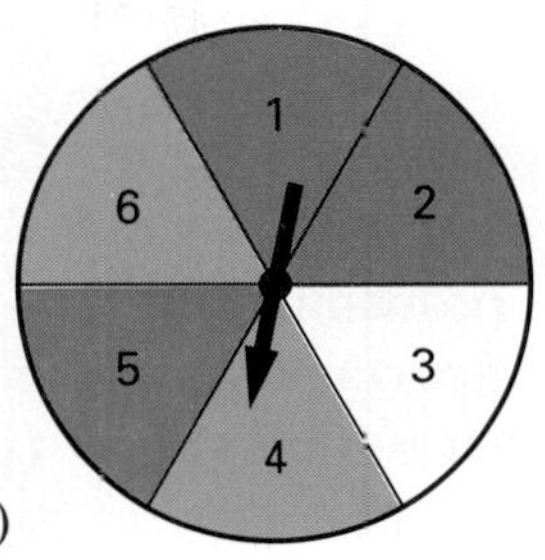

Use a tree diagram to find the number of possible outcomes in each sample space.

6. tossing a penny and then spinning a spinner like the one at the right
7. tossing a dime and tossing a number cube

Use the counting principle to find the number of possible outcomes.

8. choosing a sandwich from 3 kinds of bread and 5 sandwich fillings
9. tossing a coin five times
10. choosing a breakfast from orange juice, tomato juice, or grapefruit juice; with oatmeal or cornflakes; and with whole wheat, rye, or pumpernickel toast
11. A bag contains 3 white marbles, 2 blue marbles, and 5 orange marbles. Brian takes one marble from the bag, replaces it, and takes another. Find P(blue, then white).

A number cube is tossed 30 times. The outcomes are recorded in this table.

Number on Cube	1	2	3	4	5	6
Number of Outcomes	5	6	4	5	7	3

12. Find P(4).
13. Find P(odd number).
14. Find P(1 or 3).

CHAPTER 13

PROBABILITY

THEME Forecasting and Choices

Most situations in everyday life involve an element of chance. The ability to determine the probability of various outcomes is important in helping people decide what to do.

In this chapter you will use a formula to determine the probability of an event.

Raindrops form in clouds when tiny water droplets join together or when larger ice crystals melt. A raindrop must contain at least 1,000 droplets for it to be heavy enough to fall.

GREATEST AVERAGE ANNUAL RAINFALLS

Continent	Rainfall (in millimeters)	Place
Oceania (Pacific islands)	11,684	Mt. Wai-'ale-'ale, Hawaii
Asia	11,430	Cherrapunji, India
Africa	10,277	Debundseha, Cameroon
South America	8,991	Quibdo, Colombia
North America	6,655	Henderson Lake, British Columbia
Europe	4,648	Crkvice, Yugoslavia
Australia	4,496	Tully, Queensland

DECISION MAKING

Using Data

1. Use the information in the chart on page 472. On the average, how much more rainfall do the residents of Henderson Lake receive than do the residents of Tully?
2. In the tropical rain forests in South America, it rains nearly every day. Each year at least 2,030 mm of rain can fall. Decide which of the following ranges best describes the probability of this area suffering from a drought in the next year.
   a. 0%–20% b. 20%–50% c. 50%–100%

Working Together

Your group's task will be to track the predictions of a particular weather forecaster in the newspaper or on TV for one week. Make a chart showing the weather that was forecast and then the actual weather that occurred for each day during the week.

Based on your data, your group should decide on the probability that your forecaster's next prediction will be accurate.

13-1 Probability

EXPLORE/ WORKING TOGETHER

Work with a partner. Toss a number cube 20 times. Keep a tally of the number of times each number lands face up.

1. Predict how many times the number 5 would land face up if you tossed the number cube 40 times.
2. Toss the number cube 40 times and record the times the number 5 lands face up.
3. Compare this number to your prediction. Were you close?

SKILLS DEVELOPMENT

When you toss a number cube, there are six possible **outcomes.** Each outcome is **equally likely** to happen.

An **event** is an outcome or a combination of outcomes.

The **probability** of an event, written $P(E)$, is a number between 0 and 1. It is usually written as a fraction in lowest terms. You can find $P(E)$ by using this formula.

$$P(E) = \frac{\text{number of favorable outcomes}}{\text{number of possible outcomes}}$$

Example 1

Find each probability. Use this spinner.

a. $P(2)$
b. $P(\text{blue})$
c. $P(10)$
d. $P(\text{blue, red, or green})$

Solution

a. $P(2) = \frac{1}{6}$ ← **one favorable outcome** / ← **six possible outcomes**

b. Three of the six sections are blue.

$$P(\text{blue}) = \frac{3}{6} = \frac{1}{2}$$

c. No section is labeled 10, so there are no favorable outcomes.

$$P(10) = \frac{0}{6} = 0$$ ← **The probability of any *impossible event* is 0.**

d. All the outcomes are favorable.

$$P(\text{blue, red, or green}) = \frac{6}{6} = 1$$ ← **The probability of any *certain event* is 1.** ◄

CHECK UNDERSTANDING

1. In Example 1c, explain what is meant by an impossible event.
2. In Example 1d, how do you know that the event is certain to occur?

Example 2

Find the probability of picking a B or a Y from this set of cards without looking.

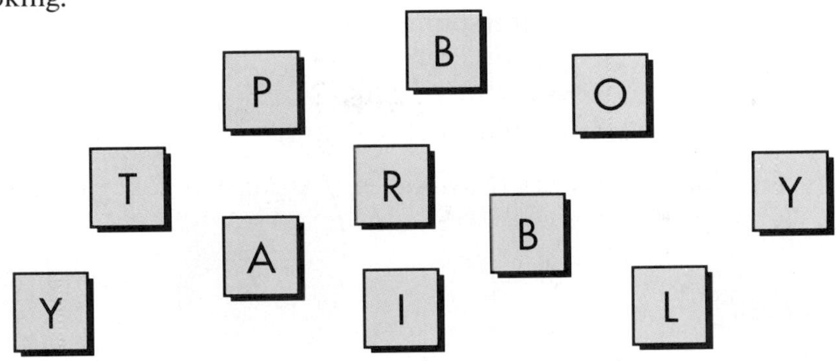

Solution

$P(\text{B or Y}) = \frac{4}{11}$ ← two Bs, two Ys; ← total number of possible outcomes ◄

Example 3

A bag contains 3 purple, 4 green, 4 blue, and 4 white marbles. What is the probability of taking a marble that is not purple out of the bag?

Solution

Find the total number of marbles.

$$3 + 4 + 4 + 4 = 15$$

There are 3 purple marbles, so subtract to find the number of *not purple* marbles.

$$15 - 3 = 12$$

Find the probability.

$P(\text{not purple}) = \frac{12}{15} = \frac{4}{5}$ ◄

TRY THESE

Find each probability. Use this spinner.

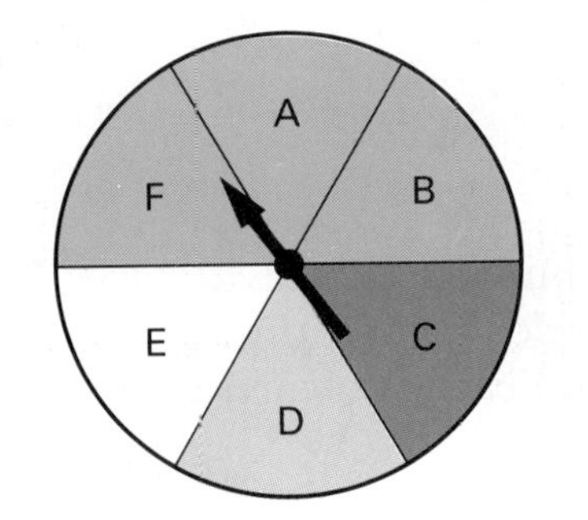

1. P(blue)
2. P(not blue)
3. P(C or D)
4. P(Z)
5. P(red, green, or blue)
6. P(A, B, C, D, E, or F)

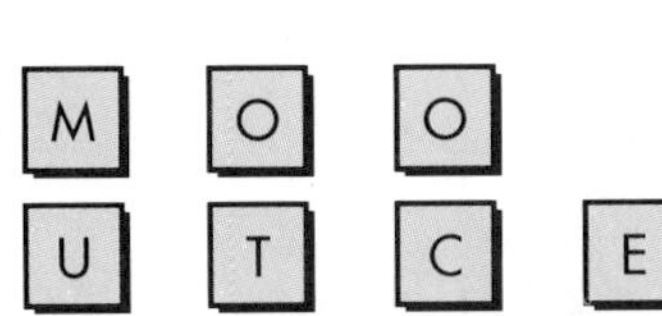

One of the cards is picked without looking. Find each probability.

7. P(O)
8. P(C, O, or E)
9. P(B)
10. P(T)

EXERCISES

PRACTICE/ SOLVE PROBLEMS

Find each probability. Use the spinner.

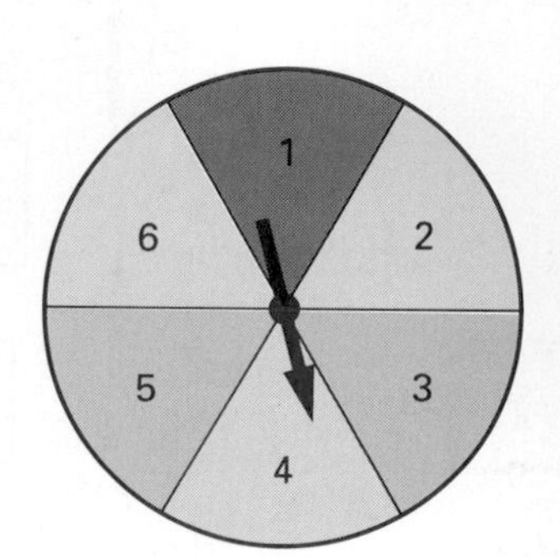

1. P(blue)
2. P(4 or 5)
3. P(even number)
4. P(number divisible by 2)
5. P(not green)
6. P(10)
7. P(not red)
8. P(6)
9. P(red)
10. P(number $>$ 0)

One of the cards is drawn without looking. Find each probability.

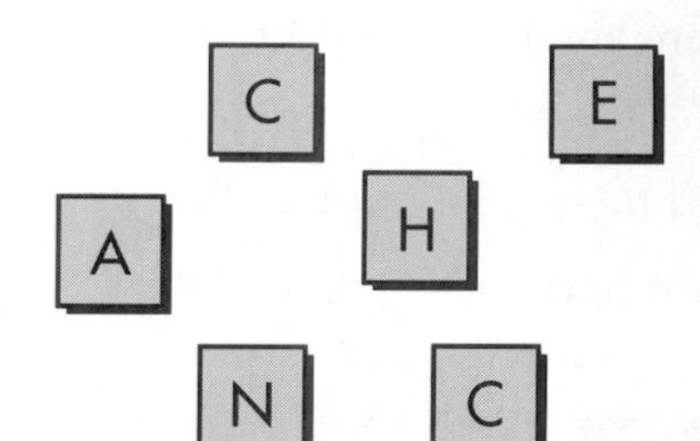

11. P(vowel)
12. P(C)
13. P(N or H)
14. P(consonant)
15. P(S)
16. P(consonant or vowel)

EXTEND/ SOLVE PROBLEMS

A number cube, with faces labeled from 1 through 6, is tossed. Find each probability.

17. P(6)
18. P(even number)
19. P(prime number)
20. P(number $>$ 10)
21. P(2, 3, 5, or 6)
22. P(composite number)

The *odds in favor of an event* can be found using this formula.

$$\text{odds in favor of } E = \frac{\text{number of favorable outcomes}}{\text{number of unfavorable outcomes}}$$

Use the formula and the spinner to find the odds in favor of each event. Write your answer as a ratio in the form *a* to *b*.

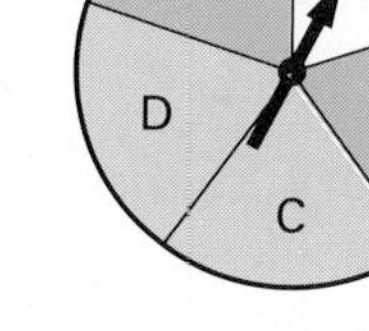

23. B

24. green

25. blue or not green

26. C or D

27. not A

28. white

29. green or A

30. B, C, or D

31. *USING DATA* Use the Data Index on page 546 to locate information that will help you answer the following question. Suppose the names of all the Mountain states were put into a hat and one name were to be drawn from the hat without looking. What is the probability that the name drawn would be that of a state larger than Utah?

THINK CRITICALLY/ SOLVE PROBLEMS

Many times you will see a probability expressed as a percent. For instance, if the probability that it will rain today is $\frac{3}{5}$, a meteorologist usually reports this as a 60% chance of rain. If an event will definitely occur, people often describe it as "100% sure."

For Exercises 32–37, choose an appropriate probability for each event from probabilities a–e.

a. 0%
b. less than 10%
c. about 50%
d. more than 90%
e. 100%

32. A name chosen at random from a telephone directory is a woman's name.

33. The next car that passes through an intersection was manufactured before 1960.

34. The sun will rise in the west tomorrow.

35. The next person you meet will be someone who sometimes watches television.

36. Water will freeze when its temperature is lowered to $-10°F$.

37. The next person you meet will be an identical twin.

Write the probability as a percent.

38. A meteorologist says that there is a 30% chance of snow tomorrow. What is the probability that it will *not* snow tomorrow?

13-2 Sample Spaces and Tree Diagrams

EXPLORE

TALK IT OVER

If you are only able to guess at the answers, what happens to your chances of getting a passing grade as the number of true–false items increases? Do you think that guessing is ever a good strategy?

Suppose that your history teacher surprises you with a true–false quiz. You haven't done the reading yet, so you have to guess at the answers.

1. List all possible ways to guess at the answer if the quiz has only one item.
2. Now suppose that the quiz has just two items. List all the possible ways you might guess at the answers.
3. What if the quiz has just three items? List all the possible ways you could guess. How can you make sure you are accounting for all the possibilities?

SKILLS DEVELOPMENT

To find a probability, it is important that you account for all outcomes. The term **sample space** is the special name we use for a set of all possible outcomes. Many times it is easy to identify a sample space. For instance, when you toss a number cube, there are six outcomes in the sample space, which you can list as follows.

$$\{1, 2, 3, 4, 5, 6\}$$

At other times, though, a sample space is not easy to identify. In such cases, you might use a **tree diagram** to picture the sample space and then count the outcomes.

Example 1

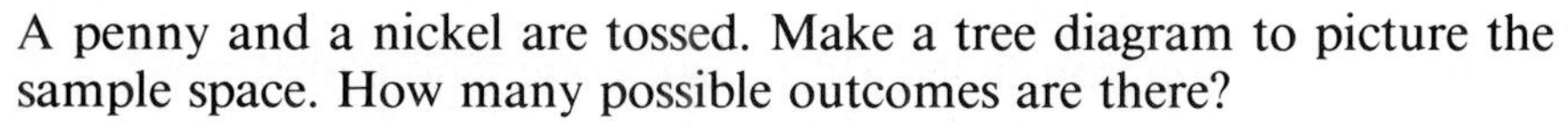

A penny and a nickel are tossed. Make a tree diagram to picture the sample space. How many possible outcomes are there?

Solution

When each coin is tossed there are two possibilities.

penny → heads or tails
nickel → heads or tails

This tree diagram pictures the outcomes.

PENNY	NICKEL		OUTCOMES
H	H	→	HH
	T	→	HT
T	H	→	TH
	T	→	TT

There are four possible outcomes. ◄

You can also use a tree diagram to find a probability.

Example 2

Three coins are tossed. Find P(exactly 2 tails).

Solution
Make a tree diagram to picture the sample space.

COIN 1	COIN 2	COIN 3		OUTCOMES
H	H	H	→	HHH
		T	→	HHT
	T	H	→	HTH
		T	→	HTT
T	H	H	→	THH
		T	→	THT
	T	H	→	TTH
		T	→	TTT

Look at the list of outcomes in the tree diagram. There are 8 possible outcomes. Only 3 of these outcomes contain *exactly* 2 tails.

HTT, THT, TTH

So $P(\text{exactly 2 tails}) = \frac{3}{8}$. ◄

TRY THESE

List each sample space.

1. tossing a coin
2. spinning this spinner

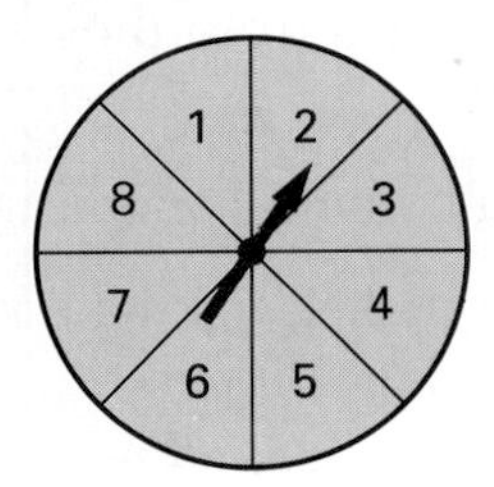

Use a tree diagram to find the number of possible outcomes in the sample space.

3. tossing four coins
4. tossing a coin and tossing a number cube

Suppose that you toss a coin and spin the spinner in Exercise 2 above. Find each probability.

5. P(tail and 5)
6. P(head and even number)

EXERCISES

PRACTICE/ SOLVE PROBLEMS

List each sample space.

1. picking a card from those shown here.
2. spinning spinner 1

Spinner 1

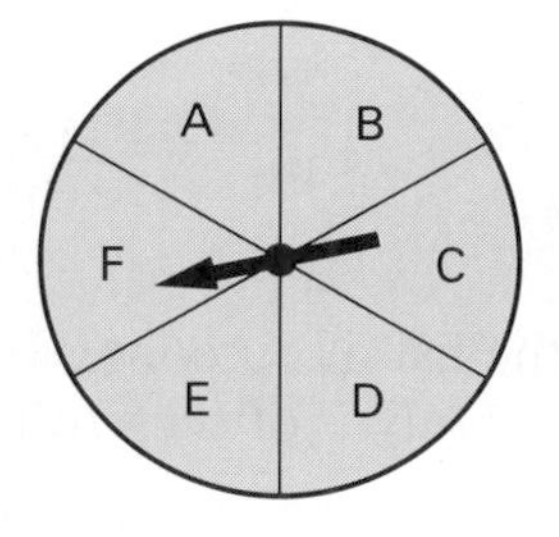

Spinner 2

Use a tree diagram to find the number of possible outcomes in each sample space.

3. tossing a penny, a nickel, and a dime
4. picking a card from those shown above and tossing a coin
5. tossing a coin and spinning spinner 1
6. tossing a nickel, tossing a number cube, and spinning spinner 1

Suppose you spin both spinner 1 and spinner 2. Find each probability.

7. P(3 and E)
8. P(2 and consonant)
9. P(odd number and vowel)
10. P(prime number and C)

MIXED REVIEW

Find the sum or difference.

1. $9 - (-12)$
2. $22 + (-7)$

Write each decimal as a fraction in lowest terms.

3. 2.75
4. 0.8
5. 3.15

Use $I = prt$ to find the total amount due.

6. $p = \$300$
$r = 5\%$
$t = 2$ years
7. $p = \$2,000$
$r = 8\%$
$t = 36$ months
8. $p = \$1,750$
$r = 11\%$
$t = 6$ months

Solve.

9. A rectangular garden measures 7.4 m by 5.5 m. What is the perimeter?

EXTEND/ SOLVE PROBLEMS

The Highlands Recreation Center has the following rules for coding their membership cards.

Rule 1: Letters and numbers may be used only once in each card.

Rule 2: A membership code must be two numbers followed by two letters.

Use the information given and list the possible codes for membership cards that can be made from the letters and numbers given for each of Exercises 11–13.

11. 4, 9, 3, A, J, R
12. 1, 6, 7, 5, P, N
13. Choose three letters and four numbers and make up your own membership codes. Make a tree diagram to show all possible outcomes in this sample space.

THINK CRITICALLY/ SOLVE PROBLEMS

For Exercises 14–15, use these three spinners.

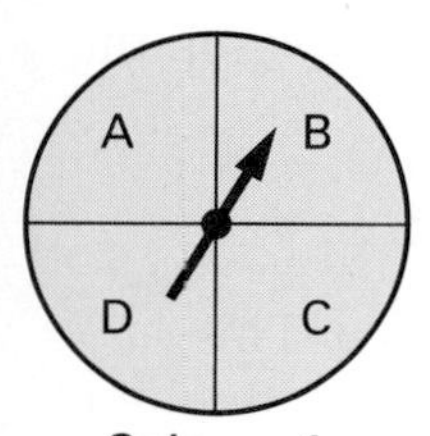

Spinner 1

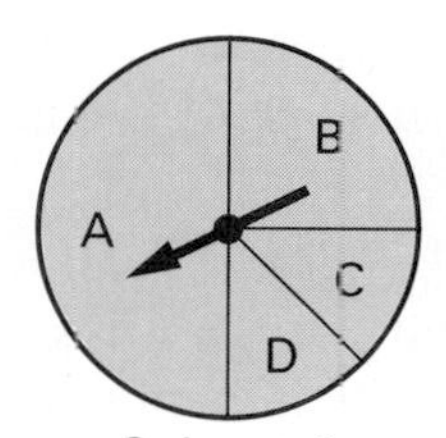

Spinner 2

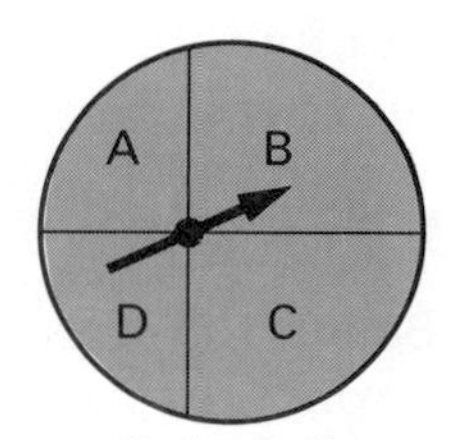

Spinner 3

14. List the sample space for each spinner.
    a. spinner 1
    b. spinner 2
    c. spinner 3
15. Find $P(\text{A})$ for each spinner.
    a. spinner 1
    b. spinner 2
    c. spinner 3
16. Compare your answers to Exercises 14 and 15. How are the spinners alike? How are they different?

13-3 The Counting Principle

EXPLORE

Suppose that you and your friend are making sandwiches for lunch. You can use either chicken salad or turkey for the filling. For bread, your choices are pita, whole wheat, and pumpernickel. How many different kinds of sandwiches can you make? (Use only one type of filling and one type of bread for each sandwich.)

SKILLS DEVELOPMENT

You have probably noticed that, as the number of choices increases, it becomes increasingly less convenient to count outcomes by using a tree diagram. Another method you can use to find the number of outcomes is the **counting principle.** Using the counting principle, you can find the number of outcomes for an activity by multiplying the number of outcomes at each stage of the activity.

TALK IT OVER

When is it not practical to use a tree diagram to show the sample space for an event?

Example 1

Benita is configuring a computer system for her home office. This means that she must choose which combination of computer, monitor, and printer to buy. The table shows the possibilities from which she can choose.

Brand of Computer	Type of Monitor	Type of Printer
Plum	color	laser
Ideal	black and white	dot matrix
Sunflower		daisy wheel
AA Tech		

How many different configurations can she choose from?

Solution

There are three stages involved in choosing a configuration: which computer to buy, which monitor to buy, and which printer to buy.

Multiply the number of choices at each stage.

$$\begin{matrix} \text{computer} & & \text{monitor} & & \text{printer} & & \\ 4 & \times & 2 & \times & 3 & = & 24 \end{matrix}$$

Benita can choose from 24 possible configurations. ◄

Example 2

In a game, a player tosses a number cube and chooses one of 26 alphabet cards. Using the counting principle, find P(prime number, T or Q).

Solution

Remember that $P(E) = \frac{\text{number of favorable outcomes}}{\text{number of possible outcomes}}$.

First find the number of possible outcomes.

number cube		alphabet cards		
6	×	26	=	156

Then find the number of favorable outcomes. There are 3 prime numbers: 2, 3, and 5. There are 2 letters: T and Q.

prime number		letters		
3	×	2	=	6

P(prime number, T or Q) $= \frac{6}{156} = \frac{1}{26}$ ◄

TRY THESE

Use the counting principle to find the number of possible outcomes.

1. choosing a stereo system from 2 kinds of receivers, 3 kinds of CD players, and 4 kinds of cassette decks
2. making a sandwich using wheat, rye, or pumpernickel bread and peanut butter or jelly as a filling
3. tossing a number cube three times
4. tossing a coin four times

Use the counting principle to find each probability.

5. A number cube is tossed three times. Find $P(6, 6, 6)$.
6. A coin is tossed five times. Find P(all heads or all tails).

EXERCISES

PRACTICE/ SOLVE PROBLEMS

Use the counting principle to find the number of possible outcomes.

1. choosing a snack from 3 desserts and 4 kinds of herb tea
2. choosing breakfast from 4 cereals, 5 juices, and 2 kinds of toast
3. choosing a pasta from spaghetti, rigatoni, or linguini and a sauce from plain, meat, mushroom, garden vegetable, or herb
4. choosing a skirt from denim, khaki, or plaid and a blouse from white, tan, red, blue, black, or yellow

Use the counting principle to find the number of possible outcomes.

5. tossing a coin three times
6. tossing a number cube two times

Use the counting principle to find the probability.

7. A coin is tossed six times. Find P(all heads).
8. A coin is tossed three times and a number cube is tossed. Find P(all tails and even number).

EXTEND/ SOLVE PROBLEMS

9. A restaurant serves 5 kinds of soup in 3 different sizes, with and without salt. How many different ways can you order soup?
10. A type of telephone is available with or without an answering machine. The phone comes in the colors white, beige, black, slate, and gray. It is available with a programmable memory for either 3, 6, 9, 12, or 20 phone numbers. How many different ways is the telephone available for sale?
11. How many different four-digit numbers can be made using only the digits 1, 2, and 3? (Assume that a digit may be repeated.)
12. How many different three-letter "words" can be made using any of the letters A through Z? (Assume that a letter may be repeated.)

THINK CRITICALLY/ SOLVE PROBLEMS

13. Three students are going to run a relay race. The track coach has to decide which student will run in each position, first, second, and third.
    a. In how many ways can the coach assign a student to the first position?
    b. Once a student has been assigned to the first position, in how many ways can the coach assign a student to the second position?
    c. Once students have been assigned to the first and second positions, in how many ways can the coach assign a student to the third position?
    d. Multiply your answers to **a, b,** and **c.** In how many ways can the coach assign the three students to the three positions?

14. Suppose that a car has room to seat four passengers.
    a. In how many ways can four people be arranged in the four seats if they all are able to drive?
    b. In how many ways can four people be arranged in the four seats if only one of them is able to drive?

▶ READ
▶ PLAN
▶ SOLVE
▶ ANSWER
▶ CHECK

Problem Solving Applications:

DECISION MAKING IN TELEVISION PROGRAMMING

Kimi Lee, a programming manager for a television station, must decide what programs will be shown between 7:00 p.m. and 9:00 p.m. on Thursdays. Four half-hour programs are available for this time period: a comedy, a game show, news, and an interview show.

Kimi decides to use a tree diagram to show all the possible ways to fill each half-hour time slot. Then she can choose the program sequence she thinks will be most popular. Here is the beginning of her tree diagram.

7:00 p.m.	7:30 p.m.	8:00 p.m.	8:30 p.m.	POSSIBLE SEQUENCE
interview	news	comedy	game	INCG
		game	comedy	INGC
	comedy	news	game	ICNG
		game	news	ICGN
	game	news	comedy	IGNC
		comedy	news	IGCN

From past popularity ratings of the station's programs, Kimi knows several things:

a. News programs are more popular before than after 8:00 p.m.
b. Many parents of young children do not wish their children to watch this particular comedy.
c. Interview shows draw the largest audience if they come just before or just after a news show.

1. Complete Kimi's tree diagram. How many different possible choices are there for programming between 7:00 p.m. and 9:00 p.m.?
2. Decide which branches of the tree diagram you would eliminate because of fact **a.** How many possible choices would then be left?
3. After making your decision in Problem 2, decide which branches of the tree diagram you would eliminate because of facts **b** and **c.** How many possible choices would then be left?
4. Make a recommendation to Kimi about one or two program schedules you think will draw the largest audience. Give reasons for your recommendation.

13-4 Independent and Dependent Events

EXPLORE

1. Suppose that a bag contains one red, one white, and one green marble. A marble is selected from the bag and replaced, and then a marble is selected again.

   a. Make a tree diagram to show the sample space.

   b. Use the tree diagram to find *P*(red, then white).

2. Suppose that a bag contains one red, one white, and one green marble. A marble is selected from the bag, but it is *not* replaced. Then a marble is selected again.

   a. Make a tree diagram to show the sample space.

   b. Use the tree diagram to find *P*(red, then white).

SKILLS DEVELOPMENT

Two events are said to be **independent** when the result of the second event does not depend on the result of the first event.

If two events X and Y are independent, the probability that both events will occur is the product of their individual probabilities.

$$P(X \text{ and } Y) = P(X) \times P(Y)$$

Example 1

A bag contains 3 red apples and 2 yellow apples. An apple is taken from the bag and replaced. Then an apple is taken from the bag again. Find *P*(red, then red).

Solution

There are 5 apples in the bag, and 3 of these are red. Because the first apple is replaced, the second selection is independent of the first.

$$P(\text{red, then red}) = P(\text{red}) \times P(\text{red})$$

$$= \frac{3}{5} \times \frac{3}{5} = \frac{9}{25}$$ ◄

Two events are **dependent** if the result of the first event *does* affect the result of the second event.

Example 2

A bag contains 3 red apples and 2 yellow apples. An apple is taken from the bag and is *not* replaced. Then an apple is taken from the bag again. Find $P(\text{red, then red})$.

Solution

Because the first apple is not replaced, the second event is dependent on the first.

On the first selection, there are 5 apples in the bag, and 3 of the apples are red.

$$P(\text{red}) = \frac{3}{5}$$

On the next selection, there are only 4 apples in the bag. Assuming that a red apple was removed, only 2 of the apples in the bag are red.

$$P(\text{red after red}) = \frac{2}{4} = \frac{1}{2}$$

Multiply the two probabilities.

$$P(\text{red, then red}) = \frac{3}{5} \times \frac{1}{2} = \frac{3}{10}$$ ◄

Write an outline in your journal explaining the difference between independent and dependent events. Be sure to include an example of each type of event.

TRY THESE

A silverware drawer contains 10 forks, 16 teaspoons, and 4 steak knives. A utensil is taken from the drawer and replaced. Then a utensil is taken from the drawer again. Find each probability.

1. $P(\text{fork, then teaspoon})$
2. $P(\text{steak knife, then fork})$
3. $P(\text{fork, then fork})$
4. $P(\text{teaspoon, then teaspoon})$

Solve.

5. A bag contains 2 red marbles, 3 white marbles, and 5 green marbles. Find $P(\text{white, then green})$ if Jenna selects two marbles from the bag without replacement.

6. A change purse contains 2 quarters, 3 dimes, and 5 pennies. Find $P(\text{quarter, then dime})$ if the coins are selected without replacement.

EXERCISES

PRACTICE/ SOLVE PROBLEMS

A game is played by spinning the spinner and then choosing a card without looking. After each turn the card is replaced. Find the probability of each event.

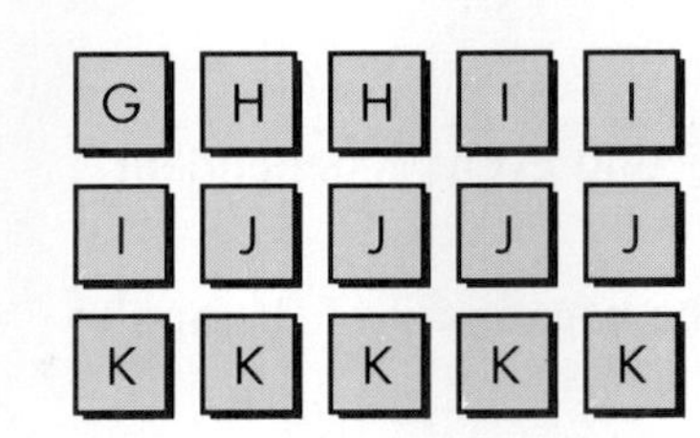

1. P(5, then K)
2. P(6, then G)
3. P(5, then G)
4. P(8, then J)
5. P(7, then I)
6. P(7, then H)
7. P(6, then J)
8. P(6, then K)
9. P(5, then H)
10. P(7, then J)
11. P(8, then G)
12. P(5, then I)

Ten cards are numbered from 1 to 10 and placed in a box. A card is taken from the box and is not replaced. Then a card is taken from the box again. Find the probability of each event.

13. P(4, then 4)
14. P(5, then odd)
15. P(6, then prime)
16. P(prime, then prime)

MIXED REVIEW

Find each product.

1. $\frac{1}{4} \times \frac{5}{16}$
2. $\frac{5}{2} \times \frac{12}{5}$

Evaluate each expression.

3. $3t - 1$ if $t = (-2)$
4. $5 - d$ if $d = (-3)$
5. $-10z$ if $z = 5$

Solve each proportion.

6. $\frac{x}{9} = \frac{35}{45}$
7. $\frac{6}{15} = \frac{m}{80}$
8. A 9-oz tube of toothpaste costs $3.55. Find the unit price to the nearest cent.

EXTEND/ SOLVE PROBLEMS

A bucket contains 8 orange golf tees and 6 blue golf tees. Kato reaches into the bucket and selects 2 tees without looking.

17. What is the probability that he selects 2 orange tees?
18. What is the probability that he selects no orange tees?
19. What is the probability that he selects one blue tee and one orange tee?

Two of these cards are chosen at random, one after another, without replacement. What is the probability of each event?

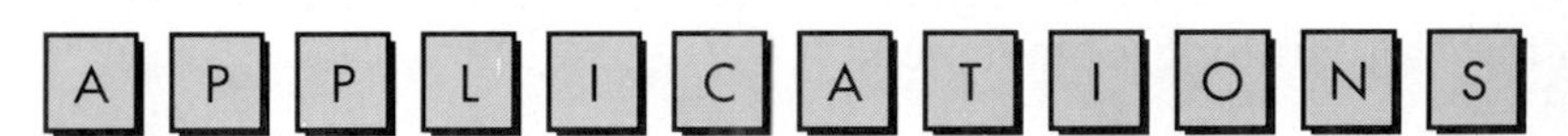

20. P(both vowels)
21. P(vowel, then consonant)
22. P(both consonants)
23. P(A, then T)

In a writing contest, 6 nonfiction stories and 4 fiction stories were chosen as finalists. Prizes were awarded on a random basis. The judges put the titles of the winning stories into a box and selected 3 winners at a time. Find each probability.

24. P(3 fiction)

25. P(3 nonfiction)

26. P(nonfiction, then fiction, then nonfiction)

27. P(fiction, then nonfiction, then fiction)

THINK CRITICALLY/ SOLVE PROBLEMS

A wallet contains three \$20 bills, one \$10 bill, and two \$1 bills. Use this information to give an example for each exercise.

28. two independent events whose probability is $\frac{1}{12}$

29. two dependent events whose probability is $\frac{1}{10}$

30. two dependent events whose probability is 0

31. two independent events whose probability is 0

13-5 Experimental Probability

EXPLORE/ WORKING TOGETHER

Work with a partner. Use a number cube with faces labeled from 1 through 6.

1. Use the probability formula to find the probability of rolling a 5 on the first roll.
2. Now roll the number cube 60 times, keeping a tally of the results of your *experiment* in a table like this.

Number on Cube	Tally
1	
2	
3	
4	
5	
6	

3. Examine the results of your experiment. How many times did the number 5 actually appear? Compare your results with those of other students in your class.

SKILLS DEVELOPMENT

The probability of an event based on the results of an experiment is called **experimental probability.**

Example 1

This table is a summary of the outcomes from an experiment in which a number cube was tossed 50 times.

Number on Cube	1	2	3	4	5	6
Number of Outcomes	7	9	10	2	11	11

Find each experimental probability.

a. $P(6)$ b. $P(3)$ c. $P(\text{even number})$ d. $P(4 \text{ or } 5)$

CHECK UNDERSTANDING

In the solution of Example 1a, what does the 11 represent in the fraction?

Solution

a. $P(6) = \frac{11}{50}$

b. $P(3) = \frac{10}{50} = \frac{1}{5}$

c. $P(\text{even number}) = \frac{22}{50} = \frac{11}{25}$

d. $P(4 \text{ or } 5) = \frac{13}{50}$ ◄

Experimental probabilities are often used when all outcomes are not equally likely. For instance, when you toss a paper cup, there generally are three ways that it could land: up, down, or on its side. So there are three outcomes, but they are not equally likely. You would expect that the cup would most often land on its side.

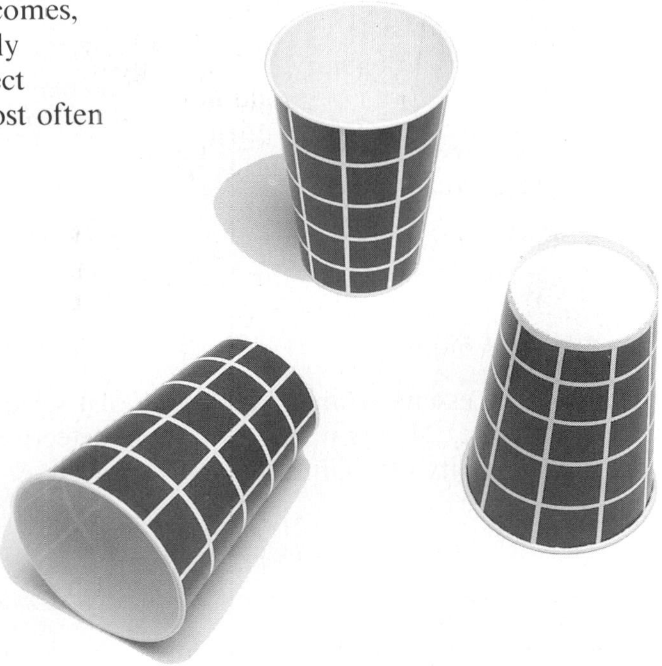

Example 2

A paper cup is tossed 50 times. The results are shown in the table. Find the experimental probability of the cup landing on its side.

Outcome	Tally
up	IIII
down	卌 卌
side	卌 卌 卌 卌 卌 卌 卌 I

Solution

Counting the tally marks, you see that the paper cup landed on its side 36 times.

$$P(\text{side}) = \frac{36}{50} = \frac{18}{25} \quad ◄$$

You often find an experimental probability using a random sample. Then you use the probability to draw conclusions about the population from which the sample was taken.

CONNECTIONS

In Chapter 1, you learned about *random sampling* as a way of collecting information from people. Example 3 shows a different way that random sampling is used. In this case, the sample is selected from a group of objects. However, the entire group is still called the *population*.

When is a sample considered a random sample?

Example 3

In a bottle factory, 1,000 bottles were selected at random. Of these, 3 were found to be defective. What is the probability of a bottle being defective?

Solution

$$P(\text{defective}) = \frac{\text{number defective in sample}}{\text{total number in sample}} = \frac{3}{1{,}000}. \quad ◄$$

TRY THESE

A number cube is tossed 20 times. The outcomes are recorded in the table. Find each experimental probability.

Number on Cube	1	2	3	4	5	6
Number of Outcomes	3	5	2	5	4	1

1. $P(1)$
2. $P(2)$
3. $P(3)$
4. $P(4)$
5. $P(5)$
6. $P(6)$
7. $P(\text{even number})$
8. $P(\text{odd number})$
9. $P(2 \text{ or } 4)$
10. On an assembly line, 5,000 randomly selected cars were tested. Of these, 24 cars were found to be defective. What is the probability of finding a car to be defective?

EXERCISES

PRACTICE/ SOLVE PROBLEMS

Jessie selected a marble from a bag 80 times. With each selection, she recorded the color and returned the marble to the bag. Use the results in the table to find each experimental probability.

Outcome	Tally	Total
striped	𝍸 𝍸 𝍸 𝍸 𝍸	25
blue	𝍸 𝍸 𝍸 𝍸 𝍸 𝍸 𝍸	35
red	𝍸 𝍸 IIII 𝍸	20

1. $P(\text{striped})$
2. $P(\text{blue})$
3. $P(\text{red})$
4. $P(\text{striped or blue})$
5. $P(\text{blue or red})$
6. $P(\text{not blue})$
7. $P(\text{striped or red})$

The table shows the number of muffins sold in the cafeteria during a recent morning. Use these results for Exercises 8–13 to find the experimental probability of each event.

blueberry	cranberry	bran	cinnamon	apple	other
22	6	10	8	4	10

8. P(blueberry)
9. P(bran)
10. P(cranberry)
11. P(not bran)
12. P(apple or cinnamon)
13. P(not blueberry)
14. Fran tossed a bottle cap 25 times. It landed up 14 times and down 11 times. Find the experimental probability that the cup lands up.
15. Bruce tossed a thumbtack 20 times. It landed point up 18 times. Find the experimental probability that a tossed tack will not land point up.
16. On a production line, 18 defective bolts were found among 2,000 randomly selected bolts. What is the probability of a bolt being defective?

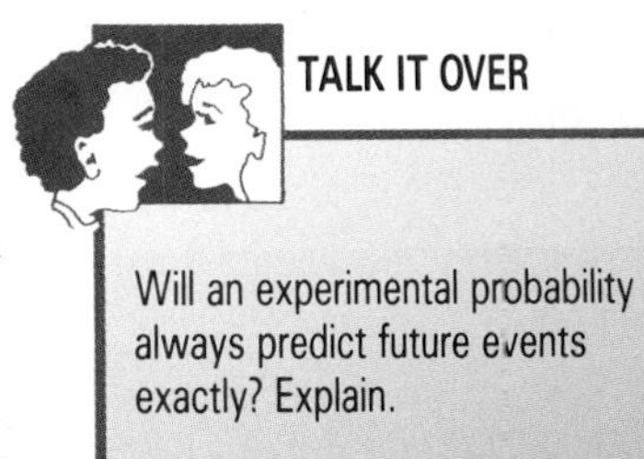

TALK IT OVER

Will an experimental probability always predict future events exactly? Explain.

EXTEND/ SOLVE PROBLEMS

17. Toss a penny and a quarter 60 times and find the experimental probability of the outcomes. Make a table to record your data.
18. Toss a number cube 100 times and find the experimental probability of each of the outcomes. Make a table to record your data.
19. A basketball team won 62 out of 120 games. What is the experimental probability that the team will win their next game?

THINK CRITICALLY/ SOLVE PROBLEMS

Make an unfair "coin" by taping a nickel and a penny together. Be sure your coin has one head and one tail.

20. Flip the coin 20 times. Use your results to predict how many tails you will get if you flip the coin 100 times.
21. Flip the coin 80 more times for a total of 100 flips. Was the prediction you made after 20 flips accurate for 100 flips?
22. If you predict the number of tails in 500 flips using your results for 100 flips, do you think your prediction will be more reliable than predicting from 20 flips? Explain.

13-6 Problem Solving Skills:

USING SAMPLES TO MAKE PREDICTIONS

Remember that a random sample is a group chosen from a large population in such a way that each member of the population has an equal chance of being chosen. As you saw in Chapter 1, random sampling is often used to poll people about their opinions or preferences. In this lesson, you will see how the responses of the random sample are used to make predictions about the entire population.

PROBLEM

Members of the Channel 3 election bureau conducted a poll before the mayoral election in Southglenn. There are 250,000 registered voters in this city. The poll-takers asked a random sample of 225 of these voters which of the three candidates they were planning to vote for. The results are shown in the table.

Candidate	Voters Planning to Vote
Miselli	35
Jackson	80
Brimmage	110

Based on this random sample, how many votes might Jackson expect to receive in the election?

SOLUTION

Use the sample results to find *P*(Jackson).

$$P(\text{Jackson}) = \frac{80}{225} = \frac{16}{45} \approx 0.356$$

You can predict how many votes Jackson will receive by finding the product of the probability and the total number of registered voters.

$$\begin{array}{ccccc} P(\text{Jackson}) & \times & \text{number of voters} & = & \text{predicted number of votes} \\ 0.356 & & 250{,}000 & & 89{,}000 \end{array}$$

So, Jackson might expect to receive about 89,000 votes in the election.

CALCULATOR

You can use your calculator to write a probability as a decimal. Just remember to divide the numerator of the fraction by the denominator.

Use a calculator to write each probability as a decimal. Round to the nearest hundredth.

1. $\frac{13}{20}$
2. $\frac{14}{25}$
3. $\frac{18}{65}$
4. $\frac{55}{75}$

PROBLEMS

A local cable company asked a random sample of 500 of their 50,000 subscribers which kind of programming they preferred. Of those polled in the survey, 150 preferred comedy shows, 80 boxing matches, 225 movies, and 45 sporting events.

1. Find *P*(boxing).
2. Find *P*(movies).
3. Find *P*(comedy).
4. Find *P*(sporting events).

Refer to the information given for Problems 1–4. If all of the subscribers were polled, how many people would you expect to select each kind of programming?

5. comedy
6. boxing matches
7. movies
8. sporting events

Of the 750,000 registered voters in Centerville, 1,000 were polled about their preferences in the upcoming election. The results of the poll are shown in this table.

Candidate	Votes
Bendel	370
Morales	325
Greenberg	155
Undecided	150

9. Based on the results of the poll, about how many of the registered voters in Centerville would you expect to be undecided?
10. Suppose that all the undecided voters vote for Morales. If all the registered voters vote, about how many votes might Morales expect to receive?
11. Suppose that Greenberg drops out of the race, and the Greenberg supporters switch their loyalty to Morales. If the undecided voters then split evenly between Bendel and Morales, about how many votes might Bendel expect to receive?
12. Would you predict the outcome of the election based on the results of this poll? Explain.

COMPUTER TIP

You may want to use this program to check your answers to Exercises 9–11. Notice how important line 20 is in giving the correct answer.

```
10 INPUT "HOW MANY
   VOTERS? " ;V: PRINT
20 LET N = V * 750
30 PRINT N;" VOTES"
```

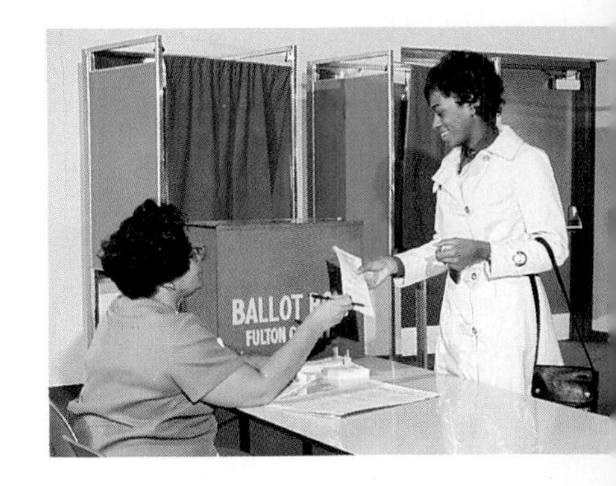

13-7 Problem Solving Strategies:
SIMULATE THE PROBLEM

Often a problem that seems complicated can be solved by relating the problem to a similar but simpler situation. This strategy is sometimes used in problems about probability. To find probabilities for a complex process you relate it to a simpler process that *simulates,* or acts in the same way as, the complex one.

PROBLEM

Dena Power is taking part in a TV game show. She has to answer 10 true-false questions. To win a prize, she must answer at least 6 of the questions correctly. Design a simulation to predict the probability of Dena winning a prize.

SOLUTION

Each of the 10 questions has two possible outcomes: either Dena answers correctly or she answers incorrectly. So, the game can be simulated by tossing 10 coins.

Let heads represent a question answered correctly and tails represent a question answered incorrectly. Toss 10 coins many times and record the outcomes in a table.

Why is the coin toss appropriate for simulating a true-false test?

What simulation would be appropriate for a multiple-choice test with 6 choices? with 4 choices?

	Number of heads (correct answers)	Number of tails (incorrect answers)
1st toss	4	6
2nd toss	7	3
3rd toss	5	5
4th toss	6	4
5th toss	7	3

On the first 5 tosses, heads showed on 6 or more coins 3 times. Based on these tosses, the probability of Dena winning a prize is given by the following.

$$P(\text{winning}) = \frac{\text{number of favorable outcomes}}{\text{number of possible outcomes}} = \frac{3}{5}$$

To obtain a more accurate estimate of the probability of Dena winning, the simulation should be continued to find the results for many more tosses.

PROBLEMS

Use this information to solve Problems 1–4.

A quality control inspector finds that 1 out of every 6 soft drink cans is not sealed properly.

1. Which of the following simulations would you use to predict the probability of getting no defective cans in a group of 7?
   a. toss 7 coins
   b. draw 7 cards from a deck
   c. toss 7 number cubes
2. Describe in detail how you would carry out the simulated experiment in Problem 1.
3. Carry out the simulation 50 times and record the results.
4. Give your estimate of the probability of getting no defective cans.

For Problems 5–11, design a simulation for the situation.

5. In the past, Frank has won 4 out of every 6 video games he has played. What is the probability that Frank will win his next three games?
6. What is the probability that, in a group of 8 people in an elevator, 5 are female?
7. Sandra guesses the answers to the 15 questions on a true-false test. What is the probability that Sandra will get 12 or more answers correct?
8. On a TV station, 7 out of every 10 commercials are for food products. What is the probability that 3 out of the next 4 commercials shown will be for food products?
9. In a school survey, 3 out of every 4 students indicated that they would prefer a rock concert to a dance. If a sample of 10 students is chosen, what is the probability that 8 or more of these students prefer a rock concert?
10. Suppose that 1 out of every 6 motorcycles has racing handlebars, and 1 out of every 2 motorcycles has a sidemarker lamp. What is the probability that the next three motorcycles you see will have neither racing handlebars nor a sidemarker lamp?
11. The entrance tests for a military college consist of two parts, an academic test and an athletic ability test. On the academic test, 4 out of every 6 applicants passes, but on the athletic test, only 1 in every 2 passes. What is the probability that both of the next two applicants will pass both tests?

COMPUTER

The random number generator built into BASIC can be used to simulate the experiments described in the text. For example, to simulate tossing a coin, the following program may be used.

```
10 INPUT "# OF TOSSES:
     ";NTOSS
20 LET H = 0: LET T = 0 :
     REM # HEADS, TAILS
30 FOR I = 1 TO NTOSS
40 LET R = RND(1) : REM
     RANDOM NUMBER
     BETWEEN 0 AND 1
50 IF R < .5 THEN LET
     H = H + 1: PRINT
     "HEADS": GOTO 70
60 LET T = T + 1: PRINT
     "TAILS"
70 NEXT I
80 PRINT "# HEADS = "; H;
     "# TAILS = "; T
```

PROBLEM SOLVING TIP

You may use any of these methods for your simulations.

a. tossing a coin
b. tossing a number cube
c. spinning a spinner

CHAPTER 13 ● REVIEW

Choose a word from the list to complete each statement.

1. The __?__ of an event is between 0 and 1 and is usually written as a fraction in lowest terms.
2. The set of all outcomes is called the __?__ .
3. Two events are said to be __?__ when the result of the second event does not depend on the result of the first event.
4. The probability of an event based on the results of an experiment is called __?__ .
5. An __?__ is an outcome or a combination of outcomes.

a. experimental probability
b. independent
c. probability
d. event
e. sample space

SECTION 13–1 PROBABILITY (pages 474–477)

► The **probability** of an event is a number between 0 and 1. You find a probability using this formula:

$$P(E) = \frac{\text{number of favorable outcomes}}{\text{number of possible outcomes}}$$

A number cube, with faces labeled from 1 through 6, is tossed.

6. Find $P(\text{number} < 5)$.
7. Find $P(\text{prime number})$.

SECTION 13–2 SAMPLE SPACES AND TREE DIAGRAMS (pages 478–481)

► The set of all outcomes is called the **sample space.**

► You can make a **tree diagram** to show a sample space.

8. Two coins are tossed. Draw a tree diagram to show the sample space. How many possible outcomes are there?

SECTION 13–3 THE COUNTING PRINCIPLE (pages 482–485)

► To find the number of possible outcomes for an activity, multiply the number of choices for each stage of the activity.

Use the counting principle to find the number of possible outcomes.

9. tossing a coin four times
10. choosing a snack from 2 desserts and 3 kinds of herb tea
11. tossing a number cube three times

SECTION 13-4 INDEPENDENT AND DEPENDENT EVENTS (pages 486-489)

► Two events are said to be **independent** when the result of the second event does not depend on the result of the first event.

► When the result of the first event does affect the result of the second event, the events are **dependent.**

12. A bag contains 3 red, 2 white, and 5 green marbles. One is taken from the bag, replaced, and another is taken. Find *P*(red, then green).

13. A change purse contains 3 quarters, 2 dimes, and 5 nickels. Find *P*(dime, then nickel) if two coins are selected from the purse without replacement.

SECTION 13-5 EXPERIMENTAL PROBABILITY (pages 490-493)

► The probability of an event based on the results of an experiment is called **experimental probability.**

A number cube is tossed 20 times. The outcomes are recorded in this table.

Number on Cube	1	2	3	4	5	6
Number of Outcomes	3	5	2	5	2	3

14. Find *P*(1).

15. Find *P*(even number).

16. Find *P*(3 or 4).

SECTIONS 13-6 AND 13-7 PROBLEM SOLVING (pages 494-497)

► You can use information from a random sample of a population to make predictions about the entire population.

► Some probability experiments are easier to understand if a simpler experiment is conducted to simulate or imitate the more difficult one.

Describe a simulation you could use to predict the probability.

17. Suppose James was on time for his piano lesson 3 days out of every 5 for the past month. What is the probability that he will be on time for his next ten lessons?

USING DATA Use the data on page 472 to answer Exercise 18.

18. Estimate the amount of rain that will fall over the next 5-year period in Mt. Wai-'ale-'ale based on the greatest amount of rain that fell one year.

CHAPTER 13 ● TEST

Find each probability. Use this spinner.

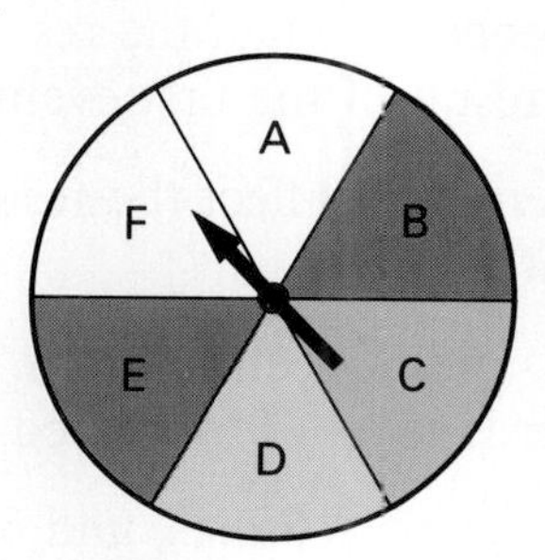

1. P(A)
2. P(E or D)
3. P(Z)
4. P(white)
5. P(A, B, C, D, E, or F)

Use a tree diagram to find the number of possible outcomes in each sample space.

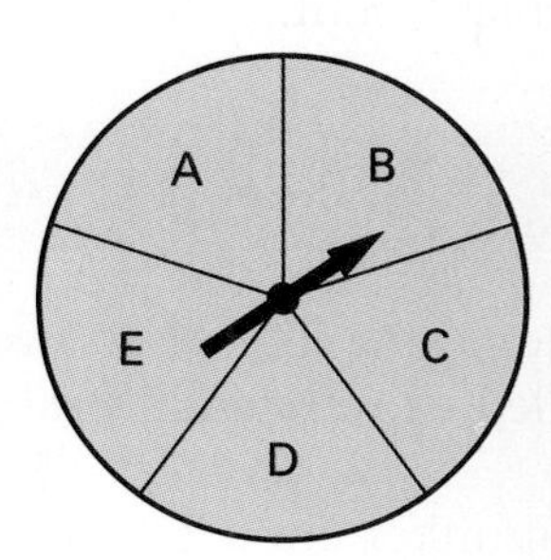

6. tossing a nickel and then tossing a number cube
7. spinning a spinner like the one shown at the right and then tossing a quarter

Use the counting principle to find the number of possible outcomes.

8. tossing a coin three times
9. choosing a stereo configuration from 2 receivers, 3 cassette decks, and 5 speakers
10. choosing a dinner consisting of either beef, lamb, fish, or chicken; with either peas or carrots; and with either milk, juice, or iced tea
11. A bag contains 1 red marble, 3 blue marbles, and 1 orange marble. Susan takes one marble from the bag, replaces it, and takes another. Find P(blue, then orange).

A number cube is tossed 30 times. The outcomes are recorded in this table.

Number on Cube	1	2	3	4	5	6
Number of Outcomes	2	4	7	3	8	6

12. Find P(5).
13. Find P(even number).
14. Find P(2 or 6).

CHAPTER 13 ● CUMULATIVE REVIEW

1. To get an average of 85 for five tests, how many total points would you need?

Find each answer.

2. $407.2 - 38.071$
3. $201.7 + 5.654$
4. 6.7×0.44
5. $3.241 \div 0.7$

6. Write two conditional statements using the following two sentences:
   You live in Dubuque.
   You live in Iowa.

Add or subtract.

7. $3\frac{3}{8} + 4\frac{5}{6}$
8. $7\frac{1}{4} - 2\frac{5}{18}$
9. $9 - 3\frac{1}{3}$
10. $\frac{7}{9} + \frac{1}{3}$

Write each answer in simplest form.

11. 8 yd − 3 yd 2 ft
12. 4 ft 3 in. − 2 ft 8 in.
13. 4 lb 8 oz + 2 lb 10 oz
14. 9 yd 2 ft × 5

Add or subtract.

15. $-65 + 8$
16. $-20 - (17)$
17. $16 + (-13)$
18. $-13 - (-18)$

Solve each equation.

19. $13 = q - 8$
20. $7 + b = -2$
21. $12 - k = -9$
22. $y + 8 = -7$

23. Write three equivalent ratios for 6:5.

24. What percent of 512 is 384?

25. One year's interest on $2,000 is $120. What is the rate of interest?

Find the unknown angle measure for each polygon.

26.

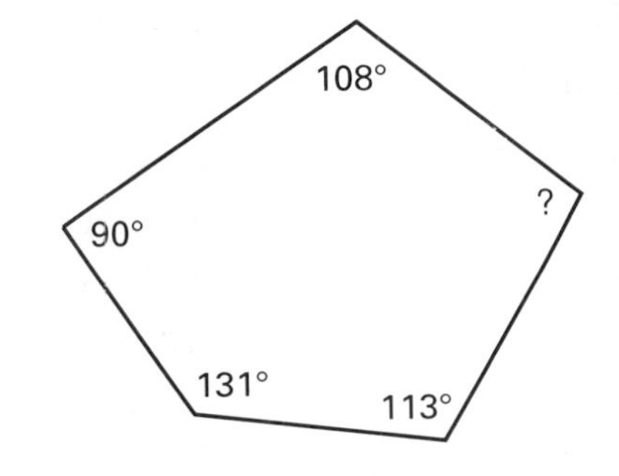

27.

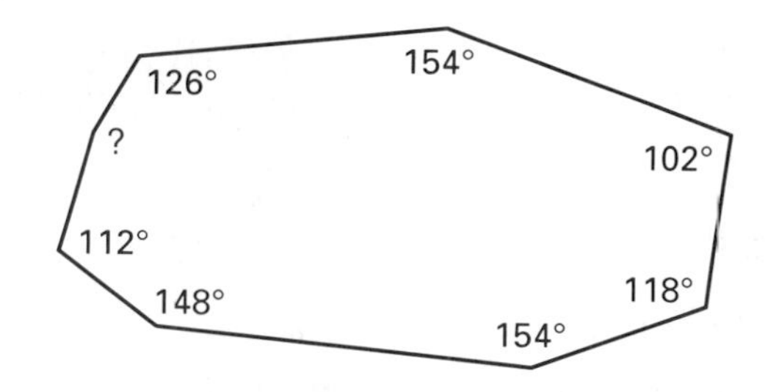

28. Line segment LM is $5\frac{1}{2}$ in. long. NO bisects LM at 0. What is the length of OM?

29. PQ bisects angle RPS. $m\angle RPS = 112°$. What is $m\angle RPQ$?

30. Find the area of a circle with a diameter of 9 cm. Round your answer to the nearest tenth.

Simplify.

31. $\frac{8n + 2}{2}$
32. $\frac{7x^2 + 14y}{7}$

33. Use the counting principle to find the number of possible outcomes for dinner with choices of 4 main courses, 4 vegetables, and 6 desserts.

1. A gasoline company wants to survey its customers. From a list of their credit card holders, they send a questionnaire to every 20th person. What type of sampling are they using?
   A. random B. clustered
   C. systematic D. convenience

2. Complete.
   $4(12 + 10) = (■ \times 12) + (4 \times 10)$
   A. 4 B. 10
   C. 12 D. 88

3. Which is a counterexample for the following statement?
   If x is a number, then x^2 is larger than x.
   A. 2 B. $\frac{1}{2}$
   C. 3 D. none of these

4. Add. $4\frac{2}{3} + 8\frac{5}{6}$
   A. $12\frac{3}{6}$ B. $13\frac{1}{6}$
   C. $13\frac{5}{6}$ D. $13\frac{1}{2}$

5. Find the perimeter of a rectangular figure whose length is 24.3 mi and whose width is 8.1 mi.
   A. 16.2 mi B. 64.8 mi
   C. 3 mi D. 196.83 mi

6. Compare.
   $3 \times (-14)$ ● $-13 \times (-3)$
   A. > B. <
   C. = D. ≈

7. Solve.
   $22 = 3r + 10$
   A. $r = 12$ B. $r = 15$
   C. $r = 7$ D. $r = 4$

8. Solve. $4(k - 9) = 24$
   A. $k = 15$ B. $k = 11$
   C. $k = 8$ D. $k = -5$

9. The scale of a drawing is 1 cm:3 m. A room in the drawing is 10 cm long. What is the length of the actual room?
   A. 13 m B. 1 m
   C. 30 cm D. 30 m

10. What percent of 900 is 270?
    A. $33\frac{1}{3}\%$ B. 25%
    C. 30% D. $333\frac{1}{3}$

11. Name the figure.
    A. triangular pyramid
    B. pentagonal prism
    C. hexagonal pyramid
    D. triangular prism

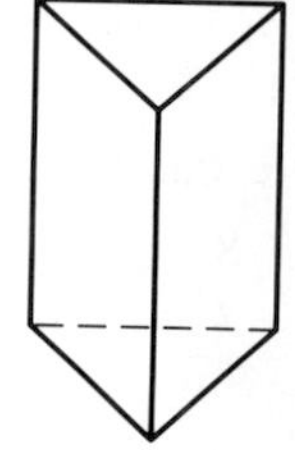

12. Find the surface area.
    A. 40 cm^2
    B. 96 cm^2
    C. 38 cm^2
    D. 76 cm^2

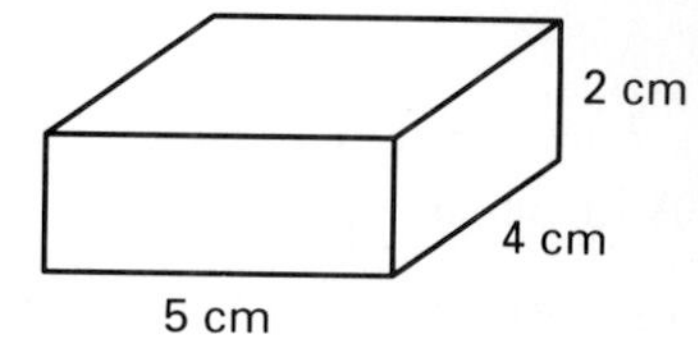

13. Simplify. $(11t^2 + 2) - (3t^2 + 2)$
    A. $8t^2$ B. $8t^4 + 4$
    C. $8t^4$ D. $14t + 4$

14. In a box of 1,000 plugs, 10 are defective. What is the probability of choosing a defective plug?
    A. 1 B. $\frac{1}{10}$
    C. $\frac{1}{100}$ D. $\frac{10}{990}$

CHAPTER 14 SKILLS PREVIEW

Use the figure at the right for Exercises 1–2.

1. Give the coordinates of points X and Y.
2. Find the slope of line segment AB.
3. From their house, Jane and Bob drove 4 blocks north and 13 blocks east to shop at the supermarket. After that, they drove 5 blocks south and 9 blocks west to deliver a birthday present to Arthur. At that point, how many blocks were they from home?

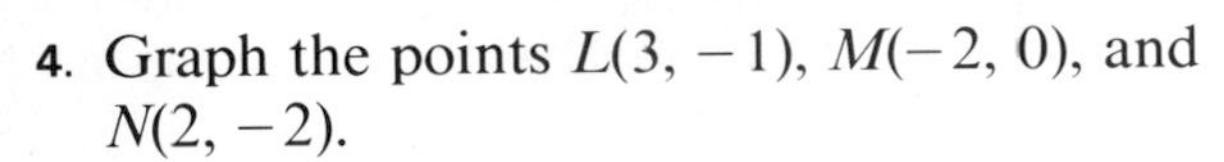

4. Graph the points $L(3, -1)$, $M(-2, 0)$, and $N(2, -2)$.
5. Graph the equation $y = x + 5$.
6. Complete the table of solutions for the equation $y = x + 6$.

$y = x + 6$

x	y	x, y
−2	■	−2, ■
0	■	0, ■
2	■	2, ■

7. At a parking lot, there is a fixed fee for using the lot in addition to the hourly rate charged for parking. Find the fixed fee and the hourly rate.

Total Amount Paid for Parking Car	\$5.50	\$8.00	\$13.00
Number of Hours Car Parked	1	2	4

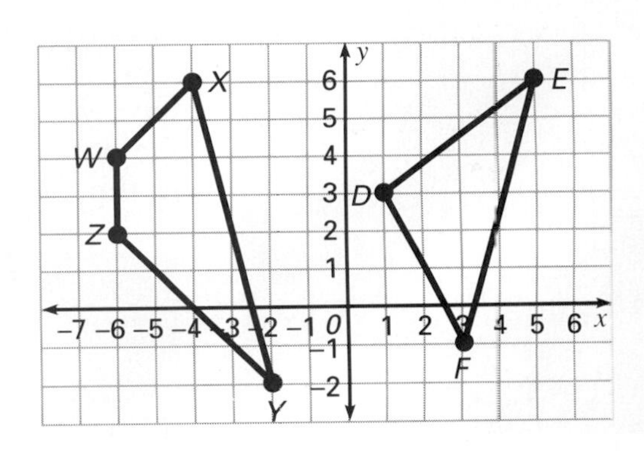

Use the figure at the right for Exercises 8–10.

8. Graph the image of $\triangle DEF$ under a translation 2 units to the right and 1 unit up.
9. Graph the image of trapezoid $WXYZ$ under a reflection across the x-axis.
10. Graph the image of $\triangle DEF$ after a 90° turn clockwise about the origin.

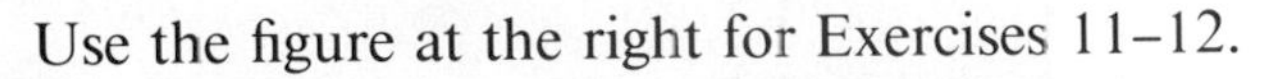

Use the figure at the right for Exercises 11–12.

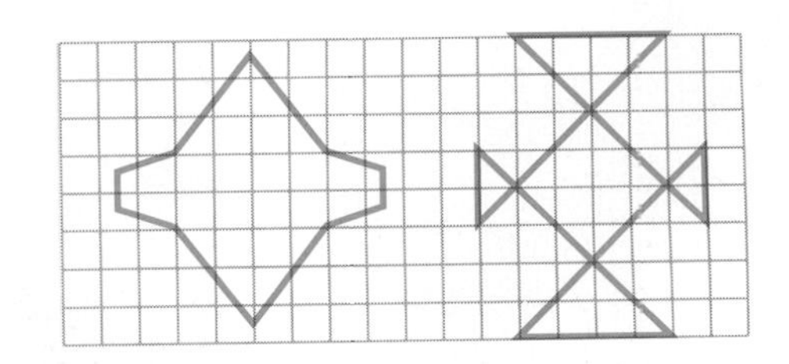

11. How many lines of symmetry does the figure on the left have?
12. What is the order of rotational symmetry for the figure on the right?

GEOMETRY OF POSITION

THEME Patterns

If you and a friend who lives on the opposite side of town planned to jog together, you would probably decide on a specific place to meet, such as the corner of Fifth and Main Streets. Because the meeting spot is at the intersection of two streets, you both would know exactly where to go. In this chapter, you will learn how to locate a specific point on a coordinate grid, using ordered pairs of numbers. These numbers are similar to the coordinates you and your friend agreed on: Fifth and Main. You will look for patterns among ordered pairs, and you will learn how to describe these patterns as mathematical functions.

Patterns can be found in the numbers of dots that make up polygons of increasingly larger sizes. These dot patterns represent numbers called **figurate numbers.** A triangular number is one type of figurate number. The first four triangular numbers are represented by these dot patterns:

triangular number →	1st	2nd	3rd	4th
number of dots →	1	3	6	10

One way to find the number of dots in a triangular number is to use the dot pattern. Another way is to use algebra. The number of dots in the nth triangular number is represented by the expression $\frac{n(n + 1)}{2}$.

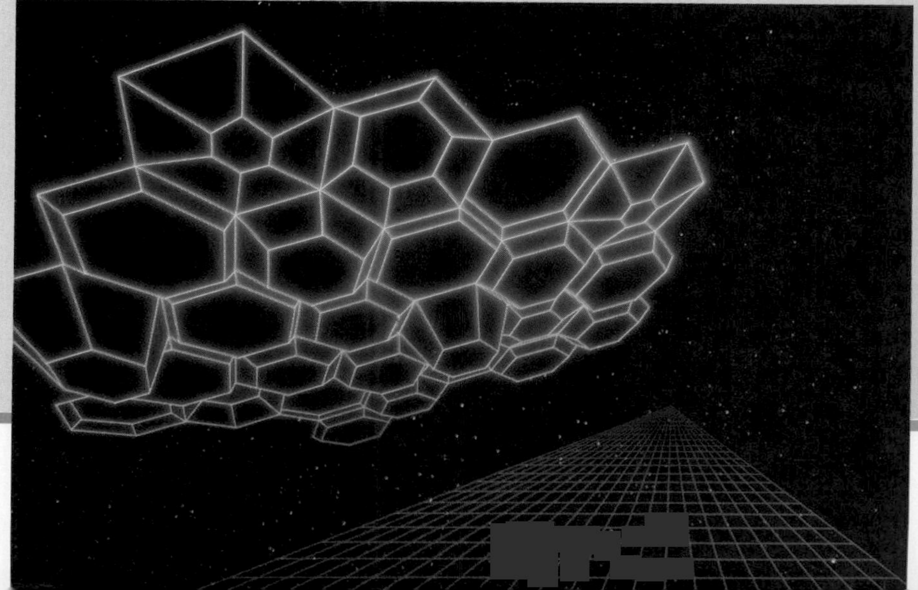

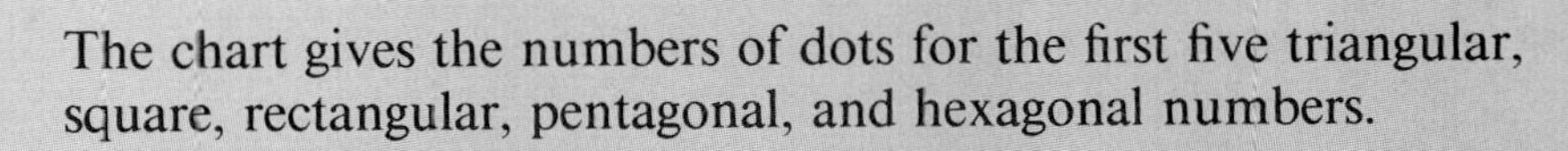

The chart gives the numbers of dots for the first five triangular, square, rectangular, pentagonal, and hexagonal numbers.

	1st	2nd	3rd	4th	5th
triangular	1	3	6	10	15
square	1	4	9	16	25
rectangular	2	6	12	20	30
pentagonal	1	5	12	22	35
hexagonal	1	6	15	28	45

DECISION MAKING

Using Data

1. Use the expression that represents the *n*th triangular number to find the 6th triangular number. Then draw a dot picture of the 6th triangular number.
2. Guess the 7th triangular number. Then use the expression to check your guess.
3. Which pentagonal number is represented by the drawing at the right?

Working Together

Make drawings of the 1st through 10th square numbers and the 1st through 10th pentagonal numbers.

14-1 The Coordinate Plane

EXPLORE

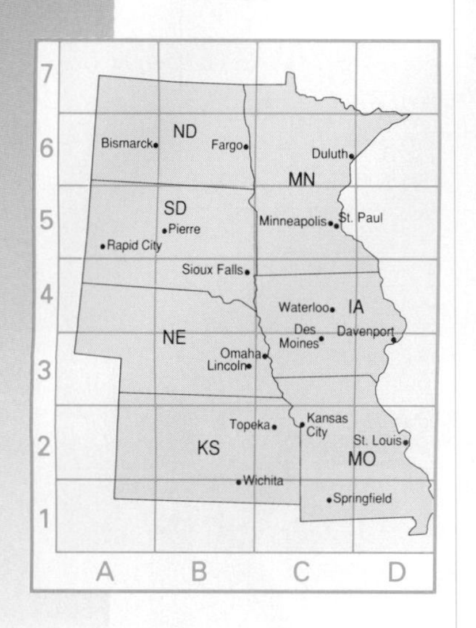

On a map, a location is specified with a letter followed by a number. For example, on the map at the left, Topeka, Kansas, is located in the box that is both above the letter C and across from the number 2, or region C2.

a. Which city is located in region C4?

b. In which region is Omaha, Nebraska, located?

c. Between which two regions would you travel if you traveled from Duluth, Minnesota, to the southwest corner of Kansas?

d. If you were to travel from St. Louis, Missouri, to Fargo, North Dakota, in which region would your trip begin? In which region would it end?

SKILLS DEVELOPMENT

The system used to identify locations on a map is much like the mathematical system called a coordinate plane. On a **coordinate plane,** two number lines are drawn perpendicular to each other. The horizontal number line is called the ***x*-axis.** The vertical number line is called the ***y*-axis.** The axes separate the plane into four regions called **quadrants.** The point where the axes cross is called the **origin.**

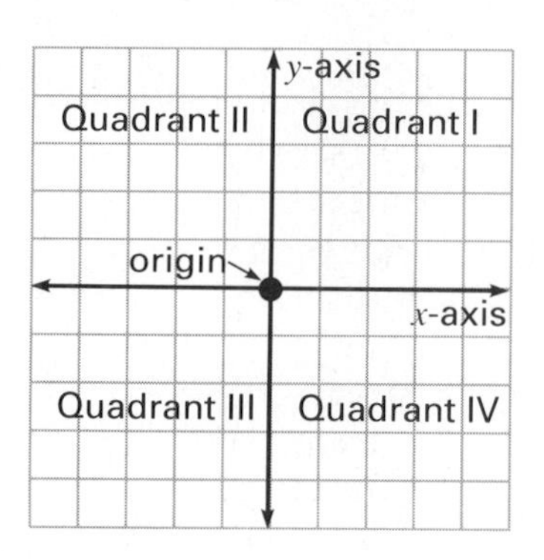

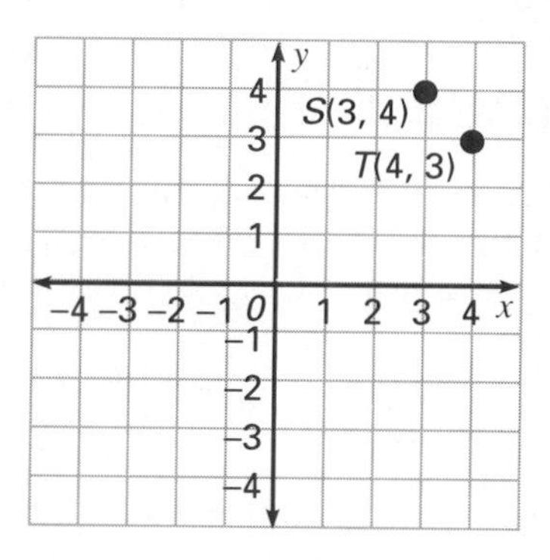

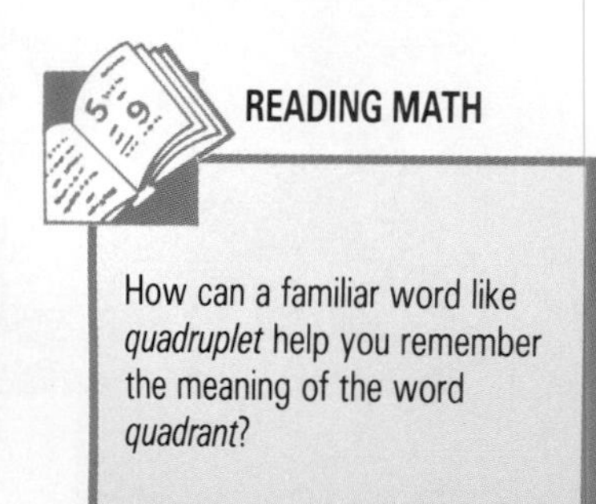

How can a familiar word like *quadruplet* help you remember the meaning of the word *quadrant?*

In the figure at the right above, point S is 3 units to the right of the origin and 4 units up from the origin. So the point is identified as $S(3, 4)$, where 3 is the ***x*-coordinate** of the point and 4 is the ***y*-coordinate.** The order of the coordinates is important because, for example, point $S(3, 4)$ is different from point $T(4, 3)$. For this reason, the coordinates for a point, such as (3, 4), are called *ordered pairs.*

When you read coordinates and graph points in the plane, you need to be aware of directions. The table shows you in which direction to move to locate a point based on the sign of each coordinate in the ordered pair.

	x-coordinate	*y*-coordinate
positive	move *right* →	move *up* ↑
negative	move *left* ←	move *down* ↓

Example 1

Give the coordinates of the point or points.

a. point D

b. two points in the third quadrant

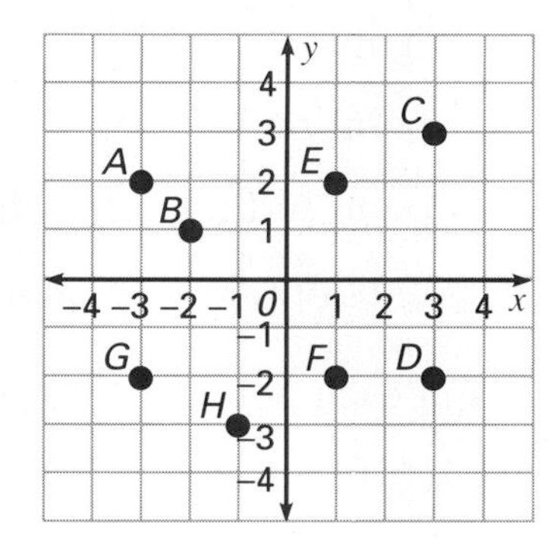

Solution

a. Point D is 3 units to the right of the origin and 2 units down from the origin. So the coordinates of D are $(3, -2)$.

b. Points G and H are in the third quadrant. Their coordinates are $G(-3, -2)$ and $H(-1, -3)$. ◄

Example 2

Graph points $X(0, 3)$, $Y(2, 0)$ and $Z(-2, -2)$.

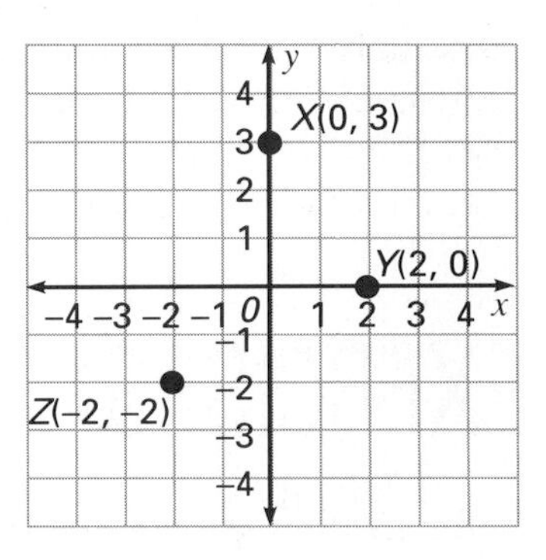

Solution

Point X is 0 units to the right or left of the origin and 3 units up.
Point Y is 2 units to the right of the origin and 0 units up or down.
Point Z is 2 units to the left of the origin and 2 units down from it. ◄

CHECK UNDERSTANDING

If the *x*-coordinate of a point is 0, what can you tell about the location of the point?

If the *y*-coordinate of a point is 0, what can you tell about the location of the point?

Example 3

Sketch the square that has a diagonal whose endpoints are $J(3, 3)$ and $L(-3, -3)$.

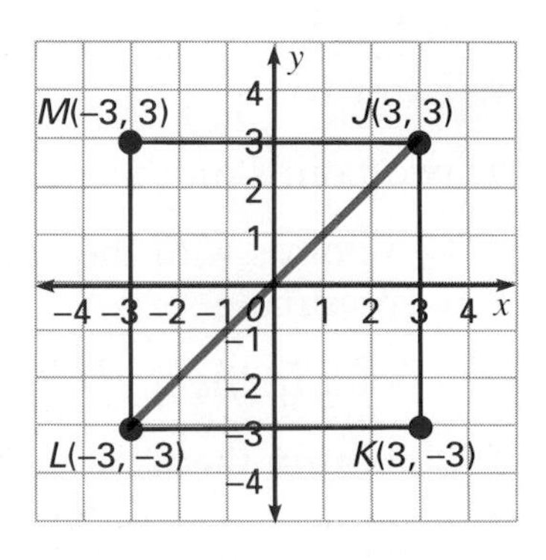

Solution

Begin by graphing $J(3, 3)$ and $L(-3, -3)$ on a coordinate plane. Point M is directly across from J and above L. So its coordinates, $M(-3, 3)$, are at a vertex of the square. Similarly, the coordinates $K(3, -3)$ are at another vertex. ◄

MIXED REVIEW

Find the answer.

1. $\frac{1}{10} + \frac{2}{3}$
2. $3\frac{1}{8} - 1\frac{2}{3}$
3. $\frac{3}{4} \times \frac{2}{9}$
4. $4^8 \div 4^3$
5. What is the square root of 169?

Solve.

6. Find the perimeter of a regular hexagon 6.3 cm on each side.
7. Find the fraction that comes next in this pattern: $\frac{1}{40}, \frac{1}{20}, \frac{1}{10}$
8. Solve the proportion: $\frac{14}{n} = \frac{3}{8}$

TRY THESE

Refer to the figure. Give the coordinates of the point or points.

1. point K
2. point D
3. point M
4. point H
5. origin
6. a point on the x-axis
7. a point in the fourth quadrant
8. two points with the same x-coordinate

Graph each point on a coordinate plane.

9. $A(4, 0)$
10. $B(0, 4)$
11. $C(0, -2)$
12. $D(3, 5)$
13. $E(5, 5)$
14. $F(-4, 2)$

Sketch each figure on a coordinate plane.

15. triangle with vertices $X(3, 2)$, $Y(5, 6)$, and $Z(2, -4)$
16. square with $C(5, -5)$ and $D(-5, 5)$ as endpoints of a diagonal

EXERCISES

PRACTICE/ SOLVE PROBLEMS

Refer to the figure. Give the coordinates of the point or points.

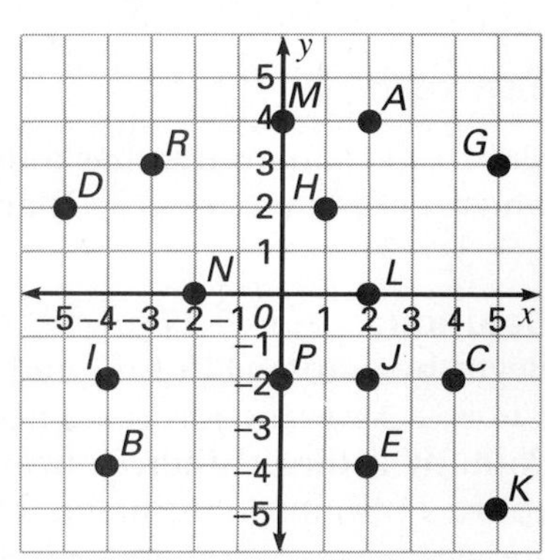

1. point A
2. point B
3. two points on the x-axis
4. two points with the same y-coordinate
5. a point in the first quadrant
6. a point in the third quadrant
7. a point whose coordinates are equal
8. a point whose coordinates are opposites

Graph each point on a coordinate plane.

9. $T(3, 2)$
10. $V(4, -1)$
11. $W(0, 4)$
12. $X(-7, 8)$
13. $Y(-4, 0)$
14. $Z(3, -11)$
15. $R(-8, -4)$
16. $O(0, 0)$

Sketch each figure on a coordinate plane.

17. rectangle with $S(3, 2)$ and $T(-3, -2)$ as endpoints of a diagonal
18. circle with center $P(3, 2)$ and radius 4 units
19. *USING DATA* Refer to the table on page 505. Draw the dot pattern for the 2nd through 4th square numbers on a coordinate plane. Give the ordered pair for the corner points of each number.
20. Which pair of points is closer together, $(-2, 7)$ and $(5, 7)$ or $(-1, 3)$ and $(-1, -1)$?
21. Find the coordinates of two points other than points on the axes that are the same distance from the origin.

EXTEND/ SOLVE PROBLEMS

Graph each set of points on a coordinate plane. Then, join the points in order, identify the geometric figure, and find its area.

22. $O(0, 0)$, $A(0, 5)$, $B(4, 5)$, and $C(4, 0)$
23. $D(-7, 1)$, $E(-3, 1)$, $F(-3, -3)$, and $G(-7, -3)$
24. Graph $L(1, 2)$. Move 3 units to the right and 2 units up to locate point M. What are the coordinates of M?
25. Graph $(1, 3)$, $(5, 7)$, $(-1, 1)$, and $(0, 2)$. Write the coordinates of four other points that fit the pattern.
26. All but one of these points suggest a pattern: $A(-7, -4)$, $B(-3, -2)$, $C(3, 1)$, $D(-5, -3)$, $E(1, 0)$, and $F(7, 4)$. Graph these points. Write the coordinates of the one point that does not fit the pattern. Tell why it does not fit the pattern.

THINK CRITICALLY/ SOLVE PROBLEMS

27. On $\overleftrightarrow{AC}$, points A and C have coordinates $(6.5, -3)$ and $(-4, 1.6)$, respectively. Find the coordinates of the midpoint of $\overline{AC}$.
28. Graph the rectangle whose vertices are $A(-2, 4)$, $B(2, 4)$, $C(2, 6)$, and $D(-2, 6)$. Now flip the rectangle over the x-axis so that it is upside down but remains the same distance from the x-axis as before. Give the coordinates of the vertices of the rectangle that results.

14-2 Problem Solving Skills:

USE A GRID TO SOLVE DISTANCE PROBLEMS

- ► READ
- ► PLAN
- ► SOLVE
- ► ANSWER
- ► CHECK

A coordinate grid can often be used to find distances.

PROBLEM

John lives 10 blocks east and 4 blocks north of the art museum. Alex lives 6 blocks west and 6 blocks south of the museum. On Saturday, John walked from his house to Alex's; then the boys went to the museum together. These are the routes they took:

To get to Alex's house, John walked 3 blocks west, 8 blocks south, 10 blocks west, 2 blocks south, and 3 blocks west.

From Alex's house, the boys walked north 4 blocks and east 6 blocks. Then Alex suddenly remembered he had left his student pass to the museum at home, so both boys retraced their steps and returned to Alex's house to get the pass. They set out again and reached the museum by walking 2 blocks east, 6 blocks north, and 4 blocks east.

How far did each boy walk between his own home and the museum?

SOLUTION

Use a grid. Let each square of the grid represent one city block. Indicate the directions north, south, east, and west on the grid.

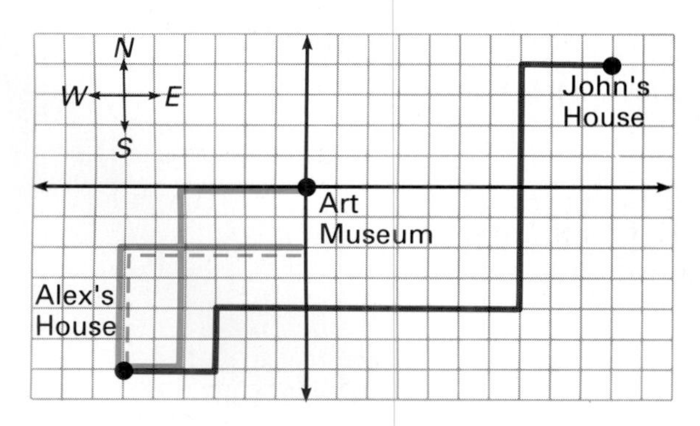

Plot the locations of the museum and each boy's house. Then draw the route John followed to reach Alex's house and the route the boys took together. Then find the total number of blocks each boy walked.

Alex walked 32 blocks.
John walked (26 + 32) blocks, or 58 blocks.

PROBLEMS

1. In a community walk-a-thon, Michele traveled the path shown on the grid at the top of page 511. Each square represents 1 block. What is the total distance Michele walked?

2. Refer to Problem 1. Assume that the grid lines represent all the sidewalks. If Michele walked only on the sidewalks, what is the shortest distance she could have walked between the start and finish points?

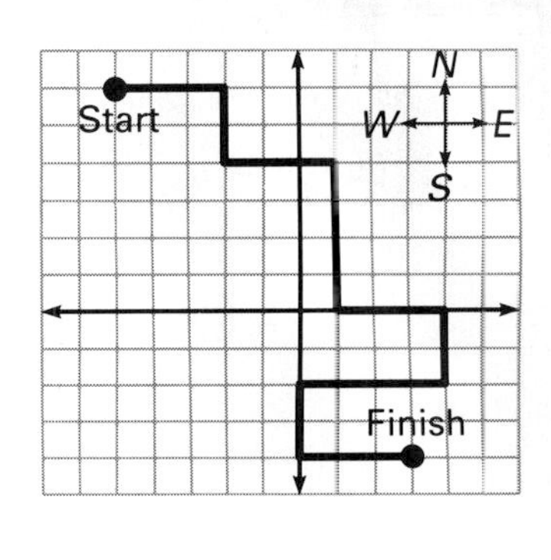

3. From her house, Marie rode her bike 4 blocks east, 4 blocks north, and 5 blocks east to return books to the library. Then she rode 7 blocks west and 3 blocks north to Angela's house. Together the girls biked 3 blocks east to visit Carla. Then Marie rode 5 blocks south and 4 blocks west to the Johnsons' house to baby-sit. How many blocks did Marie ride her bike from her house until she reached the Johnsons' house? How many blocks from home was she while she baby-sat?

4. Every Thursday, Kara leaves work and drives 6 mi west, 3 mi south, 5 mi east, 1 mi south, and 3 mi east to get to her exercise class. At about the same time, Kara's husband leaves his work, which is 4 mi due west of Kara's office, and has a friend drive him to baseball practice. They drive 3 mi west, 4 mi south, 5 mi east, 1 mi south, and 2 mi east. After her class, Kara drives the most direct route possible to pick up her husband at practice. How far and in what direction does she drive?

5. One spring, a bear came out of its cave to look for food. After traveling 20 mi due south, 20 mi due west, then 20 mi due north, the bear arrived back at its cave. What was the color of the bear?

PROBLEM SOLVING TIP

For Problem 5, think carefully about where the bear must be in order to leave its cave and then return to it by going in only three directions.

6. *USING DATA* Use the Data Index on page 546 to find the maximum speeds of animals. Traveling at maximum speed, an African elephant runs north from one spot for 5 min (one twelfth of an hour), then west for 5 min. Starting at the same spot, a Mongolian gazelle runs west for 5 min, then north for 2 min, where it stops at a watering hole. How far is the elephant from the watering hole after 10 min of running?

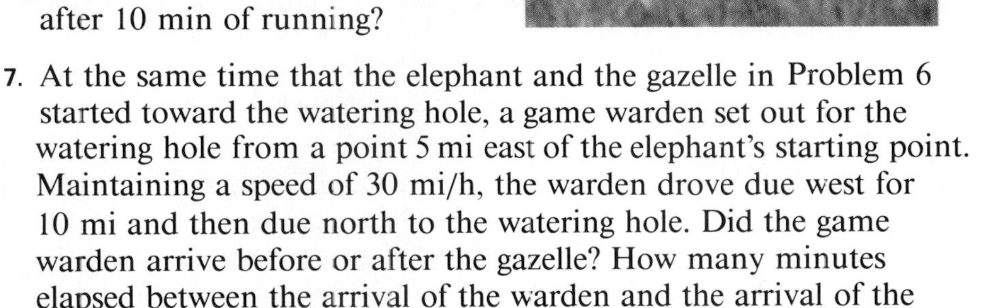

7. At the same time that the elephant and the gazelle in Problem 6 started toward the watering hole, a game warden set out for the watering hole from a point 5 mi east of the elephant's starting point. Maintaining a speed of 30 mi/h, the warden drove due west for 10 mi and then due north to the watering hole. Did the game warden arrive before or after the gazelle? How many minutes elapsed between the arrival of the warden and the arrival of the gazelle?

JUST FOR FUN

Suppose you can travel only north or east along these grid lines.

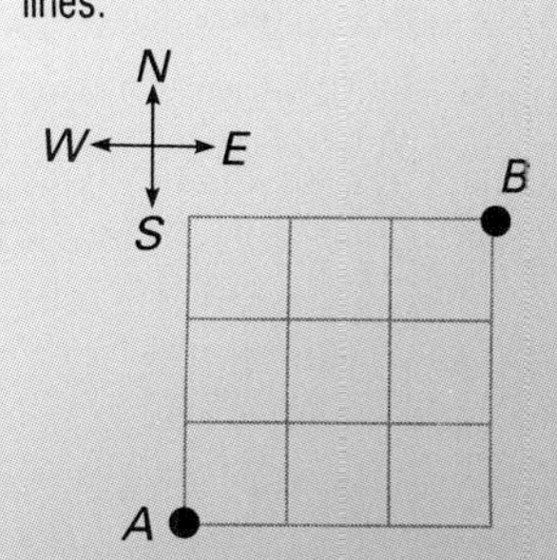

How many different routes can you take to get from *A* to *B*?

14-3 Graphing Linear Equations

EXPLORE

Read this equation: $2x + 5 = 21$
What is the solution? Is there another solution? Explain.

Now try to solve this equation: $x + y = 14$
What solution did you get?
Can you solve this equation another way? Explain.
How many solutions do you think there are to this equation?

SKILLS DEVELOPMENT

You know how to find the solution of some equations with one variable. Other equations, however, have two variables. Such an equation has an infinite number of solutions. Each of the solutions can be represented by an ordered pair.

To find some of the solutions of an equation with two variables, begin by choosing a value for the first variable, x. Substitute that value into the equation and solve to find the corresponding value of y. Do this for at least three values of x. Make a table to keep track of the ordered pairs that are solutions of the equation.

TALK IT OVER

Why do you think you are instructed to use at least three different values of x when you graph a linear equation?

Example 1

Find three solutions of the equation $y = x - 3$.

Solution
To begin solving $y = x - 3$, choose three values for x. It usually is a good idea to choose a negative number, zero, and a positive number for these values.

Let x equal -2.

$y = x - 3$
$y = -2 - 3$
$y = -5$

Let x equal 0.

$y = x - 3$
$y = 0 - 3$
$y = -3$

Let x equal 3.

$y = x - 3$
$y = 3 - 3$
$y = 0$

Write the solutions in a table.

$y = x - 3$

x	y	(x, y)	
-2	-5	$(-2, -5)$	← When x is -2, y is -5.
0	-3	$(0, -3)$	← When x is 0, y is -3.
3	0	$(3, 0)$	← When x is 3, y is 0. ◄

Since a solution of an equation in two variables is an ordered pair, you can graph it as a point on the coordinate plane. The set of all points whose coordinates are solutions of an equation is called the **graph of the equation.** When these points lie in a straight line, the equation is called a **linear equation.**

Example 2

Graph the equation $y = -3x + 2$.

Solution

Find at least three solutions of the equation, using reasonable values for x. The table shows that the values chosen for x are -2, 0, and 2.

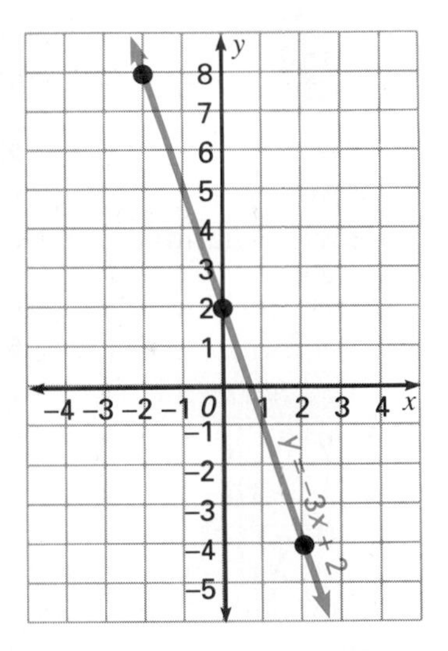

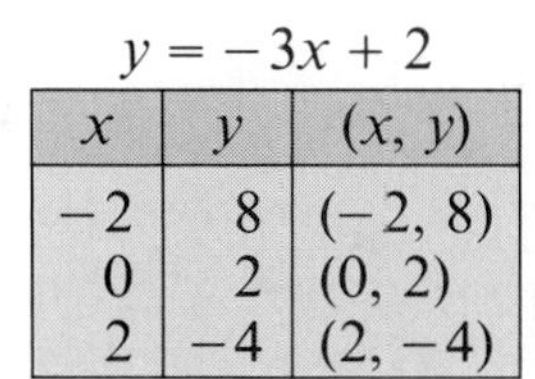

$y = -3x + 2$

x	y	(x, y)
-2	8	$(-2, 8)$
0	2	$(0, 2)$
2	-4	$(2, -4)$

Graph the ordered pairs. Draw a line through the points.

All the points along the line are solutions of the equation $y = -3x + 2$. ◄

COMPUTER TIP

Some computer software acts as a function grapher and draws graphs quickly and accurately. In most versions of this type of software, when you enter a function and values for the variables, the computer will then graph the function for those values. If you have access to this type of software, you may want to use it to investigate the graphs of functions.

When you find solutions of an equation in two variables, the y-value of the ordered pair depends on the x-value that you choose. So we say that *y is a function of x,* and the relationship between y and x is called a **function.**

In everyday life, one quantity often depends on another. Many of these relationships can be represented by mathematical functions. The graph of the equation that represents the function is called the **graph of the function.**

Example 3

Daneesha walks on a treadmill at an average rate of 4 mi/h. The distance she walks is a function of the amount of time she walks. Write an equation that represents this function. Then graph the function.

Solution

Let x = the amount of time and let y = the distance. So $y = 4x$.
Remember: The distance formula is distance = rate × time.

Make a table of solutions.

$y = 4x$

x	y	(x, y)
1	4	$(1, 4)$
2	8	$(2, 8)$
3	12	$(3, 12)$

Use ordered pairs to graph the function. ◄

CHECK UNDERSTANDING

In Example 3, the graph does *not* extend to the left of the x-axis. What is the reason for this?

TRY THESE

For each equation, make a table of three solutions.

1. $y = 4x$
2. $y = 5x - 1$
3. $y = 2x + 2$

For each equation, complete the table of solutions. Then graph the equation.

4. $y = 2x - 1$

x	y	(x, y)
−2	−5	(−2, −5)
−1	−3	(−1, −3)
0	−1	(0, −1)
1	■	(1, ■)

5. $y = -3x$

x	y	(x, y)
−1	■	(−1, ■)
0	■	(0, ■)
1	■	(1, ■)
2	■	(2, ■)

6. $y = x + 4$

x	y	(x, y)
−2	■	(−2, ■)
0	■	(0, ■)
2	■	(2, ■)
4	■	(4, ■)

7. Chicken cutlets cost $3 per pound. This is expressed by the formula $y = 3x$, where y is the cost and x is the number of pounds. Complete the table of solutions for the equation that represents the function. Then graph the function.

x	y	(x, y)
0	■	(0, ■)
1	■	(1, ■)
2	■	(2, ■)
3	■	(3, ■)

EXERCISES

PRACTICE/ SOLVE PROBLEMS

For each equation, make a table of three solutions.

1. $y = x + 4$
2. $y = 3x - 2$
3. $y = x - 1$
4. $y = -2x - 4$
5. $y = 2x + 4$
6. $y = -2x$

For each equation, complete the table of solutions. Then graph the equation.

7. $y = 4x$

x	y	(x, y)
−1	■	(−1, ■)
0	■	(0, ■)
1	■	(1, ■)
2	■	(2, ■)

8. $y = -2x + 1$

x	y	(x, y)
−3	■	(3, ■)
−1	■	(1, ■)
0	■	(0, ■)
2	■	(2, ■)

9. $y = \frac{1}{2}x + 5$

x	y	(x, y)
−4	■	−4, ■
−2	■	−2, ■
0	■	0, ■
2	■	2, ■

10. A person's age five years from now (y) is a function of the person's present age (x). Write an equation that represents this function. Then graph the function.

EXTEND/ SOLVE PROBLEMS

Write an equation for each graph.

11.

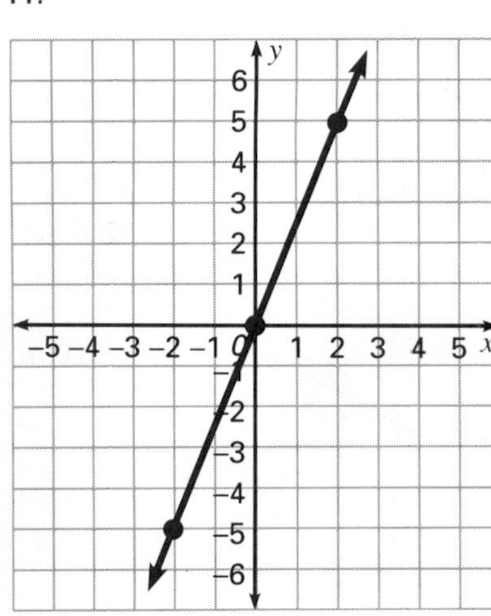

12.

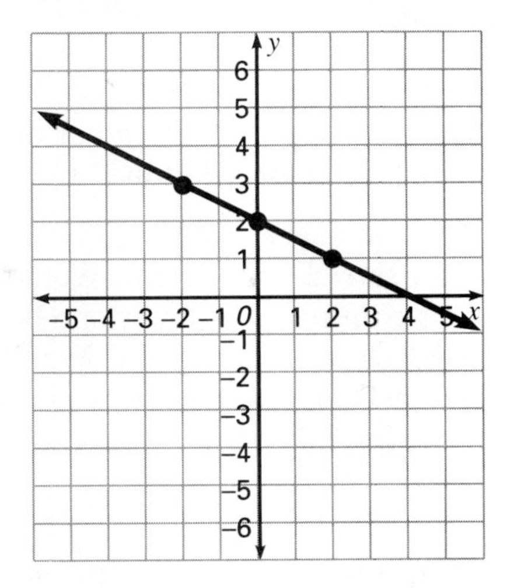

13. On a coordinate plane, graph the ordered pairs $(-2, -1)$ and $\left(1, \frac{1}{2}\right)$. Draw a line through the points. Make a table of these ordered pairs and three others that lie along the line you drew. Write the equation for the line.

THINK CRITICALLY/ SOLVE PROBLEMS

14. Look back at the graphs of the linear equations with which you have been working. Draw some conclusions about which equations are related to the graphs that slant upward and which are related to the graphs that slant downward.

14-4 Working with Slope

EXPLORE

When did you last climb up or down a slope? Describe the slope.

With your finger, trace the slope of the hill shown at the right. Would a skier coming down the slope be skiing a distance of 326 m? Would the skier be skiing a distance of 621 m? Explain.

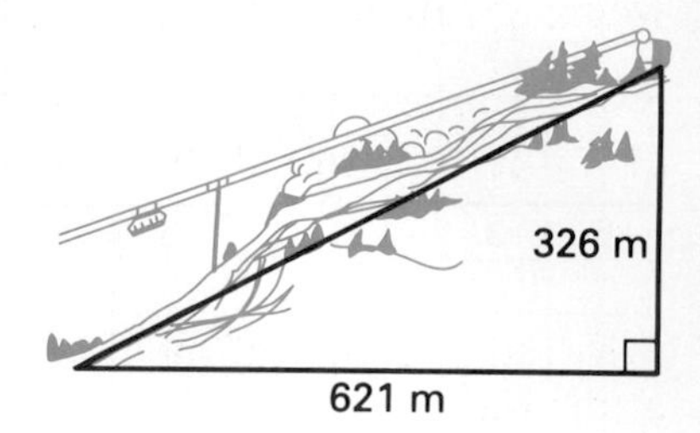

SKILLS DEVELOPMENT

TALK IT OVER

Name some instances in which someone might need to know the slope of something. Why might it be helpful to use numbers in these instances?

As you move from one point to another along a straight line, the change in position has two components. There is a vertical change in position, called the **rise,** and a horizontal change, called the **run.** The ratio of the rise to the run is called the **slope** of the line.

$$\text{slope} = \frac{\text{rise}}{\text{run}}$$

You can find the slope of a line graphed on a coordinate plane by counting the units of change between the coordinates of two points on the line.

Example 1

Find the slope of the line graphed at the right.

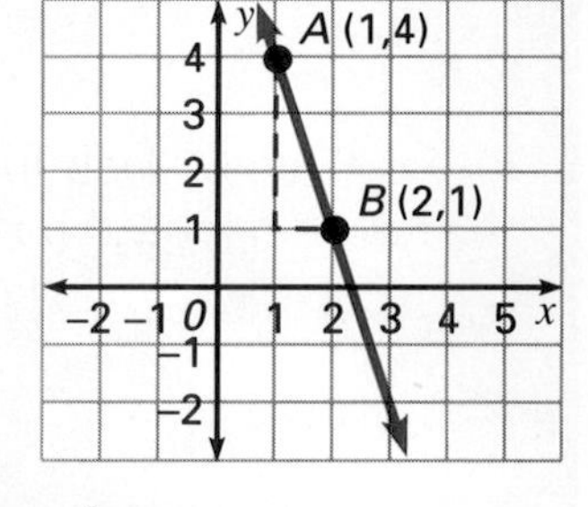

Solution

Choose two points on the line, such as $A(1, 4)$ and $B(2, 1)$. Find the number of units of change in moving from point A to point B.

$$\text{slope} = \frac{\text{rise}}{\text{run}} = \frac{\text{3 units down}}{\text{1 unit right}} = \frac{-3}{1} = -3 \quad \blacktriangleleft$$

A move *down* or to the *left* is represented by a negative number.

If you are given the coordinates of two points on a line, you can find the slope of the line by using this formula.

$$\text{slope} = \frac{\text{rise}}{\text{run}} = \frac{\text{difference of } y\text{-coordinates}}{\text{difference of } x\text{-coordinates}}$$

Example 2

Find the slope of the line that passes through the points $M(0, 2)$ and $N(-2, -1)$.

Solution

$$\text{slope} = \frac{\text{rise}}{\text{run}} = \frac{\text{difference in } y\text{-coordinates}}{\text{difference in } x\text{-coordinates}}$$

Be sure to subtract the *y*-coordinates and the *x*-coordinates in the same order.

$$\text{slope} = \frac{2-(-1)}{0-(-2)} = \frac{3}{2} \quad \blacktriangleleft$$

When a linear equation is written with y alone on one side of the equals sign, we say it is written in the form $y = mx + b$. With the equation written in this form, the coefficient of x, which is m, is the slope of the line. The other piece of information that we can read from this equation is that the constant term, b, is the y-intercept of the line. The **y-intercept of a line** is the y-coordinate of the point where it intersects the y-axis.

Since you can use $y = mx + b$ to find the slope and the y-intercept of a line, it is called the **slope-intercept form** of an equation.

Example 3

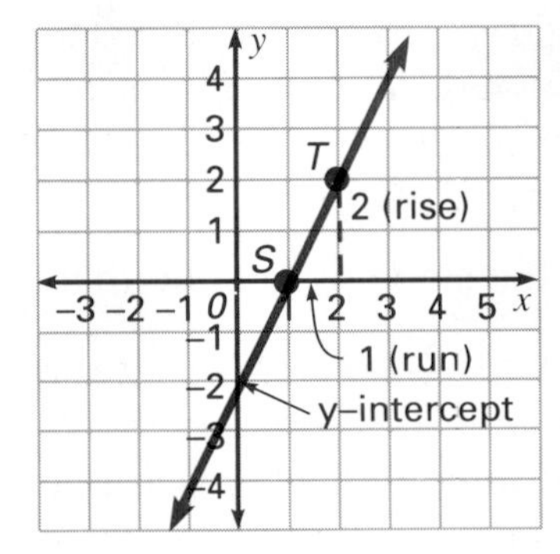

Write an equation of the line graphed at the right.

Solution
Find the slope by counting units of change from point S to point T.

$\text{slope} = \frac{\text{rise}}{\text{run}} = \frac{2}{1} = 2$

The slope of line ST is 2.

The line intersects the y-axis at $(0, -2)$, so the y-intercept is -2.

An equation of the line is $y = 2x - 2$. ◄

CHECK UNDERSTANDING

In Example 3, line ST slants upward from left to right and has a slope of 2. In what direction do you think a line with a slope of -2 would slant?

TRY THESE

Find the slope of each line on the coordinate plane at the right.

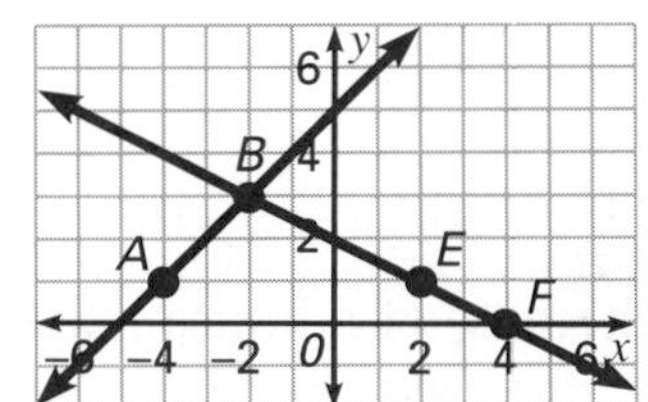
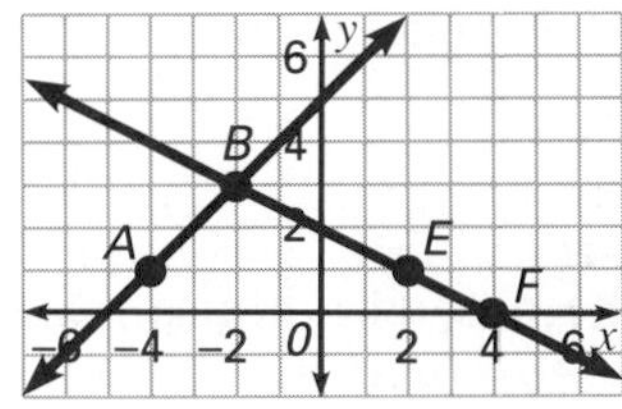

1. line AB
2. line EF

Find the slope of the line that passes through each pair of points.

3. $A(4, 5)$ and $B(3, 7)$
4. $J(1, 6)$ and $K(3, 9)$
5. $R(-5, -7)$ and $S(8, -2)$
6. $M(-3, 4)$ and $N(-8, 1)$
7. $X(6, -5)$ and $Y(9, -3)$
8. $P(9, -1)$ and $Q(0, 4)$

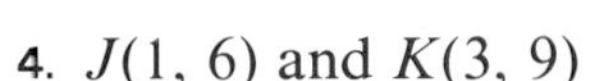

Write an equation of each line graphed at the right.

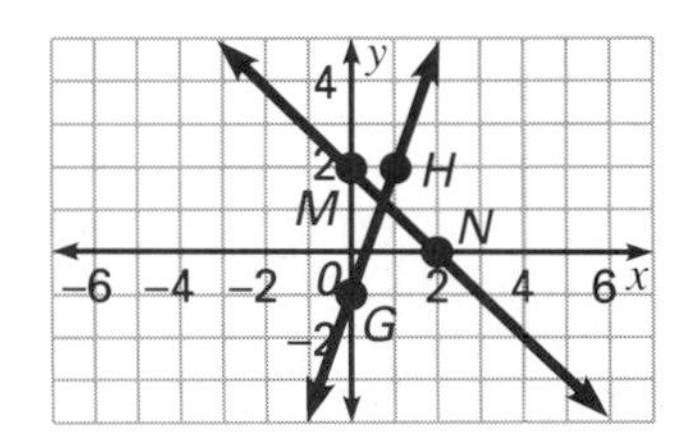

9. line GH
10. line MN

COMPUTER

You may use this program to check your answers to Exercises 3–8 and for other exercises in which you are asked to find the slope.

```
10 INPUT "ENTER THE
   COORDINATES OF THE
   FIRST POINT: ";X1, Y1:
   PRINT
20 INPUT "ENTER THE
   COORDINATES OF THE
   SECOND POINT: ";X2, Y2:
   PRINT
30 PRINT "THE SLOPE IS
   ";Y2 – Y1;"/";X2 – X1
```

EXERCISES

PRACTICE/ SOLVE PROBLEMS

Find the slope, or pitch, of each roof.

1. 

2. 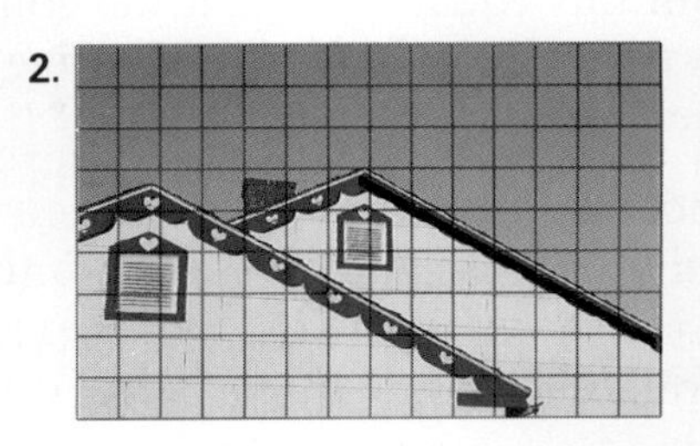

Find the slope of the line that passes through each pair of points.

3. $X(-4, 2)$ and $Z(3, -2)$
4. $M(3, 2)$ and $N(1, 1)$

CHECK UNDERSTANDING

Do you think all lines with the same y-intercept have the same slope?

Do you think all lines with the same slope have the same y-intercept? Explain.

Write an equation of each line.

5. 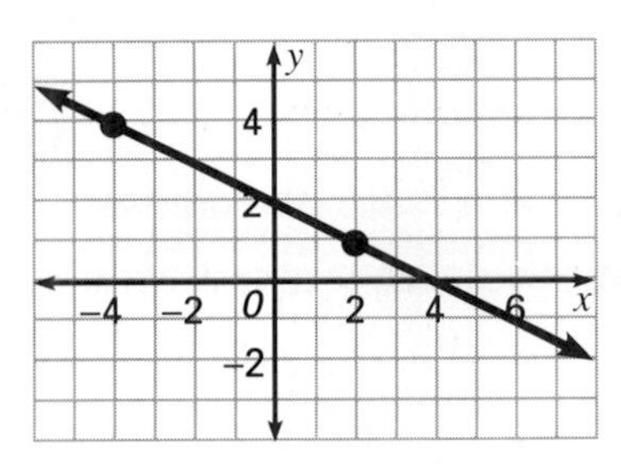

6. 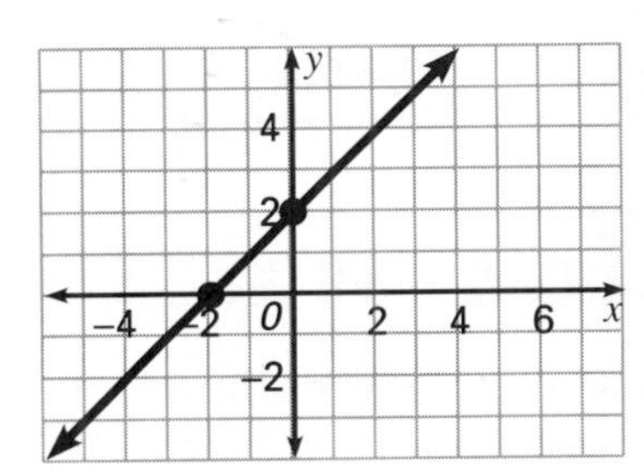

7. The roof of a house rises vertically 3 ft for every 12 ft of horizontal distance. What is the slope, or pitch, of the roof?

EXTEND/ SOLVE PROBLEMS

Use the coordinate plane at right. Find the slope of each line segment.

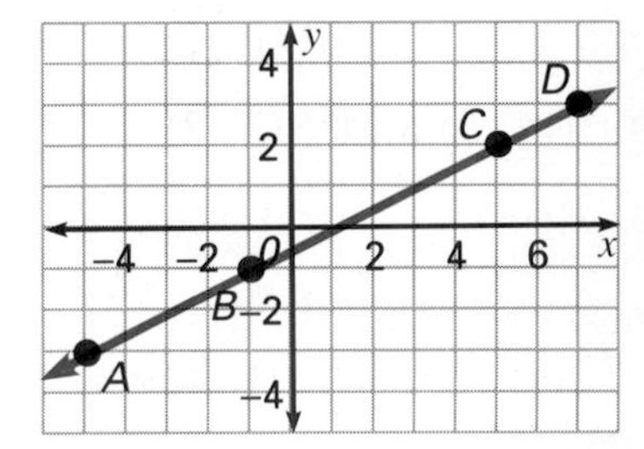

8. $\overline{AB}$
9. $\overline{BC}$
10. $\overline{CD}$
11. $\overline{AD}$
12. Compare the slopes in Exercises 8–11. What do you notice?

THINK CRITICALLY/ SOLVE PROBLEMS

13. What is an equation of the line that has a y-intercept of 2 and is parallel to the line whose equation is $y = \frac{1}{2}x + 1$?
14. Write an equation of a line that has a slope of 0.

- ► READ
- ► PLAN
- ► SOLVE
- ► ANSWER
- ► CHECK

Problem Solving Applications:

INCLINED PLANES

You may already know that the inclined plane is a simple machine that is helpful in moving objects. The effort needed to push an object up an inclined plane to a given point is less than the effort needed to raise the object the vertical distance to that point. For example, in the figure below, it would be easier to push the block of wood to the point at which it appears than to raise the block vertically to that point.

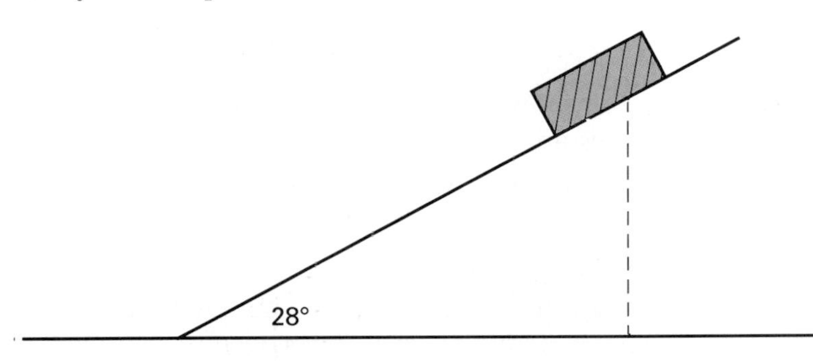

Get a block of wood. Use a smooth plank as an inclined plane.

1. Slant the inclined plane so that the block of wood can be positioned on it without sliding.
2. Position the block on the inclined plane.
3. Slowly incline the plane until the block begins to slide.
4. Record the angle of the plane at the moment when the block begins to slide.
5. Record the slope of the plane when the sliding begins.
6. Repeat steps 1–5 two more times. Record your data in a chart.

	1st trial	2nd trial	3rd trial
angle of plane			
slope of plane			

7. From your chart, how are the values of the angles and slopes related?
8. Repeat steps 1–5 using planes of different materials, such as cardboard or steel, and blocks of different sizes and materials, such as plastic or foam. Perform repeated trials using these materials. How do your data using these materials differ from the data you collected using wooden materials?

CALCULATOR

If you graph the equation $y = 6x + 3$ on a graphing calculator, the graph will show you that the line crosses the x-axis at some point between -1 and 0. The line crosses the y-axis at some point near 3.

If you change the viewing rectangle so that x goes from -1 to 1 and y goes from -1 to 4, you can use the TRACE feature to find the intercepts more accurately. The TRACE feature lets you put the cursor on the intersection points and read the x- and y-values along the bottom of the screen.

With the viewing rectangle described, you see that the y- intercept is about 3 and the x-intercept is about -0.5. Using even more refined viewing rectangles shows that the x-intercept is -0.5 and the y-intercept is 3.

Use a graphing calculator to find the x- and y-intercepts of each of these lines.

1. $y = 2x - 4$
2. $y = x + 3.4$
3. $y = -2x + 5$
4. $y = 3x$

14-5 Problem Solving Strategies:

MAKE A MODEL

► READ
► PLAN
► SOLVE
► ANSWER
► CHECK

Sometimes a problem is easier to solve if you make a model to help you solve it. There are many types of models that you can use, such as equations, sketches, scale drawings, three-dimensional constructions, and computer simulations. When a problem involves a linear function, a model that may be especially helpful is a graph of the function.

PROBLEM

Suppose you supervise employees who make sales calls over the telephone. The employees are paid a flat fee each week for making a minimum of 15 calls. In addition to their flat fee, they receive a fixed commission for each item they sell.

The chart shows the amount of the paycheck and the number of items sold by each of four employees. What is the flat fee and the amount of commission per item that each employee receives?

Number of Items Sold	10	25	30	15
Amount of Paycheck	$300	$675	$800	$425

SOLUTION

Model the situation on a coordinate plane. Label the x-axis "Number of Items Sold." Label the y-axis "Amount of Paycheck."

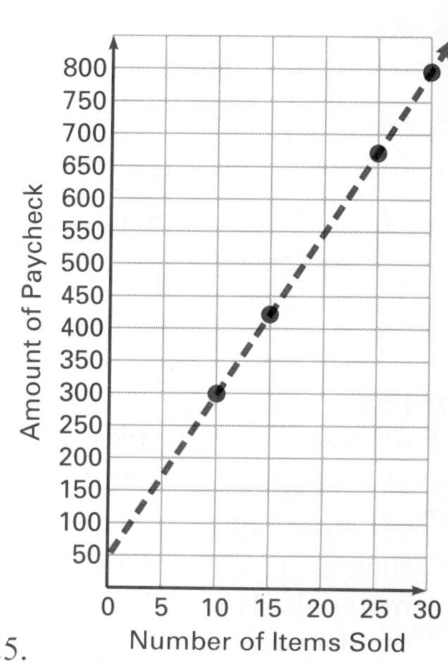

When you graph the data from the table as four points on the coordinate plane, the points lie along a straight line. The equation of that line is $y = 25x + 50$. The y-intercept, 50, represents the amount each employee receives if 0 items are sold. So the flat fee is $50. The slope of the line, 25, represents the increase in the paycheck for each 1 item sold. So the amount of commission per item is $25.

PROBLEMS

Make a model to solve each problem. Use the information in the chart provided with each problem to help you model the situation.

1. Admission to a high school soccer game is charged on a per-vehicle basis. For each vehicle, there is a fixed parking fee. Then an additional amount is charged for each student in the vehicle. What is the parking fee? What is the amount charged per student?

Amount Charged for Vehicle	$13	$7	$31
Number of Students per Vehicle	5	2	14

2. A telephone company charges a fixed fee for making a service call plus an hourly rate for each hour the service person spends making the call. Find the fixed fee and the hourly rate for each service call.

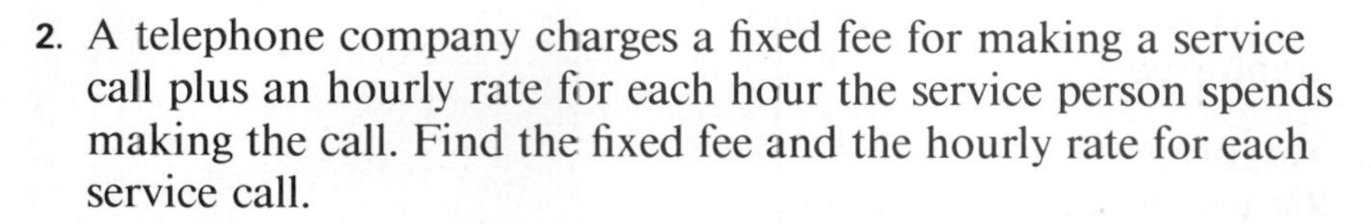

Amount Charged for Call	$85	$45	$125
Number of Hours per Call	3	1	5

3. A baby sitter charges a fixed fee for the first hour and thereafter charges an hourly rate depending on the number of children being cared for. Find the fee for the first hour and the hourly rate per child for a four-hour baby-sitting job.

Amount Charged for Four-Hour Job	$6.50	$11.50	$16.50
Number of Children	1	3	5

4. A health club charges its members a one-time fee when they first join. Thereafter each member is charged annual dues. Find the amount of the one-time fee and the amount of dues each member must pay per year.

Total Amount Paid by Member	$600	$1,050	$1,500
Number of Years of Membership	1	4	7

5. A store charges a fixed amount to ship each package and an added charge per pound. Find the fixed shipping fee and the cost per pound.

Cost of Shipping Package	$21.50	$32.00	$42.50
Number of Pounds	4	7	10

14-6 Translations

EXPLORE/ WORKING TOGETHER

Work with a partner. You will need a piece of graph paper, a ruler, and one of the small yellow square tiles from a set of algebra tiles.

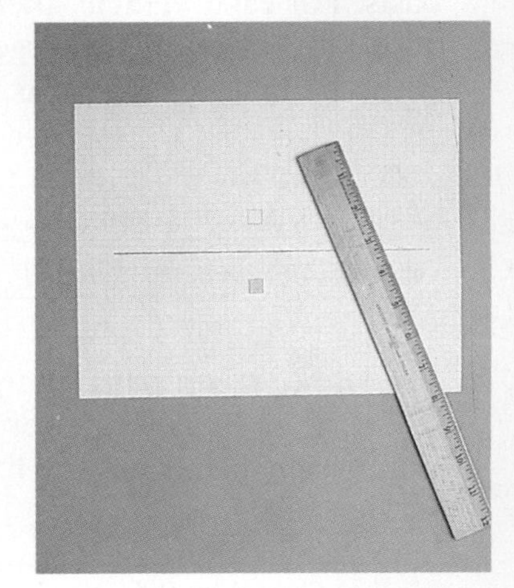

a. Use the ruler to draw a vertical rule down the middle of the graph paper.

b. Place the small yellow tile anywhere on the graph paper to the left of the vertical rule. Position the tile so that it exactly covers four of the squares on the graph paper.

c. You or your partner should hold the tile steady while the other traces it.

d. Now, slowly slide the tile horizontally along the grid lines until it is across the vertical rule and is positioned the same distance from the rule as it was in its starting position.

e. Trace the tile in its new position.

f. Draw an arrow from the first tracing to the second.

How many units did you slide the tile from its original position to its new position?

How did the tile change as you slid it? How did it stay the same?

SKILLS DEVELOPMENT

A **translation,** or *slide,* of a figure produces a new figure that is exactly like the original. As a figure is translated, you imagine all its points sliding along a plane the same distance and in the same direction. So the sides and angles of the new figure are equal in measure to the sides and angles of the original, and each side of the new figure is parallel to the corresponding side of the original.

A move like a translation is called a **transformation** of the figure. The new figure is called the **image** of the original, and the original is called the **preimage** of the new figure.

Example 1

Graph the image of the point $C(-5, 3)$ under a translation 4 units to the right and 3 units down.

Solution

Add 4 to the x-coordinate and subtract 3 from the y-coordinate.

$C(-5, 3) \rightarrow C'(-5 + 4, 3 - 3)$
$\rightarrow C'(-1, 0)$
$\uparrow$

Read C' as "C prime." ◄

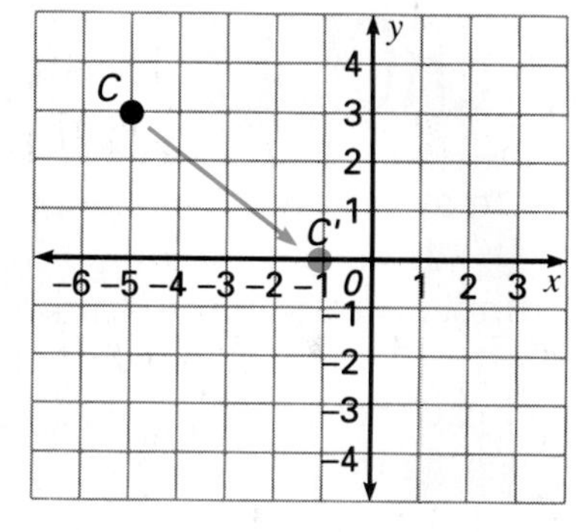

Example 2

Graph the image of $\triangle ABC$ with vertices $A(-3, 3)$, $B(-5, -1)$, and $C(-2, 0)$ under a translation 5 units to the right and 4 units up.

Solution

Add 5 to the x-coordinate of each vertex. Add 4 to the y-coordinate of each vertex.

$A(-3, 3) \rightarrow A'(-3 + 5, 3 + 4)$
$\rightarrow A'(2, 7)$

$B(-5, -1) \rightarrow B'(-5 + 5, -1 + 4)$
$\rightarrow B'(0, 3)$

$C(-2, 0) \rightarrow C'(-2 + 5, 0 + 4)$
$\rightarrow C'(3, 4)$ ◄

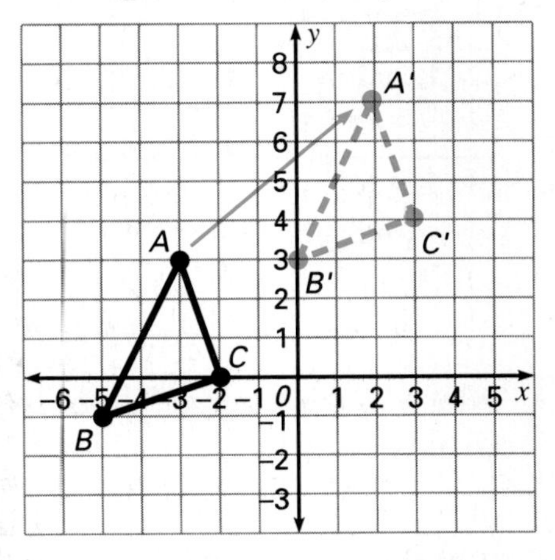

COMPUTER

You can use this program to find the coordinates of the image points in Example 2.

```
10 FOR V = 1 TO 3
20 INPUT "ENTER THE
   COORDINATES OF THE
   VERTEX: ";X, Y: PRINT
30 LET X = X + 5: LET
   Y = Y + 4
40 PRINT "THE IMAGE IS
   (";X;", ";Y;")": PRINT
50 NEXT V
```

You can use the program for a different translation of a triangle by changing the numbers in line 30. For a translation of a different type of polygon, let the V in line 10 go from 1 to the number of vertices.

TRY THESE

1. Graph the image of the point $M(3, -3)$ at the right under a translation 5 units to the left and 6 units down.

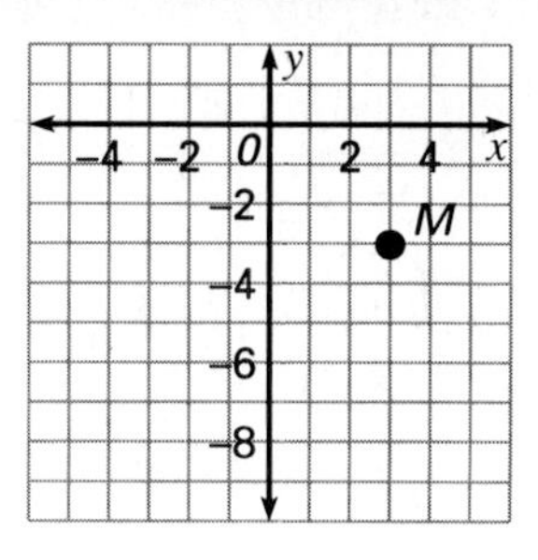

2. Graph the image of $\triangle DEF$ at the right under a translation 0 units to the right or left and 3 units down.

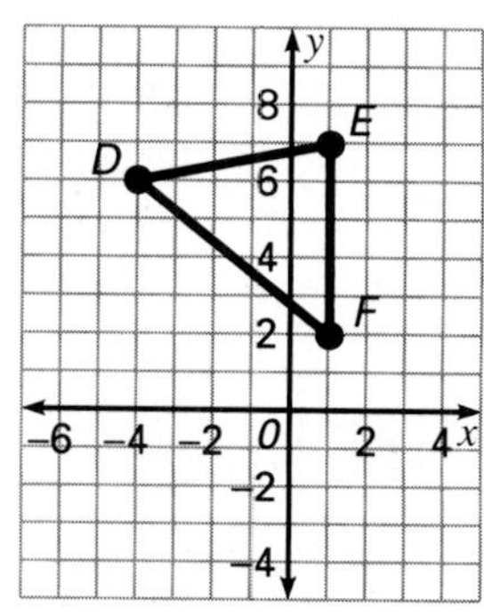

EXERCISES

PRACTICE/ SOLVE PROBLEMS

1. On a coordinate plane, graph trapezoid $DEFG$ with vertices $D(-4, 3)$, $E(6, 3)$, $F(4, -3)$, and $G(-2, -3)$. Then graph its image under a translation 4 units left and 0 units up or down.

Copy each set of figures on a coordinate plane. Then graph the image of each figure under the given translation.

2. 5 units right and 3 units up

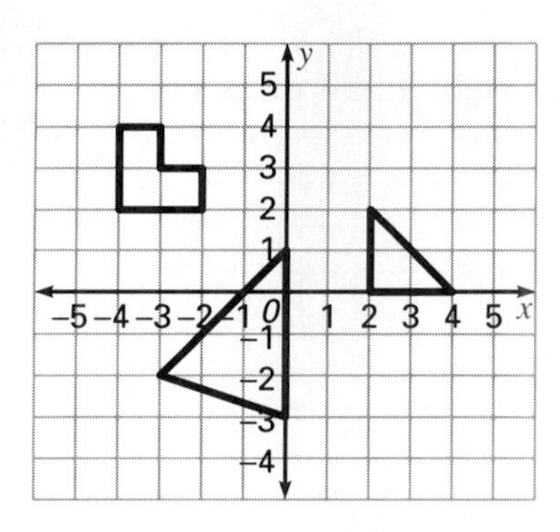

3. 1 unit left and 4 units down

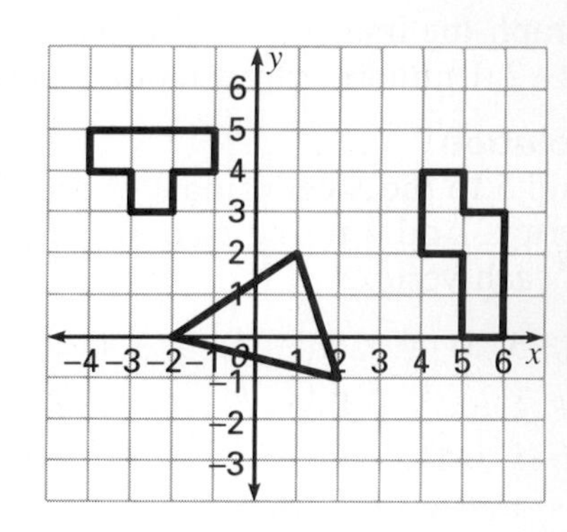

4. Which of the lettered figures are translations of the shaded figure? Give reasons for your answers.

Copy each figure onto graph paper. Then graph the image of each figure under a translation 6 units to the left and 3 units up.

5.

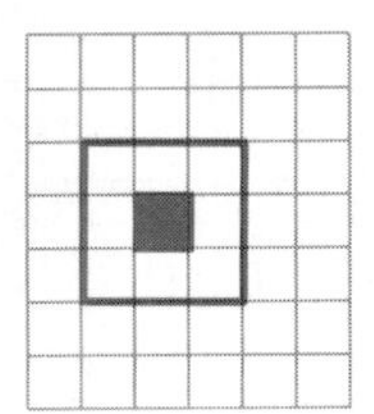

6.

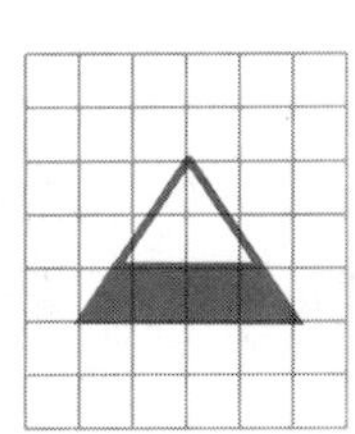

7.

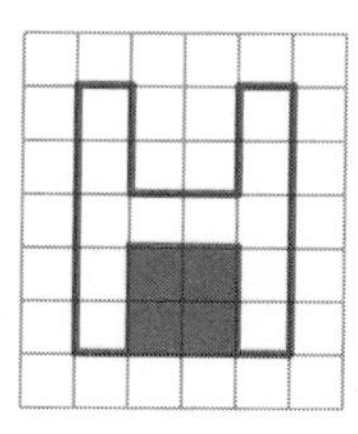

8.

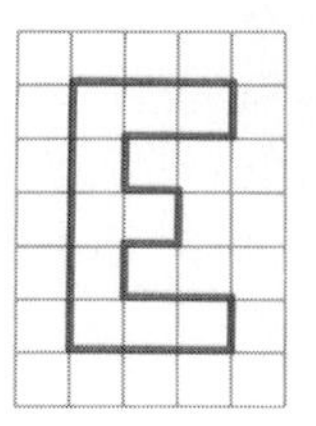

MIXED REVIEW

Find the answer.

1. $-8 - (-5)$
2. $0.007 - 0.32$
3. $36 \div 1.5$
4. $4{,}000 \times 0.366$
5. $0 - (-6)$

Solve.

6. Change 345 cm to meters.
7. Find the volume of a cube that is 8 yd long on each edge.
8. In a chemistry class of 30 students, 4 students are chosen at random to attend a conference. Does anyone have at least a 50% chance of being chosen?

EXTEND/ SOLVE PROBLEMS

Determine the direction and number of units of the translation for each preimage and its image.

9.
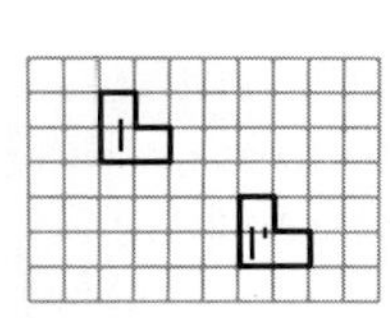

10.
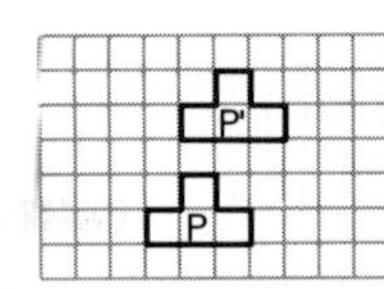

11.
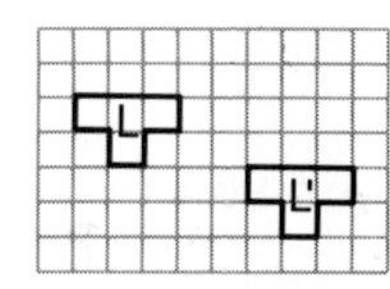

Translations were used to make each design. Decide how each was done. Then copy the design on graph paper.

12.
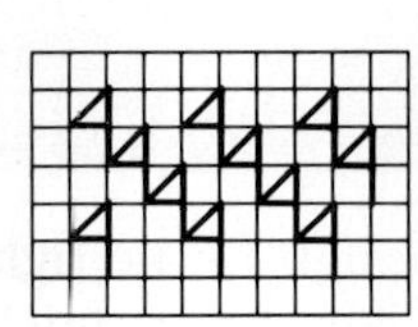

13.
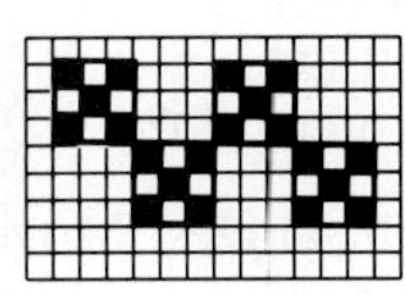

14.
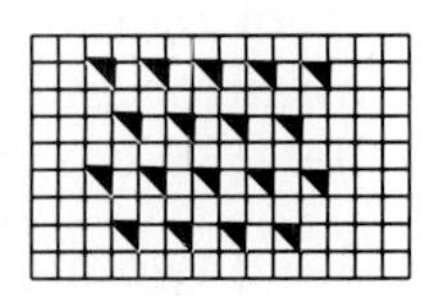

15. Make your own design with translations on graph paper.

THINK CRITICALLY/ SOLVE PROBLEMS

16. Describe the translations needed to move box *B* out of the warehouse to position *A*.

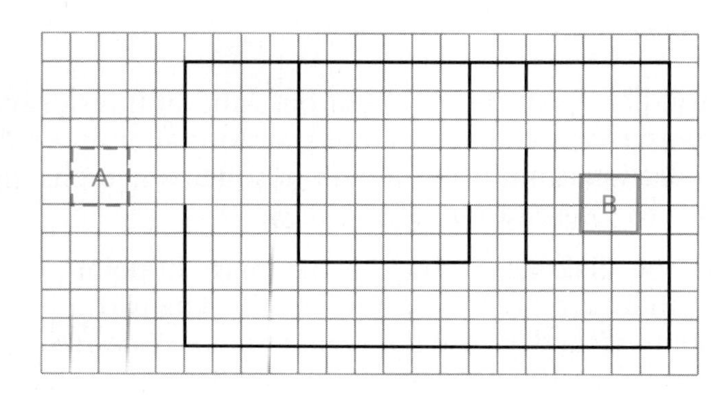

A **tessellation** is a pattern in which identical copies of a figure fill a plane so that there are no gaps or overlaps. The computer program is written in Logo. The program will create a tessellation.

17. Describe the tessellation the program will create.
18. Modify the program so that the tessellation is composed of squares.
19. Modify the program so that the tessellation is composed of parallelograms.

```
TO FIGURE
  FD 10 RT 90
  FD 5 RT 90
  FD 10 RT 90
  FD 5 RT 90
END

TO SLIDE
  PU BK 10 BK
END

TO COLUMN
  REPEAT 2 [FIGURE SLIDE]
END

TO TESS
  REPEAT 4 [COLUMN PU
  FD 20 RT 90 FD 5 LT 90 PD]
END
```

14-7 Reflections

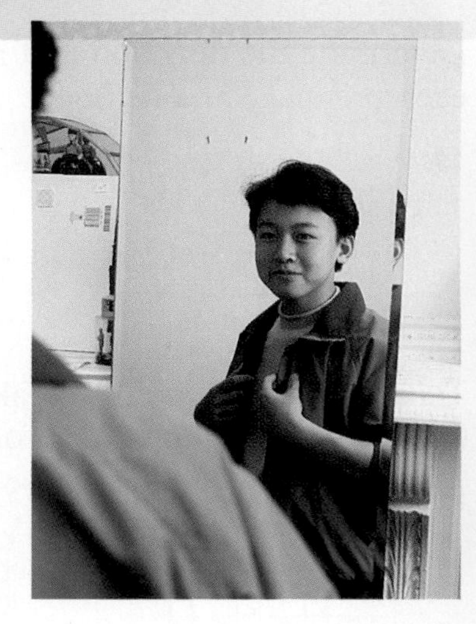

EXPLORE

When you think of reflections, you probably think of mirrors.

a. In an ordinary mirror, your image seems to be reversed from right to left. Look into a mirror and raise your right hand. What does your image do?

b. Hold two mirrors at right angles. Move close enough to the right angle so that you see just one image of yourself. How does this image differ from what you see in just one mirror?

SKILLS DEVELOPMENT

You have just learned about one kind of transformation—a translation. Another kind of transformation is a reflection. A **reflection** is a transformation in which a figure is *flipped,* or *reflected,* over a *line of reflection.*

In this diagram, point P' is the image of point P under a reflection across line EF. The two points, P and P', are the same distance from $\overleftrightarrow{EF}$, and if line PP' were drawn, it would be perpendicular to $\overleftrightarrow{EF}$.

The x- and y-axes can be used as reflection lines for figures drawn on a coordinate plane.

Example 1

Graph the image of $\triangle PQR$ with vertices $P(-6, 4)$, $Q(-5, -2)$ and $R(-2, -4)$ under a reflection across the y-axis.

Solution

Multiply the x-coordinate of each vertex by -1.

$P(-6, 4) \rightarrow P'(-6 \times (-1), 4)$
$\rightarrow P'(6, 4)$

$Q(-5, -2) \rightarrow Q'(-5 \times (-1), -2)$
$\rightarrow Q'(5, -2)$

$R(-2, -4) \rightarrow R'(-2 \times (-1), -4)$
$\rightarrow R'(2, -4)$ ◄

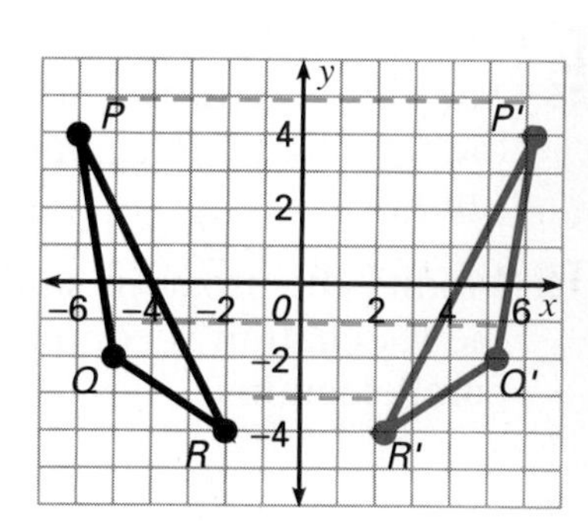

Example 2

Find the image of $\triangle STU$ with vertices $S(-2,2)$, $T(3,7)$, and $U(4,2)$ under a reflection across the x-axis.

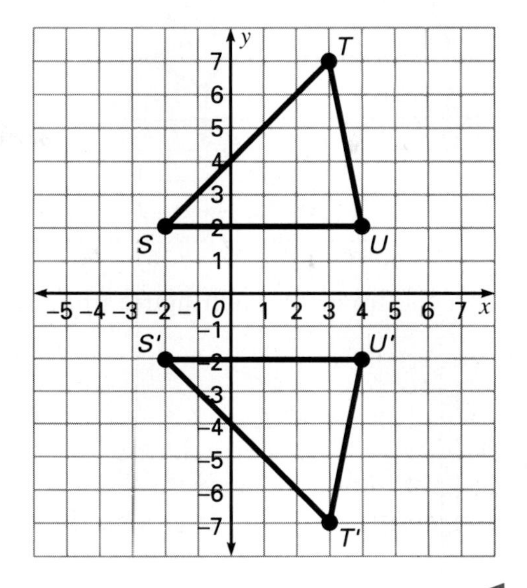

Solution

Multiply the y-coordinate of each vertex by -1.

$S(-2,2) \rightarrow S'(-2, 2 \times (-1))$
$\rightarrow S'(-2,-2)$

$T(3,7) \rightarrow T'(3,7 \times (-1))$
$\rightarrow T'(3,-7)$

$U(4,2) \rightarrow U'(4, 2 \times (-1))$
$\rightarrow U'(4,-2)$

◀

TRY THESE

Give the coordinates of the image of each point under a reflection across the given axis.

1. $(4,-5)$; y-axis
2. $(-6,-1)$; x-axis
3. $(7,9)$; x-axis
4. $(-3,8)$; y-axis
5. $(-2,0)$; y-axis
6. $(-2,0)$; x-axis

EXERCISES

PRACTICE/ SOLVE PROBLEMS

Graph the point on a coordinate plane. Then graph the image of the point under a reflection across the given axis.

1. $(4, 1)$; x-axis
2. $(-2, 5)$; x-axis
3. $(7, -2)$; y-axis
4. $(10, 5)$; y-axis
5. $(-8, -7)$; x-axis
6. $(5, 0)$; y-axis

Graph the image of each figure under a reflection across the y-axis.

7. 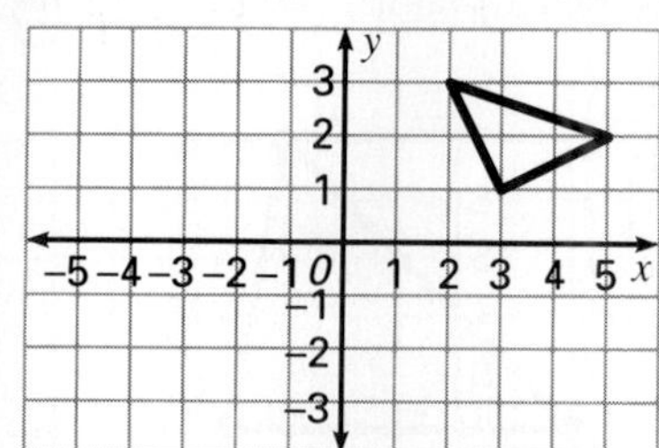

8. 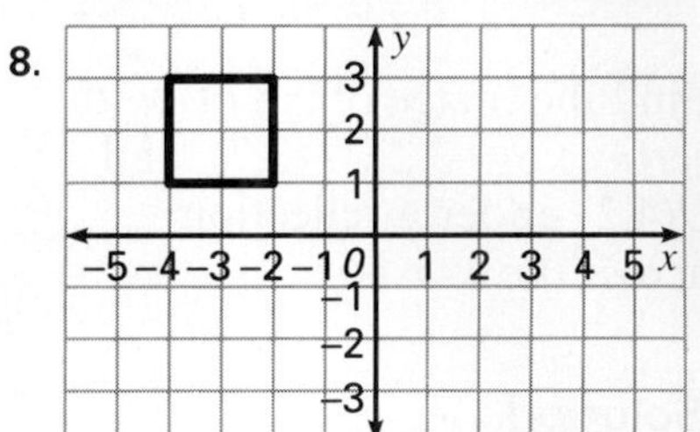

TALK IT OVER

Under what circumstances would the reflection image of a point on the coordinate plane have the same x-coordinate as its preimage?

Graph the image of each figure under a reflection across the x-axis.

9. 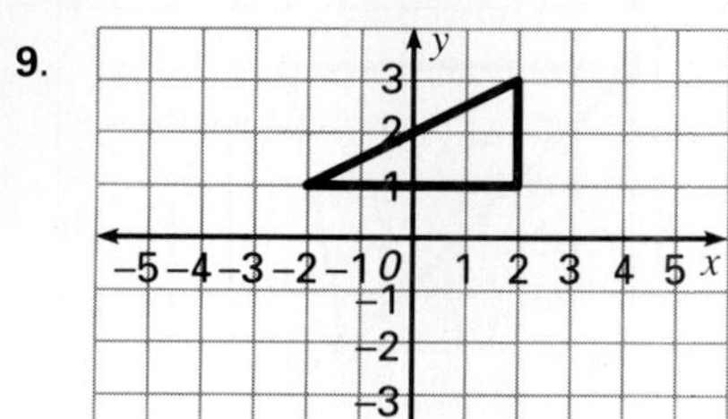

10. 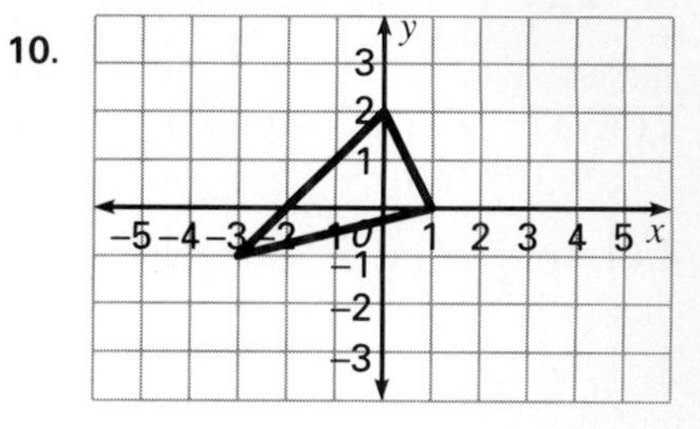

EXTEND/ SOLVE PROBLEMS

11. The letter A is a reflection of itself across the vertical line shown. Which other letters have this property?

12. The letter H is a reflection of itself in two ways. Which other letters have this property?

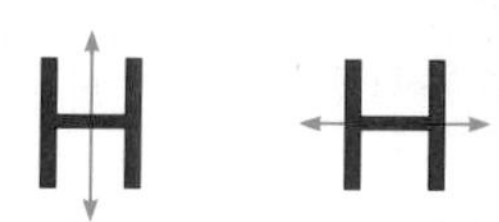

Triangle ABC is reflected in the line GK as shown.

13. Which line segments are equal in length?

14. Which angles are right angles?

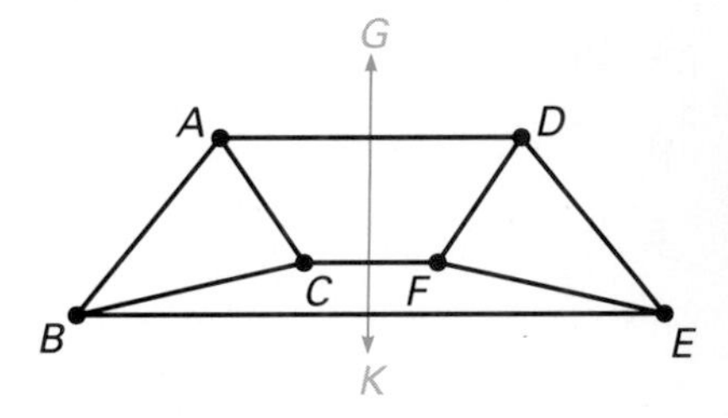

THINK CRITICALLY/ SOLVE PROBLEMS

In Exercises 15–17, $\triangle XYZ$ is drawn on a coordinate plane. The vertices are $X(2, 2)$, $Y(2, 5)$, $Z(6, 2)$.

15. Is the order of the vertices of $\triangle XYZ$ clockwise or counterclockwise? Find the reflection image $\triangle X'Y'Z'$ across the x-axis. Is the order of the vertices clockwise or counterclockwise?

16. Find the lengths of the sides of both triangles XYZ and $X'Y'Z'$. (Hint: You will need to use the Pythagorean Theorem to find the length of one side in each triangle.)

17. Think of your answers to Exercises 15 and 16. Which properties of a figure change during a reflection? Which stay the same?

► READ
► PLAN
► SOLVE
► ANSWER
► CHECK

Problem Solving Applications:

REFLECTIONS IN BILLIARDS

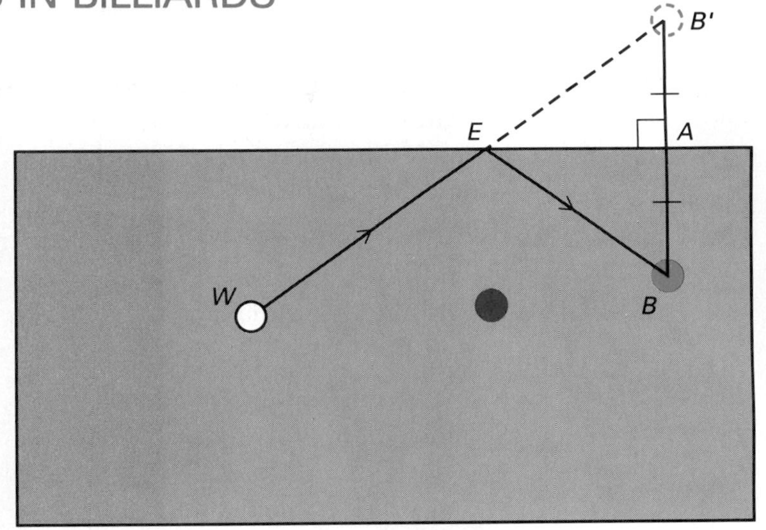

Reflections are important to the game of billiards. Often in this game, you must try to hit the cue ball so that it misses one object ball but gently taps another. To do this, you can deflect your shot off a wall of the table as shown in the diagram. In this case, the object ball, B, is blue, and the cue ball, W, is white. To miss the red ball, obtain the image of B by using the edge of the billiard table as the reflection line.

The distance $AB = AB'$, and $m\angle EAB = 90°$.

To make the shot, aim at the image ball, B'.

For each diagram, decide which edge of the billiard table should be the reflection line. Then copy the positions of the balls and draw a diagram so that the white ball will bounce off the reflection line to hit the blue ball.

1.

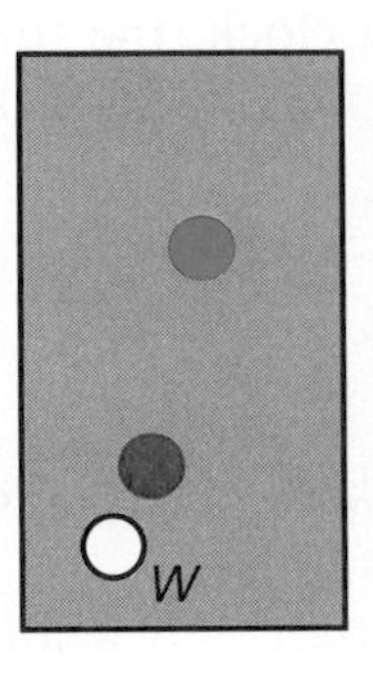

2.

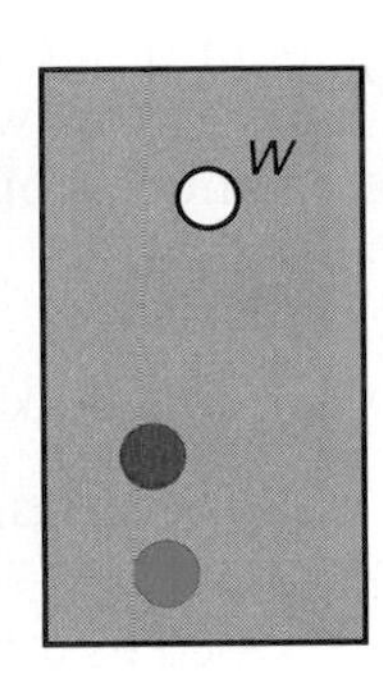

3.

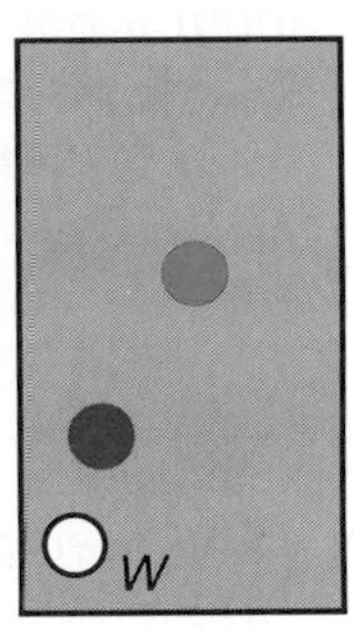

4. A *two-cushion shot* is a shot that uses two different reflecting edges. Copy the figure at the right and show how to use a two-cushion shot to make the white ball hit the blue ball.

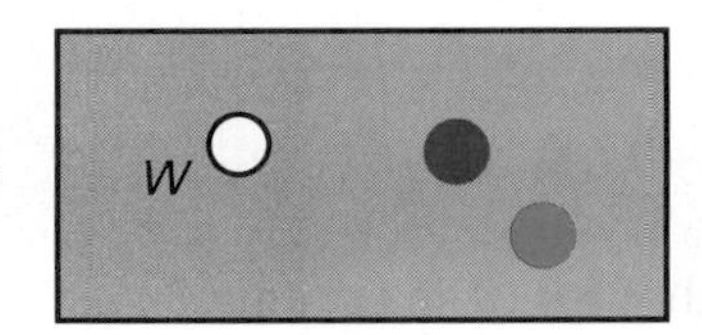

14-8 Rotations

EXPLORE

The hands of a clock turn, or rotate, around the point at which they are attached in the center of the clockface.

a. What fractional part of a full turn does the minute hand make in 30 minutes?

b. When the minute hand has moved 90° forward, or clockwise, from 3:25, to what number does it point?

c. What degree of rotation of the minute hand will set a clock back by 15 minutes?

SKILLS DEVELOPMENT

You have learned about two kinds of transformations: translations and reflections. A third type of transformation is a rotation.

A **rotation** is a transformation in which a figure is *turned,* or *rotated,* about a point. The movement of the hands of a clock, the turning of the wheels of a bicycle, and moving windshield wipers on a car are just a few examples of a rotation about a point.

CONNECTIONS

A continuous spinning motion in a single direction is called *rotary motion.* To measure rotary motion, try this. Turn a bicycle upside down and mark one tire with chalk. Spin the wheel and count the number of complete turns in one second. From this number and the circumference of the wheel you can compute approximately how far the wheel would roll along the ground in one minute.

To describe a rotation, you need three pieces of information:

- the point about which the figure is rotated (called the *center of rotation,* or *turn center*)
- the amount of turn expressed as a fractional part of a whole turn or in degrees (called the *angle of rotation*)
- the direction of rotation—either *clockwise* or *counterclockwise*

Example 1

Draw the rotation image of the flag when it is turned 90° clockwise about a turn center, T.

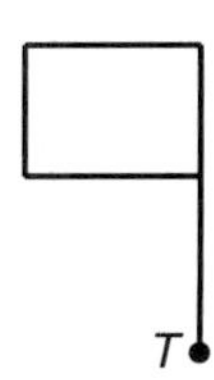

Solution

Copy the flag onto dot paper. Label point T. Then trace the flag and point T onto a sheet of tracing paper.

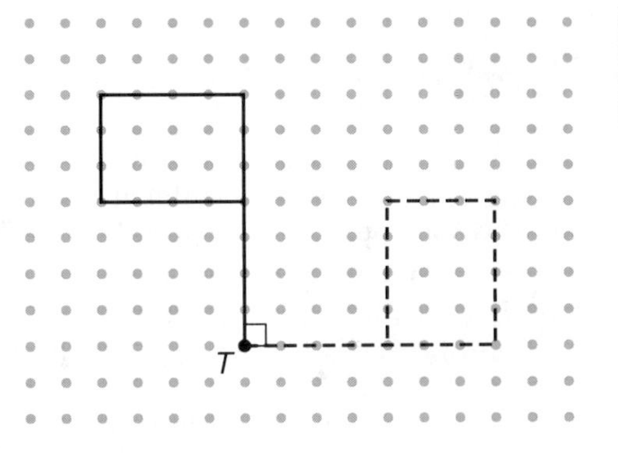

Hold your pencil point on the tracing at point T. Turn the tracing one-quarter turn, or 90°.

Remove the tracing paper and copy the image onto the dot paper. ◄

There are 360° at the center of a circle. That is why we give the measure of a complete turn as 360°. How many degrees are there in a half turn? a quarter turn? a three-quarter turn?

Example 2

Find the image of $\triangle PQR$ with vertices $P(3, 3)$, $Q(2, 1)$, and $R(5, 1)$ after a turn of 180° clockwise about the origin.

Solution

Multiply both the x-coordinate and the y-coordinate of each vertex by -1.

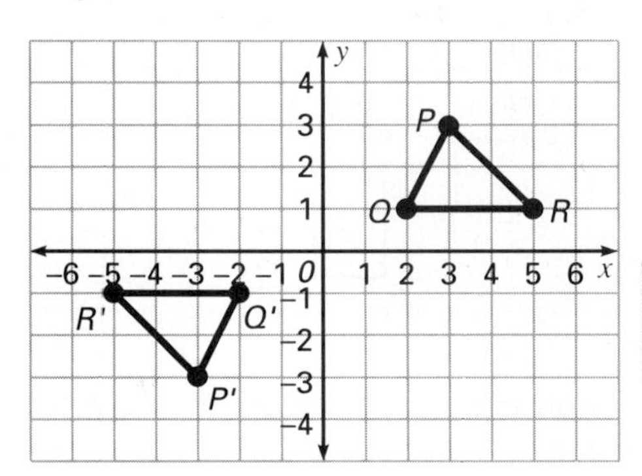

$P(3, 3) \rightarrow P'(-1 \times 3, -1 \times 3)$
$\rightarrow P'(-3, -3)$

$Q(2, 1) \rightarrow Q'(-1 \times 2, -1 \times 1)$
$\rightarrow Q'(-2, -1)$

$R(5, 1) \rightarrow R'(-1 \times 5, -1 \times 1)$
$\rightarrow R'(-5, -1)$ ◄

MATH IN THE WORKPLACE

How can an airplane manufacturer design the wing of a jet on a computer? Rosemary Chang does the kind of work that makes this possible. Ms. Chang is an expert on geometric surfaces. Her job is to write math equations that can be used to translate numerical data into computer programs. Armed with one of these programs, the computer designer can create a three-dimensional image of the jet's wing. This image can then be translated, reflected, and rotated on the computer screen, enabling the design to be evaluated from many points of view.

Ms. Chang has a bachelor's degree from New York University and a doctorate in applied mathematics from Brown University.

TRY THESE

1. Copy the figure on dot paper. Then draw the image of the figure when it is turned 270° clockwise about a turn center, T.

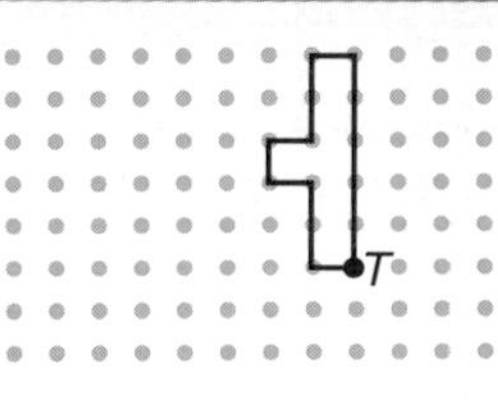

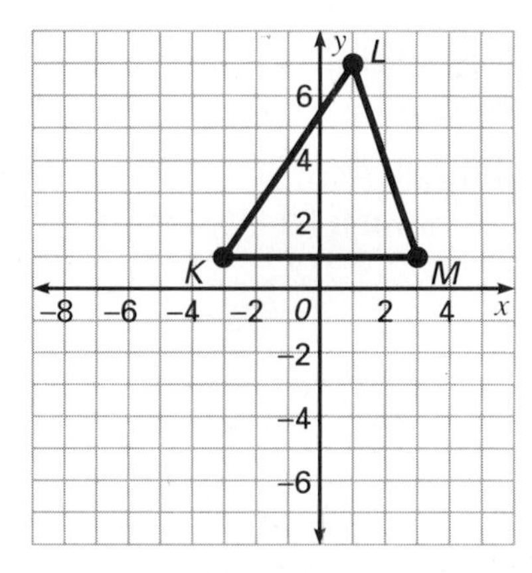

2. Graph the image of $\triangle KLM$ after a turn of 180° clockwise about the origin.

EXERCISES

PRACTICE/ SOLVE PROBLEMS

Each drawing shows a line segment, $\overline{AB}$, and its rotation image, $\overline{A'B'}$, about a turn center, T. What is the angle of rotation in a clockwise direction?

1.

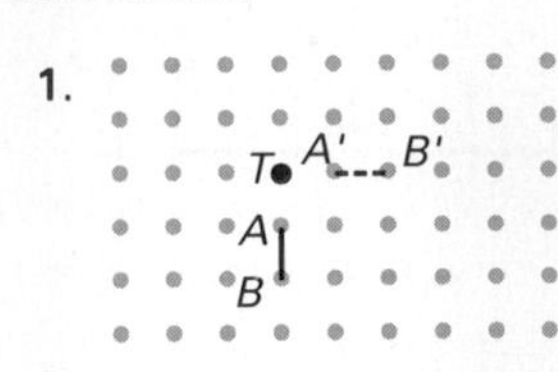

2.

3.

TALK IT OVER

The design is made entirely of straight lines. But parts of the design appear curved. What causes the curves to appear?

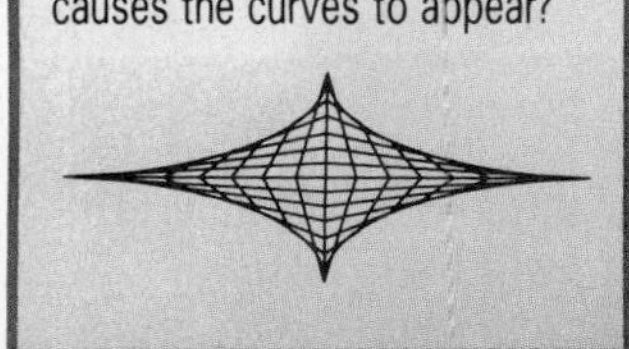

A figure has been rotated about a turn center, T. Describe each angle and direction of rotation.

4.

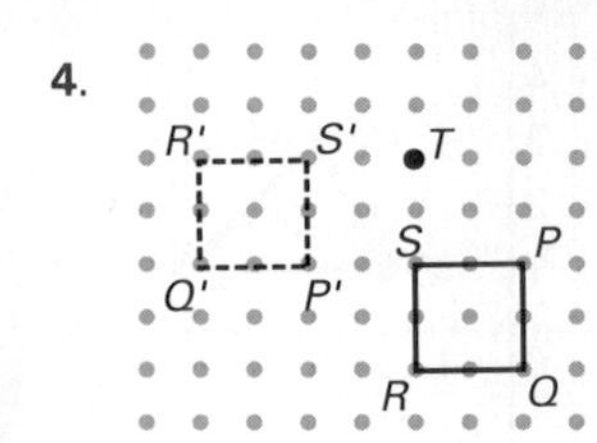

5.

6.

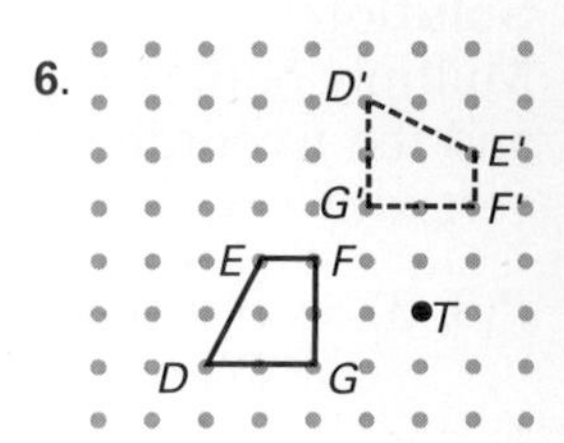

Graph the image of each figure after a 90° turn clockwise about the origin.

7.

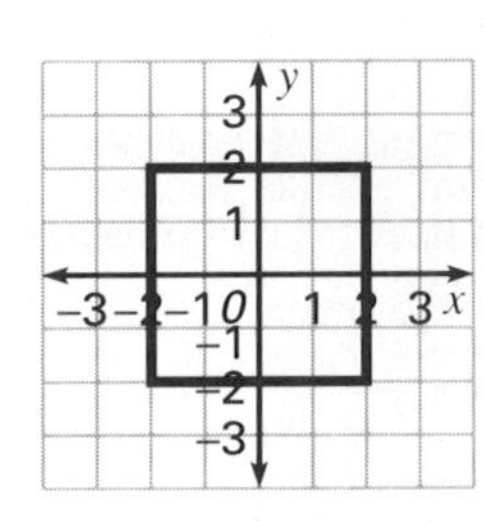

8.

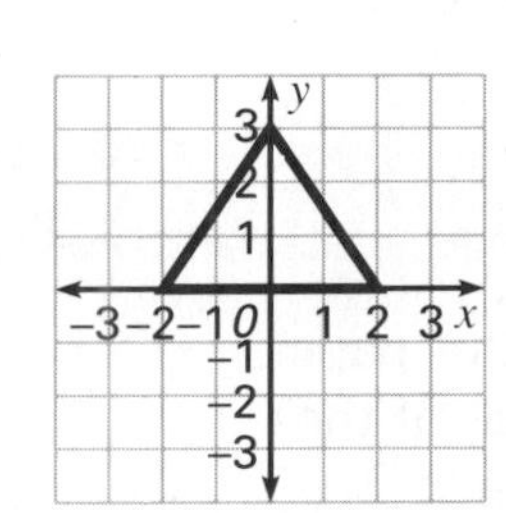

9.

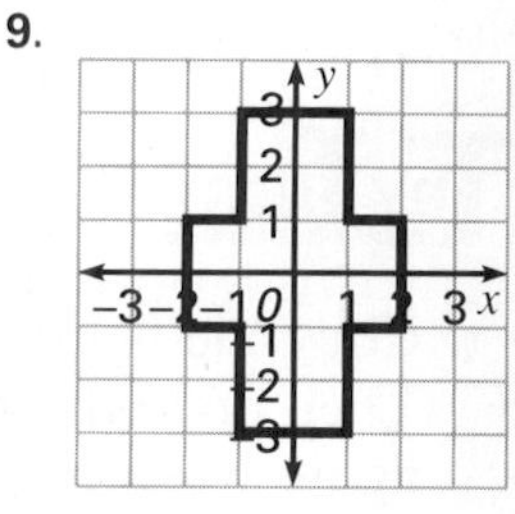

EXTEND/ SOLVE PROBLEMS

PROBLEM SOLVING TIP

Making a tracing of a figure can help you find the rotation image. Be sure to mark the point used as the center of rotation on your tracing.

Copy each figure. Then draw its image after the given rotation about point T.

10. quarter-turn counterclockwise

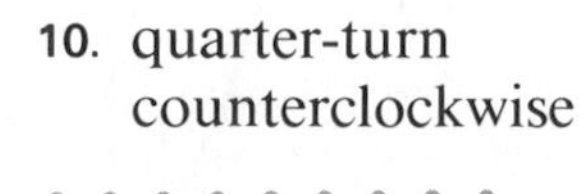

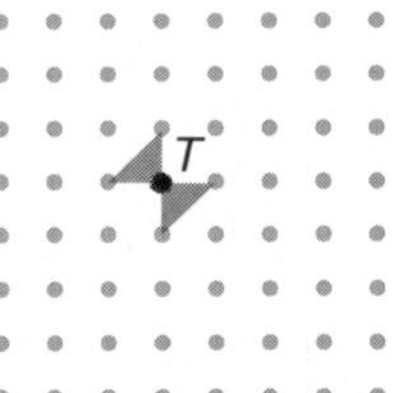

11. half-turn clockwise

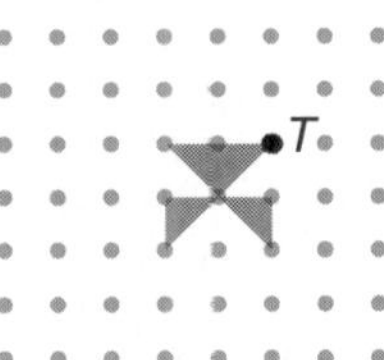

12. complete turn clockwise

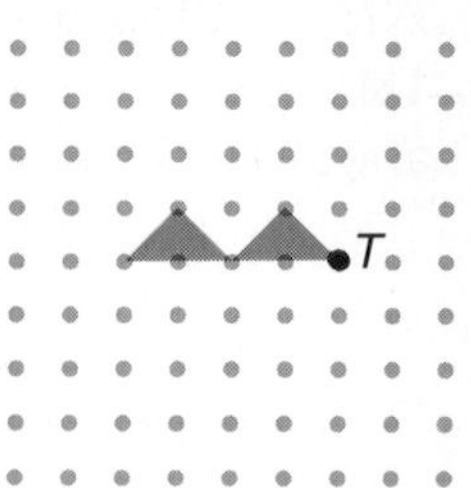

13. quarter-turn counterclockwise

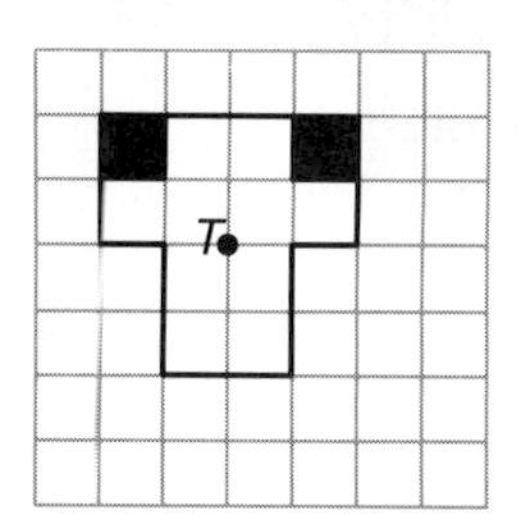

14. half-turn clockwise

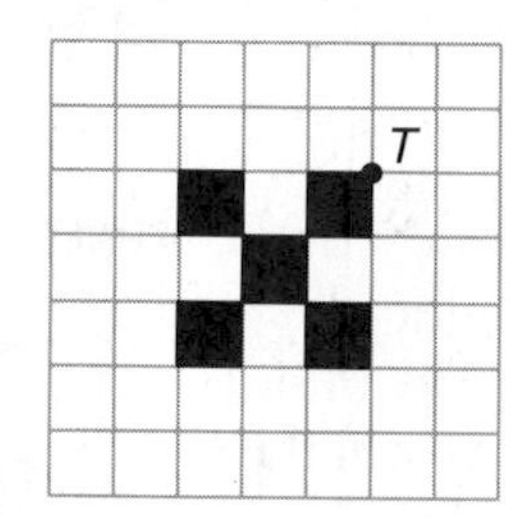

15. three-quarter-turn clockwise

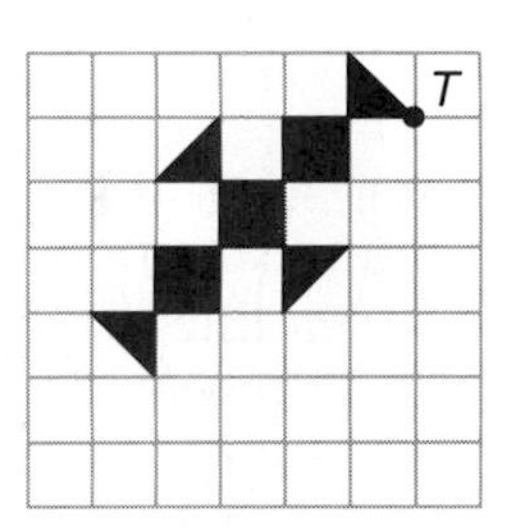

16. Each letter of the message below has been rotated either 90° or 180° clockwise or counterclockwise. The turn centers are shown by •. What is the message?

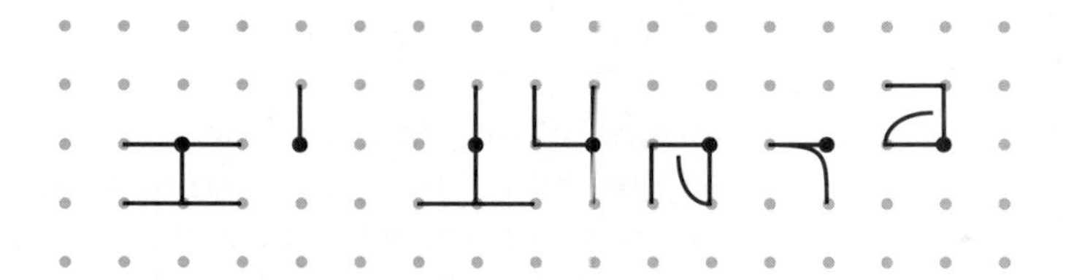

THINK CRITICALLY/ SOLVE PROBLEMS

Triangle XYZ has vertices $X(2, 0)$, $Y(5, 0)$, and $Z(2, 4)$.

17. Is the order of the vertices of $\triangle XYZ$ clockwise or counterclockwise? Rotate $\triangle XYZ$ 90° clockwise to obtain the image $\triangle X'Y'Z'$. Are the vertices of $\triangle X'Y'Z'$ in clockwise or counterclockwise order?

18. Find the lengths of the sides of both triangle XYZ and $X'Y'Z'$. (Hint: You will need to use the Pythagorean Theorem to find the length of one side of each triangle.)

19. Find the areas of $\triangle XYZ$ and $\triangle X'Y'Z'$.

20. Generalize from the results of Exercises 17–19. What properties of a figure do you think change during a rotation? What properties do you think stay the same?

READING MATH

Many mathematical terms have other, related meanings. Give the meaning of the underlined word in each sentence. (You might need to refer to a dictionary.)

1. There was a remarkable <u>transformation</u> in his appearance since the last time I saw him.
2. Her success is a <u>reflection</u> of all her hard work.
3. The meaning of the word was lost in the <u>translation</u> from Spanish to English.
4. There were five players in the baseball team's pitching <u>rotation</u>.

Solve.

21. If a moth sat on the edge of a 12-in. phonograph record rotating at a speed of $33\frac{1}{3}$ revolutions per minute, how far would it travel in 3 min?

22. Refer to Exercise 21. How many inches per minute did the moth travel?

14-9 Working with Symmetry

EXPLORE/ WORKING TOGETHER

Look around the classroom. Focus on any small object. Imagine a *vertical* line drawn down the middle of that object. Think about whether such a line would divide the object into two halves that match exactly.

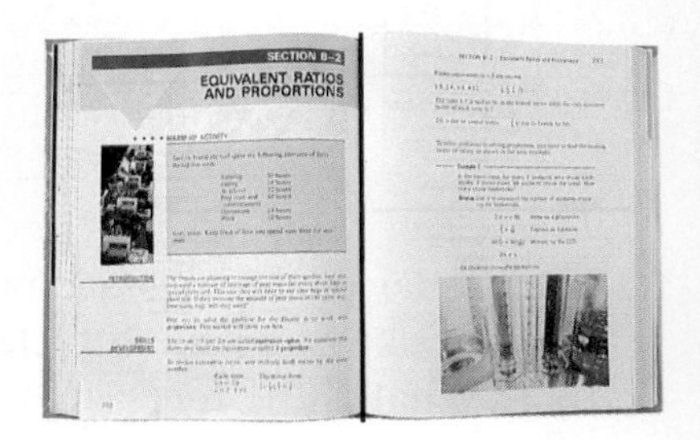
SECTION 8-3
EQUIVALENT RATIOS AND PROPORTIONS

Focus on another object. This time, imagine a *horizontal* line imposed on the object. Do you think such a line would divide the object exactly in half?

Make three lists of things found in the classroom or at home that you think could be divided into halves by a single straight line—those that could be divided by a vertical line, those that could be divided by a horizontal line, and those that could be divided by a line that is neither vertical nor horizontal.

Compare your lists. Could more than one line divide any of the things you listed? Explain.

SKILLS DEVELOPMENT

If you could fold this isosceles triangle along the line PQ, one side would fit exactly over the other. So, the triangle is said to have **line symmetry,** and line PQ is called a **line of symmetry.**

P

Q

MATH: WHO, WHERE, WHEN

Many properties of reflections find a particularly striking application in the Japanese art of paper folding. Called *origami* in Japanese, paper folding is used to make such traditional figures as the crane and the tortoise (symbols of good fortune), the carp (symbol of persistency), and the frog (symbol of love). Books with diagrams and instructions for folding these traditional paper decorations were published as early as the first quarter of the eighteenth century.

Example 1

Trace each figure and draw all the lines of symmetry.

a.

b.

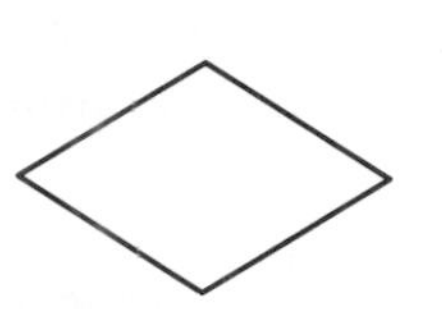

c.

Solution

a.

1 line of symmetry

b.

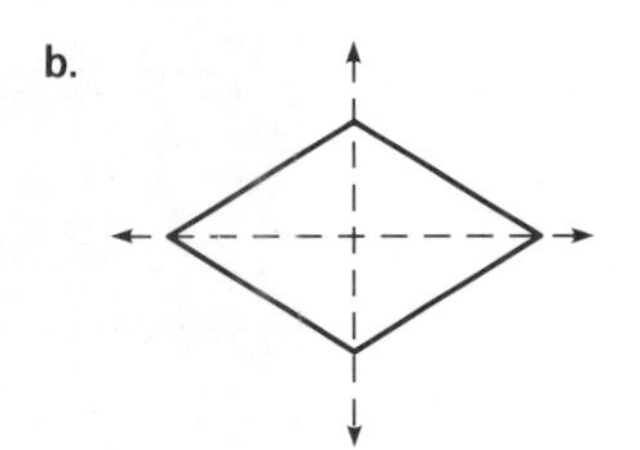

2 lines of symmetry

c.

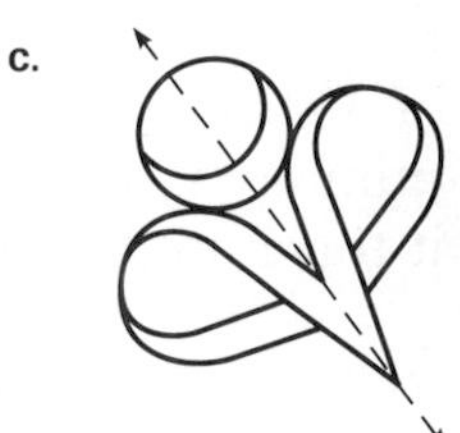

1 line of symmetry ◄

The parallelogram does not have line symmetry. However, if you could place the tip of a pencil at point A and turn this parallelogram 180° clockwise, it would fit exactly over its original position. So, the parallelogram is said to have **rotational symmetry.** Its **order of rotational symmetry** is 2, since the parallelogram would fit over its original position 2 times in the process of a complete turn.

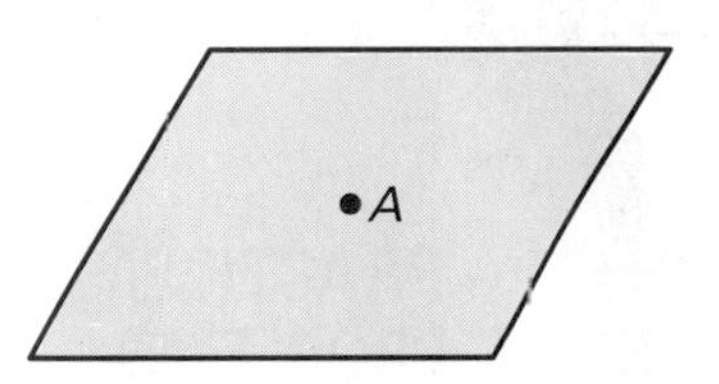

Example 2

Give the order of rotational symmetry for each figure.

a.

b.

c.

Solution

a.

This figure fits over its original position 4 times during a complete turn, so the order of rotational symmetry is 4.

b.

This figure fits over its original position 2 times during a complete turn, so the order of rotational symmetry is 2.

c.

This figure fits over its original position 6 times during a complete turn, so the order of rotational symmetry is 6. ◄

TRY THESE

Trace each figure and draw all the lines of symmetry. If a figure has no lines of symmetry, write *None.*

1.

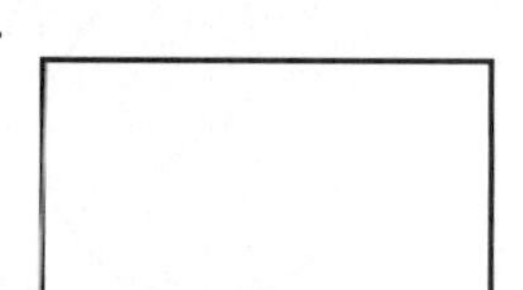

2.

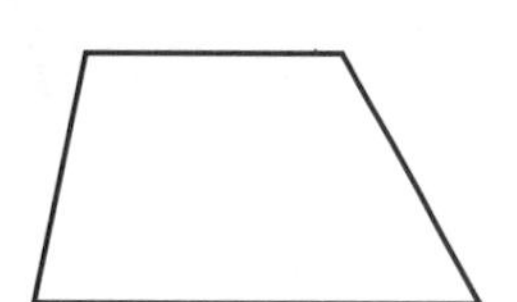

3.

Give the order of rotational symmetry for each figure.

4.

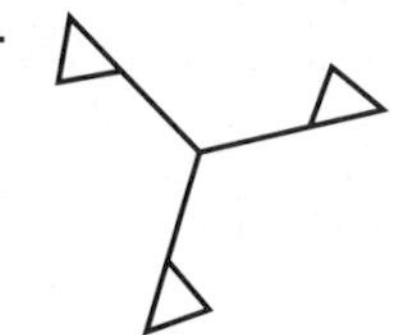

5.

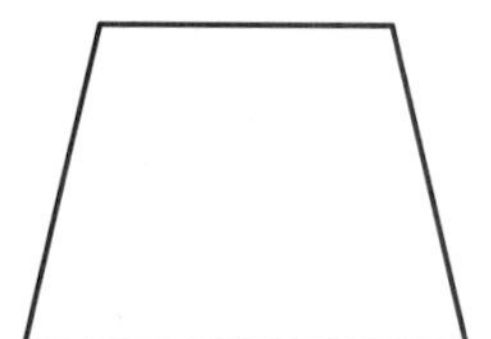

6.

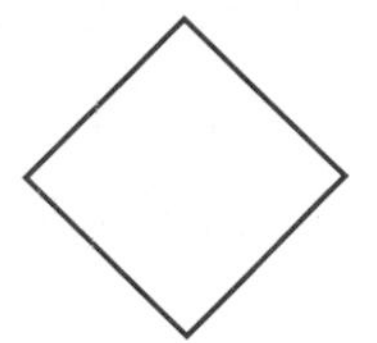

EXERCISES

PRACTICE/ SOLVE PROBLEMS

Tell whether the dashed line is a line of symmetry. (Trace the figure if you need to).

1.
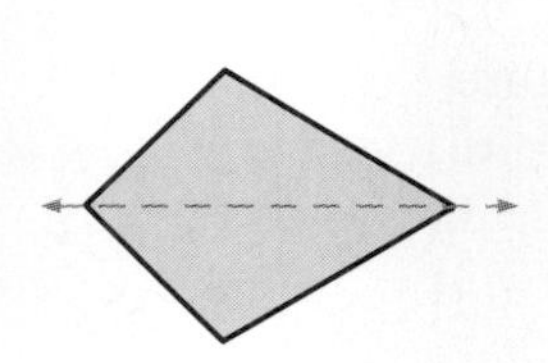

2.
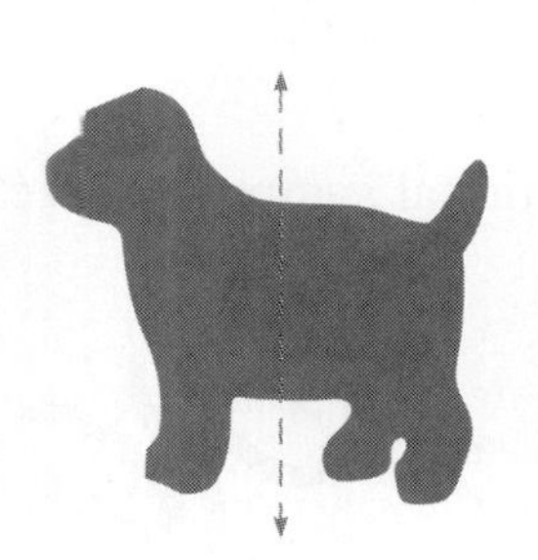

3.
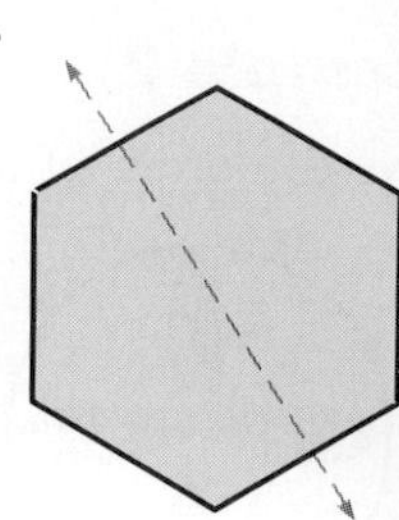

4.

5.
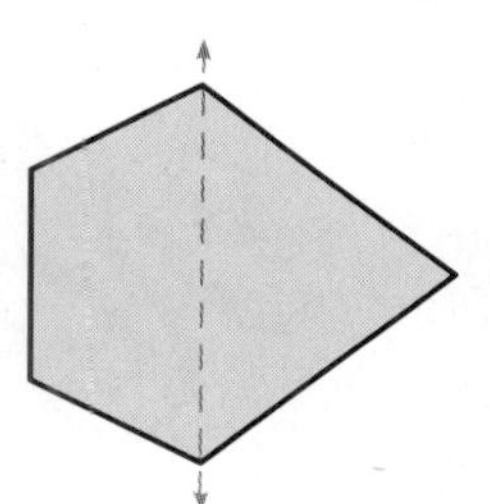

6.

Give the order of rotational symmetry for each figure.

7.
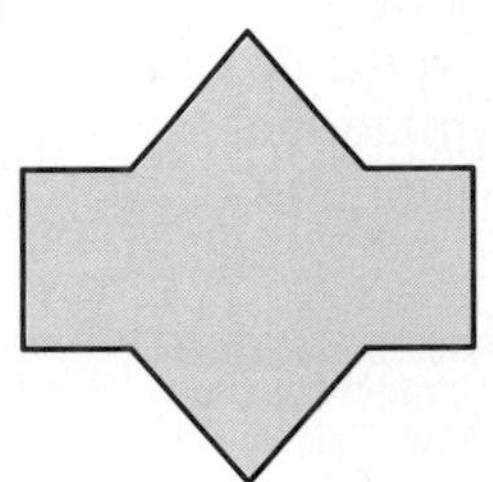

8.
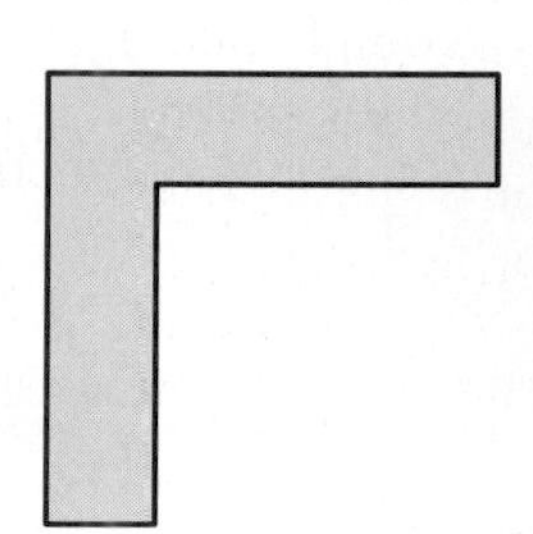

9.
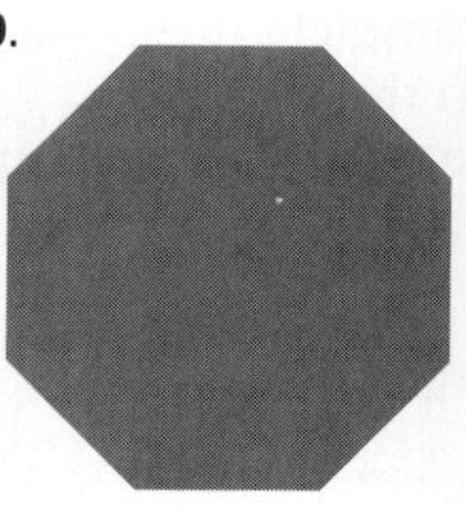

10.

11.

12.

EXTEND/ SOLVE PROBLEMS

Draw a triangle with the given number of lines of symmetry.

13. 1 **14.** 3 **15.** 0

Trace each drawing on dot paper or graph paper. Then complete the figure to show how it would appear in a mirror held vertically along the line of symmetry.

16.

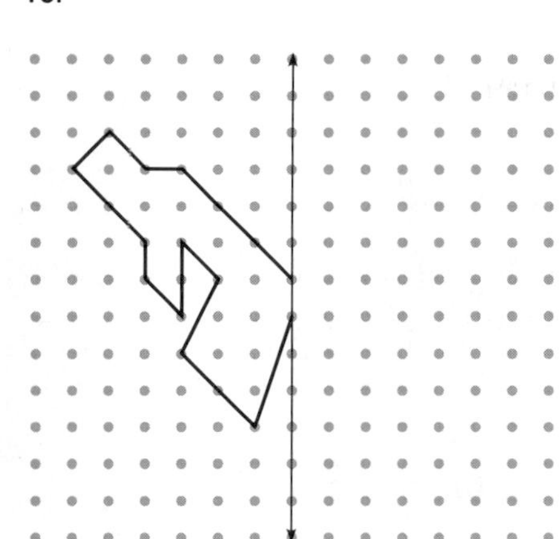

17.

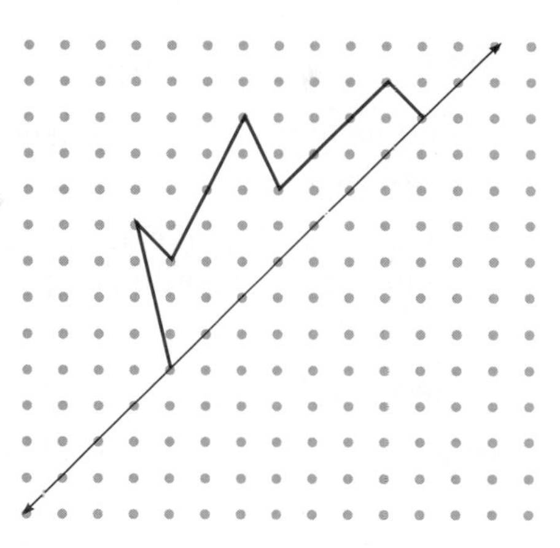

THINK CRITICALLY/ SOLVE PROBLEMS

This figure, drawn on graph paper, has rotational symmetry of order 4. Use graph paper to construct a figure having rotational symmetry of the order indicated.

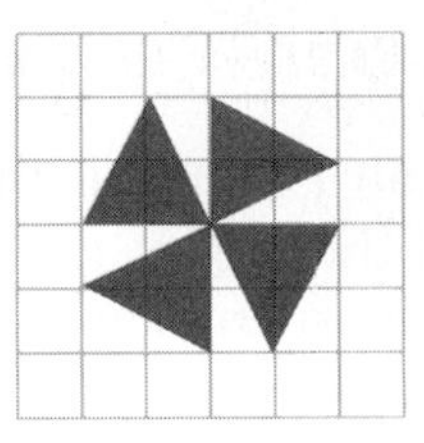

18. order 3

19. order 6

20. Some of the letters of the alphabet, when capitalized, have horizontal line symmetry. Others have vertical line symmetry. Still others have both. (Some have no symmetry at all!) Determine which of the capital letters below have horizontal line symmetry. Then, write a sentence using those letters alone. (You may use each letter more than once.)

Think about which capital letters have rotational symmetry. Write step-by-step directions that someone who is not in your class could use to determine the rotational symmetry of the letters.

CHAPTER 14 ● REVIEW

1. The movement of the hands of a clock is an example of a(n) __?__ about a point.
2. A line of __?__ works something like an ordinary mirror.
3. If the solutions to an equation form a straight line when they are graphed, the equation is __?__.
4. Because points graphed on a coordinate plane are always identified by first the x-coordinate and then the y-coordinate, (x, y) is called a(n) __?__.
5. A(n) __?__ can also be called a slide.
6. For a line segment or a straight line, the __?__ is the ratio of the rise to the run.

a. ordered pair
b. slope
c. reflection
d. translation
e. rotation
f. linear

SECTIONS 14–1 AND 14–3 LINEAR EQUATIONS (pages 506–509, 512–515)

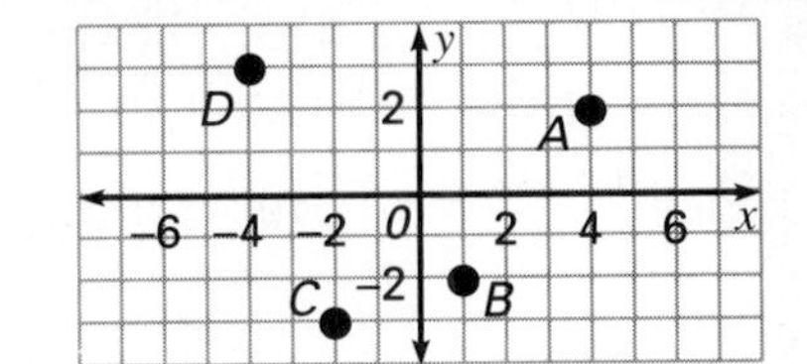

► When graphing an ordered pair (x, y), x is the horizontal distance to the left or right of the origin; y is the vertical distance up or down from the origin. The notation $A\ (x, y)$ means that point A has x-coordinate x and y-coordinate y.

► A **linear equation** in two variables has many solutions, each of which can be represented by an ordered pair. When the points that correspond to these ordered pairs are graphed on a coordinate plane, they lie along a straight line.

7. What are the coordinates of points A, B, C, and D?
8. On graph paper, graph the points $E(4, -3)$ and $F(-8, -2)$ on a coordinate plane.
9. Graph the equation $y = 2x - 3$.

SECTIONS 14–2 AND 14–4 PROBLEM SOLVING (pages 510–511, 516–519)

► A grid can often be used to find distances.

► To find the slope of a line, choose any two points on the line. Then find the difference in the y-coordinates and divide by the difference in the x-coordinates.

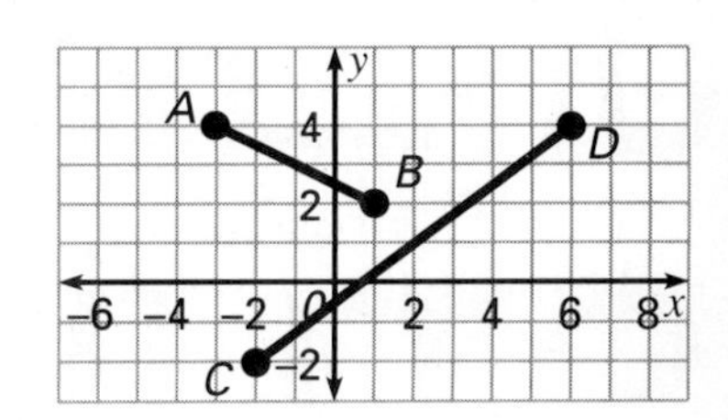

10. Henry runs from the gym 3 mi north, 5 mi west, 2 mi east, and 3 mi south. He then jogs the rest of the way back to the gym. How far does he jog?
11. Find the slope of each line segment.

SECTIONS 14–6 AND 14–7 TRANSLATIONS AND REFLECTIONS (pages 522–529)

► A **translation** is a transformation in which all the points that make up a figure slide exactly the same distance and in the same direction.
► A **reflection** is a transformation in which a figure is flipped over a line of reflection.

12. Copy the figures onto graph paper. Then graph the image of each figure under a translation 3 units to the right and 3 units up.

13. Graph the image of each figure under a reflection across the given line.

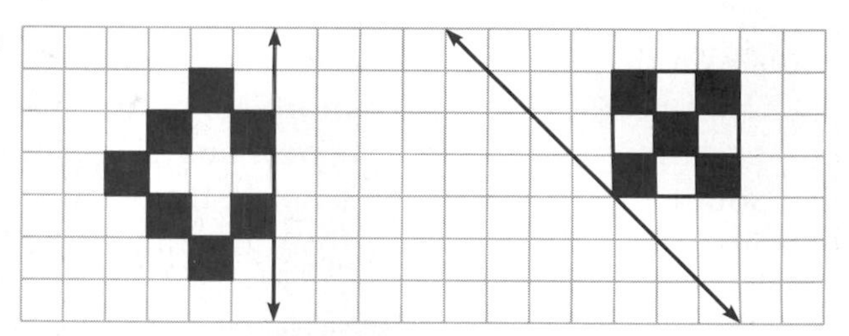

SECTIONS 14–8 AND 14–9 ROTATIONS AND SYMMETRY (pages 530–537)

► A **rotation** is a transformation in which a figure is turned, or rotated, about a point.
► A figure has **line symmetry** if, when you fold it along a line, one side fits exactly over the other.
► A figure has **rotational symmetry** if, when the figure is turned about a point, the figure fits exactly over its original position at least once before a complete turn has been made.

14. Give the number of lines of symmetry for each figure at the right.

15. Give the order of rotational symmetry for each figure at the right.

16. What is the order of rotational symmetry of the figure shown at the right?

SECTION 14–5 PROBLEM SOLVING STRATEGY: MAKE A MODEL (pages 520–521)

► Making a model is often helpful in solving problems.
► Types of models include equations, sketches, and graphs.

17. A window washer charged a fixed fee in addition to a charge per window washed. Find the flat fee and the charge per window.

Total Amount Charged	\$13.50	\$16.50	\$24.00
Number of Windows Washed	3	5	10

USING DATA Refer to the table on page 505.

18. Draw the dot pattern for the 5th square number on a coordinate plane. Give the ordered pair for the corner points.

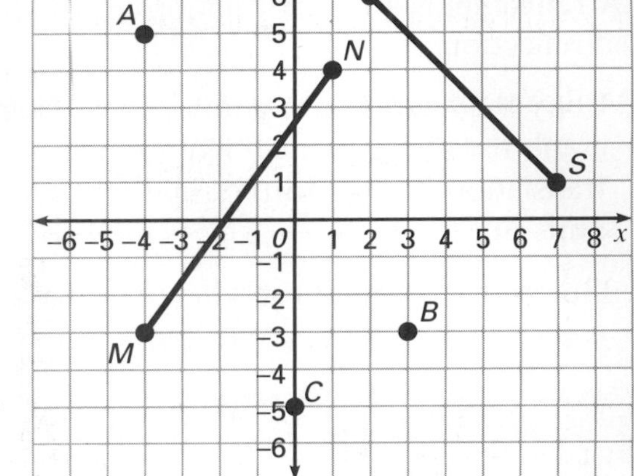

Use the figure at the right for Exercises 1–2.

1. Give the coordinates of points A, B, and C.
2. Find the slope of line segments MN and RS.
3. From the mall, Paula drove 6 mi east and 2 mi north to her grandmother's house. To go home, she drove 3 mi west and 2 mi south. How far is Paula's house from the mall? In what direction from the mall is her house?
4. Graph the points $A(4, -2)$, $B(-1, 5)$, $C(3, 0)$, $D(-1, -1)$, and $E(0, 4)$.
5. Graph the equation $y = 3x - 2$.
6. A mail-order store charges handling costs, which consist of a fixed amount for filling an order plus a charge per pound of merchandise shipped. Find the fixed fee and the cost per pound.

Total Handling Costs	\$5.50	\$7.50	\$10.50
Number of Pounds	3	5	8

7. Complete the table of solutions for the equation $y = 2x + 1$.

$y = 2x + 1$

x	y	x, y
-1	■	-1, ■
■	1	■, 1
1	■	1, ■

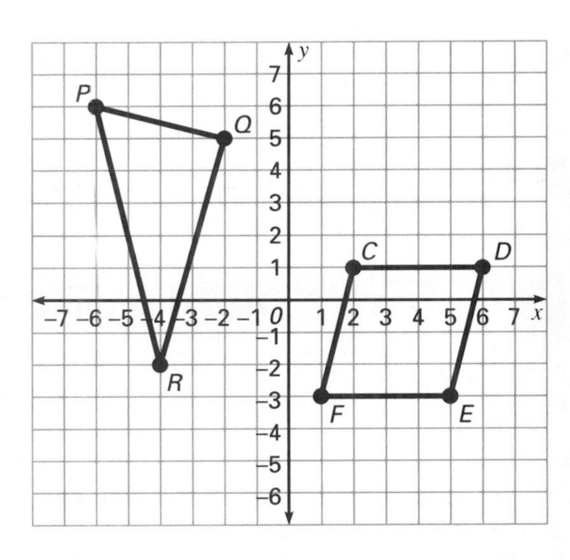

Use the figure at the right for Exercises 8–10.

8. Graph the image of parallelogram $CDEF$ under a translation 2 units to the left and 3 units up.
9. Graph the image of $\triangle PQR$ under a reflection across the x-axis.
10. Graph the image of $\triangle PQR$ after a 90° turn counterclockwise about the origin.

Use the figure at the right for Exercises 11–12.

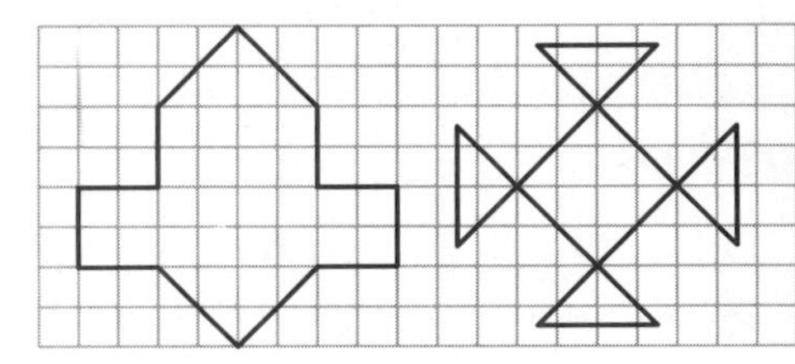

11. How many lines of symmetry does the figure on the left have?
12. What is the order of rotational symmetry for the figure on the right?

1. Members of a cooking class whose last names begin with the letter F were asked to name their favorite dessert. What type of sampling was used?

Use the bar graph for Exercises 2–5.

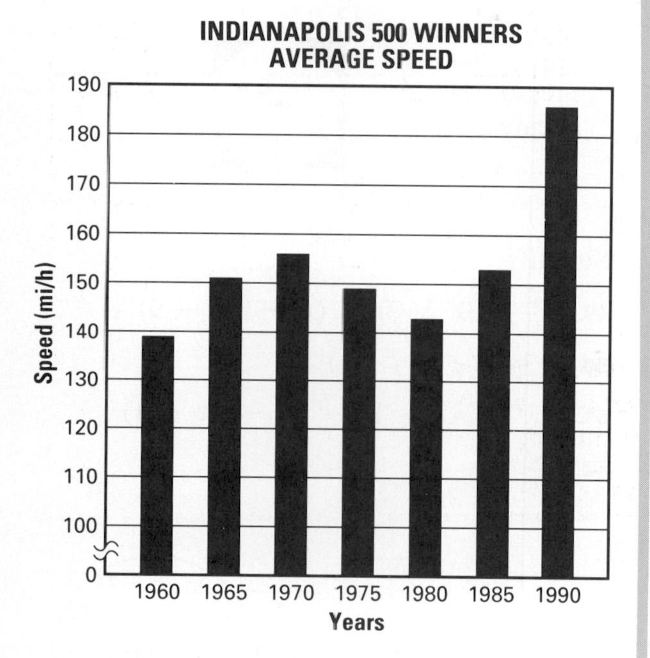

2. In 1975, about how fast was the winner's speed?

3. During which years was the average speed at least 155 mi/h?

4. Describe the scale of the speed intervals.

5. About how much faster did the winner in the 1990 race go than the winner in the 1960 race?

A consumer watch group recorded the prices of different brands of apple juice: \$1.25 \$1.18 \$1.29 \$1.32 \$1.12 \$1.20 \$1.25

6. Find the mean, median, and mode of the prices.

7. Which measure is the best indicator of the price of apple juice? Why?

Simplify.

8. $5 \times 4 + (8 - 3)$
9. $6 \times 3^2 - 2$
10. $(6 \times 4) \div (2 + 6)$
11. $3^2 + 5 \times 6 \div 2$

12. Is the argument *valid* or *invalid?*

If the number is 15, then the number is divisible by 5.
The number is divisible by 5.
Therefore, the number is 15.

Multiply or divide.

13. $1\frac{1}{2} \times 2\frac{1}{4}$
14. $1\frac{3}{8} \div 1\frac{1}{4}$
15. $\frac{7}{10} \times \frac{5}{8}$
16. $3\frac{1}{2} \div \frac{7}{8}$

Complete.

17. 0.07 L = ■ mL
18. 0.095 m = ■ mm
19. 3 km = ■ m
20. 8,000 g = ■ kg

Divide.

21. $60 \div (-15)$
22. $-49 \div (-7)$
23. $-48 \div 8$
24. $80 \div (-5)$

Solve each equation.

25. $d + 7 = 2$
26. $3(y - 4) = 27$
27. $-35 = 4x + 1$
28. $b - 19 = -17$

Solve.

29. $\frac{27}{15} = \frac{x}{20}$
30. $\frac{x}{13} = \frac{3}{2.6}$
31. $\frac{7}{3} = \frac{35}{x}$
32. $\frac{10.5}{x} = \frac{5}{10}$

33. The scale of a drawing is 1 in.:4 ft. What is the actual length of a room if the length in the scale drawing is $3\frac{1}{2}$ in.?

Find the percent of increase or decrease. Round to the nearest tenth.

34. Original amount: \$200
New amount: \$158

35. Original amount: \$35
New amount: \$49

Find the interest and the amount.

36. Principal: \$900
Rate: 12%/year
Time: 24 months

37. Principal: \$1,200
Rate: 5.5%/year
Time: 4 years

Find the area for each figure.

38.

39.

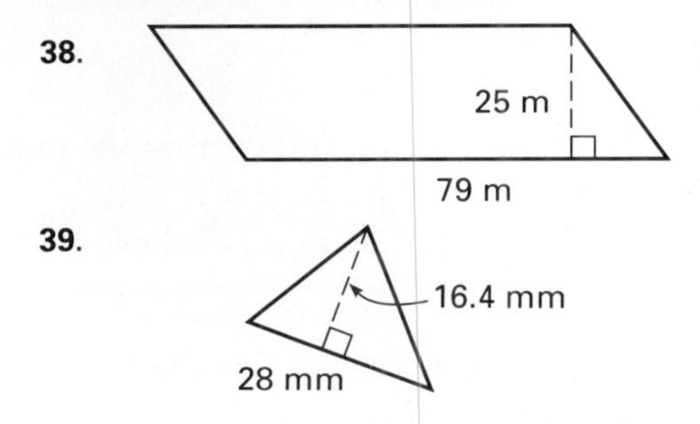

Use the figure below. $\overleftrightarrow{PT} \parallel \overleftrightarrow{UX}$. Find the measure of each angle.

40. $m\angle 5$

41. $m\angle 7$

42. $m\angle 9$

43. $m\angle 4$

44. $m\angle 12$

P 1 2 12 T
45° 3 14 15
U 4 5 8 9 X
6 7 10 11

45. Trace $\overleftrightarrow{CD}$. Then construct a perpendicular to $\overleftrightarrow{CD}$ from point B.

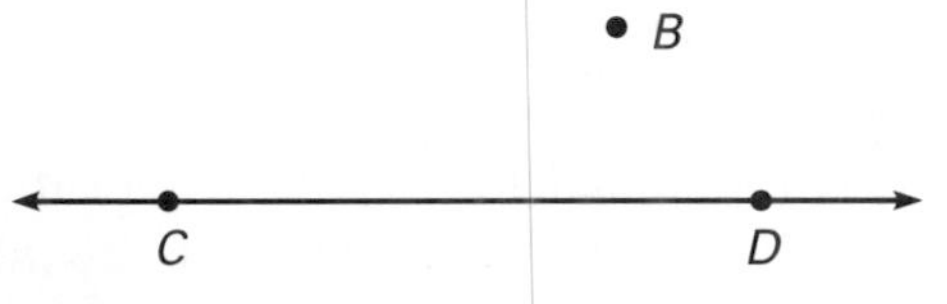

Find the unknown length in each triangle. Round to the nearest tenth.

46. **47.**

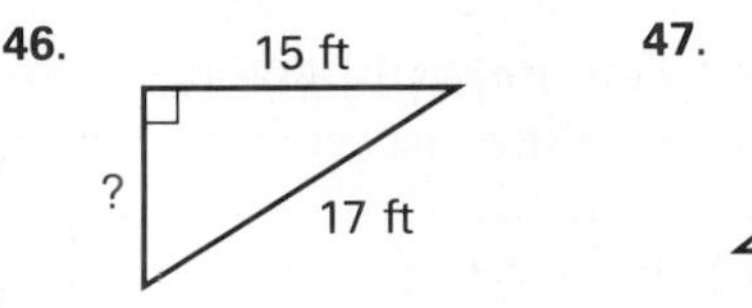

48. Find the area of the shaded region to the nearest tenth.

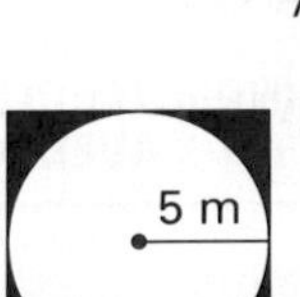

Simplify.

49. $(x^2 + 3x - 7) + (x^2 - 6x - 9)$

50. $-3y(4 + 7y - 3)$

51. $(2x^2 - 4x + 1) - (x^2 - 2x + 1)$

52. $\dfrac{25c^6d}{5c^2}$

Find each probability.

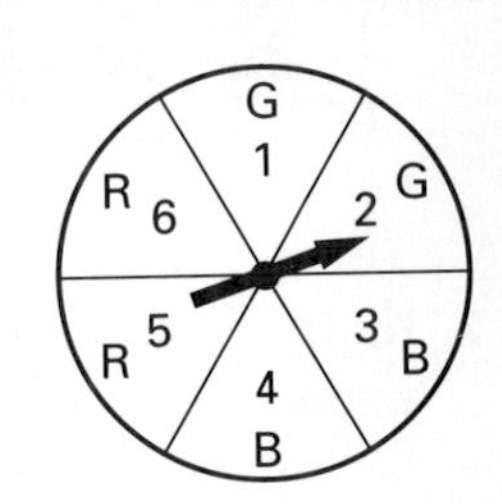

53. P (B)

54. P (R or G)

55. P (7)

56. P (not R)

57. Refer to the figure below. Find four points that will make a square whose diagonals lie along the axes. Then give the coordinates of these four points. (Call them A, B, C, and D). Is there only one such square?

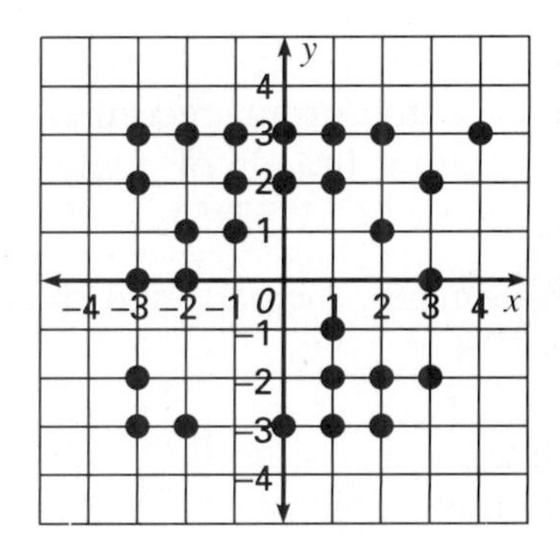

Use the bar graph for Exercises 1–4.

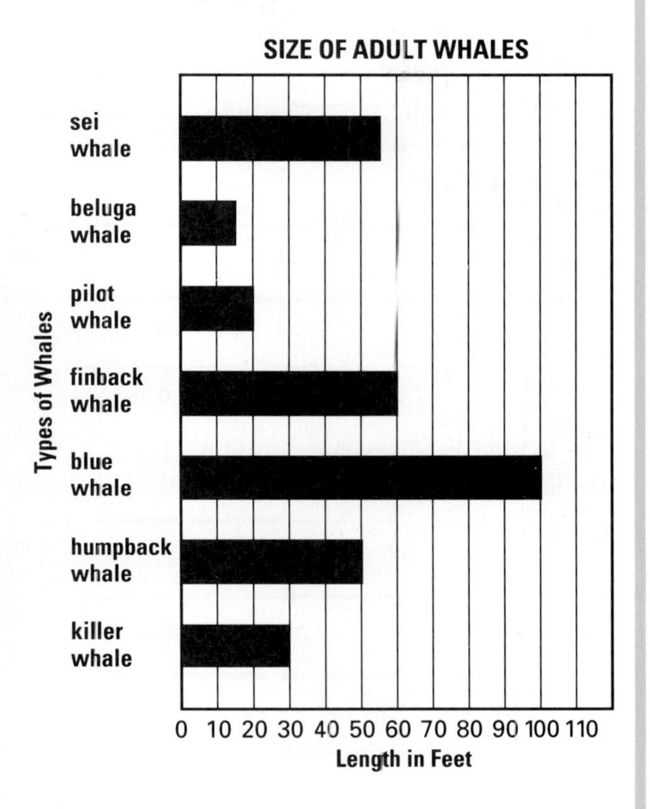

1. Which whale measures 20 feet?

   A. killer B. sei
   C. beluga D. pilot

2. How many whales are at least 55 feet long?

   A. 4 B. 2
   C. 3 D. 1

3. What is the difference in length between the largest whale and the smallest whale?

   A. 85 feet B. 70 feet
   C. 80 feet D. 75 feet

4. About how many times greater in length is the finback whale than the beluga whale?

   A. 2 times B. 4 times
   C. 3 times D. 5 times

5. What is the median for 212, 198, 157, 321, 166, 104, 158, 177?

   A. 186.6 B. 217
   C. 171.5 D. none of these

6. What is the mean for 31, 25, 39, 29, 27, 30, 33, 28, 28?

   A. 28 B. 30
   C. 11 D. none of these

7. What is the variable expression for a number divided by 9?

   A. $n \div 9$ B. $9 \div n$
   C. $\frac{9}{n}$ D. $\frac{1}{9n}$

8. Simplify. $3 \times (4 + 5) - 7$

   A. 6 B. 2
   C. 189 D. 20

9. Simplify. $7(k - 6)$

   A. $7k - 13$ B. $k - 42$
   C. $7k - 42$ D. none of these

10. Which is a counterexample for the following statement?

    If a number is divisible by 4, then it is divisible by 12.

    A. 12 B. 36 C. 32 D. 60

11. Subtract. $10 - 3\frac{2}{7}$

    A. $6\frac{5}{7}$ B. $7\frac{2}{7}$
    C. $6\frac{2}{7}$ D. $7\frac{5}{7}$

12. Multiply. $7 \times 2\frac{2}{3}$

    A. $17\frac{1}{3}$ B. $18\frac{2}{3}$
    C. $14\frac{1}{3}$ D. $\frac{8}{21}$

13. Divide. $1\frac{3}{4} \div 4\frac{3}{8}$

    A. $4\frac{1}{2}$ B. $2\frac{2}{5}$
    C. $\frac{2}{5}$ D. $2\frac{1}{2}$

14. Add. $\begin{array}{r} 2 \text{ ft } 4 \text{ in.} \\ + \quad 9 \text{ in.} \\ \hline \end{array}$

A. 2 ft 3 in. B. 3 ft 1 in.
C. 2 ft 1 in. D. none of these

15. Subtract. $\begin{array}{r} 6 \text{ gal } 2 \text{ qt} \\ -2 \text{ gal } 3 \text{ qt} \\ \hline \end{array}$

A. 3 gal 2 qt B. 3 gal 1 qt
C. 4 gal 1 qt D. none of these

16. Multiply. $\begin{array}{r} 3 \text{ ft } 5 \text{ in.} \\ \times \quad 3 \\ \hline \end{array}$

A. 10 ft 5 in. B. 9 ft 8 in.
C. 10 ft 3 in. D. 9 ft 5 in.

17. What is the perimeter of a rectangular figure whose length is 7.45 dm and whose width is 25 dm?

A. 32.45 dm B. 64.9 dm^2
C. 39.9 dm^2 D. 64.9 dm

18. Add. $19 + (-13)$

A. 6 B. 32
C. -6 D. -32

19. Subtract. $20 - (-5)$

A. -25 B. 25
C. 15 D. -15

20. Multiply. -18×3

A. 54 B. -21
C. -45 D. -54

21. Divide. $-20 \div (-5)$

A. 4 B. 15
C. -4 D. -25

22. Solve. $-15 = 3 + n$

A. $n = 18$ B. $n = -18$
C. $n = 12$ D. $n = -12$

23. Solve. $15b = -60$

A. $b = 5$ B. $b = 45$
C. $b = -4$ D. $b = -5$

24. Solve. $12 - 11s = 45$

A. $s = -33$ B. $s = 3$
C. $s = 22$ D. $s = -3$

25. Which is the correct solution for $x \leq 2$?

A. 2 1 0

B. 2 1 0

C. 2 1 0

D. 2 1 0

26. Which is the correct solution for $p > -5$?

A. -6 -5 -4

B. -6 -5 -4

C. -6 -5 -4

D. -6 -5 -4

27. Which is the unit rate for $\frac{\$108}{13.5 \text{ h}}$?

A. \$7.50/h B. \$9/h
C. \$5/h D. none of these

28. Which pair of ratios are equivalent?

A. $\frac{18}{12}, \frac{12}{18}$ B. $\frac{3}{4}, \frac{18}{24}$
C. $\frac{9}{3}, \frac{36}{15}$ D. $\frac{4}{9}, \frac{16}{27}$

29. Solve. $\frac{15}{18} = \frac{x}{6}$

A. 90 B. 5 C. 18 D. 12

30. Ellen made a scale drawing of her garden. The width of the garden in her drawing is $5\frac{1}{4}$ in. The actual length is 21 ft. What scale did Ellen use for her drawing?

A. $\frac{1}{2}$ in.:1 ft B. $\frac{1}{4}$ in.:1 ft
C. $1\frac{1}{2}$ in.:$1\frac{1}{2}$ ft D. none of these

31. The Marricks won 18 of their last 24 games. What percent of the games did the team win?

A. 133% B. 55%
C. 75% D. 60%

32. What is the amount of interest earned in one year on $900, if the rate of interest is 9.5% per year?

A. $85.40 B. $85.50
C. $85.55 D. none of these

33. Which is the alternate interior angle for $\angle 5$? $\overleftrightarrow{EF} \parallel \overleftrightarrow{GH}$

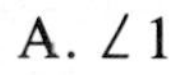

A. $\angle 1$
B. $\angle 7$
C. $\angle 4$
D. $\angle 8$

34. Find the sum of the angle measures of the polygon.

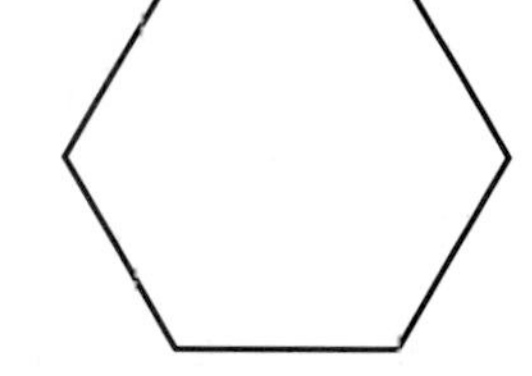

A. 540°
B. 720°
C. 1,080°
D. none of these

35. Find the volume to the nearest tenth.

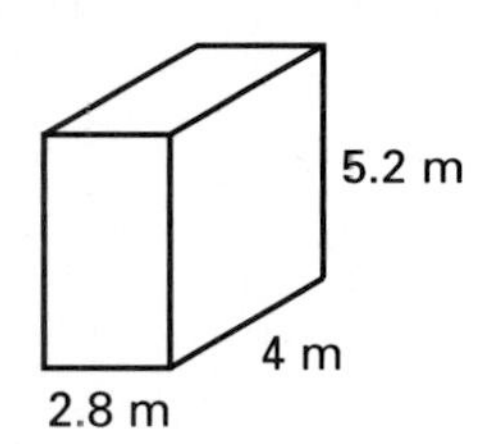

A. 49.6 m³
B. 12.1 m³
C. 58.2 m³
D. 182.9 m³

36. Find the surface area. Round to the nearest tenth.

4 ft
8 ft

A. 108.5 ft²
B. 301.4 ft²
C. 150.7 ft²
D. 125.6 ft²

37. Simplify. $(3c + 3) + (-6c + 8)$

A. $-3c + 11$ B. $-9c^2 + 11$
C. $9c^2 + 11$ D. $3c + 11$

38. Find the missing factor.
$-9x^3 = (3x)\,(?)$

A $3x$ B. $3x^2$ C. $-3x$ D. $-3x^2$

39. Which is the greatest common factor of the monomials? $4y^4z$, $4yz$, $8y^3z$?

A. $4y^3z$ B. $4yz$
C. $4y^2z$ D. yz

40. Simplify. $\frac{-48s^5t^6}{4s^4t^3}$

A. $12s^9t^9$ B. $-12st^9$
C. $-12st^3$ D. $12st^9$

41. How many possible outcomes are there for making a sandwich with 3 kinds of bread, 5 fillings, and 3 dressings?

A. 15 B. 9 C. 45 D. 11

42. A bag contains 3 red pens and 3 green pens. A pen is taken from the bag and replaced. Then a pen is taken from the bag again. Find P (red, then green).

A. $\frac{1}{4}$ B. $\frac{3}{10}$ C. $\frac{1}{2}$ D. $\frac{1}{9}$

43. In a light bulb factory, 400 bulbs were selected at random. Of these 2 were defective. What is the probability of a bulb being defective?

A. $\frac{1}{400}$ B. $\frac{1}{100}$ C. $\frac{1}{200}$ D. $\frac{1}{800}$

44. The coordinate pair $(3,-2)$ means

A. the x coordinate is -2
B. the y coordinate is 3
C. both A and B are true
D. neither A nor B is true

45. Find the slope of the line that passes through point $A(2,4)$ and point $B(-2,-4)$.

A. 2 B. 4 C. 0 D. 6

How to Use the Data File

The *Data File* contains interesting information presented in tables and graphs for your use in solving problems. The data is organized into the following categories: animals, architecture, arts and entertainment, astronomy/space science, earth science, ecology, economics, health and fitness, sports, and United States. You will need to refer to this material often as you work through this book. Whenever you come upon the words "USING DATA" at the beginning of an exercise, refer to the *Data Index* to help you locate the information you will need to complete the exercise. For a quick review of commonly used symbols, formulas, and other information, you may wish to refer to the last part of the *Data File,* Useful Mathematical Data.

DATA INDEX

Animals

MAXIMUM SPEEDS OF ANIMALS

The data on this topic are notoriously unreliable because of the many inherent difficulties of timing the movement of most animals—whether running, flying, or swimming—and because of the absence of any standardization of the method of timing, of the distance over which the performance is measured, or of allowance for wind conditions.

The most that can be said is that a specimen of the species below has been timed to have attained as a maximum the speed given.

mi/h		mi/h	
219.5	Spine-tailed Swift	37	Dolphin
180*	Peregrine Falcon	36	Dragonfly
120*	Golden Eagle	35	Flying Fish
96.29	Racing Pigeon	35	Rhinoceros
88	Spurwing Goose	35	Wolf
70†	Cheetah	33	Hawk Head Moth
60	Pronghorn Antelope	32	Giraffe
60	Mongolian Gazelle	32	Guano Bat
57	Quail	30	Blackbird
57	Swordfish	28	Grey Heron
53	Partridge	25	California Sea Lion
45	Red Kangaroo	24	African Elephant
45	English Hare	23	Salmon
40	Red Fox	22.8	Blue Whale
40	Mute Swan	22	Leatherback Turtle
38	Swallow	22	Wren
		20	Monarch Butterfly

*Stooping
†Unable to sustain a speed of over 44 mi/h over 500 yards.

HOW FAST DO INSECTS FLY?

	Wingbeats per second	Flight speed (mi/h)
White butterfly	8–12	4–9
Damselfly	16	2–4
Dragonfly	25–40	16–34
Cockchafer beetle	50	17
Hawkmoth	50–90	11–31
Hoverfly	120	7–9
Bumblebee	130	7
Housefly	200	4
Honeybee	225	4–7
Mosquito	600	0.6–1.2
Midge	1,000	0.6–1.2

LONGEST RECORDED LIFE SPANS OF SOME ANIMALS

Animal	Years	Animal	Years
Marion's tortoise	152	Lobster	50
Deep-sea clam	100	Cow	40
Killer whale	90	Domestic pigeon	35
Blue whale	90	Domestic cat	34
Fin whale	90	Dog (Labrador)	29
Freshwater oyster	80	Budgerigar	28
Cockatoo	70	Sheep	20
Condor	70	Goat	18
Indian elephant	70	Rabbit	18
Ostrich	62	Golden hamster	10
Horse	62	House mouse	6
Chimpanzee	50	Housefly	0.2
Termite	50		

TOP 25 AMERICAN KENNEL CLUB REGISTRATIONS

Breed	Rank	1989	Rank	1988
Cocker Spaniels	1	111,636	1	108,720
Labrador Retrievers	2	91,107	2	86,446
Poodles	3	78,600	3	82,600
Golden Retrievers	4	64,269	4	62,950
German Shepherd Dogs	5	58,422	5	57,139
Rottweilers	6	51,291	7	42,748
Chow Chows	7	50,150	6	50,781
Dachshunds	8	44,305	9	41,921
Beagles	9	43,314	8	41,983
Miniature Schnauzers	10	42,175	10	41,558
Shetland Sheepdogs	11	39,665	12	38,730
Yorkshire Terriers	12	39,268	13	36,040
Shih Tzu	13	38,131	11	38,829
Pomeranians	14	32,109	14	30,516
Lhasa Apsos	15	28,810	15	30,194
Chihuahuas	16	24,917	17	23,487
Pekingese	17	22,986	22	20,134
Boxers	18	22,037	20	20,604
Siberian Huskies	19	21,875	18	21,430
Doberman Pinschers	20	21,782	16	23,928
Basset Hounds	21	21,517	19	21,423
English Springer Spaniels	22	20,911	21	20,238
Collies	23	18,227	23	18,931
Dalmatians	24	17,488	27	14,109
Boston Terriers	25	15,355	24	14,988

GESTATION, LONGEVITY, AND INCUBATION OF ANIMALS

Longevity figures were supplied by Ronald T. Reuther. They refer to animals in captivity; the potential life span of animals is rarely attained in nature. Maximum longevity figures are from the Biology Data Book, 1972. Figures on gestation and incubation are averages based on estimates by leading authorities.

Animal	Gestation (days)	Average longevity (years)	Maximum longevity (yr, mo)
Ass	365	12	35-10
Baboon	187	20	35-7
Bear: Black	219	18	36-10
Grizzly	225	25	—
Polar	240	20	34-8
Beaver	122	5	20-6
Buffalo (American)	278	15	—
Bactrian camel	406	12	29-5
Cat (domestic)	63	12	28
Chimpanzee	231	20	44-6
Chipmunk	31	6	8
Cow	284	15	30
Deer (white-tailed)	201	8	17-6
Dog (domestic)	61	12	20
Elephant (African)	—	35	60
Elephant (Asian)	645	40	70
Elk	250	15	26-6
Fox (red)	52	7	14
Giraffe	425	10	33-7
Goat (domestic)	151	8	18
Gorilla	257	20	39-4
Guinea pig	68	4	7-6
Hippopotamus	238	25	—
Horse	330	20	46
Kangaroo	42	7	—
Leopard	98	12	19-4
Lion	100	15	25-1
Monkey (rhesus)	164	15	—
Moose	240	12	—
Mouse (meadow)	21	3	—
Mouse (dom. white)	19	3	3-6
Opossum (American)	14-17	1	—
Pig (domestic)	112	10	27
Puma	90	12	19
Rabbit (domestic)	31	5	13
Rhinoceros (black)	450	15	—
Rhinoceros (white)	—	20	—
Sea lion (California)	350	12	28
Sheep (domestic)	154	12	20
Squirrel (gray)	44	10	—
Tiger	105	16	26-3
Wolf (maned)	63	5	—
Zebra (Grant's)	365	15	—

Incubation time (days)	
Chicken	21
Duck	30
Goose	30
Pigeon	18
Turkey	26

Architecture

THE LONGEST BRIDGE SPANS (as of 1988)

Bridge	Country	Year Built	Span (meters)	Span (feet)
Akashi-Ohashi*	Japan	1988	1,780	5,840
Humber	England	1981	1,410	4,626
Verrazano Narrows	United States	1964	1,298	4,260
Golden Gate	United States	1937	1,280	4,200
Mackinac	United States	1957	1,158	3,800
Bosphorus	Turkey	1973	1,074	3,524
George Washington	United States	1931	1,067	3,500
Tagus River	Portugal	1966	1,013	3,323
Forth	Scotland	1964	1,006	3,300
Severn	England/Wales	1966	988	3,240

*Scheduled completion information

THE TALLEST BUILDINGS IN THE WORLD (as of 1990)

Building	City	Year Built	Stories	Height (meters)	Height (feet)
Sears Tower	Chicago	1974	110	443	1,454
World Trade Center, North	New York	1972	110	417	1,368
World Trade Center, South	New York	1973	110	415	1,362
Empire State	New York	1931	102	381	1,250
Bank of China Tower	Hong Kong	1988	72	368	1,209
Amoco	Chicago	1973	80	346	1,136
John Hancock	Chicago	1968	100	344	1,127
Chrysler	New York	1930	77	319	1,046

LAND VEHICULAR TUNNELS IN U.S. (over 2,000 feet in length)

Name	Location	Length (feet)
E. Johnson Memorial	I-70, Col.	8,959
Eisenhower Memorial	I-70, Col.	8,941
Allegheny (twin)	Penna. Turnpike	6,072
Liberty Tubes	Pittsburgh, Pa.	5,920
Zion Natl. Park	Rte. 9, Utah	5,766
East River Mt. (twin)	Interstate 77, W. Va.-Va.	5,412
Tuscarora (twin)	Penna. Turnpike	5,400
Kittatinny (twin)	Penna. Turnpike	4,660
Blue Mountain (twin)	Penna. Turnpike	4,435
Lehigh	Penna. Turnpike	4,379
Wawona	Yosemite Natl. Park	4,233
Big Walker Mt.	Route I-77, Va.	4,229
Squirrel Hill	Pittsburgh, Pa.	4,225
Fort Pitt	Pittsburgh, Pa.	3,560
Mall Tunnel	Dist. of Columbia	3,400
Caldecott	Oakland, Cal.	3,371
Cody No. 1	U.S. 14, 16, 20, Wyo.	3,202
Kalihi	Honolulu, Ha.	2,780
Ft. Cronkhite	Sausalito, Cal.	2,690
Memorial	W. Va. Tpke. (I-77)	2,669
Cross-Town	178 St., N.Y.C.	2,414
F. D. Roosevelt Dr.	81–89 Sts., N.Y.C.	2,400
Dewey Sq.	Boston, Mass.	2,400
Battery Park	N.Y.C.	2,300
Battery St.	Seattle, Wash.	2,140
Big Oak Flat	Yosemite Natl. Park	2,083

NOTED RECTANGULAR STRUCTURES

Structure	Country	Length (meters)	Width (meters)
Parthenon	Greece	69.5	30.9
Palace of the Governors	Mexico	96	11
Great Pyramid of Cheops	Egypt	230.6	230.6
Step Pyramid of Zosar	Egypt	125	109
Temple of Hathor	Egypt	290	280
Cleopatra's Needle (base)	England*	2.4	2.3
Ziggurat of Ur (base)	Middle East	62	43
Guanyin Pavilion of Dule Monastery	China	20	14
Izumo Shrine	Japan	10.9	10.9
Kibitsu Shrine (main)	Japan	14.5	17.9
Kongorinjo Hondo	Japan	21	20.7
Bakong Temple, Roluos	Cambodia	70	70
Ta Keo Temple	Cambodia	103	122
Wat Kukut Temple, Lampun	Thailand	23	23
Tsukiji Hotel	Japan	67	27

*Gift to England from Egypt

Parthenon

PENNY SIZES OF NAILS

Penny Size	Length (in centimeters)
2-penny	2.5 cm
3-penny	3.125 cm
4-penny	3.75 cm
6-penny	5 cm
8-penny	6.25 cm
10-penny	7.5 cm
12-penny	8.125 cm
16-penny	8.75 cm

Mosque of Omar

HOUSING UNITS—SUMMARY OF CHARACTERISTICS AND EQUIPMENT, BY TENURE AND REGION: ONE RECENT YEAR

[In thousands of units, except as indicated. Based on the American Housing Survey]

Item	Total housing units	Sea-sonal	Year-round Units							
			Occupied							Vacant
			Total	Owner	Renter	North-east	Mid-west	South	West	
Total units	**99,931**	**3,182**	**88,425**	**56,145**	**32,280**	**18,729**	**22,142**	**30,064**	**17,490**	**8,324**
Percent distribution	100.0	3.2	88.5	56.2	32.3	21.2	25.0	34.0	19.8	8.3
Units in structure:										
Single family detached	60,607	1,834	55,076	46,703	8,373	9,368	14,958	19,984	10,766	3,697
Single family attached	4,514	64	4,102	2,211	1,890	1,431	814	1,212	645	349
2–4 units	11,655	134	10,217	1,996	8,221	3,324	2,515	2,426	1,952	1,304
5–9 units	5,134	73	4,372	344	4,029	903	967	1,408	1,094	689
10–19 units	4,558	99	3,760	261	3,500	818	766	1,360	817	699
20–49 units	3,530	146	2,913	287	2,627	904	603	651	756	470
50 or more units	3,839	135	3,230	438	2,792	1,540	661	583	446	474
Mobile home or trailer	6,094	698	4,754	3,906	848	440	860	2,440	1,014	642

Arts and Entertainment

LONGEST BROADWAY RUNS	
Show	**Performances**
1. A Chorus Line (1975–90)	6,137
2. Oh! Calcutta (1976–89)	5,959
3. 42nd Street (1980–89)	3,486
4. Grease (1972–80)	3,388
5. Cats (1982–)	3,269
6. Fiddler on the Roof (1964–72)	3,242
7. Life with Father (1939–47)	3,224
8. Tobacco Road (1933–41)	3,182
9. Hello, Dolly! (1964–71)	2,844
10. My Fair Lady (1956–62)	2,717
11. Annie (1977–83)	2,377
12. Man of La Mancha (1965–71)	2,328
13. Abie's Irish Rose (1922–27)	2,327
14. Oklahoma! (1943–48)	2,212
15. Pippin (1971–77)	1,944
16. South Pacific (1949–54)	1,925
17. Magic Show (1974–78)	1,920
18. Deathtrap (1978–82)	1,792
19. Gemini (1977–81)	1,788
20. Harvey (1944–49)	1,775
21. Dancin' (1978–82)	1,774
22. La Cage aux Folles (1983–87)	1,761
23. Hair (1968–72)	1,750
24. The Wiz (1975–79)	1,672
25. Born Yesterday (1946–49)	1,642

ALL-TIME TOP TELEVISION PROGRAMS

	Program	Date	Network	Households (000)
1	M*A*S*H	2/28/83	CBS	50,150
2	Dallas	11/21/80	CBS	41,470
3	Roots Pt. VIII	1/30/77	ABC	36,380
4	Super Bowl XVI	1/24/82	CBS	40,020
5	Super Bowl XVII	1/30/83	NBC	40,480
6	Super Bowl XX	1/26/86	NBC	41,490
7	Gone With The Wind—Pt. 1	11/7/76	NBC	33,960
8	Gone With The Wind—Pt. 2	11/8/76	NBC	33,750
9	Super Bowl XII	1/15/78	CBS	34,410
10	Super Bowl XIII	1/21/79	NBC	35,090
11	Bob Hope Christmas Show	1/15/70	NBC	27,260
12	Super Bowl XVIII	1/22/84	CBS	38,800
12	Super Bowl XIX	1/20/85	ABC	39,390
14	Super Bowl XIV	1/20/80	CBS	35,330
15	ABC Theater (The Day After)	11/20/83	ABC	38,550
16	Roots Pt. VI	1/28/77	ABC	32,680
16	The Fugitive	8/29/67	ABC	25,700
18	Super Bowl XXI	1/25/87	CBS	40,030
19	Roots Pt. V	1/27/77	ABC	32,540
20	Ed Sullivan	2/9/64	CBS	23,240
21	Bob Hope Christmas Special	1/14/71	NBC	27,050
22	Roots Pt. III	1/25/77	ABC	31,900
23	Super Bowl XI	1/9/77	NBC	31,610
23	Super Bowl XV	1/25/81	NBC	34,540
25	Super Bowl VI	1/16/72	CBS	27,450

Source: A. C. Nielsen estimates, Jan. 30, 1960 through Apr. 17, 1989, excluding unsponsored or joint network telecasts or programs under 30 minutes long.
Ranked by percent of average audience.

OPTICAL ILLUSIONS

THE RECORDING INDUSTRY ASSOCIATION OF AMERICA
1990 Sales Profile
Released September 1991

PERCENTAGE OF DOLLAR VALUE BY TYPE OF MUSIC

Type of Music	1986	1987	1988	1989	1990
Rock	46.8	47.2	46.2	42.9	37.4
Pop	14.2	12.9	15.2	14.4	13.6
Urban Contemporary	10.1	11.6	13.3	14.0	18.3
Country	9.7	9.5	7.4	6.8	8.8
Classical	5.9	2.6	3.5	4.3	4.1
Jazz	0.4	5.2	4.7	5.7	5.2
Gospel	3.0	3.9	2.5	3.1	2.4
Other	5.6	6.6	6.0	8.0	9.3

PERCENTAGE OF DOLLAR VALUE BY CONFIGURATION

Configuration	1986	1987	1988	1989	1990
Cassette	57.4	59.3	60.4	50.4	48.4
Cassette Singles	0	0	1.0	2.7	2.6
LP's	28.4	18.2	13.9	8.3	4.3
7" Singles	1.5	0.8	3.4	0.5	0.3
12" Singles	2.1	1.5	2.2	1.6	0.7
CD's 3" Singles	0	0	0	0.4	1.1
CD's 5" Full	10.1	19.8	18.6	35.9	42.5

TOTAL U.S. SALES IN BILLIONS OF DOLLARS—1986 THROUGH 1990	1986	1987	1988	1989	1990
	4.6	5.5	6.2	6.4	7.5

ALL-TIME TOP 25 MOVIES
Source: *Variety*, January, 1990

Title	Total $ Revenue
1. E.T. the Extra-Terrestrial; 1982	228,618,939
2. Star Wars; 1977	193,500,000
3. Return of the Jedi; 1983	168,002,414
4. Batman; 1989	150,500,000
5. The Empire Strikes Back; 1980	141,600,000
6. Ghostbusters; 1984	130,211,324
7. Jaws; 1975	129,549,325
8. Raiders of the Lost Ark; 1981	115,598,000
9. Indiana Jones and the Last Crusade; 1989	115,500,000
10. Indiana Jones and the Temple of Doom; 1984	109,000,000
11. Beverly Hills Cop; 1984	108,000,000
12. Back to the Future; 1985	104,408,738
13. Grease; 1978	96,300,000
14. Tootsie; 1982	96,292,736
15. The Exorcist; 1973	89,000,000
16. The Godfather; 1972	86,275,000
17. Superman; 1978	82,800,000
18. Rain Man; 1989	86,000,000
19. Close Encounters of the Third Kind; 1977/1980	82,750,000
20. Three Men and a Baby; 1987	81,313,000
21. Who Framed Roger Rabbit; 1988	81,244,000
22. Beverly Hills Cop II; 1987	80,857,776
23. The Sound of Music; 1965	79,748,000
24. Gremlins; 1984	79,500,000
25. Top Gun; 1986	79,400,000

Revenue figures are in absolute dollars, reflecting actual amounts received by the distributors (estimated for movies in current release). Ticket price inflation favors recent films, but older films have the advantage of numerous reissues adding to their totals.

MULTIMEDIA AUDIENCES—SUMMARY: 1987

[**In percent, except as indicated.** As of **spring**. For persons 18 years old and over. Based on sample and subject to sampling error; see source for details]

	Television viewing and coverage	Television prime time viewing and coverage	Cable viewing and coverage	Radio listening and coverage	Newspaper reading and coverage
Total	**91.5**	**77.1**	**45.1**	**86.2**	**85.7**
18–24 years old	88.0	68.5	45.5	93.9	82.6
25–34 years old	90.2	75.8	47.5	93.7	86.2
35–44 years old	90.8	78.3	51.0	91.3	88.2
45–54 years old	91.9	77.6	45.5	87.3	87.4
55–64 years old	94.5	82.9	44.6	79.1	88.1
65 years old and over	94.9	80.8	33.9	66.2	81.7

Source: Mediamark Research Inc., New York, NY, *Multimedia Audiences*, Spring 1988. (Copyright.)

Astronomy/ Space Science

PLANETARY DIAMETERS AND DISTANCES FROM SUN

Planet	Equatorial Diameter (km)	Average Distance from Sun (km)
Mercury	4,880	58,000,000
Venus	12,100	108,000,000
Earth	12,756	150,000,000
Mars	6,780	228,000,000
Jupiter	142,800	778,000,000
Saturn	120,000	1,427,000,000
Uranus	50,800	2,870,000,000
Neptune	48,600	4,497,000,000
Pluto	2,200	5,900,000,000

PLANETARY DAYS AND YEARS

Planet	Length of Day (in Earth Hours)	Length of Year (in Earth Years; 365.26 days = 1 year)
Mercury	1406.4	0.2409
Venus	5832	0.6152
Earth	23.934	1.00
Mars	24.623	1.88
Jupiter	9.842	11.86
Saturn	10.23	29.46
Uranus	23.0	84.01
Neptune	22.0	164.8
Pluto	153.0	248.4

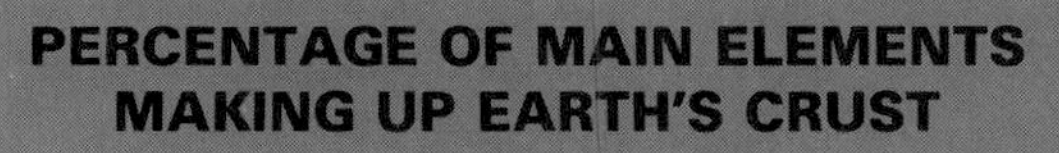

PERCENTAGE OF MAIN ELEMENTS MAKING UP EARTH'S CRUST

Element	Percentage	Element	Percentage
Oxygen	47%	Calcium	4%
Silicon	28%	Magnesium	2%
Aluminum	8%	Sodium	3%
Iron	5%	Potassium	3%

LARGEST MOONS OF THE SOLAR SYSTEM

Moon	Planet	Diameter (mi)
Titan	Saturn	3,500
Triton	Neptune	3,300
Callisto	Jupiter	3,220
Ganymede	Jupiter	3,200
Io	Jupiter	2,310
Moon	Earth	2,160
Europa	Jupiter	1,950

LARGE TELESCOPES OF THE WORLD

Type	Location	Size of Lens[1]
Refractor	Yerkes Observatory, Wisconsin	102 cm
	Lick Observatory, California	91 cm
	Paris Observatory, Meudon, France	84 cm
Reflector	W. M. Keck Telescope, Mauna Kea, Hawaii	10 m*
	Special Astrophysical Observatory, Zelenchukskaya, USSR	6 m
	Hale Telescope, Mount Palomar, California	5 m
	Cerro Tololo Inter-America Observatory, Chile	4 m
	Kitt Peak National Observatory, Arizona	4 m
	Mount Stromlo Observatory, Australia	3.8 m
	European Southern Observatory, Chile	3.6 m
	Lick Observatory, California	3 m
	McDonald Observatory, Texas	2.7 m
	Hale Observatory, Mt. Wilson, California	2.5 m

[1] for refractor telescopes, size is diameter of objective lens; for reflector telescopes, size is diameter of mirror *multiple mirrors

ORBITAL PERIODS OF SELECTED COMETS

Name	Orbital Period (yr)	First Observed
Encke	3.30	A.D. 1786
Giacobini-Zinner	6.42	1900
Biela	6.62	1772
Whipple	7.42	1933
Wolf	8.43	1884
Temple-Tuttle	33.18	1866
Halley	76.03	B.C. 239

SPACE SHUTTLE SYSTEM EXPENDITURES BY NASA IN CONSTANT (1982) DOLLARS: 1972 TO 1987

[In millions of dollars. For year ending Sept. 30. These data cannot be used alone to estimate the total shuttle cost. Only operating expenditures correspond to an annual cost in the economic sense, and even those estimates require adjustment for the three-year period over which the cost of a shuttle flight is incurred and reimbursements are received.]

Year	Total	D.D.T. and E.[1]	Construction	Production	Operations	Year	Total	D.D.T. and E.[1]	Construction	Production	Operations
1972	40	—	40	—	—	1980	2,751	1,339	42	907	463
1973	58	—	58	—	—	1981	2,724	1,041	13	1,093	578
1974	325	216	109	—	—	1982	2,932	894	20	1,283	735
1975	1,543	1,407	136	—	—	1983	3,014	—	26	1,635	1,354
1976	2,619	2,541	79	—	—	1984	3,127	—	73	690	2,364
1977	2,258	2,089	60	109	—	1985	2,636	—	42	1,376	1,218
1978	2,062	1,898	103	61	—	1986	3,311	—	—	1,504	1,807
1979	2,251	1,713	42	496	—	1987	5,806	—	—	3,841	1,965

— Represents zero. [1]Design, development, testing, and evaluation.

Source: U.S. Congress, Congressional Budget Office, estimate March 1985 and U.S. National Aeronautics and Space Administration, *NASA Budget Estimates for Fiscal Year 1989.*

Earth Science

HIGHEST AND LOWEST CONTINENTAL ALTITUDES

Source: National Geographic Society, Washington, D.C.

Continent	Highest point	Feet of elevation	Lowest point	Feet below sea level
Asia	Mount Everest, Nepal-Tibet	29,028	Dead Sea, Israel-Jordan	1,312
South America	Mount Aconcagua, Argentina	22,834	Valdes Peninsula, Argentina	131
North America	Mount McKinley, Alaska	20,320	Death Valley, California	282
Africa	Kilimanjaro, Tanzania	19,340	Lake Assai, Djibouti	512
Europe	Mount El'brus, USSR	18,510	Caspian Sea, USSR	92
Antarctica	Vinson Massif	16,864	Unknown	...
Australia	Mount Kosciusko, New South Wales	7,310	Lake Eyre, South Australia	52

From *The World Almanac and Book of Facts,* 1991.

SOME NOTABLE TORNADOES IN U.S. SINCE 1960

Date	Place	Deaths
1960 May 5, 6	SE Oklahoma, Arkansas	30
1965 Apr. 11	Ind., Ill., Oh., Mich., Wis.	271
1966 Mar. 3	Jackson, Miss.	57
1966 Mar. 3	Mississippi, Alabama	61
1967 Apr. 21	Ill., Mich.	33
1968 May 15	Midwest	71
1969 Jan. 23	Mississippi	32
1971 Feb. 21	Mississippi delta	110
1973 May 26–27	South, Midwest (series)	47
1974 Apr. 3–4	Ala., Ga., Tenn., Ky., Oh.	350
1977 Apr. 4	Ala., Miss., Ga.	22
1979 Apr. 10	Tex., Okla.	60
1980 June 3	Grand Island, Neb. (series)	4
1982 Mar. 2–4	South, Midwest (series)	17
1982 May 29	So. Ill.	10
1983 May 18–22	Texas	12
1984 Mar. 28	N. Carolina, S. Carolina	67
1984 Apr. 21–22	Mississippi	15
1984 Apr. 26	Okla. to Minn. (series)	17
1985 May 31	N.Y., Pa., Oh., Ont. (series)	90
1987 May 22	Saragosa, Tex.	29
1990 June 2–3	Midwest, Great Lakes	13

MEASURING EARTHQUAKES

The energy of an earthquake is generally reported using the Richter scale, a system developed by American geologist Charles Richter in 1935, based on measuring the heights of wave measurements on a seismograph.

On the Richter scale, each single-integer increase represents 10 times more ground movement and 30 times more energy released. The change in magnitude between numbers on the scale can be represented by 10^x and 30^x, where x represents the change in the Richter scale measure. Therefore, a 3.0 earthquake has 100 times more ground movement and 900 times more energy released than a 1.0 earthquake.

Richter scale	
2.5	Generally not felt, but recorded on seismometers.
3.5	Felt by many people.
4.5	Some local damage may occur.
6.0	A destructive earthquake.
7.0	A major earthquake. About ten occur each year.
8.0 and above	Great earthquakes. These occur once every five to ten years.

From *The Universal Almanac* © 1990.

SIZE AND DEPTH OF THE OCEANS

Ocean	Mi²	Greatest Depth-ft
Pacific	63,800,000	36,161
Atlantic	31,800,000	30,249
Indian	28,900,000	24,441
Arctic	5,400,000	17,881

RECORD HOLDERS FROM THE EARTH SCIENCES

Traditionally, the study of early life by way of its remains—which are called fossils—has been the province of earth scientists, rather than biologists. This is because fossils were the first known way to group rocks of the same age together, since it was assumed that at a given time in the past certain life-forms existed and later became extinct. Earth itself is thought to be about 4.5 billion years old.

By Size

This group of record holders includes a couple of ancient life-forms but concentrates more on those features of Earth that are studied by earth scientists.

Record	Record holder	Size or distance
Longest dinosaur	*Seismosaurus,* who lived about 150 million years ago in what is now New Mexico	100–120 ft long
Largest island	Greenland, also known as Kalaallit Nunnaat	About 840,000 mi²
Largest ocean	Pacific	64,186,300 mi²
Deepest part of ocean	*Challenger* Deep in the Marianas Trench in Pacific Ocean	35,640 ft, or 6.85 mi
Greatest tide	Bay of Fundy between Maine and New Brunswick	47.5 ft between high and low tides
Largest geyser	Steamboat Geyser in Yellowstone Park	Shoots mud and rocks about 1,000 ft in air
Longest glacier	Lambert Glacier in Antarctica (upper section known as Mellor Glacier)	At least 250 mi
Longest known cave	Mammoth Cave, which is connected to Flint Ridge Cave system	Total mapped passageway of over 300 mi
Largest canyon on land	Grand Canyon of the Colorado, in northern Arizona	Over 217 mi long, from 4 to 13 mi wide, as much as 5,300 ft deep
Highest volcano	Cerro Aconagua in the Argentine Andes	22,834 ft high
Most abundant mineral	Magnesium silicate perovskite	About $\frac{2}{3}$ of planet, it forms Earth's mantle

By Age

Record	Record holder	Age in years
Oldest rocks	Zircons from Australia	4.4 billion
Oldest fossils	Single-celled algae or bacteria from Australia	3.5 billion
Oldest slime molds (bacteria)	Slime molds—also called slime bacteria—that gathered together to form multicellular bodies for reproduction	2.5 billion
Oldest petroleum	Oil from northern Australia	1.4 billion
Oldest land animal	Millipede? Known only from its burrows	488 million
Oldest fish	*Sacabambasis,* found in Bolivia	470 million
Oldest land plant	Moss or algae	425 million
Oldest insect	Bristletail (relative of modern silverfish)	390 million
Oldest reptile	"Lizzie the Lizard," found in Scotland	340 million
Oldest bird	*Protoavis,* fossils found near Post, Texas	225 million
Oldest dinosaur	Unnamed dinosaur the size of ostrich	225 million

SOME PRINCIPAL RIVERS OF THE WORLD

River	Length (miles)
Amazon	4,000
Arkansas	1,459
Columbia	1,243
Danube	1,776
Ganges	1,560
Indus	1,800
Mackenzie	2,635
Mississippi	2,340
Missouri	2,540
Nile	4,160
Ohio	1,310
Orinoco	1,600
Paraguay	1,584
Red	1,290
Rhine	820
Rio Grande	1,900
St. Lawrence	800
Snake	1,038
Thames	236
Tiber	252
Volga	2,194
Zambezi	1,700

SIZE OF THE CONTINENTS

Continent	Mi²
Asia	17,297,000
Africa	11,708,000
North America	9,406,000
South America	6,883,000
Antarctica	5,405,000
Europe	3,835,000
Australia	2,968,000

Ecology

FUEL ECONOMY AND CARBON DIOXIDE

The better your car's gas mileage, the less carbon dioxide your car will emit into the environment. As stated earlier, cars and light trucks emit one-fifth of all carbon dioxide in the United States, one of the main causes of the greenhouse effect.

Here are estimates of the amount of carbon dioxide emissions over a single car's lifetime, courtesy of the Energy Conservation Coalition, a project of Environmental Action:

ACID RAIN TROUBLE AREAS

Source: National Acid Precipitation Assessment Program

Area	No. of lakes over 10 acres	No. acidified	%
Adirondacks			
Southwest Lakes	450	171	38
New England			
Seaboard Lowlands Lakes	848	68	8
Highland Lakes	3,574	71	2
Appalachia			
Forested Lakes	433	43	10
Forested Streams	11,631	1,396	12
Atlantic Coastal Plain			
Northeast Lakes	187	21	11
Pine Barrens Streams	675	378	56
Other Streams	7,452	745	10
Florida			
Northern Highland Lakes	522	329	63
Northern Highland Streams	669	187	28
Eastern Upper Midwest			
Low Silica Lakes	1,254	201	16
High Silica Lakes	1,673	50	3

CONTENTS OF GARBAGE CANS

Our garbage cans are filled with a diverse mix (by weight):

- paper and paperboard $\frac{9}{25}$
- yard wastes $\frac{19}{100}$
- glass $\frac{2}{25}$
- metals $\frac{9}{100}$
- food $\frac{9}{100}$
- plastics $\frac{2}{25}$
- wood/fabric $\frac{1}{25}$
- rubber and leather $\frac{3}{100}$
- textiles $\frac{1}{50}$
- other $\frac{4}{250}$
- household hazardous waste $\frac{1}{250}$

From: *Nontoxic, Natural, & Earthwise* by Debra Lynn Dood.

QUANTITIES OF HAZARDOUS WASTE BURNED IN U.S.

Incinerator Type	Number of Facilities	Quantity of Hazardous Waste Burned, lbs/yr
Commercial incinerators	17	1.3 billion (1)
Captive/on-site incinerators	154	2.3 billion (1)
Cement kilns	25–30	1.8 billion (2)
Aggregate kilns	6	1.2 billion (2)
Boilers/other furnaces	900 +	1.0 billion (2)
Total	1,100 +	7.6 billion

(1) Highum 1990. Data from review of states' capacity assurance plans.
(2) Holloway 1990. Data from U.S. EPA'S Office of Solid Waste and Emergency Response.

AIR AND WATER POLLUTION ABATEMENT EXPENDITURES IN CONSTANT (1982) DOLLARS: 1975 TO 1986

[In millions of dollars]

Year	AIR							WATER				
	Total	Mobile sources[1]			Stationary sources			Total[4]	Industrial		Public sewer systems[5]	
		Total	Cars	Trucks	Total[2]	Industrial						
						Facilities	Operations[3]		Facilities	Operations[3]	Facilities	Operations[3]
1975	20,768	9,324	7,419	1,904	11,444	6,669	4,348	22,840	4,200	2,950	8,997	3,428
1978	23,774	11.877	8,928	2,949	11,900	5,652	5,711	26,631	4,277	3,934	10,090	4,392
1979	24,462	11,757	8,469	3,288	12,706	5,969	6,108	26,470	4,013	4,222	9,758	4,583
1980	24,744	11.764	8,818	2,946	12,980	5,946	6,304	24,647	3,725	4,081	8,942	4,694
1981	25,850	13,401	10,564	2,837	12,448	5,446	6,299	21,984	3,259	4,180	6,882	4,880
1982	24,961	13,464	10,530	2,934	11,496	5,086	5,675	21,199	3,080	4,022	6,148	5,156
1983	26,367	15,581	12,274	3,307	10,785	4,104	5,990	21,543	2,811	4,509	5,551	5,475
1984	28,591	17,561	13,481	4,081	11,030	4,115	6,260	23,257	2,900	4,795	6,387	5,648
1985	29,524	18,704	14,217	4,488	10,819	3,929	6,342	24,770	2,941	5,042	6,990	5,946
1986, prel.	30,401	19,535	14,830	4,704	10,867	3,881	6,425	26,201	2,852	5,279	7,707	6,567

[1] Excludes expenditures to reduce emissions from sources other than cars and trucks. [2] Includes other expenditures not shown separately. [3] Operation of facilities. [4] Includes nonpoint sources not shown separately. [5] Includes expenditures for private connection to sewer systems, by owners of animal feedlots, and by government enterprises.

Source: U.S. Bureau of Economic Analysis, *Survey of Current Business,* May 1988.

Economics

MONEY AROUND THE WORLD

In the United States, the basic monetary unit is the dollar and the chief fractional unit is the penny. One dollar = 100 pennies. Unless noted otherwise, the basic monetary unit equals 100 chief fractional units for the countries listed below.

Country	Basic Monetary Unit	Chief Fractional Unit	Coin and Paper Denominations
Australia	dollar	cent	100, 50, 20, 10, 5, and 2 dollar notes; 2 and 1 dollar coins; 50, 20, 10, 5, 2, and 1 cent coins
Canada	dollar	cent	1,000, 500, 100, 50, 20, 10, 5, 2, and 1 dollar notes; 1 dollar coin; 25, 10, 5, and 1 cent coins
China	yuan	jiao (10 jiao) fen (1,000 fen)	10, 5, 2, and 1 yuan notes; 5, 2, and 1 jiao notes; 1 yuan coin; 5, 2, and 1 jiao coins; 5, 2, and 1 fen coins
France	franc	centime	500, 200, 100, 50, and 20 franc notes; 10, 5, 2, and 1 franc coins; 50, 20, 10, and 5 centime coins
India	Rupee	paise	1,000, 500, 100, 40, 20, 10, 5, 2, and 1 rupee notes; 2 and 1 rupee coins; 50, 25, and 20 paise coins
Mexico	peso	centavo	50,000, 20,000, 10,000, 5,000, 2,000, 1,000, and 500 peso notes; 500, 200, 100, 50, 20, 10, 5, 2, and 1 peso coins
Netherlands	gulden	cent	1,000, 250, 100, 50, 25, 10, and 5 gulden notes; 5, 2.5, and 1 gulden coins; 25, 10, 5 cents
Sudan	pound	piasters, milliemes (1,000 milliemes)	50, 20, 10, 5, and 1 pound notes; 50 and 25 piaster notes; 50, 10, 5, and 2 piaster coins; 10, 5, 2, and 1 millieme coins

SEIGNIORAGE OF COIN AND SILVER BULLION MADE IN THE UNITED STATES

Seigniorage is the difference between the monetary value of a coin and the cost of making a coin.

Fiscal Year	Dollars
1/1/35–6/30/65 cumulative	2,525,927,763.84
1968	383,141,339.00
1970	274,217,884.01
1972	580,586,683.00
1974	320,706,638.49
1980	662,814,791.48
1983	477,479,387.58
1984	498,371,724.09
1986	692,445,674.57
1988	470,409,480.20
1/1/35–6/30/88 cumulative	13,727,374,271.32

EARLY MONETARY SYSTEMS

Pacific Islands (New Britain, San Cristobel)

10 coconuts = 1 string white whales' teeth
10 strings of white teeth = 1 string of red whales' teeth or 1 dog's tooth
10 strings of red teeth = 50 porpoise teeth
500 porpoise teeth = 1 wife of good quality
1 "marble" (shell) ring = 1 good pig

THE SHRINKING VALUE OF THE DOLLAR

The average retail cost of certain foods in selected years, 1890–1975 (to the nearest cent)

	5 lb flour	1 lb round steak	1 qt milk	10 lb potatoes
1890	$0.15	$0.12	$0.07	$0.16
1910	0.18	0.17	0.08	0.17
1930	0.23	0.43	0.14	0.36
1950	0.49	0.94	0.21	0.46
1970	0.59	1.30	0.33	0.90
1975	0.98	1.89	0.45	0.99

How to Read the Exchange Rates Table →

The first two columns show how many U.S. dollars are needed to equal one unit of another country's currency. For example, on Thursday, one British pound could be exchanged for $1.6045 U.S. The last two columns show the value of one U.S. dollar in another country. For example, on Thursday, $1.00 U.S. had the same value as 0.6232 British pound.

EXCHANGE RATES

Thursday, July 11, 1991

The New York foreign exchange selling rates below apply to trading among banks in amounts of $1 million and more, as quoted at 3 p.m. Eastern time by Bankers Trust Co. and other sources. Retail transactions provide fewer units of foreign currency per dollar.

	U.S. $ equiv.		Currency per U.S. dollar	
Country	Thurs.	Wed.	Thurs.	Wed.
Argentina (Austral)	.0001010	.0001010	9902.00	9902.00
Australia (Dollar)	.7667	.7670	1.3043	1.3038
Austria (Schilling)	.07746	.07846	12.91	12.75
Bahrain (Dinar)	2.6525	2.6525	.3770	.3770
Belgium (Franc)	.02648	.02682	37.76	37.28
Brazil (Cruzeiro)	.00320	.00320	312.34	312.34
Britain (Pound)	1.6045	1.6225	.6232	.6163
Canada (Dollar)	.8705	.8715	1.1487	1.1473
Chile (Peso)	.002945	.002944	339.51	339.71
China (Renmimbi)	.186567	.186567	5.3600	5.3600
Colombia (Peso)	.001753	.001751	570.38	571.00
Denmark (Krone)	.1409	.1425	7.0949	7.0163
Ecuador (Sucre)	.000965	.000965	1036.00	1036.00
Finland (Markka)	.22656	.22911	4.4138	4.3647
France (Franc)	.16085	.16260	6.2170	6.1500
Germany (Mark)	.5451	.5516	1.8345	1.8130
Greece (Drachma)	.004995	.005062	200.20	197.55
Hong Kong (Dollar)	.12877	.12882	7.7660	7.7625
India (Rupee)	.04167	.04167	24.00	24.00
Indonesia (Rupiah)	.0005123	.0005123	1952.00	1952.00
Ireland (Punt)	1.4579	1.4765	.6859	.6773
Israel (Shekel)	.4276	.4291	2.3386	2.3302
Italy (Lira)	.0007337	.0007419	1363.01	1347.85
Japan (Yen)	.007211	.007220	138.67	138.50
Jordan (Dinar)	1.4535	1.4535	.6880	.6880
Kuwait (Dinar)	z	z	z	z
Lebanon (Pound)	.001110	.001110	901.00	901.00
Malaysia (Ringgit)	.3588	.3588	2.7870	2.7870
Malta (Lira)	2.9028	2.9028	.3445	.3445
Mexico (Peso)	.0003305	.0003305	3026.00	3026.00
Netherland (Guilder)	.4842	.4918	2.0654	2.0335
New Zealand (Dollar)	.5620	.5627	1.7794	1.7771
Norway (Krone)	.1410	.1410	7.0937	7.0937
Pakistan (Rupee)	.0410	.0410	24.40	24.40
Peru (New Sol)	1.2231	1.2231	.82	.82
Philippines (Peso)	.03724	.03724	26.85	26.85
Portugal (Escudo)	.006389	.006351	156.51	157.46
Saudi Arabia (Riyal)	.26660	.26660	3.7510	3.7510
Singapore (Dollar)	.5690	.5700	1.7575	1.7545
South Africa (Rand)	.3454	.3452	2.8950	2.8968
South Korea (Won)	.0013805	0013805	724.35	724.35
Spain (Peseta)	.008702	.008787	114.92	113.80
Sweden (Krona)	.1508	.1525	6.6308	6.5576
Switzerland (Franc)	.6285	.6365	1.5910	1.5710
Taiwan (Dollar)	.037397	.037355	26.74	26.77
Thailand (Baht)	.03891	.03891	25.70	25.70
Turkey (Lira)	.0002300	.0002309	4347.01	4330.02
United Arab (Dirham)	.2723	.2723	3.6730	3.6730
Uruguay (New Peso)	.000500	.000500	2000.00	2000.00
Venezuela (Bolivar)	.01803	.01803	55.46	55.46

z—Not quoted.

STATE GENERAL SALES AND USE TAXES, JULY 1988

State	Percent rate	State	Percent rate	State	Percent rate
Alabama	4	Kentucky	5	Ohio	5
Arizona	5	Louisiana	4	Oklahoma	4
Arkansas	4	Maine	5	Pennsylvania	6
California	4.75	Maryland	5	Rhode Island	6
Colorado	3	Massachusetts	5	South Carolina	5
Connecticut	7.5	Michigan	4	South Dakota	4
D.C.	6	Minnesota	6	Tennessee	5.5
Florida	6	Mississippi	6	Texas	6
Georgia	3	Missouri	4.225	Utah	5.09375
Hawaii	4	Nebraska	4	Vermont	4
Idaho	5	Nevada	5.75	Virginia	3.5
Illinois	5	New Jersey	6	Washington	6.5
Indiana	5	New Mexico	4.75	West Virginia	6
Iowa	4	New York	4	Wisconsin	5
Kansas	4	North Carolina	3	Wyoming	3
		North Dakota	5.5		

NOTE: Alaska, Delaware, Montana, New Hampshire, and Oregon have no statewide sales and use taxes.
Source: Information Please Almanac questionnaires to the states and Tax Foundation, Inc.

Health and Fitness

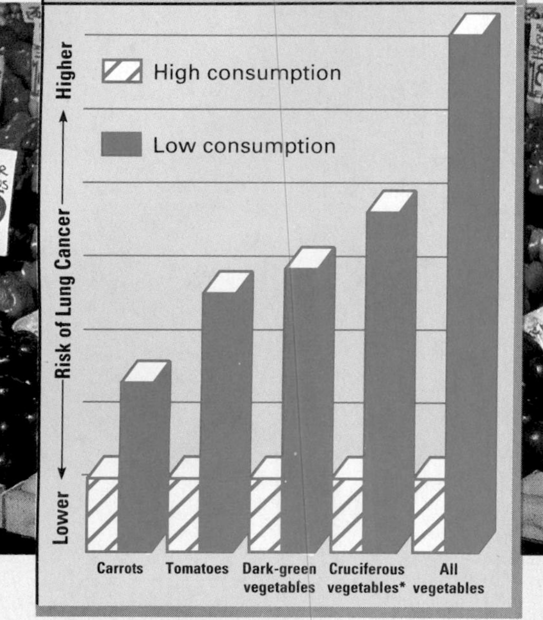

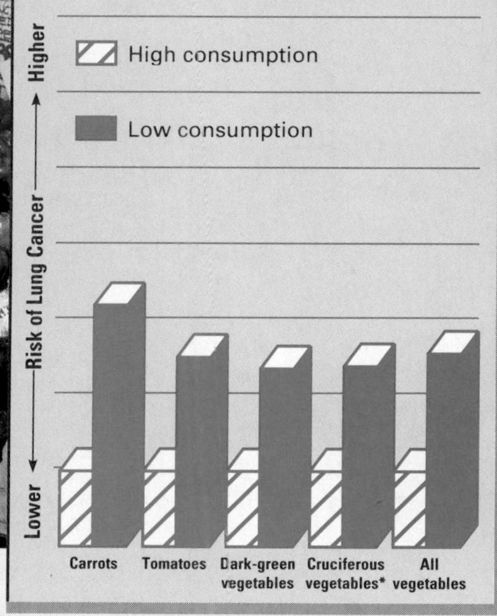

*Cruciferous vegetables are vegetables in the cabbage family, such as cabbage, broccoli, cauliflower, and Brussels sprouts.

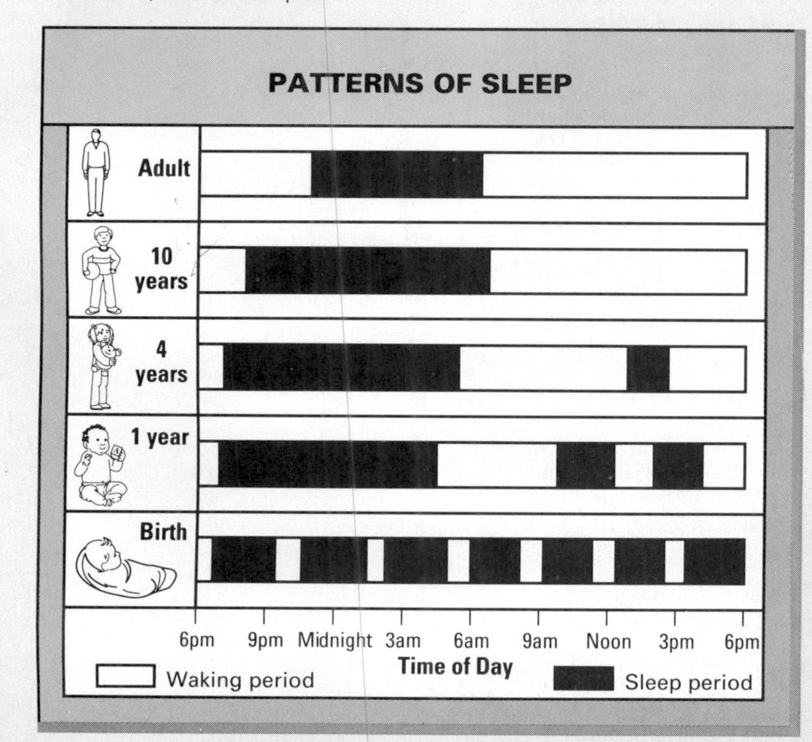

HOW MUCH WALKING IS DONE ON THE JOB?

Job	Miles Walked per Day
Hospital nurse	5.3
Security officer	4.2
City messenger	4.0
Retail salesperson	3.5
Waiter/waitress	3.3
Hotel employee	3.2
Doctor	2.5
Advertising rep.	2.4
Architect	2.3
Secretary	2.2
Newspaper reporter	2.1
Banker	2.0
Accountant	1.8
Teacher	1.7
Lawyer	1.5
Housewife	1.3
Magazine editor	1.3
Radio announcer	1.1
Dentist	0.8

LIFE EXPECTANCY

Life Expectancy at Birth (years)

80
75
70
65
60
55
50
45
40

80.4
78.4
77.8
74.6
73.2
65.9
57.3
50.7

73.5
71.5
70.1
67.0
66.8
61.6
55.4
47.9

1900 1920 1940 1960 1970 1980 1987 2000

Year of Birth

Men

Women

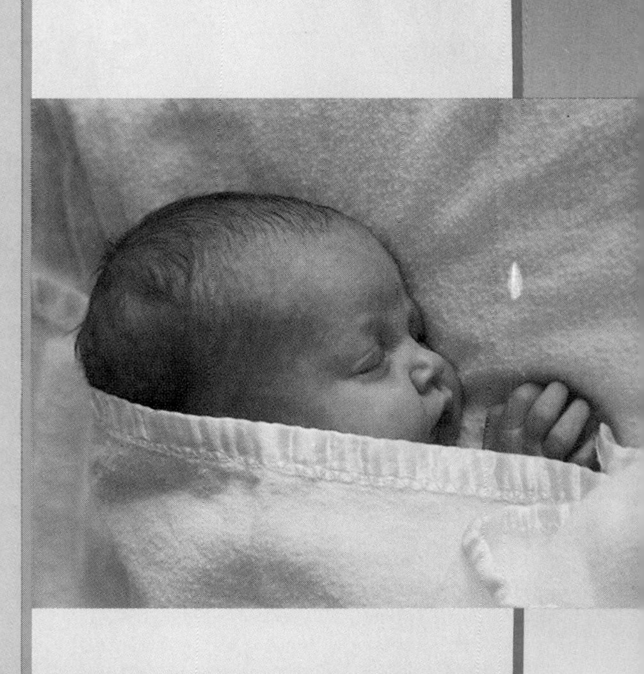

IMMUNIZATION SCHEDULE FOR CHILDREN

	DTP	Polio	TB Test*	Measles	Mumps	Rubella	Hib/ Conjugate	Tetanus-Diphtheria
2 months	✓	✓						
4 months	✓	✓						
6 months	✓							
12–15 months			✓					
15 months				✓	✓	✓	✓	
15–18 months	✓	✓						
4–6 years	✓	✓						
5–21 years				✓	✓	✓		
14–16 years								✓

*Only in high-prevalence populations.

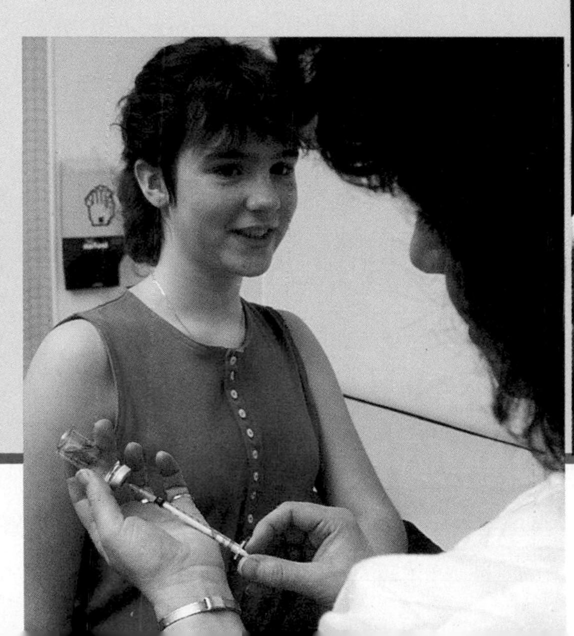

Sports

BASEBALL STADIUMS

Team	Stadium	Seating Capacity
ATLANTA BRAVES	Atlanta—Fulton County Stadium	52,003
BALTIMORE ORIOLES	Memorial Stadium	54,017
BOSTON RED SOX	Fenway Park	34,182
CALIFORNIA ANGELS	Anaheim Stadium	64,593
CHICAGO CUBS	Wrigley Field	39,600
CHICAGO WHITE SOX	Comiskey Park	44,087
CINCINNATI REDS	Riverfront Stadium	52,392
CLEVELAND INDIANS	Cleveland Stadium	74,483
DETROIT TIGERS	Tiger Stadium	52,416
HOUSTON ASTROS	Astrodome	45,000
KANSAS CITY ROYALS	Royals Stadium	40,625
LOS ANGELES DODGERS	Dodger Stadium	56,000
MILWAUKEE BREWERS	Milwaukee County Stadium	53,192
MINNESOTA TWINS	Hubert H. Humphrey Metrodome	55,883
MONTREAL EXPOS	Olympic Stadium	59,149
NEW YORK METS	Shea Stadium	55,300
NEW YORK YANKEES	Yankee Stadium	57,545
OAKLAND A's	Oakland Coliseum	49,219
PHILADELPHIA PHILLIES	Veterans Stadium	64,538
PITTSBURGH PIRATES	Three Rivers Stadium	58,727
SAN DIEGO PADRES	Jack Murphy Stadium	58,433
SAN FRANCISCO GIANTS	Candlestick Park	58,000
SEATTLE MARINERS	Kingdome	58,150
ST. LOUIS CARDINALS	Busch Stadium	54,224
TEXAS RANGERS	Arlington Stadium	43,508
TORONTO BLUE JAYS	Skydome	53,000

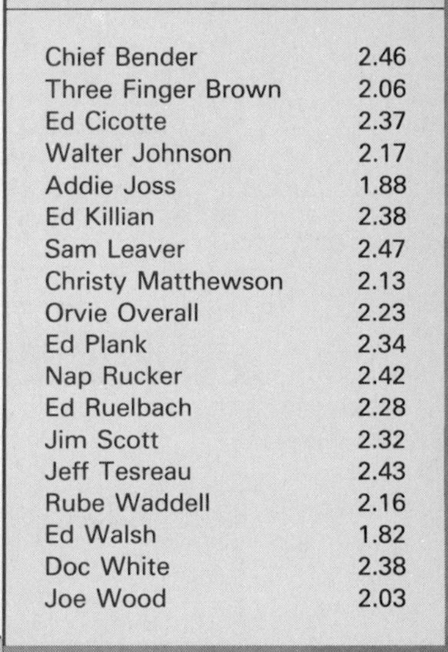

LIFETIME EARNED RUN AVERAGE
(minimum 1,500 innings pitched)

Pitcher	ERA
Chief Bender	2.46
Three Finger Brown	2.06
Ed Cicotte	2.37
Walter Johnson	2.17
Addie Joss	1.88
Ed Killian	2.38
Sam Leaver	2.47
Christy Matthewson	2.13
Orvie Overall	2.23
Ed Plank	2.34
Nap Rucker	2.42
Ed Ruelbach	2.28
Jim Scott	2.32
Jeff Tesreau	2.43
Rube Waddell	2.16
Ed Walsh	1.82
Doc White	2.38
Joe Wood	2.03

NUMBER OF CALORIES BURNED BY PEOPLE OF DIFFERENT WEIGHTS

Exercise	Calories Burned per Hour		
	110 lb	154 lb	198 lb
Martial arts	620	790	960
Racquetball (2 people)	610	775	945
Basketball (full-court game)	585	750	910
Skiing—cross country (5 mi/h)	550	700	850
downhill	465	595	720
Running—8-min mile	550	700	850
12-min mile	515	655	795
Swimming—crawl, 45 yd/min	540	690	835
crawl, 30 yd/min	330	420	510
Stationary bicycle—15 mi/h	515	655	795
Aerobic dancing—intense	515	655	795
moderate	350	445	540
Walking—5 mi/h	435	555	675
3 mi/h	235	300	365
2 mi/h	145	185	225
Calisthenics—intense	435	555	675
moderate	350	445	540
Scuba diving	355	450	550
Hiking—20-lb pack, 4 mi/h	355	450	550
20-lb pack, 2 mi/h	235	300	365
Tennis—singles, recreational	335	425	520
doubles, recreational	235	300	365
Ice skating	275	350	425
Roller skating	275	350	425

WORLD SPEED SKATING RECORDS (Ice Skating)

Distance	min:sec	Name and Nationality	Place	Date
MEN				
500 m	36.23*	Nick Thometz (USA)	Medeo, USSR	Mar 26, 1987
	36.45	Jens-Uwe May (E Ger)	Calgary, Canada	Feb 14, 1988
1000 m	1.12.05	Nick Thometz (USA)	Medeo, USSR	Mar 27, 1987
1500 m	1.52.06	Andre Hoffman (E Ger)	Calgary, Canada	Feb 20, 1988
3000 m	3.59.27	Leo Visser (Neth)	Heerenveen, Netherlands	Mar 19, 1987
5000 m	6.47.01	Leo Visser (Neth)	Heerenveen, Netherlands	Feb 14, 1987
10,000 m	13.48.20	Tomas Gustafson (Swe)	Calgary, Canada	Feb 21, 1988
WOMEN				
500 m	39.10	Bonnie Blair (USA)	Calgary, Canada	Feb 22, 1988
1000 m	1.18.11	Karin Kania (*nee* Enke) (GDR)	Calgary, Canada	Dec 5, 1987
1500 m	1.59.30	Karin Kania (GDR)	Medeo, USSR	Mar 22, 1986
3000 m	4.11.94	Yvonne van Gennip (Neth)	Calgary, Canada	Feb 23, 1988
5000 m	6.43.59	Yvonne van Gennip (Neth)	Calgary, Canada	Feb 28, 1988
10,000 m	15.25.25	Yvonne van Gennip (Neth)	Heerenveen, Netherlands	Mar 19, 1988

*Unofficial (represents an average speed of 30.87 mi/h).

THE SPORTS WITH THE MOST INJURIES

Percentage of All Sports Injuries

- Baseball 2.3%
- Track 2.7%
- Soccer 3.2%
- Basketball 5.0%
- Football 63.9%

OLYMPIC RECORD TIMES FOR 400-m FREESTYLE SWIMMING (minutes)

YEAR	1924	1928	1932	1936	1948	1952	1956	1960	1964	1968	1972	1976	1980	1984	1988
MALE	5:04.2	5:01.6	4:48.4	4:44.5	4:41.0	4:30.7	4:27.3	4:18.3	4:12.2	4:09.0	4:00.27	3:51.93	3:51.31	3:51.23	3:46.25
FEMALE	6:02.2	5:42.8	5:28.5	5:26.4	5:17.8	5:12.1	4:54.6	4:50.6	4:43.3	4:31.8	4:19.44	4:09.89	4:08.76	4:07.10	4:03.85

SIZES AND WEIGHTS OF BALLS USED IN VARIOUS SPORTS

Type	Diameter (cm)	Average Weight (g)
Baseball	7.6	145
Basketball	24.0	596
Croquet ball	8.6	340
Field hockey ball	7.6	160
Golf ball	4.3	46
Handball	4.8	65
Soccer ball	22.0	425
Softball, large	13.0	279
Softball, small	9.8	187
Table tennis ball	3.7	2
Tennis ball	6.5	57
Volleyball	21.9	256

United States

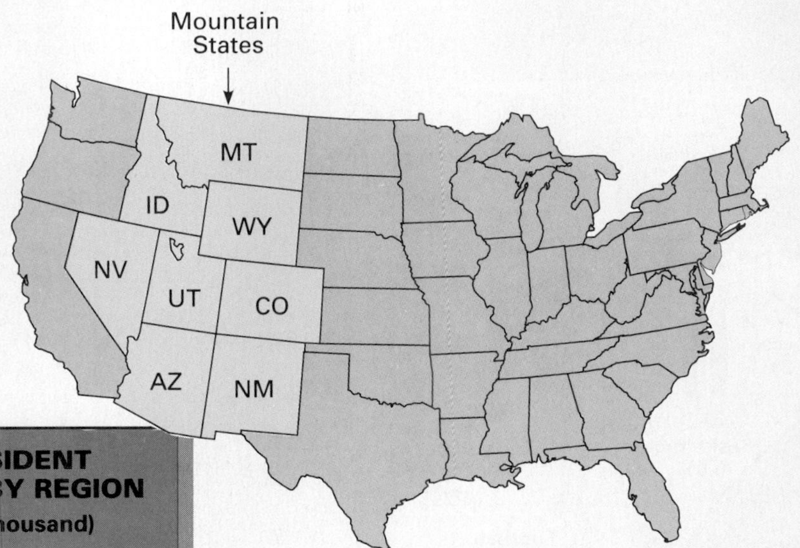

UNITED STATES RESIDENT POPULATION CHANGE BY REGION
(to the nearest hundred thousand)

	1980	1990
Northeast	49,100,000	50,800,000
Midwest	58,900,000	59,700,000
South	75,300,000	85,400,000
West	43,200,000	52,800,000

UTAH AND THE OTHER MOUNTAIN STATES

In area, Utah ranks 7th among the Mountain states (shaded) and 11th among all 50 states.

SOME ENDANGERED MAMMALS OF THE UNITED STATES

Source: U.S. Fish and Wildlife Service, U.S. Interior Department; as of April 15, 1990

Common name	Scientific name	Range
Ozark big-eared bat	*Plecotus townsendii ingens*	Mo., Okla., Ariz.
Brown or grizzly bear	*Ursus arctos horribilis*	48 conterminous states
Columbian white-tailed deer	*Odocoileus virginianus leucurus*	Wash., Ore.
San Joaquin kit fox	*Vulpes macrotis mutica*	Cal.
Southeastern beach mouse	*Peromyscus polionotus phasma*	Fla.
Ocelot	*Felis pardalis*	Tex., Ariz.
Southern sea otter	*Enhydra lutris hereis*	Wash., Ore., Cal.
Florida panther	*Felis concolor coryi*	La., Ark. east to S.C., Fla.
Utah prairie dog	*Cynomys parvidens*	Ut.
Morro Bay kangaroo rat	*Dipodomys heermanni morroensis*	Cal.
Carolina northern flying squirrel	*Glaucomys sabrinus coloratus*	N.C., Tenn.
Hualapai Mexican vole	*Microtus mexicanus hualpaiensis*	Ariz.
Red wolf	*Canis rufus*	Southeast to central Tex.

FARMS—NUMBER AND ACREAGE, BY STATE: 1980 AND 1988

[1988 data preliminary. Based on 1974 census definition of farms and farmland]

State	Farms (1,000)		Acreage (mil.)		Acreage per farm	
	1980	1988	1980	1988	1980	1988
U.S.	**2,433**	**2,159**	**1,039**	**999**	**427**	**463**
Alabama	59	49	12	11	207	224
Alaska	(z)	1	2	1	3,378	2,123
Arizona	8	8	38	37	5,080	4,506
Arkansas	59	47	17	15	280	319
California	81	78	34	33	417	417
Colorado	27	27	36	34	1,358	1,234
Connecticut	4	4	(z)	(z)	117	119
Delaware	4	3	1	1	186	197
Florida	39	40	13	13	344	325
Georgia	59	49	15	13	254	265
Hawaii	4	4	2	2	458	443
Idaho	24	23	15	14	623	609
Illinois	107	83	29	29	269	345
Indiana	87	72	17	16	193	228
Iowa	119	107	34	34	284	313
Kansas	75	69	48	48	644	694
Kentucky	102	99	15	15	143	146
Louisiana	37	35	10	10	273	271
Maine	8	8	2	2	195	192
Maryland	18	16	3	2	157	147
Massachusetts	6	6	1	1	116	111
Michigan	65	58	11	11	175	193
Minnesota	104	94	30	30	291	319
Mississippi	55	43	15	14	265	314
Missouri	120	113	31	30	261	9
Montana	24	23	62	61	2,601	2,605
Nebraska	65	55	48	47	734	856
Nevada	3	2	9	9	3,100	3,667
New Hampshire	3	3	1	1	160	158
New Jersey	9	7	1	1	109	112
New Mexico	14	14	47	45	3,467	3,333
New York	47	40	9	9	200	213
North Carolina	93	70	12	11	126	150
North Dakota	40	33	42	40	1,043	1,243
Ohio	95	84	16	16	171	186
Oklahoma	72	69	35	33	481	478
Oregon	35	37	18	18	517	488
Pennsylvania	62	56	9	8	145	150
Rhode Island	1	1	(z)	(z)	87	96
South Carolina	34	27	6	5	188	200
South Dakota	39	35	45	44	1,169	1,278
Tennessee	96	94	14	13	142	136
Texas	189	156	138	132	731	846
Utah	14	13	12	11	919	850
Vermont	8	7	2	2	226	223
Virginia	58	49	10	10	169	196
Washington	38	38	16	16	429	421
West Virginia	22	21	4	4	191	176
Wisconsin	93	82	19	18	200	215
Wyoming	9	9	35	35	3,846	4,000

z represents less than 500 farms or 500,000 acres.

1990 POPULATION AND NUMBER OF REPRESENTATIVES, BY STATE

TOTAL POPULATION 249,632,692

State	Apportionment Population	Number of Representatives Based on the 1990 Census	Change from 1980 Apportionment
U.S. TOTAL*	249,022,783	435	
Alabama	4,062,608	7	–
Alaska	551,947	1	–
Arizona	3,677,985	6	+1
Arkansas	2,362,239	4	–
California	29,839,250	52	+7
Colorado	3,307,912	6	–
Connecticut	3,295,669	6	–
Delaware	668,696	1	–
Florida	13,003,362	23	+4
Georgia	6,508,419	11	+1
Hawaii	1,115,274	2	–
Idaho	1,011,986	2	–
Illinois	11,466,682	20	–2
Indiana	5,564,228	10	–
Iowa	2,787,424	5	–1
Kansas	2,485,600	4	–1
Kentucky	3,698,969	6	–1
Louisiana	4,238,216	7	–1
Maine	1,233,223	2	–
Maryland	4,798,622	8	–
Massachusetts	6,029,051	10	–1
Michigan	9,328,784	16	–2
Minnesota	4,387,029	8	–
Mississippi	2,586,443	5	–
Missouri	5,137,804	9	–
Montana	803,655	1	–1
Nebraska	1,584,617	3	–
Nevada	1,206,152	2	–
New Hampshire	1,113,915	2	–
New Jersey	7,748,634	13	–1
New Mexico	1,521,779	3	–
New York	18,044,505	31	–3
North Carolina	6,657,630	12	+1
North Dakota	641,364	1	–
Ohio	10,887,325	19	–2
Oklahoma	3,157,604	6	–
Oregon	2,853,733	5	–
Pennsylvania	11,924,710	21	–2
Rhode Island	1,005,984	2	–
South Carolina	3,505,707	6	–
South Dakota	699,999	1	–
Tennessee	4,896,641	9	–
Texas	17,059,805	30	+3
Utah	1,727,784	3	–
Vermont	564,964	1	–
Virginia	6,216,568	11	+1
Washington	4,887,941	9	+1
West Virginia	1,801,625	3	–1
Wisconsin	4,906,745	9	–
Wyoming	455,975	1	–

*Total population, not including the District of Columbia.

Useful Mathematical Data

SYMBOLS

Symbol	Meaning	Symbol	Meaning
$=$	is equal to	%	percent
$\neq$	is not equal to	$0.\overline{3}$	repeating decimal
$\cong$	is congruent to	π	pi: $\approx \frac{22}{7}$ or 3.14
$\sim$	is similar to	$\sqrt{}$	square root
$\approx$	is approximately equal to	$\lvert x \rvert$	absolute value of x
$\parallel$	is parallel to	$\overleftrightarrow{AB}$	line AB
$\perp$	is perpendicular to	$\overline{AB}$	line segment AB
$>$	is greater than	$\overrightarrow{AB}$	ray AB
$<$	is less than	$\angle ABC$	angle ABC
$\geq$	is greater than or equal to	$\llcorner$	right angle
$\leq$	is less than or equal to	$^\circ$	degrees
( )	parentheses: "Do this operation first."	$A \rightarrow A'$	point A maps onto point A'
3^2	exponent (2); base (3)	$(1, -2)$	coordinates of a point where $x = 1$ and $y = -2$

PRIME NUMBERS TO 499

2	3	5	7	11	13	17	19
23	29	31	37	41	43	47	53
59	61	67	71	73	79	83	89
97	101	103	107	109	113	127	131
137	139	149	151	157	163	167	173
179	181	191	193	197	199	211	223
227	229	233	239	241	257	263	269
271	277	281	283	293	307	311	313
317	331	337	347	349	353	359	367
373	379	383	389	397	401	409	419
421	431	433	439	443	449	457	461
463	467	479	487	491	499		

FORMULAS/USEFUL EQUATIONS

Geometric

l = length $\quad w$ = width
b = base $\quad h$ = height
s = side $\quad r$ = radius
d = diameter
B = area of the base of a three-dimensional figure

Perimeter: P

Rectangle: $P = 2 \times l + 2 \times w$
Square: $P = 4 \times s$
Parallelogram: $P = 2 \times (s_1 + s_2)$
 or $P = 2 \times s_1 + 2 \times s_2$
Rhombus: $P = 4 \times s$
Regular Polygon with n sides: $P = n \times s$
Triangle: $P = s_1 + s_2 + s_3$
Circle (Circumference): $C = \pi d$ or $C = 2\pi r$

Area: A

Rectangle: $A = l \times w$
Parallelogram: $A = b \times h$
Trapezoid: $A = \frac{1}{2} \times h \times (b_1 + b_2)$
Triangle: $A = \frac{1}{2} \times b \times h$
Circle: $A = \pi r^2$
Square: $A = s^2$

Surface Area: SA

Rectangular Prism:
 $SA = 2 \times (l \times w + l \times h + w \times h)$
Cube: $SA = 6 \times s^2$
Cylinder: $SA = 2\pi rh + 2\pi r^2$
Sphere: $SA = 4\pi r^2$
Cone: $SA = \pi rs + \pi r^2$

Volume: V

Rectangular Prism:
 $V = l \times w \times h$ or $V = B \times h$
Triangular Prism: $V = B \times h$
Cube: $V = s^3$
Cylinder: $V = \pi r^2 \times h$
Sphere: $V = \frac{4}{3}\pi r^3$
Rectangular Pyramid: $V = \frac{1}{3} \times B \times h$
Cone: $V = \frac{1}{3} \times \pi r^2 \times h$

Pythagorean Theorem

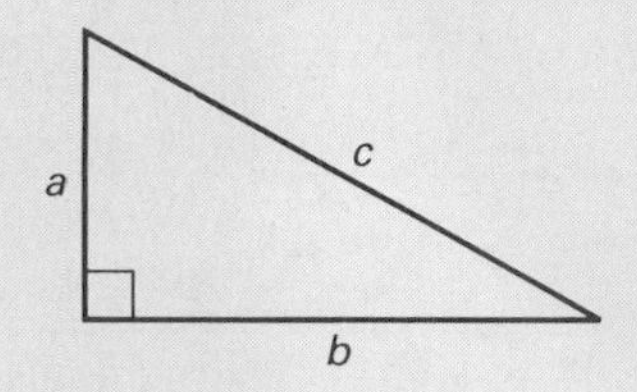

In a right triangle, as shown,
 $a^2 + b^2 = c^2$ or $c = \sqrt{a^2 + b^2}$

Temperature

C: degrees Celsius
F: degrees Fahrenheit

$C = \frac{5}{9}(F - 32)$

$F = \frac{9}{5}C + 32$

Distance

d: distance
r: rate
t: time

$d = r \times t$
$r = \frac{d}{t}$
$t = \frac{d}{r}$

Interest

I: interest
p: principal
t: time (in years)
A: total amount
r: rate

$I = p \times r \times t$
$A = p + p \times r \times t$

Laws of Exponents

$a^m \times a^n = a^{m+n}$
$(a^m)^n = a^{m \times n}$
$\frac{a^m}{a^n} = a^{m-n}$

MEASUREMENT: METRIC UNITS

Length

1 centimeter (cm) = 10 millimeters (mm)
10 centimeters or
100 millimeters = 1 decimeter (dm)
10 decimeters or
100 centimeters = 1 meter (m)
1,000 meters = 1 kilometer (km)

Area

100 square millimeters (mm^2) = 1 square centimeter (cm^2)
10,000 square centimeters = 1 square meter (m^2)
100 square meters = 1 are (a)
10,000 square meters = 1 hectare (ha)

Volume

1,000 cubic millimeters (mm^3) = 1 cubic centimeter (cm^3)
1,000 cubic centimeters = 1 cubic decimeter (dm^3)
1,000,000 cubic centimeters = 1 cubic meter (m^3)

Capacity

1,000 milliliters = 1 liter (L)
1,000 liters = 1 kiloliter (kL)

Mass

1,000 milligrams (mg) = 1 gram (g)
1,000 grams = 1 kilogram (kg)
1,000 kilograms = 1 metric ton (T)

Temperature

0°C = freezing point of water
37°C = normal body temperature
100°C = boiling point of water

MEASUREMENT: CUSTOMARY UNITS

Length		
12 inches (in.)	=	1 foot (ft)
3 feet or 36 inches	=	1 yard (yd)
1760 yards or 5280 feet	=	1 mile (mi)
6,076 feet	=	1 nautical mile
Area		
144 square inches (in.^{2})	=	1 square foot (ft^2)
9 square feet	=	1 square yard (yd^2)
4,840 square yards	=	1 acre (A)
Volume		
1,728 cubic inches (in.^{3})	=	1 cubic foot (ft^3)
27 cubic feet	=	1 cubic yard (yd^3)
Capacity		
8 fluid ounces (fl oz)	=	1 cup
2 cups	=	1 pint (pt)
2 pints	=	1 quart (qt)
4 quarts	=	1 gallon (gal)
Weight		
16 ounces (oz)	=	1 pound (lb)
2,000 pounds	=	1 ton (T)
Temperature		
32°F	=	freezing point of water
98.6°F	=	normal body temperature
212°F	=	boiling point of water

Metric/Customary Comparisons

5 centimeters is about the same length as 2 inches.
1 meter is slightly longer than 1 yard.
5 kilometers is about the same length as 3 miles.

Selected Answers

CHAPTER 1 EXPLORING DATA

Section 1-1, pages 6-7

Exercises **1.** advantage: these are the people who play the games; disadvantage: may name only those games available in that arcade **3.** advantage: convenience, some or all of these people may play arcade games; disadvantage: some or all of these people may not play arcade games **5.** The interviewer could ask every tenth person who passes. This should be done on several different days. **7.** Answers will vary. **9.** Answers will vary. **11.** Answers will vary.

Problem Solving Applications **1.** 2 **3.** sports and music videos; mysteries and soap operas; mysteries and news; music videos and comedies **5.** Answers will vary.

Section 1-2, pages 10-11

Exercises **1.** 36 **3.** 2; 1; 0

5. 2|5 represents 2.5 mi.

Stem	Leaves
0	8
1	8 7 5 3 3 1
2	5 4 3 2 1 0
3	5 3 2
4	2 0
5	3

7. 2|3 represents 230 voters.

Stem	Leaves
1	5 5
2	9 7 0 9 3 4 9 9 8 3 9
3	9 2 5 4 8 0 5 0
4	6 4 0 2 1 1 1 9 1
5	8 2 3 3 2 3
6	2 0 2 7 9

9. Answers will vary.

Problem Solving Applications

1.

Age at Inauguration		Age at Death
	9	00
	8	3508
	7	3891470128
4912054811	6	785463707034
265141561455046201478777	5	36687
3279689	4	96

3. youngest–T. Roosevelt, 42; oldest–Ronald Reagan, 69 **5.** Answers will vary. **7.** Answers will vary.

Section 1-3, pages 14-15

Exercises **1.** rock **3.** 200 **5.** Answers will vary. **7.** 8 **9.** 60

11. The graph should show the following: Kelly, 6 symbols; Green, $6\frac{1}{2}$ symbols; Currie, 3 symbols; Smith, 4 symbols; Charron, $4\frac{3}{4}$ symbols **13.** Smith **15.** excellent **17.** 170 19. Answers will vary. **21.** Answers will vary.

Section 1-4, page 17

Problems **1.** 20 classes **3.** \$315 **4.** \$937

Section 1-5, pages 18-19

Problems Answers will vary. Samples are given. **1.** 500 or 1,000 (performances) **3.** 400 or 500 (camps) **5.** 0.50 or 1.00 (dollars) **7.** 25 or 30 (millions of yd^3)

Section 1-6, pages 22-23

Exercises **1.** male, 15–18 **3.** 800 **5.** 2,200

7.

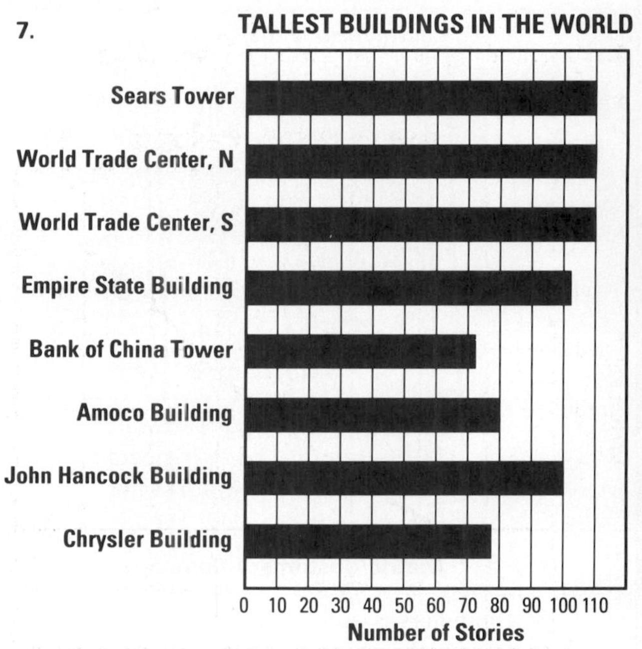

9. Hudson, Mediterranan, Arctic, Black **11.** Answers will vary. **13.** about $2\frac{1}{2}$ times lower **15.** about $14\frac{1}{2}$ h

Section 1-7, pages 24-25

Problems **1.** 1955; 1959 **3.** about 4 times **5.** tobogganing

Section 1-8, pages 28-29

Exercises **1.** \$125 **3.** 1985

5.

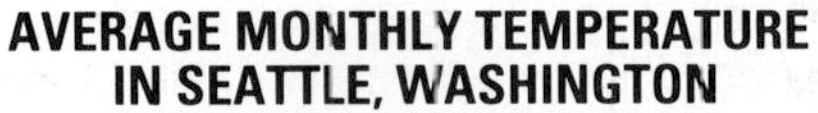

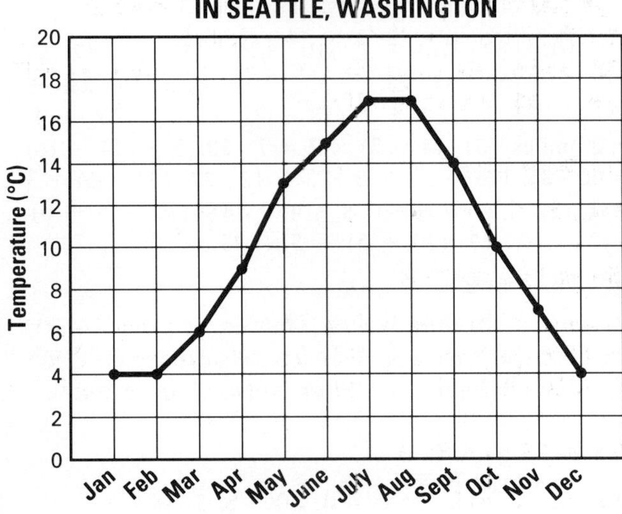

7. 30 kg; 45 kg **9.** from birth to 3 mo; 15 kg

11.

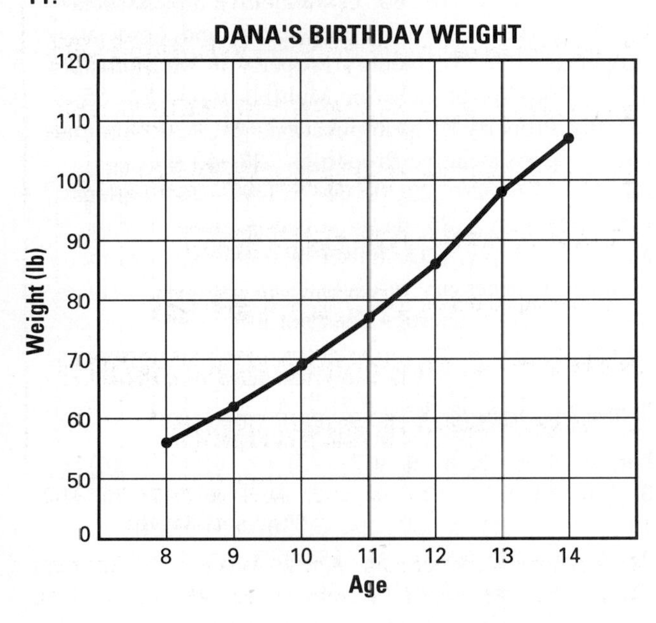

13. between her twelfth and thirteenth birthdays **15.** about 25 calories **17.** The data could be used in helping to determine the proper amount of exercise needed to gain, lose, or maintain weight. **19.** 79 years

Section 1-9, pages 31-33

Exercises **1.** 9.7; 9.05; 8.7 and 9.2 **3.** mean **5.** mean or median **7.** \$69,620.83; \$62,250; none **9.** 8.8; 9.1; 7.0 **11.** 191 **13.** 5,730 **15.** 344 **17.** 9,442 **19.** Answers will vary. **21.** Each would be 5 min longer **23.** mean; median **25.** 159 lb **27.** The mean and the median are the same. **29.** no **31.** yes

CHAPTER 2 EXPLORING WHOLE NUMBERS AND DECIMALS

Section 2-1, pages 44-45

Exercises **1.** 30 **3.** 700 **5.** 1,000

	Actual	Estimate
7.	5,972	6,000
9.	8,448	8,000
11.	4,652	5,000
13.	29	30
15.	81	80
17.	61	60

19a. 16,000 people **b.** 2,000 people **21.** 45,000; C **23.** 11,000; A **25.** 9,000; C **27.** \$498,000,000 **29.** Answers will vary. Sample: Yes. Rounded to the nearest hundred million, 300,000,000 + 700,000,000 + 700,000,000 = 1,700,000,000. This is close to the answer, so the answer is reasonable.

Estimate:

31. a. $\frac{90 + 80 + 80 + 90 + 80}{5} = 84$; yes

b. median: 85; mode: 75

33. b

Section 2-2, pages 48-49

Exercises **1.** $\frac{n}{2}$ **3.** $6n$ **5.** $n + 7$ **7.** $10n$ **9.** 7 **11.** 3 **13.** 3 **15.** 3 **17.** 27 **19.** 31 **21.** 10 **23.** 4 **25.** 0 **27.** 448 **29.** 3 **31.** 2 **33.** w+9 **35.** $25p$ **37a.** $c \div 8$ or $\frac{c}{8}$ **b.** 33 **39a.** $65 + d$ **b.** 79 **41.** 3 **43.** 11 **45.** 6 **47.** 5 **49.** 25 **51.** 5 **53a.** $3d + 4$ **b.** 79 **55a.** $x + 2$ **b.** $x - 2$ **c.** $x + 1$ **d.** $x - 1$ **e.** $x + x + 1$ **f.** $x + x + 2$ **57.** Answers will vary.

Section 2-3, pages 50-51

Problems **1.** subtract **3.** add; 211 people **5.** multiply; 42 cents **7.** add; 38,321 worms **9.** divide; 8 people

11. divide; 12 cartons **13.** multiply, multiply; 7,200 breaths **15.** 100 coconuts **17.** 50 porpoise teeth

Section 2-4, pages 53–55

Exercises **1.** 187.84 **3.** 1,145.45 **5.** 4,377.9 **7.** 806.646 **9.** $66.65

11.

1890	$0.50
1910	$0.60
1930	$1.16
1950	$2.10
1970	$3.12
1975	$4.31

13. $24.80 **15.** 9,3,6 **17.** 8,6,6,7 **19.** 2,2,5 **21.** $61.51

Problem Solving Applications **1a.** $370.08 **b.** $231.00 **c.** $221.31 **3.** $274.13 **5.** 2,500 pesos **7.** $14.57

Section 2-5, pages 57–59

Exercises **1.** B **3.** B **5.** 26 **7.** 86 **9.** 28.8 **11.** 20.7 **13.** 23.53 **15.** B **17.** B **19.** 0.124 **21.** 0.121 **23.** 0.12 **25.** 1.5 **27.** $6.92 **29.** 326.7 **31.** 1.805 **33.** 3.30 **35.** $49, 583.98 **37.** 132.42 **39.** $6,833.36 **41.** True

Section 2-6, pages 62–63

Exercises **1.** 3^3 **3.** 6^4 **5.** 16 **7.** 36 **9.** 6.25984×10^5 **11.** 8.92×10^2 **13.** 63,900 **15.** 300 **17.** No; $9 \neq 8$ **19.** 85 **21.** 89 **23.** 248 **25.** 87 **27.** 62,748,517 **29.** 1,452 **31.** 2^5 **33.** 7^3 **35.** 7 hours

Problem Solving Applications **1.** b **3.** c **5.** 1.39×10^{11}; 139,000,000,000 **7.** 1.0941899×10^{20} **9.** 3.6379788×10^{26} **11.** 2.39245×10^{20}

Section 2-7, pages 66–67

Exercises **1.** 10^{29} **3.** 5^8 **5.** n^6 **7.** m^{12} **9.** 10^2 **11.** 2^3 **13.** b^8 **15.** y^2 **17.** 4^{560} **19.** 5^{63} **21.** a^{30} **23.** n^{49} **25.** 10^5 cm **27.** 4^{150} **29.** 8^{27} **31.** 7^{12} **33.** 12^{25} **35.** 8^{30} **37.** 10^6 **39.** $<$ **41.** $=$ **43.** $=$ **45.** 3.5 shows 10 times more ground movement than 2.5. **47.** 5.5 shows 1,000 times more ground movement than 2.5. **49.** 15 **51.** 4 **53.** $e = 2254$ **55.** $g = 1$ **57.** $i = 30$

Section 2-8, pages 70–71

Exercises **1.** 21 **3.** 81 **5.** 1 **7.** 62

9. 28
First key sequence:
4 [×] 4 [=] [MR]
2 [×] 6 [+] [MR] [=]
Second key sequence:
4 [x^y] 2 [+] 2 [×] 6 [=]

11. $(10 + 3 \times 10) \div 4$, $10 **13.** 1 **15.** 15 **17.** 35 **19.** $25 - (6 \times 4)$ **21.** $5^2 \times (32 - 28) \div 20$ **23.** $-$ **25.** $+$ **27.** $\div$ **29.** $4 + (18 + 22 + 10 + 26) + 4 = 23$ miles **31.** $(4 - 3) \times 7 = 7$ **33.** $5 \times (24 - 16) = 40$ **35.** $(16 + 2^2) \div 5 \times 3 = 12$ **37.** $(32 - 6) \div 3 = 1$ **39.** $7 + 3 + -1 \times 1 = 9$ **41.** $(10 + 5 + 5) \div 2 = 10$ **43.** $(30 + 5) \div 5 + 23 = 30$

45. $3 + 5 \times 4 - 1 = 22$
or $3 + (5 \times 4) - 1 = 22$
$(3 + 5) \times 4 - 1 = 31$
$(3 + 5) \times (4 - 1) = 24$
$3 + 5 \times (4 - 1) = 18$

Section 2-9, pages 74–75

Exercises **1.** 8.5 **3.** 963 **5.** 2.58 **7.** 0 **9.** 2 **11.** 6 **13.** 60; Identity Property of Addition **14.** 73; Commutative and Associative Properties **15.** 113; Associative Property **16.** 60; Commutative and Associative Properties **17.** 81; Commutative and Associative Properties **18.** 24; Identity Property of Multiplication **19.** 0; Property of Zero for Multiplication **20.** 55; Identity Property of Multiplication **21.** 820; Commutative and Associative Properties **23.** 40 socks **25.** 720 **27.** 320 **29.** 0 **31.** $21.99 **33.** Explanations may vary. **35.** True. It can be used only to add and multiply. **37.** False. You can use a property as often as necessary to evaluate an expression since the use of the properties does not affect the value of an expression.

Section 2-10, pages 78–79

Exercises **1.** 6 **3.** 8 **5.** $9(7 - 3)$ **7.** $14(12 + 16)$ **9.** 130 **11.** 648 **13.** 594 **15.** $24 + 6b$ **17.** $7d + 7$ **19.** $3f - 6$ **21.** 5,580 **23.** 2,730 **25.** 3,150 **27.** 6,480 **29.** 5,940 **31.** 154,380 **33.** 22,055 **35.** 98,892 **37.** 316,100 **39.** $25 \times 198 = 25(200 - 2) = 5{,}000 - 50 = 4{,}950$ packages **41.** Answers will vary. **43.** Yes **45.** True. It is used only with multiplication over addition or subtraction. **47.** True. $ab - ac = ba - ca$. The order of the variables as factors in a product does not affect the value of the product.

Section 2-11, pages 80–81

Problems **1.** a; 0.32, 0.40 **3.** b; 10.765, 11.876 **5.** Subtract 0.2; 0.9, 0.7 **7.** Add 4, subtract 0.1; 8.7; 12.7 **9.** Subtract 1, add 0.20; 7.74, 6.74

11. a. $38 **b.** $258

J	F	M	A	M	J	J	A	S	O	N	D
5	8	11	14	17	20	23	26	29	32	35	38

13. 240 miles

1st	2nd	3rd	4th	5th	6th	7th	8th
9	15	21	27	33	39	45	51

15a. 34, 55, 89 **b.** Answers will vary.

CHAPTER 3 REASONING LOGICALLY

Section 3-1, pages 91–93

Exercises **1.** Lines l and m are parallel. **3.** Line segments *DE* and *FG* have the same length. **5.** Viewed from one perspective, the face of an old woman is seen. From another perspective, the side of the head of a young woman is seen. **7.** Answers will vary.

Problem Solving Applications **1–5.** Only 5 is possible. **7.** The connection between the top front cube and the bottom cube at back is impossible.

Section 3-2, pages 96–97

Exercises **1.** 4 **3.** odd **5.** even **7.** The result is always 5. **9.** The digit will always be 9. **11.** If the original number is abc, the final number is abc, abc. **13.** Answers will vary. This conjecture is always true. **15.** yes; no

Section 3-3, pages 100–101

Exercises **1.** If the number is divisible by 9, then the number is divisible by 3. If the number is divisible by 3, then the number is divisible by 9. **3.** valid **5.** valid **7.** If a number is even, then it is divisible by 2. **9.** If a creature is a whale, then it is a mammal. **11.** valid **13.** If a number is a whole number, then it is a rational number.

15. If an animal is a fish, then it is a sea creature.
A shark is a fish.
Therefore, a shark is a sea creature.

17. If a canary is well fed, then it will sing loud.
If a canary sings loud, then it is not melancholy.
Therefore, if a canary is well fed, then it is not melancholy.

Section 3-4, page 103

Problems **1.** 30 students play hockey and basketball. 30 − 12 = 18 **3.** 33 **5.** fall, 8; winter, 7; spring, 9 **7.** 20 **9.** 144

Section 3-5, page 105

Problems **1.** Tillie **3.** Sue married Tom; Betty married Harry; DeeDee married Bill. **5.** Armand is the potter. Belinda is the violinist. Colette is the writer. Dimitri is the painter.

Section 3-6, pages 107–109

Exercises **1.** $2.50 **3.** 20 posts **5.** Fill the 13-gallon container and from it fill the 5-gallon container. You now have 8 gallons left in the 13-gallon container. Pour it into the 11-gallon container for safekeeping. Repeat the process with the 13- and 5-gallon containers to divide up the remaining water. You will end up with 8 gallons of water in each of the 11-, 13-, and 24-gallon containers.

7.

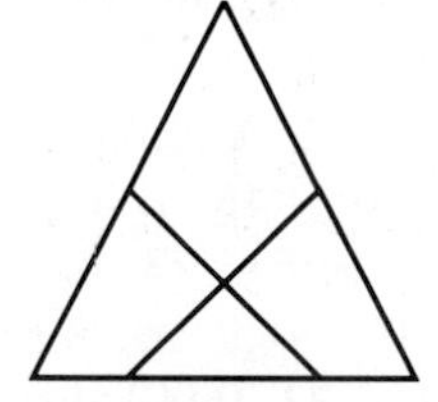

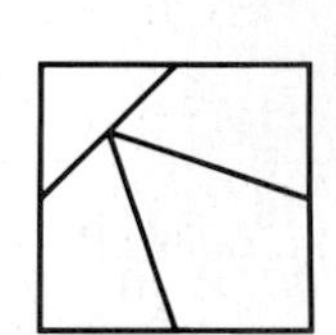

9. 27 cubes **11.** 6 cubes **13.** 8 cubes

15.

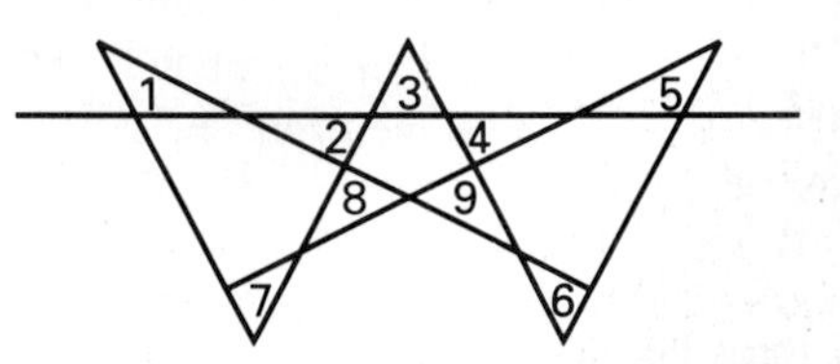

17. He first takes the goat across and returns alone. Next, he takes the cabbage across and returns with the goat. Then he leaves the goat and takes the wolf across and returns alone to get the goat. At no time are the potential enemies alone.

CHAPTER 4 EXPLORING FRACTIONS

Section 4-1, pages 120–121

Exercises **1.** $2 \times 2 \times 2 \times 13$, or $2^3 \times 13$ **3.** $2 \times 3 \times 5 \times 5$, or $2 \times 3 \times 5^2$ **5.** 2 **7.** 3 **9.** 2 **11.** 8 **13.** 28 **15.** 48 **17.** 28 **19.** 180 **21.** 12th person **23.** 5, 7 **25.** 37, 41, 43, 47, 53, 59, 61 **27.** 2 **29.** $3ab$ **31.** $3a^6b^3$ **33.** 60 **35.** $12a^2b^2$ **37.** $6a^2b^2$ **39.** 2 friends **41.** It is odd. **43.** It is the product of two numbers.

Section 4-2, pages 124–125

Exercises **1.** Samples: $\frac{2}{6}$, $\frac{3}{9}$ **3.** Samples: $\frac{10}{24}$, $\frac{15}{36}$ **5.** Samples: $\frac{6}{9}$, $\frac{2}{3}$ **7.** Samples: $\frac{15}{18}$, $\frac{10}{12}$ **9.** $\frac{1}{2}$ **11.** $\frac{1}{3}$ **13.** $\frac{9}{10}$ **15.** $\frac{2}{3}$ **17.** $<$ **19.** $=$ **21.** $<$ **23.** $\frac{1}{3}$ **25.** $\frac{3}{4}$ **27.** $\frac{1}{3}$, $\frac{2}{3}$, $\frac{5}{6}$ **29.** $\frac{21}{50}$ **31.** $\frac{3}{50}$ **33.** $\frac{51}{100}$ **35.** $\frac{9}{100}$ **37a.** $\frac{1}{4}$ **b.** $\frac{3}{4}$ **39.** $\frac{3}{4}$ **41.** $\frac{3}{5}$ **43.** 1

Section 4-3, pages 128–129

Exercises **1.** $\frac{7}{10}$ **3.** $\frac{233}{250}$ **5.** $\frac{29}{500}$ **7.** $8\frac{1}{20}$ **9.** 0.9 **11.** 0.417 **13.** 0.25 **15.** 0.375 **17.** 1.4 **19.** 4.875 **21.** $0.8\overline{3}$ **23.** $0.\overline{5}$ **25.** $0.\overline{8}$ **27.** $0.\overline{3}$ **29.** $>$ **31.** $>$ **33.** Christine and her coach are both correct since $\frac{60}{75} = \frac{4}{5}$ 0.8 **35.** 0.25, $\frac{3}{7}$, 0.561, 0.781, $\frac{4}{5}$, $\frac{7}{8}$ **37.** $>$ **39.** Mona **41.** Otis **43.** Answers may vary. Sample: $\frac{3}{4}$, $\frac{6}{8}$, $\frac{9}{12}$, $\frac{75}{100}$ **45.** between 0.8 and 0.832

Section 4-4, pages 132–133

Exercises **1.** $\frac{1}{64}$ **3.** $\frac{1}{9}$ **5.** 8^{-6} **7.** 2^{-2} **9.** 0.0071 **11.** 0.0000091302 **13.** 2.5×10^{-3} **15.** 1.96×10^{-12} **17.** 10^{-15} **19.** $\frac{1}{1{,}048{,}576}$ **21.** $\frac{1}{2{,}562{,}890{,}625}$ **23.** 6.2×10^{-5}, 6.27×10^{-5}, 7.3×10^{-4}, $\frac{1}{1000}$ **25.** 8.668×10^{-8}, 10^{-7}, $\frac{1}{10^6}$, 8.886×10^{-6} **27.** 0.001 does not belong. It is equal to $\frac{1}{1000}$, not $\frac{1}{10{,}000}$. **29.** equal **31.** Picometer. A picometer $= 10^{-12}$, which is greater than 10^{-15}. **33.** 5^{-3} **35.** 2^{-6} or 4^{-3} or 8^{-2} **37.** $\frac{1}{2^2}$ **39.** $6.2193 \times 10{,}000$ **41a.** neutron **b.** 2×10^{-30} kg **43.** $>$ **45.** $<$

Section 4-5, pages 136–137

Exercises **1.** $\frac{2}{3}$ **3.** 18 **5.** $5\frac{17}{56}$ **7.** $6\frac{3}{10}$ **9.** $1\frac{1}{3}$ **11.** 20 **13.** $1\frac{23}{25}$ **15.** $3\frac{3}{26}$ **17.** 11 hours **19.** $34\frac{29}{50}$ tons **21.** 36 students **23.** 8 **25.** $\frac{1}{4}$ **27–29.** Exercise 29 cannot be true because the product of $\frac{3}{5} \times 24$ is a mixed number, not a whole number.

Problem Solving Applications

1. DOUGH
$\frac{1}{8}$ c warm water
$\frac{1}{2}$ t dry yeast
$\frac{1}{4}$ t sugar
$1\frac{3}{8}$ c all-purpose flour
$\frac{3}{4}$ t salt
$\frac{3}{8}$ c cold water
$1\frac{1}{2}$ T olive oil

3. CHEESE TOPPINGS
$\frac{3}{8}$ lb sliced or grated mozzarella cheese
$\frac{1}{6}$ to $\frac{1}{4}$ c grated Parmesan cheese

5. SAUCE
2 c tomato sauce
2 6-oz can tomato paste
2 T olive oil
4 T water
2 t oregano
1 t salt
1 t sugar
$\frac{1}{4}$ t crushed dried hot red chili peppers

7. 18 pizzas **9.** The amount of each ingredient should be multiplied by $1\frac{1}{2}$. **11.** Unless the recipe is very large, it usually is difficult to divide and still measure accurately. Janet would have a hard time measuring $\frac{1}{24}$ t, so in this case it is probably not a reasonable endeavor.

Section 4-6, pages 139–141

Exercises **1.** $\frac{3}{4}$ **3.** $\frac{1}{5}$ **5.** $\frac{1}{24}$ **7.** $\frac{11}{15}$ **9.** $2\frac{1}{10}$ **11.** $9\frac{1}{5}$ **13.** $8\frac{13}{24}$ **15.** $4\frac{5}{12}$ **17.** $6\frac{7}{8}$ lb **19.** $1\frac{1}{12}$ **21.** $1\frac{7}{24}$ **23.** Answers may vary. Sample answer: Yard wastes, glass, metals, textiles, other, and household hazardous waste total $\frac{40}{100} = \frac{4}{10}$ **25.** 4 h 10 min **27.** 10 h 25 min **29.** 7 **31.** $\frac{1}{8}$ **33.** $\frac{1}{4}$

Problem Solving Applications **1.** eighth note **3.** eighth note **5.** sixteenth note **7.** quarter note **9.** half note **11.** quarter note; quarter note; half note; half note **13.** quarter note; eighth note; quarter note

Section 4-7, page 143

Problems **1.** R **3.** R **5.** NR **7.** R **9.** R **11.** no **13.** no **15.** No. Answers may vary. Sample answers:

6 balloons and 12 favors; 8 balloons and 8 favors
17. yes **19.** about 4 km

Section 4-8, page 145

Problems **1.** 58 lb **3.** 1,350 h **5.** 277.2 mi **7.** $19.82
9. $16.56 **11.** $317.32

CHAPTER 5 MEASUREMENT AND GEOMETRY

Section 5-1, pages 156–157

Exercises **1.** 2 **3.** 7,000 **5.** 1, 1 **7.** 15, 1 **9.** fl oz
11. gal **13.** oz **15.** 8 yd **17.** b **19.** a **21.** 9,700 lb
23. 48 steps **25.** 12 pint jars

Section 5-2, pages 160-161

Exercises **1.** 8,000 **3.** 2,000 **5.** 4.783 **7.** 4 **9.** m
11. L **13.** kg **15.** 7,000 mL **17.** km **19.** cm
21. g **23.** kg **25.** L **27.** 5,013 m **29.** 1,250 mL

31.

Birthday	16	17	18	19	20	21
Grams of Gold	10	20	40	80	160	320
Value of Gold	$200	$400	$800	$1,600	$3,200	$6,400

$ 200
400
800
1,600
3,200
+6,400
$12,600

Section 5-3, pages 164–165

Exercises **1.** 11 lb 5 oz **3.** 5 yd 2 ft **5.** 9 c 4 fl oz
7. 2 yd 2 ft **9.** 3 **11.** 320 **13.** 175 **15.** 650
17. Yes; 2 lb = 32 oz and 3 × 12 oz = 36 oz
19. 4 gal 2 qt **21.** 2 ft 6 in **23.** $4\frac{2}{3}$ ft **25.** $12\frac{1}{2}$ qt
27. $4\frac{3}{4}$ lb **29.** 10 lb 12 oz **31.** 4 oz

Section 5-4, pages 167–169

Exercises **1.** A: 14 ft B: 24 ft C: 28 ft D: 28 ft E: 42 ft
F: 16 ft G: 24 ft H: 50 ft I: 62 ft **3.** 90 yd **5.** 165 ft
7. 468 m **9.** 214 m **11.** 83.4 m **13.** 424 mm or
42.4 cm **15.** $7\frac{2}{3}$ yd **17.** 4.9 m **19.** 12 m
21. 12 m **23.** 17 m, 5.9 m, 5.9 m

Section 5-[illegible]

Problems **1a.** [illegible]
5. 66 ft **7.** 40 [illegible]
$116.35; dining roo[illegible]
bedroom 2, $71.60; liv[illegible]
23 ft **13.** $51\frac{3}{4}$ ft

Section 5-6, pages 174–175

Exercises **1.** 59.7 cm **3.** 42.4 ft **5.** 132 [illegible]
7. 76.6 mm **9.** 30 cm; 94.2 cm **11.** 1,00[illegible]
3,140 in **13.** 22.3 m **15.** 31.4 cm **17.** 37.6[illegible]
19. circumference of circle; perimeter is 222 mm,
circumference is 348.5 mm **21.** 314 ft **23a.** 125.5 m
b. 251 m **c.** 2,510 m **25.** 301 cm

Section 5-7, page 177

Problems **1.** 16 thumbtacks **3.** 19 thumbtacks
5. 22 thumbtacks **7.** 24 thumbtacks **9.** 1 photograph **11.** 9 photographs **13.** 25 photographs
15. 49 photos **17.** 33 yd 1 ft

CHAPTER 6 EXPLORING INTEGERS

Section 6-1, pages 188–189

Exercises **1.** −6 and 6 **3.** −2 and 2 **5.** 3 and −3
7. 34 **9.** 17 **11.** > **13.** < **15.** < **17.** 4,278

19. and 21.

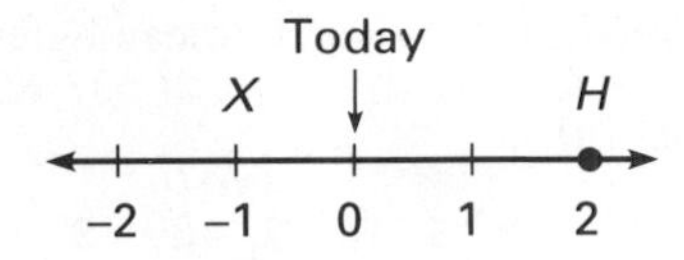

23. −13, −9, 0, 7, 14 **25.** −15, −8, 8, 9, 14
27. False. |7| = |−7|, but 7 > −7. **29.** True.
|6| > |4|, and 6 > 4.

Problem Solving Applications **1.** −10°C to −1°C **3.** +35°C to +40°C **5.** 100°C **7.** 85°F **9.** 0°C **11.** 198°F
13. 80°F **15.** Less. By comparing the thermometer scales, one can see that the equivalent temperature, in degrees Celsius, is about 56°C. **17.** Honolulu, Key West, (tie between Hartford and Boston), Madison, (tie between Atlanta and Helena)

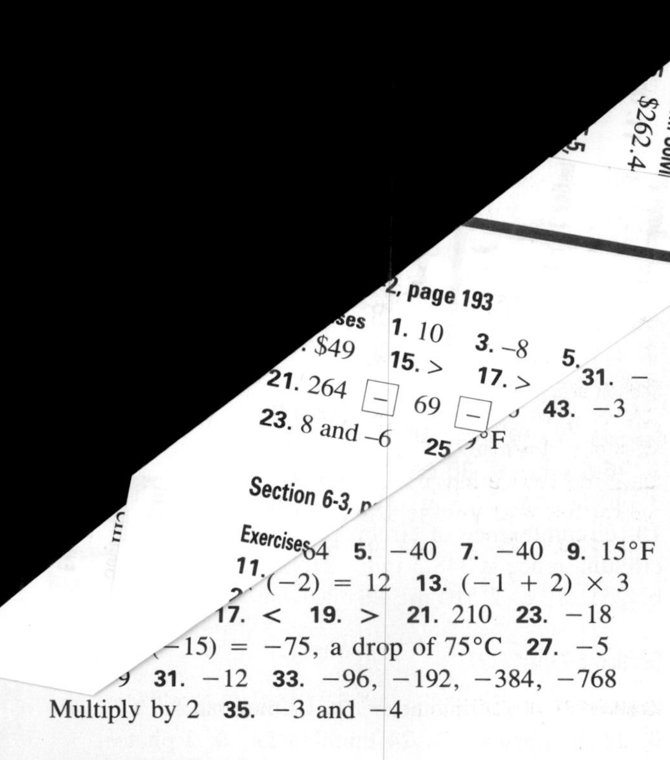

...4 **5.** −40 **7.** −40 **9.** 15°F **11.** ... (−2) = 12 **13.** (−1 + 2) × 3 ... **17.** < **19.** > **21.** 210 **23.** −18 ... (−15) = −75, a drop of 75°C **27.** −5 ...9 **31.** −12 **33.** −96, −192, −384, −768 Multiply by 2 **35.** −3 and −4

Section 6-5, pages 203–205

Exercises **1.** −9 **3.** 4 **5.** −5 **7.** 7 **9.** 7 **11.** 2 **13.** −5 **15.** 12 **17.** decreased 2°F **19.** −8 **21.** −21 **23.** 56 **25.** 64 **27.** about 6.7 times deeper **29.** = **31.** > **33.** < **35.** −5 **37.** 3 **39.** decreased 2.5°F **41.** \$1.70 **43.** 2 **45.** 50 **47.** −22 **49.** −, × **51.** ×, × **53.** ×, ÷ **55.** No. For example, the answer to 6 ÷ 5 is not an integer. **57.** 4 and −4 **59.** Answers may vary. Sample answer is given. Because a zero quotient times a non-zero divisor equals a dividend of zero, but a non-zero quotient times zero as a divisor will not equal any dividend except zero. So zero as a divisor is meaningless.

Section 6-6, pages 206–207

Problems **1.** −10 **3.** −1 **5.** 18 **7.** 189 **9.** −12 **11.** 8 **13.** −\$31 **15.** −540 calories **17.** \$40

Section 6-7, pages 210–211

Exercises **1.** Answers may vary. Sample answers: $\frac{7}{1}$, $\frac{14}{2}$ **3.** Answers may vary. Sample answers: $\frac{10}{3}$, $\frac{20}{6}$ **5.** B **7.** C

9. Intervals chosen for the number line may vary. $-2\frac{2}{3}$ will fall between −3 and −2; $-\frac{2}{3}$ will fall between −1 and 0; $1\frac{1}{3}$ will fall between 1 and 2; $3\frac{2}{3}$ will fall between 3 and 4.

... chosen for the number line may vary. ... between −2 and −1; $-\frac{1}{5}$ will fall ... and 0; $\frac{2}{5}$ will fall between 0 and 1; ... between 1 and 2.

... $\frac{7}{10}$ **15.** $\frac{7}{12}, \frac{2}{3}, \frac{3}{4}$ **17.** Answers will vary. ... answers are $\frac{28}{100}$ and $\frac{56}{200}$. **19.** $-0.6, -\frac{18}{30}$ ... > **23.** > **25.** = **27.** > **29.** $3\frac{1}{5}$ or 3.2 **31.** $-32\frac{1}{8}$ **33.** $-\frac{3}{4}$ **35.** $-\frac{3}{4}$ **37.** false: $1.5 - 2.5 = -1$ **39.** true **41.** false: $\frac{3}{4} + -\frac{3}{4} = 0$

Section 6-8, pages 212–213

Problems **1.** 36 mi, 81 mi, 182.25 mi; sample strategy: draw a picture **3.** the twelfth week; sample strategy: find a pattern **5.** 28 games; 45 games; 300 games; sample strategy: find a pattern **7.** 13 mi; 6.5 mi; 3.25 mi; sample strategy: find a pattern **9.** \$1,200; \$4,800; \$614,400; sample strategy: find a pattern

CHAPTER 7 EQUATIONS AND INEQUALITIES

Section 7-1, pages 224–225

Exercises **1.** open **3.** true **5.** true **7.** 4 **9.** −8 **11.** −12 **13.** 1 **15.** $8\frac{1}{2}$ **17.** 48 **19.** 2 **21.** 8 **23.** −5 **25.** 4 **27.** 28 points **29.** $\frac{1}{2}$ **31.** 1 **33.** 1.5 **35.** 27 nickels **37.** \$72 **39.** 3, −3 **41.** Addie Joss

Section 7-2, pages 228–229

Exercises **1.** add 5 **3.** add 8 **5.** add $\frac{4}{5}$ **7.** 3 **9.** 8 **11.** 4 **13.** $-1\frac{3}{4}$ **15.** 0 **17.** 13 **19.** 9 **21.** −4 **23.** −5.3 **25.** 33.273 **27.** 138,086 **29.** −17°C **31.** Answers may vary. Sample: $x + 3 = 7$ **33.** $x = \pm 3$

Section 7-3, page 231

Problems **1.** c **3.** 58 **5.** \$63 **7.** 20 points

Section 7-4, pages 233–235

Exercises **1.** divide by −8 **3.** multiply by 2 **5.** multiply by 16 **7.** 57 **9.** 392 **11.** 33 **13.** 0.6 **15.** $-\frac{4}{3}$ **17.** 2.3 **19.** $8\frac{3}{4}$ **21.** 2.5 **23.** −18 **25.** −0.4 **27.** 21.3 **29.** 108.24 **31.** −3.6162 **33.** 4.7 **35.** \$0.85 **37.** There is exactly one solution: $n = 0$. **39.** There are an infinite number of solutions, since the product of 0 and any number is 0 and $0 = 0$. **41.** The equation becomes $0 = 0$.

Problem Solving Applications **1.** $12 + x = 14\frac{1}{2}$; $+2\frac{1}{2}$ **3.** $2x = 300$; 150 shares **5.** $\frac{180}{120} = x$; $-1\frac{1}{2}$

Section 7-5, pages 238–239

Exercises **1.** Add 3 **3.** Subtract 8 **5.** 2 **7.** 24 **9.** −42 **11.** 1.3 **13.** 77 **15.** 7 **17.** $\frac{M}{4} + 6 = 15$ **19.** 10 **21.** −2.6 **23.** 229 **25.** 5 **27.** Answers may vary: Sample: $2f - 5 = 1$

Section 7-6, pages 242–243

Exercises **1.** −8 **3.** 10 **5.** $\frac{5}{4}$ **7.** 30 **9.** 16 **11.** $-\frac{2}{3}$ **13.** $4\frac{9}{14}$ h **15.** 3 **17.** $4\frac{1}{2}$ **19.** $-\frac{2}{3}$ **21.** $\frac{1}{2}$ **23.** They are equal; one or the other or both are 0.

Problem Solving Applications **1.** $d = -3$; foal **3.** $p = 1$; squab **5.** $t = 4$; chick

Section 7-7, pages 246–247

Exercises **1.** 24 cm **3.** 11,200 cm³ **5.** 152 chirps per minute **7.** 36 runs **9.** $t = \frac{I}{pr}$ **11.** $i = A - p$ **13.** $C = K - 273$

15. $A = \frac{1}{2}bh$
$2A = bh$
$\frac{2A}{b} = h$

17. $y = mx + b$
$y - b = mx$
$\frac{y - b}{m} = x$

19a. \$95 **b.** \$105 **c.** \$129

Section 7-8, pages 248–249

Problems **1.** \$0.22 **3.** 6:45 a.m. **5.** 19 **7.** \$9,440

Section 7-9, pages 252–253

Exercises **1.** unlike **3.** like **5.** like **7.** like **9.** like **11.** $5s + 5t$ **13.** $10y - 4$ **15.** $6a$ **17.** $9n + 2x$ **19.** $-6x + 3xy$ **21.** $11s + 9p + 7s + 6k$; $18s + 9p + 6k$ **23.** 7 **25.** $\frac{5}{2}$ **27.** −1 **29.** $4x$ **31.** $18a - 1$ **33.** $700k$ **35.** $15n$ **37.** $10a + 12b + 14c$ **39.** $1{,}485x - 81y$ **41.** $4(x + y) + 3(2x - y)$; $10x + y$ **43.** −1 **45.** −32 **47a.** $n + 3$ **b.** $4n + 12$ **c.** $4n + 6$ **d.** $2n + 3$ **e.** 3 **f.** The variable n is eliminated from the calculations in the final step, leaving 3 no matter what the value of n.

Section 7-10, pages 256–257

Exercises **9.** yes; no; no; yes **11.** no; no; yes; yes **13.** yes; yes; yes; no **15.** $r > 300$ **17.** $c < 25$ **19.** $x > 0$ (but answers may vary) **21.** $x = 2$ (but answers may vary) **23.** $f \geq 8$ **25.** $p \geq 50{,}000$ **31.** true **33.** true **35.** false

Section 7-11, pages 260–261

Exercises **1.** $w < 4$ **3.** $x \leq -6$ **5.** $y \leq -3$ **7.** $t > 2$ **9.** $m > -7$ **11.** $x > -2$ **13.** $w > -1$ **15.** $n \geq -2$ **17.** $n \geq -6$ **19.** $y \geq -4$ **21.** $x \geq 12$ **23.** $e \leq 0$ **25.** 15 min **27.** Answers will vary. Sample: Anthony plans to continue going to school until he's twice his present age, then to travel for 4 years. He will still be under 30 years of age. How old is he? **29.** 3, 4, 5, 6, 7, 8 **31.** none

CHAPTER 8 EXPLORING RATIO AND PROPORTION

Section 8-1, pages 272–273

Exercises **1.** 6:9; $\frac{6}{9}$ **3.** 8 to 1; 8:1 **5.** 1 to 3; 1:3 **7.** $\frac{5}{2}$ **9.** $\frac{5}{1}$ **10.** $\frac{16}{3}$ **11.** $\frac{7}{4}$ **13.** Part A: 15; 20; 25 Part B: 6; 12 **15.** $\frac{6}{25}$ **17.** $\frac{13}{100}$ **19.** $\frac{6}{11} > \frac{5}{13}$ **21.** $\frac{4}{9} < \frac{7}{12}$

Problem Solving Applications **1.** Possible answer: The average is a comparison of two numbers. Even though Thomas had fewer at-bats than Brett, the ratio of his at-bats to his hits is about the same as Brett's. **3.** .750

Section 8-2, pages 276–277

Exercises **1.** 50 mi/h **3.** 55 words/min **5.** \$5/lb **7.** \$65/room **9.** 500 km/day **11.** 55 mi/h **13.** 1,498 km in 7 h **15.** 144 students in 3 buses **17.** \$3.96 for 36 oz **19.** 48 mi/gal; about 0.02 gal/mi

Problem Solving Applications **1.** \$0.02/sheet **3.** \$0.73/lb **5.** the half gallon

Section 8-3, pages 280–281

Exercises **1.** $\frac{12}{14}$; $\frac{18}{21}$; $\frac{24}{28}$ **3.** 15:20; 6:8; 9:12 **5.** 10 to 6; 15 to 9; 20 to 12 **7.** $\frac{16}{18}$; $\frac{24}{27}$; $\frac{32}{36}$ **9.** 6 to 20; 9 to 30; 12:40 **11.** no **13.** no **15.** no **17.** yes **19.** yes; 7 to 10 = 28 to 40 **21.** no; $\frac{6}{9} = \frac{2}{3}$; $\frac{24}{45} = \frac{8}{15}$ **23.** 9 to 10 **25.** 3:4 **27.** 6:17:12 **29.** Possible answer: 6:17:12; 60:170:120

Section 8-4, pages 284–285

Exercises **1.** yes **3.** yes **5.** no **7.** $n = 8$ **9.** $x = 18$ **11.** $b = 1$ **13.** $n = 4$ **15.** $y = 16$ **17.** 60 laps **19.** $m = 52$ **21.** $m = 35$ **23.** $x = 45$ **25.** $x = 11$ **27.** $a = 15$ **29.** $y = 1.2$ **31.** 70 yd/min **33.** Answers will vary. Possible answer: $\frac{9}{27} = \frac{35}{105}$ **35.** $w = 60$ ft; $l = 160$ ft

Section 8-5, page 287

Problems **1.** 64 statuettes **3.** 104 photographs **5.** 92 students **7.** 45 sculptures

Section 8-6, pages 289–291

Exercises **1.** 18 m **3.** 15 in. **5.** 11 cm **7.** 1.4 m **9.** 1 m:2.5 mm **11.** 1 in.:10 mi **13.** 1 in.:49 mi **15.** 1 in.:5 ft **17.** Answers will vary.

Problem Solving Applications **1.** 3.75 mi **3.** 22.5 mi **5.** school and post office **7.** 220 km **9.** $1\frac{1}{4}$ in. **11.** 4.6 cm **13.** They both are drawings made so actual dimensions can be determined by using a proportion.

Section 8-7, page 293

Problems **1.** 17 frames **3.** 6 ways **5.** 8 odd numbers **7.** 18 ways

CHAPTER 9 EXPLORING PERCENT

Section 9-1, pages 304–305

Exercises **1.** 0.5 **3.** 0.17 **5.** 0.085 **7.** 0.881 **9.** 0.135 **11.** 0.0575 **13.** $\frac{11}{20}$ **15.** $\frac{1}{5}$ **17.** $\frac{1}{250}$ **19.** $\frac{7}{1000}$ **21.** $\frac{3}{800}$ **23.** $\frac{13}{80}$ **25.** 20% **27.** 79% **29.** 1.5% **31.** 0.5% **33.** $12\frac{1}{2}\%$ **35.** $16\frac{2}{3}\%$ **37.** 12% **39.** 75% **41.** 87.5% **43.** $41\frac{2}{3}\%$ **45.** 15% **47.** 0.05, 5% **49.** 0.4, 40% **51.** 75% **53.** 45% **55.** no **57.** no **59.** $33\frac{1}{3}\%$

Section 9-2, pages 308–309

Exercises **1.** 69.6 **3.** 18.6 **5.** 33 **7.** 63 **9.** 490 **11.** 20 **13.** 48 **15.** 27 **17.** 8.5 **19.** 16.6 **21.** $46.20 **23.** 200 **25.** 40 **27.** 37 or 38 **29.** 845 **31.** 987 **33.** 162.9 **35.** 0.15 **37.** 0.021 **39.** $184.47 **41.** $3,000 to $3,200 **43.** 26% of $7,000 = $1,820 **45.** $5,032 to $5,320. The lowest value is found by subtracting 26% of $6,800 from $6,800; the highest value is found by subtracting 24% of $7,000 from $7,000

Section 9-3, pages 312–313

Exercises **1.** 25% **3.** 34% **5.** 10% **7.** 35% **9.** 100% **11.** 70% **13.** 70%; 69% **15.** $24{,}000x = 6{,}500$; $x = 0.0625 \approx 6.3\%$ $24{,}000y = 22{,}500$; $y = 0.9375 \approx 93.8\%$ They save about 6.3% and spend about 93.8%. **17.** It is halved to 3.1%

Problem Solving Applications **1.** 56.4%; 60.4%; 55.1%; 59.4%; 55.6% **3.** 48.7%; 57.9%; 53.6%; 59.1%; 56.1%; 53.1%; 51.4% **5.** Esiason, Krieg, Marino, Eason, O'Brien

Section 9-4, pages 316–317

Exercises **1.** 10% **3.** 10% **5.** 40% **7.** 25% **9.** 18% **11.** 15% **13.** 15% **15.** 30% **17.** 24% increase **19.** 5% increase **21.** 120% **23.** Northeast, about 3.5% increase (least); West, about 22.2% increase (greatest)

Section 9-5, pages 320–321

Exercises **1.** 156 **3.** 88 **5.** 83 **7.** 25 **9.** 325 **11.** 600 **13.** 72,000 seats **15.** 100 cm^3 **17.** 600 cm^3 **19.** 300 cm^3 **21.** Nevada **23.** Enter 8.5 for R in Line 20.

Section 9-6, page 323

Problems **1.** mental math; $15.25 **3.** calculator; $16.90 **5.** mental math; 67 **7.** calculator; $382.00 or $382.02, depending on method used

Section 9-7, pages 326–327

Exercises **1.** D = $15.05 S = $135.45 **3.** D = $127.65 S = $562.34 **5.** D = $64.44 S = $446.65 **7.** $102.50 **9.** $68.09 **11.** $17.38 **13.** the banjo **15.** No. The 35% reduction represents a greater saving. The additional 15% is based on a sale price lower than the original price.

Problem Solving Applications **1.** $2,550 **3.** 3% **5.** $5,400 **7.** $629.55 **9.** $1,375

Section 9-8, pages 330–331

Exercises **1.** $10.25 $420.25 **3.** $590.63 $2,165.63 **5.** $810 $3,810 **7.** $53.08 $558.58 **9.** $348.75 $1,123.75 **11.** $1,313 $11,413 **13.** $2,240 **15.** $738 **17.** $60.94 **19.** $30,100 **21.** $900 **23.** $1\frac{1}{2}$ y **25.** 12% **27.** 10% **29.** 10.5% = $141,440 over 4 years 3% = $137,892.34 over 4 years

Section 9-9, pages 332–333

Problems **1.** Possible strategy: Work backward; original salary: \$237; current salary: \$236 **3.** Possible strategy: Use logical reasoning; 9 movies **5.** Possible strategy: Guess and check. There are 12 pentominoes. **7.** Possible strategy: Make a table; 10 days **9.** Possible strategy: Work backward; 960 figurines

11. $\begin{array}{r} 5 \\ \times 5 \\ \hline 25 \end{array}$ $\quad$ $\begin{array}{r} 6 \\ \times 6 \\ \hline 36 \end{array}$

CHAPTER 10 GEOMETRY OF SIZE AND SHAPE

Section 10-1, pages 344–345

Exercises **1.** GH **3.** $\angle PQR$ or $\angle RPQ$ or $\angle Q$ **5.** PQ **7.** PR **9.** pentagon **11.** quadrilateral **13.** Answers may vary. At least 4 of these: AC, AB, BC, CA, CB, BA **15.** BA, BC **17.** b **19.** a **21.** f **23.** c **25.** false **27.** false **29.** true **31.** 8 vertices in each combination

Section 10-2, pages 348–349

Try These **9.** $\angle IGH$ **10.** $\angle FDE$ and $\angle IGH$ **11.** $\angle ABC$ **12.** $\angle FDE$ and $\angle ABC$

Exercises **1.** 25°, acute **3.** 140°, obtuse **5.** 115°, obtuse **7.** 30°, acute **15.** $\angle LMP$ and $\angle PMN$ **17.** $\angle LMP$ and $\angle PMN$ **19.** acute, 56° **21.** $\angle CAB$, $\angle EAD$, $\angle EAF$, $\angle GAH$, $\angle IAJ$ **23.** $\angle GAD$, $\angle DAB$, $\angle BAI$, $\angle IAG$, $\angle EAC$, $\angle CAJ$, $\angle JAH$, $\angle HAE$ **25.** false **27.** $m\angle C = m\angle B + 90°$

Section 10-3, pages 352–353

Exercises **1.** 30° **3.** 150° **5.** 110° **7.** 70° **9.** 95° **11.** 95° **13.** $m\angle 1 = 22°$; $m\angle 2 = 68°$; $m\angle 3 = 68°$; $m\angle 3 = 68°$; $m\angle 4 = 90°$ **15.** $m\angle QRN = 50°$; $m\angle PRN = m\angle MRQ = 130°$ **17.** $AB \parallel CD$ because $\angle BAC$ and $\angle DCE$ are corresponding angles.

Section 10-4, page 357

Exercises **1.** acute **3.** obtuse **5.** isosceles **7.** scalene **9.** 57° **11.** true **13.** true **15.** true

Section 10-5, pages 360–361

Exercises **1.** $3 \times 180° = 540°$ **3.** $5 \times 180° = 900°$ **5.** $7 \times 180° = 1{,}260°$ **7.** 3.73 cm **9.** has no diagonals

13.

number of sides	3	4	5	6	7	8	9
number of diagonals	0	2	5	7	14	20	27

15. true **17.** false **19.** Yes.

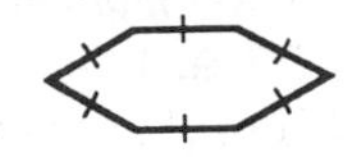

Section 10-6, pages 364–365

Exercises **1.** A polyhedron with two parallel congruent bases that are shaped like triangles. **3.** A right prism having six congruent square faces. **5.** octagonal pyramid **7.** rectangualr prism **9.** hexagonal pyramid **11.** Answers will vary. **13.** Answers will vary.

	Polyhedron	Number of Faces (F)	Number of Vertices (V)	Number of Edges (E)	$F + V - E$
15.	rectangular prism	6	8	12	2
17.	triangular pyramid	4	4	6	2
19.	pentagonal pyramid	6	6	10	2

21. Built in the shape of a pentagon, the Pentagon Building is one of the largest office buildings in the world. **23.** $E = 2t$ **25.** triangular prism

Section 10-7, pages 366–367

Problems **1a.** Answers will vary. **b.** 78° **3a.** Answers will vary. **b.** 9 cm **5a.** Answers will vary. **b.** 59° **7.** false **9.** true

Section 10-8, pages 368–369

Problems **1-2.**

3. rectangle **5.** square **7.** $\frac{1}{2}$ **9.** 12 ways **11.** 2 quarters + 5 dimes, or 3 quarters + 1 dime + 3 nickels **13.** 6 ft

Section 10-9, page 373

Problem Solving Applications

1. The following are the central angles for the graph: water 144°; fat approximately 122°; and protein approximately 94°.

3. The following are the central angles for the graph: hamburger 117°; kidney beans 63°; tomato sauce approximately 34°; tomato paste approximately 34°; seasoning approximately 14°; and water approximately 99°.

Section 10-10, pages 376-377

Exercises **1.** Line segment should be 3.5 cm **3.** Line segment should be 5 cm. **5.** The line of intersection should be 90° and cut half way between Point Q and Point R. **7-9.** Line K should intersect CD at 90°.

13. The distance to each vertex is the same. **15.** the distances are the same.

CHAPTER 11 AREA AND VOLUME

Section 11-1, pages 388-389

Exercises **1.** 20 cm² **3.** 56 cm² **5.** perimeter, 22 in.; area, 20 in.² **7.** 50 cm³ **9.** 96 cm³ **11.** 18 cm³ **13.** The manufacturer can pack 12 square boxes in each carton. **15.** 38 ft² **17.** 24 cm² **19.** 23 cm² **21.** 30 in.² **23.** 11 m³ **25.** 2 bags; Explanations may vary. Sample: Since the area of the region is 3 ft², 2 bags will be needed. **27.** 5 **29.** 34 unit³

Section 11-2, pages 392-393

Exercises **1.** 360 cm² **3.** 1,280 m² **5.** 4,156 ft² **7.** 497 ft² **9.** 26.5 cm² **11.** 600.25 ft² **13.** $594; area = 72 yd² **15.** 320 in.²

Problem Solving Applications **1.** 468 in.² **3.** 324 in.² **5.** 324 in.² **7.** 444 in.² **9.** 336 in.² **11.** 60 in.²

Section 11-3, pages 396-397

Exercises **1.** 900 cm² **3.** 8.06 m² **5.** 24 cm² **7.** 36 mm² **9.** 60 in.² **11.** 500 cm² **13.** 56 yd² **15.** 250 ft² **17.** 18 cm **19.** For Exercise 17, the picture should be of a parallelogram with an indicated length of 18 cm and an indicated altitude of 10 cm **21.** 12.48 cm² **23.** 38.605 m² **25.** 25.8 m² **27.** 209 m² **29.** height, length of the base **31.** 32 cm²

Section 11-4, pages 398-399

Problems **1.** P **3.** A **5.** $P = 2l + 2w$; 43.8 m **7.** $P = 4 \times s$; 1,700 cm **9.** $A = l \times w$; $6,969.60 **11.** 3.120 ft² **13.** 140 ft **15.** 14,020 ft²

Section 11-5, pages 402-403

Exercises **1.** 324 **3.** 441 **5.** 3.6 **7.** 4.41 **9.** $\frac{1}{25}$ **11.** $\frac{64}{81}$ **13.** 16 **15.** −10 **17.** −0.7 **19.** $\frac{2}{9}$ **21.** 3.5 **23.** 9.2 **25.** 6.5 **27.** 2.5 **29.** 12.369 **31.** 31.241 **33.** 10.050 **35.** 28.548 **37.** 4.8 cm **39.** 1,156 **41.** 0.04 **43.** 0.64 **45.** 0.000064 **47.** 8 **49.** 24 **51.** $916.76 **53.** 3.9 s

Section 11-6, pages 406-407

Exercises **1.** 9 **3.** 31.2 **5.** 10 **7.** 34 **9.** 20 cm **11.** 25.5 cm **13.** 30 ft **15.** 3.7 **17.** 5.7 **19.** 10.2 **21.** 23.75 **23.** 2.8 in. **25.** 8.2 m **27.** 14.7 m **29.** 36.2 m **31.** 9.05 m ≈ 9.1 m

Section 11-7, pages 410-411

Exercises **1.** 24 in.³ **3.** 648 cm³ **5.** 149.5 cm³ **7.** 4,704 in.³, 5,040 in.³ D has the greater volume. **9.** 5.52 cm³ **11.** 1,404 in.³ **13.** 13,568.3 m³ **15.** $66.82 **17.** 2 trips **19.** $V = \frac{1}{2}s^3$; 0.864 m³ **21.** 1,890,000 cm³ **23.** 1,890 goldfish **25.** 8,763.3 cm³ or 8.7633 L

Section 11-8, pages 413-415

Exercises **1.** 452.2 cm² **3.** 706.5 ft² **5.** 201.0 ft² **7.** 301.6 in.² **9.** C **11.** B **13.** 379.9 cm² **15.** square **17.** about 1.3 yd² **19.** 153.9 in.² **21.** 201 in.² **23.** 6¢ per in.² **25.** large with two choices, Exercise 24 **27.** 314 in.² **29.** 201.0 ft² **31.** 36.1 in.² **33.** 99.3 yd² **35.** 16,717.2 cm²

Section 11-9, pages 417-419

Exercises **1.** 113 cm³ **3.** 2,703 cm³ **5.** r = 3 ft, V = 226 ft³ **7.** d = 13.8 in., V = 628 in.³ **9.** 7,235 in.³ **11.** Answers will vary. **13.** 2,735 cm³ **15.** 401,593 cm³

Problem Solving Applications **1.** 238.5 km³ **3.** 122,844.2 km³ **5.** A = 1,707.4 km³, B = 1,429.2 km³, $A > B$ **7.** 0.3 km²

Section 11-10, pages 422-423

Exercises **1.** 65.3 cm³ **3.** 7,234.6 ft³ **5.** 19.3 ft³ **7.** 2,143.6 in.³ **9.** 1,316.4 ft³ **11.** 3,052.1 m³ **13.** 1,316.4 mm³ **15.** 198.3 cm³ **17.** 2,142.6 m³ **19.** 5,128.7 cm³ **21.** 8.2 m³ **23.** 473.7 m³ **25.** 510.3 yd³ **27.** 237.6 cm³/1,097.5 cm³; 21.6% **29.** The volume is multiplied by 4.

Section 11-11, pages 426-427

Exercises **1.** 222 cm² **3.** 11 m² **5.** 377 ft² **7.** 314 in.² **9.** 312 in.² **11.** 184 in.² **13.** 808.8 cm² **15.** 289.4 cm² **17.** 181.37 cm² **19.** 912 in.²; 984 in.²; 864 in.² **21.** 803.84 cm²

Section 11-12, page 429

Problems **1.** base 10 in.; height 12 in.; 10 **3.** 6 cm **5.** base 40 yd; height 120 yd **7.** 12 cm

CHAPTER 12 POLYNOMIALS

Section 12-1, pages 440–441

Exercises **1.** $4n^4 + n^3 + 2n^2 + 7n$ **3.** $5z^2 + (-5z) + (-4z) + 4$ **5.** $5n^5 + 2n^3 + (-8n)$ **7.** $3a + (-3)$ **9.** $-5c + (-2)$ **11.** $12r^2 + r$ **13.** $12s^2 + (-2s)$ **15.** $-9h^3 + 22h^2 + (-2)$ **17.** $3s^3 + (-2s^2) + (-5s) + 7$ **19.** $5x^2y$ **21.** $3mn$ **23.** $3z^2 + (-11x^2y^2)$ **25.** $2x^2 + (-x) + 5xy$ **27.** $a^3b + (-2a^2b) + 3ab + 4ab^2 + 8ab^3$ **29.** 2 **31.** 1 **33.** false **35.** None of the other terms are like terms.

Section 12-2, pages 444–445

Exercises **1.** $6y --3$ **3.** $7m - 7$ **5.** $-2a^2 - 1$ **7.** $2a^2 - 2a - 3$ **9.** $5x$ **11.** $4z - 7$ **13.** $3 + k$ **15.** $5p^2 - 2p + 14$ **17.** $17m^3 - m^2 - 5m$ **19.** $3x - y$ **21.** $3a + 3b$ **23.** $13a + b$ **25.** $-x - 6xy$ **27.** $8r^2t - 2rt^2 - 4rt$ **29.** $3a + b$ **31.** $-3z + 2y$ **33.** $9f - 11g - 4$ **35.** $6r^2t^2 + 10rt^2$ **37.** $-a^2bc + 5abc^2 + 2abc - ab^2c$ **39.** xy **41.** $5x + y$ **43.** $4x + y$ **45.** They are opposites of each other. **47.** The sign of each term in the difference is opposite when the order of polynomials is reversed in a subtraction.

Section 12-3, pages 448–449

Exercises **1.** $6ab$ **3.** $-12km$ **5.** $6kp$ **7.** $7cd$ **9.** $4a^2$ **11.** $7y^5$ **13.** $-24p^3$ **15.** $-15k^6$ **17.** $9x^2$ **19.** $-8d^9$ **21.** $81y^3$ **23.** $36c^4$ **25.** $-64b^{12}$ **27.** $6mp$ **29.** $12a^6b^3$ **31.** $-3p^5r^5$ **33.** $x^{21}y^6$ **35.** $4x^{14}y^{30}$ **37.** $36p^{10}r^8$ **39.** $24x^2$ **41.** $\frac{27}{64}a^3$ **43.** $9p^3$ **45.** Any values such that st is negative.

Section 12-4, pages 451–453

Exercises **1.** $2x^2 + 6x^3$ **3.** $6a^2 + 2a$ **5.** $2a^2 + 10a - 2$ **7.** $-3f^2 + 6f + 24$ **9.** $-r^4 + 9r^2 - 6r$ **11.** $4x^3 - 8x^2 + 6x$ **13.** $10x^2 - 5x$ **15.** $6x + 2$ **17.** $-4p + 12$ **19.** $5a - 7b$ **21.** $-x^4 - 2x^3 + 2x^2 + 2x$ **23.** $c^4 - 3c^3d - 6c^2d^2$ **25.** $-2r^4 + 6r^3s^2 - 8rs^3$ **27.** $c(25t + 24) = 25ct + 24c$ **29.** Ganymede **31.** $24m^2 - 12m$ **33.** $x - y$ **35.** $x - y$ **37.** $w + z$

Section 12-5, pages 454–455

Problems **1.** $5n$, $6n$; guess and check **3.** four; $1 + x^6$; $x + x^5$; $x^2 + x^4$; $x^3 + x^3 = 2x^3$; make an organized list **5.** 100, 5,625; solve a simpler problem **7.** 44 minutes fast; make an organized list **9.** $24x^2$; draw a picture **11.** Lums: 6; Mads: 16; Nogs: 14; Ools: 8; use logical reasoning **13.** 204 squares; find a pattern.

Section 12-6, pages 457–459

Exercises **1.** $4y$ **3.** $5ab$ **5.** $5s^2t$ **7.** d **9.** a **11.** c **13.** $9(1 - 2x)$ **15.** $3b(4a - b)$ **17.** $6x(4x - y)$ **19.** $y(x + xy - y)$ **21.** $7xy(1 - 2xy + 4z)$ **23.** $-3a^2$ **25.** $4y^2$ **27.** $\frac{1}{3}xyz(1 - y - xz)$ **29.** $4r(s + u) + 4t(s - u)$ **31.** $x^2(8x - 9)$

Problem Solving Applications **1.** $8a$ **3.** $20n + 2$ **5.** $n(2n + 6) = 2n^2 + 6n$ **7.** $SA = 2\pi r(h + r)$ **9.** $r^2(\pi - 2)$

Section 12-7, page 461

Problems **1.** $n(n + 1) + 3(n + 1) = (n + 3)(n + 1)$ **3.** $(a + 2)(a - 3)$ **5.** $(r + 2)(r + 6)$ **7.** $(c + 1)(c^2 + c + 5)$ **9.** Use the chunk $(m + 1)$. Let $m + 1 = 5$. Then $m = 5 - 1 = 4$ **11.** Since $(y + 2)^2 = 5^2$ write these two equations: $y + 2 = 5$ and $y + 2 = -5$. Then $y = 3$ and $y = -7$.

Section 12-8, pages 464–465

Exercises **1.** $5c$ **3.** $4y$ **5.** $-4p$ **7.** y **9.** cd **11.** ab^2 **13.** $a + 6$ **15.** $3a - 6b$ **17.** $2x^2 - 3x + 1$ **19.** $3abc - 5c$ **21.** $-6b^8$ **23.** $\frac{x}{y}$ **25.** $\frac{z^2}{3}$ **27.** $-6q^4rs$ **29.** $8a^2 - 4b + 6ab^2$ **31.** $2z - 3x + 4xy - 6zx^2y^2$ **33.** $5q$ **35.** $9a$ **37.** $5m^2n^3$ **39.** $90xy^2z^2$ **41.** $x^2 - 4$

CHAPTER 13 PROBABILITY

Section 13-1, pages 475–477

Exercises **1.** $\frac{1}{3}$ **3.** $\frac{1}{2}$ **5.** $\frac{1}{2}$ **7.** $\frac{5}{6}$ **9.** $\frac{1}{6}$ **11.** $\frac{1}{3}$ **13.** $\frac{1}{3}$ **15.** 0 **17.** $\frac{1}{6}$ **19.** $\frac{1}{2}$ **21.** $\frac{2}{3}$ **23.** 1 to 4 **25.** 3 to 2 **27.** 4 to 1 **29.** 3 to 2 **31.** $\frac{3}{4}$ **33.** b **35.** e **37.** b

Section 13-2, pages 479–481

Exercises **1.** (M,A,T,H,F,U,N) **3.** 8 possible outcomes **5.** 6 possible outcomes **7.** $\frac{1}{18}$ **9.** $\frac{2}{9}$ **11.** 36 possible outcomes **13.** 72 possible outcomes **15a.** $\frac{1}{4}$ **b.** $\frac{1}{2}$ **c.** $\frac{1}{4}$

Section 13-3, pages 483–485

Exercises **1.** 12 **3.** 15 **5.** 8 **7.** $\frac{1}{64}$ **9.** 30 **11.** $3 \times 3 \times 3 \times 3 = 81$ **13a.** 3 ways **b.** 2 ways **c.** 1 way **d.** $3 \times 2 \times 1 = 6$; 6 ways

Problem Solving Applications **1.** 24 possible choices **3.** Answers will vary.

Section 13-4, pages 488–489

Exercises **1.** $\frac{1}{30}$ **3.** $\frac{1}{150}$ **5.** $\frac{3}{50}$ **7.** $\frac{4}{75}$ **9.** $\frac{1}{75}$ **11.** $\frac{2}{75}$ **13.** 0 **15.** $\frac{2}{45}$ **17.** $\frac{4}{13}$ **19.** $\frac{24}{91}$ **21.** $\frac{35}{132}$ **23.** $\frac{1}{66}$ **25.** $\frac{1}{6}$ **27.** $\frac{1}{10}$ **29.** pulling out a $20 bill, not replacing it, and then pulling out a $10 bill **31.** pulling out a $10 bill, replacing it, and then pulling out a $5 bill

Section 13-5, pages 492–493

Exercises **1.** $\frac{5}{16}$ **3.** $\frac{1}{4}$ **5.** $\frac{11}{16}$ **7.** $\frac{9}{16}$ **9.** $\frac{1}{6}$ **11.** $\frac{5}{6}$ **13.** $\frac{19}{30}$ **15.** $\frac{1}{10}$ **17.** Answers will vary. **19.** $\frac{31}{60}$ **21.** Answers will vary.

Section 13-6, page 495

Problems **1.** $\frac{4}{25}$ **3.** $\frac{3}{10}$ **5.** 15,000 **7.** 22,500 **9.** 112,500 **11.** 333,750

Section 13-7, page 497

Exercises **1–11.** Answers will vary.

CHAPTER 14 GEOMETRY OF POSITION

Section 14-1, pages 508–509

Exercises **1.** (2,4) **3.** $N(-2,0)$; $L(2,0)$ **5.** $A(2,4)$; $G(5,3)$; $H(1,2)$ **7.** $B(-4,-4)$

9. **11.** **13.** **15.**

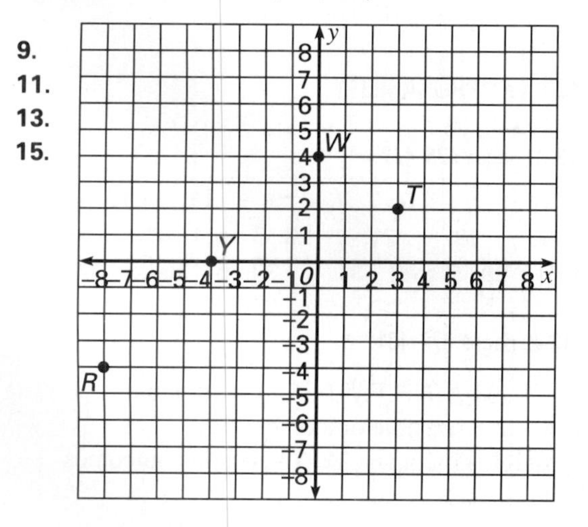

17.

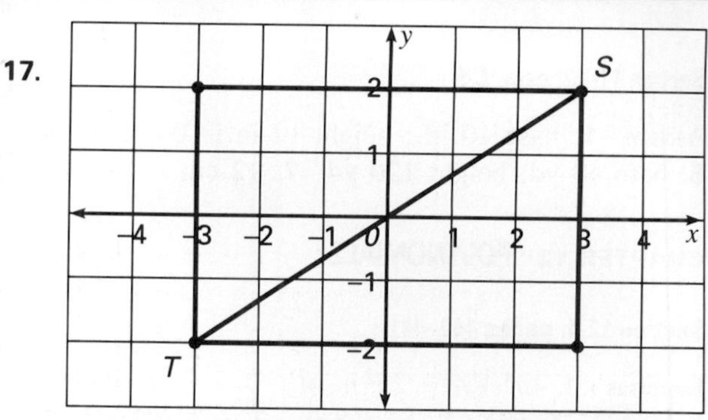

19. Answers will vary. **21.** Answers will vary. A possible solution is (3,2) and (−3,−2).

23. The area is 16 square units.
25. (–2,0); (2,4); (3,5); (4,6) **27.** (1.25, –0.7)

Section 14-2, pages 510–511

Problems **1.** 26 blocks **3.** 35 blocks; 3 blocks **5.** white **7.** after; 17 min

Section 14-3, pages 514–515

Exercises **1.** Answers will vary. **3.** Answers will vary. **5.** Answers will vary. **7.** −4; −4; 0; 0; 4; 4; 8; 8; The straight line that includes points (−1,−4), (1,4), and (2,8) **9.** 3; 3; 4; 4; 5; 5; 6; 6; The straight line that includes points (−4,3), (−2,4), and (2,6) **11.** $y = 2\frac{1}{2}x$ **13.** $y = \frac{1}{2}x$; The straight line that includes points (–2,–1), $(-1, -\frac{1}{2})$, and $(1, \frac{1}{2})$

Section 14-4, pages 518–519

Exercises **1.** 1 **3.** $-\frac{4}{7}$ **5.** $y = -\frac{1}{2}x + 2$ **7.** $\frac{1}{4}$ **9.** $\frac{1}{2}$ **11.** $\frac{1}{2}$ **13.** $y = \frac{1}{2}x + 2$

Problem Solving Applications **1–8:** Answers will vary.

Section 14-5, page 521

Problems **1.** Parking Charge = $3.00
Ticket Charge = $2.00

Amount charged for vehicle	$13	$7	$31
Number of students per vehicle	5	2	14

3. Flat fee = $4.00 Hourly = $2.50 per child

Amount charged for 4-hour job	$6.50	$11.50	$16.50
Number of children	1	3	5

5. Fixed fee = $7.50 cost per lb = $3.50

Cost of shipping package	$21.50	$32.00	$42.50
Number of pounds	4	7	10

Section 14-6, pages 524-525

Exercises **1.**

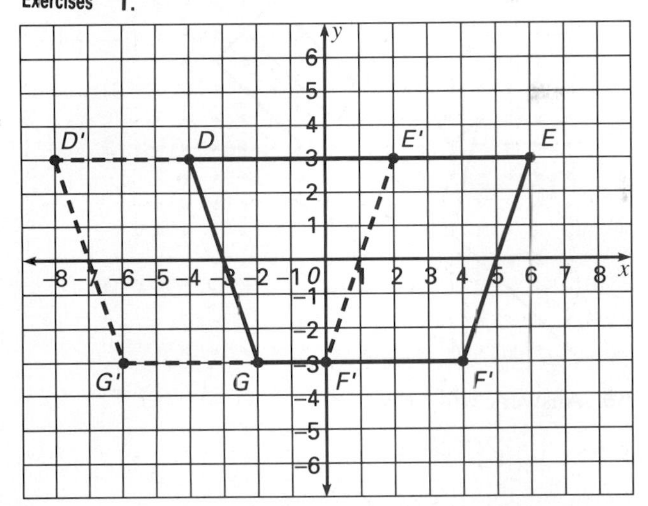

3. The coordinate for the top right corner of the T-shaped figure is (5, −1). For the translated figure, the top right corner would be at the coordinate (−2,1).

The coordinates for the vertices of the triangle are (1,2), (2, −1), and (−2,0). For the translated figure, the coordinates would be at coordinates (0, −2), (1, −5), and (−3, −4).

The coordinate for the top right corner of the remaining figure is (6,4). For the translated figure, the top right corner would be at coordinate (5,0).

5. Answers will vary. If for example, the top right corner of the figure was on coordinate (5,4) the translated figure would have the coordinate (0,7).
7. Answers will vary. If for example, the top right corner of the figure was on coordinate (7, −1) the translated figure would have the coordinate (1,2).
9. 4 right, 3 down **11.** 5 right, 2 down **13.** Various descriptions are possible. For example, for each small square in the design, a corresponding small square is positioned 6 grid squares to the right. **15.** Answers will vary. **17.** tessellating rectangles. **19.** Answers will vary.

Section 14-7, pages 527-529

Exercises **1.** (4, −1) **3.** (−7, −2) **5.** (−8,7) **7.** Coordinates for vertices of reflected image: (−2,3), (−5,2), (−3,1) **9.** Coordinates for vertices of reflected image: (2, −1), (2, −3), and (−2, −1)
11. *H, I, M, O, T, U, V, W, X, Y* **13.** $\overline{AC}$ and $\overline{DF}$; $\overline{AB}$ and $\overline{DE}$; $\overline{BC}$ and $\overline{EF}$ **15.** clockwise; counterclockwise **17.** The order of the vertices changes. The lengths of the sides stay the same.

Problem Solving Applications

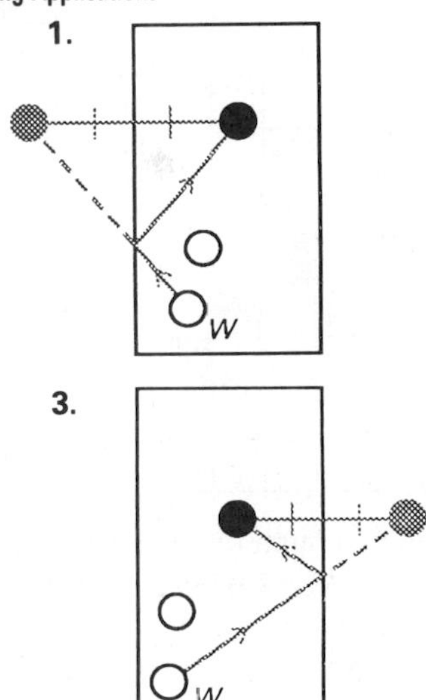

Section 14-8, pages 532-533

Exercises **1.** 270° **3.** 180° **5.** 180° counterclockwise or 180° clockwise **7.** The coordinates of the rotated figures are (2, 2), (−2, 2), (−2, −2) and (2, −2). **9.** The two top coordinates of the rotated figure are (1, 2) and (−1, 2). The two bottom coordinates of the rotated figure are (1, −2) and (−1, −2).

11.

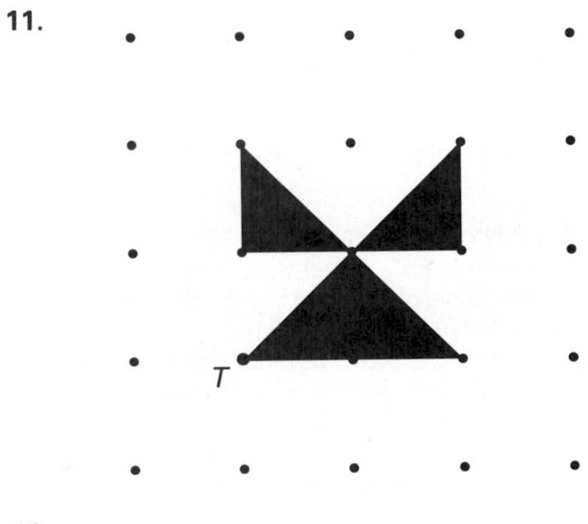

13.

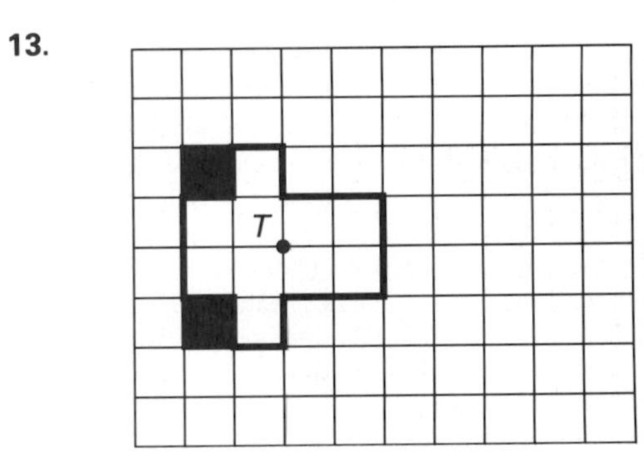

15.

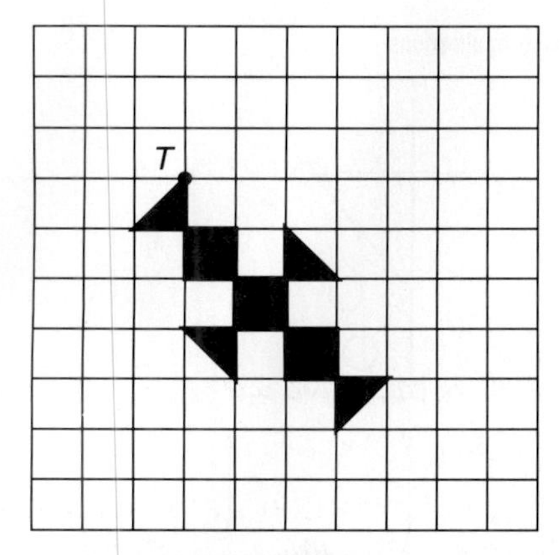

17. counterclockwise; counterclockwise **19.** Both equal 6 square units. **21.** 3,678 in., or 314 ft

Section 14-9, pages 536-537

Exercises **1.** yes **3.** yes **5.** no **7.** 2 **9.** 8 **11.** 4 **13.** isosceles triangle **15.** scalene triangle

17.

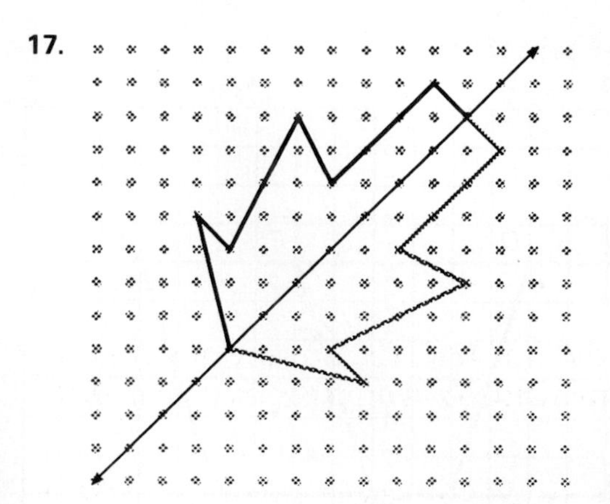

19. Answers will vary.

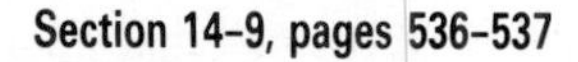

Prevention's Giant Book of Health Facts, © 1991, Rodale Press, Inc., Emmaus, PA, "Vegetables and Risk of Lung Cancer in Men," "Patterns of Sleep," p. 562; "Life Expectancy," p. 563; "Number of Calories Burned by People of Different Weights," p. 564; "The Sports with the Most Injuries," p. 565.
Reader's Digest Book of Facts, © 1987 The Reader's Digest Association, Inc., Pleasantville, NY, "How Fast Do Insects Fly?" and "Longest Recorded Life Spans of Some Animals," p. 548.
The Recording Industry Association of America, "The Recording Industry Association of America 1990 Sales Profile," p. 553.
The Universal Almanac, © 1990 Andrews and McMeel, Kansas City, MO, "Measuring Earthquakes," p. 556.
U.S. Bureau of the Census, *Statistical Abstract of the United States: 1989* (109th edition) Washington, DC, 1989, "Housing Units — Summary of Characteristics and Equipment, by Tenure and Region: One Recent Year," p. 551; "Multimedia Audiences — Summary 1987," p. 553; "Space Shuttle Expenditures by NASA in Constant Dollars: 1972 to 1987," p. 555; "Air and Water Pollution Abatement Expenditures in Constant Dollars: 1975 to 1986," p. 559; "Farms — Number and Acreage, by State: 1980 and 1988," p. 567.
U.S. Bureau of the Census, "1990 Population and Number of Representatives, by State," p. 567.
The World Almanac and Book of Facts, 1991, © 1990 Pharos Books, "Gestation, Longevity, and Incubation of Animals," and "Top 25 Kennel Club Registrations," p. 549; "All-Time Top Television Programs," p. 552; "All-Time Top 25 Movies," p. 553; "Highest and Lowest Continental Altitudes," p. 556; "Acid Rain Trouble Areas," p. 558; "Immunization Schedule for Children," p. 563; "Some Endangered Mammals of the United States," p. 566.

Glossary

A

absolute value (p. 186) The distance of any number, x, from zero on the number line. Represented by $|x|$.

acute angle (p. 347) An angle measuring between 0° and 90°.

acute triangle (p. 354) A triangle having three acute angles.

addition property of opposites (p. 191) When opposites are added, the sum is always zero.

adjacent angles (p. 347) Have a common vertex and a common side but have no interior points in common.

alternate exterior angles (p. 351) Pairs of angles that are exterior to the lines and on alternate sides of the transversal.

alternate interior angles (p. 351) Pairs of angles that are interior to the lines but on alternate sides of the transversal.

altitude of a triangle (p. 377) Segment from a vertex of a triangle perpendicular to the opposite side or to a line containing that side.

angle (p. 342) A plane figure formed by two rays having a common endpoint.

area (p. 384) The amount of surface covered by a plane figure. Area is measured in square units.

associative property (p. 72) If three or more numbers are added or multiplied, the numbers can be regrouped without changing the result. For example, $4 + (6 + 5) = (4 + 6) + 5$; $4 \times (6 \times 5) = (4 \times 6) \times 5$.

average (p. 30) See *mean*.

axes (p. 506) The perpendicular lines used for reference in a coordinate plane.

B

back-to-back stem-and-leaf plots (p. 11) See *stem-and-leaf plot*.

bar graph (p. 20) A means of displaying statistical information in which horizontal or vertical bars are used to compare quantities.

base (of a power) (p. 60) The repeating factor in a power. For example, in 2^3 the base is 2.

binomial (p. 438) A polynomial with two terms.

bisector of an angle (p. 371) A ray that divides the angle into two equal adjacent angles.

C

Celsius (p. 189) See *metric units*.

chunking (p. 460) The process of collecting several pieces of information and grouping them together as a single piece of information.

circle (p. 172) The set of all points in a plane the same distance from a given point called the center.

circumference (p. 172) The distance around a circle.

closed set (p. 208) If an operation is performed on two numbers in a given set and the result is also a member of the set, then the set is closed, with respect to that operation (closure property).

clusters (p. 9) See *stem-and-leaf plot*.

cluster sampling (p. 4) The members of the population are chosen at random from a part of the population and are then polled in clusters, not individually.

coefficient (p. 438) A number by which a variable or group of variables is multiplied. For example, in $7ab^2$ the coefficient is 7.

collinear points (p. 342) Two or more points lying in a straight line.

combining like terms (p. 250) Process using the distributive property to simplify algebraic expressions.

commission (p. 327) An amount of money that is a percent of a sale. The percent is called the *commission rate.*

common factors (p. 119) The factors that are the same for a given set of numbers are the common factors. For example, a common factor of 12 and 18 is 6.

common multiple (p. 119) A number that is a multiple of two different numbers is a common multiple of those two numbers.

commutative property (p. 72) If two numbers are added or multiplied, the operations can be done in any order. For example, $4 \times 5 = 5 \times 4$; $5 + 4 = 4 + 5$.

compass (p. 370) An instrument used to draw circles and arcs and to transfer measurements.

compatible numbers (p. 43) Numbers that are easy to compute mentally.

complementary angles (p. 347) Two angles whose sum measures 90°.

composite numbers (p. 118) A whole number greater than one having more than two factors.

conditional statement (p. 98) One made from two simple sentences. It has an *if* part and a *then* part. The *if* part is called the *hypothesis* and the *then* part is called the *conclusion.*

cone (p. 363) A three-dimensional figure with one circular base and one vertex. The line segment from the vertex perpendicular to the base is the *altitude.* If the endpoint opposite the vertex is the center of the base, the figure is a *right cone.*

conjecture (p. 94) (1) A conclusion reached as part of the process of inductive reasoning.
(p. 366) (2) As a problem solving skill, a guess that seems reasonable based on knowledge available.

constant (p. 438) A monomial that does not contain a variable.

construction (p. 370) A drawing of a geometric figure made by using a compass and a straightedge.

convenience sampling (p. 4) The population is chosen only because it is readily available.

coordinate for a point (p. 506) A number associated with a point on a number line or an ordered pair of numbers associated with a point on a grid.

coordinate plane (p. 506) Two perpendicular number lines forming a grid. The horizontal number line is called the x-axis. The vertical number line is called the y-axis.

coplanar points (p. 342) Points lying on the same plane.

corresponding angles (p. 351) Angles that are in the same position relative to the transversal and the lines.

counterexample (p. 98) A counterexample shows that a conditional statement is false because it satisfies the hypothesis but not the conclusion.

cross-products (p. 282) For $\frac{a}{b}$ and $\frac{c}{d}$, the cross products are ad and bc. In a proportion cross products are equal.

cube (p. 362) A right prism with six faces that have the same size and shape.

cubed (p. 60) A number multiplied by itself two times. For example, 10^3 is read as "10 cubed."

customary units (p. 154) Units used in the customary system of measurement. Frequently used customary units include the following: length—inch (in.), feet (ft), yard (yd), and mile (mi); capacity—cup (c), pint (pt), quart (qt), and gallon (gal); weight—ounce (oz), pound (lb), and ton (T); and temperature—degrees Fahrenheit.

cylinder (p. 363) A three-dimensional figure having two parallel and congruent circular bases.

►D

data (p. 2) Statistical information most often gathered in numerical form. The word data is the plural form of the Latin word *datum.*

deductive reasoning (p. 99) A process of reasoning in which a conclusion is drawn from a conditional statement and additional information.

degree (p. 189, p. 346) A common unit of measurement for angles or temperature.

degree of a polynomial (p. 446) The greatest degree of any monomial within the polynomial.

denominator (p. 122) In the fraction $\frac{2}{3}$, the number 3 is the denominator. The denominator tells the total number of equal parts into which a whole has been divided.

dependent events (p. 487) Two events such that the outcome of the first affects the outcome of the second. For example, drawing a first card and then a second card without replacing the first.

diagonal (p. 358) A line segment joining two non-adjacent vertices of a polygon.

diameter (p. 172) A line segment passing through the center of a circle and having endpoints on the circle.

discount (p. 324) The difference between the regular selling price of an item and its sale price.

distributive property (p. 76) If one factor in a product is a sum, multiplying each addend by the other factor before adding does not change the product.

►E

endpoint (p. 342) A point at the end of a segment or ray.

equation (p. 222) A mathematical statement that two numbers or expressions are equal.

equiangular (p. 355) Having angles of the same measure.

equilateral triangle (p. 355) A triangle with all three sides having the same length and all angles the same measure.

equivalent fractions (p. 122) Fractions that represent the same amount.

equivalent ratios (p. 278) Two ratios that represent the same comparison. For example, 2:5, 4:10, 6:15 are equivalent ratios.

estimate (p. 42) An approximation of an answer or measurement.

evaluate (p. 46) To find the value of an algebraic expression.

event (p. 474) A set of one or more outcomes.

experimental probability (p. 490) The probability of an event based on the results of an experiment.

exponent (p. 60) A number showing how many times the base is used as a factor. For example, in 2^3 the exponent is 3.

exponential form (p. 60) A number written with a base and an exponent. For example, $2 \times 2 \times 2 \times 2 = 2^4$.

►F

factor (p. 118) Any number multiplied by another number to produce a product.

factoring (p. 118, p. 456) Finding two or more factors of a number or a polynomial.

Fahrenheit (p. 189) See *customary units.*

formula (p. 170) A specific equation giving a rule for relationships between quantities.

frequency table (p. 7) In a frequency table a tally mark is used to record each response. The total number of tally marks for a given response is the frequency of that response.

front-end estimation (p. 53) Estimate with the front-end digits only and then adjust the estimate by approximating the values of the other digits.

function (p. 513) A relation in which each member of the domain is paired with one, and only one, number of the range.

►G

gaps (p. 9) See *stem-and-leaf plot.*

graph of the equation (p. 512) The set of all points whose coordinates are solutions of an equation.

graph of the function (p. 513) A graph of an equation that represents the function.

greatest common factor (GCF) (p. 119) The greatest integer that is a factor of two or more numbers.

►H

hypotenuse (p. 404) The side opposite the right angle in a right triangle.

►I

identity property of addition (p. 73) The sum of any number and 1 is that number.

identity property of multiplication (p. 73) The product of any number and 1 is that number.

independent events (p. 486) Events such that the outcome of the first does not affect the outcome of the second. For example, tossing a coin and rolling a number cube.

inductive reasoning (p. 94) Logical reasoning where a conclusion is made based on a set of specific examples. The conclusion is called a *conjecture.*

inequality (p. 254) A mathematical sentence involving one of the symbols $<$, $>$, $\leq$, or $\geq$.

integers (p. 186) The set of whole numbers and their opposites.

interest (p. 328) Amount that is paid for the use of money over a period of time.

intersecting lines (p. 350) Lines that have exactly one point in common.

inverse operations (p. 202, p. 226) Operations that undo each other such as addition and subtraction or multiplication and division.

irrational number (p. 255) Numbers such as π that are non-terminating non-repeating decimals.

isosceles triangle (p. 355) A triangle having at least two sides the same length and at least two angles the same measure.

►L

laws of exponents (p. 64) Rules that govern how to perform operations with numbers in exponential form. Examples are: To multiply numbers with the same base, add the exponents. To divide numbers with the same base, subtract the exponents.

least common denominator (LCD) (p. 123) The least common multiple of the denominators of fractions.

least common multiple (LCM) (p. 119) The smallest number that is a multiple of two or more numbers.

legs of a triangle (p. 404) In a right triangle, the two sides that are not the hypotenuse.

like fractions (p. 138) Fractions having the same denominator. For example, $\frac{5}{8}$ and $\frac{1}{8}$.

line (p. 342) A set of points that extends without end in two opposite directions. Two points determine a line.

linear equation (p. 512) An equation that has two variables and many solutions, all of which lie in a straight line when they are graphed on a coordinate grid.

line graph (p. 26) A means of displaying changes in statistical information using points and line segments to represent the data.

line of symmetry (p. 534) A line on which a figure can be folded, so that when one part is reflected over that line it matches the other part exactly.

line plot (p. 7) In a line plot, an *X* is made to record each response. The *X*'s may be stacked one on top of another until all the data are recorded.

line segment (p. 342) The set of points containing two endpoints and all points between them.

line symmetry (p. 534) A figure has line symmetry if a line drawn through it divides the figure into two matching parts.

lowest terms (p. 123) A fraction in which the numerator and denominator do not have a common factor other than one.

►M

mean (p. 30) The sum of a set of numbers divided by the number of data in that set. Also known as *average.*

measures of central tendency (p. 30) Statistics used to analyze data. These measures are the *mean,* the *median,* and the *mode.*

median (p. 30) In a list of data ordered from least to greatest or greatest to least, the middle number or the *mean* of the middle two numbers.

metric units (p. 158, p. 189) Units used in the metric system of measurement. Frequently used metric units include the following: length—meter (m), millimeter (mm), centimeter (cm), and kilometer (km); liquid capacity—liter (L) and milliliter (mL); mass—kilogram (kg) and gram (g); and temperature—degrees Celsius.

midpoint (p. 374) The midpoint of a line segment is the point that separates it into two line segments of equal length.

mode (p. 30) In a list of data, the number or item occurring most frequently.

monomial (p. 438) An expression that is either a single number, a variable, or the product of a number and one or more variables.

multiple (p. 119) The product of any number and another whole number.

►N

negative integers (p. 186) Integers that are less than zero.

numerator (p. 122) In the fraction $\frac{3}{7}$, the number 3 is the numerator. The numerator tells how many of the equal parts of a whole are being considered.

►O

obtuse angle (p. 347) Any angle measuring between 90° and 180°.

obtuse triangle (p. 354) A triangle having one obtuse angle.

open sentence (p. 222) A sentence that contains one or more variables. It can be true or false, depending upon what values are substituted for the variables. A value of the variable that makes an equation true is called a *solution* of the equation.

opposites (p. 186) Two integers having a sum of zero. For example −27 and 27.

ordered pair (p. 506) Two numbers named in a specific order.

order of operations (p. 68) Rules followed to simplify expressions.

origin (p. 506) The point of intersection of the x-axis and y-axis in a coordinate plane.

outcomes (p. 474) The possible result of an experiment. For example, tossing a coin.

outliers (p. 9) See *stem-and-leaf plot.*

►P

parallel lines (p. 350) Lines lying on the same plane that do not intersect.

parallelogram (p. 359) A quadrilateral having both pairs of opposite sides parallel.

percent (p. 302) *Per one hundred.* A ratio that compares a number to 100.

perfect square (p. 400) The exact square of any other number or polynomial. For example, 4 is a perfect square.

perimeter (p. 166) The distance around a figure.

perpendicular bisector (p. 374) The perpendicular bisector of a line segment is the line, ray, or line segment that is perpendicular to a line segment at its midpoint.

perpendicular lines (p. 350) Intersecting lines that form right angles.

pi (π) (p. 172) The ratio of the circumference of a circle to its diameter.

pictograph (p. 12) A graph that uses pictures or symbols to represent data. The *key* identifies the number of data items represented by each symbol.

plane (p. 342) A flat surface that extends without end in all directions.

point (p. 342) A specific location in space.

polygon (p. 343) A closed plane figure that is formed by joining three or more line segments at their endpoints. Each line segment joins exactly two others and is called a *side of the polygon.* Each point where two sides meet is a *vertex.*

polyhedron (p. 362) Plural: polyhedra. A three-dimensional figure in which each surface is a polygon. The surfaces of a polyhedron are called its *faces*. Two faces meet, or intersect, at an *edge*. Three or more edges intersect at a *vertex.*

polynomial (p. 438) The sum of two or more monomials. Each monomial is called a *term* of the polynomial. A polynomial is in *standard form* when its terms are in order from greatest to least powers of one of the variables.

positive integers (p. 186) Integers that are greater than zero.

power (p. 60) A number that can be expressed using an exponent. Read 2^4 as "2 to the fourth power."

power rule (p. 65) To raise an exponential number to a power, multiply the exponents. For example, $(3^2)^4 = 3^{2 \times 4} = 3^8$.

prime factorization (p. 118) A composite number expressed as a product of prime numbers.

prime number (p. 118) A whole number greater than one having exactly two distinct factors, one and the number itself.

principal (p. 328) The amount of money on which interest is paid or the amount of money borrowed at interest.

prism (p. 362) A polyhedron with two *bases* that are parallel and are the same size and shape.

probability (p. 474) The ratio of the number of favorable outcomes to the total number of possible outcomes.

property of zero for multiplication (p. 73) The product of any number and zero is zero.

proportion (p. 282) Two equivalent ratios set up to make a mathematical sentence.

protractor (p. 346) An instrument used to measure angles.

Pythagorean Theorem (p. 404) In a right triangle the sum of the squares of the length of legs is equal to the square of the length of the hypotenuse. $a^2 + b^2 = c^2$

pyramid (p. 363) A three-dimensional figure having three or more triangular faces and a polygonal base.

► Q

quadrant (p. 506) One of the four regions formed by the axes of the coordinate plane.

quadrilateral (p. 359) A polygon having four sides.

► R

radius (p. 172) A line segment having one endpoint at the center of the circle and the other endpoint on the circle.

random sampling (p. 4) Each member of the population is given an equal chance of being selected.

range (p. 30) The difference between the greatest and least number in a set of numerical data.

rate (p. 274) A ratio that compares two different kinds of quantities.

ratio (p. 270) A comparison of two numbers, represented in one of the following ways: 2 to 5, 2 out of 5, 2:5, or $\frac{2}{5}$.

rational number (p. 208) Any number that can be expressed in the form $\frac{a}{b}$, where a is any integer and b is any integer except 0.

ray (p. 342) A part of a line having one endpoint and extending without end in one direction.

real numbers (p. 255) The set of irrational and rational numbers together.

reciprocals (p. 134) Two numbers are reciprocals when their product is 1.

rectangle (p. 359) A parallelogram having four right angles.

reflection (p. 526) A transformation in which a figure is flipped or reflected over a line of reflection.

regular polygon (p. 359) A polygon all of whose sides are equal and all of whose angles are equal.

repeating decimal (p. 127) A decimal in which a digit or group of digits repeats. A bar above the group of repeating decimals is used to express the repeating decimal. For example, $\frac{2}{3} = 0.\overline{6}$.

rhombus (p. 359) A parallelogram with all sides the same length.

right angle (p. 347) An angle having a measure of 90°.

right triangle (p. 354) A triangle having a right angle.

rotation (p. 530) A transformation in which a figure is *turned* or *rotated.*

rotational symmetry (p. 535) A figure has rotational symmetry if when the figure is turned about a point the figure fits exactly over its original position at least once during a complete rotation.

rounding (p. 42) Expressing a number to the nearest ten, hundred, thousand and so on.

►S

sale price (p. 324) The purchase price less the discount.

sample space (p. 478) The set of all possible outcomes of an event.

sampling (p. 4) See *cluster sampling, convenience sampling, random sampling,* and *systematic sampling.*

scale drawing (p. 288) A drawing that represents a real object. All lengths in the drawing are proportional to actual lengths in the object. The ratio of the size of the drawing to the size of the actual object is called the *scale* of the drawing.

scalene triangle (p. 355) A triangle with no sides the same length and no angles the same measure.

scientific notation (p. 61) A notation for writing a number as the product of a number between 1 and 10 and a power of 10.

simplify (p. 250) To simplify an expression is to perform as many of the indicated operations as possible.

slide (p. 522) See *translation.*

slope (p. 516) The ratio of the change in y to the change in x, or the ratio of the change in the *rise* (vertical change) to the *run* (horizontal change).

slope-intercept form (p. 517) A linear equation in the form of $y = mx + b$ where m is the slope of the graph of the equation and b is the y-intercept.

solution (p. 222) A replacement set for a variable that makes a mathematical sentence true.

sphere (p. 363) A three-dimensional figure consisting of points that are the same distance from the center.

square (geometric) (p. 359) A parallelogram with four right angles and all sides the same length.

square (numeric) (p. 60, p. 400) The product of a number and itself.

squared (p. 60) A number multiplied by itself. For example, 10^2 is read as "10 squared."

square root (p. 400) One of two equal factors of a number.

statistics (p. 2) The field of mathematics involving the collection, analysis, and presentation of data.

stem-and-leaf plot (p. 8) A means of organizing data in which certain digits are used as *stems,* and the remaining digits are used as *leaves. Outliers* are data values that are much greater than or much less than most of the other values. *Clusters* are isolated groups of values. *Gaps* are large spaces between values. A *back-to-back stem-and-leaf plot* displays two sets of data simultaneously.

straight angle (p. 347) An angle having a measure of 180°.

supplementary angles (p. 347) Two angles whose sum measures 180°.

surface area (p. 424) The sum of the areas of all the faces of a three-dimensional figure.

survey (p. 4) A means of collecting data for the analysis of some aspect of a group or area.

systematic sampling (p. 4) After a population has been ordered in some way, the members of the population are chosen according to a pattern.

►T

terminating decimal (p. 126) A decimal in which the only repeating digit is 0.

terms (p. 250) The parts of an expression separated by addition or subtraction signs. In the expression $2a + 3b + 4a - 5b^2$, the terms $2a$ and $4a$ are *like terms* because they have identical variable parts. The terms $3b$, $4a$, and $5b^2$ are *unlike terms* because they have different variable parts.

time (p. 328) Refers to the period of time during which the principal remains in a bank account. It also refers to the period of time that borrowed money has not been paid back.

transformation (p. 522) A way of moving a geometric figure without changing its size or shape. Prior to the move the figure is called a *preimage*. After the move, the figure is called an *image*.

translation (p. 522) A change in the position of a figure such that all the points in the figure slide exactly the same distance and in the same direction at once.

transversal (p. 351) A line that intersects two or more lines.

trapezoid (p. 359) A quadrilateral having one and only one pair of parallel sides.

tree diagram (p. 478) A diagram that shows all the possible outcomes of an event.

triangle (p. 343) A polygon having three sides.

trinomial (p. 438) A polynomial with three terms.

►U

unit price (p. 277) A ratio comparing the price of an item to the unit of its measure.

unit rate (p. 274) A rate that has a denominator of 1 unit.

►V

variable expressions (p. 46) Expressions that contain at least one variable.

variables (p. 46) Placeholders in mathematical expressions or sentences.

Venn diagram (p. 102) A means of showing relationships between two or three different classes of things. In a Venn diagram, each class is represented by a circular region inside a rectangle. The rectangle is called the *universal set*.

vertex (p. 342) The common endpoint of two rays that form an angle. The rays are called the *sides* of the angle. Plural form: *vertices*.

vertical angles (p. 350) Angles of the same measure formed by two intersecting lines.

volume (p. 387) The number of cubic units needed to fill a space.

►X

x-axis (p. 506) Horizontal number line.

x-coordinate (p. 506) The first number in an ordered pair.

►Y

y-axis (p. 506) Vertical number line.

y-coordinate (p. 506) The second number in an ordered pair.

y-intercept (p. 517) The y-intercept of a line is the y-coordinate of the point where it intersects the y-axis.

Index